H. B. Griffiths
P. J. Hilton

A Comprehensive Textbook of Classical Mathematics

A Contemporary Interpretation

Springer-Verlag
New York Heidelberg Berlin

H. B. Griffiths
University of Southampton
Southampton, S09 5 NH
England

P. J. Hilton
Case Western Reserve University
Cleveland, Ohio 44106
USA

AMS Subject Classifications: 00-01, 00A05

Library of Congress Cataloging in Publication Data

Griffiths, Hubert Brian.
A comprehensive textbook of classical mathematics.

Bibliography: p.
Includes index.
1. Mathematics—1961- I. Hilton, Peter
John, joint author. II. Title.
QA37.2.G75 1978 510 78-15692

Originally published in 1970 by Van Nostrand Reinhold Co.

Printed in the United States of America.

9 8 7 6 5 4 3 2 1

ISBN 0-387-90342-9 Springer-Verlag New York
ISBN 3-540-90342-9 Springer-Verlag Berlin Heidelberg

To Catherine and Margaret

Introduction to the Springer-Verlag Edition

This edition does not differ in content from its predecessor, but whereas the first edition appeared in 1970, the book is now addressed to an audience at a period when mathematics itself, the study of mathematics, and attitudes to both, have all undergone very profound and rapid change—especially compared with the years during which the book was conceived and written. An important (but by no means the sole) influence on these changes has been the increasing penetration into Mathematics programs, of information theory, numerical analysis, and computer science. We excluded such topics from the book, because we did not have any special contribution to make to their discussion—although we explicitly recognised their interest and importance. Today further areas of mathematics have come into prominence largely as a result of the key role played by the computer in society and in education. But, although these subjects have become of very great importance, they have not displaced the more classical subjects which form the content of this book. Moreover there are still, of course, vast areas of application of mathematics—for example, differential equations—which centre more on the classical subjects than on the more modern ones. Thus we would claim that the mathematics treated in this book has not lost its relevance, either within mathematics or with a view to applications to the real world; it still has to be learned, even though mathematics has grown apace and economic realities are forcing greater stress on applications than a decade ago. Indeed, we ourselves were pointing out applications throughout the book as we wrote it.

Our next remarks are directed at American readers. When this book was written, it was not our thought that the distribution of the book would be largely confined to Europe and Canada (but this seems to have happened as a result of certain unpredictable changes of policy in the publishing house which originally handled our book). Now a significant difference between practice in America and the European system, is that at American universities, undergraduate courses are always accompanied by assigned textbooks, whereas in the European system, texts may be recommended, but very rarely form an integral part of the course. Thus we included for American readers, on page xviii of the original Introduction (page **xviii** in this edition), a list of five courses which can be constructed by suitable selection of material from the text. Each of these courses could find a home in an American university or college program. Thus for example, Course 2 (Algebra) incorporates the arithmetic of the integers, linear algebra, an introduction to group theory, the theory of polynomial functions and polynomial equations, and some Boolean algebra. It could be supplemented, of course, by material from other chapters. Again, Course 5 (Calculus) discusses the differential and integral calculus more or less from the beginnings of these theories, and proceeds through functions of several real variables, functions of a complex variable, and topics of real analysis such as the implicit function theorem.

We would, however, like to make a further point with regard to the appropriateness of our text in course work. We emphasized in the Introduction to the original edition that, in the main, we had in mind the reader who had already met the topics once and wished to review them in the light of his (or her) increased knowledge and mathematical maturity. We therefore believe that our book could form a suitable basis for American graduate courses in the mathematical sciences, especially those prerequisites for a Master's degree. It is the view of many who are currently seriously considering the question of the appropriate education for students of the mathematical sciences entering the American job market now or in the foreseeable future,[1] that the mathematics which they will use in industry is not likely to be in itself of a very sophisticated technical nature, but that such students would need to have a very mature attitude towards that mathematics in order to use it and adapt it in an effective and flexible manner. Thus we contend that a second, perhaps more systematic, look at undergraduate mathematics might be the most appropriate type of course for students entering the job market with a Master's degree.

Similar considerations, with appropriate naming of University degrees, are wholly relevant to the European system also, where too many University courses force advanced, abstract Mathematics, on to students who possess a totally inadequate (because neglected) foundation of basic, Classical Mathematics. We might also argue that a course from this book will be valuable for the student going on to become a professional mathematician, but for such a student there would naturally be a need of very substantial supplementation.

The present edition, then, is identical with the original except for the elimination of errors in the original text which have been drawn to our attention by readers. To those readers we are, of course, extremely grateful. A few additional calculus exercises have also been added.

[1]This has been argued, for example, by members of the United States National Research Council Committee on Applied Mathematics Training, of which one of us (P.J.H.) is chairman.

INTRODUCTION

1. Origins and Purpose of the Book

This book owes its origin to a course of lectures given by the authors, with the aid of some of their colleagues, while members of the mathematics department of the University of Birmingham, England. The course was designed for grammar school teachers of mathematics working in the Birmingham area. Although it included individual lectures on the uses of 'modern mathematics' in engineering, the theory of intelligence, statistics, and computer science, our primary purpose was to convey something of the climate of thought in a contemporary mathematics department of a university, and to discuss its relation with school mathematics. We hoped that our mutual discussions would help both schools and universities to prepare their pupils better for careers in mathematics, with an eye on the growing shortage of mathematicians. The lectures took place on Wednesday afternoons during term time and the sessions lasted between two and three hours each. Such co-operation (and in-service training) was highly unusual at the time (1961–2) but fortunately is now less rare. We would like to say here how much we gained from the give and take of our arguments with the teachers in our audience; such contacts are likely to be of immense value to the universities, not only with regard to mathematics.

It was certainly not possible to provide in the course a comprehensive coverage of the material relevant to sixth-form and first-year undergraduate courses in 'pure' mathematics. We therefore decided to take certain topics selected from classical mathematics and, within these topics, to treat the underlying concepts and theories from a modern standpoint. In referring to a 'modern' standpoint we do not wish to imply that we subscribe to any particular view of how to teach mathematics today. Rather, we intend to convey the impression that we tried to treat these topics from the point of view of working mathematicians, adopting the language and standards of present-day mathematics. The same intention informs this book.

However, the book is, naturally, much more than a record of the course. Although we have not included any topics which did not find mention in the course, nevertheless we have given much more attention to each topic than was possible—or would have been desirable—in the course itself. Our hope all along in the writing of the book has been that students (of any age!) who have reached roughly the level of completing one year of specialist mathematical study at a university may be able to take up this book and read it without requiring further instruction through lectures or course work (see p. xviii).

The topics themselves are more or less represented by the titles of the various parts of the book, with the exception that Parts I, II, and VIII may all be subsumed as treating the topic *Set Theory and Foundations.* Thus, apart from this topic, we discuss Arithmetic; Geometry; Algebra; Number Systems; Calculus. We do not imply by this choice that no other parts of mathematics are appropriate to a reader at the level at which he may wish to review the topics which we treat here, but we chose them as the central part of our course because they formed a background from classical (pure) mathematics which teachers in schools have in common with teachers in universities. It has not been our primary intention to offer readers their *first* contact with the concepts and ideas we treat. In the spirit of the course of lectures itself, we wish to encourage the reader to look at rather familiar ideas a second time, with a view to fitting them into the framework of present-day, contemporary mathematical thought; and we hope thus to enable the reader to see how certain key ideas recur again and again and give a real unity to apparently separate parts of his early mathematical experience. Perhaps we might believe, in certain cases, that what we have written is the best way to introduce and develop an idea, but this forms no part of our claim nor of our motive in writing the book.

Now, in selecting our topics, we have been very much aware of the importance of presenting probability and statistics, and information theory, at this stage; and we would certainly hope that a student would by this time have some acquaintance with numerical analysis and computer science. Also, we fully recognise the vital role which computers are playing in our society, and we appreciate the effect that role must surely have on the whole of education; nevertheless, we have not taken account of it in this book. To have done so would not only have further lengthened an already outsize text; it would also have led to a very different organization of the material. Further, these excluded topics are not so likely to be *already* familiar to the reader as those included; moreover if they are familiar, then it is reasonably likely that the treatment accorded them on the reader's first introduction is not so very different from that which we ourselves would give. (It seems a good rule for authors to remain silent if they have nothing new to say.) However, we have often emphasized the algorithmic nature of certain techniques, because it is usually helpful, when presenting a topic, to get a student to pretend that he is clarifying the ideas ready to explain them to a computing machine. (This is only an extension of the notion that the best way of understanding Pythagoras's Theorem is to pretend that one is Pythagoras himself, searching—and struggling—for a proof.) It is our hope that the informed reader will recognize where the availability of computers can lead to an insightful approach to a piece of mathematics and, if he is a teacher, will communicate this awareness to his students.

2. Mode of Presentation

Our style of presentation reflects our intention to provide a review (in the sense of a survey starting from a familiar background) of the material treated. We present a great deal of theory and we present it as a definitive body of knowledge. Naturally we are concerned, often very explicitly, with the question of motivation for the introduction of a new concept, for a generalization, or for attributing significance to a theorem; but we adopt a didactic style in the actual enunciations of definitions and results, and we apparently leave the student little area of choice for the development of his own approach to the material. We do not forget, however, that 'the most important existence theorem in mathematics is the existence of people'. And our didacticism is not, we repeat, due to any prejudice against 'discovery methods' in mathematics; nor does it imply that we fail to recognize that the creative aspect of mathematics is at least as important as the use of mathematics for systematizing our knowledge. But, where a topic has already become familiar through use and applications at a less sophisticated level (or, in the language of *Goals for School Mathematics*†, at the *premathematical* level), it remains to elucidate the nature of the mathematical ideas involved and to *organize* the knowledge gained. This requires, of course, a systematic style of presentation and a careful organization of the material to bring out its interrelations with other parts of mathematics, so that the pattern of thought may the more clearly emerge. This was, essentially, the rationale underlying the course as originally given, and the transfer from the lecture room to the printed page has tended to emphasize the style in which the topics have been treated. We very much hope that teachers will use our material as a quarry, so to speak, of material from which they can select topics for treatment by bold classroom techniques, along the lines of the remarkable ATM book *Some Lessons in Mathematics* (listed as [36a] in the Bibliography at the end of this book). Before they can effect this transformation they need to have the mathematical material in 'professional' form, and that form is what we intend the book to have.

On the other hand, mathematical ideas are not communicated from mathematician to mathematician in extremely precise form; indeed, it is often true that a published article, while perfectly precise and formally correct, may fail to convey the essential nature of the result in question just because the author has neglected to include in his article an informal description of the motivation underlying the problem or his chosen line of solution. The implication for mathematical exposition would seem to be to use the 'spiral approach' recommended in the *Goals* pamphlet. Thus ideas should first be introduced informally, and precise definitions and proofs should later be

† The report of the Cambridge Conference on School Mathematics, published by Houghton Mifflin (1964). This report should be read by all interested in mathematics education.

provided for the key concepts, after which informality becomes once more appropriate. We have attempted to implement this principle. Moreover, in conformity with the spiral approach, we believe that topics should reappear in any well-planned syllabus (if we may, for brevity, use such an authoritarian phrase); typically a topic, treated in the first place for its own sake, will be reconsidered later as part of a broader generalization or as exemplifying a feature somewhat too subtle to be understood and appreciated at the initial exposure. Where a new concept first makes its appearance we discuss it informally and intuitively; we then introduce a degree of precision unusual in a text at this level in order to isolate and identify the concept in question; and, once we have been precise, we take the view that we and the reader have both earned the right to be informal in the interest of the free flow of ideas and of thought. Since we regard this book principally as a review, we proceed somewhat rapidly through the first stage to the point where we feel it appropriate to adopt a precision of language and notation; nevertheless we hope that we pass sufficiently rapidly to the less formal style to avoid the charge of pedantry.

3. The Spiral Approach versus Pedantry: Notational Difficulties

While we certainly do not wish to be rigid we also believe that, in departing from precise terminology, the student should know just what it is from which he is departing! The appearance of pedantry will perhaps be particularly conspicuous in matters of notation. At first we insist, for example, that a function is simply denoted by a symbol 'f' and that we must (for practical computational reasons) specify the domain and range of f; moreover, after the introductory 'chat', a function first appears in its precise form as a set, suitably restricted, of ordered pairs taken from an appropriate Cartesian product. We completely repudiate the attitude that so formalistic a view should be maintained when one comes to *work* with functions, as in the differential calculus (say); and we adopt and encourage more conventional and traditional notation and terminology such as '$f(x)$', 'dy/dx', 'y is a function of x'. Similarly we emphasize the role of the identity function, calling it Id or even 1, irrespective of its domain, so that the reader may get used to thinking of it as a very important and ubiquitous mathematical concept; but it may, and usually should, become a very inconspicuous element in most calculations. Similarly we lay great stress initially on the composition of functions, but here an extra purpose is served. For we believe that a real confusion is created unless the student understands at the outset the nature of function-composition and the distinction between such a composition and the important product operation in the ring of real-valued functions. The notation f^{-1} is really dangerous for the function inverse to f (with respect to function-composition), when the student knows that $x^{-1} = 1/x$. It is not possible to work with a notation that is entirely free of ambiguity—such a notation would be intolerably cumbersome; but at least a teacher must do his duty to his student by explaining precisely—at the appropriate turn on the spiral—where the danger of misunderstanding lies, and why a potentially misleading notation is adopted. In this case, to avoid early misunderstanding, we use the symbol $f^\flat$ for f^{-1} until we feel the time is ripe to revert to f^{-1}.

We have, rather reluctantly, continued the traditional practice of writing the function symbol to the *left* of the variable (thus, we write $f(x)$). We are well aware of the advantages to be gained by writing the function symbol on the right, but we have thought it prudent not to introduce notational innovations beyond those which may be regarded as standard, if not universal, practice. Undoubtedly it would be most unusual to find the right-hand convention adopted in a text at this level and we do not wish to present the reader with a further, and somewhat gratuitous, difficulty in bridging the gap between this book and others he may encounter before, during, or immediately after contact with our terminology and notation. Nevertheless, since we have placed considerable emphasis on flexibility, it would be retrograde not to recommend to the reader that he prepare himself for the use of either convention. At the risk of underlining the obvious, let us point out that the notational problem comes most clearly into focus when discussing *composition* of functions (in Chapter 2). For, given

$$S \xrightarrow{f} T \xrightarrow{g} U,$$

the composite function from S to U seems to demand the symbol $f \circ g$ or, simply, fg. However, with the left-hand convention, we are forced to write $g \circ f$, or gf; for, if x is in S, then the image of x under the composite function is $g(f(x))$. Certainly most mathematicians use the left-hand convention (this, of course, accounts for our own choice in this book), but it should be remembered that the convention arose before so much emphasis was placed, as it is today, on composition of functions. It is often an amusing experience to see how different mathematicians, giving a lecture, cope with the anomaly referred to above. Many, in writing 'fg', actually write the 'g' first (that is, first in time!) and the 'f' afterwards! Some even go as far as to read the symbol 'fg' as 'gee eff'!

In the spirit of the spiral approach we do not always recommend a single point of view towards a given topic, nor even a single designation for a particular concept. Thus, for example, we describe both the Cauchy-sequence procedure and the Dedekind-section procedure for completing the rationals to the reals. One of us believes that the former has greater algebraic appeal and the latter greater geometric appeal, while the other believes the opposite: but both agree that it would be folly to insist here on a single viewpoint. Similarly we describe a function $f: S \to T$ variously as *invertible* or as an *equivalence* if it admits a function $g: T \to S$, such that gf is the identity on S and fg the identity on T. Again, the term 'invertible' brings out the algebraic flavor while the term 'equivalence' brings out the set-theoretic flavor. In other cases, too, we try to cater for the various uses of a concept and the varying tastes of students of mathematics; and to prepare them for reading texts which use differing terminology, notations and founts of type.

Particular notational features are, of course, always explained in the text when they are introduced. However a few should be mentioned here in the introduction. We use the symbol ■ to indicate that a proof has been completed; whereas we use the symbol □ to indicate that no proof is being given of the statement thus adorned (although a proof exists!). The neologism 'iff' is used to mean 'if and only if', since this phrase is of such frequent occurrence; but we often use the symbols '$\Rightarrow$', '$\Leftarrow$', and '$\Leftrightarrow$' to denote 'only if', 'if' 'if and only if', respectively. Also we frequently use the existential quantifiers $\exists x$ and $\forall x$ to mean 'there is an x' and 'for all x'. We lay stress on the grammatical skill necessary for negating them fluently, and on the flexibility necessary for the efficient use of any symbolic abbreviations.

4. Unification in Mathematics

Of course, we would not wish to be taken too literally when we disclaimed, earlier, the intention of introducing new ideas to the reader. New formulations of familiar ideas and techniques are bound to involve new ideas. But such new ideas as we do treat in the text are unifying ideas, taken from set theory, algebra or topology; by mastering them the reader should be better able to grasp the meaning of arguments carried out in special cases. This quest for unification, whose purpose is partly aesthetic, has the practical aim of helping to bring masses of detailed knowledge under control, to aid our memories. It has led us to include a chapter on categories and functors. These notions, recently introduced into mathematics, have been found very useful by mathematicians in order to provide a common framework and language for vastly different branches of mathematics (e.g., set theory, group theory, topology). However, as always in the development of mathematics, once a new concept is born, a theory grows up round this concept. Category theory is a very exciting, young branch of mathematics; here, however, we have resisted the temptation to go into this theory, contenting ourselves with describing how its basic concepts enrich our understanding of familiar parts of mathematics and render explicit those links between these parts which previously were apprehended only intuitively. Our aim, therefore, does not go beyond that of introducing categorical language.

5. Organization of the Text

The text is organized as follows. The book is divided into eight Parts; each part is divided into chapters (numbered consecutively through the book);

and each chapter is divided into sections. Although each part is prefaced by an explanatory introduction, we summarize their contents here, to clarify the relationships of the parts with the whole. The first Part, *The Language of Mathematics*, is different in kind from the rest. As its title suggests, this part is designed to establish the language in which the remainder of the book will be communicated,† and thus requires some discussion at this stage. This part is, very literally, prerequisite for the further parts, whereas the latter deal with various areas of mathematics and we expect that many students will choose to read particular parts and neglect others—or, at least, to read the various parts in an order different from that in which they are presented.

Naturally, since the *leitmotiv* of the book is the very unity of mathematics, familiarity with certain ideas from, say, arithmetic will be necessary to understand the Part devoted to algebra; and cross-referencing of propositions and examples serves to bind the text into a whole. But the order of presentation of the material is largely arbitrary and the student is encouraged to make choices in the interest of maintenance of appetite. We must emphasize that our approach is *not* always the most economical, and in a strictly logical order, of the sort favoured by many mathematics lecturers in universities. But we believe that most students (including the aforesaid lecturers) learn in a highly non-linear way, and it is only at a certain level of fluency that there comes a desire to see a subject laid out in a linear order. It is *far easier* to present lectures in a ruthless linear way than by the approach of the spiral technique; but if the object of teaching is to communicate, rather than to give aesthetic satisfaction to the expositor, then we must be prepared to put pedagogical techniques above mere logic and the demands of the examination system. Having been influenced by the mathematical unity and order brought to us by Bourbaki, we must also take note of the communicatory implications of Piaget. In any case, a wave of children will soon be in the universities who have been subjected to the freer teaching approaches of the several new mathematics projects; and the universities must be ready for them, and adaptable to them.

Part II is concerned with Set Theory itself; that is, with set-theory as a mathematical discipline rather than with set-theory as providing the medium of mathematical discourse. The material of this part (for example, the study of permutations) may well strike the reader as unnecessarily pedantic, concerned largely to underline the obvious, but in this part we have tried to render explicit the underlying assumptions that so often remain implicit—even unnoticed—in elementary mathematical reasoning. On the other hand, the set-theory we present is 'naïve', and we reserve a more sophisticated attitude for Chapter 39 in Part VIII.

Part III is concerned with arithmetic. This part consists of a re-examination of the elementary but basic concepts of arithmetic from a more general

† Let us repeat, however, that the language will become increasingly colloquial as familiarity with the concepts increases.

standpoint. Thus we introduce algebraic structures (abelian groups, rings, etc.) in order to exhibit the nature of those properties of the integers that render valid the traditional treatment of such topics as factorization and prime numbers. We then get into a position from which we can greatly enlarge the scope of application of the theory; and in Chapter 13 we give certain applications which may well be less familiar to the student at this stage.

Part IV is concerned with the geometry of three-dimensional space. We place this part immediately after that on arithmetic because of our feeling that counting and measurement are the two basic aspects of mathematics and constitute between them the rationale for our number system. Our treatment is broadly algebraic and we make heavy use of vectors. However, we do discuss axiomatic questions and, in Chapter 17, we present the real and the complex projective plane and use them to illustrate Klein's view of the nature of geometry.

Part V is concerned with algebra. Of course, algebra cannot be regarded as distinct from arithmetic; the two overlap very substantially. Our pragmatical view has been to regard arithmetic as that part of algebra directly inspired by, and applicable to, the study of the integers; and correspondingly, we treat in this part those aspects of algebra which are of more general applicability (e.g., to geometry). Many of the ideas of this part were involved in our discussions in Part IV (and Part III); in this part these ideas are rendered quite precise and explicit and studied from the point of view of the elucidation of algebraic structure. In Chapter 21 we are concerned with the interplay of algebraic structure and an order relation in a given set; and this leads us to present Dedekind's definition of the real numbers, and a brief treatment of Boolean algebra which takes up aspects arising in Part II (Chapter 8).

Part VI is concerned with the systematic development of the number system and with the topological aspects of real n-dimensional space $\mathbb{R}^n$, the set of n-tuples of real numbers. Here, more conspicuously than anywhere else in the book, the spiral approach is in evidence. For, having presumed familiarity with the number system all the way through, we now proceed to construct it! Chapter 24 is devoted principally to the construction of the real numbers from the rationals, and it is perhaps the most difficult chapter in the entire book, because the rigorous treatment follows hard upon the motivational introduction there, and little time is allowed for digestion. Moreover, the construction (via Cauchy sequences) is presented from the algebraic point of view of ideal theory in commutative rings, so that we are assuming the student has absorbed the necessary algebraico-arithmetic background. There are those who argue that it is a pedagogic error to present the construction of the reals; their point of view is that one should introduce, axiomatically, a complete, Archimedean-ordered field, and that the completion process should be introduced later in a discussion of general metric spaces. We hope that our pedantic treatment of the construction will enable the student, when the time comes, to provide easily, for himself, the generalization to arbitrary metric spaces; and we believe that, for many students, this will be their only contact with the completion process. We do not take the view that students who are not continuing through to their degree as specialists in pure mathematics should remain all their lives in ignorance of how the real numbers are constructed from the rationals; mathematicians have a duty to such students, too! In Chapter 25 we embark on a genuine study of the topology of $\mathbb{R}^n$. We hope that, in this way, the reader will see that certain basic ideas (like continuity of functions and limits of sequences) are really topological in nature, and that he will thereby be led to understand these ideas better. He will also, we hope, acquire a meaningful introduction to a subject which plays a role in very many parts of mathematics today and is itself a very active and extensive research area. Part of this chapter first appeared in the journal *Mathematics Teaching* (and as the pamphlet [45]) and we are indebted to the editor for permission to use it here.

Part VII is concerned with the differential and integral calculus of one and several variables. A comprehensive treatment of this topic would require a substantial book in itself—as the plethora of textbooks devoted to the calculus attest. Thus in this part, more than in any other, we have not aimed at being self-contained, but have been content to refer to other texts for proofs of standard theorems. Even apart from the desire to keep the size of the present work within bounds, we have also had in mind that our principal innovation is in our notation and point of view, and that we are in essential agreement with other authors as to what are the important features of a course in the calculus at this level. However, we should add that we regard the question of point of view as absolutely crucial in the effective teaching of the calculus. Undoubtedly many students have foundered in their mathematics course on their first contact with the calculus; and many others, having learnt the calculus as a set of rules of procedure, have been severely confounded on being exposed to a course in mathematical analysis which has seemed to them to do nothing but call in question everything they have hitherto learnt in the calculus. It has been our aim to present the calculus as a part of the natural mathematical development based on the two foundations of algebra and the real numbers, but without forgetting its geometrical inspiration. It is misleading, however, to disguise the fact that the calculus is a difficult mathematical discipline, so that to present even the 'elementary' theory in its entirety, with full proofs, would be to put even the most talented student to great strain. Thus, in omitting certain proofs, we do not merely rest our case on their availability elsewhere; we genuinely do not believe that most students at this stage are ready to grapple with them. Here we invoke the spiral approach once more, viewing the student at this level as ready for the ideas, needing the techniques and able to apply them, but not yet mature enough mathematically for a rigorous demonstration of all the theorems. It is surely common ground that preciseness is possible without completeness of treatment; in eschewing the aim of completeness, we insist on accuracy and clarity of statement with honesty of presentation, so that the student is in no doubt as to what has been proved and what has not. We would add, parenthetically, that the most painful procedure in mathematical education, for student and teacher, is that of unlearning; so that honestly and openly to omit proofs is far better than to give basically dishonest arguments which must subsequently be repudiated. Of course, plausible arguments can often be important, provided their crudity is honestly described. At any level it is a major task in designing a mathematical curriculum to combine the spiral approach with a minimization of the amount of unlearning; and we will be content if we have even approximately fulfilled this exacting criterion.

Part VIII, like Part I, should have been exempted from the general statement made earlier about the role of each part in the book as a whole. It is really in the nature of an appendix, containing two chapters which, while related as to content, fulfil distinct roles in the book as a whole. Chapter 38, devoted to categories and functors, is, in a sense, a continuation of Part I, enriching the basic language of mathematics so that the interconnections between different branches of mathematics can be discussed as freely as the content of any particular branch. We have already indicated, earlier in this introduction, our reasons for including this chapter. The final chapter is, likewise, an extension of Part II, leading the reader to a deeper understanding of the preoccupations of mathematical logicians in examining the nature and properties of the systems which constitute the raw material of mathematics. Of course, we are only able to give a preview of this fascinating area of present-day research, sufficient, however, we hope, to convince the student that, whatever mathematics is, it is not the quest for absolute certainty.

6. The Exercises

In the belief that almost any mathematical book is rendered more intelligible —if not more readable—thereby, we have included many exercises in the text. These are in sets, numbered consecutively throughout each chapter: thus 21 Exercise 3(ii) refers to the second in the third set of exercises of Chapter 21. We do not intend to convey the impression that these exercises are included just so that the book might be regarded as suitable for some school or university course; its suitability for such use is for others to say. In any case, many of the exercises are highly open-ended, serving to show that mathematical activity is not a matter of cleverly answering half-hour questions posed by someone else: this widely-held impression is fostered by the traditional frozen examination system, and it is hard to eradicate from students who have been successful in that system. We have tried to show that mathematics abounds with questions where not only are the answers unknown, but we are not even sure what the questions are! Other exercises are standard tests to give practice in understanding the theory, and we hope we have included enough to rebut the charge that 'modern mathematics' is all theory and gives no practice in problem-solving. But there is a great need to add good problems to the common stock, and we urge the reader to cultivate the art of making up his own exercises: we ourselves have certainly acquired the greater part of our understanding of mathematics from our own stumbling attempts to ask and answer our own mathematical questions (at the levels of both research and class-teaching). A fundamental exercise is to set problems suitable for one's friends (who may include one's students of course), not with the object of demonstrating one's own cleverness, but with the hope of illuminating some aspect of mathematics.

To help the reader find out more about parts of mathematics which may interest him (perhaps through the exercises), we have included a fairly substantial bibliography, and references to it are made in the text by quoting author(s) and number, thus Hilbert [57]. We particularly recommend the reader to read about the way in which great mathematicians have worked, for example as explained in Bell [10], and about the way the subject has developed through its long history. Above all, he should remember that he is part of that history, and that the great mathematicians are related to him by the community of a common craft; he should not be too humble to say to himself 'Go, and do thou likewise'.

7. Acknowledgements

We wish to thank our publishers, Van Nostrand Reinhold Company, for their immense patience and constant encouragement during the long years of gestation of the manuscript, and for their unfailing courtesy throughout the period from submission of the manuscript to its publication. Professor J. F. P. Hudson read the whole text critically, from a mathematical point of view, and he did a very thorough job. We are grateful to him for pointing out errors, inconsistencies, and obscurities; any remaining ones are of course the responsibility of the authors.

It is also appropriate to acknowledge the valuable guidance of those who have taught us the mathematics we know. We hope to have conveyed to the reader something of the attitude of mind which they imparted to us.

Southampton H. B. G.
Ithaca, N.Y. P. J. H.
1969

SOME 'PACKAGES' OF DIRECTED READING

To guide readers interested in following particular topics (for example, through a course of lectures) we suggest the following packages and indicate the prerequisites for each. These assume always that Part I has been given a first, even if not detailed, reading. While no other material is really required, we assume for the purposes of illustration and exercises that the reader is familiar with the following topics from Algebra, Analytic Geometry and Calculus. Thus, in algebra he should be able to: manipulate algebraic formulae, factorize integers and simple polynomials, solve pairs of simultaneous equations, know about decimals, logarithms and indices and possibly a little of complex numbers. In analytic geometry he should know about graphs, gradients, circles and simple conics as well as the basic ideas of plane trigonometry. In calculus he should be able to differentiate and integrate simple expressions involving the trigonometric exponentials and logarithmic functions. The reader is also expected to know a little about deductive arguments such as the idea of proof by contradiction. If there are gaps in his knowledge of these topics they will in any case be filled during the course of reading the book as he can find out by looking at the index.

Possible 'packages' of topics, or 'courses' through the book are then:

Course 1. *Naïve Set Theory and Logic.* Chapters 1–8 and possibly Chapter 39. No prerequisites except for illustrative purposes.

Course 2. *Algebra.* Chapters 9–13, 18–21 with possibly Chapters 8, 23, or 38. Prerequisites: Knowledge of factorization of quadratics, the splitting of integers into factors; roots of equations, complex numbers.

Course 3. *Geometry.* Chapters 14–20.
Prerequisites: Familiarity with the idea of doing co-ordinate geometry, and with elementary trigonometry.

Course 4. *Geometry and Topology.* Chapters 14–17 and Chapter 25.
Prerequisites: As for Course 3, with intuitive ideas of continuity plus the notion of group.

Course 5. *Calculus.* Chapters 26–30; deeper, Chapters 31–37.
Prerequisites: Definitions of vector space and ring, with preferably a previous manipulative course in calculus.

Chapters 23 and 24 could be included in any of the Courses. Various topics could be omitted or added from other packages depending on the depth required, but each package covers approximately 120 pages (plus Part I) and thus corresponds to about one semester of an undergraduate course, i.e., approximately 30 lectures, but much will naturally depend on the amount of detail omitted or retained.

TEXTUAL CONVENTIONS

Notation such as '8.5.4' refers to item 4 of Section 5 of Chapter 8; but '8.5' refers to Section 5 of Chapter 8.

A line down the left-hand margin of a page indicates material not essential to a first reading.

Exercises are indicated by the method described on p. xvi, but the chapter number is omitted in the chapter in which an exercise occurs. Those of greater technical difficulty are indicated by the symbol §. Hints are indicated by curly brackets, and commentary by square ones.

Numbers in square brackets refer to the books listed in the Bibliography on pp. 617–622.

In addition to the Index (of words), there is an index of special symbols listed for the reader's convenience. In particular, this should enable him to locate quickly any set of axioms (such as those for a group) that he may need to recall.

CONTENTS

PART I
THE LANGUAGE OF MATHEMATICS

PART V
ALGEBRA

Part VI
NUMBER SYSTEMS AND TOPOLOGY

Part VII
CALCULUS

ADDITIONAL TOPICS IN THE CALCULUS

Part VIII
FOUNDATIONS

Part I

THE LANGUAGE OF MATHEMATICS

Contemporary mathematics is expressed in a language which has gradually evolved, and which a student must learn. The language was found to be necessary for conveying thoughts about subtle ideas which had previously been vague or non-existent. Remember that mathematics does not exist until it is communicated between people; a supple language is necessary for a teacher trying to clarify a difficulty for a pupil, and for research mathematicians arguing with each other, to make sure that a new result is properly proved without mistakes. The contemporary language manipulates such basic notions as *sets*, *functions*, and *relations*, and describes constructions and relationships involving these entities. Our concern is to get the reader fluent in the working form of this language, and to show how abbreviated—even slang—forms are introduced as substitutes for longer pedantic forms. When resolving mathematical difficulties, we must always be able to use the pedantic, precise form, where the slang might be ambiguous and lead to errors.

Once the basic entities are described by the language, our curiosity makes us ask natural questions about the entities and to manipulate them. The beginner tends to regard only the manipulations as mathematics, whereas the linguistic task of formulating the questions (and describing the entities) is just as much a part of mathematics. Thus, there is a convention for laying out the work; there are the *definitions*, followed by *theorems* (often labelled also by such words as 'proposition', 'lemma', 'corollary', depending on taste) and their *proofs*. A definition describes a new entity in terms of previously defined entities, and is *intended to be taken literally*. We then know *exactly* what we are talking about. A theorem is a statement giving an answer to questions raised about these entities—certainly not always a complete answer; while the proof of a theorem records the manipulations necessary for convincing us that the theorem is a correct statement.

In an expository text such as this, there is also a good deal of commentary about the reasons for making the definitions, the significance of a theorem, or the structure of a proof. Although we have tried to separate out such commentary, the reader must be on the watch to distinguish between the mathematics and the commentary: he will improve with experience.

The expository technique just described will persist throughout the book. In this first part, however, the subject-matter is confined to the elements of set-theory (as it is called), with commentary using illustrations from other, more familiar, parts of mathematics. Thus, Chapter 1 introduces *sets* with the associated ideas of subsets, and inclusion. In Chapter 2 we introduce *functions*, and discuss their composition and inversion, together with the question of counting. Sets and functions are united in Chapter 3, with the construction of the *Cartesian product* of sets. Subsets of a Cartesian product are called *relations*, and two kinds of these are of great importance: equivalence relations (discussed in Chapter 4) and ordering relations (discussed in Chapter 5). Our discussion of the latter includes the method of mathematical induction.

Once armed with the language of this part, the reader can then begin any of the remaining parts, since these are essentially elaborations, in the same spirit, about the special kinds of sets, functions, and relations appropriate to arithmetic, algebra, geometry, and calculus.

Chapter 1

DESCRIPTIVE THEORY OF SETS

1.1 Notion of Set

The notion of a 'set', or collection of objects, is basic, both in our daily lives and in mathematics. As we grow up, we become aware of collections of toys, groups of people, families of relatives, heaps of sand, classes of school-children, mobs of rioters, and whole lists of collective nouns. In mathematics, we use the word 'set' most often, with synonyms like 'class', 'family', 'aggregate', and 'collection'; but we reserve the word 'group' for a technical name in algebra. And whereas housewives speak of a set of china (which must, for example, include a milk jug), while philatelists speak of a set of stamps (which must consist of certain stamps and no others), we lay down no such condition in mathematics. To us, the housewife still has a *set* of china utensils even after she has smashed the sugar bowl, and any number of stamps at all will for us form a set. We have a notation for describing sets: for example,

1.1.1 $$D = \{x \mid x \text{ is a dog}\}$$
$$C = \{x \mid x \text{ is Cleopatra, Queen of Egypt (50 B.C.)}\}$$
$$\mathbb{I} = \{x \mid 0 \leqslant x \leqslant 1\}$$

are (abbreviated) descriptions of sets, and in general we may write

1.1.2 $$\{x \mid \phi(x)\}$$

(*read:* 'the set of all x such that $\phi(x)$ holds'), where $\phi(x)$ is some statement one can make about x. But what is x? To exclude irrelevant objects, like elephants in a discussion about dogs, we first specify a **universe of discourse**, which will be a set taken to be constant throughout any particular discussion. Denote this set by $\mathscr{U}$. Thus, if $\mathscr{U}$ is the set of all mammals, living or dead, then the first set D in 1.1.1 denotes the set of all dogs, living or dead; but had we laid down that $\mathscr{U}$ was to be the set of all breathing organisms at Garden Party X on July 4th, 1962, then D would consist of a (fairly small) set of dogs. Here, too, we might also simply list the set as, say,

$$D = \{\text{Rover, Bob}\}.$$

If we keep the same universe of discourse $\mathscr{U}$, the second set C contains no members at all, and is an example of the **empty** set. If we change $\mathscr{U}$ to include all Queens of Egypt in the year 50 B.C., then C has (to our knowledge) just one member. It is obvious on many counts that we must then distinguish between the set C and the woman Cleopatra; and in general between any object x and the set $\{x\}$ whose only member is x. We call such a set a **singleton.**

Again, the third set in 1.1.1 is not completely known until we have stipulated from which universe $\mathcal{U}$ our numbers x must come, although the notation at least suggests that we are only interested in numbers there.

In elementary mathematics, we commonly take $\mathcal{U}$ to be one of the following sets:

1.1.3
- $\mathbb{Z}$, the set of all **integers,** $0, \pm 1, \pm 2, \ldots$;
- $\mathbb{N}$, the set of all **natural numbers,** $1, 2, 3, \ldots$;
- $\mathbb{Z}^+$, the set of all **non-negative integers,** $0, 1, 2, \ldots$;
- $\mathbb{Q}$, the set of all **rational numbers** (i.e., fractions of the form p/q, where p, q lie in $\mathbb{Z}$, and $q \neq 0$;
- $\mathbb{R}$, the set of all **real numbers** (i.e., all possible decimals, terminating or not), which we sometimes think of as a straight line of geometrical points, particularly as the x-axis in co-ordinate geometry;
- $\mathbb{R}^2$, the set of all points (x, y) in the plane of co-ordinate geometry.

[The sets listed in 1.1.3 will all be described with greater precision later: see especially Part VI.]

Then, for example, 'the line $ax + by = c$' in geometry is the set

$$l = \{(x', y') | ax' + by' = c\}$$

where $\mathcal{U}$ is understood to be $\mathbb{R}^2$.

Here, $\phi(P)$ in 1.1.2 is the statement '$ax' + by' = c$' about the object $P = (x', y')$. Similarly for 'the circle $x^2 + y^2 = a^2$'. Indeed, all 'loci' in geometry are merely sets of the form $\{P | \psi(P)\}$ and we need no talk of 'points which move'. The common lack of explicit distinction between the set $\{P | \psi(P)\}$ and the sentence '$\psi(P)$' is often a source of difficulty to students of traditional textbooks. *Once pointed out*, however, it can sometimes be practical and convenient to use the shorter form while meaning the longer. This is an example of what Bourbaki calls 'abuse of language' in his treatise [17].

1.2 Inclusion

It is customary to denote sets by capital letters $A, B, X, \ldots$, and their 'members' or 'elements' or 'points' by lower-case letters $a, b, x, \ldots$. The facts that x lies in X, or y does not, are recorded as

$$x \in X, \qquad y \notin X$$

(*read*: 'x belongs to X; y does not belong to X') and a form of the Greek letter 'epsilon' is used because it is the first letter of an appropriate Greek word.

In particular, our universe $\mathscr{U}$ having been fixed, then for all x (in $\mathscr{U}$ understood), $x \in \mathscr{U}$. If A, B are sets (with all their members in $\mathscr{U}$), we write

1.2.1 $$A \subseteq B \quad \text{or} \quad B \supseteq A$$

to mean that if $x \in A$ then $x \in B$ (thus B **contains** A, or A **lies in** B); A is then a **subset** of B.

From the list 1.1.3, for example, we can write

1.2.2 $$\mathbb{N} \subseteq \mathbb{Z}^+, \qquad \mathbb{Z}^+ \subseteq \mathbb{Z}, \qquad \mathbb{Z} \subseteq \mathbb{Q}, \qquad \mathbb{Q} \subseteq \mathbb{R}.$$

Notice that the direction of the curved inclusion symbol $\subseteq$ corresponds to that of the inequality sign $\leqslant$ between numbers. This correspondence is no accident, because $\subseteq$ behaves like $\leqslant$ in other ways, as we shall show in a moment. First however a warning: numbers satisfy the **law of trichotomy**, i.e., given $x, y \in \mathbb{R}$, then either $x \leqslant y$, or $y \leqslant x$, whereas, for example, the sets A, B of odd integers and even integers satisfy neither $A \subseteq B$ nor $B \subseteq A$.

As to the similarities of behaviour between $\subseteq$ and $\leqslant$, we have, obviously, for any $A \subseteq \mathscr{U}$,

1.2.3 $$A \subseteq A,$$

and we therefore say that 'inclusion is reflexive' (cf. 4.2.1).

We next observe the fact that $\subseteq$, like $\leqslant$, is **transitive**: that is, for any subsets X, Y, Z of $\mathscr{U}$,

1.2.4 *If* $X \subseteq Y$ *and* $Y \subseteq Z$, *then* $X \subseteq Z$.

Proof. We have to show that, given $x \in X$, then $x \in Z$. But since $x \in X$, then $x \in Y$ (because $X \subseteq Y$), and then $x \in Z$ (because $Y \subseteq Z$) as required. ∎

Thus we are justified in writing 1.2.2 as

$$\mathbb{N} \subseteq \mathbb{Z}^+ \subseteq \mathbb{Z} \subseteq \mathbb{Q} \subseteq \mathbb{R}.$$

1.3 Venn Diagrams

A convenient aid in thinking about sets is the 'Venn diagram' (Fig. 1.1), in which areas in the plane $\mathbb{R}^2$ are used to represent sets; the rectangle in

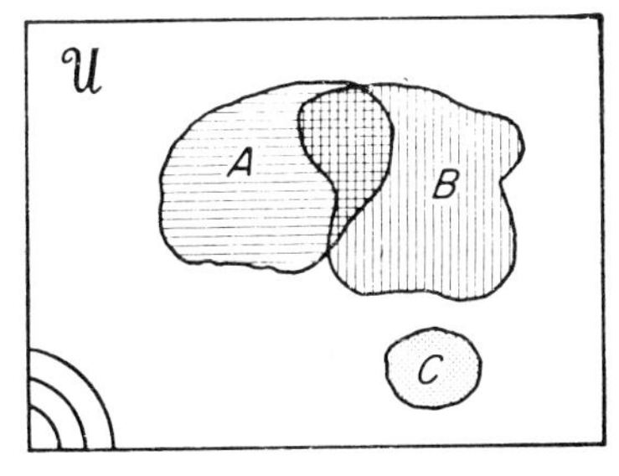

Fig. 1.1

Fig. 1.1 represents $\mathscr{U}$, and the areas enclosed by the curves represent sets A, B, C, etc. We shall often see how a Venn diagram can suggest a formal argument; for example, the three quadrants in the corner of the rectangle in Fig. 1.1 suggest at once the transitivity of inclusion (1.2.4), while the areas A, B, C show how the law of trichotomy may fail. However, care must be taken, since the diagram may represent too special a case.

1.4 Equality

We can go no further without saying when two sets X, Y (in $\mathscr{U}$) are equal, since they might have different definitions of the form 1.1.2, and yet contain the same elements. [Bertrand Russell gives in his *Introduction to Mathematical Philosophy* (see Newman [93], p. 539) the example of the set of all men, which can be described 'as featherless bipeds, or (more correctly) by the traits by which Swift delineates the Yahoos'.] We therefore define equality of sets by:

1.4.1 DEFINITION. $X = Y$ whenever $X \subseteq Y$ and $Y \subseteq X$.

(If only the first inclusion holds, but $X \neq Y$, then X is called a **proper** subset of Y.) Definition 1.4.1 corresponds to the provable *theorem* for numbers, that $x = y$ if and only if $x \leqslant y$ and $y \leqslant x$. It is a reasonable definition of equality, since it allows us to conclude that (i) $X = X$, {using the reflexivity of inclusion 1.2.3}, (ii) if $X = Y$, then $Y = X$ {by symmetry in 1.4.1}, and (iii) if $X = Y$ and $Y = Z$, then $X = Z$ {using the transitivity of inclusion 1.2.4}.

Many statements of mathematics can be expressed in the form that two sets, X, Y are equal; and the reader will observe numerous instances in this book. The proof that $X = Y$ then has two halves, as required by 1.4.1: one half must prove that $X \subseteq Y$, and the other half that $Y \subseteq X$, and the separate proofs may be quite different in spirit. Consider the following elementary example.

1.4.2 EXAMPLE. Let X denote the directrix† $x = -a$ of the parabola† $y^2 = 4ax$, and let Y denote the set of all intersections of perpendicular pairs of tangents to the parabola. Let us prove that $X = Y$. The first half, $Y \subseteq X$, is straightforward: two such tangents, at the points $(at^2, 2at)$, $(as^2, 2as)$, have equations $yt = x + at^2$, $ys = x + as^2$, and intersect at $x = ast = -a$ since $st = -1$ (so $s \neq t$). Thus $Y \subseteq X$. But to prove that $X \subseteq Y$, we have the more difficult task of showing that, if $P \in X$ then (i) exactly two tangents from P to the parabola *exist*, and (ii) they are perpendicular. Once seen, of course, the difficulty is soon resolved: if $P = (-a, h)$ then it lies on a tangent if and only if there exists a real number t such that $at^2 - ht - a = 0$; and since $a \neq 0$ and $h^2 + 4a^2 > 0$ there are two such numbers—proving (i)—with

† We are using the customary abbreviated forms, instead of the longer $\{(x, y) | x = -a\}$, $\{(x, y) | y^2 = 4ax\}$.

product $-a/a = -1$—which proves (ii). Thus $X \subseteq Y$; so the earlier inclusion enables us to conclude that $X = Y$, by 1.4.1. ■

In 'loci problems' of this type, it is not unusual to be offered solutions, in examinations and textbooks, which supply only one of the two required inclusions. The omission arises in the first place because it is not made clear that two sets are being proved equal. Sometimes, however, the problem reduces to that of eliminating t from a pair of equations $x = f(t)$, $y = g(t)$, to obtain a locus $h(x, y) = 0$, and there may be points on the latter which are *not* of the previous parametric form; thus only one inclusion is possible and the requirements of the question were wrongly stated.

Exercise 1

(i) Prove that, if the number k is non-zero, then the lines in $\mathbb{R}^2$, with equations $ax + by + c = 0$ and $k \cdot (ax + by + c) = 0$, are identical. Which set is described by the second equation, when $k = 0$?
(ii) Examine some theorem you know in analytic geometry, about loci, to see if its proof can be expressed using the kind of argument given in Example 1.4.2.

1.5 The Power Set

Many problems in mathematics concern the analysis of a certain set X; and a powerful approach is that of breaking up X into simpler subsets, analysing these, and reassembling them to get a picture of X. We therefore now consider ways of forming new sets from old. Let then $\mathscr{U}$ be a fixed universe of discourse. We form another universe called the **power set** $\wp\mathscr{U}$ of $\mathscr{U}$, which shall consist of *all* the subsets of $\mathscr{U}$. [The strange name comes from property 6.3.1 below.] This idea of a set, whose members are themselves sets, may at first surprise the reader; but he is already familiar with such sets in daily life, because of such examples as the United Nations, whose elements are sets (nations) of people, the Trades Union Congress (a set of brotherhoods of employees), consortia of construction firms, and so on. Returning to $\wp\mathscr{U}$, note that since $\mathscr{U}$ is a subset of itself, then $\mathscr{U} \in \wp\mathscr{U}$. Moreover, the empty set (mentioned above) is in $\wp\mathscr{U}$, a fact which seems at first sight strange. But let us denote the empty set by the Danish symbol $\varnothing$, defined for example by

$$\varnothing = \{x | x \in \mathscr{U} \text{ and } x \neq x\};$$

clearly no x can lie in $\varnothing$. We now have a small theorem.

1.5.1 THEOREM. *If A is any subset of $\mathscr{U}$, then $\varnothing \subseteq A$. Hence $\varnothing \in \wp\mathscr{U}$.*

Proof. We have to show, by definition of inclusion in 1.2.1, that given $x \in \varnothing$, then $x \in A$, i.e., there is no x in $\varnothing$ which fails the test of belonging to A. Since there *is* no x in $\varnothing$ at all, no x fails (or, for that matter, passes either), so 1.2.1 holds. ■

1.5.2 COROLLARY (of proof). *The empty set is unique.* For, the proof works for any A at all, in particular for an A_0 with no members, so $\varnothing \subseteq A_0$; but all we used about $\varnothing$ was that it had no members, so the proof shows equally well that $A_0 \subseteq \varnothing$. Hence $\varnothing = A_0$ by 1.4.1, and this justifies our speaking of 'the' empty set. ■

Exercise 2

Prove that, if $\mathscr{U}$ has n elements, then $\mathfrak{p}\mathscr{U}$ has 2^n elements. What is $\mathfrak{p}\varnothing$? [Use the theory of combinations here; another method is given in Chapter 6.]

Now consider the following mediaeval problem.

1.5.3 EXAMPLE. A ferryman must transport a wolf, a goat and a cabbage across a river, and can carry only one passenger besides himself. How must he proceed, knowing that the wolf cannot be left alone with the goat, nor the goat with the cabbage?

Solution. Let the universe $\mathscr{U}$ be the set {F, W, G, C}, the letters denoting the obvious objects. We then can represent the subsets of $\mathfrak{p}\mathscr{U}$ by points in the plane (Fig. 1.2) and think of these points as representing possible 'states' of

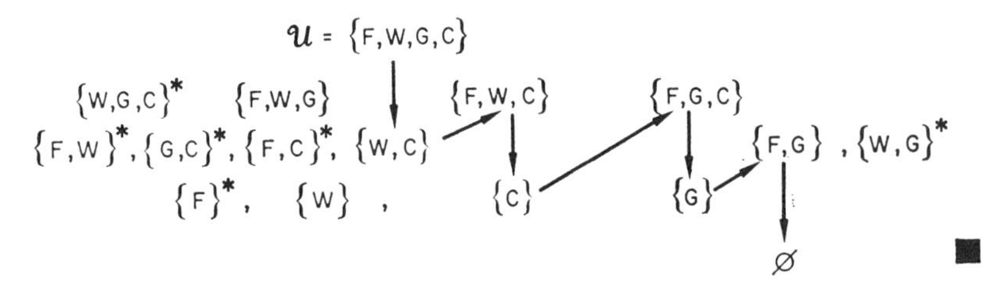

Fig. 1.2

the initial bank of the river. We eliminate those not allowed (such as {W, G}), and join by an arrow any state S to another S' if S' can be obtained from S by one allowed move by the ferryman. Not all arrows are shown in Fig. 1.2; stars denote eliminated states. We then look for a continuous route from the initial state $\mathscr{U}$, to the final state $\varnothing$, (shown by the heavy line), and this gives the ferryman his schedule. Another schedule exists, and we leave the reader to find it. ■

Exercise 3

Given a flask containing 8 litres of water, and two empty flasks of capacity 5 and 3 litres, derive a 'pouring schedule' for obtaining two flasks each with 4 litres.

1.6 Union and Intersection

Next, let A, B be subsets of $\mathscr{U}$. Then we form two new subsets of $\mathscr{U}$; their **union** and **intersection**, and denoted respectively by $A \cup B$, $A \cap B$.

[As a mnemonic, remember that the 'cup' symbol $\cup$ is like the 'u' in 'union'. The symbol $\cap$ is called a 'cap'. Older books use $+$ and $\cdot$ respectively, following George Boole, the originator (see Bell [10], Chapter 23).] These are defined by the rules

1.6.1 $$A \cup B = \{x | x \in A \quad \text{or} \quad x \in B\},$$

1.6.2 $$A \cap B = \{x | x \in A \quad \& \quad x \in B\},$$

where the word 'or' is taken in the weak sense, customary in mathematics, of 'and/or'. Thus, in the appropriate Venn diagram (Fig. 1.1), $A \cup B$ is all the larger shaded area, and $A \cap B$ the smaller. Also, $A \cap C = B \cap C = \varnothing$ and we say that A and C are **disjoint** or **non-overlapping**. Notice that we always have $A \cup A = A = A \cap A$, and that if $A \subseteq B \subseteq D$ and $A \subseteq C \subseteq D$, then

1.6.3 $$A \subseteq B \cap C \subseteq B \cup C \subseteq D.$$

1.6.4 EXAMPLE. Let l, l' denote the lines $ax + by + c = 0$, $a'x + b'y + c' = 0$ in the plane $\mathbb{R}^2$. Then

(i) *$l \cup l'$ is the locus $(ax + by + c)(a'x + b'y + c') = 0$.* To see this, let the locus be L; we then have to prove that $l \cup l' = L$; so, applying 1.2.4, we prove $l \cup l' \subseteq L$ first. If $(p, q) \in l \cup l'$ then $(p, q) \in l$ or $(p, q) \in l'$, so either $ap + bq + c = 0$ or $a'p + b'q + c' = 0$, and in any case $(ap + bq + c) \cdot (a'p + b'q + c') = 0$, so $(p, q) \in L$. Thus $l \cup l' \subseteq L$. Conversely, given $(p, q) \in L$, then $(ap + bq + c)(a'p + b'q + c') = 0$, so at least one of the factors is zero, say $ap + bq + c = 0$. Hence $(p, q) \in l \subseteq l \cup l'$, so $L \subseteq l \cup l'$. By 1.4.1 then, $l \cup l' = L$, proving (i).

Secondly, we prove

(ii) *If $ab' - a'b \neq 0$, then $l \cap l'$ is a singleton and conversely; while*
(iii) *if $ab' = a'b$, then either $b'c = c'b$ and $l = l' = l \cap l'$,* or *$b'c \neq c'b$ and $l \cap l' = \varnothing$.* [This says that, geometrically, either l and l' intersect in a single point, or $l = l'$, or l is parallel to l'.]

Proof. There are three mutually exclusive possibilities:

(1) $ab' - a'b \neq 0$;
(2) $ab' = a'b$ and $b'c = bc'$;
(3) $ab' = a'b$ and $b'c \neq bc'$.

If (1) holds, we can solve the simultaneous equations $ax + by + c = 0 = a'c + b'y + c'$ for (x, y), to give a solution $(p, q) \in l \cap l'$. Further, if $(r, s) \in l \cap l'$ then (r, s) is also a solution of the simultaneous equations, so $(r, s) = (p, q)$ since $ab' - a'b \neq 0$. Hence $l \cap l'$ is the singleton $\{(p, q)\}$. If (2) holds, either $a \neq 0$, $a' \neq 0$ or $b \neq 0$, $b' \neq 0$ [Why?]. Assume $b \neq 0$, $b' \neq 0$. Then an equation of l is $0 = b' \cdot (ax + by + c) = ba'x + bb'y + bc' = b \cdot (a'x + b'y + c')$, which is an equation also of l', so $l \subseteq l'$. Similarly $l' \subseteq l$, so $l = l'$. With (3) we must have $l \cap l' = \varnothing$; otherwise there exists $(u, v) \in l \cap l'$, so $au + bv + c = 0 = a'u + b'v + c'$, whence we can multiply the equations by b', b respectively to obtain $b'c = bc'$ (since $ab' = a'b$), contradicting (3). We have proved, then, that the cases (1), (2), (3) imply, respectively, that $l \cap l'$ is a singleton, that $l = l'$, and that $l \cap l' = \varnothing$. Hence, if conversely $l \cap l'$ is known to be a singleton, then (2) and (3) must be false since a singleton is neither a whole line [Why?] nor the empty set. Hence, since the possibilities are mutually exclusive, (1) holds. Thus (1) holds *iff*† $l \cap l'$ is a singleton; similarly, (2) and (3) hold iff $l \cap l' = l = l'$ or $l \cap l' = \varnothing$ respectively. This proves (ii) and (iii). ■

Exercise 4

Let P, Q be any loci (e.g., conics) in the plane, of the forms

$$P = \{(x, y) | p(x, y) = 0\}, \qquad Q = \{(x, y) | q(x, y) = 0\},$$

where p, q are, say, polynomials in x and y. Prove that

$$P \cup Q = \{(x, y) | p(x, y) \cdot q(x, y) = 0\}$$
$$P \cap Q \subseteq \{(x, y) | \lambda \cdot p(x, y) + \mu \cdot q(x, y) = 0\}$$

for any numbers $\lambda, \mu \in \mathbb{R}$.

1.6.5 EXAMPLE (Lewis Carroll). The following three premisses are given, and we are to deduce a conclusion.

(1) My saucepans are the only things I have that are made of tin;
(2) I find all your presents very useful;
(3) None of my saucepans is of the slightest use.

Solution. Let $\mathscr{U}$ be the set of all my things, A those made of tin, B my saucepans, C my useful things, and D those which were presents from you. Then (1), (2), and (3) translate as: $A = B$; $D \subseteq C$, $B \cap C = \varnothing$. But $B \cap D \subseteq B \cap C$ (since $D \subseteq C$), so $B \cap D \subseteq \varnothing$, whence $B \cap D = \varnothing$. Therefore $A \cap D = \varnothing$, i.e., none of your presents to me is made of tin. The corresponding Venn diagram is shown in Fig. 1.3. (See also Exercise 5 below.)

1.6.6 EXAMPLE. (This gives a foretaste of Part III and the ideas here will be treated in detail there.) Let n, m denote any natural numbers (i.e., $n, m \in \mathbb{N}$). It is of particular interest when the fraction n/m is not merely a rational number, but an *integer*; so $n = mk$, say, where $k \in \mathbb{N}$. In this case, we say that m **divides** n, m is a **factor** of n, or n is **divisible** by m. Then divisibility is *transitive*; i.e., if m divides n (so $n = mk$) and n divides r (so $r = nt$, $t \in \mathbb{N}$) then m divides r (since $r = (mk)t = m(kt)$ and $kt \in \mathbb{N}$). If $n \in \mathbb{N}$ and $n > 1$, then n is called **prime**, provided its only factors are itself and 1. The first

† As explained in the Introduction, iff means 'if, and only if'.

few primes are 2, 3, 5, 7, ... A **common factor** of m, n is any member of $\mathbb{N}$ which divides each of m and n; a **common multiple** of m, n is any member of $\mathbb{N}$ which is divisible by each of m and n. The **highest common factor,** denoted in this section by (m, n), is that common factor which is divisible by every other common factor of m and n; the **least common multiple**, denoted by $[m, n]$ is that common multiple which divides every other common multiple of m and n. Primes have the property (see Section 12.2) that if a prime p divides a product mn, then p divides m or n (or both).

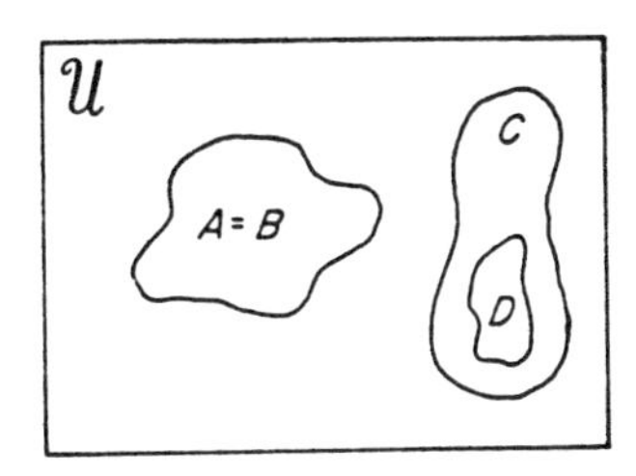

Fig. 1.3

Now let $A(n)$ denote the set of all primes which divide n. Thus

$$A(1) = \varnothing, \qquad A(2) = \{2\}, \qquad A(12) = \{2, 3\} = A(6),$$

and if p is a prime, $A(p)$ is the singleton $\{p\}$. Since divisibility is transitive, then

(i) *if q divides n, then $A(q) \subseteq A(n)$.*

Hence, if q, r are any common factor or multiple, respectively, of m, n, then

(ii) $A(q) \subseteq A(m) \cap A(n)$; $A(m) \cup A(n) \subseteq A(r)$.

But if the prime p lies in $A(m) \cap A(n)$, then p is a common factor of m, n, so p divides the highest common factor (m, n); thus

$$A(m) \cap A(n) \subseteq A((m, n)).$$

Using the first inclusion in (ii), with $q = (m, n)$, and applying the definition 1.4.1 of equality of sets, we obtain

(iii) $A((m, n)) = A(m) \cap A(n)$.

Similarly, if in (ii) we take $r = [m, n]$, then any p in $A(r)$ must lie in either $A(m)$ or $A(n)$; otherwise the integer r/p would be a common multiple of m, n by the property of primes noted above; so r/p is divisible by the least common multiple, r, and this is impossible since $p > 1$. Hence

$$A([m, n]) \subseteq A(m) \cup A(n),$$

so by the second inclusion in (ii) and by 1.2.4

(iv) $A([m, n]) = A(m) \cup A(n)$.

Next, define $B(n)$ to be the set of all prime powers which divide n; thus since $p^0 = 1$ for any prime p,

$$B(1) = \{1\}, \qquad B(2) = \{1, 2\}, \qquad B(12) = \{1, 2, 2^2, 3\}.$$

It follows from the 'fundamental theorem of arithmetic' 12.3.2 that given $B(n)$ we can reconstruct n by simply multiplying together all the highest powers in $B(n)$. Thus if $B(n) = B(m)$, then $n = m$, or (put another way) distinct integers n, m correspond to distinct sets $B(n)$, $B(m)$. Arguments similar to those given above establish analogues of the statements (i)–(iv), and the reader is invited to write them out in detail. In particular, the equations

$$B((m, n)) = B(m) \cap B(n), \qquad B([m, n]) = B(m) \cup B(n),$$

taken with the above remark about reconstructing n from $B(n)$, suggest one process for constructing (n, m) and $[n, m]$, given n and m.

The reader will notice a kind of duality between statements about multiples and factors, corresponding to dual statements about cups and caps. These matters will be discussed further, later on, in Chapters 8 and 21.

1.7 The Complement

Given $A, B \subseteq \mathscr{U}$, we define a new set, $A - B$, by

1.7.1 $$A - B = \{x | x \in A \text{ and } x \notin B\}.$$

Clearly, $A - B \subseteq A$, and $(A - B) \cap B = \varnothing$. If, in particular, $A = \mathscr{U}$ then $\mathscr{U} - B$ is called the **complement** of B in $\mathscr{U}$, and denoted by $\complement B$; since it is relative to $\mathscr{U}$, we ought to embody $\mathscr{U}$ in the notation, but this is unnecessary if $\mathscr{U}$ is constant throughout a discussion. There is a general theory of the complement, as of the cup and cap symbols, and this will be treated formally in Chapter 8. Meanwhile, we state certain of the rules, postponing their proofs until Chapter 8; and the reader should acquire an informal familiarity with them, by using Venn diagrams, and particularly by using them to formulate mathematical results in a succinct way (as in the conclusions of Examples 1.6.5 and 1.6.7). It is only in certain specialized applications that one needs complicated algebraic manipulations with sets; but the theory has an intrinsic interest, once this basic familiarity is acquired.

The $\cup$ and $\cap$ operations are related by the complement, as expressed in the **rules of Pierce and De Morgan:**

1.7.2 $$\complement(A \cup B) = \complement A \cap \complement B, \qquad \complement(A \cap B) = \complement A \cup \complement B$$

for all $A, B \in \mathfrak{p}\mathscr{U}$. We may express these laws by saying that 'complementation interchanges cups and caps'. Now, if $x \notin \complement A$, then $x \in A$, whence

1.7.3 $$\complement(\complement A) = A.$$

Consider now the additional rules, all of which are intuitively obvious from Venn diagrams. Thus, for any sets X, Y, $Z \in \mathfrak{p}\mathscr{U}$ we have the **associative** laws:

1.7.4 $X \cup (Y \cup Z) = (X \cup Y) \cup Z; \qquad X \cap (Y \cap Z) = (X \cap Y) \cap Z;$

so we can omit brackets and write $X \cup Y \cup Z$, $X \cap Y \cap Z$ without ambiguity. Similarly for more sets $X_1, X_2, \ldots, X_n$. We have, too, a **distributive** law:

1.7.5 $$X \cap (Y \cup Z) = (X \cap Y) \cup (X \cap Z),$$

corresponding to $a \cdot (b + c) = a \cdot b + a \cdot c$ in ordinary Arithmetic; and the **commutative** law:

1.7.6 $$X \cup Y = Y \cup X, \qquad X \cap Y = Y \cap X,$$

together with the **idempotent** laws:

1.7.7 $$X \cup X = X = X \cap X$$

(which of course have no analogues in ordinary arithmetic). Observe that the laws 1.7.4, 1.7.6, and 1.7.7 go in pairs, in that each equation in a pair gives the other if we interchange $\cup$ and $\cap$. This should raise the suspicion that 1.7.5 should have the mate

1.7.8 $$X \cup (Y \cap Z) = (X \cup Y) \cap (X \cup Z),$$

unlike arithmetic again; and this we can *deduce* easily as follows, where for brevity we write X' for $\complement X$. By 1.7.3, $X = (X')'$, so

$$\begin{aligned} X \cup (Y \cap Z) &= (X')' \cup ((Y')' \cap (Z')') \\ &= (X')' \cup (Y' \cup Z')' && \text{by 1.7.2,} \\ &= (X' \cap (Y' \cup Z'))' && \text{by 1.7.2,} \\ &= ((X' \cap Y') \cup (X' \cap Z'))' && \text{by 1.7.5,} \\ &= (X' \cap Y')' \cap (X' \cap Z')' && \text{by 1.7.2,} \\ &= (X'' \cup Y'') \cap (X'' \cup Z'') && \text{by 1.7.2,} \\ &= (X \cup Y) \cap (X \cup Z) && \text{by 1.7.3.} \end{aligned}$$ ■

Exercise 5

(i) Use the method of the last proof, to prove *one* of the equations in each of the pairs 1.7.4, 1.7.6, 1.7.7, given the other equation.
(ii) Prove from the rules that

$$\complement[(X_1 \cup X_2 \cup X_3 \cup X_2 \cup X_1) \cap (X_1 \cap X_2 \cap X_3 \cap X_2 \cap X_1)] = \complement X_1 \cup \complement X_2 \cup \complement X_3.$$

(iii) Prove that $\complement \varnothing = \mathscr{U}$, $\complement \mathscr{U} = \varnothing$.
(iv) Prove that $A - B = A \cap \complement B$. When is $(A - B) \cup B$ equal to B? Can we have $B = \complement B$?
(v) If $\mathscr{U} = \mathbb{Z}$, what is $\complement X$ when X is the set of all (a) even integers, (b) odd integers, (c) positive integers?

Use the method of Example 1.6.5 to establish the indicated conclusions in the following problems.
(vi) *Premisses.* Nobody takes *The Times* unless he is well-educated. No hedgehog can read. Those who cannot read are not well-educated.
Conclusion. No hedgehog takes *The Times.*
(vii) *Premisses.* All babies are illogical. Nobody is despised who can manage a crocodile. All illogical persons are despised.
Conclusion. No babies can manage a crocodile.
(viii) *Premisses.* No boys under 12 are admitted to this school as boarders. All industrious boys have red hair. No day-boy learns Greek. None but those under 12 are idle.
Conclusion. None but red-haired boys learn Greek in this school.
(ix) Do the problem of the April Fool's Dance (Exercise 3, p. 73 of Whitesitt [135]).

See also the collection of problems (from which some of these are taken) invented by Lewis Carroll, and included in the book *Symbolic Logic* [31] by Rev. C. L. Dodgson (Carroll's real name, used by him when writing on mathematics).

1.8 Quantifiers

There is a tradition in classical elementary algebra of distinguishing between an *identity* (for example, $(x - 1)(x + 1) = x^2 - 1$) and an *equation* (for example, $2x - 1 = 3$). This distinction is overcome in many modern treatments by talking of *open sentences.* Thus either of the examples quoted is an open sentence, and the game is to test, for each replacement of x by an element from $\mathscr{U}$, the universe of discourse, whether the resulting assertion is true or false. Indeed, the point that an open sentence only becomes a proposition when x is so replaced is often emphasized by using some formal **place-holder** (like $\square$) instead of x, so that open sentences take the form

$$(\square - 1)(\square + 1) = \square^2 - 1$$

$$2\square - 1 = 3$$

$$\square + \triangle = \triangle + \square$$

and so on. It is at least plain, in the terminology of open sentences, that an identity is simply one which is true when any legitimate substitution is made for the place-holder. However, as we proceed in mathematics we find ourselves frequently changing our domain of discourse; for example, our notion of number begins with the natural numbers 1, 2, 3, . . ., but soon we wish to speak of rational numbers, real numbers, complex numbers, and so on. Thus we need a more flexible formalism and this we now introduce.

Let $\psi(P)$ be a proposition or statement about the element P (of the set S) which is meaningful whatever element P is taken; see 1.1 for examples. Then the symbol

$$\underset{P \in S}{\forall} P, \psi(P),$$

or more briefly, if the set S may be understood,

$$\forall P, \psi(P)$$

is to be read 'for all P, $\psi(P)$ holds'. It asserts the truth of $\psi(P)$ whatever element of S is represented by P; every P has property ψ. Thus we have an 'identity', in the old-fashioned terminology, provided S is regarded as the universe of discourse. The symbol $\forall$ is called a **quantifier.**

Now, associated with any statement ψ there is the **negation,** $\neg\psi$, of that statement (*read*: 'not ψ'). Formally, the negation $\neg\psi$ is true if and only if ψ itself is false. There is, of course, an extremely close connection between the negation of statements and the complement of sets, expressed by

1.8.1 $$\complement\{P \mid \psi(P)\} = \{P \mid \neg\, \psi(P)\}.$$

Let us examine the negation of the statement $\forall P, \psi(P)$. In commonsense terms we are denying that $\psi(P)$ is true for every element P in S, and thus we are asserting the existence of at least one element P in S such that $\psi(P)$ is false, that is, such that $\neg\psi(P)$ is true. For example, to demonstrate the falsity of the statement 'Every elephant has five legs' it suffices to produce an elephant with four legs (or even with six!). In the general situation, then, we arrive at the formula

1.8.2 $$\neg(\forall P, \psi(P)) = \exists P, \neg\psi(P)$$

where the symbol $\exists$, which is also called a quantifier, is to be read as 'there exists' or 'for some'. Indeed, replacing $\neg\psi$ by ϕ (so that† $\psi = \neg\phi$) we may rewrite 1.8.2 as

1.8.3 $$\exists P, \phi(P) = \neg(\forall P, \neg\phi(P))$$

and regard the right-hand side of 1.8.3 as a *definition* of the left. Thus a standard method of proving that $\phi(P)$ is true for some P is to assume that $\neg\phi$ is true for all P and arrive at a contradiction. The interesting and, indeed, extraordinary feature of this method is that it provides no clue as to *which* elements P do, in fact, render $\phi(P)$ true. A proof of this kind is called an *existence proof,* in contrast with one which allows one to find, or, at least, to approximate to an element P rendering $\phi(P)$ true. [There is a school of mathematical logic, called Intuitionism and founded by the great Dutch mathematician L. E. J. Brouwer, who died in 1966, which questions the validity of pure existence proofs.]

Of course, 1.8.3 may be further rewritten as

1.8.4 $$\neg(\exists P, \phi(P)) = \forall P, \neg\phi(P),$$

or

1.8.5 $$\neg(\exists P, \neg\psi(P)) = \forall P, \psi(P).$$

† The equals sign should be taken, with propositions, to mean 'has the same meaning as'. A fuller discussion is given in 8.1.

1.8.6 The form 1.8.5 is of some importance since it symbolizes the principle of the **counter-example.** That is to say, if you assert that $\psi(P)$ is true for all P, then I refute you by demonstrating the existence of an element P such that $\psi(P)$ is false; and P is called a counter-example to your assertion since your position is then untenable. On the other hand, I do not refute you—even if I undermine your confidence—by demonstrating the existence of a flaw in your reasoning. Such a demonstration is neither necessary nor sufficient for invalidating your assertion. By contrast, it is remarkable how often one hears a discussion like this: A asserts that all Italian women are passionate, B says he knows an Italian woman who is emotionally cold, and A retorts that 'that is only one example'. *One* example is often enough!

1.8.7 The quantifiers $\forall$, $\exists$ may, of course, exist in combination. Thus

$$\forall P, \exists Q, \qquad P + Q = 10$$

is an assertion which is true if the domain of discourse is $\mathbb{Z}$, false if the domain of discourse is $\mathbb{N}$ (P, Q could, in principle, belong to different domains of discourse). Its negation is worked out thus:

$$\begin{aligned} &\neg(\forall P, \exists Q, P + Q = 10) \\ &\quad = (\exists P, \neg(\exists Q, P + Q = 10) \\ &\quad = (\exists P, \forall Q, \neg(P + Q = 10) \\ &\quad = \exists P, \forall Q, P + Q \neq 10. \end{aligned}$$

Exercise 6

(i) Express symbolically, using quantifiers, the following statements:

(a) 12 is not the largest integer.
(b) There is no largest integer.
(c) Between any two rational numbers lies a third.
(d) Every continuous transformation of a circular region into itself leaves a point of the region unmoved.

(ii) Negate the assertions (as in 1.8.7):

(a) $\forall P, P \in A \cup B$ (without using the symbol $\notin$)
(b) $\forall P, \exists Q, \psi(P, Q)$.

(iii) Show how, in the discussion of Italian women, above, A might gradually modify (or 'refine') his statement until B can give no counter-example. Need the refined version be necessarily true?
(iv) Discuss the philosophical belief that 'the great propositions are those whose negations are also true'.
(v) Show that, if S consists of just two elements A, B, then $\forall_{P \in S} P, \psi(P)$ means '$\psi(A)$ and $\psi(B)$' while $\exists_{P \in S} P, \psi(P)$ means '$\psi(A)$ or $\psi(B)$'. Rewrite 1.8.4, 1.8.5 in this case, without quantifiers. Use mathematical induction (Chapter 5) to extend these

results to the case when S is a finite set of n elements. [N.B.: remember that 'or' is not mutually exclusive.]

1.8.8 It is worth making the grammatical point that before negating a sentence with quantifiers, these must be brought as near as possible to the beginning. Thus, in ordinary mathematical speech we often say '$\psi(x)$ is true for all $x \in A$', but we should *write* '$(\forall x \in A)\psi(x)$'. As an example, consider the negation of the statement R: 'The set A is contained in the set B'. Then $\neg R$ may be written $\neg\{(\forall x \in A)x \in A \;.\!\Rightarrow\!.\; x \in B\}$—where '$Q \;.\!\Rightarrow\!.\; R$' means 'If Q, then R'—and now this is, by 1.8.2,

$$(\exists x \in A)x \in A \;.\&.\; \neg (x \in B),$$

i.e., some x in A is not in B. Observe also that our interpretation of $\neg(Q \;.\!\Rightarrow\!.\; R)$ is '$Q \;.\&.\; \neg R$', i.e., 'Q holds and (yet) R is false'; cf. 8.5.

Chapter 2

FUNCTIONS: DESCRIPTIVE THEORY

2.1 The Notion of Function

During the course of mathematical history, the notion of a 'function' has gradually broadened from its early appearance; at first functions were simple algebraic or trigonometrical 'expressions' or formulae, which expressed a relationship between the 'dependent variable' and the 'independent variable', these variables being numbers. A much more general and precise notion is now necessary for doing mathematics, and we first give a 'working' definition, with a preciser discussion in 3.4. Thus, a function involves three things, a set A called the **domain**, a set B called the **range**, and a rule, denoted by $f: A \rightarrow B$, according to which we are told how to assign, to each $a \in A$, a unique $b \in B$. When A and B are fixed throughout a discussion, we often abbreviate '$f: A \rightarrow B$' to 'f'. We write $b = f(a)$ and call b the **value** of f at the **argument** a. The equation $b = f(a)$ is *not* read 'b equals function a', as advocated in many traditional books, but as 'b equals f of a' (to emphasise that the function in question is f, and not g, or sine or log). Traditional books speak, e.g., of 'the function $y = x^2$'; but in some parts of mathematics, this usage leads to confusion. In certain contexts, however, it is safe to use 'the function x^2' as an abbreviation for 'the function whose value at x is x^2' (and the domain should then be given). Our notion of function is what traditional writers would call 'single-valued'; *we never* deal with 'multi-valued functions'. The following examples will illustrate the generality of this notion of function.

2.1.1 EXAMPLE. Let A, B be the sets of humans and of males, respectively. Then $f: A \rightarrow B$, given for each $a \in A$ by $f(a) =$ father of a, is a function, since we have a perfectly definite way of describing a's father (using biology) even if we do not know his name. Thus 'father' is a function, while on the other hand 'child' is not. That is, the rough rule given by $g(b) =$ child of b, is *not* a function, because b might not have any children, or b may have more than one child. These defects in g are described technically as 'g is not defined at b', and 'g is not single-valued', respectively. The first defect can be remedied by replacing B with the subset B_0 of all males having children, and the second by one of two devices. Either we can pick a particular child e.g., take $g_1(b) =$ eldest child of b, or we can replace A by $\mathfrak{p}A$, and take $g_2(b) =$ set of all children of b. These remarks should suffice to counter the common saying that 'y is a function of x, so x is a function of y'; the saying, of course, is true under certain conditions (see Theorem 2.9.1).

2.1.2 EXAMPLE. Let A be the set of points in space occupied by an opaque body, and let B be the set of points on a wall, occupied by the shadow of the body. Let $f: A \to B$ be given by $f(a) =$ shadow of a. This example is the source of certain terminology, as we shall see below.

2.1.3 EXAMPLE. Let A, B be any non-empty sets, and let b_0 denote a fixed element of B. Then we have a function $g: A \to B$, defined by $g(a) = b_0$ for each $a \in A$. Such a function is called a **constant** function (for example, a function like '$y = 2$' in calculus), and we often indicate the dependence of g on b_0 by calling g 'the constant function $b_0: A \to B$.'

2.1.4 EXAMPLE. In elementary calculus, particularly, it is traditional not to bother about the domain of a function, but this custom is strongly to be deprecated. Questions like 'differentiate $\log \cos x$' are all very well, but the bald answer '$-\tan x$' is unsatisfactory, since, for example, (avoiding complex numbers) $-\tan \pi$ makes sense, whereas $\log \cos \pi$ does not; we must know where the graph *is* before we can work out its gradient! We therefore urge the reader to be practical and acquire the habit of asking for the domain as an essential preliminary to sketching roughly the graph of the function. Thus, still avoiding complex numbers, we shall mean by 'the domain of $\log \cos x$' the largest set of $x \in \mathbb{R}$ for which $\log \cos x$ is defined (i.e., makes sense). Here, the domain is the set $\{x | \cos x > 0\}$ and this is the family of intervals:

$$I_n = \{x | (4n - 1)\pi/2 < x < (4n + 1)\pi/2\}, \qquad (n \in \mathbb{Z}).$$

[If $x = (2n + 1)\pi/2$, then $\cos x = 0$ and a schoolboy might say that $\log \cos x$ is $-\infty$, which still 'makes sense'. But $-\infty \notin \mathbb{R}$, and we regard $\log t$ as undefined if $t \leqslant 0$; in fact the domain of log is $\mathbb{R}_+ = \{t | t > 0\}$ as explained in Chapter 31.] If complex numbers are allowed, the necessity for knowing the domain is just as great to avoid 'multi-valued functions'.

Exercise 1

(i) Using your knowledge of elementary calculus, calculate the domains of the following functions: 2, $\sin x$, $\cos x$, $\tan x$, $\tan \sin x$, $\exp(\sqrt{(1 - \tan^2 x)})$, $\cos \log x$.
(ii) Use quantifiers to negate the statement '$f: A \to B$ is constant'.

2.1.5 We would also say, at this point, that several synonyms for the word 'function' are used, depending on the branch of mathematics. The commonest synonyms are 'transformation', 'correspondence', 'operator', and 'mapping'. The last occurs particularly in geometry, because to draw on paper a map of a country A, is to assign some point of the paper to each point of A; of course certain extra conditions have to be observed if the map is to be of interest. Indeed, the different branches of mathematics arise precisely because of the kinds of extra condition imposed on the functions which occur.

2.2 Equality of Functions

Just as we had to *define* equality of sets in 1.4, we shall now *define* equality of functions. Taking first a familiar example, consider the exponential function exp: $\mathbb{R} \to \mathbb{R}_+$ whose domain is the set of all real numbers, and whose range is the set $\mathbb{R}_+ = \{x | x \in \mathbb{R} \text{ and } x > 0\}$: exp (x) is also denoted by e^x. Now it is shown in books on calculus that exp (x) can be defined in different ways; for example, two alternatives are

$$p(x) = \sum_{n=0}^{\infty} x^n/n!, \qquad q(x) = \lim_{n\to\infty} (1 + x/n)^n.$$

Here the *formulae* representing $p(x)$ and $q(x)$ are quite different, and so are the numerical routines for calculating them with a given x. Nevertheless, when all computing is done, the rule for p assigns to each $x \in \mathbb{R}$ the same number as that for q. We therefore say that p and q denote the *same* function (namely exp). Similar remarks apply, for example, to the function which assigns to each customer in a store the amount payable by him; on leaving, the customer and clerk may have different ways of computing a value which is what the customer finally agrees to pay. More generally, we *define* equality of functions as follows.

2.2.1 DEFINITION. Two functions $f: A \to B$, $g: C \to D$ are said to be **equal**, whenever $A = C$, and $B = D$, while for each $a \in A$, $f(a) = g(a)$.

Here, the reader may object, and say that, for example, he regards as equal the functions $f: \mathbb{N} \to \mathbb{N}$, $g: \mathbb{N} \to \mathbb{Q}$ given by $f(n) = g(n) = n$. However, g can be 'halved', in the sense that we can form $\frac{1}{2}g: \mathbb{N} \to \mathbb{Q}$ where $(\frac{1}{2}g)(n) = \frac{1}{2}n$. This construction is not possible with f without altering its range. We shall state some conventions and allow abuses of language later (see 2.6.1). However, Definition 2.2.1 is reasonable because it allows us—together with 1.2.4 for sets—to conclude that in 2.2.1, $f = f$; $f = g$ iff $g = f$; and $f = g$, together with $g = h: E \to F$ imply $f = h$.

2.3 The Image

When a function $f: A \to B$ is constructed, further study often shows that the values $f(a)$ are in some proper subset of B. In calculus, for example, this subset can tell us the absolute maxima and minima of f. In view of its importance, we give this subset the special name 'image of f'. More precisely, we make the following definition.

2.3.1 DEFINITION. The **image** of $f: A \to B$ is the subset of B consisting of all $b \in B$ for which $b = f(a)$ for some $a \in A$. It is denoted by Im (f) or $f(A)$. Clearly, equal functions must have the same image, but not conversely.

Exercise 2

(i) Having calculated the domain D of each function f in Exercise 1(i), calculate $f(D)$, given that sine and cosine have images† $\langle -1, 1\rangle$. Find the domain of $\log\cos x + \cos\log x$, and show that its image I lies in $\langle -\infty, 1\rangle$. [It seems to be a difficult problem to decide whether or not $1 \in I$.]
(ii) Use quantifiers to negate the statements '$f\colon A \to B$ and $g\colon C \to D$ are equal' and 'Im (f) = Im (g) but $A = C$ and $B = D$'.

2.4 Injections, Surjections, and Equivalences

We next pick out three important classes of functions as follows, giving all three definitions before the examples.

2.4.1 DEFINITION. We say that $f\colon A \to B$ is **onto**, or **surjective**, or‡ a **surjection** whenever $f(A) = B$.
A second class of functions, as important as the surjective ones, consists of the class of *injective* functions, defined as follows.

2.4.2 DEFINITION. The function $g\colon A \to B$ is **one-one** or **injective** or‡ an **injection** whenever, for any distinct elements a_1, a_2 in A, $g(a_1) \neq g(a_2)$ in B. (Thus, distinct arguments have distinct values.)
An equivalent formulation, often used in proofs to show that a function g is one-one is:

2.4.3 g is one-one iff, whenever $g(a_1) = g(a_2)$ then $a_1 = a_2$.

2.4.4 DEFINITION. The function $h\colon A \to B$ is an **equivalence** or **bijective** or‡ a **bijection** iff h is both injective and surjective.

Exercise 3

(i) It is usual to denote the set of *all* possible functions $f\colon A \to B$ by B^A (because if A, B are finite with a, b elements respectively, then B^A has b^a elements; see 7.1). Let I, S, E, denote respectively the subsets of B^A consisting of injections, surjections and equivalences. Prove that $E = I \cap S$.
(ii) If a map of a country is regarded as a function (as in 2.1.5), is this function one-one, or onto, or either?
(iii) Let $\mathcal{U}$ denote a universe. Prove that the function $\mathbf{C}\colon \mathfrak{p}\mathcal{U} \to \mathfrak{p}\mathcal{U}$ (which assigns to each $A \in \mathfrak{p}\mathcal{U}$, the complement $\mathbf{C}(A)$) is an equivalence.
(iv) When is a constant function (2.1.3) injective, surjective, or bijective?

† We use the notation $\langle a, b\rangle = \{x \mid a \leqslant x \leqslant b\}$, $\langle a, b\rangle = \{x \mid a < x \leqslant b\}$ and similar obvious variants.
‡ We make no rule about which word to use. The nouns 'surjection', 'injection', 'bijection' are of French origin (cf. 'jeter sur'), and have the useful associated adjectives 'surjective', etc. The properties denoted by 'one-one' and 'onto' are more self-evident; and 'similarity' is sometimes used for 'equivalence'.

(v) In Example 1.6.6 the assignments $n \to A(n)$, $n \to B(n)$ are each functions with domain $\mathbb{N}$. Describe the image of each, and show that the second function is one-one, while the first is not.
(vi) Use quantifiers to negate the statements 'f is injective', 'f is surjective', 'f is bijective'.

2.5 Examples

In Example 2.1.1, $f(A)$ is the set of all fathers, and since $f(A) \neq B$, f is not onto. Nor is f injective, because different humans may have the same father. On the other hand, the functions g_1, g_2 are one-one but not onto. In Example 2.1.2, f is onto, since B was given to be the shadow of A. Also f is not one-one, except in special circumstances depending on the thickness of A and its position relative to the sun and wall.

2.5.1 EXAMPLE. Let A denote the set of all people belonging to a union affiliated to the T.U.C., and let B be the T.U.C.—a set of sets. Then $f: A \to B$, defined by: $f(a) =$ *the union to which a belongs*, is a function, which is obviously not one-one. But f is onto, if we assume the most likely definition of 'affiliated'.

2.5.2 EXAMPLE. If A, B are finite sets, we can display functions $f: A \to B$ graphically as in Fig. 2.1, where for each $a \in A$, we have joined a to $f(a)$ by an arrow.

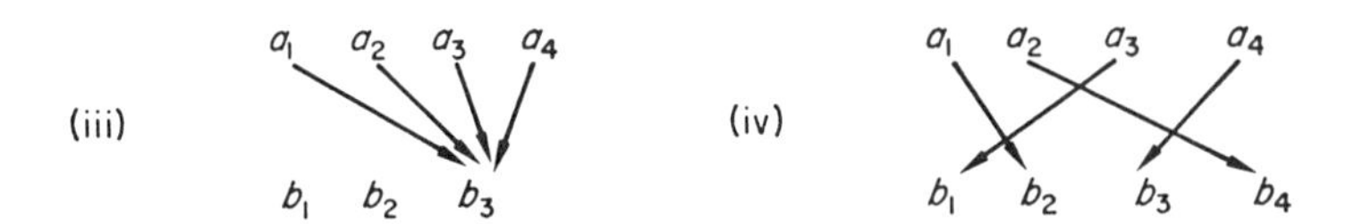

Fig. 2.1 (i) f is onto. (ii) f is one-one. (iii) f is not onto, not one-one. (iv) f is an equivalence.

This method of display is essentially the same as giving the function by means of a finite table, as in (say) books of log tables (where logarithms are given for a finite set of arguments). Thus, in Fig. 2.1(ii) for instance, we could specify f by the list

$$(a_1) = b_3, \qquad f(a_2) = b_1, \qquad f(a_3) = b_2.$$

Exercise 4

Describe the set B^A (notation of Exercise 3(i)), i.e., find all functions $f: A \to B$, when $A = \{a_1, a_2, a_3, a_4\}$, $B = \{b_1, b_2, b_3\}$; and state which (if any) are one-one, which are onto, and which are equivalences. Verify that, here, B^A has 3^4 elements.

2.5.3 EXAMPLE. In the projective plane (see Chapter 17), let P denote a point on the conic Σ, and let l denote a line not through P. Regarding Σ and l as sets of points, define a function $f: \Sigma \to l$ by the following rule. Let the line PQ—which shall be the tangent at P if $P = Q$—cut l in R. Then $f(Q)$ shall be R; certainly f is a function, and by the axioms of projective geometry, it is both one-one and onto, and hence an equivalence. It is called a 'one-one correspondence' in textbooks, and such books often say 'to each Q corresponds an R, and to each R a Q'; but it is important here that $R = f(Q)$, and not $g(Q)$ for some different function g.

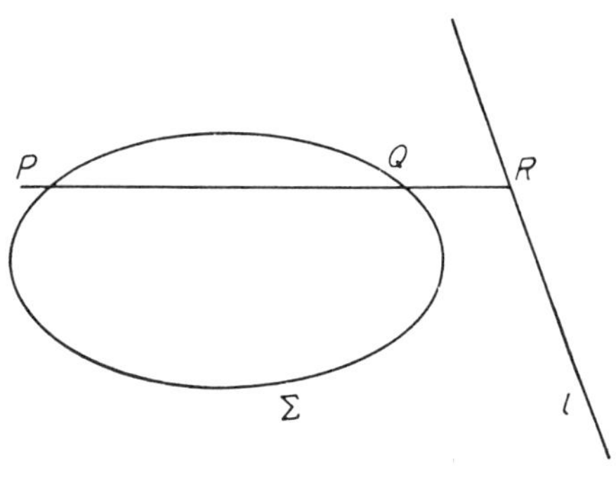

Fig. 2.2

2.5.4 EXAMPLE. An **involution** $h: l \to l$ of the line l in Example 2.5.3 is defined to be a function such that, if $h(R) = S$, then $h(S) = R$, for each $R \in l$. This amounts to saying that

(i) $h(h(R)) = R$, *for all* $R \in l$

or, technically, that 'h is of period 2'. Then h is an equivalence.

Proof. If $R_1 \neq R_2$, yet $h(R_1) = h(R_2)$, then $h(h(R_1)) = h(h(R_2))$, so $R_1 = R_2$ by (i), a contradiction, showing that h is one-one. And given $R \in l$, then $R = h(h(R))$ (by (i)), so $R = h(S)$ where $S = h(R)$; thus h is onto. Therefore h is both one-one and onto, so h is an equivalence. ■

2.5.5 EXAMPLE. For any set A, the function $\mathrm{id}_A: A \to A$ given by $\mathrm{id}_A(a) = a$ for each $a \in A$, is an equivalence, called the **identity function on** A. In the special case when A is the set $\mathbb{R}$ of real numbers, id_A is the function confusingly called '$y = x$' in traditional textbooks.

2.5.6 EXAMPLE. The logarithmic function, $\log: \mathbb{R}_+ \to \mathbb{R}$ in 2.1.4 is an equivalence, because distinct numbers have distinct logs (so log is one-one) and every x is the log of some y, viz., of e^x (so log is onto).

2.5.7 EXAMPLE. Similarly, the exponential function $\exp: \mathbb{R} \to \mathbb{R}_+$ is an equivalence ($\exp(x) = e^x$).

2.5.8 EXAMPLE. Let C denote the set of all functions $f: \mathbb{R} \to \mathbb{R}$ which have an infinity of derivatives f', f'', etc. [Thus, as explained in Chapter 29, the number

$$f'(x) = \lim_{h \to 0} f \frac{(x + h) - f(x)}{h}$$

exists for each $x \in \mathbb{R}$, so f' is a new function $f': \mathbb{R} \to \mathbb{R}$, called the **derivative** of f: and similarly for the higher derivatives $f'', f''', \ldots$.] Then the rule, which tells us how to assign to each $f \in C$ another function (namely f') in C, is itself a function but with domain C, and often denoted by $D: C \to C$; $D(f) = f'$. It is usual to call D an **operator**, in view of its property of linearity:

$$D(\lambda \cdot f + \mu \cdot g) = \lambda \cdot Df + \mu \cdot Dg$$

for all $f, g \in C$, and numbers $\lambda, \mu \in \mathbb{R}$; where $\lambda \cdot f + \mu \cdot g$ denotes that function whose value at x is $\lambda \cdot f(x) + \mu \cdot g(x)$.

A second operator on C is the function $Q: C \to C$, given by $Q(f) = g$, where g is that member of C whose value at $x \in \mathbb{R}$ is the definite integral

2.5.9 $$g(x) = \int_0^x f(t)\, dt.$$

Notice that C can be defined analogously for any interval

$$\langle a, b \rangle = \{x \in \mathbb{R} | a < x < b\},$$

and it is not 'just' a set but has other features which make it what is technically known as a 'real algebra'. If $\mathscr{J}$ denotes the set of all intervals of the form $\langle a, b \rangle$, then the assignment of $C(j)$ to each $j \in \mathscr{J}$, is itself a *function* C with domain $\mathscr{J}$ and range the family of all real algebras. The detailed discussion of the image of C is part of the subject called 'functional analysis'. (See also Chapter 26.)

2.6 Notation and Abuse of Language

Our next remarks are concerned with notation and some points arising therefrom, which may strike the reader initially as pedantic. First, if, in the equation 2.5.9 for $g(x)$, we had written $Q(f)$ for g (which is what g stands for), then we would obtain the cumbersome formula $Q(f)(x)$. It is usual, therefore, often to omit certain parentheses, and to write Qf instead of $Q(f)$. However, if f

itself had to be replaced by a complicated formula, we might then write $Q(f)|_x = g(x)$. In certain branches, the notation xf is also used for $f(x)$; e.g., when $n \in \mathbb{N}$, factorial n is written $n!$ $(=1 \cdot 2 \cdot \cdots \cdot n)$. After this practical matter, consider the following more subtle one, which takes up the remark following 2.2.1.

2.6.1 Suppose that $f\colon A \to B$ is a given function, and that C is a set such that $f(A) \subseteq C \subseteq B$. Then we have a new function $g\colon A \to C$ defined by setting $g(a) = f(a)$ for each $a \in A$. For brevity, we often refer to the functions as 'f' and 'g', respectively, and even regard them as equal. But an ambiguity can occur: for suppose $C = f(A) \neq B$. Then using the brief forms we would say 'f is not onto', whereas 'g is onto'. Thus two 'equal' things have different properties! In some contexts, however, such ambiguities do not lead to confusion, and the brevity is worthwhile, so there we allow the abuse of language '$f = g$'. When the distinction has to be made, we may denote g by $f\colon A \to C$, which is still a recognizably different formula from $f\colon A \to B$, and so avoids the memory-strain that 'regarding' things can impose.

2.6.2 The same conventions apply to the use of the identity function (2.5.5). Thus, if $C \subseteq B$, there are three functions, $\mathrm{id}_C\colon C \to C$, $\mathrm{id}_B\colon B \to B$, and the **inclusion map** $j\colon C \to B$ given by $j(c) = c$. We shall often refer to all three as 'the identity function id on C'.

The upshot of this discussion is that we allow convenient abbreviations in contexts where ambiguities are not serious, now that we have the language to resolve these ambiguities when required. These logical distinctions must be grasped, since they are essential in many, even quite elementary, branches of mathematics.

2.6.3 EXAMPLE. Let $\mathbb{R}^+ = \{x \in \mathbb{R} | x \geqslant 0\}$, and let us temporarily denote by $\mathrm{Sq}\colon \mathbb{R} \to \mathbb{R}$ the function such that $\mathrm{Sq}\,(x) = x^2$ for each $x \in \mathbb{R}$. Then Sq is not one-one, since $\mathrm{Sq}\,(1) = \mathrm{Sq}\,(-1)$, yet $1 \neq -1$ (for example). Also, the image, $\mathrm{Sq}\,(\mathbb{R})$, is $\mathbb{R}^+$, so taking account of the above remarks, we see that the function $\mathrm{Sq}\colon \mathbb{R} \to \mathbb{R}^+$ is onto. If we *restrict* Sq to $\mathbb{R}^+$ (that is, we consider the function $f\colon \mathbb{R}^+ \to \mathbb{R}^+$ given by $f(x) = x^2$), then it is known that f is an equivalence (see 28.6). We call f the **restriction** of f to $\mathbb{R}^+$, and write

(i) $f = \mathrm{Sq}|\mathbb{R}^+$.

2.6.4 More generally, for any function $g\colon A \to B$, if $C \subseteq A$, we define the **restriction of** g **to** C to be the function $h\colon C \to B$, given by

$$h(c) = g(c) \quad \text{for each} \quad c \in C;$$

and we indicate the dependence of h on C and g, by using the notation

(i) $h = g|C$.

2.6.5 EXAMPLE. The sine function, sin: $\mathbb{R} \to \mathbb{R}$ (explained, e.g., in 29.8) is not onto, because in particular $2 \neq \sin x$ for any $x \in \mathbb{R}$. Its image is known to be the interval

$$A = \langle\!\langle -1, 1 \rangle\!\rangle.$$

Nor is sine injective, because $\sin 0 = \sin \pi$, yet $0 \neq \pi$. If, however, we restrict to the interval $B = \langle\!\langle -\pi/2, \pi/2 \rangle\!\rangle$, then $f = \sin|B$ is known to be an equivalence $f: B \to A$. It is customary still to denote f by sine; and ambiguity rarely occurs in this context.

Exercise 5

(i) Given a function $f: X \to Y$, suppose that A, B, C are subsets of X. Prove that, if $B \subseteq C$, then $f(B) \subseteq f(C)$,

$$f(A \cup B) = f(A) \cup f(B), \qquad f(A \cap B) \subseteq f(A) \cap f(B);$$

and find an example to show that the last inclusion cannot always be replaced by equality.

(ii) Show that, if $\mathscr{U}$ is a non-empty universe, the function $s: \mathscr{U} \to \mathfrak{p}\mathscr{U}$, which assigns to each $x \in \mathscr{U}$ the singleton $\{x\}$, is one-one but not onto. [We can therefore regard Im (s) as a copy of $\mathscr{U}$, and thus regard $\mathscr{U}$ as a subset of $\mathfrak{p}\mathscr{U}$. We say that s **embeds** $\mathscr{U}$ **in** $\mathfrak{p}\mathscr{U}$.]

(iii) Given any subset $D \subseteq Y$, define $f^{\flat}(D)$, the **inverse image of** D **under** $f: X \to Y$, by

$$f^{\flat}(D) = \{x | x \in X \text{ and } f(x) \in D\}.$$

[Thus we have a function $f^{\flat}: \mathfrak{p}Y \to \mathfrak{p}X$; other notations exist, but we avoid them for the present.] When f is sin: $\mathbb{R} \to \mathbb{R}$, and D is the singleton $\{\frac{1}{2}\}$, show that $\sin^{\flat}\{\frac{1}{2}\} = \{x | x = \pi/6 + 2k\pi \text{ or } x = 5\pi/6 + 2k\pi, k \in \mathbb{Z}\}$, while $\sin^{\flat}\{2\} = \varnothing$. Do the analogous problems when f is replaced respectively by $\sin|\langle\!\langle 0, \pi/6 \rangle\!\rangle$ and $\sin|\langle\!\langle 0, \pi/7 \rangle\!\rangle$. [The relationship between $\sin^{\flat}\{x\}$ and $\sin^{-1} x$, suspected by the reader, is discussed below, in 2.9.6.]

(iv) With a general $f: X \to Y$, and $D \subseteq Y$ prove that, if also $E \subseteq Y$, then

$$f^{\flat}(D \cup E) = f^{\flat}(D) \cup f^{\flat}(E); \qquad f^{\flat}(D \cap E) = f^{\flat}(D) \cap f^{\flat}(E).$$

(v) Show by examples that, in (iv),

(1) $f^{\flat}(D)$ can be empty, although $D \neq \varnothing$;
(2) $f(f^{\flat}(D)) \subseteq D$ } without necessarily having equality.
(3) $f^{\flat}(f(A) \supseteq A$ }
(4) If $f^{\flat}(f(A)) = A$ for all $A \subseteq X$, prove that f is one-one, and prove the converse of this result.
(5) If $f(f^{\flat}(D)) = D$ for all $D \subseteq Y$, prove that f is onto, and prove the converse of this result.
(6) Prove that $f^{\flat}(Y) = X$, $f^{\flat}(Y - D) = X - f^{\flat}(D)$.

2.7 Composition of Functions

The reasons mentioned in 1.5 for assembling new sets from old apply also to functions. From two given functions

$$f: A \to B, \qquad g: B \to C$$

we form a new function denoted by

$$g \circ f: A \to C$$

called the **composite** of g and f (*read*: 'g circle f'), and defined by the rule that

2.7.1 $$(g \circ f)(a) = g(f(a)), \quad \textit{for all} \quad a \in A.$$

In calculus, $g \circ f$ is often called a 'function of a function', but the reader should by now be able to see why this is a misleading description. Notice also that $f \circ g$ will not even be defined, if $A \neq C$.

2.7.2 EXAMPLES. On using the id function of Example 2.5.5, the equation (i) of Example 2.5.4 shows that $h \circ h = \mathrm{id}_l$. If $f: A \to B$ is the 'father' function of Example 2.1.1, then $(f \circ f)(a)$ denotes the paternal grandfather of a. The function $f: A \to B$ in 2.6.1 is $j \circ g$. In 2.6.4, the restriction $g|C$ of $g: A \to B$, is $g \circ k$, where $k: C \to A$ is the inclusion map. Given any $f: A \to B$, we have

2.7.3 $$f \circ \mathrm{id}_A = f = \mathrm{id}_B \circ f;$$

thus id functions behave like a unity element in algebra. For this reason we shall often use the symbol **1** to denote an identity function (with or without a subscript, as explained in 2.6.2).

2.7.4 EXAMPLE. Consider the functions $\log: \mathbb{R}_+ \to \mathbb{R}$ and $\sin: \mathbb{R} \to \langle\!\langle -1, 1 \rangle\!\rangle$. Then $\sin \circ \log: \mathbb{R}_+ \to \langle\!\langle -1, 1 \rangle\!\rangle$ has at each $x \in \mathbb{R}_+$ the value $\sin(\log x)$, whereas $\log \circ \sin(x)$ is not even defined for all x [e.g., when $x = 3\pi/2$]; and when it *is* defined, $\sin(\log x) \neq \log(\sin x)$, in general.

2.7.5 Thus, the operation of forming $g \circ f$ from g and f is not **commutative.** On the other hand, it is **associative**, i.e., given functions†

$$A \xrightarrow{f} B \xrightarrow{g} C \xrightarrow{h} D$$

then we have a theorem:

2.7.6 THEOREM. $$h \circ (g \circ f) = (h \circ g) \circ f: A \to D$$

which says that the order of bracketing is not important. To avoid clumsy formulae, we shall often omit the composition symbol and write gf for $g \circ f$. Thus we can write each function in 2.7.6 (once it is proved!) without ambiguity as

$$hgf: A \to D.$$

Proof of 2.7.6. By Definition 2.2.1 it suffices to prove that, for each $a \in A$, $(h(gf))(a) = ((hg)f)(a)$; but, using 2.7.1 four times, we have

$$(h(gf))(a) = h((gf)(a)) = h(g(f(a))) = (hg)(f(a)) = ((hg)f)(a),$$

so 2.7.6 follows. ■

† This method of displaying the functions is commonly used, instead of the longer

$$f: A \to B, \qquad g: B \to C, \qquad h: C \to D.$$

The reader who is unsure should insert the composition symbol at the appropriate places in the above equations, for practice.

2.7.7 EXAMPLE. The function traditionally denoted by $e^{\sin x^2}$ can be broken down, using the notation following 2.7.4, into

$$\mathbb{R} \xrightarrow{\text{Sq}} \mathbb{R}_+ \xrightarrow{\sin} \langle\!\langle -1, 1 \rangle\!\rangle \xrightarrow{\exp} \langle\!\langle e^{-1}, e \rangle\!\rangle,$$

where we have taken the smallest relevant range each time. Thus (retaining the composition symbol, for punctuation!) we have

$$e^{\sin x^2} = (\exp \circ \sin \circ \text{Sq})(x)$$

and the *name* of the function is $\exp \circ \sin \circ \text{Sq}: \mathbb{R} \to \langle\!\langle e^{-1}, e \rangle\!\rangle$.

2.8 Composition of Injections, etc.

Let us now state two results, which tell us about the composition of injections, surjections, and equivalences. We leave their proofs to the reader. [See also Exercise 6 below.]

2.8.1 THEOREM. *If the functions $f: A \to B$, $g: B \to C$ are both one-one, then so is $g \circ f$. If they are both onto, so is $g \circ f$.* □

[Hence, in Example 2.7.7, the function $\exp \circ \sin \circ \text{Sq}: \mathbb{R} \to \langle\!\langle e^{-1}, e \rangle\!\rangle$ is onto; it is not one-one. Why?]

2.8.2 COROLLARY. *If in 2.8.1, f and g are both equivalences, so is $g \circ f$.* ■ [See also 2.9.9.]

2.9 The Inversion Theorem

We now give a theorem, the 'Inversion Theorem', which is basic in many branches of mathematics. It is a precise version of the dogma 'If b is a function of a, then a is a function of b'. In each application, more is usually required. Thus, one often wants to say something like: 'If b is a linear, or continuous or differentiable, or projective, function of a, then a is the *same kind* of function of b'. The Inversion Theorem enables one to get started before doing the extra work necessary for the stronger conclusion. (See, for example, Theorem 29.9.2.) Since we prefer to emphasize functions rather than variables, we use again the id function and denote it by $\mathbf{1}$, as explained after 2.7.3.

2.9.1 (**The Inversion Theorem.**) *Suppose that the function $f: A \to B$ is an equivalence. Then there exists a function*† *$g: B \to A$, such that*

$$\text{(i) } g \circ f = \mathbf{1}_A, \text{ (ii) } f \circ g = \mathbf{1}_B.$$

† In fact, g is an equivalence: see 2.9.2.

Proof. Given $b \in B$, there exists $a \in A$ such that $f(a) = b$, since f is onto. Moreover, if any other $a' \in A$ is such that $f(a') = b$, then $a = a'$ since f is one-one. Thus, given b, the element a is unique; and therefore we denote a by $g(b)$, obtaining a function $g\colon B \to A$. Then by 2.7.1, $(f \circ g)(b) = f(g(b)) = f(a)$, since $a = g(b)$ by definition of g; but $f(a) = b$ by definition of a, so (omitting now the composition symbol) $fg(b) = f(a) = b = \mathbf{1}_B(b)$ by definition of $\mathbf{1}_B$. Since this equation holds for every $b \in B$, then $f \circ g = \mathbf{1}_B$, by Definition 2.2.1. This establishes (ii) of 2.9.1. To prove (i) of 2.9.1, we have, for each $a \in A$,

$$\begin{aligned} gf(a) &= g(f(a)) && \text{(by 2.7.1)} \\ &= g(b), && \text{say,} \end{aligned}$$

where $b = f(a)$. But then $a = g(b)$, by definition of g, so

$$gf(a) = g(b) = a = \mathbf{1}_A(a)$$

by definition of $\mathbf{1}_A$. This holds for each $a \in A$, so $g \circ f = \mathbf{1}_A$, by Definition 2.2.1; and (i) of 2.9.1 is established. ■

2.9.2 THEOREM. *Given functions $f\colon A \to B$, $g\colon B \to A$ satisfying equations* (i) *and* (ii) *of* 2.9.1, *then both f and g are equivalences.*

Proof. The hypotheses are symmetrical in f and g, so it suffices to prove that g is an equivalence. Thus we split the proof into two parts, the first to show that g is one-one, the second to show that g is onto.

For the first it suffices by 2.4.3 to prove that if $b_1, b_2 \in B$, and if $g(b_1) = g(b_2)$, then $b_1 = b_2$. Now if $g(b_1) = g(b_2)$, then $f(g(b_1)) = f(g(b_2))$, i.e., $(fg)b_1 = (fg)b_2$ (by 2.7.1); so $\mathbf{1}_B(b_1) = \mathbf{1}_B(b_2)$ by (ii) of 2.9.1; so $b_1 = b_2$ by definition of $\mathbf{1}_B$, as required.

To prove that g is onto, let $a \in A$ be given. By Definitions 2.3.1 and 2.4.1, we must find some $b \in B$, such that $g(b) = a$.

But

$$\begin{aligned} a &= \mathbf{1}_A(a) && \text{(by definition of } \mathbf{1}_A) \\ &= gf(a) && \text{(by (i) of 2.9.1)} \\ &= g(f(a)) && \text{(by 2.7.1);} \end{aligned}$$

so $a = g(b)$, where the element b required is $f(a)$. Thus g is onto. Hence g is an equivalence. ■

The two sections of the last proof form also a proof of the following remark, which we label as a corollary.

2.9.3 COROLLARY. *If, given a function $f\colon A \to B$, there exists a function $g\colon B \to A$ such that $g \circ f = \mathbf{1}_A$, then g is onto; if instead, $f \circ g = \mathbf{1}_B$, then g is one-one; and if both, then f and g are both equivalences.* ■

This remark rexpresses the notion of an equivalence in an algebraic manner (cf. the discussion of inverses in Chapter 3) and enables algebraic techniques to be applied to proofs about functions. Such an application is exemplified in the proof of the following theorem, in which we do not need to refer to the elements of the sets involved; the flavour is quite different from that of the *proof* of 2.9.2 and is in fact a proof about 'multiplicative systems' in algebra. (See 3.6 below.)

2.9.4 THEOREM. *The function* $g: B \to A$ *in* 2.9.1 *is unique. More precisely, if* $f: A \to B$, $g: B \to A$ *are functions known to satisfy* (ii) *of* 2.9.1, *then any other function* $h: B \to A$ *satisfying* 2.9.1(i) *is equal to* g.

Proof. We shall compute hgf in two ways. Thus with the given data, replace f by h in 2.7.3 to get

$$\begin{aligned} h = h \circ \mathbf{1}_B &= h(fg) && \text{(by 2.9.1(ii) for } g) \\ &= (hf)g && \text{(by associativity, 2.7.6)} \\ &= \mathbf{1}_A g \end{aligned}$$

since h is given to satisfy 2.9.1(i). But $\mathbf{1}_A g = g$, substituting $g: B \to A$ for f in 2.7.3. Hence $h = g$. ■

Exercise 6

(i) Let $f: A \to B$, $g: B \to C$ be functions. Prove that if $g \circ f$ is one-one, then f is one-one, and that if $g \circ f$ is onto, then g is onto. Give examples to show that g need not be one-one and f not onto, respectively.
(ii) Give examples of functions $f: A \to B$, $g: B \to A$, such that $g \circ f = \mathbf{1}_A$, with g not one-one.
(iii) Give examples where $f \circ g = \mathbf{1}_B$ with g not onto.
(iv) Let f, g, h be functions such that $g \circ f = \mathbf{1}$, $h \circ g = \mathbf{1}$. Show that f, g, h are equivalences and $f = h$.

Since the function g in 2.9.1 is unique and depends on f, we display this dependence by writing

2.9.5 $$g = f^{-1}: B \to A;$$

it is called the **inverse function** *of* f. This notation is natural if we write (i) and (ii) of 2.9.1 in the form $gf = \mathbf{1} = fg$. Whenever such equations hold, we say that f **has an inverse** or f is **invertible**, and Theorems 2.9.1, 2.9.2 assert that f *is an equivalence if and only if* f *is invertible.*

2.9.6 EXAMPLE. Let us now consider the relationship, foreshadowed in Exercise 5(iii), between the functions $\sin^\flat: \mathfrak{p}B \to \mathfrak{p}A$ and $\sin: A \to B$ where $A = \langle\!\langle -\pi/2, \pi/2 \rangle\!\rangle$ and $B = \langle\!\langle -1, 1 \rangle\!\rangle$. As stated before, $\sin: A \to B$ is an equivalence, and hence it is invertible, with inverse the familiar function $\sin^{-1}: \langle\!\langle -1, 1 \rangle\!\rangle \to \langle\!\langle -\pi/2, \pi/2 \rangle\!\rangle$. The latter is also an equivalence by Theorem 2.9.2; so for each $t \in B$, $\sin^{-1} t$ is a (single) point in A. Taking singletons, we have by *definition* of $\sin^\flat$

$$\sin^\flat \{t\} = \{\sin^{-1} t\}.$$

If now we regard B as a subset of $\flat B$ (as in Exercise 5(ii)) then we must likewise regard $\{\sin^{-1} t\} \in \flat A$ as $\sin^{-1} t \in A$. Thus the last equation tells us that $\sin^{-1}$ is the *restriction* (in the sense of 2.6.4) of $\sin^\flat$ to B, i.e.,

$$\sin^{-1} = \sin^\flat|B.$$

Quite generally, it turns out (see Exercise 7(ii) below) that for any equivalence $f: X \to Y$ we may identify $f^{-1}: Y \to X$ with the restriction $f^\flat|Y$ of $f^\flat: \flat Y \to \flat X$, when Y is regarded as a subset of $\flat Y$; thus $f^{-1} = f^\flat|Y$. For this reason, $f^\flat$ itself is commonly denoted by f^{-1} *even when f is not invertible.* We advise the inexperienced reader to use the notation $f^\flat$ until he can confidently avoid the pitfalls.

2.9.7 CAUTION. When A is a set of numbers, do not confuse $f^{-1}(x)$ with $(f(x))^{-1}$ (the reciprocal of $f(x)$). With the first we have taken f, then its inverse, and evaluated at x; with the second we have taken f, evaluated at x, and then taken the reciprocal. Inverting and evaluating f are two operations which do not commute. As an example, $\sin^{-1}(1) = \pi/2 = 1.57\cdots$, while $(\sin 1)^{-1} = 1.18\cdots$.

We may conveniently summarize the results 2.9.1–2.9.4 in the following theorem, which we call the **Algebraic Inversion Theorem.**

2.9.8 THEOREM. *A function $f: A \to B$ is an equivalence iff $\exists g: B \to A$ satisfying the equations $gf = \mathbf{1}_A$, $fg = \mathbf{1}_B$. Such a g is then unique; it is the inverse of f.*

As an application, suppose that $f: A \to B$, $g: B \to C$ are invertible; then $g \circ f: A \to C$ is an equivalence by 2.8.2; so it has an inverse. Let us prove that, in this case

2.9.9 $$(g \circ f)^{-1} = f^{-1} \circ g^{-1}: C \to A.$$

Before beginning the proof we remark that we are really establishing that the composition of two invertible functions is again invertible. Thus the proof is formal, or algebraic, and does not rely on the nature of functions but only on their laws of composition 2.7.3, 2.7.6.

Proof. We leave the reader to supply reasons for the following equalities:

$$\begin{aligned}(f^{-1}g^{-1})(gf) &= f^{-1}(g^{-1}g)f \\ &= f^{-1}1f = f^{-1}f = 1.\end{aligned}$$

Similarly, $(gf)(f^{-1}g^{-1}) = 1$. Hence gf has $f^{-1}g^{-1}$ as an inverse; and by 2.9.4, an inverse is unique if it exists, so we are forced to write

$$(g \circ f)^{-1} = f^{-1} \circ g^{-1}.$$ ■

Exercise 7

(i) Given functions $A \xrightarrow{f} B \xrightarrow{g} C$ (not necessarily equivalences), prove that the composite of $g^\flat: \flat C \to \flat B$ and $f^\flat: \flat B \to \flat A$ is $(g \circ f)^\flat: \flat C \to \flat A$, and $(g \circ f)^\flat = f^\flat \circ g^\flat$. Prove also that $(\mathbf{1}_A)^\flat$ is the identity on $\flat A$. Hence prove that if f is invertible, so is $f^\flat$. {*Hint*: consider $(ff^{-1})^\flat$ and $(f^{-1}f)^\flat$. See also 6.2.9.}

(ii) Following Example 2.9.6, suppose we regard A, B as subsets of $\flat A$, $\flat B$, as suggested in Exercise 5(ii). Show that if $f\colon A \to B$ is an equivalence, then its inverse is the restriction to B of $f^\flat\colon \flat B \to \flat A$. More precisely, to avoid the 'regarding' process let $s_A\colon A \to \flat A$ denote the embedding of A in $\flat A$, and similarly for s_B. Show that the diagram 'commutes', i.e., that $f^\flat \circ s_B = s_A \circ f^{-1}$. This equation is the precise form of the equation $f^\flat|B = f^{-1}$.

$$\begin{array}{ccc} B & \xrightarrow{f^{-1}} & A \\ {\scriptstyle s_B}\downarrow & & \downarrow{\scriptstyle s_A} \\ \flat B & \xrightarrow[f^\flat]{} & \flat A \end{array}$$

2.9.10 EXAMPLE. In Examples 2.5.6, 2.6.7 the functions $\log\colon \mathbb{R}_+ \to \mathbb{R}$, $\exp\colon \mathbb{R} \to \mathbb{R}_+$ are inverses of each other, because of the equations

$$\log(\exp x) = x \qquad (x \in \mathbb{R})$$
$$\exp(\log x) = x \qquad (x \in \mathbb{R}_+),$$

so $\log \circ \exp = \mathrm{id}$, $\exp \circ \log = \mathrm{id}$, where, in the spirit of 2.6.2, we omit the subscripts $\mathbb{R}$ and $\mathbb{R}_+$ from the first and second occurrences of id, respectively†. In 2.6.3, we remarked that the function $\mathrm{Sq}|\mathbb{R}^+$ of 2.6.3(i) is an equivalence: $\mathbb{R}^+ \to \mathbb{R}^+$. Hence, by the Inversion Theorem, it has an inverse, usually called 'the positive square root'; and of course

$$\mathrm{Sq}(\sqrt[+]{x}) = x = \sqrt[+]{(\mathrm{Sq}(x))}.$$

In Example 2.5.8, the 'fundamental theorem of calculus' (30.1.2) gives $D \circ \underset{\sim}{Q} = \mathrm{id}_C$; but $\underset{\sim}{Q} \circ D \neq \mathrm{id}_C$, since $(\underset{\sim}{Q} \circ D)(f)$ is that function whose value at x is

$$\int_0^x f'(t)\,dt = f(x) - f(0),$$

which is not $f(x)$ unless $f(0) = 0$. Thus, with this exception, $(\underset{\sim}{Q} \circ D)(f) \neq f$ for any $f \in C$, so $\underset{\sim}{Q} \circ D \neq \mathrm{id}_C$.

2.10 Equivalent Sets

It is convenient to abbreviate the statement '$f\colon A \to B$ is an equivalence' to‡

2.10.1 $$f\colon A \approx B.$$

To express the fact that there exists *some* equivalence between A and B we write

2.10.2 $$A \approx B$$

saying that A is **equivalent** to B.

† Another excuse is: 'for typographical reasons', (because of the expense of setting up mathematical type). Notice that we write id or 1 for the identity function.

‡ When writing symbols like $\sim$ or $\approx$, lecturers often emit the sound 'twiddles'.

The statement 'f is an equivalence' is often, and naturally, confused with 'A is equivalent (to B)'. To avoid this confusion we will often say 'f is invertible', or 'f is a bijection', instead of 'f is an equivalence'; we recall Theorem 2.9.8.

Exercise 8

Prove the following:

$$A \approx A;$$

$$\text{if} \quad A \approx B \quad \text{then} \quad B \approx A;$$

$$\text{if} \quad A \approx B \quad \text{and} \quad B \approx C, \quad \text{then} \quad A \approx C.$$

2.11 Counting

Let $\mathbb{N}_n$ denote the subset $\{1, 2, \ldots, n\}$ of $\mathbb{N}$, consisting of the first n natural numbers; $\mathbb{N}_0$ is therefore empty. Then to say of a set A that we have counted it and found it to have n members, is to say that we have constructed an equivalence

$$f\colon A \approx \mathbb{N}_n;$$

for the act of counting consists in pairing off each element in A with one in $\mathbb{N}_n$, until A is exhausted. The last integer to be paired is n. Thus we make the definition:

2.11.1 DEFINITION. We write $\#A = n$ (read: the **number of elements** in A is n) iff there exists a bijection $f\colon A \to \mathbb{N}_n$. Thus $A = \varnothing$ iff $\# A = 0$. If there does not exist such a bijection, we write

$$\#A = \infty,$$

and† call A **infinite**; otherwise A is **finite.**

Caution. The fact that we speak of 'the' number of elements in A, implies that A cannot have both n elements and m elements, with $n \neq m$. Common-sense, and experience with counting, both combine to assure us that this view is reasonable. However, experience also reminds us of occasions when we have inaccurately counted a set twice and obtained different answers; and we have no experience at all of counting really large sets, like the total number of possible games of chess having fewer than 50 moves. To set these doubts at rest we shall, in Chapter 7, *prove*:

2.11.2 THEOREM. *It is impossible for bijections*

$$h\colon A \approx \mathbb{N}_n, \qquad k\colon A \approx \mathbb{N}_m, \qquad n \neq m,$$

to exist simultaneously.

† Note that we do not define the symbol ∞ on its own.

Hence the integer $\#A$ *is uniquely defined*. Assuming this result temporarily, then, we have:

2.11.3 PROPOSITION. *Two finite sets are equivalent, iff they have the same number of elements.*

Proof. Suppose $f: A \approx \mathbb{N}_n$ and $g: A \approx B$. Then $g^{-1}: B \approx A$ (by 2.9.1); so $fg^{-1}: B \approx \mathbb{N}_n$ by 2.8.2. Hence $\#A = n = \#B$. Conversely, if $\#A = \#B = n$, then there exist bijections $p: A \approx \mathbb{N}_n$, $q: B \approx \mathbb{N}_n$; so $q^{-1}p: A \approx B$. ∎

Remarks. The proof uses the method, common in real life, of counting the set B by comparing it with a known set A. Also, we have shown that it is impossible for B to be simultaneously infinite and equivalent to the set A. The method of proof shows too that if h, k existed in 2.11.2, then $hk^{-1}: \mathbb{N}_m \approx \mathbb{N}_n$, $n \neq m$, and it is basically this statement that we prove to be impossible in 7.1.3. We shall also prove there that *every subset of a finite set is finite;* this statement is not trivial, since it must be shown to hold by using Definition 2.11.1, and not other usages of the word 'finite'. A consequence is that, for each finite universe $\mathscr{U}$, $\#$ is a *function*: $\mathfrak{p}\mathscr{U} \to \mathbb{Z}^+$, where $\mathbb{Z}^+$ is the set $\mathbb{N} \cup \{0\}$ of non-negative integers.

Exercise 9

(i) Prove that $\mathbb{N}$ contains proper subsets B (i.e., $B \neq \mathbb{N}$) such that $B \approx \mathbb{N}$.
(ii) If X is a set such that $X \approx \mathbb{N}$, then X is said to be **denumerably** (or **countably**) **infinite**. Prove that the following sets are denumerably infinite: the even integers, any infinite subset of $\mathbb{N}$, the rational numbers, the set of all points (p, q) in the plane such that $p, q \in \mathbb{N}$. {In the last, count off the points in a diagram like Fig. 5.3 by going along successive diagonals like the one shown there.}
§(iii) Use the same diagram to show that if $X_n \approx \mathbb{N}$, $n = 1, 2, \ldots$, then $\bigcup_{n=1}^{\infty} X_n \approx \mathbb{N}$. Hence prove that the set of all polynomials with integer coefficients is equivalent to $\mathbb{N}$. Then show that the set of all **algebraic numbers** (i.e., real roots of such polynomials) is denumerably infinite.

Let us now obtain some formulae for counting finite subsets of a finite universe $\mathscr{U}$. First, since $\mathbb{N}_0 \approx \mathbb{N}_0 = \varnothing$, then, as remarked in 2.11.1,

2.11.4 $$\#\varnothing = 0.$$

The following result is used constantly in ordinary life, as a Venn diagram shows, but we must prove it from the definition of $\#$.

2.11.5 THEOREM. *If $A, B \subseteq \mathscr{U}$, and $A \cap B = \varnothing$, then*

$$\#(A \cup B) = \#A + \#B.$$

Proof. Let $\#A = n$, $\#B = m$. The formula 2.11.5 obviously holds when n or m is zero, by 2.11.4; hence suppose $0 < n$ and $0 < m$. By Definition 2.11.1, there exist bijections

$$a: A \approx \mathbb{N}_n, \qquad b: B \approx \mathbb{N}_m$$

and we construct a bijection $c: A \cup B \approx \mathbb{N}_{n+m}$ by the rules:

$$c(x) = \begin{cases} a(x) & \text{if } x \in A, \\ b(x) + n & \text{if } x \in B. \end{cases}$$

We leave the reader to verify that the function c is in fact one-one and onto; the fact that $A \cap B = \varnothing$ ensures no ambiguity in the definition of c. But then, by Definition 2.11.1, the existence of c gives

$$\begin{aligned} \#(A \cup B) = \#\mathbb{N}_{n+m} &= n + m \\ &= \#A + \#B. \end{aligned}$$

and 2.11.5 follows. ■

2.11.6 COROLLARY. *If $X \subseteq A \subseteq \mathscr{U}$, then* (by using the notation $A - X$ explained in 1.7):

(i) $\#A = \#X + \#(A - X)$
(ii) $\#X \leqslant \#A$.

{A is the union of the two disjoint sets X, $A - X$, so (i) is an application of 2.11.6; (ii) follows from (i), since $\#(A - X) \geqslant 0$.} ■

We can now prove a generalization of 2.11.6 which is again fairly obvious from a Venn diagram; it reduces to 2.11.5 when $A \cap B = \varnothing$, by 2.11.4.

2.11.7 THEOREM. *If $A, B \subseteq \mathscr{U}$, then*

$$\#(A \cup B) = \#A + \#B - \#(A \cap B).$$

Proof.

$$\begin{aligned} A \cup B &= [(A - (A \cap B)) \cup (A \cap B)] \cup [(A \cap B) \cup (B - (A \cap B))], \\ &= (A - (A \cap B)) \cup (A \cap B) \cup (B - (A \cap B)), \end{aligned}$$

applying the rule 1.7.7 with $X = A \cap B$. Now $A \cup B$ has been expressed as the union of three subsets, any pair of which are disjoint. Thus, we can apply 2.11.5 twice to get

$$\begin{aligned} \#(A \cup B) &= \#(A - (A \cap B)) + \#(A \cap B) + \#(B - (A \cap B)) \\ &= [\#A - \#(A \cap B)] + \#(A \cap B) + [\#B - \#(A \cap B)], \end{aligned}$$

using 2.11.6(i). Rearrangement gives 2.11.7. ■

Notice the purely algebraic nature of this proof, which is different from that of the proof of 2.11.5. In a purely algebraic manner, also, we can extend 2.11.7 to a union of more sets, as follows.

2.11.8 THEOREM.

$$\#(A \cup B \cup C) = \#A + \#B + \#C - \#(A \cap B) - \#(B \cap C) - \#(C \cap A) + \#(A \cap B \cap C),$$

and more generally

$$\#(A_1 \cup \cdots \cup A_n) = \sum_{i=1}^{n} \#A_i - \sum_{i<j} \#(A_i \cap A_j) + \sum_{i<j<k} \#(A_i \cap A_j \cap A_k) + \cdots + (-1)^{n-1} \#(A_1 \cap A_2 \cap \cdots \cap A_n),$$

where the rth term is $(-1)^{r-1}$ times the sum of all integers of the form

$$\#(A_{i_1} \cap A_{i_2} \cap \cdots \cap A_{i_r}), \qquad i_1 < i_2 < \cdots < i_r \leqslant n.$$

Proof. The proof for n sets requires the method of finite induction, discussed in Chapter 5, and we leave it to the reader, since the main step occurs already when $n = 3$, and uses the distributive law 1.7.5. Thus, for three sets A, B, C we have

$$\begin{aligned} \#(A \cup B \cup C) &= \#(A \cup (B \cup C)) \\ &= \#A + \#(B \cup C) - \#(A \cap (B \cup C)) \qquad \text{(by 2.11.7)} \\ &= \#A + [\#B + \#C - \#(B \cap C)] - \#((A \cap B) \cup (A \cap C)), \end{aligned}$$

applying 2.11.7 to $B \cup C$, and the distributive law 1.7.5 to $A \cap (B \cup C)$. If now we apply 2.11.7 to the final term of the last equality, we can put it equal to $\#(A \cap B) + \#(A \cap C) - \#((A \cap B) \cap (A \cap C))$. Now $(A \cap B) \cap (A \cap C) = B \cap (A \cap A) \cap C = A \cap B \cap C$, using the commutative, associative and idempotent laws for $\cap$; hence

$$\#((A \cap B) \cap (A \cap C)) = \#(A \cap B \cap C),$$

so assembling our equations gives 2.11.8. ■

Further results are obtained in Chapter 8.

2.11.9 EXAMPLE. A survey showed that on a certain day in winter, 1000 houses were heated and used gas, oil, or coal as fuels. The numbers using these fuels respectively were 265, 51, 803; and 287, 843, 919 householders respectively recalled using gas or oil, oil or coal, and gas or coal during the day. Show that this information contains a discrepancy.

Solution. If we take $\mathscr{U}$ to be the 1000 houses, and let A, B, C denote respectively those using gas, oil, and coal, then $\mathscr{U} = A \cup B \cup C$, where

$$\#A = 265, \qquad \#B = 51, \qquad \#C = 803.$$

Also, $\#(A \cup B) = 287$, so, by 2.11.7,

$$\#(A \cap B) = 265 + 51 - 287 = 29,$$

and similarly, $\#(B \cap C) = 11$, $\#(C \cap A) = 149$. Hence, by 2.11.8,

$$\begin{aligned} 1000 &= \#(A \cup B \cup C) \\ &= 265 + 51 + 803 - (29 + 11 + 149) + \#(A \cap B \cap C) \end{aligned}$$

so $\#(A \cap B \cap C) = 1189 - 1119 = 70 > \#B$, which contradicts the fact that $A \cap B \cap C \subseteq B$. The given figures cannot therefore be correct. (See also Birkhoff–MacLane [16], Ch. XI, for other problems of this type.)

Exercise 10

(i) If $A \approx A'$, $B \approx B'$, and $A \cap B = \varnothing = A' \cap B'$, prove that $A \cup B \approx A' \cup B'$.
(ii) Make up, and solve, a problem like that of 2.11.9.

Chapter 3

THE CARTESIAN PRODUCT

3.1 Pairs and Products

When we say in analytic geometry that a point has co-ordinates (x, y), the *order* in which x and y occur, in the symbol (x, y), is important: $(1, 2) \neq (2, 1)$. For this reason we call (x, y) an **ordered pair.** Moreover, x and y come from sets; in this case $x, y \in \mathbb{R}$. This idea can be generalized† as follows. Let $\mathscr{U}$ be a universe. We can then form ordered pairs (u, v) where $u, v \in \mathscr{U}$; two such ordered pairs (u, v), (x, y) are **equal** iff $u = x$ and $v = y$. It is therefore important to distinguish between the ordered pair (x, y), and the **unordered** pair $\{x, y\}$ consisting of the set whose elements are x and y (so $\{x, y\} = \{y, x\}$). Also we can, without confusion, call x the **first co-ordinate** and y the **second co-ordinate** of (x, y).

Once armed with the concept of ordered pair, we construct the *Cartesian product* $A \times B$ of subsets $A, B \subseteq \mathscr{U}$ as follows.

3.1.1 DEFINITION. The **Cartesian product** $A \times B$ of A and B is the set of all ordered pairs (a, b) with $a \in A$ and $b \in B$. In symbols

$$A \times B = \{(a, b) | a \in A \ \& \ b \in B\}.$$

Of course, our universe is now $\mathscr{U} \times \mathscr{U}$, which can be thought of as containing (a copy of) $\mathscr{U}$: see the comment following 3.2.4.

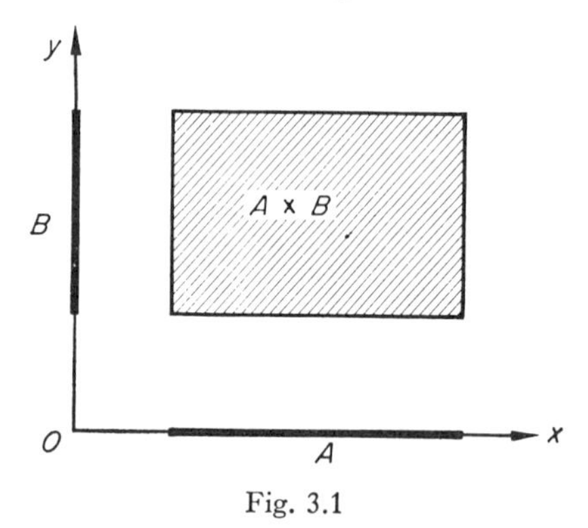

Fig. 3.1

† Our procedure here is open to logical objection, which we correct in 3.5; we have adopted it on pedagogical grounds.

The notation (a, b) might be taken to mean the hcf of a and b when $a, b \in \mathbb{N}$ (see 1.6.7), but the meaning is usually clear from the context.

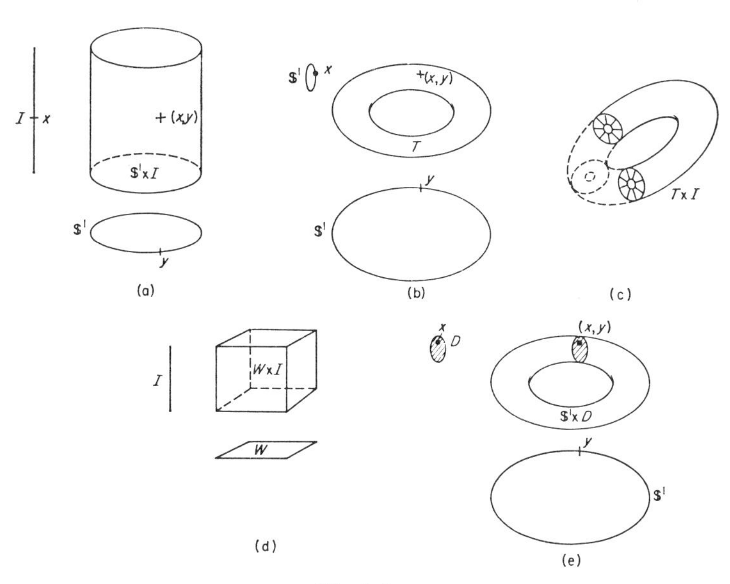

Fig. 3.2

3.1.2 EXAMPLE. The plane $\mathbb{R}^2$ of analytic geometry (defined in 1.3) was the first historical example of a Cartesian product, since

$$\mathbb{R}^2 = \mathbb{R} \times \mathbb{R};$$

hence the superscript 2, and the name 'Cartesian', after Descartes†, who invented analytic geometry. More generally, if A, B are intervals in $\mathbb{R}$, and and if we consider A, B respectively to lie in the x-axis and the y-axis (each a copy of $\mathbb{R}$), then $A \times B$ is a rectangular area (see Fig. 3.1). Other geometrical examples are the cylinder ($\approx \mathbb{S}^1 \times I$, where $\mathbb{S}^1$ is a circle and I an interval), the torus $T \approx \mathbb{S}^1 \times \mathbb{S}^1$, the 'thick' torus $T \times I$, the solid cube $W \times I$, where W is a square disc, and the solid torus $\mathbb{S}^1 \times D$ where D is a circular disc (see Fig. 3.2). These products are discussed from a topological point of view in Example 25.4.1.

3.2 Algebraic Properties

From Fig. 3.3, we are led to write down three algebraic properties of the Cartesian product. Proofs of these will be given in Chapter 8. The properties are:

3.2.1 $$A \times (B \cup C) = (A \times B) \cup (A \times C)$$

3.2.2 $$A \times (B \cap C) = (A \times B) \cap (A \times C)$$

† Bell [10], Ch. 3.

which we describe by saying that $\times$ is **distributive** over $\cup$ and $\cap$. Now if, in Fig. 3.3, B and C were pulled apart so as to become disjoint, then

3.2.3 $$A \times \varnothing = \varnothing,$$

an equation which can be *proved* immediately from the definitions. Also, if $\{z\}$ denotes a singleton, then the correspondence $a \to (a, z)$ is† easily seen to be an equivalence

3.2.4 $$A \approx A \times \{z\}.$$

This result enables us to regard a universe $\mathscr{U}$ as being contained in $\mathscr{U} \times \mathscr{U}$; for if we select a fixed element $u_0 \in \mathscr{U}$, then $\mathscr{U} \times \{u_0\} \subseteq \mathscr{U} \times \mathscr{U}$, so we can (by abuse of language) identify $\mathscr{U}$ with the copy $\mathscr{U} \times \{u_0\}$. This is a good copy, given by an equivalence $u \to (u, u_0)$ like that of 3.2.4. (If $\mathscr{U} = \varnothing$, then $\mathscr{U} = \mathscr{U} \times \mathscr{U}$, by 3.2.3.)

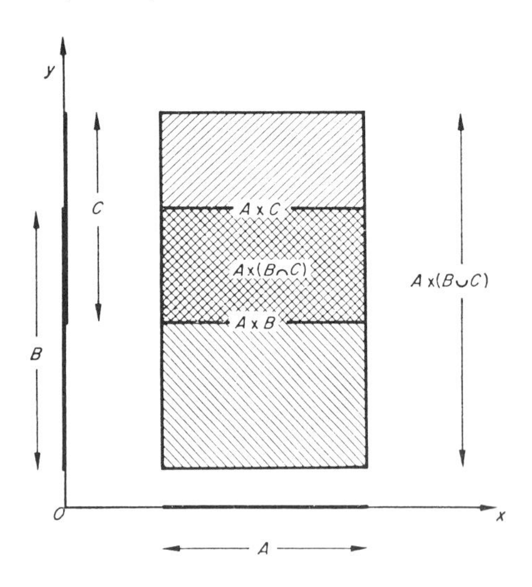

Fig. 3.3

Exercise 1

(i) Prove that the function $(a, b) \to (b, a)$ is an equivalence: $A \times B \approx B \times A$.
(ii) Prove that $A \times (B \times C) \neq (A \times B) \times C$ in general, but that the function $(a, (b, c)) \to ((a, b), c)$ is an equivalence: $A \times (B \times C) \approx (A \times B) \times C$. Find an inverse of this function. (Each set is equivalent to one denoted without ambiguity by $A \times B \times C$; see problem (v) below.)
(iii) Suppose $f: A \approx A'$, $g: B \approx B'$. Show that $h: A \times B \approx A' \times B'$, where $h(a, b) = (fa, gb)$.
§(iv) Prove that, if A and B are finite sets, then $\#(A \times B) = \#A\,\#B$. {Write B as $\{b_1, \ldots, b_{n-1}\} \cup \{b_n\}$ and use 3.2.4, 3.2.1, with induction on n.}

† We here adopt a common abbreviation, since 3.2.4 gives the range and domain of the correspondence.

(v) Prove that $A \times B$ is equivalent to the set of all functions $f: \mathbb{N}_2 \to A \cup B$, such that $f(1) \in A, f(2) \in B$.
(vi) Similarly, let $A \times B \times C$ denote the set of all functions $f: \mathbb{N}_3 \to A \cup B \cup C$, such that $f(1) \in A, f(2) \in B, f(3) \in C$. Prove that $A \times B \times C \approx A \times (B \times C)$.
(vii) If $A \subseteq X$ and $B \subseteq Y$, prove that $A \times B \subseteq X \times Y$.
(viii) The functions $p_1: X \times Y \to X$, $p_2: X \times Y \to Y$, given by $p_1(x, y) = x$, $p_2(x, y) = y$, are called the **projection functions**. Assuming $X \neq \varnothing \neq Y$, prove that they are onto. Let x_0, y_0 be fixed elements in X, Y respectively, and following 3.2.4, define the **injections** $i_1: X \to X \times Y$, $i_2: Y \to X \times Y$ by $i_1(x) = (x, y_0)$, $i_2(y) = (x_0, y)$. Prove that they are one-one, and that $p_\alpha \circ i_\alpha$ is the identity ($\alpha = 1, 2$) while $p_1 \circ i_2$ and $p_2 \circ i_1$ are the constant functions x_0 and y_0 respectively. When $X = Y = \mathbb{R}$ and $x_0 = y_0 = 0$, $i_1 \circ p_1$ and $i_2 \circ p_2$ are the projections on the x- and y-axes. To what geometrical operations do $i_2 \circ p_1$, $i_1 \circ p_2$ correspond here? (If $X \neq Y$, these maps are not defined.)
(ix) A, B are disjoint sets, and S, T are subsets of $A \cup B$ such that $S \cap A = T \cap A$ and $S \cap B = T \cap B$. Prove that $S = T$. Hence, if f denotes the function such that $f(S) = (S \cap A, S \cap B)$ prove that

$$f: \mathfrak{p}(A \cup B) \approx \mathfrak{p}A \times \mathfrak{p}B.$$

(x) X and Y are finite sets. Denote by $\mu(A)$ the set of functions $f: A \to Y$, for each $A \subseteq X$ where $\mu(A) = \varnothing$ if $A = \varnothing$. Given A and the function $f: X \to Y$ ($A \neq \varnothing$), assign to f the pair $(f|A, f|(X - A)) \in \mu(A) \times \mu(X - A)$. Show that this assignment is an equivalence: $\mu(X) \approx \mu(A) \times \mu(X - A)$. Hence prove that, given $f: A \to Y$, there are exactly $m(X - A)$ functions $g: X \to Y$, such that $g(a) = f(a)$ for each $a \in A$ (where $m(B) = \#\mu(B)$). Deduce that

$$m(X) = m(A) \cdot m(X - A),$$
$$m(A \cup B) = m(A) \cdot m(B)/m(A \cap B)$$

for each $A, B \subseteq X$. Hence calculate $m(A)$ in terms of $\#A, \#Y$.
(xi) In $A \times A$, the subset $\Delta = \{(a, a) | a \in A\}$ is called the **diagonal**. Show that the function $g: A \to A \times A$, given by $g(a) = (a, a)$, is one-one with image Δ. Thus g embeds A in $A \times A$. Compare the embedding following 3.2.4.

3.3 The Graph of a Function

In Example 3.1.2 where A, B are intervals in $\mathbb{R}$, suppose we have a function $f: A \to B$, and plot its graph. This means that we form the subset of $A \times B$, consisting of all ordered pairs (a, b) such that $a \in A$, and $b = f(a)$. Thus the graph of f is a subset G of $A \times B$, with the special properties that:

3.3.1 *given $a \in A$, there exists $b \in B$ such that $(a, b) \in G$;*

3.3.2 *if (a_1, b_1) and (a_2, b_2) are in G, and if $a_1 = a_2$, then $b_1 = b_2$.*

In Fig. 3.4, we show a subset C of $A \times B$ which is not the graph of any function since there are, for example, the three points (a, b_1), (a, b_2), and (a, b_3) in C. The above description of a graph can be extended to general functions, with arbitrary domain and range. Thus, we define the **graph** of any function

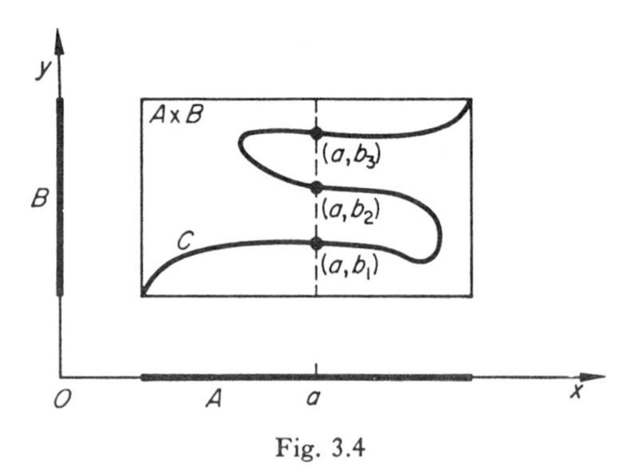

Fig. 3.4

$f: X \to Y$ to be that subset G of $X \times Y$ consisting of all pairs $(x, f(x))$, as x runs through X. The properties 3.3.1, 3.3.2 still hold for G.

Exercise 2

(i) Show that the circle $x^2 + y^2 = 1$ in $\mathbb{R}^2$ is not the graph of any function, but that it is the union of the graphs of two functions.
(ii) Let S be the circle in (i). Define $f: S \to \mathbb{R}$ by $f(x, y) = x$. Describe the graph of f in $S \times \mathbb{R}$.
(iii) With the same circle S, describe the graph of id: $S \to S$ on the torus $S \times S$.
(iv) Prove that for any sets A, B, the graph of $g: A \to B$ is equivalent to A.
(v) Prove that, for any set A, the graph of $\text{id}_A: A \to A$ is the diagonal (see Exercise 1(x)).

3.4 The Notion of Function, again

We shall now use the language of sets and Cartesian products to express the notion of 'function', thus clarifying our earlier use of the vague notion of a 'rule' or 'law'. From 3.3, it is clear that all the information about a function $f: X \to Y$ is given by the graph G, so nothing (except perhaps convenience) is lost by discarding f in favour of G. Since we already have the notions of subset and Cartesian product, we can avoid saying what a 'law' is (in the earlier working definition of the function concept) by framing the following definition, based on 3.3.1 and 3.3.2.

3.4.1 DEFINITION. Given $X, Y \subseteq \mathscr{U}$, then a **function**, with **domain** X and **range** Y, is defined to be any subset G of $X \times Y$ satisfying:

(i) every $x \in X$ is the first co-ordinate of some $(x, y) \in G$;
(ii) if (x, y_1) and (x, y_2) lie in G, then $y_1 = y_2$.

We cannot *prove* that this definition agrees with the earlier one given prior to 2.1.1, because of the undefined term 'law'. However, assuming that we can recognize a 'law' when we see one, we can with the same degree of certainty reconcile 3.4.1 with the earlier definition by the following observation. Given $x \in X$, then 3.4.1 guarantees that there is exactly one pair $(x, y) \in G$; so the 'rule', which assigns to x the second co-ordinate of (x, y), is a function (old definition), and its graph is G.. If we denote y by $G(x)$, it becomes natural to indicate the lack of symmetry in the definition by writing $G: X \to Y$ to denote G in 3.4.1. And now we can define a 'law' or 'rule' as a *function* (new definition)!

At the risk of pedantry we point out that a function, according to 3.4.1, is a subset G of $X \times Y$, *with its subset structure.* That is, it is a triple (X, Y, G). The point here is that if $X \supseteq X_0$, $Y \supseteq Y_0$ and G is a subset of $X_0 \times Y_0$, then it is also a subset of $X \times Y$. If, further, G is a function and $X = X_0$, then (X, Y_0, G) is again a function, but, according to our viewpoint, a different function from (X, Y, G).

Notice that the definition of equality in 2.2.1 now becomes a provable *theorem*, since we have already agreed on a definition (1.2.4) of equality of sets and so can *deduce* 2.2.1 from 1.2.4 and 3.4.1 (see Exercise 3(iii) below).

In practice, when specifying some function $h: A \to B$, we continue to use the method of Chapter 2, in which we use a form: 'Let h be given by $h(a) = \cdots$'; for this amounts to the same as giving all the ordered pairs $(a, h(a))$ as a runs through A. When we wish to emphasize this set of ordered pairs, we shall speak of the *graph* of h—a redundant but convenient expression.

Exercise 3

(i) In definition 3.4.1, what additional properties must G have to be (a) one-one, (b) onto, (c) a bijection? In (c) show that the description of G is symmetrical in X and Y. Hence give another proof of the Inversion Theorem 2.9.1.
(ii) Given functions $f, g: A \to B$ (thought of as graphs), when are $f \cap g$ and $f \cup g$ graphs of functions?
(iii) Prove the statement in the text, that 2.2.1 is deducible from 1.2.4 and 3.4.1.

3.4.2 EXAMPLE. Suppose we ask: what *sort* of object is denoted by the 'cup' symbol in $A \cup B$? And similarly for 'cap', and for the '+' denoting the sum of two integers in the set $\mathbb{Z}$ of 1.1.3. The concept of Cartesian product provides us with the language. For, if $\mathscr{U}$ is the universe of discourse, then we have a function

$$j: \mathfrak{p}\mathscr{U} \times \mathfrak{p}\mathscr{U} \to \mathfrak{p}\mathscr{U}$$

defined by $j(A, B) = A \cup B$, for each $A, B \in \mathfrak{p}\mathscr{U}$. Thus $\cup$ is nothing more than a function; similarly

$$\cap: \mathfrak{p}\mathscr{U} \times \mathfrak{p}\mathscr{U} \to \mathfrak{p}\mathscr{U}$$

is the function whose value at (A, B) is written $A \cap B$ instead of $\cap((A, B))$. And $+$ is the function

$$+: \mathbb{Z} \times \mathbb{Z} \to \mathbb{Z}$$

whose value at (n, m) is written $n + m$ instead of $+(n, m)$. Earlier writers used the word 'operation' to describe $\cup$, $\cap$, $+$ and similar objects, and the name is still in use, especially for functions with simple properties.

In case the domain or range of a function is empty, we have the following result, in which the graph concept is particularly helpful.

3.4.3 THEOREM. *Given the function $f: X \to Y$, then if X is non-empty, so is Y. If X is empty, so is f* (i.e., $f = \varnothing \subseteq X \times Y = \varnothing$).

Proof. If X is non-empty, there is at least one element in X. Therefore by condition (i) of 3.4.1, there must be at least one $y \in Y$ (such that $(x, y) \in f$). Hence $Y \neq \varnothing$. If $X = \varnothing$, then $X \times Y = \varnothing$, by 3.2.3. Therefore the *only* subset of $X \times Y$ in this case is $\varnothing$; and $\varnothing$ is a function since it satisfies conditions (i) and (ii) of 3.4.1 'vacuously', i.e., $\varnothing$ contains no (x, y) which violates these conditions. ■

Exercise 4

(i) When $X = \varnothing$, prove that the unique function $f: X \to Y$ is one-one, and that f is also a bijection if $Y = \varnothing$. Use 1.4.1 to show that f is then the identity function on $\varnothing$!

(ii) Given functions $f: A \to B$, $p: X \to Y$, define† a function $f \times p: A \times X \to B \times Y$ by $(f \times p)(a, x) = (f(a), p(x))$. Prove that $f \times p$ is one-one [onto] if f and p are one-one [onto]. Given also functions $g: B \to C$, $q: Y \to Z$, prove that $(g \times q) \circ (f \times p) = (g \circ f) \times (q \circ p)$. Hence show that if f and p are invertible, then so is $f \times p$, and $(f \times p)^{-1} = f^{-1} \times p^{-1}$. Investigate $f^\flat \times p^\flat$ in the general case.

3.5 Ordered Pairs again

We have just seen how to build the notion of function on those of subset and Cartesian product. The latter depends on the notion of ordered pair, which was defined above somewhat intuitively. (We did not say what (x, y) *was*.) However, this, too, can be defined by using only the notion of subset, at the cost of a certain complexity (which we preferred not to dwell on, when first introducing the idea to the reader). The point of this approach is to avoid non-mathematical statements like 'x appears first', and possible philosophical difficulties about time, about how an element can 'appear', and about references to our method of writing from left to right.

Let then $\mathscr{U}$ be a universe of discourse, and let $x, y \in \mathscr{U}$. Then the ordered pair (x, y) is a member of the universe $\mathfrak{p}(\mathfrak{p}\mathscr{U})$ and defined as follows.

3.5.1 DEFINITION. The **ordered pair** (x, y) is a set, given by

$$(x, y) = \{\{x\}, \{x, y\}\},$$

† Here is an example where it is easier *not* to think of a function as a graph!

consisting of the singleton $\{x\}$ and the set $\{x, y\}$. [Thus $(x, y) \in \mathfrak{p}(\mathfrak{p}\mathscr{U})$.] The **first co-ordinate** of (x, y) is x, the **second co-ordinate** is y.

The reader here might object, and say 'I never thought of an ordered pair like that'. We simply challenge him to express *what* he thought, in equally basic language. Further, as a proof of the pudding, we prove the following result, which tells us that our previous *usage* of ordered pairs has been correct, even if we did not say (or know) what an ordered pair was. A mathematician is content with such a statement; it might, however, worry a philosopher.

3.5.2 THEOREM. *The ordered pairs (x, y), (u, v), with $x, y, u, v \in \mathscr{U}$ are equal iff $x = y$ and $u = v$.*

Proof. Analogously to 3.5.1,

$$(u, v) = \{\{u\}, \{u, v\}\}.$$

Hence, if $(x, y) = (u, v)$, the two singletons must be equal (by definition of equality of sets), so $x = u$. Therefore

$$\{x, y\} = \{u, v\} = \{x, v\},$$

so $y = v$. Conversely, if $x = u$ and $y = v$, then clearly $(x, y) = (u, v)$. ■

Observe that, if $A, B \in \mathfrak{p}\mathscr{U}$, and $a \in A$, $b \in B$, then $(a, b) \in \mathfrak{p}(\mathfrak{p}\mathscr{U})$, so $A \times B \in \mathfrak{p}(\mathfrak{p}(\mathfrak{p}\mathscr{U}))$. This makes $A \times B$ seem fearsomely complex, but we do not worry: we continue to use $A \times B$ as we did before, and regard the results of this section as justification. However, we must realize that, in Example 3.1.2, the torus is *equivalent* to $\mathbb{S}^1 \times \mathbb{S}^1$, not *equal* to it, and similarly for the other sets of that example. Once realized, the abuse of language 'the torus is $\mathbb{S}^1 \times \mathbb{S}^1$' is permitted—indeed encouraged!

Exercise 5

Show how statements of the form $(\forall P)(\forall Q)\psi(P, Q)$ can be written as $\forall R, \phi(R)$; and similarly for the quantifier $\exists$.

3.6 Multiplicative Systems

We now take up the line of thought in Example 3.4.2, that is, when we have a function of the form $f\colon A \times A \to A$, as in algebra. It is customary then to write f 'multiplicatively', denoting $f(a, a')$ by aa', and to call (A, f)—or simply A, if f is understood—a 'multiplicative' system. However, the reader should beware of supposing that this 'multiplication' has any necessary connection with the multiplication of ordinary arithmetic: for example A might be a set of functions, and 'multiplication' aa' might mean composition $a \circ a'$. We shall consider in this section a few general facts about multiplicative systems which are applied repeatedly in many apparently different situations; indeed they already have been touched on, in 1.5, 1.7, and 2.7, concerning the algebra of sets and that of functions, and they apply also to such things as matrices. The treatment may perhaps strike the reader as aridly abstract, but, because of the several possible concrete interpretations of the results, we prefer an abstract presentation.

It often happens that a multiplicative system A possesses a **neutral element** e, that is to say an element having the property:

3.6.1 $\quad$ *For all* $a \in A, \quad ae = a = ea.$

For example, if f is addition of integers, so that f is $+: \mathbb{Z} \times \mathbb{Z} \to \mathbb{Z}$, then the neutral element is the integer zero; if f is multiplication of integers then the neutral element is 1; if f is union of sets, so that f is $\cup: \mathfrak{p}\mathcal{U} \times \mathfrak{p}\mathcal{U} \to \mathfrak{p}\mathcal{U}$, then e is $\varnothing$. Here as elsewhere the reader must acquire the habit of 'casting roles': just as we move from arithmetic to algebra in school by 'allowing x to denote any number', we are here allowing A to denote any multiplicative system, and we shall introduce the reader to a stock of such systems as we proceed through the book. We continually need

3.6.2 PROPOSITION. *A neutral element in A is unique, if it exists.*

Proof. If e' were a second neutral element, then putting $a = e'$ in 3.6.1 we have

$$e'e = e' = ee';$$

while replacing e in 3.6.1 by e' and putting $a = e$ we have

$$ee' = e = e'e,$$

whence $e = e'$ and e is unique. ■

Thus when we have found *one* element in A which acts as a neutral element, we need look no further.

Exercise 6

(i) In the system $(\mathbb{Z}, f)$ where $f(a, b) = \lambda a + \mu b$ for fixed $\lambda, \mu \in \mathbb{Z}$, show that there exists a neutral element only if $\lambda = \mu = 1$, and then this element is 0.
(ii) By considering $\cup: \mathfrak{p}\mathcal{U} \times \mathfrak{p}\mathcal{U} \to \mathfrak{p}\mathcal{U}$, prove that $\varnothing$ is unique.
(iii) Show that $\mathrm{id}_X: X \to X$ is the only function $g: X \to X$ with the property that for all functions $p: X \to X$, $pg = p = gp$. [Here, with 2.7.3 in mind, we 'cast roles' by considering the multiplicative system M of all functions $X \to X$, where ab now means the composition $a \circ b$.]
(iv) Let A, B be multiplicative systems with neutral elements e, e' respectively. Show that $A \times B$ can be turned into a multiplicative system in which (e, e') is the neutral element.

Particularly important multiplicative systems are the 'associative' ones; we call A **associative** iff

3.6.3 $\quad$ *For all* $a, b, c \in A, \quad a(bc) = (ab)c.$

For example, addition and multiplication of numbers are each associative operations. With subsets of $\mathcal{U}$, $\cup$ and $\cap$ are associative; and the system M

of all functions $X \to X$ in Exercise 6 (iii) is associative, by 2.7.6. On the other hand, the integers do not form an associative system when 'multiplication' means ordinary subtraction, nor do the rational numbers under† ordinary division. The force of 3.6.3 is to allow us to ignore the order of the bracketing.

Now suppose that A has a neutral element e. It may then happen that a given $a \in A$ has a **multiplicative inverse**, that is to say, there exists $a_0 \in A$ such that

3.6.4 $$aa_0 = e = a_0a.$$

Such an a_0 is unique (given a) if A is associative.

3.6.5 PROPOSITION. *Let A be an associative multiplicative system with neutral element. If a_0, a_1 are multiplicative inverses of $a \in A$, then $a_0 = a_1$.*

Proof. (Cf. 2.9.4.) By the first half of 3.6.1,

$$\begin{aligned} a_1 = a_1e &= a_1(aa_0) \\ &= (a_1a)a_0 \quad \text{by associativity} \\ &= ea_0 = a_0 \quad \text{by the second half of 3.6.1.} \end{aligned}$$

Thus $a_0 = a_1$. ■

This unique inverse of a (when it exists) is denoted by a^{-1} and a is called **invertible** (in A). In examples of associative systems, once we have found *an* inverse for a, we need look no further. *We henceforth assume our systems associative.*

3.6.6 COROLLARY. *If a is invertible in A, so is a^{-1}; and $(a^{-1})^{-1} = a$.*

Proof. If we read 3.6.4 from right to left, the equations tell us that a is an inverse for a_0; thus a is *the* inverse of a^{-1}, whence $a = (a^{-1})^{-1}$. ■

3.6.7 COROLLARY. *If $a, b \in A$ have inverses, so has ab, and $(ab)^{-1} = b^{-1}a^{-1}$.*

The proof is word for word the same as that of 2.9.9. ■

These results can be generalized; see for example the ATM pamphlet [9] or Zassenhaus [138]. See also 18.2.2, and Chapter 38.

Exercise 7

(i) Show that e is invertible in A.
(ii) Show that in the multiplicative system $(\mathfrak{p}\mathscr{U}, \cup)$, the only invertible element is $\varnothing$; while in $(\mathfrak{p}\mathscr{U}, \cap)$ the only invertible element is $\mathscr{U}$.
(iii) What are the invertible elements in $\mathbb{Z}$ with ordinary multiplication?
(iv) Show that the invertible elements in a multiplicative system form a group.
(v) Let A, B be multiplicative systems. Turn $A \times B$ into a multiplicative system by setting

$$(a_1, b_1)(a_2, b_2) = (a_1a_2, b_1b_2).$$

† Here 'under' is a common technical abbreviation for 'when "multiplication" means'.

Show that $A \times B$ is associative iff A and B are. Find the invertible elements in $A \times B$ in terms of those of A and B.

Given multiplicative systems A, B, consider a function $h: A \to B$, such that, for all $a_1, a_2 \in A$,

3.6.8 $$h(a_1a_2) = h(a_1)h(a_2).$$

Then h is called a **homomorphism**† (note that the product on the left is in A, that on the right in B). Such functions are said to 'preserve the structure' since they map products to products; and a powerful method of studying a system such as A is to find a known B and investigate homomorphisms of A in B or of B in A. If h happens also to be invertible, then $h^{-1}: B \to A$ exists (see 2.9.5); we prove

3.6.9 PROPOSITION. *h^{-1} is a homomorphism.*

Proof. Given $b_1, b_2 \in B$, we must (by 3.6.8) show that

$$h^{-1}(b_1b_2) = h^{-1}(b_1)h^{-1}(b_2).$$

Since h is invertible it is onto, so there exist $a_1, a_2 \in A$, such that $h(a_i) = b_i$, $i = 1, 2$. Then

$$\begin{aligned} h^{-1}(b_1b_2) = h^{-1}(h(a_1)h(a_2)) &= h^{-1}(h(a_1a_2)) && \text{by 3.6.8,} \\ &= a_1a_2 && (h^{-1} \circ h = \mathrm{id}_A), \\ &= h^{-1}(b_1)h^{-1}(b_2). \quad \blacksquare \end{aligned}$$

Thus if h is invertible as a function, it is invertible as a homomorphism. If h is an invertible homomorphism, it is called an **isomorphism**‡, and A and B are then **isomorphic** (differing 'only in notation'). An important problem of mathematics is to make a collection of standard multiplicative systems, no two isomorphic, such that any other is demonstrably isomorphic to one in the collection. This problem is largely unsolved, even with the restriction that all members of the collection should have, say, 1000 elements. On the other hand, many such 'structure' theorems are known, and several will be discussed later in this book.

Exercise 8

(i) Use Exercise 4(ii) to show that the following diagram 'commutes' (i.e. $g \circ (h \times h) = h \circ f$) iff h is a homomorphism. Use Exercise 7(v) to show that $h \times h$ is a homomorphism. Prove that the multiplication on $A \times B$ of Exercise 7(v) is the only one for which the projection functions in Exercise 1(vii) are homomorphisms. (Here f, g are the multiplications on A, B respectively.)

† From Greek *homo* = 'same', and Greek *morphe* = 'structure'.
‡ *iso-* is Greek for 'equal'.

$$\begin{array}{ccc} A \times A & \xrightarrow{f} & A \\ {\scriptstyle h\times h}\downarrow & & \downarrow{\scriptstyle h} \\ B \times B & \xrightarrow[g]{} & B \end{array}$$

(ii) Let $h: A \to B$ be a homomorphism. Show that if A is an associative multiplicative system with neutral element e, so is $h(A)$, with neutral element $h(e)$; and that if h is onto, then $h(e)$ is the neutral element of B.
(iii) If, further, $h(e)$ is the neutral element of B, prove that h maps invertible elements in A to invertible elements in B.
(iv) Show that $\complement: (\mathfrak{p}\mathscr{U}, \cup) \to (\mathfrak{p}\mathscr{U}, \cap)$ is an isomorphism, which maps the neutral element of the first system to that of the second.
(v) For each $a \in A$, denote by $L_a: A \to A$ the function such that $\forall x \in A, L_a(x) = ax$. Show that if a is invertible, then L_a is invertible, and find an example to show that L_a is not necessarily a homomorphism. Show that A is associative iff $(\forall a, b)L_a \circ L_b = L_{ab}$.
(vi) Let $R_a: A \to A$ denote the function such that $(\forall x \in A)R_a(x) = xa$. Show that A is associative iff $(\forall b, c)R_c \circ R_b = R_{bc}$; and also, A is associative iff $(\forall a, c)L_a \circ R_c = R_c \circ L_a$. [The last equation tells us that the two functions $L_a, R_c: A \to A$ 'commute'.]
(vii) Let A be associative and let B denote the multiplicative system of all functions $A \to A$ (with 'multiplication' taken as composition). Show that the function $L: A \to B$, where $L(a) = L_a$, is a homomorphism. If A has a neutral element e, show that L is one-one. Show also that $L_e = \mathrm{id}_A$.

[Here, L **embeds** A in the 'bigger' system B.]

(viii) Describe B when A is $\mathbb{Z}_2$ or $\mathbb{Z}_3$, and describe the image of A under the embedding L.

Chapter 4
RELATIONS

4.1 What Is a Relation?

We often have to say, of two objects x, y, that they are 'related', in some way. For example, if $x, y \in \mathbb{R}$ we might have '$x < y$'; if x, y are integers we might have '$x - y$ is exactly divisible by 3', and in ordinary conversation we might have x 'better than' y, x 'more beautiful than' y, x 'like' y, x 'a brother of' y, and so on. These are all instances of a sentence of the form 'xRy'—x 'is in the relation R to' y; but, while we may know what the above specific instances of an R mean, what do we mean in general by a *relation* R? On reflection we can give a definition as follows. We would in principle know all we need know about a relation R, if we knew *all* pairs x, y such that xRy. Thus, we use the same idea as in Definition 3.4.1 and make the following definition.

4.1.1 DEFINITION. Given $X, Y \subseteq \mathscr{U}$, then a **relation** with **domain** X and **range** Y is defined to be any subset R of $X \times Y$. If $(x, y) \in R$ we write as well xRy; and if $X = Y$ we say that R is a **relation on** X.

Since R in 4.1.1 is a set, it may well be given in the form 1.1.2, as $R = \{(x, y) | \psi(x, y)\}$, where $\psi(x, y)$ is a statement about x and y. Just as sets are sometimes not distinguished from their defining propositions, it is common to call ψ a 'relation' also, and then R is called the **graph** of ψ. In practice, one wishes to know a convenient ψ rather than its graph; for example, it is necessary to know when a new baby becomes a brother, in order to decide whether or not to add to the list of all brother-pairs, since this list is changing in time.

4.1.2 EXAMPLE. If $X = \{\text{Cain, Abel}\}$, then the relation of brotherhood on X is the whole of $X \times X$, provided brothers are simply required to have the same parents and be of the male sex. This is not, however, standard usage!

Exercise 1

(i) Let $X = \{\text{Lear's three daughters}\}$, $Y = \{\text{King Lear}\}$. Show that here the relation 'is a daughter of' has graph $X \times Y$.
(ii) Let X denote the *dramatis personae* of the Theban plays of Sophocles. Enumerate, on $X \times X$, the sets corresponding to the relations 'is a child of', 'is a parent of'.

4.2 The RST Conditions

By 3.4.1, a function $f: X \to Y$ is just a special kind of relation. The most useful relations, other than functions, are those where $X = Y$ and which satisfy one or other of the conditions 4.2.1 below. Let then R be a relation on a set X. We call R **reflexive**, **symmetric**, or **transitive**, according as R satisfies the appropriate one of the following **RST conditions**:

4.2.1
- **Reflexivity.** For all $x \in X$, xRx.
- **Symmetry.** Whenever xRy, then also yRx.
- **Transitivity.** If xRy and yRz, then xRz.

R is **strict** iff *it is never* reflexive (i.e., for no x is xRx).

4.2.2 EXAMPLE. The relation† of equality, $x = x$, satisfies all three of the conditions 4.2.1; it is not strict. The relation of inclusion, $X \subseteq Y$, between subsets of $\mathscr{U}$, is reflexive (by 1.2.3) and transitive (by 1.4.1) but clearly not symmetric. The relation 'not greater than', on the set of integers is also reflexive and transitive, but not symmetric; while the relation 'less than' is transitive and strict only. The relation specified by 'x loves y' on the set of humans is often not reflexive, while its (frequent!) lack of symmetry and transitivity is the basis of a large part of poetry and drama. [Strictly speaking, a relation R is not reflexive iff there exists an x such that xRx is false. But we allow the abuse of language, meaning by 'R is often not reflexive' that 'for a lot of x, xRx is false'. Similar remarks apply to our later use of 'frequent'.]

4.3 Linear Graphs

When X is finite, we can picture a strict relation R on X by means of a **linear graph**, as follows. Let $X = \{x_1, \ldots, x_n\}$, and let $P_1, \ldots, P_n$ be n distinct points in space $\mathbb{R}^3$. For each $i, j = 1, \ldots, n$, we join P_i to P_j by an arrow λ_{ij} directed from P_i to P_j iff x_iRx_j; no two of the arrows λ_{ij}, λ_{rs} are to intersect except possibly at end-points. The points P_i and arrows λ_{ij} are called the 'nodes' and 'edges', respectively, of the graph. We have already constructed such a linear graph in Example 1.5.3, where X was the set of 'states' there described, and sRs' means 'state s' is obtainable in one allowed move from state s'. If the relation R on X is not strict, then we can represent the fact that x_iRx_i by adding a loop at P_i. This would occur with the ferryman problem (1.5.3), if to leave a state unaltered is an allowed move.

In Fig. 4.1, we illustrate the linear graph corresponding to the relation $\subset$ of strict inclusion on the set $\mathfrak{p}X$, $X = \{a, b, c\}$. We have denoted the nodes of the graph by the corresponding elements of $\mathfrak{p}X$; and some arrows have been omitted, with the understanding that $\subset$ is transitive (e.g., the arrow from $\varnothing$

† In commentary of this kind we allow a certain imprecision of the same kind as saying 'the function $f(x)$'.

to $\{c, a\}$). The crossings of arrows could be avoided by curving the arrows in space.

For a general theory of linear graphs see Berge [13]; also 5 Exercise 1(xiii). As to their value as a means of displaying information vividly, an interesting development of the technique can be found in the book of Papy [100]; here, a picture is used to convey the essence of an entire proof, and much work is needed to develop these techniques further. The reader has only to see some of the diagrams in a book like Eilenberg–Steenrod [35], to see how diagrammatic methods of proof can help the understanding of complicated mathematics.

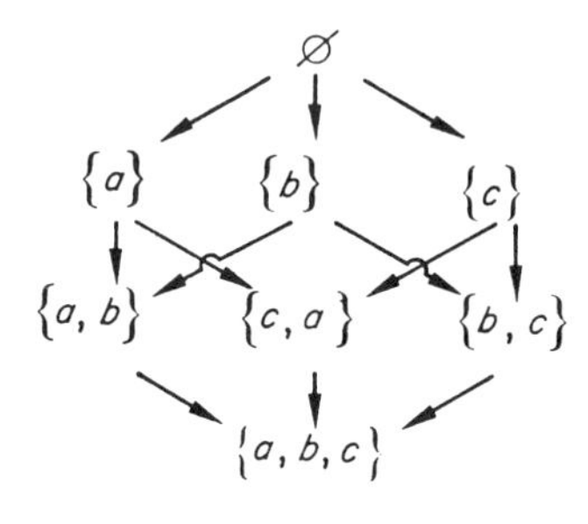

Fig. 4.1

Exercise 2

(i) What is the relation to which a family tree corresponds?

(ii) With the line 'Your servant's servant is your servant, madam' (*Twelfth Night*, III, 1, 99), Shakespeare implies that the relationship of being a servant is transitive. Do you agree with him?

(iii) Consider the set X of characters in a Shakespeare play (e.g., *A Midsummer Night's Dream*, or *Othello*). Construct a linear graph corresponding to the relations 'x loves y', 'x hates y', 'x kills y', 'x is killed by y'. In case of difficulty in deciding, of a given pair, whether or not a relationship is fulfilled, modify the relation or make it more precise if you can. Would you say that *A.M.N.D.* was less 'hate-full' than *Othello* from their graphs? Could you formulate relations of this kind on the set of all plays? The graph might tell when one play had a 'more complex' structure than another. See if you can formulate this notion more precisely. Do the same for musicals, or for some symphonies that you know.

(iv) Formulate Grimm's Laws concerning sound changes in language, in terms of functions and relations.

(v) Analyse the legends in Frazer's *Golden Bough* after the manner of (iii).

(vi) Let R, S be relations with domain X, and range Y. Prove that if R, S are both reflexive, then $R \cap S$ is reflexive, and similarly for the properties 'symmetric', and 'transitive'. Show by counter-examples where necessary, that similar statements do not always hold for $R \cup S$. Define $\hat{R}$, the **inverse** of R, to be $\{(x, y) | (y, x) \in R\}$ and discuss the reflexivity, symmetry, and transitivity of $\hat{R}$ in terms of the corresponding properties for R. If R is a function, relate $\hat{R}$ and R^{-1}. (Cf. 3 Exercise 3(i).)

(vii) If R is a relation on X, prove that $R \cap (A \times A) = R_A$ is a relation on A, when $A \subseteq X$ (R_A is said to be **induced** by R on A). Investigate the RST properties of R_A in terms of those of R. {Can R_A be RST while R is not?}
(viii) Show how to represent a function $f: A \to B$ by means of a linear graph whose nodes are $A \cup B$ with edges joining a to $f(a)$. If f is invertible, describe the linear graph of f^{-1} in terms of that of f, and hence give a quick plausible argument for the inversion theorem. {See Fig. 2.1.}

4.4 Orderings

Two kinds of relation are of especial importance, namely the *order* relations and the *equivalence* relations. We shall deal with order relations first.

4.4.1 DEFINITION. Let R be a relation on a set X. If R is transitive, and satisfies the condition

(i) *xRy and yRx together imply $x = y$ $(x, y \in X)$*,

then R is said to be an '**order relation**' or '**ordering**' on X, and X is '**ordered**' by R. The pair (X, R) is an **ordered set.**

4.4.2 EXAMPLE. The relation of equality is an order relation.

4.4.3 EXAMPLE. If $\mathscr{U}$ is a universe of discourse, then the relation $\subseteq$ is an order relation on $\mathfrak{p}\mathscr{U}$, by 1.4.1 and 1.2.4. The strict relation $\subset$ is also an order relation on $\mathfrak{p}\mathscr{U}$, with (i) of 4.4.1 'vacuously' satisfied (we never have $A \subset B$ and $B \subset A$ for any $A, B \in \mathfrak{p}\mathscr{U}$, so (i) is never violated).

4.4.4 EXAMPLE. The usual alphabetical ordering in a dictionary is an order relation on the set of words listed in the dictionary. It is based on the order relation inherent in the alphabet.

4.4.5 EXAMPLE. The set $\mathbb{Z}$ of positive and negative integers is ordered by $\leqslant$; and also by the strict relation $<$, for the same reason as in 4.4.3. Similarly, the natural numbers $\mathbb{N}$, the rationals $\mathbb{Q}$ and the real numbers $\mathbb{R}$ are ordered by $\leqslant$ and $<$; these relations on $\mathbb{R}$ induce those on the subsets $\mathbb{Q}$ and $\mathbb{Z}$. They are used, also, to obtain orderings in physics, as for example, when one point p of a region G is said to be hotter than another, q. Here, the temperature function $\theta: G \to \mathbb{R}$ is used, to say that p is hotter than q when $\theta(q) < \theta(p)$. Similarly when we say that a note is louder, or higher, than another; temperature is replaced by decibels or frequency, respectively.

Exercise 3

Let (X, R), (Y, S) be ordered sets. Define a relation V on $X \times Y$ by setting $(x_1, y_1)V(x_2, y_2)$ iff either x_1Rx_2 and $x_1 \neq x_2$ or $x_1 = x_2$ and y_1Sy_2. Show that V is an ordering; it is called a **dictionary** (or **lexicographical**) **ordering**, by analogy with Example 4.4.4, which it generalizes.

The orderings in Examples 4.4.4, 4.4.5 have the additional property that they satisfy the *law of trichotomy*:

4.4.6 DEFINITION. The ordered set (X, R) is **linearly** ordered or **totally** ordered iff it satisfies the **law of trichotomy**, that, given $x, y \in X$, then $x = y$ or xRy or yRx.

4.4.7 EXAMPLE. The ordered set $(\mathfrak{p}\{a, b, c\}, \subset)$, whose linear graph is Fig. 4.1, is not linearly ordered, since (for example), given the two elements $\{a\}$, $\{b\}$, none of the three possibilities in 4.4.6 is satisfied.

Exercise 4

(i) Is an army, ordered by rank, a linearly ordered set? How about the set of generals, ordered by success? [Here there is a preliminary difficulty, that of defining the orderings.]

(ii) Let X denote the set of objects in a large store, ordered by cost; write xCy when x is cheaper than y. Prove that the function $f: X \to \mathbb{Z}^+$, where $f(x) =$ cost of x to the nearest penny, is an *order-preserving* function $f: (X, C) \to (\mathbb{Z}^+, \leqslant)$; i.e., show that if xCy then $f(x) \leqslant f(y)$. Need f be onto or one-one?

The ordering in a telephone directory, and the ordering $\leqslant$ on $\mathbb{N}$ have an even stronger property than that of a linear ordering; they are *well-orderings*:

4.4.8 DEFINITION. The ordered set (X, R) is **well-ordered** iff every non-empty subset S has a **least** member, i.e., some $s_0 \in S$ is such that s_0Rs for every $s \in S$.

Observe that the definition tells us *two* things about S: first that some s_0 exists with s_0Rs, and second that s_0 lies not just in X, but in S itself. The last remark about $\mathbb{N}$ is so important, that we record it as

4.4.9 $\qquad$ *$\mathbb{N}$ is well-ordered by $\leqslant$.*

4.4.10 EXAMPLE. If we let $\mathbb{Q}$ denote the set of all rationals in $\mathbb{R}$, $\mathbb{Q}$ is *not* well-ordered. For, if

$$S = \{q | q \in \mathbb{Q} \ \&\ 0 < q\}$$

then S is certainly non-empty (since $\mathbb{N} \subseteq S$), and yet *there exists no $q_0 \in S$ such that $q_0 \leqslant q$ for every $q \in S$.* For if q_0 were such, then $0 < \frac{1}{2}q_0 < q_0$, contradicting the 'least' property of q_0. Of course $0 \leqslant q$ for each $q \in S$, but $0 \notin S$.

Exercise 5

(i) Let $T = \{q | q \in \mathbb{Q} \ \&\ 2 \leqslant q^2\}$. Show that T is non-empty but has no least member. {Prove that if $r/s = q_0$ were such a least member ($r, s \in \mathbb{N}$), then $(kr - 1)/ks = q' \in T$ for a sufficiently large integer k; and obtain a contradiction, since $q' < q_0$. If we assume that $\sqrt{2} \in \mathbb{R}$ is not rational (see 24 Exercise 1(i)), then q_0 would have to be $\sqrt{2}$, which is not even in $\mathbb{Q}$ let alone T. The same example shows that $(\mathbb{R}, \leqslant)$ is not well-ordered; for here, $\sqrt{2}$ certainly exists in $\mathbb{R}$, but it is not rational, and so cannot lie in T.}

(ii) Prove that every subset of a well-ordered set is well-ordered.

4.4.11 EXAMPLE. (A quibble.) $(\mathbb{Z}^+, <)$ is *not* well-ordered; for $\mathbb{Z}^+$ itself is non-empty, with first member zero; but it is not true that $0 < n$ for every $n \in \mathbb{Z}^+$ (viz., take $n = 0$). Similarly, no strictly ordered non-empty set is well-ordered. (Our 'error' was to take $<$ instead of $\leqslant$.)

The fact that $(\mathbb{N}, \leqslant)$ is well-ordered immediately implies

4.4.12 THEOREM. *Let A be any subset of $\mathbb{Z}^+$. Then $(A, \leqslant)$ is well-ordered.* ■

Exercise 6

(i) Prove 4.4.12 in detail, and also show by displaying a suitable set A, that $(\mathbb{Z}, \leqslant)$ is *not* well-ordered. [This example has been used by Russell [111] to show the fallacy in the mediaeval Scholastic 'first cause' argument for the existence of God.]

(ii) Investigate the dictionary ordering V in Exercise 3. Is it linearly ordered, or well-ordered, when R and S are?

(iii) Prove that a well-ordered set is linearly ordered. Hence prove that the empty subset of an ordered set is both well-ordered and linearly ordered.

A prime consequence of 4.4.9 is the following result, to be discussed more fully in Chapter 5.

4.4.13 (**The Induction Theorem.**) *Let $A \subseteq \mathbb{N}$ be a subset with the propertiesl*

(i) $1 \in A$,

(ii) *if $n \in A$, then $n + 1 \in A$.*

Then $A = \mathbb{N}$.

Proof. By 1.4.1, we must prove that $\mathbb{N} \subseteq A$ and we accomplish this by the method of *reductio ad absurdum*. If $\mathbb{N} \nsubseteq A$ then there exists some $m \in \mathbb{N} - A$. Hence $\mathbb{N} - A$ is a non-empty set; and since $\mathbb{N}$ is well-ordered by $\leqslant$, then $\mathbb{N} - A$ possesses a least member, say m_0. That is, $m_0 \in \mathbb{N} - A$ and for any $m \in \mathbb{N} - A$, $m_0 \leqslant m$. Now, $m_0 > 1$ by property (i) above; so $m_0 = 1 + k$, say, when $k \in \mathbb{N}$. But then $k < m_0$, so $k \in A$. Hence, by property (ii) above, $m_0 = 1 + k \in A$. Thus, $m_0 \in A \cap (\mathbb{N} - A) = \varnothing$, which is impossible. This contradiction has arisen from the assumption that $\mathbb{N} \nsubseteq A$, which assumption must therefore be false. Hence† $\mathbb{N} \subseteq A$. ■

4.5 Equivalence Relations

We shall return to the topic of well-orderings in Chapter 5. Meanwhile,. let us now consider those relations called 'equivalence' relations. These are defined as follows.

† Observe that we have assumed of $\mathbb{N}$ only that it is non-empty, well-ordered by $\leqslant$, and that there is a one-one function $S: \mathbb{N} \to \mathbb{N} - \{1\}$ (with $S(n)$ *denoted* by $n + 1$) such that $n < S(n)$ for each $n \in \mathbb{N}$. It was shown by Dedekind that this is all one needs to know about $\mathbb{N}$ to be able to deduce all its properties. See Chapter 23.

4.5.1 DEFINITION. Let R be a relation on X, which is reflexive, symmetric, and transitive. Then R is called an **equivalence relation.**

After 2.10.2 we warned the reader not to confuse 'equivalence relation', with 'function which is an equivalence'. Consider, however, the following example which shows that the relation of being equivalent sets *is* an equivalence relation.

4.5.2 EXAMPLE. Let $\mathscr{U}$ denote a universe of discourse; if $A, B \in \mathfrak{p}\mathscr{U}$ and A and B are equivalent, we write $A \approx B$ as in 2.10.2. This is a relation on $\mathfrak{p}\mathscr{U}$, and we assert that it is an *equivalence relation.* For, if in 2.10.1 we take $B = A$, $f = \mathrm{id}_A$, we get reflexivity, $A \approx A$; symmetry follows from the inversion theorem and 2.9.5, because then $f^{-1}: B \approx A$ exists and is invertible; and transitivity follows from 2.8.2, because if $A \approx B$, $B \approx C$, then there are bijections $f: A \approx B$, $g: B \approx C$, so $g \circ f: A \approx C$ is invertible and $A \approx C$.

4.5.3 EXAMPLE. In the set $\mathbb{Z}$ of all integers let n be a fixed integer such that $n > 0$. Define the relation C_n of 'congruence mod n' on $\mathbb{Z}$ by setting xC_ny or more conventionally

$$x \equiv y (\mathrm{mod}\ n)$$

(*read:* x is congruent to y modulo n) iff n divides $x - y$ without remainder. Thus $15 \equiv 0 (\mathrm{mod}\ 5)$ and $15 \equiv 1 (\mathrm{mod}\ 2)$. Then C_n is an equivalence relation, because (a) $x - x = 0$ and n divides 0, so C_n is reflexive; (b) if n divides $x - y$, then it divides $y - x$, so C_n is symmetric; and (c) if n divides $x - y$ and $y - z$, then n divides their sum, $x - z$, so C_n is transitive. By (a), (b), and (c), it is therefore an equivalence relation (one of the earliest known to mathematicians).

4.5.4 EXAMPLE. Let A denote the set of all people living in Britain on the night of the last census. Let R denote the relation on A given by: xRy iff x slept under the same roof as y on the night in question. Disallowing sleep-walkers, R is an equivalence relation on A.

The reason why we consider equivalence relations is that for some purposes we wish to ignore some distinguishing characteristic between objects, and lump together those objects differing only with respect to that characteristic†. In Example 4.5.3 we ignore multiples of n, and in Example 4.5.4 we ignore individuality to the extent of a household, and divide up the population, not into humans, but into households. In physics, this is forced upon us by the coarseness of measuring apparatus; we observe 'blobs' of space and time, not individual points and instants. This idea has been used by Zeeman, for example; see his article 'Topology of the brain and visual perception' in the symposium [37].

† Cf. the saying, of certain families whose members all possess a marked facial resemblance: 'If you've seen one, you've seen the lot'.

More precisely, an equivalence relation R on a set X 'partitions' X into 'cosets' in the following way. For each $x \in X$ define the **coset** xR (the **coset of** x **modulo** R) by

4.5.5 $$xR = \{y | y \in X \;\&\; xRy\}.$$

4.5.6 EXAMPLE. In Example 4.5.3, if $j \in \mathbb{Z}$, then jC_n is the set of all integers which leave the same remainder, after division by n, as does j. Thus, $0C_2$ and $1C_2$ are respectively the sets of all even, and all odd, integers.

Before commenting further on Definition 4.5.5, we shall make some deductions from it. Thus, since xRx, then $x \in xR$; since R is symmetric, then $y \in xR$ iff $x \in yR$. Even more:

4.5.7 *If* xRy, *i.e., if* $y \in xR$, *then* $xR = yR$.

Proof. Let $z \in xR$. Then xRz, by 4.5.5. In particular, $y \in xR$, so xRy, and therefore yRx, since R is symmetric. Therefore yRx and xRz, so yRz, since R is transitive. Hence $z \in yR$, by 4.5.5. This proves that $xR \subseteq yR$. But, by the remark preceding 4.5.7, we know that $x \in yR$, since $y \in xR$. Therefore the same argument as before proves $yR \subseteq xR$. Hence $yR = xR$ by Definition 1.4.1. ∎

4.5.8 *Two cosets*, xR, yR, *intersect* iff *they are equal.*

Proof. Since R is reflexive, $x \in xR$, so $xR \neq \varnothing$. Hence, *if* $xR = yR$, then xR and yR intersect (i.e., $xR \cap yR \neq \varnothing$). Conversely, if $xR \cap yR \neq \varnothing$, then some $z \in xR \cap yR$; so $z \in xR$, whence $xR = zR$ (by 4.5.7). Similarly $zR = yR$, so $xR = yR$. ∎

4.5.9 These results show that a given coset xR can be described in many ways; to be precise, xR is yR for each $y \in xR$. We say that we have made the **choice of representative** x or y, according as we denote this one coset by xR, or yR, respectively. It is therefore often necessary to show that an argument about a coset xR is independent of the choice of representative x, and therefore of the particular description xR of that coset. This part of the argument is called **checking independence of representatives**. Occasionally there is a compelling choice of representative; for instance, in Example 4.5.6, we often pick as representative of jC_n the *smallest* positive member. Thus $10C_3 = 1C_3 = (-2)C_3$, and we choose the form $1C_3$.

4.5.10 Returning to 4.5.8, we see that R splits X into a family of subsets of X, each being a coset modulo R, and any two of them are either equal or completely disjoint. This family is denoted by X/R (*read*: X **mod** R), and the construction of this **quotient set** is a method of making new sets from old.

Exercise 7

On the set A of all non-zero integers, define xRy to mean that $xy > 0$. Assuming the usual rules for operating with inequalities, prove that R is an equivalence relation on A, and that A/R consists of exactly two equivalence classes.

In the general case, as in Exercise 7, X is the *union* of the cosets xR—and we write

4.5.11 $X = xR \cup \cdots \cup yR \cup \cdots$, or $X = \bigcup_{x \in X} xR = \{y | \exists x, y \in xR\}$—

meaning that every element y of X lies in some one, at least, of the subsets xR. Of course, this usage of the union symbol is not confined to subsets which are cosets, nor to the case when they are equal or disjoint. Notice that it is sufficient to take the union $\cup$ over a set of *coset representatives.*

Exercise 8

(i) Prove that the definition of union, in 4.5.11, agrees with that of 1.6.1 when there are no more than two subsets. [Compare 1 Exercise 5(iv).]
(ii) Let $B = \{x | x \in \mathbb{R} \;\&\; \sin x > 0\}$. Prove that

$$B = \bigcup_{n \in \mathbb{Z}} B_n, \qquad B_n = \langle 2n\pi, (2n+1)\pi \rangle.$$

The way in which X has been expressed in 4.5.11, *with all the distinct subsets xR disjoint,* is called a **partitioning** of X. A fuller discussion of this notion is given in 4.6 below; for the moment, we conveniently summarize the above theory of an equivalence relation R on X in a theorem:

4.5.12 THEOREM. *An equivalence relation partitions its domain.* ■
(A form of converse is given in 4.6.4.)

4.5.13 EXAMPLE. In Example 4.5.4, the population A is split into separate households, these being the cosets for the relation R defined in the example. In Example 4.5.3, the set $\mathbb{Z}$ of integers is split into exactly n cosets; using the particular choice of representatives mentioned in 4.5.9, these are $1C_n, 2C_n, \ldots, nC_n$ where, for each integer $j \in \{1, \ldots, n\}$, the coset jC_n (often denoted by $[j]_n$ or $[j]$) is

$$[j]_n = \{k | k \in \mathbb{Z} \;\&\; j \equiv k \bmod n\}.$$

(Thus j and k leave the same remainder after division by n.) The resulting quotient set is written $\mathbb{Z}_n$. There are many other examples throughout the book

Exercise 9

(i) Let $f: X \to Y$ be a function. Define a relation R_f on X by: $x_1 R_f x_2$ iff $f(x_1) = f(x_2)$. [We say that R is **induced** by f.] Prove that R_f is an equivalence relation on X, and that the coset xR_f is $f^{\flat}\{f(x)\}$ (where $f^{\flat}: \mathfrak{p}Y \to \mathfrak{p}X$ is the function of 2 Exercise 5(iii)). What are the cosets for the function $s: \mathbb{R}^2 \to \mathbb{R}$ when $s(x, y) = x^2 + y^2$? Prove that, here, $X/R_s \approx \mathbb{R}^+ (= \{x | x \geqslant 0\})$.
(ii) On $\mathbb{R}^2$, define a relation L by: $(x, y)L(u, v)$ iff $x - y = u - v$. Show that L is an equivalence relation on $\mathbb{R}^2$, and that the coset mod L of (x_1, y_1) is the whole line $y = x + (y_1 - x_1)$. Thus L splits $\mathbb{R}^2$ into the system of all lines of gradient 1. Show also that L is the relation R_f of problem (i), when f is taken to be $f: \mathbb{R}^2 \to \mathbb{R}$, with $f(x, y) = y - x$. Prove that $\mathbb{R}^2/L \approx \mathbb{R}^1$.
(iii) The set L of all straight lines in the plane $\mathbb{R}^2$ contains the subset X of all lines through the origin; and the x-axis $\mathbb{R}$ is in X. Define relations P, I on L by:

lPl' iff $l = l'$ or l is parallel to l'; and lIl' iff either $l \cap l' \cap \mathbb{R} \neq \varnothing$ or lPl' and $lP\mathbb{R}$. Prove that P and I are equivalence relations on L, $L - \{\mathbb{R}\}$, respectively, and that each coset lP is equivalent to $\mathbb{R}$. Prove also that

$$L/P \approx X, \qquad (L - \mathbb{R}P)/I \approx \mathbb{R}.$$

4.6 Partitionings

The notation 4.5.11 requires a little more language to discuss it effectively. The system X/R of cosets xR of the equivalence relation R is an example of an *indexed family of subsets of* X:

4.6.1 DEFINITION. An **indexed** family of subsets of $\mathscr{U}$ is a function $X\colon A \to \mathfrak{p}\mathscr{U}$, where A is a set called the **indexing set** of the family. The notation $\{X_\alpha\}_{\alpha \in A}$ is common; we merely write X_α for the value $X(\alpha)$ of the function X. The **union** of the X_α is a subset of $\mathscr{U}$, denoted by

$$\bigcup_{\alpha \in A} X_\alpha$$

and is defined to be the set of all $x \in \mathscr{U}$, for which x lies in some one X_α (at least). The indexing is **strict** iff X is one-one.

4.6.2 EXAMPLE. The cosets of X mod R in 4.5.11 form a family indexed by X; the indexing is strict iff R is equality. The sets $\{B_n\}$ in Exercise 8(ii) form a family indexed strictly by $\mathbb{Z}$.

We now have the language to describe a converse of Theorem 4.5.12. Let us first say what we mean by a 'partitioning' of a set X; the theory leading to 4.5.12 gave an (informal) example.

4.6.3 DEFINITION. A family $\{X_\alpha\}_{\alpha \in A}$ **partitions** a set X, provided,

(i) $X = \bigcup_{\alpha \in A} X_\alpha$;

(ii) *for any* $\alpha, \beta \in A$, *either* $X_\alpha = X_\beta$ *or* $X_\alpha \cap X_\beta = \varnothing$.

The family $\{X_\alpha\}_{\alpha \in A}$ is then called a **partitioning** of X.

If we ignore the cement in a brick wall, a pavement, or a mosaic, the bricks or stones form a partitioning; for this reason, partitionings in geometry are sometimes called pavings. There should be no confusion between the notations $\{X\}$ (for a singleton) and $\{X_\alpha\}_{\alpha \in A}$ for an indexed family.

Given a partitioning $\{X_\alpha\}_{\alpha \in A}$ as in 4.6.3, define an equivalence relation R on X by: $x_1 R x_2$ iff x_1 and x_2 lie in the same set X_α. We say that R is **induced** by the partitioning, and leave the reader to prove that R is an equivalence relation on X, such that each coset xR is one of the sets X_α. Thus, we have a form of converse of 4.5.12:

4.6.4 THEOREM. *A partitioning* $\{X_\alpha\}_{\alpha \in A}$ *induces an equivalence relation* R *on* $X = \bigcup_{\alpha \in A} X_\alpha$; *and* $X/R = \{X_\alpha\}_{\alpha \in A}$. ■

4.7 The Quotient Map

Again, let R be an equivalence relation on X. We define a function

4.7.1 $$\nu: X \to X/R$$

called the **quotient map**, or **natural map**, by taking $\nu(x)$ to be the coset xR, for each $x \in X$. Clearly ν is onto, and for each element $xR \in X/R$, the inverse image $\nu^\flat\{xR\}$ is the subset $xR \subseteq X$. Thus we have a converse to the situation of Exercise 9(i), where we obtained an equivalence relation R_f from a given function $f: X \to Y$; and R here is R_ν (induced by ν) in that notation. We shall now prove that the image, $\mathrm{Im}\,(f)$, of f satisfies:

4.7.2 THEOREM. *There is an equivalence*† $X/R_f \approx \mathrm{Im}\,(f)$.

Proof. The definition of R_f is: $x_1 R_f x_2$ iff $f(x_1) = f(x_2)$. Thus, for every x in the coset $x_1 R_f$, the value $f(x)$ is constant and equal to $f(x_1)$. This constant value we define to be $j(x_1 R_f)$, thus obtaining a function

$$j: X/R_f \to \mathrm{Im}\,(f); \qquad j(xR_f) = f(x).$$

Then j is onto, because every $y \in \mathrm{Im}\,(f)$ is of the form $f(x_1)$, for some $x_1 \in X$, and $f(x_1) = j(x_1 R_f)$. To see that j is one-one, suppose that $x_1 R_f \neq x_2 R_f$. Then

(i) $$j(x_1 R_f) = f(x_1) \neq f(x_2) = j(x_2 R_f),$$

because if $f(x_1) = f(x_2)$, then $x_1 R_f x_2$, so x_1, x_2 lie in the same coset, contrary to the fact that the unequal cosets $x_1 R_f$, $x_2 R_f$ are disjoint (by 4.5.8). Thus, j is one-one by (i); hence i is the required equivalence. ■

4.7.3 NOTE. The procedure in the last proof, where we defined $j(x_1 R)$ to be $f(x_1)$, and remarked that $f(x_1) = f(x)$ for each $x \in x_1 R_f$, is common in proofs about cosets. It is an example of the process of 'checking independence of representatives', mentioned in 4.5.9; and if the details are routine, they are often omitted. They are essential, of course, because j would not be a function on X mod R_f if it were false that $f(x_1) = f(x)$ whenever $x \in x_1 R_f$.

Exercise 10

(i) Consider the diagram

$$\begin{array}{ccccc} X & \xrightarrow{g} & \mathrm{Im}\,(f) & \xrightarrow{i} & Y \\ & {\scriptstyle\nu}\searrow & \nearrow{\scriptstyle j} & & \\ & & X/R_f & & \end{array}$$

where $g(x) = f(x)$ for each $x \in X$, i is the inclusion map (see 2.6.2), ν is induced by R_f as in 4.7.1, and j is the equivalence constructed in the proof of 4.7.2. Prove

† In 4.7.1 this equivalence is an equality: $X/R_\nu = \mathrm{Im}\,(\nu)$, since ν is onto.

that the triangle is 'commutative', i.e., that $g = j \circ \nu$, and that the arbitrary function $f: X \to Y$ is 'factored' into $i \circ j \circ \nu$, where i is an injection, j is a bijection and ν is a surjection. Note that i, j, ν all depend on f.
(ii) Show that the equivalence in Exercise 9(i) follows from Theorem 4.7.2.
(iii) Let X be an object (e.g., a locomotive or a country), and let Y be a model of X (e.g., a scale drawing or map). Regarding X and Y as sets of points, in what sense is Y equivalent to X/R, for some equivalence relation R?
(iv) Let $\{X_\alpha\}_{\alpha \in A}$ be as in Definition 4.6.1. Show that

$$\bigcup_{\alpha \in A} X_\alpha = \left\{ y \in \mathcal{U} \,\middle|\, \underset{\alpha \in A}{\exists}\ \alpha,\ y \in X_\alpha \right\}.$$

Define $\bigcap_{\alpha \in A} X_\alpha$ to be the set of all $x \in \mathcal{U}$ such that x lies in every $X_\alpha (\alpha \in A)$. Prove that

$$\complement\left(\bigcup_{\alpha \in A} X_\alpha\right) = \bigcap_{\alpha \in A} (\complement X_\alpha)$$

using 1.8.2. [See also 5 Exercise 1(v).]

The reader will find the ideas of this section applied extensively in Chapter 24, and elsewhere throughout the book. In particular, the diagram of Exercise 10(i) plays an important role in group theory (see 18.9.6).

Chapter 5

MATHEMATICAL INDUCTION

5.1 Physical and Mathematical Induction

Further progress depends on our being able to use the method of mathematical induction to construct proofs. Let us first explain this method and its philosophy.

In ordinary life, we often use the method of 'physical induction'; for example, on the basis of a large number of observations, we expect that every human being we meet will have one nose, in the centre of his face. Strictly speaking, we have no right to be surprised if the next person we meet should have two noses, or one nose badly off centre. In physics, laws are often propounded after a very small number of experiments, although of course such laws are expected to be good predictors and they are usually discarded when they fail. In pure mathematics, however, words like 'large' or 'small' have little meaning when applied to infinite sets like $\mathbb{N}$ or $\mathbb{Z}$; and although it is true that we often use the method of physical induction to *guess* a result, the result itself does not become part of pure mathematics until a 'genuine' proof is found, i.e., one which follows according to certain rules from allowed hypotheses. Thus, for example, there is the conjecture named after Goldbach, that every even number >2 is the sum of two odd primes. By the standards of ordinary life or of physics, the evidence is overwhelming, because every even number up to 10^8 has certainly been tested and found to satisfy the conjecture. Nevertheless, we do not regard such tests as a proof of the general hypothesis, for 10^8 is an insignificant number compared with 'most' integers, or even when compared with the estimated number 10^{79} of protons and electrons in the universe. Now Goldbach's conjecture can be expressed in the form: 'Given $n \in \mathbb{N}$, then $2n + 2$ can be expressed as the sum of two primes', or briefly 'Given $n \in \mathbb{N}$, then $E(n)$', where $E(n)$ denotes the statement '$2n + 2$ can be expressed as the sum of two primes'. Empirical evidence shows the truth of the statement:

5.1.1 $$E(1) \mathbin{\&} E(2) \mathbin{\&} \cdots \mathbin{\&} E(10^8),$$

whereas we want to prove a statement expressible roughly as

5.1.2 $$E(1) \mathbin{\&} E(2) \mathbin{\&} \cdots \mathbin{\&} E(n) \mathbin{\&} \cdots.$$

The difference between 5.1.1 and 5.1.2 is like that between summing a finite set of numbers, and summing an infinite series; and just as we have to make conventions about the latter, we are going to make one about what we will regard as a proof of an infinite conjunction of the form 5.1.2. A convention *must* be made, since we cannot give a separate proof for *every* $n \in \mathbb{N}$ that $E(n)$ holds; indeed, it is unlikely that we could even give in our lifetime a separate proof for every $n \leqslant 10^{250}$.

We shall use the fact that $\mathbb{N}$ is well-ordered, as explained in 4.4.9. No proof of well-ordering was given there, since the only 'proofs' known must assume the Principle of Induction, which we are about to derive from well-ordering! (See Exercise 1(xviii), below.) The reader should observe that we have not said what $\mathbb{N}$ *is*, in full detail, but only what we need to be able to *do* with $\mathbb{N}$.

Let us therefore formulate, and prove, the Principle of Mathematical Induction. It will apply to propositions of the form

5.1.3 '*Given* $n \in \mathbb{N}$, *then* $P(n)$ *is true*',

where $P(n)$ is a statement about the integer n. Observe that if $\mathbb{N}$ is replaced in 5.1.3 by a finite set, say $\{1, 2, 3\}$, then 5.1.3 means '$P(1)$ & $P(2)$ & $P(3)$', a proof of which could be given by proving $P(1)$, then $P(2)$, then $P(3)$. A statement like 5.1.3 can then be regarded as an infinite conjunction of the form 5.1.2. We now state:

5.1.4. **(The Principle of Mathematical Induction)** *Suppose* (i) $P(1)$ *is true;*† (ii) *the truth of* $P(n)$ *implies that of* $P(n + 1)$, *for any* $n \in \mathbb{N}$.

Then, 5.1.3 *is true.*

Proof. The statement is so strongly reminiscent of that of the induction theorem (4.4.13) that we naturally see if it can be deduced therefrom. But then, all we need do is to let $A \subseteq \mathbb{N}$ be the set of all $n \in \mathbb{N}$ for which $P(n)$ is true. The suppositions (i) and (ii) in 5.1.4 tell us that A satisfies conditions (i) and (ii) of 4.4.13. Therefore, by the induction theorem, $A = \mathbb{N}$; so $P(n)$ is true for all $n \in \mathbb{N}$. ■

5.2 A Bad Custom

Consider an example of a proof using the induction principle (many other such proofs will be found in this book). Elementary books on algebra often contain illustrations of the principle, applied to finite series; for example, they would let $P(n)$ denote the sentence

$$\sum_{m=1}^{n} m^2 = n(n + 1)(2n + 1)/6$$

† Here, as always in pure mathematics, 'true' means 'provable'; cf. the discussion in Chapter 38 and in 8.3.

(which† is a statement about n with verb 'equals'). Then, they argue, $P(1)$ is '$1 = 1\cdot(1+1)\cdot(2\cdot 1+1)/6$' which is true, and

$$\sum_{m=1}^{n+1} m^2 = \sum_{m=1}^{n} m^2 + (n+1)^2$$
$$= n(n+1)(2n+1)/6 + (n+1)^2 \qquad \text{(by } P(n)\text{)},$$

so
$$\sum_{m=1}^{n+1} m^2 = (n+1)(n+2)(2(n+1)+1)/6,$$

which equality is the sentence $P(n+1)$. The books then often go on: 'Thus $P(1)$ is true, therefore $P(2)$, $P(3)$, and so on; hence $P(n)$ is always true'. This last sentence gives the impression that, after all, they have resorted to physical induction; whereas all they need have said is: 'So the hypotheses (i) and (ii) {of 5.1.4} are satisfied, and therefore $P(n)$ holds for every n, by mathematical induction'.

Before considering some problems, let us make some remarks about inductive *definitions*.

5.3 The Method of Inductive Definition

We can use the induction theorem (4.4.13) for making definitions, as well as for the construction (5.1.4) of proofs. It lies beyond the scope of this book to formulate a general description of the 'method of inductive definition', but we give here two illustrations of the use of the method, and other uses are made throughout this book.

5.3.1 What does the symbol $\sum_{m=1}^{n} a_m$ mean, when the a_m's are given numbers?

Traditional writers often say that $\sum_{m=1}^{n} a_m$ means $a_1 + a_2 + \cdots + a_n$; but what do the dots mean? Usually these can be inferred from the context, but not always (e.g., 'Sum the series $1 + 4 + 9 + \cdots$': is the fourth term $16 (a_n = n^2)$ or $22 (a_{n+1} = 2a_n + a_{n-1})$?). If we wanted an electronic computer to work out $\sum_{m=1}^{n} a_m$, it would need to be told exactly what to do when it came to the dots. So we programme it to compute a function $f\colon \mathbb{N} \to \mathbb{R}$, where

5.3.2 $$f(1) = a_1, \qquad f(n+1) = f(n) + a_{n+1},$$

and the numbers a_i are part of the data given to the computer in advance. Assuming that the computer can distinguish between symbols, and can add

† The symbol $\sum_{m=1}^{n}$ is defined in 5.3.2 below; that is, $\sum_{m=1}^{n} a_m = f(n)$.

and remember them, then the equations 5.3.2 enable it to work out $f(n)$ $(= \sum_{m=1}^{n} a_m)$ for any $n \in \mathbb{N}$. For the set A of those n for which it can compute $f(n)$ satisfies conditions (i) and (ii) of the induction theorem, so $A = \mathbb{N}$. The equations in 5.3.2 are said to **define f inductively**, or to be an **inductive definition** of f. Since the integer m does not appear in 5.3.2, it is called a 'dummy suffix' in the symbol $\sum_{m=1}^{n} a_m$, so $\sum_{m=1}^{n} a_m = \sum_{t=1}^{n} a_t$ (for example).

The symbol $\mathbb{N}_n$ in 2.11 was defined informally, using dots, as the subset $\{1, 2, \ldots, n\}$ of $\mathbb{N}$. We can define $\mathbb{N}_n$ inductively for each $n \in \mathbb{N}$ by

5.3.3 $$\mathbb{N}_1 = \{1\}, \qquad \mathbb{N}_{n+1} = \mathbb{N}_n \cup \{n + 1\}.$$

Thus, a computer assigned to print out the membership of (say) $\mathbb{N}_{4516}$ could work successively, using its own construction of $\mathbb{N}_{3129}$ (say) in the second equation to get $\mathbb{N}_{3130}$, and so onwards to $\mathbb{N}_{4516}$.

When the replacing of dots is a routine matter for human beings, the dots are often substituted for the inductive definition which the reader is expected to give. The use of dots often makes formulae easier to comprehend, but, again, only for human readers. And in more advanced work, the formal definition by induction *must* be given, especially when that definition has only just been created by an author.

Exercise 1

(i) Using the inductive definition of $\sum_{m=1}^{n}$, prove that $\sum_{m=1}^{n} (\lambda a_m + \mu b_m) = \lambda \sum_{m=1}^{n} a_m + \mu \sum_{m=1}^{n} b_m$ when $\lambda, \mu \in \mathbb{R}$. If $a_m = 1$ for each m, prove that $\sum_{m=1}^{n} a_m = n$. Prove that the sum of an even(odd) number of odd numbers is even(odd). Hence prove that the number of people who have shaken hands an odd number of times is even.

(ii) Use the method of inductive definition to define the symbol $\prod_{m=1}^{n} a_m$, which (using dots) means $a_1 \times a_2 \times \cdots \times a_n$ in the two cases where (1) the a's are numbers, and $\times$ means ordinary multiplication; and (2) the a's are sets and $\times$ denotes Cartesian product (cf. 3 Exercise 1(v)). In case (1), prove from the definition that $\prod_{m=1}^{n} a_m = \prod_{r=1}^{n} a_r$; $\prod_{m=1}^{n} a_m b_m = \prod_{m=1}^{n} a_m \cdot \prod_{m=1}^{n} b_m$; and if $a_m = a$ for all a, then $\prod_{m=1}^{n} a_m = a^n$, where a^n is defined inductively by $a^1 = a$, $a^{n+1} = a^n \cdot a$.

(iii) Use induction to show that, when $n \in \mathbb{N}$:

$$n^3 + 2n \equiv 0 \pmod 3; \qquad 5^{2n} - 1 \equiv 0 \pmod{24};$$

$$10^n + 3.4^{n+2} + 5 \equiv 0 \pmod 9; \qquad 2n \leqslant 2^n;$$

if n is odd and positive, then $n(n^2 - 1) \equiv 0 \pmod{24}$.

(iv) Given an integer k between 0 and n, how would you give inductive definitions of the symbols $\sum_{m=0}^{n} a_m$, $\prod_{m=0}^{n} a_m$ so that you could *prove* that

$$\sum_{m=0}^{n} a_m = \sum_{m=0}^{k} a_m + \sum_{m=k+1}^{n} a_m; \qquad \prod_{m=0}^{n} a_m = \prod_{m=0}^{k} a_m \cdot \prod_{m=k+1}^{n} a_m,$$

(where the a's are numbers)? Also, prove that

$$\sum_{r=0}^{n} a_r \cdot \sum_{s=0}^{n} b_s = \sum_{\substack{t=0 \\ r+s=t}}^{n} a_r b_s,$$

first defining the right-hand sum.

(v) Frame definitions of $\bigcup_{r=k}^{n} A_r$, $\bigcap_{r=k}^{n} A_r$, when the A_r are sets, analogous to your definitions in (iv), and prove analogues of the equations in (ii) (case (1)) and in (iv). Use induction also to establish the equations

$$B \cap \bigcup_{r=k}^{n} A_r = \bigcup_{r=k}^{n} (B \cap A_r), \qquad B \cup \bigcap_{r=k}^{n} A_r = \bigcap_{r=k}^{n} (B \cup A_r);$$

$$\complement\Big(\bigcup_{r=k}^{n} A_r\Big) = \bigcap_{r=k}^{n} \complement A_r, \qquad \complement\Big(\bigcup_{r=k}^{1} A_r\Big) = \bigcap_{r=k}^{n} \complement A_r;$$

and prove that if $A_r = A$, all r, then

$$\bigcup_{r=k}^{n} A_r = A = \bigcap_{r=k}^{n} A_r.$$

Compare this definition of union with that of 4.6.1, by letting $J = \{1, \ldots, n\}$, and proving that

$$\bigcup_{r \in J} A_r = \bigcup_{r=1}^{n} A_r.$$

Hence formulate a definition of $\bigcap_{\alpha \in A} X_\alpha$ for an indexed family $\{X_\alpha\}_{\alpha \in A}$, and prove analogues of the above results.

(vi) Extend 3 Exercise 1(iv) as follows. Fix $n \in \mathbb{N}$ and let S denote the set of all functions $f\colon \mathbb{N}_n \to \bigcup_{r=1}^{n} A_r$ such that $f(i) \in A_i$ for each $i \in \mathbb{N}_n$. Prove that there is an equivalence: $S \approx \prod_{r=1}^{n} A_r$, where the second set is the Cartesian product defined in (i), case (2), above.

(vii) Give a careful proof, by induction on k, of the following proposition (the **Pigeonhole principle**): 'If a box contains $k + 1$ objects, of k different colours, then exactly two of the objects are of the same colour'.

(viii) Given a positive integer x, let x^* denote the sum of all the digits in x (in the scale of 10). Define a sequence $x_1, x_2, \ldots, x_n, \ldots,$ of integers by the rules:

$$x_1 = x, \qquad x_{n+1} = (x_n)^*.$$

Show that, after some n, $x_{n+1} = x_n$ and $0 \leqslant x_n < 10$, while $x \equiv x_n \pmod 9$. [This is connected with the process of 'casting out the nines'; cf. 10.1.]

(ix) A set S of positive integers is **bounded above** iff there exists $h \in \mathbb{N}$, such that $j \leqslant h$ for each $j \in S$. Prove that then S has a greatest member, provided $S \neq \varnothing$.
(x) Prove by induction that given any positive integer n, some power of 2 is greater than n. Hence show that an integer $k = k(n)$ exists such that

$$2^k \leqslant n < 2^{k+1},$$

and therefore that $k(n) + k(m) \leqslant k(nm) \leqslant k(n) + k(m) + 1$ for any positive integers n, m.
(xi) Prove the following alternative forms of the principle of mathematical induction.

(1) s is a positive integer, and a sequence $A_s, A_{s+1}, A_{s+2}, \ldots$, of propositions is such that (a) For every integer $r \geqslant s$, $A_r \Rightarrow A_{r+1}$; (b) A_s is true.
Then A_n is true for every integer $n \geqslant s$.
(2) $A_1, A_2, A_3, \ldots$ is a sequence of propositions such that (a) For every positive integer r, the truth of $A_1, A_2, \ldots, A_r$ implies the truth of A_{r+1}, (b) A_1 is true.
Then A_n is true for every positive integer n.

(xii) What is wrong with the following 'proof' that any two positive integers are equal? First, a definition: If a and b are two unequal positive integers, we define max (a, b) to be a or b, whichever is greater; if $a = b$ we set max $(a, b) = a = b$. Thus max $(3, 5) =$ max $(5, 3) = 5$, while max $(4, 4) = 4$. Now let A_n be the statement: 'If a and b are any two positive integers such that max $(a, b) = n$, then $a = b$'.
(a) Suppose A_r to be true. Let a and b be any two positive integers such that max $(a, b) = r + 1$. Consider the two integers

$$\alpha = a - 1$$
$$\beta = b - 1;$$

then max $(\alpha, \beta) = r$. Hence $\alpha = \beta$, for we are assuming A_r to be true. It follows that $a = b$; hence A_{r+1} is true.
(b) A_1 is obviously true, for if max $(a, b) = 1$, then since a and b are by hypothesis positive integers they must both be equal to 1. Therefore, by mathematical induction, A_n is true for every n.
Now if a and b are any two positive integers whatsoever, denote max (a, b) by r. Since A_n has been shown to be true for every n, in particular A_r is true. Hence $a = b$!!
(xiii) Let G be a linear graph (see 4.3) with vertices P_i and edges e_i. A **route** in G is a set of edges $e_1, e_2, \ldots, e_m$, in order, where each e_i meets the next; and the route is **unicursal** if all the e's are distinct. G is **connected** provided any pair P_i, P_j can be joined by the edges of a route; and if there is a unicursal route which visits all vertices, then G is **unicursal**. Prove that the graphs shown in Fig. 5.1 are connected and not unicursal. Define the order $w(P_i)$ of a vertex P_i to be the number of edges having P_i as end-point. Prove that G is unicursal if, and only if, the number of 'odd' vertices P (i.e., with $w(P)$ odd) is zero or two. {Hint: use induction on the number of edges of a connected graph; the simplest graphs are .—. and '.'. Try a unicursal route from one odd node to another. It either

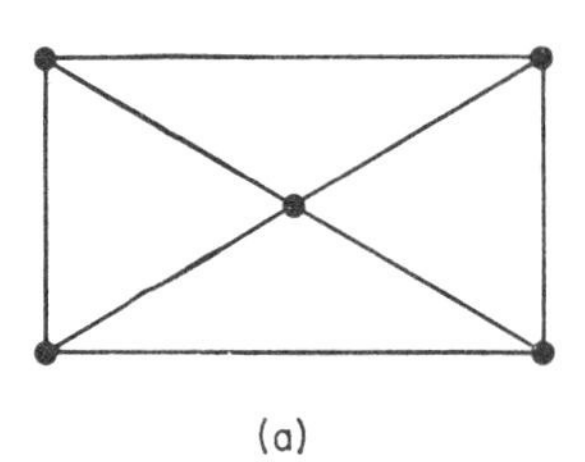

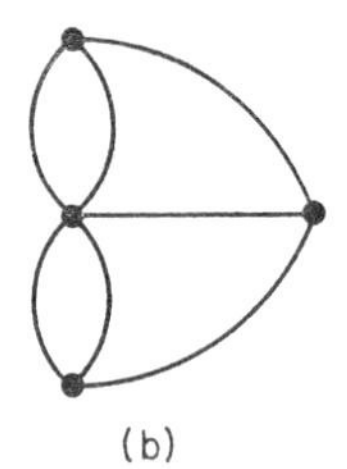

Fig. 5.1

arrives or contains a loop. Remove it, or the loop, and use induction, since the number of odd nodes is still either zero or two.} Prove also by induction on k that if the number of odd nodes is $2k$, then k unicursal routes will cover the graph.

[The graph in Fig. 5.1(b) arises in connection with the **problem of the Koenigsberg bridges** first solved by Euler (1707–1783; see Bell [10], Ch. 9); the question was, to find a tour of Koenigsberg (if possible) which crossed each of its seven bridges once and only once. Figure 5.2 shows a map of Koenigsberg, a town in

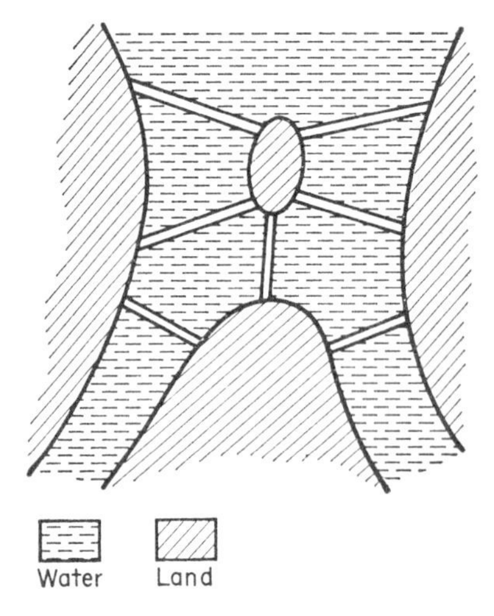

Fig. 5.2

East Prussia, consisting of two islands standing in the river Pregel, with bridges as shown. Euler showed that a route is possible iff the graph in Fig. 5.1(b) is unicursal; but since there are four odd nodes, the required route is impossible.]

(xiv) Write $\prod_{i=1}^{n} (x + a_i) = \sum_{r=0}^{n} S_r x^{n-r}$. Prove that S_r is the sum of all products of the form $a_{i_1} \cdots a_{i_r}$, $(i_1 < \cdots < i_r)$, $r \geqslant 1$; $S_0 = 1$.

(xv) In the equation $\sum_{r=1}^{n} r = n(n+1)/2$, the right-hand side suggests half a rectangular area of dots above the broken diagonal line in an array like that of Fig. 5.3. Use induction to show that the successive diagonals have 1, 2, 3, ..., points on them, and hence establish the original equation.

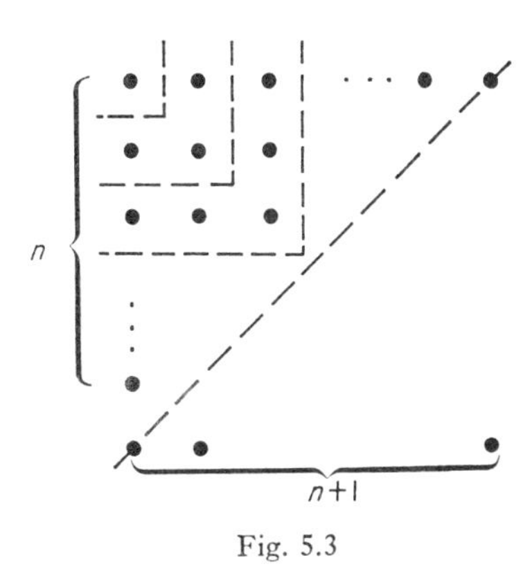

Fig. 5.3

A number T_n of the form $n(n+1)/2$ is called **triangular**, while n^2 is **square**. Prove that $T_n + T_{n+1} = (n+1)^2$, and illustrate in a picture. Find a three-dimensional model for the formula:

$$\sum_{r=1}^{n} r^2 = n(n+1)(2n+1)/6,$$

using the fact that the volume of a tetrahedron is equal to one-sixth of its base times its height. Can you find a four-dimensional model for the formula

$$\sum_{r=1}^{n} r^3 = \left(\sum_{r=1}^{n} r\right)^2?$$

On the other hand, use induction to show that the right-angled corners, shown in Fig. 5.3, can successively mark off $1^3, 2^3, \ldots,$ points, to fill up squares of side $1, 3, \ldots, \sum r$.

(xvi) For each $n \in \mathbb{N}$ define $n!$ inductively by:

$$0! = 1, \qquad (n+1)! = (n+1)\cdot n!$$

[Thus, we have a function, $!\colon \mathbb{Z}^+ \to \mathbb{N}$, and an example of writing xf for $f(x)$. For typographical reasons, we do not use the old 'angle' notation.] Prove that $2^{n-1} \leqslant n!$, if $n \in \mathbb{N}$. Given $r = 0, 1, \ldots, n$, define $\binom{n}{r}$ by

$$\binom{n}{r} = n!/r!\,(n-r)!,$$

and prove from the formula that, if $r > 0$,

$$\binom{n+1}{r} = \binom{n}{r-1} + \binom{n}{r}.$$

Hence, use induction on n to prove the **binomial theorem for positive integral index**:

$$(a+b)^n = \sum_{r=0}^{n} \binom{n}{r} a^r b^{n-r}, \qquad (a, b \in \mathbb{R}).$$

and the **theorem of Leibniz** in calculus:

$$D^n(fg) = \sum_{r=0}^{n} \binom{n}{r} D^r f \cdot D^{n-r} g, \qquad Df = df/dx,$$

for real-valued functions on an interval of $\mathbb{R}$, assuming that all the necessary derivatives exist. (About G. Leibniz (1646–1716), see Bell [10], Ch. 7.)

(xvii) Suppose that, in the set $\mathbb{N}$ of positive integers, there is assigned to each $n \in \mathbb{N}$, a finite set $S_n \subseteq \mathbb{N}$. Say that S_p 'straddles' S_q whenever one contains an integer between the first and last integers of the other. Suppose further that (i) if S_p straddles S_q, then one includes the other and extends it at each end; (ii) if n is the greatest element of some S_r, $n + 1$ is not the least member of any S_s. Prove that, for an infinity of integers $j \in \mathbb{N}$, no S_n contains j.

{Prove that each S_n lies in only a finite number of S_p; call the biggest such S_p 'maximal'. Then show that no two maximal S's straddle each other.}

(xviii) Prove the induction theorem (4.4.13) assuming the principle of mathematical induction (5.1.4). This is the first step towards deriving the well-ordering of $\mathbb{N}$ from 5.1.4.

(xix) In the plane, there are n non-parallel lines drawn so that no three intersect at a single point. Use induction to calculate the number of regions into which the lines divide the plane. Do the analogous problem, considering n planes in space.

(xx) Justify the accountant's method of adding m rows of n numbers; that is, he checks that the totals along the m rows have the same sum as do the totals down the n columns.

§(xxi) Prove the general associative law in an associative multiplicative system that a product $a_1 \cdots a_n$ can be computed using any form of bracketing. (The difficulty lies in formulating 'any form of bracketing' precisely, using induction.)

Part II

FURTHER SET THEORY

Part I, *The Language of Mathematics*, contained material which is often described today as 'set theory'. We adopted the given title in order to emphasize that we were concerned to establish the language in which mathematical statements may be made, mathematical dialogue conducted and mathematical ideas exchanged and developed. Having given the reader his mathematical vocabulary, we are now in a position to talk about the various areas of mathematics and to exhibit their special and their common features. Indeed we are proposing to do just that throughout the remainder of this book (except for Chapter 38, *Categories and Functors*, which is again concerned with language). Thus in this Part we apply mathematical language to set theory itself. Perhaps, then, the most important initial point to make about this Part is the negative one that it is not essential reading for what follows (as Part I was) but constitutes a study in itself which may be omitted by the reader who would prefer to get straight on to the later topics of the book.

On the positive side, this Part is concerned with set theory as a mathematical discipline. Our treatment is, of course, deliberately naïve from the standpoint of present-day logic; nevertheless it will strike many readers as pedantically fussy. This paradox serves only to emphasize the gulf between intuitive concepts and rigorous formulations; both, of course, have their place in the growth of mathematics and there is no question of making any extramathematical value judgment as to their importance—that is not our concern. We do believe, however, that in moving towards more rigorous mathematics we may not only clarify certain familiar ideas but also bring unfamiliar and traditionally 'advanced' topics within the ambit of the less experienced reader.

The primary object of Chapter 6 is to clarify the theory of permutations and combinations through a rather careful study of the set of functions from one set to another. Chapter 7 begins by clarifying (or, perhaps for some, obfuscating) the counting process in the elementary context of finite sets, in order to explain the nature of transfinite cardinals and ordinals and their arithmetic. This theory, essentially due to the great German mathematician, Georg Cantor (1845–1918, see Bell [10], Ch. 29), is surely one of the most revolutionary developments in human thought, throwing out the ancient Aristotelian axiom that 'the whole is greater than the part'. Chapter 8 develops the parallelism between the algebra of sets and the propositional calculus—or, as we might less precisely say, between set theory and logic. The common notion here is that of a Boolean algebra, a type of algebraic system which, while having wide application (not just to set theory and logic but to switching theory, for example) was traditionally not presented to any but the mathematical specialist simply because it is not concerned with numbers! Nowadays, however, many schoolteachers are introducing Boolean algebra to their classes, and we show here and in Chapter 21 how the subject develops beyond the mere axioms.

Chapter 6

SETS OF FUNCTIONS

6.1 The Set B^A

When A, B are sets, the set B^A of all functions $f: A \to B$ is mentioned at various places throughout this book. We here intend to count it, when A, B are finite, and discuss its relationships with other sets, such as the set of all shuffles of A. Now, there are two useful general principles for counting a finite set X:

(i) we find a known set Y, already counted, and show that $X \approx Y$;
(ii) we find some equivalence relation R on X, partition X into cosets mod R as in Theorem 4.5.12, count each coset, and add up; often the cosets all have the same number of elements.

We need to recall from 2.6 the notion of the restriction $f|X_0$, of a function $f: X \to Y$, when $X_0 \subseteq X$. Also, by 3.4.3, $B^\varnothing = \{\varnothing\}$ for all B, including $B = \varnothing$; and $\varnothing^A = \varnothing$ if $A \neq \varnothing$. From now on, therefore, we consider B^A only when A and B are non-empty. Our first theorem shows that if X splits into simpler sets so does B^X (although they split in different ways).

6.1.1 THEOREM. *For any* (possibly infinite) *sets A, B, C* such that $A \cap C = \varnothing$ *there is an equivalence*

$$B^{A \cup C} \approx B^A \times B^C.$$

Proof. We must first define a function

$$\theta: B^{A \cup C} \to B^A \times B^C$$

and prove it to be one-one and onto. Given $f \in B^{A \cup C}$, define $\theta(f)$ to be the pair of functions $(f|A, f|C)$. This pair clearly lies in $B^A \times B^C$.

To prove that θ is one-one, suppose $\theta(f) = \theta(g)$, so (by definition of equality of ordered pairs)

(i) $f|A = g|A, \qquad f|C = g|C.$

Any $x \in A \cup C$ lies either in A, or in C, but not both, since $A \cap C = \varnothing$. Hence, if $x \in A$, then $f(x) = g(x)$, by (i), and similarly if $x \in C$. Thus, $f(x) = g(x)$ for each $x \in A \cup C$, so, by definition of equality of functions, $f = g$. Therefore θ is one-one.

To prove that θ is onto, let the pair $(h, k) \in B^A \times B^C$ be given. We must find $f \in B^{A \cup C}$ such that $\theta(f) = (h, k)$, i.e., such that

(ii) $f|A = h, \qquad f|B = k.$

These equations suggest the definition of f: define $f(x)$ to be $h(x)$ or $k(x)$ according as x lies in A or C—there is no ambiguity, since $A \cap C = \varnothing$. By construction the equations (ii) hold, so $\theta(f) = (h, k)$ and θ is onto. (In fact the function $(h, k) \to f$ is an inverse of θ.) ■

Now consider the case of the simplest possible 'exponent'.

6.1.2 THEOREM. *If C is a singleton $\{c\}$, then*

$$B^{\{c\}} \approx B.$$

Proof. Define $\psi: B^{\{c\}} \to B$ by $\psi(f) = f(c)$ for each $f \in B^{\{c\}}$. We leave the reader to prove that ψ is one-one and onto. ■

6.1.3 THEOREM. *If A is finite, say $A = \{a_1, \ldots, a_n\}$, then*

$$B^A \approx B^{\{a_1\}} \times B^{\{a_2\}} \times \cdots \times B^{\{a_n\}}.$$

Proof. By induction on n; if $n = 1$, then 6.1.3 is clear, and if $n > 1$, then

$$B^A = B^{\{a_1, \ldots, a_{n-1}\} \cup \{a_n\}} \approx B^{\{a_1, \ldots, a_{n-1}\}} \times B^{\{a_n\}},$$

applying 6.1.1. Thus 6.1.3 follows by finite induction on $\#A$. ■

6.1.4 COROLLARY. $B^{\{a_1, \ldots, a_n\}} \approx B \times B \times \cdots \times B$ (n times).

Proof. Use induction on n, with 6.1.2 and 3 Exercise 1(iv). ■

The following corollary gives the reason for the choice by Cantor of the notation B^A.

6.1.5 COROLLARY. *If A and B are both finite, then*

$$\#(B^A) = (\#B)^{\#A}.$$

Proof. By 6.1.4, $B^A \approx B \times B \times \cdots \times B$ (n times), where $n = \#A$, so

$$\#(B^A) = (\#B)^n$$

by 3 Exercise 1(iv), using induction on n. ■

These results will be used as a starting point, in 7.2, for the arithmetic of cardinal numbers.

Exercise 1

(i) If $C \subseteq A$, show that the function $r: B^A \to B^C$ given by $r(\phi) = \phi | C$, is onto. If $\theta \in B^C$, show that the inverse image $r^\flat\{\theta\} \approx B^{A-C}$, and hence give a second proof of 6.1.5.

(ii) Each member of a set of n people votes for just one of a set of m candidates, where $0 < m < n$. Show that the number of possible results of the voting is m^n. If $m = 2$, for how many of these results will a particular candidate be elected? What if $m = 3$?

6.2 Mappings of B^A

It follows from 6.1.5 that, if $B \approx D$, and $A \approx C$, then (all sets being finite) $\#(B^A) = \#(D^C)$; so there must be an equivalence $B^A \approx D^C$, by 2.11.1. Let us analyse the nature of this equivalence in terms of the given two. As is so often the case, it is best first to be more general, so we do not here assume our sets $A, B, \ldots,$ to be finite; and we first suppose merely that there is a function $f: B \to D$ (not necessarily an equivalence). We shall now construct a new function, denoted by $f^A: B^A \to D^A$ and said to be 'induced' by f.

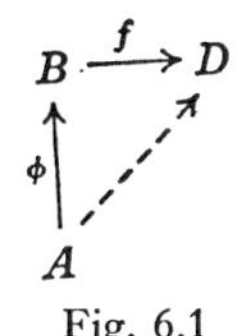

Fig. 6.1

The definition of f^A is suggested by the diagram (Fig. 6.1): given $\phi \in B^A$, then the dotted arrow from A to D can be labelled $f \circ \phi$. That is, we define $f^A(\phi)$ to be $f \circ \phi$; f^A is clearly a function from B^A to D^A. By 2.7.3, we already have a small theorem:

6.2.1 PROPOSITION. *If $B = D$ and $f = \mathrm{id}_B$, then f^A is the identity on B^A.* ■

Suppose next that a further function $g: D \to E$ is given. Then we have the diagram of Fig. 6.2, where g induces a function $g^A: D^A \to E^A$ constructed like

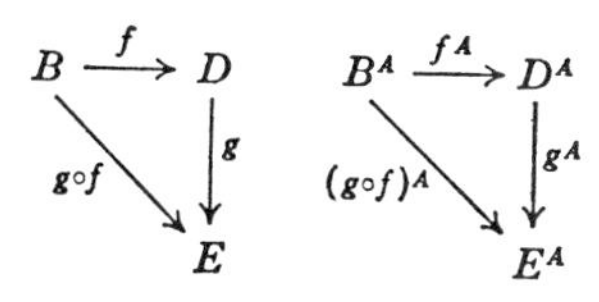

Fig. 6.2

f^A, so for each $\theta \in D^A$, $g^A(\theta) = g \circ \theta$ and we have by composition a function $g^A \circ f^A: B^A \to E^A$, as well as a function $(g \circ f)^A: B^A \to E^A$ induced by the function $g \circ f: B \to E$. One therefore guesses that the following theorem holds:

6.2.2 THEOREM. $$g^A \circ f^A = (g \circ f)^A.$$

Proof. The two functions have the same domain B^A and range E^A. If $\phi \in B^A$, then by definition (and omitting now the composition symbol)

$$(gf)^A(\phi) = (gf)\phi$$

while

$$(g^A f^A)(\phi) = g^A(f^A(\phi)) = g^A(f\phi) = g(f\phi).$$

Hence, by associativity of composition (2.7.6), $(gf)^A\phi = (g^Af^A)(\phi)$, and so, by definition of equality of functions,

$$(gf)^A = g^Af^A. \quad \blacksquare$$

6.2.3 COROLLARY. *If $f: B \to D$ is an equivalence, so is $f^A: B^A \to D^A$.*

Proof. By the inversion theorem (2.9.1 and 2.9.5), there exists the function $f^{-1}: D \to B$. Hence, applying 6.2.2,

$$f^A(f^{-1})^A = (\mathbf{1}_D)^A, \qquad (f^{-1})^Af^A = (\mathbf{1}_B)^A$$

where $\mathbf{1}_B$, $\mathbf{1}_D$ denote identity maps; and then by 6.2.1, $(\mathbf{1}_D)^A$, $(\mathbf{1}_B)^A$ are the identity maps respectively on D^A, B^A. Therefore, by the algebraic inversion theorem 2.9.8, $f^A: B^A \to D^A$ is an equivalence, with inverse $(f^{-1})^A$; thus

$$(f^A)^{-1} = (f^{-1})^A. \quad \blacksquare$$

Exercise 2

(i) Prove that if $f: B \to D$ is an injection, so is $f^A: B^A \to D^A$; and if f is a surjection, so is f^A.
(ii) If the dotted arrow in Fig. 6.1 is labelled α, and if $\alpha = f^A(\phi)$, calculate ϕ in terms of α and f, when $f: B \to D$ is an equivalence. Hence prove directly that if f is an equivalence, so is f^A.
(iii) What happens to f^A when A is empty?

6.2.4 We now proceed similarly with a function $h: C \to A$. Then h induces a function $h_B: B^A \to B^C$ (note the reversal of direction!) in a way suggested by the diagram (Fig. 6.3), in which the dotted arrow might be taken as $\phi \circ h$.

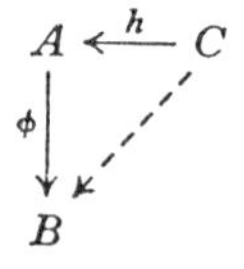

Fig. 6.3

Thus, we define $h_B(\phi)$ to be $\phi \circ h$, and obtain a function $B^A \to B^C$. [A more logical notation is B^h, but it disturbs the feeling that capitals denote sets rather than functions; the reader may accept ${}_Bh$ as a compromise, and as a mnemonic to remember that h is on the right of ϕ in the definition of h_B given.] Proofs similar to those of 6.2.1–6.2.3 yield the following analogues, and we leave the details to the reader.

6.2.5 *If $\mathbf{1}: A \to A$ is the identity map, so is $\mathbf{1}_B: B^A \to B^A$.* $\blacksquare$

6.2.6 *Given functions $F \xrightarrow{k} C \xrightarrow{h} A$, then* (note the reversal of order):

$$(h \circ k)_B = k_B \circ h_B: B^A \to B^F. \quad \blacksquare$$

The algebraic inversion theorem (2.9.8) implies:

6.2.7 COROLLARY. *If $h: C \to A$ is invertible, so is $h_B: B^A \to B^C$.* ■
If now we couple 6.2.3 and 6.2.7 we obtain immediately the

6.2.8 THEOREM. *Given bijections $f: B \approx D$, $h: C \approx A$, then there exists an equivalence $B^A \approx D^C$ compounded of the bijections $B^A \xrightarrow{u} B^C \xrightarrow{v} D^C$ where $u = h_B$, $v = f^C$.* ■

[The apparent asymmetry in the choice of equivalence $B^A \approx D^C$ is removed by Exercise 3(ii) below.]

The statement 6.2.8 is valid for finite or infinite sets, and is much more informative (in the finite case) than the statement: 'If $\#B = \#D$, $\#C = \#A$, then $\#(B^A) = \#(D^C)$'.

6.2.9 REMARK. We often meet pairs of equations like 6.2.1 with 6.2.2, or 6.2.5 with 6.2.6, calling them **covariant** or **contravariant** to emphasize that the order is preserved in the first case, and reversed in the second. For example, in 2 Exercise 7(i), we have contravariant equations for $f^\flat$. The proofs of 6.2.3, 6.2.7 are standard in form, occurring always in the presence of covariant or contravariant rules, and they may conveniently be recalled and summarized by saying 'by covariance' and 'by contravariance', respectively. See Chapter 37 for a fuller discussion.

Exercise 3

(i) In Fig. 6.3, suppose $h_B(\phi_1) = h_B(\phi_2)$. Prove that $\phi_1|H = \phi_2|H$, where $H = \text{Im}(h)$. Prove that h_B is an injection or surjection according as h is a surjection or injection, respectively. Prove directly that if h is a bijection, then so is h_B.
(ii) Given functions $f: B \to D$, $h: C \to A$, show that the diagram of Fig. 6.4 is 'commutative', i.e., $h_D \circ f^A = f^C \circ h_B$.

$$\begin{array}{ccc} B^A & \xrightarrow{f^A} & D^A \\ {\scriptstyle h_B}\downarrow & & \downarrow{\scriptstyle h_D} \\ B^C & \xrightarrow{f^C} & D^C \end{array}$$

Fig. 6.4

(iii) Let $I(A, B)$, $S(A, B)$ denote the subsets of B^A consisting of injections and surjections respectively. Prove that

$$f^A(I(A, B)) \subseteq I(A, D), \quad \text{if } f \text{ is one-one, and } f \in I(B, D);$$
$$f^A(S(A, B)) \subseteq S(A, D), \quad \text{if } f \text{ is onto, and } f \in S(B, D).$$

Show that equality occurs in each case if f is invertible.

(iv) When A, B are finite in (iii), prove that

$$I(A, B) = \varnothing \quad \text{if} \quad \#A > \#B; \qquad S(A, B) = \varnothing \quad \text{if} \quad \#A < \#B;$$
$$\text{and when} \quad \#A = \#B, \qquad I(A, B) = S(A, B).$$

Would the last equality hold if A is not finite and $A \approx B$?
(v) Similarly prove that

$$h_B(I(C, B)) \subseteq I(A, B) \quad \text{if } h \text{ is onto}, \qquad \text{and} \quad h \in S(C, A),$$
$$h_B(S(C, B)) \subseteq S(A, B) \quad \text{if } h \text{ is one-one}, \quad \text{and} \quad h \in I(C, A);$$

with equality in each case when f is invertible.
(vi) Show that when f, h are bijections in (ii), then there are equivalences $I(A, D) \approx I(C, B)$, $S(A, D) \approx S(C, B)$. Name these equivalences explicitly.
(vii) Let $E(A, B) = I(A, B) \cap S(A, B)$; it is the set of all bijections $A \to B$. Prove that when f, h are equivalences, then

$$f^A(E(A, B)) = E(A, D), \qquad h_B(E(C, B)) = E(A, B);$$

so f and h induce an equivalence $E(A, D) \approx E(C, B)$.

6.3 The Case When $\#B = 2$

We continue with two famous results about the size (so to speak) of $\mathfrak{p}A$, the family of subsets of a set A.

6.3.1 THEOREM. *If B has exactly two elements, then*

$$B^A \approx \mathfrak{p}A.$$

Proof. By Corollary 6.2.3 we may take $B = \{0, 1\}$. We must construct a function $\psi: B^A \to \mathfrak{p}A$, and then prove it to be an equivalence. Therefore, given $f \in B^A$, we want to define $\psi(f)$, a subset of A; so we choose to define

$$\psi(f) = \{x | x \in A \; \& \; f(x) = 0\}.$$

(We might equally well have used 1 instead of 0.) We leave to the reader the easy proof that if the sets of zeros, $\psi(f)$ and $\psi(g)$, are equal then $f = g$. Thus ψ is one-one.

To prove that ψ is onto, we must show that, given $X \in \mathfrak{p}A$, then there is a function $h \in B^A$ such that $\psi(h) = X$. The h required is quite obvious; we take $h: A \to B$ to be that function which has the constant value 0 on X and the value 1 everywhere else. By definition then, $\psi(h) = X$, so ψ is onto; since ψ is also one-one, then

$$\psi: B^A \approx \mathfrak{p}A. \quad ■$$

6.3.2 COROLLARY. *If A is finite then $\#\mathfrak{p}A = 2^{\#A}$.*

Proof. By 6.1.5 and 6.3.1. ■

Exercise 4

Suppose B and C each have two elements. The proof of 6.3.1 constructs a bijection $\psi: B^A \approx \mathfrak{p}A$, and similarly there is a bijection $\theta: C^A \approx \mathfrak{p}A$. Also there are two possible bijections $B \to C$ inducing (by 6.2.8) bijections $\alpha, \beta: B^A \to C^A$. Is either of α or β equal to $\theta^{-1} \circ \psi$?

The following very beautiful theorem is due to Cantor himself; when A is finite, it follows from Corollary 6.3.2, of course.

6.3.3 THEOREM. *For no set A is there an equivalence between A and $\mathfrak{p}A$.*

Proof. Suppose the theorem to be false. Then there would exist a set A and a bijection

(i) $\theta: A \approx \mathfrak{p}A$.

Now A is not empty, because $\mathfrak{p}\varnothing = \{\varnothing\}$ and a singleton cannot be equivalent o $\varnothing$. Given $X \subseteq A$, then $X = \theta(x)$, say, since θ is onto; and x either lies in X or not. Consider the subset T of A given by

$$T = \{x | x \in A \;\&\; x \notin \theta(x)\};$$

then $T = \theta(t)$, say, since θ is onto. We now ask whether or not $t \in T$. If $t \in T$, then by definition, $t \notin \theta(t) = T$; and if $t \notin T$, then $t \in \theta(t)$, so $t \in T$ since $T = \theta(t)$. We have therefore shown that $t \notin T$ and *also* $t \in T$, which is an impossibility. This contradiction can only have arisen from our initial supposition that the statement 6.3.3 is false. This statement must therefore be true, so the theorem is proved. ■

Note. The proof made no use of the one-one property of θ in (i) and so we can say at once:

6.3.4 THEOREM. *There is no surjection $A \to \mathfrak{p}A$.* ■

This result is important in the theory of cardinals; see 7.2.

Of course, there is at least one injection $A \to \mathfrak{p}A$, namely that given by

$$g(a) = \{a\}$$

as in 2 Exercise 5(ii). Thus we can regard A as being 'embedded' in $\mathfrak{p}A$. One might ask: can there be an injection $\mathfrak{p}A \to A$? The answer is 'No', by the Schroeder–Bernstein theorem in 7.4 below.

Exercise 5

(i) Prove that, for all integers $n \geqslant 0$, $n < 2^n$.
(ii) When is $A \approx B^A$?

6.4 Shuffles, Permutations, and the Set $I(A, B)$

There is always a chapter on 'permutations and combinations' in the traditional books on algebra. Most readers find it difficult, partly because of the wordy nature of the problems and the methods of solution, partly because the material itself is difficult. We hope here to demonstrate how the language we have now available will help eliminate the first difficulty. A prominent part is played by what the traditional books call a 'permutation of r things from n, attention being paid to order', and the problem of counting all such permutations. On reflection, we see that a **permutation** *of r things from a set A of n things, is nothing other than an injection* $f: \mathbb{N}_r \to A$, the first object selected being $f(1)$, the second $f(2)$, and so on. The objects selected from A form a subset $B = \text{Im}(f)$; if we keep B and change the order, we change f to some injection g with the same image, $\text{Im}(g) = B$. And if we 'pay no attention to order', all the injections f with image B give one '**combination** of n things from A, r at a time'. We shall therefore count the number $\#I(\mathbb{N}_r, A)$ of injections (it is traditionally denoted by ${}_nP_r$) and the number ${}_nC_r$ of combinations; and we adopt the second of the two methods indicated at the beginning of 6.1. It is useful to recall the 'factorial function', $!: \mathbb{Z}^+ \to \mathbb{N}$, whose value at n is traditionally denoted by $n!$ and defined inductively by $0! = 1$, $n! = n(n-1)!$ (so $n! = n(n-1)\cdots 2\cdot 1$).

6.4.1 THEOREM. ${}_nP_r = \#I(\mathbb{N}_r, \mathbb{N}_n) = n!/(n-r)!, \qquad (0 \leqslant r \leqslant n).$

Proof. If $\#A = n$, then (since there exists an equivalence $A \approx \mathbb{N}_n$) there is an equivalence $I(\mathbb{N}_r, A) \approx I(\mathbb{N}_r, \mathbb{N}_n)$ by Exercise 3(iv). Hence

$${}_nP_r = \#I(\mathbb{N}_r, A) = \#I(\mathbb{N}_r, \mathbb{N}_n).$$

Also, if $r = 0$ then $I(\mathbb{N}_0, A) = I(\varnothing, A) = \{\varnothing\}$ by 3.4.3. Thus ${}_nP_0 = 1$ as required.

Suppose then that $1 \leqslant r \leqslant n$. We shall construct a surjection

$$\nu: I(\mathbb{N}_r, \mathbb{N}_n) \to I(\mathbb{N}_{r-1}, \mathbb{N}_n)$$

with the property that for each $\phi \in I(\mathbb{N}_{r-1}, \mathbb{N}_r)$,

(i) $\#\nu^{\flat}\{\phi\} = n - r + 1$;

thus $I(\mathbb{N}_r, \mathbb{N}_n)$ is the union of the (obviously mutually disjoint) sets $\nu^{-1}(\phi)$ as ϕ runs through $I(\mathbb{N}_{r-1}, \mathbb{N}_r)$. Since there are ${}_nP_{r-1}$ such sets, one for each ϕ, then by (i)

$$\begin{aligned} {}_nP_r &= \#I(\mathbb{N}_r, \mathbb{N}_n) = (n - r + 1)\cdot {}_nP_{r-1} \\ &= (n-(r-1))(n-(r-2))\cdots(n-(r-k))\cdot {}_nP_{r-k}, \end{aligned}$$

using induction on $k = 1, \ldots, r$. But ${}_nP_0 = 1$, proved above, whence

$${}_nP_r = (n-(r-1))(n-(r-2))\cdots(n-1)n = n!/(n-r)!.$$

It remains to define the function satisfying (i), so let $\theta \in I(\mathbb{N}_r, \mathbb{N}_n)$, and define $\nu(\theta)$ to be the restriction $\theta|\mathbb{N}_{r-1}$; $\nu(\theta)$ is one-one since θ is, so $\nu(\theta) \in I(\mathbb{N}_{r-1}, \mathbb{N}_n)$. To settle (i), it suffices to observe that if $\theta|\mathbb{N}_{r-1} = \phi$, then $\theta(r)$ can be only one of the $n - r + 1$ elements x in $\mathbb{N}_n - \theta(\mathbb{N}_{r-1})$; while to each such x corresponds just one injection θ of $\mathbb{N}_r$ with $\theta|\mathbb{N}_{r-1} = \phi$ and $\theta(r) = x$. Plainly ν is onto, that is, every $\phi \in I(\mathbb{N}_{r-1}, \mathbb{N}_n)$ appears as $\nu(\theta)$ for some θ; so the theorem is proved. ■

Exercise 6

(i) If $\#A = n$, define a relation R on $I(\mathbb{N}_r, A)$ by: $\theta R\theta'$ if $\theta|\mathbb{N}_{r-1} = \theta'|\mathbb{N}_{r-1}$ and show that R is an equivalence relation. Show that, if $A = \mathbb{N}_n$, then the cosets θR (see 4.5.5) of R are the sets $\nu^{\flat}(\phi)$ of the last proof, and that there is an equivalence $I(\mathbb{N}_r, A)/R \approx I(\mathbb{N}_{r-1}, A)$.
(ii) Let $\alpha: A \to B$ be an injection. By Exercise 3(iii), the function $\alpha_*: I(\mathbb{N}_r, A) \to I(\mathbb{N}_r, B)$ is one-one, where $\alpha_*(\theta) = \alpha \circ \theta$. If R' denotes the relation on $I(\mathbb{N}_r, B)$ corresponding to R in (i), show that $(\alpha_*\theta)R'(\alpha_*\theta')$ when $\theta R\theta'$; and $\alpha_*(\theta R) \subseteq (\alpha_*\theta)R'$, with equality if α is invertible.
(iii) In a game of 'Bingo', each player receives a matrix with ten columns and five rows. Half the entries in the matrix are blank and the rest are distinct integers j, where $0 \leqslant j \leqslant 100$. The integer j in the kth row of the ith column increases with k, and satisfies $10(i - 1) \leqslant j \leqslant 10i$. How many such matrices are there?

In many situations of ordinary life, we often think of the members of a set A as occupying a set P of 'positions', e.g., people in seats, books on shelves, coats on hooks, cards in a pack, and so on. The elements of P are thought of as fixed, while those of A change their positions, as in a shuffle of a pack of cards. Consider a particular distribution of the elements of A among the places in P; it defines an injection $f: A \to P$, where $f(a)$ is the position in P occupied by $a \in A$, and f is one-one, since two a's cannot occupy the same position. Such an injection f is usually called an 'arrangement' or 'rearrangement' of A in the positions P; the set of all such is $I(A, P)$. If $\#A = n$ and $\#P = p$, then by definition there exist bijections $\alpha: \mathbb{N}_n \to A$, $\beta: \mathbb{N}_p \to P$; hence by Exercise 3(vi), there exists a bijection

$$\gamma_*: I(A, P) \to I(\mathbb{N}_n, \mathbb{N}_p)$$
$$\text{(in fact } \gamma_*(f) = \beta^{-1}f\alpha\text{)}.$$

Hence by 2.11.1 and 6.4.1 we have proved:

6.4.2 THEOREM. *The number of arrangements of n things in p places is ${}_pP_n$.* ■

If $n = p$, then $I(A, P) = E(A, P)$ (by Exercise 3(iv) and (vii)). An element of $E(A, P)$ is usually called a 'shuffle' (relative to P). If $P = A$, an element of $E(A, A)$ is called a **permutation** of A, and we write for brevity

6.4.3 $$E(A, A) = \text{Perm } A.$$

Hence, again by 2.11.1, we obtain

6.4.4 THEOREM. *The number of shuffles of n things in n places is $n!$. This also the number of permutations of n things.* ■

Permutations are related to the theory of groups, and the reader will need to know something of Chapter 18 before he can attempt the following exercise, except for (ix), which does not involve the group concept.

Exercise 7

(i) Prove that Perm A in 6.4.3 is a group with respect to the composition operation $\circ$. When $A = \mathbb{N}_n$, Perm A is called the **symmetric group on n symbols.** Prove $f \to f^{-1}$ is a bijection of Perm A onto itself but not an isomorphism if $n > 2$.
(ii) If $\theta: A \approx B$, then by Exercise 3(vii), there is an equivalence Perm $A \approx$ Perm B. Prove that more is true, however, in that the function θ_*: Perm $A \to$ Perm B given by $\theta_*(\alpha) = \theta\alpha\theta^{-1}$, is an isomorphism of the *groups* (Perm A, $\circ$), (Perm B, $\circ$). (Thus $\theta_*(\alpha \circ \beta) = \theta_*\alpha \circ \theta_*\beta$.)
(iii) Let x denote a fixed element of A, and let S_x^A denote the subset of Perm A consisting of all α such that $\alpha(x) = x$. (We call x a **fixed point** of α.) Prove that S_x^A is a subgroup of Perm A. (It is called the **stabilizer** of x in Perm A.) With the notation of (ii), let $y = \theta(x) \in B$. Show that $\theta_*|S_x^A$ is an isomorphism of S_x^A onto S_y^B.
(iv) In the last problem, write S_x instead of S_x^A, let $z \in A$, and let $c: A \to A$ denote that permutation which interchanges x and z but leaves all other elements of A fixed. Prove that $c \circ c$ is the identity, and that the correspondence $\alpha \to c\alpha c$ is an isomorphism $S_x \to S_z$. Hence prove that S_x and S_z are conjugate subgroups of Perm A, and that each is isomorphic to Perm $\mathbb{N}_{n-1}$, where $n = \#A > 0$.
(v) Show that, if $g, h \in$ Perm A, then $gx = hx$ iff $h^{-1} \circ g \in S_x$.
(vi) In the set A, the **orbit** O_x of $x \in A$ 'under the action of the group $G =$ Perm A' is the set $\{gx | g \in \text{Perm } A\}$. Show that $f: G/S_x \to O_x$ is a bijection, where $f[g] = gx$ and $[g] = gS_x$ is a coset of g. Hence give a proof of 6.4.4, using induction on n and the isomorphism $S_x \approx$ Perm $\mathbb{N}_{n-1}$.
(vii) With A as before, let $A_0 = \{a(1), \ldots, a(k)\} \subseteq A$. Let $h: S_{a(1)} \cap S_{a(2)} \cap \cdots \cap S_{a(k)} \to$ Perm $(A - A_0)$ denote the function which assigns to each α the permutation $\alpha|(A - A_0)$. Prove that h is invertible.
(viii) Let $\alpha \in$ Perm A. Since A is finite, then (Perm A, $\circ$) is a finite group, so α has a period p, $1 \leqslant p \leqslant n!$, such that p divides $n!$. Define a relation V on $I(\mathbb{N}_r, A)$ by: fVg whenever $f = \alpha^k \circ g$, for some k with $0 < k < p$. Prove that V is an equivalence relation, and calculate $\#I(\mathbb{N}_r, A)/V$ when $r = n$.
(ix) Let $\#A = n$, $\#B = m \geqslant n$. Count $I(A, B)$ by the following argument. Fix $a \in A$ and define a relation C on $I(A, B)$ by: fCg whenever $f(a) = g(a)$. Prove that C is an equivalence relation. At a, all the elements of the coset fC have the same value $f(a) \in B$. Hence prove that there are just m cosets. If $g \in fC$, prove that $g|A - a$ lies in† $I(A - a, B - g(a))$ and hence that the correspondence $g \to g|A - a$ is a bijection: $fC \approx I(A - a, B - g(a))$. Show therefore that ${}_mP_n = m \cdot {}_{m-1}P_{n-1}$, and so calculate ${}_mP_n$ inductively.

6.5 Combinations

As mentioned earlier, two permutations $f, g \in I(\mathbb{N}_r, A)$, where $r \leqslant \#A = n$ are said to yield the same combination, whenever Im $(f) =$ Im (g). (Thus f and g pick out the same subset of A, and we are ignoring the *order* in which

† We often write $A - a$ for the correct $A - \{a\}$.

they pick it out.) More precisely, let us define a relation T on $I(\mathbb{N}_r, A)$ by setting fTg whenever $\mathrm{Im}(f) = \mathrm{Im}(g)$. Clearly, T is reflexive, symmetric and transitive, and hence T is an equivalence relation. It partitions $I(\mathbb{N}_r, A)$ into cosets fT; and each coset is what we mean by 'a combination of r things from A'. Traditionally, the notations ${}_nC_r$, ${}^nC_r$, $\binom{n}{r}$ are used to denote $\#(I(\mathbb{N}_r, A)/T)$, the 'number of combinations of r things from A'.

6.5.1 THEOREM $\binom{n}{r} = n!/(n-r)!\,r!$.

Proof. When $r = 0$, $I(\mathbb{N}_0, A) = \{\varnothing\}$, so the only coset modulo T is $\{\varnothing\}$ itself. Hence $\#(I(\mathbb{N}_0, A)/T) = 1 = n!/0!\cdot n!$, as required. Suppose then that $r > 0$.

We already know that $\#I(\mathbb{N}_r, A) = {}_nP_r = n!/(n-r)!$, so by 4.5.12 our result will follow if we can prove that each coset fT in $I(\mathbb{N}_r, A)$ has $r!$ elements. To accomplish this, fix one element, $h \in fT$. Then define a function

$$\theta: fT \to \mathrm{Perm}\,\mathbb{N}_r$$

by: $\theta(g) = g^{-1} \circ h$, for each $g \in fT$. We see that this is a valid definition, as follows. We have the diagram $\mathbb{N}_r \xrightarrow{h} X \xrightarrow{g^{-1}} \mathbb{N}_r$, where X is the common image for all $g \in fT$ (by definition of T); g^{-1} exists, since $g: \mathbb{N}_r \to X$ is one-one by definition of $I(\mathbb{N}_r, A)$, and g is onto X since $X = \mathrm{Im}(g)$.

It is left to the reader to prove that θ is invertible. Hence by 2.11.1, $\#(fT) = \#\,\mathrm{Perm}\,\mathbb{N}_r = r!$. As stated above, this suffices to complete the proof. ■

Exercise 8

(i) Prove that $\binom{n-1}{r} + \binom{n-1}{r-1} = \binom{n}{r}$, both from the algebraic definition of $\binom{n}{r}$ and by an argument about combinations which select a specified element. Hence form **Pascal's table**

	0	1	2	3	4	...	r
1	1	1					
2	1	2	1				.
3	1	3	3	1			.
4	1	4	6	4	1		.
5	1	5	10	10	5		
6	1	6	15	20	15		
⋮							
n	1	n		...			$\binom{n}{r}$

to work out $\binom{n}{r}$ by iteration. Note that the nth column gives the set of coefficients in the binomial expansion of $(a + b)^n$. (See 5 Exercise 1(xvi).)

(ii) Estimate the size of $n!$ by comparing $\sum_{r=1}^{n} \log r$ and $\int_1^n \log x \, dx$. [The resulting formula is due to Stirling, and says that $n!$ is approximately $\sqrt{(2\pi)}n^{n+\frac{1}{2}}e^{-n}$ with error $<1/400n\%$. See Courant [74], Vol. 1, p. 361.]

(iii) Prove that there is an equivalence between the set $I(\mathbb{N}_r, A)/T$ (defined prior to 6.5.1) and the family $\mathfrak{p}_r A$ of all subsets $X \subseteq A$ with $\#X = r$.

(iv) Prove that the number of sets of r integers $i_1, \ldots, i_r$ between 1 and n, such that $i_1 < i_2 < \cdots < i_r$, is $\binom{n}{r}$.

(v) Using the data and notation of Exercise 7(iii), let S denote the set of all $\alpha \in \operatorname{Perm} A$ with no fixed point. Prove that

$$S = \operatorname{Perm} A - (S_{a(1)} \cup S_{a(2)} \cup \cdots \cup S_{a(n)})$$

where $A = \{a(1), \ldots, a(n)\}$. Hence, using (iv) and 2.11.8 prove that

$$\#S = n! - \sum_{r=1}^{n} (-1)^{r-1}\binom{n}{r}\cdot(n - r)!$$

(vi) Let X be a finite set with subset Y. Suppose Y is defined by means of some property ψ, so that $Y = \{x | x \in X, \ \& \ \psi(x) \textit{ holds}\}$.

We define the **probability** p that $x \in X$ shall have property ψ, to be the fraction

$$p = \#Y/\#X.$$

(Thus $0 \leqslant p \leqslant 1$.) In (v), let p denote the probability that α, from Perm A, has no fixed points. Calculate the error η when we write $p = \mathrm{e}^{-1} + \eta$ and e is the base of natural logarithms.

(vii) In Exercise 6(iii) suppose that each of n players receives a matrix of the kind described there. Integers j between 0 and 100 are called out in random order, and if k is called, then any player having k in his matrix crosses it out. The first player to have crossed out all non-blank spaces in his matrix has won. Calculate a player's probability of winning, as a function of n.

(viii) A typist types seven letters, addresses seven envelopes accordingly, and inserts the letters in the envelopes at random. Show that the probability of her putting every letter in a wrong envelope is about 1/3.

(ix) Let X be a set of the form $X = X_1 \cup X_2$ where $X_1 \cap X_2 = \varnothing$. Define a relation W on $I(X, \mathbb{N}_m)$ (where $\#X = n \leqslant m$) by: fWg whenever $f(X_i) = g(X_i)$ for each i. Prove that W is an equivalence relation. Fixing h in a coset fW, define $\theta: fW \to \operatorname{Perm} X_1 \times \operatorname{Perm} X_2$ by: $\theta(g) = (k|X_1, k|X_2)$, where $k = h^{-1} \circ g$. Prove that θ is a bijection, and hence that

$$\#(I(X, \mathbb{N}_m)/W = {}_mP_n/m_1!\, m_2! \quad (\text{where } m_i = \#X_i).$$

Extend this result to the case when $X = X_1 \cup X_2 \cup \cdots \cup X_n$ where the X_i are mutually disjoint. Hence calculate the 'number of different ways of arranging m objects, when m_1 are of one kind, m_2 of another kind, and so on'.

(x) Let G be a group, and X a set. G is said to *act* on X if there is a function $G \times X \to X$ (and we write $(g, x) \to gx$) such that $1 \cdot x = x$ and $g_1(g_2x) = (g_1g_2)x$ for all $g_1g_2 \in G$ and $x \in X$. For example, Perm A in Exercise 7, acted on A, and the *orbit* $O_x \subseteq X$ and *stabilizer* $S_x \subseteq G$ are defined as there. Prove that S_x is a subgroup of G, and if $y = gx$, then $S_x = g^{-1}S_yg$, and $O_x = O_y$. Prove that the orbits are the cosets xR of a suitable equivalence relation R on X, and if X and G are finite then $\#X = \sum \#O_{x(i)}$, where we take one $x(i)$ from each orbit. Hence obtain the results of Exercise 7(vi), (viii), (ix), and of exercise (ix) above (e.g., in (ix) above take X here to be $I(X, \mathbb{N}_m)$ there, G to be Perm $X_1 \times$ Perm X_2, so R becomes W there; in the proof of Theorem 6.5.1, X is $I(\mathbb{N}_r, A)$, G is Perm A, and R is T). The orbit formula leads to many results in combinatorial mathematics. (See also Herstein [56], p. 71.)

6.6 The Set $S(A, B)$

Let us now consider the more difficult problem of finding the number

$$e_{nm} = \#S(A, B)$$

of surjections $A \to B$, where $n = \#A$, $m = \#B$. Of course, if $n < m$ then $e_{nm} = 0$; if $n = m$ then $S(A, B) = E(A, B)$, so $e_{nn} = n!$; and if $m = 1 \leqslant n$, then $S(A, B)$ consists of just one (constant) function, so $e_{n1} = 1$. Suppose then that $n > m > 1$. By Exercise 3(v), $S(A, B) \approx S(\mathbb{N}_n, B)$, and we count $S(\mathbb{N}_n, B)$.

6.6.1 THEOREM. $$e_{nm} = m \cdot (e_{n-1,m} + e_{n-1,m-1}).$$

Proof. Define a function $r: S(\mathbb{N}_n, B) \to B^{\mathbb{N}_{n-1}}$ by $r(f) = f|\mathbb{N}_{n-1}$; it need not occur that $r(f)$ is onto, but certainly

(i) $S(\mathbb{N}_{n-1}, B) \subseteq \text{Im}\,(r)$.

Let $B = \{b_1, \ldots, b_m\}$. Now if $g \in S(\mathbb{N}_{n-1}, B)$, then $g = r(f_i)$ for exactly those m functions f_i for which $f_i|\mathbb{N}_{n-1} = g$ while $f_i(n) = b_i$. Hence

(ii) *if* $g \in S(\mathbb{N}_{n-1}, B)$, *then* $\#r^{\flat}(g) = m$.

By (i), $\text{Im}\,(r)$ is a disjoint union of $S(\mathbb{N}_{n-1}, B)$ and its complementary set X; thus $S(\mathbb{N}_n, B) = V \cup W$, where $V = r^{\flat}(S(\mathbb{N}_{n-1}, B))$, $W = r^{\flat}(X)$ and $V \cap W = \varnothing$. Hence

(iii) $\#S(\mathbb{N}_n, B) = \#V + \#W$.

To count W, suppose $f \in W$. Then $r(f)$ can fail to lie in $S(\mathbb{N}_{n-1}, B)$ only because n is the sole element mapped by f to $f(n)$. Thus we can express W as a disjoint union

(iv) $W = W_1 \cup W_2 \cup \cdots \cup W_m$

where

$$W_i = \{f | f \in W \text{ and } f(n) = b_i\}.$$

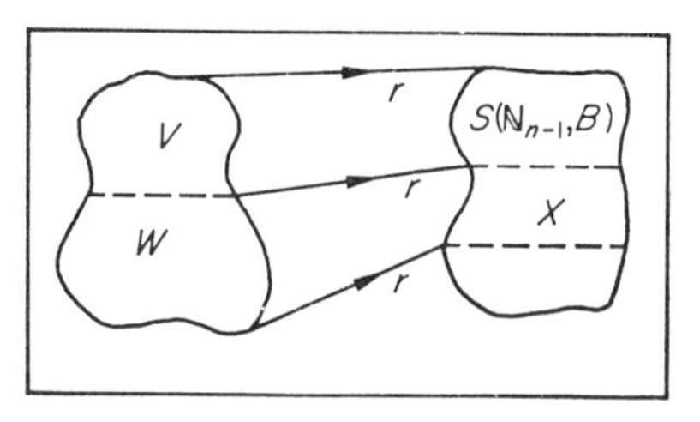

Fig. 6.5

Now, if $f \in W_i$, then $r(f)$ is a surjection $\mathbb{N}_{n-1} \to (B - b_i) = B_i$, so $r|W_i$ is a function $r_i: W_i \to S(\mathbb{N}_{n-1}), B_i)$. We leave the reader to prove that r_i is a bijection; so by (iv)

$$\#W = m \cdot e_{n-1,m-1}.$$

By definition of V and by (ii), $\#V = m \cdot e_{n-1,m}$. Therefore 6.6.1 follows now from (iii). ■

The equation 6.6.1 enables us to compute e_{nm} inductively, at any rate for given n, m. Thus, for example, we obtain from it

$$e_{n+1,1} = 1 \cdot (e_{n1} + e_{n0}) = e_{n1}$$

since $S(A, \varnothing) = \varnothing$ if $A \neq \varnothing$. Hence by induction $e_{n+1,1} = \cdots = e_{11} = 1$, which was already calculated above. To calculate $e_{n+1,n}$ we could argue from first principles (see Exercise 9(ii) below) or use $e_{nn} = n!$ and Theorem 6.6.1 to prove inductively that $e_{n+1,n} = n!\binom{n+1}{2}$.

We can also proceed systematically to build a tableau, as follows, in which the blanks denote zeros; and we enter e_{nm} in the nth row, as m times the sum of the two entries above and to the left in the previous row.

	1	2	3	4	5	6	7	$\cdots$	m
1	1								
2	1	2!							
3	1	6	3!						
4	1	14	36	4!					
5	1	30	150	240	5!				
6	1	62	540	1560	1800	6!			
7	1	126	1806	8400	16,800	15,120	7!		$\cdots$
$\vdots$									$\vdots$
n		$\cdots$		$\cdots$		$\cdots$		$\cdots$	e_{nm}

Exercise 9

(i) If B has just two elements and $\#A \geqslant 1$, show that exactly two members of B^A are not onto. Hence prove that $e_{n,2} = 2^n - 2$.
(ii) Calculate $e_{n+1,n}$ directly by setting up a bijection $S(A, B) \approx C \times \text{Perm}\, B$, where C denotes the set of combinations of pairs from A, and $\#A = n + 1$, $\#B = n$. {If $f \in S(A, B)$, then there is just one pair x, y in A such that $f(x) = f(y)$, and then $f|(A - x)$ is a bijection onto B.}

Chapter 7

COUNTING, AND TRANSFINITE ARITHMETIC

7.1 Counting

We shall now prove Theorem 2.11.2, namely that there is no pair of bijections $A \approx \mathbb{N}_m$, $A \approx \mathbb{N}_n$, if $n \neq m$. Recall that the point of doing this is to allow us to talk of 'the' number, $\#A$, of elements of a finite set A. Several of the steps in the proof of the latter will seem to the reader to consist of proving the obvious. We emphasize our concern to show that they follow from earlier definitions. They are thought to be obvious since we use them (without proof) from childhood; and this is an excellent reason for scrutinizing them carefully, to prevent possible prejudices of one generation from passing to the next. We will also, of course, consider more sophisticated counting procedures.

Let us begin by defining the set $\mathbb{N}_n$ without using dots (see 5.3.3). We take $\mathbb{N}$ as known, and then use the order relation $\leqslant$ on $\mathbb{N}$ to define

7.1.1 $$\mathbb{N}_n = \{i \mid i \in \mathbb{N} \quad \text{and} \quad 1 \leqslant i \leqslant n\}.$$

Thus, if $n = 0$, there is no $i \in \mathbb{N}$ with $1 \leqslant i \leqslant 0$, so $\mathbb{N}_0 = \varnothing$. If $n \geqslant 0$, then

7.1.2 $$\mathbb{N}_{n+1} = \mathbb{N}_n \cup \{n + 1\},$$

and the definition also shows that if $1 \leqslant r < n$, then $\mathbb{N}_r \subset \mathbb{N}_n$.

As was remarked in 2.11.4, it suffices, in order to prove Theorem 2.11.2, to prove:

7.1.3 THEOREM. *If $m \neq n$, then it is impossible that $\mathbb{N}_m \approx \mathbb{N}_n$.*

Proof. We shall prove this by induction, and we first reformulate it slightly, into a statement about n alone; this is justified since either $m \leqslant n$ or $m \geqslant n$. Thus, we prove:

(i) *Given $n \in \mathbb{N}$, then if $1 \leqslant m < n$, there is no injection $\mathbb{N}_n \to \mathbb{N}_m$.*

(Even more so, there can then be no equivalence $\mathbb{N}_n \approx \mathbb{N}_m$.) The statement (i) is of the form '$(\forall n)Q(n)$' where $Q(n)$ is a statement about n, since m is a 'dummy variable'. We are ready, therefore, to use induction on n.

$Q(1)$ cannot be denied, so assume $n > 1$, and make the inductive hypothesis that $Q(n-1)$ holds. Suppose that $Q(n)$ were false, so that there exists an injection $\psi: \mathbb{N}_n \to \mathbb{N}_m$ for some $m < n$, with $1 \leqslant m$. Since $1 \neq 2$ in $\mathbb{N}_n$, $\psi(1) \neq \psi(2)$ in $\mathbb{N}_m$, so $1 < m$ and $\mathbb{N}_{m-1}$ exists. Let $\psi(n) = x$ in $\mathbb{N}_m$. There is a bijection $\theta: \mathbb{N}_m \approx \mathbb{N}_m$ which interchanges x and m and leaves fixed all other elements of $\mathbb{N}_m$ (see Fig. 7.1); if $x = m$, then θ is the identity on $\mathbb{N}_m$.

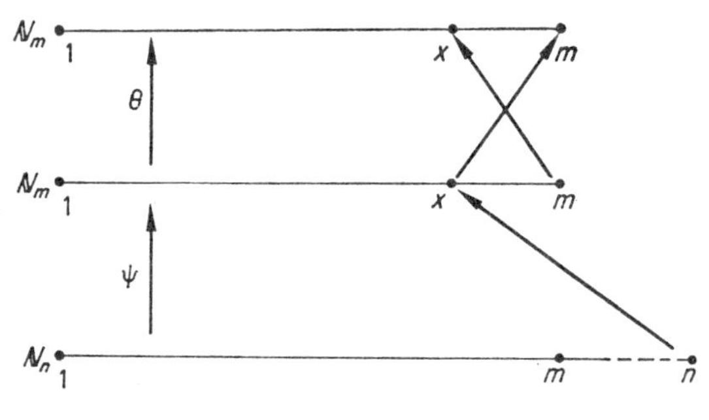

Fig. 7.1

Then $\theta \circ \psi: \mathbb{N}_n \to \mathbb{N}_m$ is an injection, ϕ, by 2.8.1. Now $\phi(n) = m$, so by 7.1.2, $\phi(\mathbb{N}_{n-1}) \subseteq \mathbb{N}_{m-1}$ since ϕ maps no element of $\mathbb{N}_{n-1}$, other than n, to m. Hence $\phi|\mathbb{N}_{n-1}$ is an injection of $\mathbb{N}_{n-1}$ into $\mathbb{N}_{m-1}$, yet $m - 1 < n - 1$ since $m < n$. Thus $Q(n-1)$ is contradicted; so $Q(n-1) \Rightarrow Q(n)$. Therefore, by the principle of induction we have $(\forall n)Q(n)$, as required. ∎

The remarks in 2.11 now show that if A is a finite set (i.e., if there is a bijection $A \approx \mathbb{N}_n$ for some $n \in \mathbb{N}_n$) then n is uniquely defined, and we write: $\#A = n$. We now prove the convenient and expected fact that:

7.1.4 THEOREM. *A finite subset of a finite set is finite.*

Proof. Let $X \subseteq Y$, where Y is finite; we must show that there exists a bijection $X \approx \mathbb{N}_m$ for some m, knowing only that there exists a bijection $f: Y \approx \mathbb{N}_n$, $n = \#Y$. It suffices, then, to show that the subset $A = f(X)$ of $\mathbb{N}_n$ is finite; for if $g: A \approx \mathbb{N}_m$ is a bijection, so is $g \circ (f|X): X \approx \mathbb{N}_m$. If $A = \varnothing$, then $g = \mathrm{id}_\varnothing: A \approx \mathbb{N}_0$; so we assume $A \neq \varnothing$. We now use induction on $\#Y$, and consider the proposition

$$P(n): \quad \textit{Every subset of } \mathbb{N}_n \textit{ is finite.}$$

We have just proved $P(0)$, and we now assume $P(n-1)$. Then we show that if $A \subseteq \mathbb{N}_n$, A is finite. To see this, let $B = A \cap \mathbb{N}_{n-1}$, so either

$$A = B \quad \text{or} \quad A = B \cup \{n\}.$$

In the first case A is finite, by $P(n-1)$. In the second, there exists a bijection $\phi: B \approx \mathbb{N}_k$ for some $k \in \mathbb{N}$, by $P(n-1)$; so we obtain a function $\psi: A \to \mathbb{N}_{k+1}$ by defining

$$\psi|B = \phi; \qquad \psi(n) = k + 1.$$

It is easily checked that ψ is a bijection, so $P(n)$ holds. Hence by the general principle of induction, $P(n)$ holds for all n, and Theorem 7.1.4 is proved. ■

7.1.5 COROLLARY. *If $X \subseteq Y$ and Y is finite, then $\#X \leqslant \#Y$.*

Proof. All that is needed is to strengthen $P(n)$ in the last proof to 'Every subset A of $\mathbb{N}_n$ is finite and $\#A \leqslant n$'; for then the integer k which arose is $\leqslant n - 1$, so $\#A \leqslant k + 1 \leqslant n$. ■

We explained in 2.11 how Theorem 7.1.3 justifies our speaking of 'the' number of things in a finite set A. To show that there are *infinite* sets X, i.e., for which no equivalence $X \approx \mathbb{N}_n$ exists, let us prove that $\mathbb{N}$ itself is infinite.

7.1.6 THEOREM. *For no $n \in \mathbb{N}$ is there an equivalence $\mathbb{N} \approx \mathbb{N}_n$.*

Proof. If $\theta: \mathbb{N} \to \mathbb{N}_n$ were such an equivalence, let $\theta' = \theta|\mathbb{N}_{n+1}$. Then $\theta': \mathbb{N}_{n+1} \to \mathbb{N}_n$ is an injection, contrary to 7.1.3(i). ■

Exercise 1

(i) Prove that $\mathbb{R}$ is infinite.
(ii) If $X \subseteq Y$ and X is infinite, prove that Y is infinite.
(iii) If A is finite, prove that $\mathfrak{p}A$ is finite.
(iv) Prove: if $\varnothing \neq A \subseteq \mathbb{N}_n$, then there is an integer $m \leqslant n$ and a permutation ϕ of $\mathbb{N}_n$ such that $\phi(A)$ is the subset $\mathbb{N}_m$ of $\mathbb{N}_n$.

7.2 Transfinite Arithmetic

What do we mean by 'the number two', by 'infinity', or by 'the set X has more elements than the set Y'? Questions such as these were investigated in the later 19th century by such logicians as Cantor, Frege, and Russell. Their studies had repercussions in almost all branches of mathematics, because the 'set theory' they developed was of universal application. One, more special, outgrowth was an extension of ordinary arithmetic called 'transfinite arithmetic', which we shall now survey.

The question 'what is a number?' is not of much interest in mathematics. Rather we need to know: is there a set of things *with the properties of* $\mathbb{N}$? The 'nature' of the things themselves is of interest only to philosophers. However, to prove the existence of a copy of $\mathbb{N}$, one must build it out of something even more basic, and Russell decided to use the ordinary material of logic—sets, grammar, and so on—as his basic material. He modelled his construction on that of Cantor for infinite sets. Thus:

7.2.1 DEFINITION. The **cardinal**, card X, of a set X, is the class of all sets Y such that $Y \approx X$.

In the language of Theorem 4.5.12 the family of all sets X is partitioned into disjoint cosets by the equivalence relation $\approx$, and *card* X is nothing other than the coset of X. Then we define certain 'integers' as follows: they are merely short names for complicated sets. Thus:

$$0 = \text{card } \varnothing, \qquad 1 = \text{card } \{0\}; \qquad 2 = \text{card } \{0, 1\};$$

and so on. As Russell puts it, '2 is the class of all twins' (and 1 the class of all singletons).

Cardinals are added and multiplied by the following rules, which are suggested by 2.11.5 and 3 Exercise 1(iv), but we emphasize that these hold for all sets, finite or infinite. Thus, we define:

7.2.2 $$\text{card } A + \text{card } B = \text{card } (A \cup B) \qquad (A \cap B = \varnothing)$$

7.2.3 $$\text{card } A \cdot \text{card } B = \text{card } (A \times B).$$

The usual procedure must be undergone, of 'checking independence of representatives' (see 4.5.9) of the cosets. For example, a prerequisite for 7.2.2 is that the classes card A, card B have members A', B' which are disjoint; and of course they have (using 3.2.4), since we may take $A' = A \times \{p\}$, $B' = \{p\} \times B$ where p is any object not in $A \cup B$. [Does such a p always exist?] And then, if $A'' \in$ card A, $B'' \in$ card B had been a second choice of disjoint representatives, then by 2 Exercise 10(i), $A' \cup B' \approx A'' \cup B''$, since $A' \approx A''$, $B' \approx B''$. Similarly, for 7.2.3, if A', B' are chosen from card A, card B, then $A \times B \approx A' \times B'$ [by 3 Exercise 1(iii)], so card $(A \times B) =$ card $(A' \times B')$.

Commutativity of multiplication follows from 3 Exercise 1(i), for then $A \times B \approx B \times A$, so card $(A \times B) =$ card $(B \times A)$, whence

$$\text{card } A \cdot \text{card } B = \text{card } (A \times B) = \text{card } (B \times A) = \text{card } B \cdot \text{card } A.$$

Distributivity follows from 3.2.1; and 0(= card $\varnothing$) is a true zero, by 3.2.3 and the fact that $A \cup \varnothing = A$.

Exercise 2

(i) Give detailed proofs, along the above lines, that the set of cardinals satisfies the axioms $\mathfrak{A}_1 - \mathfrak{A}_8$ (see 9.2) of a ring with unit 1, but without subtraction being always possible.

(ii) It is customary to use the Hebrew letter $\aleph$ (pronounced 'aleph') to denote many cardinals. In particular, we define $\aleph_0 =$ card $\mathbb{N}$.

Prove that, for any integer $n \in \mathbb{N}$,

$$n + \aleph_0 = \aleph_0 = n \cdot \aleph_0 = \aleph_0^n.$$

{Hint: use 2 Exercise 9(ii), and assume the ordinary properties of integers.}

(iii) For any cardinal a, show that $a + a = 2 \cdot a$.

(iv) Denote card ($\mathbb{R}$) by $\mathbf{c}$ [because $\mathbb{R}$ was called by Cantor the 'Continuum']. Prove some facts about $\mathbf{c}$ as follows. Let $X = \mathbb{R} - \{\text{all even integers}\}$. Find a bijection $f: \mathbb{R} \approx X$ such that $f(x) = x$ if x is not an integer, and $f(\mathbb{Z})$ lies in the odd integers. Hence prove that $\mathbf{c} = \mathbf{c} + \aleph_0$, and hence that if $n \in \mathbb{N}$ then $\mathbf{c} + n = \mathbf{c}$.
(v) Prove that, if I is the interval $\{x|0 < x < 1\}$ in $\mathbb{R}$, then $\mathbf{c} = \text{card } I$. Use (iv) to prove that $I \approx I \cup \{0\} \approx I \cup \{0, 1\} \approx \mathbb{R}^+$ and

$$\mathbf{c} + \mathbf{c} = \mathbf{c} = n \cdot \mathbf{c} \qquad (n \in \mathbb{N})$$
$$= \aleph_0 \cdot \mathbf{c}.$$

(vi) Every number a such that $0 \leqslant a < 1$ has a decimal expansion

$$a = 0.a^{(1)}a^{(2)} \cdots a^{(n)} \cdots,$$

which is unambiguously defined, unless there are recurring nines (e.g., $\frac{1}{10} = 0.1000 \cdots = 0.0999 \cdots$). Show that the set of a with recurring nines is countably infinite and hence that the set Δ of 'unambiguous' a has cardinal $\mathbf{c}$. {Recall from 2 Exercise 9(ii) that a set X is countably infinite iff $X \approx \mathbb{N}$.}
(vii) Prove that $\mathbf{c} \neq \aleph_0$, by Cantor's 'diagonal' process: suppose that Δ in (vi) could be enumerated as a sequence $a_1, a_2, \ldots, a_n, \ldots$. Then a_n has the unique decimal expansion $a_n = 0.a_n^{(1)}a_n^{(2)} \cdots a_n^{(n)} \cdots$. Now look at the **diagonal elements** $a_n^{(n)}$ and form a new number $x = 0.x^{(1)} \cdots x^{(2)} \cdots {}^{(n)}$ in Δ, where

$$x^{(n)} = \begin{cases} 7 & \text{if} \quad a_n^{(n)} \neq 7 \\ 3 & \text{if} \quad a_n^{(n)} = 7, \end{cases}$$

(other choices could be made). Show that x has no place in the supposed enumeration of Δ, and that this contradiction implies $\mathbf{c} \neq \aleph_0$.

[A consequence of this result is *the existence of transcendental numbers*. Recall from 2 Exercise 9(iii) that the set A of algebraic numbers is countable. The diagonal process explained above shows that $\mathbb{R}$ contains numbers not in A, and such numbers are called **transcendental**. Cantor's proof of their existence is not constructive, since it gives no test for deciding whether or not a given number like π or e is transcendental. If the a's in the construction above are all the algebraic numbers, then the x constructed there is transcendental but its properties are difficult to work out. J. Liouville (1809–1892) gave a direct construction for making some transcendental numbers; see Courant–Robbins [25], p. 105. For example, Liouville showed that $\sum_{n=0}^{\infty} 10^{-(n!)}$ is transcendental.]

Notice that with the cardinal numbers we already have an 'arithmetic' built on very basic notions: a child can recognize twins before he can count, he can form the union of his pile of toys with a disjoint pile before he can add, and he is aware of rectangular block patterns before he can multiply. Yet few adults can specify the relationship, between the 'five' in 'five toes' and in 'five shillings'! They simply lack the language.

For the purpose of building the ordinary integers, we now select a subset of the cardinals, called the 'finite' cardinals. Cantor observed a basic distinction among the sets we meet in ordinary mathematical experience, namely, that 'infinite' sets, unlike finite ones, are equivalent to proper subsets of themselves (e.g., $\mathbb{N} \approx E$, where E is the set of even integers, through the correspondence $n \to 2n$). We give then his definition, which can be reconciled with the definition we gave in 7.1.4.

7.2.4 DEFINITION. Card A is **finite** iff A is not equivalent to a proper subset of A. The set A is then **finite.** Otherwise, A and card A are **infinite.**

Russell then showed that the set F of all finite cardinals has all the properties [see the Peano axioms in 23.1] that we require of $\mathbb{N}$. Further, if A is finite in this sense, then card A turns out to be what we denoted by $\#A$ in 2.11.1, provided we take $\mathbb{N}$ to *be* F itself; and the two definitions of 'finite' agree. Our definition of $\#A$ in 2.11.1 took $\mathbb{N}$ itself as basic, and used the subsets $\mathbb{N}_n$ as 'models', as standard n-element sets. This is connected with the fact that mathematical logicians still differ as to whether mathematics is part of logic (as Russell believed), or vice versa. Since *something* has to be taken as basic building material, taste and philosophy will govern an individual's choice.

7.3 The Order Relation in Transfinite Arithmetic

Our remaining discussion of cardinals is now from Cantor's point of view†, of having an 'arithmetic' that makes sense for infinite cardinals. In particular, we shall want to be able to compare two cardinals as to 'size', thus distinguishing between infinite sets.

First, consider the problem of exponentiation, of defining b^a when a, b are cardinals. If a is finite, then we could use induction on a, to say that b^a is the product (in the sense of 7.2.3) of b with itself, taken a times. But what if a is infinite? Here we generalize from Corollary 6.1.5 and make the definition:

7.3.1 DEFINITION. We define $(\text{card } B)^{\text{card } A}$ to be card (B^A).

When A is finite, then this is just repeated multiplication, by Corollary 6.1.4. The following exercise shows how the 'rules of indices' are obeyed.

Exercise 3

Prove that

(i) $(\text{card } B)^{(\text{card } A + \text{card } C)} = (\text{card } B)^{\text{card } A} \cdot (\text{card } B)^{\text{card } C}$ {use 6.1.1};
(ii) $(\text{card } B)^0 = 1$, $(\text{card } B)^1 = \text{card } B$;
(iii) $((\text{card } B)^{\text{card } A})^{\text{card } C} = (\text{card } B)^{\text{card } A \cdot \text{card } C}$.

{*Hint:* To prove that $(B^A)^C \approx B^{A \times C}$, assign to each function $f: A \times C \to B$ the function $\theta f \in (B^A)^C$ as follows. θf is to be a function $g: C \to B^A$, so $g(c)$ has to be a function $A \to B$; take it to be that function $A \to B$ whose value at $a \in A$ is $f(a, c)$. (Thus θ 'regards f as a function of c only' for each a.) Now prove this assignment to be invertible.}

† Technically, 'within naïve set theory'.

(iv) For all cardinals a, b, c prove that $(ab)^c = a^c b^c$.

{*Hint:* for sets A, B, C, use 3 Exercise 4(ii) to prove $(A \times B)^C \approx A^C \times B^C$.}

To define inequality of cardinals, we make a definition as follows, based on judging sets in real life to be 'smaller' than others.

7.3.2 DEFINITION. We say that the cardinal of A is **less than or equal to** the cardinal of B and write card $A \leqslant$ card B iff there is an injection $j: A \to B$. To prove independence of representatives in this definition, suppose A', B' had been chosen instead to represent card A, card B; then there exist equivalences $\alpha: A \approx A'$, $\beta: B \approx B'$. If there is an injection $k: A' \to B'$, then by 2.8.1, $\beta^{-1}k\alpha$ is an injection $A \to B$; this checks that the definition is independent of representatives. ■ Of course we have not shown that card $A \leqslant$ card B is an order relation, so the form of words in the definition is a little premature. But see Exercise 4(i).

Exercise 4

(i) Use 2.8.1 again to prove that $\leqslant$, as here defined, is reflexive and transitive.
(ii) If card $A \leqslant$ card B and card A is infinite, prove card B is infinite.
(iii) How do Definitions 7.2.4, 7.3.2 clarify the Aristotelian axiom that 'the whole is greater than the part'?

To express *strict* inequality, we write card $A <$ card B, iff

$$\text{card } A \leqslant \text{card } B \quad \text{and} \quad \text{card } A \neq \text{card } B.$$

Thus, for example, Theorems 6.3.2 and 6.3.3 give the important result, that for all B,

7.3.3 $$\text{card } B < 2^{\text{card } B},$$

which shows that *to every cardinal, there is a greater one.*

Exercise 5

(i) If card $A \leqslant$ card B, prove that for any cardinal X, $X +$ card $A \leqslant X +$ card B, and (card A)$\cdot X \leqslant$ (card B)$\cdot X$.
(ii) Show that if $a < b$, we need not have $x + a < x + b$.
(iii) Prove that $\aleph_0 < \mathbf{c}$.
(iv) Let $\mathbb{I} = \{x | 0 \leqslant x \leqslant 1\} \subseteq \mathbb{R}$, let $A = \{0, 1, \ldots, 9\}$ and let $\theta: A^{\mathbb{N}} \to \mathbb{I}$ be the function such that $\theta(f)$ is the number with decimal expansion $0.f(1)f(2)\cdots f(n)\cdots$. Show that θ is onto, and that for all $x \in \mathbb{I}$, $\theta^{\flat}\{x\}$ has either one or two elements, according as $x \in \Delta$ (see Exercise 2(vi)) or not. Hence show that $10^{\aleph_0} = \mathbf{c}$.
(v) Let m be an integer $\geqslant 2$. Repeat the last exercise, when A is replaced by the set $\{0, 1, 2, \ldots, m\}$, and show that $m^{\aleph_0} = \mathbf{c}$. {When $m = 2$, for example, the decimal expansion $\sum_{n \in \mathbb{N}} f(n)/10^n$ is replaced by the 'binary' expansion $\sum_{n \in \mathbb{N}} f(n)/2^n$.}
(vi) Prove that $\mathbf{c}^2 = \mathbf{c}$. {Use Exercise 3(iv).}
§(vii) Investigate Peano's mapping of $\mathbb{I}$ onto the unit square: for each $x \in \mathbb{I}$ choose a decimal representation $x = 0.x_1x_2\cdots$, (with recurring nines if necessary) and

map x to $(0.x_1x_3x_5\cdots, 0.x_2x_4x_6\cdots)$. Note that, for example, $0.x_1x_3x_5\cdots$ may end in recurring zeros, but this still represents unambiguously a number in $\mathbb{I}$. Is the mapping continuous?
(viii) Prove that $\mathbf{c} = \aleph_0^{\aleph_0}$.
(ix) Let a be an infinite cardinal. Show that $\aleph_0 \leqslant a$ by using induction to choose a sequence $x_1, \cdots, x_n, \cdots$, from A where card $A = a$. Hence show that $\aleph_0 + a = a$.

If the definition 7.3.2 of inequality is to be any good, it ought to be not only reflexive and transitive but also antisymmetric. We would also wish it to satisfy the law of trichotomy 4.4.6. We need then to be able to prove two things, and the reader may already have hoped for the first of them, because it would have made some of the earlier exercises simpler. Thus we need to prove

7.3.4 THEOREM. *If* card $A \leqslant$ card B *and* card $B \leqslant$ card A, *then*

$$\text{card } A = \text{card } B;$$

7.3.5 THEOREM. *Either* card $A \leqslant$ card B *or* card $B \leqslant$ card A (*and not both* (*by* 7.3.4) *unless* card A = card B.

7.3.6 Theorem 7.3.4 is essentially just the famous **Schroeder–Bernstein theorem**; see 7.3.7 below. We have to prove that if there exist injections $u: A \to B$, $v: B \to A$, then there exists an equivalence $A \approx B$. Much more formidable, however, is a proof of Theorem 7.3.5, since we have somehow to find an injection $A \to B$ (or the other way round), given *any* sets A, B. To solve this problem, Cantor had to develop the theory of ordinal numbers, and we discuss it further in 7.4 below.

Let us first settle 7.3.4 by proving:

7.3.7 THEOREM (Schroeder–Bernstein). *Given injections* $u: A \to B$, $v: B \to A$, *there exists a bijection* $A \approx B$.

Proof. We shall say that x in A has a 'parent' in B if $\exists z \in B$ such that $x = v(z)$; and if z similarly has a 'parent' y in A, then we say that x has a 'parent' in A, and $x = vu(y)$. Now trace the 'ancestry' of x; it might go back for ever, or cease in A (perhaps at x itself), or cease in B. More precisely, it might happen that

(i) $(\forall n \in \mathbb{Z}^+)\exists y \in A$ and $x = (vu)^n(y)$; or
(ii) there is a largest $n \in \mathbb{N}$, and $y \in A$, such that $x = (vu)^n y$ but y is parentless in B (and hence also in A); or
(iii) there is a largest $n \in \mathbb{Z}^+$, and $z \in B$, such that $x = (vu)^n v(z)$ but z is parentless in A (and hence also in B).

Let A_1, A_2, A_3 respectively denote the subsets of A consisting of elements of types (i), (ii), (iii); and let B_1, B_2, B_3 in B be defined similarly. Then

$$A = A_1 \cup A_2 \cup A_3, \qquad A_1 \cap A_2 = \varnothing = A_2 \cap A_3 = A_3 \cap A_1,$$

and similarly for B. It is easily verified that

$$u(A_1) = B_1, \qquad u(A_2) = B_3, \qquad v(B_2) = A_3.$$

Let $p = u|A_1$, $q = u|A_2$, $r = v|B_2$; each is an injection since u, v are, and hence each is a *bijection*

(iv) $p\colon A_1 \approx B_1; \qquad q\colon A_2 \approx B_3; \qquad r\colon B_2 \approx A_3.$

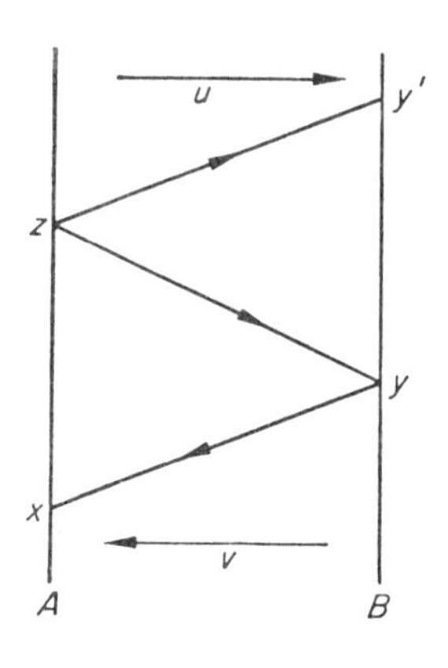

Fig. 7.2

We now use the method employed in proving Theorem 2.11.5 to piece together the functions p, q, r^{-1}, obtaining a function $w\colon A \to B$ given by

$$w|A_1 = p; \qquad w|A_2 = q; \qquad w|A_3 = r^{-1}.$$

By (iv), w is a bijection $A \approx B$, as required. ■

This settles the problem of establishing 7.3.4, and we now sketch a way of dealing with 7.3.5; for a fuller treatment see Halmos [50] or Hausdorff [55].

7.4 The Axiom of Choice

Many proofs in mathematics involve choosing an element from a set X; but examples in real life already display a difficulty (apart from ethical questions of 'fairness'). For instance, how does one choose one electron from a pair of electrons? The notion of a set is not defined precisely, and is so general†, that we have to carry with it a belief that if X is any conceivable (non-empty) set then we can in principle choose an element from X. The difficulty is worse when we have a whole family $\{X_\alpha\}_{\alpha \in A}$ of non-empty sets, where A is infinite; and suppose that we wish to choose one element x_α from each X_α. Thus we want a **choice function** $c\colon A \to \bigcup_{\alpha \in A} X_\alpha$ such that $\forall \alpha$,

† See also the Russell paradox, 39.2.1.

$c(\alpha) \in X_\alpha$, and if we believe that the set† of all such functions is non-empty, then we still have the problem of 'choosing' one. Ideally we would like a precise prescription for such a choice which a computer could follow; but most mathematicians are content with believing that a choice exists 'in principle'. In fact, they hide the difficulties behind the existential quantifier, with the usage '$X \neq \varnothing$, so $\exists x \in X$ and *that* x is the one we imagine to be picked out'. Nevertheless, it is important to state the belief, so that future mathematical errors may be traced back to it if necessary; and it does lead to some conclusions which are surprising relative to our acquaintance with the finite world.

Thus, we shall say that we can 'choose' an element from a non-empty set 'by the **axiom of choice**', and we shall say that we can choose a choice function $c: A \to \bigcup_{\alpha \in A} X_\alpha$ 'by the **strong** axiom of choice'. For the case when A is finite and $\#A = n > 1$, we can use induction on n and bypass the strong axiom of choice by using the steps:

(i) Choose a bijection $p: A \approx \mathbb{N}_n$;
(ii) Choose a choice function $c: \mathbb{N}_{n-1} \to X_1 \cup \cdots \cup X_{n-1}$ [that is, let us repeat, an element of $X_1 \times \cdots \times X_{n-1}$ (see 3 Exercise 1(v), (vi)];
(iii) Choose $x \in X_n$ and define $c(n)$ to be x;

whence cp is a choice function $A \to \bigcup_{m \in A} X_m$, picked out by the proof.

Exercise 6

(i) Fill in the details of the last proof.
(ii) Specify (if you can) a choice of one element from a set X when X consists of (a) a pair of shoes, (b) a pair of socks, (c) a group, (d) the set of all unfamiliar transcendental numbers, (e) the set of maxima of a given continuous function $\langle 0, 1\rangle \to \mathbb{R}$, (f) $\mathfrak{p}A$ where A is a given set.

Now, if a non-empty set X is well-ordered (4.4.8), we can choose an element x from X by stipulating that x shall be the *first* member of X in the well-ordering. This suggests that we should try to well-order every set Y, to be able to make choices; but we would still need to choose one of the (possibly many) well-orderings which Y may possess among the subsets of $\mathfrak{p}(Y \times Y)$. And nobody has ever succeeded in prescribing a well-ordering (in a finite usable formula) for such a 'familiar' set as $\mathbb{R}$.

We will state a lemma which is a useful tool, and which is implied by the axiom of choice. It is due to the contemporary mathematician M. Zorn. First we need some definitions. Let $(X, \leqslant)$ be an ordered set as in 4.4.1, and let $A \subseteq X$.

7.4.1 DEFINITION. An element x in the ordered set $(X, \leqslant)$ is **maximal** with respect to the subset A of X, provided there is no $y \in A$ such that $x < y$. A maximal element x is called an **upper bound** for A in X if, in fact, $z \leqslant x$ for every $z \in A$. If, further, $x \leqslant y$ for all upper bounds y of

† This set is usually called the **Cartesian product** of the family $\{X_\alpha\}_{\alpha \in A}$.

A in X, then x is a **least upper bound** for A in X. Plainly a least upper bound, if it exists, is unique. We abbreviate the least upper bound to lub.

For example, let X be the set of points shown in Fig. 7.3, where $u < v$ means that u is lower down the page than v. Then each of the points 6, 7, 8 is maximal with respect to X itself but none is an upper bound for X; while each of 4, 5 is maximal with respect to $A = \{1, 2, 3, 4\}$ but 4 is an upper bound for A in X, yet 5 is not. Neither 4 nor 5 is maximal with respect to $B = \{3, 4, 5, 6\}$, and hence neither of them can be an upper bound for B in X.

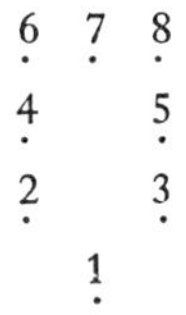

6 7 8
4 5
2 3
1

Fig. 7.3

Now the subset A in 7.4.1 is itself partially ordered by $\leqslant$, and we are interested in those A which are *totally* ordered by $\leqslant$; that is (see 4.4.6) if $x, y \in A$, then either $x \leqslant y$ or $y \leqslant x$ (but not both, unless $x = y$, by 4.4.1(i)).

7.4.2 DEFINITION. $(X, \leqslant)$ is **inductively** ordered provided that every totally ordered subset A possesses a lub in X.

For example, the set X of rational numbers x with $0 \leqslant x < 1$, in their natural ordering, is not inductively ordered even though this ordering of X is total; for X has no maximal element with respect to itself. On the other hand, if we regard this set X as a subset of $(\mathbb{Q}, \leqslant)$ then 1 is maximal (in $\mathbb{Q}$), with respect to X and it is also the lub for X in $\mathbb{Q}$. But $\mathbb{Q}$ is not inductive since it possesses no maximal element.

We next quote the following result whose proof does not require any choices; for the (fairly long) proof we refer the reader to Hall and Spencer [47], 277.

7.4.3 PROPOSITION. *Let $(X, \leqslant)$ be a non-empty inductively ordered set, and let $f\colon X \to X$ be a function such that, $\forall x \in X$, $x \leqslant f(x)$. Then f has a fixed point y (i.e., $y = f(y)$) in X.* □

If now we assume the strong axiom of choice, we have:

7.4.4 ZORN'S LEMMA. *Every non-empty inductively ordered set has a maximal element.*

Proof. To get a contradiction, suppose that $(X, \leqslant)$ is an inductively ordered set without a maximal element. Then to each $x \in X$, $\exists y \in X$ with $x < y$. Thus the set R_x, consisting of all y with $x < y$, is non-empty. Now pick a choice function $c\colon X \to \bigcup_{x \in X} R_x \subseteq X$, using the strong axiom of choice. Then, by 7.4.3, c has a fixed point $y = c(y) \in R_y$, contrary to the condition that $y \notin R_y$. This contradiction establishes the lemma. ■

For our purposes it suffices to give two applications of Zorn's lemma; but many other applications exist, especially in algebraic parts of mathematics.

7.4.5 THEOREM. *Every non-empty set X has a well-ordering.*

Proof. Let W denote the family of all pairs (A, R) where $A \subseteq X$ and R is a well-ordering on A (so $R \subseteq A \times A$). W is not empty; for $X \neq \varnothing$ and $\forall a \in X$, the singleton $\{a\}$ is well-ordered by the relation of equality. Order W by the rule: $(A, R) \leqslant (B, S)$ provided (i) $A \subseteq B$; (ii) $S|A$ is R (i.e., S agrees on A with R); and (iii) if $a \in A$, $b \in B$, and bSa, then also $b \in A$. Plainly $(W, \leqslant)$ is an ordered set; *we claim that $(W, \leqslant)$ is inductively ordered.*

For let $V = \{(A_\alpha, R_\alpha)\}_{\alpha \in I}$ be a subset of W, which is totally ordered by $\leqslant$; we must first find an upper bound for V in W. To do this, let $A = \bigcup_{\alpha \in I} A_\alpha$, and we find a well-ordering R of A as follows.

If $x, y \in A$, then $\exists \alpha, \beta \in I$ such that $x \in A_\alpha$, $y \in A_\beta$. Since V is totally ordered, we may assume $(A_\alpha, R_\alpha) \leqslant (A_\beta, R_\beta)$, so $A_\alpha \subseteq A_\beta$ by (i) above, whence x and y lie in A_β. Now R_β *well-orders* A_β by definition, so just one of $xR_\beta y$, $yR_\beta x$ holds, say the first. Define R by setting xRy to mean $xR_\beta y$; by (ii) above, this is independent of the choice of α, β. Observe that by construction

(iv) $\forall \alpha \in I, \qquad R|A_\alpha = R_\alpha.$

It is easy to see that R totally orders A. To show that R well-orders A we have to find, for each non-empty $Y \subseteq A$, a first element in the sense of R. By definition of A, $\exists \alpha \in I$ such that $A_\alpha \cap Y \neq \varnothing$, since $Y \neq \varnothing$. Since R_α well-orders A_α, there is a first element $u \in A_\alpha \cap Y$ (in the sense of R_α), and we claim that u is the first element of Y in the sense of R. Thus we must show that given $y \in Y$, uRy; since $uR_\alpha y$ if $y \in A_\alpha$, we may assume that $y \in A_\beta$ for some $\beta \in I$, and $y \notin A_\alpha$. Hence $A_\beta \nsubseteq A_\alpha$, so in the totally ordered set V we must have $(A_\alpha, R_\alpha) \leqslant (A_\beta, R_\beta)$; thus $A_\alpha \subseteq A_\beta$. Since R_β well-orders A_β and $u \in A_\alpha \subseteq A_\beta$, then either $uR_\beta y$ or $yR_\beta u$. If the latter held, then by (iii) above, y would lie in A_α; thus $uR_\beta y$, whence by (iv) uRy as required. This proves that $(A, R) \in W$; we now invite the reader to show that $(A_\alpha, R_\alpha) \leqslant (A, R)$ for all $a \in I$, by checking (i), (ii), and (iii) above. Thus (A, R) is an upper bound for V in W, and the same kind of checking shows that $(A, R) \leqslant$ any other upper bound (B, S). [Observe that $A_\alpha \subseteq \bigcup_{\alpha \in I} A_\alpha = A \subseteq B$.] Therefore (A, R) is the lub for V in W, so it is verified that $(W, \leqslant)$ is inductively ordered.

Now apply Zorn's lemma to $(W, \leqslant)$, to obtain a maximal element (B, S). We claim that $B = X$; otherwise $\exists x \in X - B$, and we can define a well-ordering T of $B' = B \cup \{x\}$ by setting $T|B = S$, with bTx for all $b \in B$ (in effect 'sticking x at the end of B'). Thus $(B, S) \leqslant (B', T)$ while $(B, S) \neq (B', T)$, so (B, S) would not be maximal. Therefore (B, S) is in fact (X, S), where S well-orders X. This completes the proof. ∎

Exercise 7

(i) In the last proof, in what sense was choice used when we remarked that $\exists\alpha: A_\alpha \cap Y \neq \varnothing$, that $y \in A_\beta$, and that $\exists x \in X - B$?
(ii) Two partially ordered sets $(X, \leqslant)$, $(Y, \leqslant)$ are called *isomorphic* if there exists a bijection $f: X \to Y$ such that f and f^{-1} are order-preserving, i.e., $x_1 \leqslant x_2$ in X iff $fx_1 \leqslant fx_2$ in Y. Show that if X is totally ordered, so is Y, and that then f^{-1} is order-preserving if f is. If X is not totally ordered and $f: X \to Y$ is an order-preserving bijection, show that f^{-1} need not preserve order. {Map the system of Fig. 7.3 to $\mathbb{N}_8$.}
(iii) Use induction to show that if in (ii) the sets X, Y are finite and $\#X = \#Y$, then there exists an isomorphism between $(X, \leqslant)$ and $(Y, \leqslant)$, provided both systems are well-ordered.

Our second application of Zorn's lemma will settle 7.3.5, as explained in 7.3.6.

7.4.6 THEOREM. *Given sets A, B, then there is either an injection $A \to B$ or an injection $B \to A$.*

Proof. Let W denote the set of all pairs (X, i) where $X \subseteq A$ and $i: X \to B$ is an injection. By 3.4.3 we may assume that neither A nor B is empty; so $W \neq \varnothing$ because W contains an element (X, i) with X a singleton $\{a\}$, and $i(a)$ a chosen $b \in B$. Order W by the rule: $(X, i) \leqslant (Y, j)$ if $X \subseteq Y$ and $j|X = i$ (so j extends i). We assert: $(W, \leqslant)$ is inductively ordered.

For, let $V = \{X_\alpha, i_\alpha\}_{\alpha \in I}$ be a totally ordered subset of W, and let us find an upper bound for V in W. Let $X = \bigcup_{\alpha \in I} X_\alpha$, and define $j: X \to B$ as follows. If $x \in X$, then $x \in X_\alpha$ for some $\alpha \in I$, so we define $j(x)$ to be $i_\alpha(x)$; we must show that if also $x \in X_\beta$ then $i_\beta(x) = i_\alpha(x)$ and so j is well-defined. But V is totally ordered, so either $(X_\alpha, i_\alpha) \leqslant (X_\beta, i_\beta)$ or vice versa, whence $i_\alpha(x) = i_\beta(x)$ by definition of $\leqslant$. Thus j is indeed a function and

(i) $\forall \alpha \in I,\ i_\alpha = j|X_\alpha$;

it is easily checked that j is an injection, so $(X, j) \in V$. To see that (X, j) is an upper bound for V in W, let $(X_\alpha, i_\alpha) \in V$; we prove that $(X_\alpha, i_\alpha) \leqslant (X, j)$ in W. But this follows by definition of $\leqslant$ in W, since $X_\alpha \subseteq X$ and (by (i)) $j|X_\alpha = i_\alpha$. Therefore (X, j) is an upper bound for V. It is obviously the lub, and $(W, \leqslant)$ is inductively ordered.

Now apply Zorn's lemma to get an element (Y, k) of W, maximal with respect to W. We cannot have *both* $Y \neq A$ and $k(Y) \neq B$, otherwise we can choose $a \in A - Y$, $b \in B - k(A)$, and define $k(a) = b$, to obtain $(A, k) < (A \cup \{a\}, k)$, contrary to the maximality of (A, k). Hence *either* $Y = A$ —whence $k: A \to B$ is the required injection—*or* $k(Y) = B$, in which case $k^{-1}: B \to Y \subseteq A$ is the required injection. This completes the proof. ■

Remark. In fact, in applying Zorn's lemma it is logically unnecessary to check that the upper bound we have found is the lub. Thus we may simplify, mildly, the proofs given above using Zorn's lemma. We do not have to make the check because (see Kelley [69]) we obtain the same concept of an inductive ordering if we remove the word 'least' from Definition 7.4.2.

Exercise 8

§(i) Use Zorn's lemma to prove that every vector space V over a field F has a basis linearly independent (over F), whether or not $\dim_F V$ is finite. {Consider the set B of all linearly independent subsets of V, and order B by inclusion.}
(ii) Let X be a set. Find a function which assigns to each subset A of X an element in A.

We conclude this chapter with a mention of ordinal numbers. In ordinary speech, these are the 'numbers' (so-called) 'first', 'second', etc. Their meaning is never made precise in the grammar books, but they have a connotation of well-ordering. This led Cantor to define an **ordinal number** α as an equivalence class $[(X, R)]$ of isomorphic well-ordered sets (X, R) (see Exercise 7(ii)). The class of ordinals is then well-ordered by saying that $\alpha \leqslant \beta$ iff there is an order-preserving injection $(X, R) \leqslant (Y, S)$ for some $(X, R) \in \alpha$, $(Y, S) \in \beta$. Just as for cardinals, this definition can be shown to be independent of representatives. Addition $\alpha + \beta$ is defined by choosing disjoint representatives $(X, R) \in \alpha$, $(Y, S) \in \beta$ and setting $\alpha + \beta = [(X \cup Y, T)]$ where every element of Y 'follows' X, i.e., $T|X = R$, $T|Y = S$ and xTy for all $x \in X$, $y \in Y$. Addition is associative but not always commutative. Abbreviations are then introduced, starting with the class of the well-ordered empty set $(\varnothing, \varnothing)$; thus $0 = [(\varnothing, \varnothing)]$, $1 = [(\{0\}, =)]$, $2 = 1 + 1$, $3 = 2 + 1$, $\ldots$, and the class of $\mathbb{N}$, with its usual ordering, is denoted by ω. Thus we have the **ordinal series**

$$0, 1, 2, \ldots, \omega, \omega + 1, \ldots, \omega + \omega, \ldots, \omega^{\omega}, \ldots$$

where exponentiation and multiplication can be defined, but multiplication need not be commutative: if $\alpha = [(X, R)]$ and $\beta = [(Y, S)]$, then we define $\alpha\beta = [(X \times Y, T)]$ where T is the 'lexicographical' ordering given in 4 Exercise 3. It turns out by using Exercise 7(iii) that the set of finite ordinals is isomorphic to that of the finite cardinals (including addition and multiplication). Also, the method of finite induction can be extended to 'transfinite induction' to define and prove statements of the form $P(\alpha)$ where α is an ordinal; but such proofs can often be avoided by using Zorn's lemma.

Cantor used the ordinals for proving many results in transfinite arithmetic; for example that if a, b are cardinals with $a \geqslant b \geqslant \aleph_0$ then $ab = a$. (See Hausdorff [55], Ch. IV, p. 71.) But he was unable to prove one particular result, whose statement has come to be called the 'continuum hypothesis'; it states that *there is no cardinal strictly between* $\aleph_0$ and $\mathbf{c}$. An interesting literature has grown round this conjecture (see Sierpinski [119]); but then the contemporary Austro-American logician Gödel* [44] proved a breathtaking theorem, which states roughly the following. For brevity let A, C respectively

* (Added in present edition) Kurt Gödel died in 1978, aged 71.

denote the strong axiom of choice, and the continuum hypothesis. Then Gödel shows: if mathematics is consistent without assuming A and C, it remains consistent if we do assume them: no new errors are introduced by incorporating A and C into our mathematics. This result was further strengthened, in 1966, by Paul Cohen; see Chapter 39.

Exercise 9

(i) Verify that addition of ordinals, defined above, is associative but that (for example) $1 + \omega = \omega \neq \omega + 1$. Construct a workable definition of $\sum_{\alpha < \sigma} x_\alpha$, a sum of ordinals x_α as α runs through all ordinals $\alpha < \sigma$.
(ii) Use the definition of multiplication of ordinals to verify that

$$2\omega = \omega + \omega \neq \omega 2 = \omega.$$

(iii) Prove that multiplication of ordinals is associative.
(iv) Show that the ordinals form an ordered set by setting $[(X, R)] \leqslant [(Y, S)]$ if there exists an order-preserving injection $(X, R) \to (Y, S)$. [In fact, the ordinals are well-ordered by this ordering.]
(v) Assign to each ordinal $[(X, R)]$ its cardinal card X. Show that this assignment preserves addition, multiplication, and order.

Chapter 8

ALGEBRA OF SETS, AND THE PROPOSITIONAL CALCULUS

8.1 Algebra of Sets

We have several times stated certain rules about operations with sets (e.g., $A \cup B = B \cup A$) without giving any formal proof. In this section, such proofs will be given; and in formulating these proofs we shall find that we have to state explicitly some grammatical conventions used in ordinary language. These conventions are expressed as rules about the construction of new sentences ('propositions') from old ones. Since many mathematical difficulties (though by no means all!) arise as difficulties of grammar, knowledge of this grammar can help a student enormously. It turns out that we shall observe a parallelism between the behaviour of sets and of the propositions (sentences) which define them; thus analogous rules for sets and propositions will be distinguished by the subscripts 's' and 'p', respectively. By a 'proposition' we simply mean any sentence, and to help discussions with propositions, we use some abbreviations: '&' denotes 'and'; 'not p' denotes the proposition obtained from a proposition p by negating the verb in its main clause (thus if p denotes 'It is raining', then 'not p' denotes 'It is not raining'). [We used $\neg p$ for 'not p' in 1.8.] Also, $p \;.\!\equiv\!.\; q$ signifies that the propositions denoted by p and q mean the same thing. We need not say what 'mean the same thing' means (even if we could!) but we give rules for the use of the symbol. [Observe that the relation is $\equiv$; the dots in '$p \;.\!\equiv\!.\; q$' are used for punctuation, instead of brackets. If q is a complicated formula, we may use $p \;.\!\equiv\!.\; q$, as in 8.1.6$_p$ below; and dots may be omitted when it is safe to do so.] The rules are:

8.1.1 *The relation* $\equiv$ *between propositions is an equivalence relation. Moreover, if* $p \;.\!\equiv\!.\; q$ *and* $r \;.\!\equiv\!.\; s$, *then*

$$p \,\&\, r \;.\!\equiv\!.\; q \,\&\, s,$$

$$p \text{ or } r \;.\!\equiv\!.\; q \text{ or } s;$$

and

$$\text{not } p \;.\!\equiv\!.\; \text{not } q.$$

Further, if S *is a set, and* $p(x)$, $q(x)$ *are propositions about* x *such that for all* $x \in S$, $p(x) \;.\!\equiv\!.\; q(x)$, *then the subsets* $\{x|p(x)\}$, $\{x|q(x)\}$ *of* S *coincide.*

Now let $\mathscr{U}$ be a universe of discourse. Then by 1.7.1, if A is a subset of $\mathscr{U}$, the complement $\complement A$ is defined by $\complement A = \{x|x \notin A\}$, or $\complement A = \{x|\text{not } (x \in A)\}$.

We now give a formal proof of the equality stated in 1.7.3:

$8.1.2_s$ $\complement(\complement A) = A.$

Proof.
$$\begin{aligned}\complement(\complement A) &= \{x \mid \text{not } (x \in \complement A)\}\\ &= \{x \mid \text{not } (\text{not } (x \in A))\}\\ &= \{x \mid x \in A\}\end{aligned}$$

by the third part of 8.1.1 and the grammatical rule that (at least in English) for any proposition p (here p is '$x \in A$'):

$8.1.2_p$ $$\text{not } (\text{not } p) \;.\!\equiv p. \quad \blacksquare$$

We have often used the **commutative laws**

$8.1.3_s$ $$\text{(i) } A \cup B = B \cup A; \qquad \text{(ii) } A \cap B = B \cap A;$$

to prove the first equality, we have

$$\begin{aligned}A \cup B &= \{x \mid x \in A \text{ or } x \in B\}\\ &= \{x \mid x \in B \text{ or } x \in A\} = B \cup A\end{aligned}$$

because of the last part of 8.1.1 and the rule that, for any propositions p, q (here p is '$x \in A$', q is '$x \in B$'):

$8.1.3_p$ $$\text{(i) } p \text{ or } q \;.\!\equiv.\; q \text{ or } p. \quad \blacksquare$$

Similarly, $A \cap B = B \cap A$ follows from

$8.1.3_p$ $$\text{(ii) } p \;\&\; q \;.\!\equiv.\; q \;\&\; p. \quad \blacksquare$$

The same kind of appeal to the definitions proves the **associative laws**

$8.1.4_s$

$$\text{(i) } A \cup (B \cup C) = (A \cup B) \cup C, \qquad \text{(ii) } A \cap (B \cap C) = (A \cap B) \cap C,$$

and the **idempotent laws**

$8.1.5_s$ $$A \cup A = A, \qquad A \cap A = A$$

from the corresponding rules for propositions:

$8.1.4_p$
$$\begin{aligned}&\text{(i) } p \text{ or } (q \text{ or } r) \;.\!\equiv.\; (p \text{ or } q) \text{ or } r,\\ &\text{(ii) } p \;\&\; (q \;\&\; r) \;.\!\equiv.\; (p \;\&\; q) \;\&\; r,\\ &p \text{ or } p \;.\!\equiv.\; p,\\ &p \;\&\; p \;.\!\equiv.\; p.\end{aligned}$$

For example we give:

Proof of 8.1.4$_s$(i)

$$\begin{aligned} A \cup (B \cup C) &= \{x \mid x \in A \text{ or } x \in B \cup C\} \\ &= \{x \mid x \in A \text{ or } (x \in B \text{ or } x \in C)\} \\ &= \{x \mid (x \in A \text{ or } x \in B) \text{ or } x \in C\} \qquad \text{by } 8.1.4_p\text{(i)} \\ &= \{x \mid x \in A \cup B \text{ or } x \in C\} \\ &= (A \cup B) \cup C. \quad \blacksquare \end{aligned}$$

Next, we demonstrate the 'rules of Pierce and De Morgan', mentioned in 1.7.2:

8.1.6$_s$

$$\text{(i) } \complement(A \cup B) = \complement A \cap \complement B,$$
$$\text{(ii) } \complement(A \cap B) = \complement A \cup \complement B,$$

using their companions for propositions:

8.1.6$_p$

(i) not (p or q) .≡. not p. & not q.

(ii) not (p & q) .≡. not p. or not q.

For example, in (ii), if we say: 'It is false that she has beautiful eyes and hair', then we mean 'It is false that she has beautiful eyes, or it is false that she has beautiful hair'. (Perhaps her eyes and hair are *both* awful!) The reader is advised to insert familiar sentences for p, q, r in the other rules we give, to test for plausibility.

Proof of 8.1.6$_s$(i). The left-hand side is

$$\begin{aligned} &\{x \mid \text{not } (x \in A \cup B)\} \\ = &\{x \mid \text{not } (x \in A \text{ or } x \in B)\} \\ = &\{x \mid \text{not } (x \in A) \ \& \text{ not } (x \in B)\} \qquad \text{by } 8.1.6_p\text{(i)} \\ = &\{x \mid x \in \complement A \ \& \ x \in \complement B\} \\ = &\complement A \cap \complement B, \qquad \text{the right-hand side of } 8.1.6_s\text{(i)}. \end{aligned}$$

We could prove 8.1.6$_s$(ii) similarly, using 8.1.6$_p$(ii) in the appropriate place, or use earlier results and give a purely algebraic proof as follows.

Proof of 8.1.6$_s$(ii). Write A' for $\complement A$; then by 8.1.2$_s$, $A = (A')'$, $B = (B')'$. Therefore if $X = A \cap B$, then $X = (A')' \cap (B')' = (A' \cup B')'$ by 8.1.6$_s$(i); so the left-hand side of 8.1.6$_s$(ii) is X', which is $((A' \cup B')')'$, i.e., $X' = A' \cup B'$ by 8.1.2$_s$ again. This proves 8.1.6$_s$(ii). ■

The importance of the laws of Pierce and De Morgan is that they introduce a 'duality' principle, namely that to every statement involving $\cup$ and $\cap$, there corresponds another, its 'dual', involving $\cap$ and $\cup$. Similarly for the rules $8.1.6_p$; other examples of this kind of duality will occur.

The last proof deduced $8.1.6_s$(ii) from $8.1.2_s$ and $8.1.6_s$(i); but its algebraic nature enables us to say more. For, had we written respectively '(not p)', '&', 'or', instead of P', $\cap$, $\cup$, and used $8.1.2_p$ and $8.1.6_p$(i) instead of $8.1.2_s$ and $8.1.6_s$(i), then the *same algebra* would give us a deduction of $8.1.6_p$(ii) from $8.1.2_p$ and $8.1.6_p$(i).

After carrying out the details of the following exercise the reader will see that if he agrees to $8.1.2_p$, then he is *forced* to agree to both $8.1.6_s$(i) and $8.1.6_s$(ii) if he agrees to one of them.

Exercise 1

(i) Deduce $8.1.6_s$(i) from $8.1.2_s$ and $8.1.6_s$(ii), and deduce $8.1.6_p$(i) from $8.1.2_p$ and $8.1.6_p$(ii), along the lines of the last proof.
(ii) What does 'not p' mean, when p denotes the proposition 'Even though the sun is shining, it is raining'?

We now come to the **distributive laws**:

$8.1.7_s$

$$\text{(i)}\ A \cap (B \cup C) = (A \cap B) \cup (A \cap C)$$
$$\text{(ii)}\ A \cup (B \cap C) = (A \cup B) \cap (A \cup C),$$

of which the first was stated in 1.7.5.

To prove them, we need the analogous rules for propositions:

$8.1.7_p$

$$\text{(i)}\ p\ \&\ (q \text{ or } r)\ . \equiv .\ (p\ \&\ q) \text{ or } (p\ \&\ r),$$
$$\text{(ii)}\ p \text{ or } (q\ \&\ r)\ . \equiv .\ (p \text{ or } q)\ \&\ (p \text{ or } r);$$

for example: if we say 'She has golden hair and blue or brown eyes', we mean 'She has golden hair and blue eyes, or golden hair and brown eyes'. The first, briefer, form is short for 'She has golden hair and (she has blue eyes or she has brown eyes)', with similar abbreviations in the second form.

Thus for example we prove $8.1.7_s$:

$$\begin{aligned} A \cap (B \cup C) &= \{x \mid x \in A\ \&\ x \in B \cup C\} \\ &= \{x \mid x \in A\ \&\ (x \in B \text{ or } x \in C)\} \\ &= \{x \mid (x \in A\ \&\ x \in B) \text{ or } (x \in A\ \&\ x \in C)\} \qquad \text{by } 8.1.7_p\text{(i)} \\ &= \{x \mid x \in A \cap B \text{ or } x \in A \cap C\} \\ &= (A \cap B) \cup (A \cap C). \quad \blacksquare \end{aligned}$$

Exercise 2

Deduce $8.1.7_s$(ii) similarly from $8.1.7_p$(ii). {Compare the proof of 1.7.8.}

It is sometimes objected that $8.1.7_p$(i) is no more basic than $8.1.7_s$(i), since some people find $8.1.7_p$(i) hard to grasp while $8.1.7_s$(i) is 'obvious from a diagram'. However, let us show how $8.1.7_p$(i) implies another proposition about sets, viz. (cf. 3.2.1):

8.1.8 $$A \times (B \cup C) = (A \times B) \cup (A \times C),$$

which therefore gives it some priority over $8.1.7_s$(i).

Proof of 8.1.8. The left-hand side is

$$\begin{aligned} &\{(x, y) | x \in A \;\&\; y \in B \cup C\} \\ &= \{(x, y) | x \in A \;\&\; (y \in B \text{ or } y \in C)\} \\ &= \{(x, y) | (x \in A \;\&\; y \in B) \text{ or } (x \in A \;\&\; y \in C)\} \qquad \text{by } 8.1.7_p(\text{i}) \\ &= \{(x, y) | (x, y) \in A \times B \text{ or } (x, y) \in A \times C\} \\ &= (A \times B) \cup (A \times C). \quad \blacksquare \end{aligned}$$

Classroom experience shows that some adults will agree at once to $8.1.7_p$(i) but not to $8.1.7_p$(ii); however, we shall now show how agreement to the first forces agreement to the second, and vice versa. We use an algebraic method, as in the proof of $8.1.6_s$(ii). Thus, to prove $8.1.7_p$(ii) from $8.1.7_p$(i), write X for 'p or (q & r)', and p' for 'not p'. Then

$$\begin{aligned} X' &\equiv .\; p' \;\&\; (q \;\&\; r)' && \text{by } 8.1.6_p(\text{i}) \\ &\equiv .\; p' \;\&\; (q' \text{ or } r') && \text{by 8.1.1 and } 8.1.6_p(\text{ii}) \\ &\equiv .\; (p' \;\&\; q') \text{ or } (p' \;\&\; r') && \text{by } 8.1.7_p(\text{i}) \\ \text{so} \quad X &\equiv .\; (X')' && \text{by } 8.1.2_p \\ &\equiv .\; (p' \;\&\; q')' \;\&\; (p' \;\&\; r')' && \text{by 8.1.1 and } 8.1.6_p(\text{i}) \\ &\equiv .\; (p \text{ or } q) \;\&\; (p \text{ or } r) && \text{by 8.1.1 and } 8.1.6_p(\text{ii}), \end{aligned}$$

which proves $8.1.7_p$(ii). Similarly, $8.1.7_p$(i) follows from $8.1.7_p$(ii). A proof with the same algebraic pattern yields $8.1.7_s$(ii) from $8.1.7_s$(i), and vice versa.

Exercise 3

(i) Give these proofs in detail (of $8.1.7_p$(i), $8.1.7_s$(ii), etc.).
(ii) Given sets A, B, define their **symmetric difference** (denoted by $A \sim B$) to be the set $(A \cup B) - (A \cap B)$. Prove that, for the subsets $A, B, C, \ldots$, of a universe $\mathscr{U}$,

$$\begin{aligned} &A \sim B = \varnothing \text{ iff } A = B; \\ &A \sim B = B \sim A, \qquad A \sim (B \sim C) = (A \sim B) \sim C: \\ &A \sim B \subseteq (A \sim C) \cup (C \sim B). \end{aligned}$$

If $A \subseteq B$, is it always true that $A \sim C \subseteq B \sim C$?
(iii) Prove that $\mathfrak{p}\mathscr{U}$ is an abelian group with respect to $\sim$, and that every element except $\varnothing$ is of order 2. {Heavy algebraic manipulation can be avoided by using the notion of **characteristic function**. Thus, for each $A \in \mathfrak{p}\mathscr{U}$, let $c_A\colon \mathscr{U} \to \mathbb{Z}_2$ denote the function such that $c_A(x) = 1$ if $x \in A$ and $c_A(x) = 0$ otherwise. Then $c_{A \cap B} = c_A c_B$, $c_{A \sim B} = c_A + c_B$, (algebra mod 2) and for any $X, Y \subseteq \mathscr{U}$, $c_X = c_Y$ iff $X = Y$.}

8.2 B-Algebras

The various rules we have been proving show that $\mathfrak{p}\mathcal{U}$ has the features of a kind of algebra. For it is an example of a set Σ with the following properties:
8.2.1 (i) *Σ is closed under two operations* $+$, $\cdot$, *which are each associative and commutative, and each distributes over the other*, i.e.,

$$x \cdot (y + z) = x \cdot y + x \cdot z, \qquad x + (y \cdot z) = (x + y) \cdot (x + z)$$

for all $x, y, z \in \Sigma$. Moreover, each is **idempotent**, i.e.,

$$x + x = x = x \cdot x \qquad \textit{for all } x \in \Sigma.$$

(ii) *There is a function $\sigma\colon \Sigma \to \Sigma$ such that $\sigma \circ \sigma = \mathrm{id}_\Sigma$ and*

$$\sigma(x + y) = \sigma x \cdot \sigma y, \qquad \sigma(x \cdot y) = \sigma(x) + \sigma(y)$$

for all $x, y \in \Sigma$.

Such a system $(\Sigma, +, \cdot, \sigma)$ will be called a **B-algebra**—a name chosen because it is almost a 'Boolean algebra' in the sense of 8.2.5 below. If we take Σ to be $\mathfrak{p}\mathcal{U}$, and $+$, $\cdot$, σ to be $\cup$, $\cap$, $\complement$ respectively, then the rules in 8.1 show that $(\mathfrak{p}\mathcal{U}, \cup, \cap, \complement)$ is a B-algebra: so also is $(\mathfrak{p}\mathcal{U}, \cap, \cup, \complement)$.

We have also seen that the set $\mathscr{S}$ of all propositions, divided into equivalence classes by the relation '$\equiv$', is a B-algebra with $+$, $\cdot$, and σ defined for cosets $[p]$, $[q]$ by

8.2.2
$$[p] + [q] = [p \text{ or } q]; \qquad [p] \cdot [q] = [p \;\&\; q]$$
$$\sigma[p] = [\text{not } p].$$

Exercise 4

Prove in detail that $\mathscr{S}$, with $+$, $\cdot$, σ thus defined, is a B-algebra.

The algebraic arguments given in 8.1, for example the proof of 8.1.6(ii), could have been written out for any B-algebra Σ, and therefore are valid for the two particular B-algebras $\mathfrak{p}\mathcal{U}$ and $\mathscr{S}$ (as well as for all other B-algebras). This was the point of demonstrating them; we were getting several theorems for the price of one.

8.2.3 EXAMPLE. A further example of a B-algebra is generated by switching-circuits; such a circuit consists of switches $P_1, P_2, \ldots, P_n$ connected in series or parallel, and with one 'input' terminal, and one 'output' terminal. Here

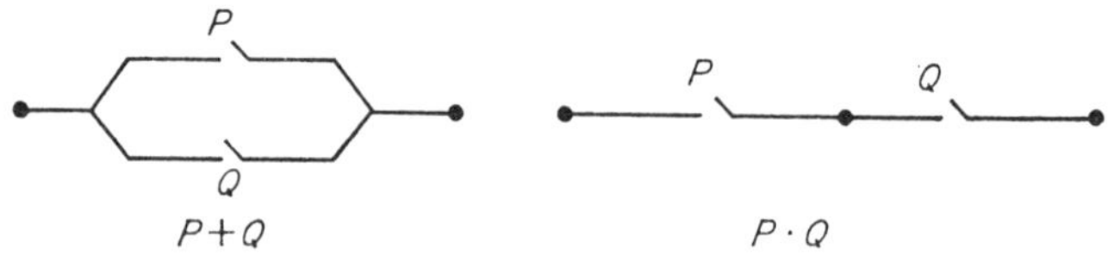

Fig. 8.1

the interest lies in whether current flows or does not flow across the terminals (e.g., in such devices as computers), and not in variations of voltage or current. We denote two switches by the same symbol, if they are always both on, or both off, simultaneously. If P and Q are two circuits, denote by $P + Q$, $P \cdot Q$ the circuits consisting of P and Q connected in parallel and in series respectively (Fig. 8.1). Two circuits of switches $(P_1, \ldots, P_n)$ are defined to be **equal** if they produce the same output for the same input, when P_i in each circuit is in the same 'state', i.e., 'on' in both, or 'off' in both. For example, Fig. 8.2 illustrates a distributive law, viz: $P + (Q \cdot R) = (P + Q) \cdot (P + R)$,

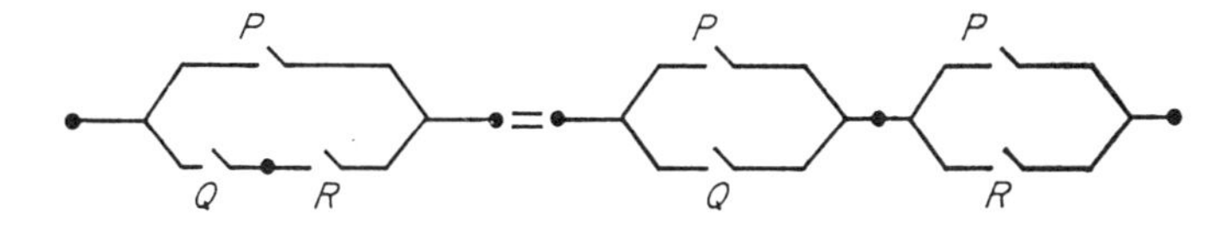

Fig. 8.2

analogous to 8.1.7. To check the validity of this equality, we form the table of Fig. 8.3 where 'l', 'r' denote, respectively, the left and right circuits of Fig. 8.2. Here '1', '0', in a column denote, respectively, that current does, or does not, flow in the device at the head of that column. Since the columns 'l', 'r' are identical, the outputs of the corresponding circuits in Fig. 8.2 are identical. This checks the equality required. We leave the reader to check that the set **SW** of all circuits is a B-algebra, when $+$, $\cdot$, represent connection in parallel and in series, and when $\sigma(C)$ is taken to be conversion of the circuit C to a device whose output is 0 or 1 according as that of C is 1 or 0. [Electrical engineers can manufacture a gadget which converts any switch C to one with the properties of $\sigma(C)$.]

P	Q	R	l	r
0	0	0	0	0
0	0	1	0	0
0	1	0	0	0
0	1	1	1	1
1	0	0	1	1
1	0	1	1	1
1	1	0	1	1
1	1	1	1	1

Fig. 8.3

Thus, theorems about B-algebras in general have interpretations which give results for **SW** in particular. See, for example, Whitesitt [135], Kemeny–Snell–Thompson [72] and Cundy–Rollett [27].

The B-algebras $\mathfrak{p}\mathscr{U}$ and **SW** have an additional feature. In $\mathfrak{p}\mathscr{U}$ there are the elements $\varnothing$ and $\mathscr{U}$, which have the properties

8.2.4 $A = A \cup \varnothing = A \cap \mathscr{U}$; $A \cup \mathscr{U} = \mathscr{U}$, $A \cap \varnothing = \varnothing$, $\complement\varnothing = \mathscr{U}$, for all $A \in \mathfrak{p}\mathscr{U}$. In **SW** there is the 'open' switch Ω which is always off, and the 'closed' switch C which is always on, and such that for any switch $P \in$ **SW**,

$$P = P + \Omega = P \cdot C;$$

$$P + C = C; \qquad P \cdot \Omega = \Omega, \qquad \sigma\Omega = C$$

$$(\text{so } \sigma(C) = \sigma^2(\Omega) = \Omega).$$

Hence, $\mathfrak{p}\mathscr{U}$ and **SW** are in fact examples of a richer kind of algebra, called 'Boolean', after G. Boole (1815–1864; Bell [10], 23). (For details and a different point of view see 21.4.)

We may define a Boolean algebra thus:

[More economical sets of axioms can be given: see for example the set in 21.4.]

8.2.5 DEFINITION. A **Boolean algebra** S is a B-algebra in which there exist two elements 0, 1, such that, for all $x \in S$,

$$x = x + 0; \qquad x + 1 = 1 = x + \sigma(x) \quad \text{and} \quad \sigma(0) = 1.$$

By operating with σ on each equation we obtain for all $y \in S$ (since $y = \sigma x$ where $x = \sigma y$, because $\sigma^2 = \mathrm{id}_\Sigma$):

8.2.6 $$y = y \cdot 1; \qquad y \cdot 0 = 0 = y \cdot \sigma(y) \quad \text{and} \quad \sigma(1) = 0.$$

Exercise 5

(i) Show that a singleton $\{x\}$ forms a Boolean algebra, if we take 0 and 1 each to be x, with $x + x = x = x \cdot x = \sigma(x)$.

(ii) Verify that $\{0, 1\}$ forms a Boolean algebra, if $\sigma(0) = 1$, $\sigma(1) = 0$, while $+$ and $\cdot$ are given by the tables

+	0	1
0	0	1
1	1	1

·	0	1
0	0	0
1	0	1

[We denote this 'standard' Boolean algebra by $\mathbb{B}_2$.]

(iii) Devise similar tables for a Boolean algebra with exactly four elements. Is there a Boolean algebra with exactly three elements?

(iv) Let X_{120} denote the set of all positive divisors of 120. Given $x, y \in X_{120}$, define $x \cap y$ to be the hcf of x and y, let $x \cup y$ be their lcm, and let $\sigma(x) = 120/x$. Show that $(X_{120}, \cup, \cap, \sigma)$ is a B-algebra but not a Boolean algebra with $\cup$ as $+$, and $\cap$ as $\cdot$. {Consider the law $x \cdot x = 0$.} Show that operations, similarly defined for X_{30}, make it into a Boolean algebra with zero and 1 taken to be 1 and 30, respectively. Why does X_{120} behave differently from X_{30}?

(v) Given a Boolean algebra $\mathbf{S} = (S, +, \cdot, \sigma, 0, 1)$ with the indicated operations, zero and unit element, prove that $(S, \cdot, +, \sigma, 1, 0)$ is also a Boolean algebra, whose zero is 1 and whose unit element is 0. (This new algebra is called the **dual**, $\mathbf{S}^*$, of $\mathbf{S}$.)

8.2.7 It is not apparent that the B-algebra $\mathscr{S}$ of propositions is Boolean, since we have to find two propositions a, b, such that the classes $[a]$, $[b]$ act respectively as a zero and one (as required by Definition 8.2.5), i.e., such that, for any $[p] \in \mathscr{S}$,

$$[p] = [p \text{ or } a] = [p \,\&\, b], \qquad [p \text{ or } b] = [b], \qquad [p \,\&\, a] = [a],$$
$$[\text{not } a] = [b].$$

We cannot consider this question further until we inquire more closely into the meaning of the relation $p \;.\!\equiv\!.\; q$ between propositions. This we do in the next section. Meanwhile we conclude the present section with a few remarks about Boolean algebras in general.

8.2.8 As in other branches of algebra, having defined an object with structure (namely, a Boolean algebra, here), we often want to compare two such objects, a known with an unknown, and therefore we define a Boolean *subalgebra* and a *homomorphism* of Boolean algebras. So, if $\mathbf{S} = (S, +, \cdot, \sigma)$, $\mathbf{T} = (T, +', \times, \tau)$ are two Boolean algebras with the indicated operations, then a **homomorphism** $\theta\colon \mathbf{S} \to \mathbf{T}$ is defined to be a function $\theta\colon S \to T$ such that, for all $x, y \in S$,

$$\theta(x + y) = \theta(x) +' \theta(y); \qquad \theta(x \cdot y) = \theta(x) \times \theta(y);$$
$$\theta(\sigma(x)) = \tau(\theta(x)), \qquad \theta(0) = 0.$$

Plainly then, $\theta(1) = 1$.

Exactly the same definition applies for a homomorphism of B-algebras, (except that we suppress $\theta(0) = 0$). If, in addition, θ is (1-1) or invertible, then θ is called a **monomorphism** or **isomorphism**, respectively, of Boolean algebras (B-algebras). If S happens also to be a subset of T, and the inclusion function $i\colon S \to T$ is a homomorphism: $\mathbf{S} \to \mathbf{T}$ then $\mathbf{S}$ is called a Boolean **subalgebra** (**B-subalgebra**) of $\mathbf{T}$.

Exercise 6

(i) Prove that the function id_S is an isomorphism $\mathbf{S} \to \mathbf{S}$.
(ii) Let $\mathbf{S}$ denote the system of positive divisors of 120 as in Exercise 5(iv), and let $\mathbf{T}$ denote that corresponding to 240. Show that the correspondence $x \to 2x$ is not a B-algebra homomorphism of $\mathbf{S}$ to $\mathbf{T}$.
(iii) The **kernel** K of a homomorphism $\theta\colon \mathbf{S} \to \mathbf{T}$ of Boolean algebras is $\theta^{\flat}\{0\}$, the set of all x such that $\theta(x) = 0$. Is K a Boolean sub-algebra of $\mathbf{S}$? Prove that if $y \in K$, then $x \cdot y \in K$ for all $x \in S$, and that $x + y \in K$ if also $x \in K$. [K is a **Boolean ideal**.]
(iv) Let A, B be sets, and $f\colon A \to B$ a function. Given $X \in \mathfrak{p}A$, then $f(X) \in \mathfrak{p}B$. Is the assignment $X \to f(X)$ a homomorphism $\mathfrak{p}A \to \mathfrak{p}B$? Prove that the function $Y \to f^{\flat}(Y)$ {see 2 Exercise 5(iii)} is a homomorphism $f^{\flat}\colon \mathfrak{p}B \to \mathfrak{p}A$. Is $f^{\flat}$ one-one, if f is? Prove that $f^{\flat}$ is an isomorphism if f is a bijection. {See 2 Exercise 7(ii).}

(v) Given a B-algebra or Boolean algebra $\mathbf{S} = (S, +, \cdot, \sigma)$, show that σ is an isomorphism of $\mathbf{S}$ with its dual, $\mathbf{S}^*$ (defined in Exercise 5(v)). [Hence every theorem true of all Boolean algebras in a family containing $\mathbf{S}$ is true of both $\mathbf{S}$ and $\mathbf{S}^*$. This is a precise statement of the 'duality' principle mentioned after 8.1.6. In particular, if we assume that one of the Pierce–De Morgan laws holds in $\mathbf{S}$, we may prove that the other holds, using *one* proof and the duality principle. Similarly, if one distributive law holds, so does the other.]
(vi) If $\theta: \mathbf{S} \to \mathbf{T}$ is an isomorphism of B-algebras, is θ^{-1}? Similarly for Boolean algebras.

8.3 The Propositional Calculus

We shall now look more closely at matters concerning propositions, since our treatment in 8.1 was somewhat informal. For example, we did not say clearly what a proposition was, nor what 'and', 'or', 'not', and 'means' were to mean. It is important to study these questions mathematically, partly out of curiosity, but especially to be able to cope with difficulties arising from problems of grammar. We must (as in all *applied* mathematics) make a model within *pure* mathematics of the system which interests us. Here, we are thinking of sentences and their formation as something in the outside world, which exist because people talk and write; and we must now think how to use the methods of pure mathematics to make a model for precise study.

Beginning tentatively, let us become physicists and suppose that we have a set S of sentences ('propositions'), which are simple enough for us to recognize them as sentences, and for us to decide the truth or falsity of each. For the moment, we do not define 'truth' and 'falsity'. If $p \in S$ is 'true', its **truth-value** is said to be T, and otherwise its **truth-value** is F; we suppose the sentences in S are either true or false and not undecidable (e.g., 'A rose is a rose' might be in S, but not 'A rose is a rose is a rose'). From the elements of S, we generate new, more complicated sentences called 'formulae', by using the† **logical connectives**:

8.3.1 $\neg$ (not), & (and), $\vee$ (or), $\to$ (if, . . . then), $\leftrightarrow$ (if and only if)

to form expressions like

$$\neg p, \qquad (\neg p \,\&\, q) \to (r \leftrightarrow s), \quad \text{etc.}$$

We do not want to consider expressions like $\vee p$, or $p\!\to$, but we temporarily avoid a rigid specification; however, our intention is to copy our usage of 'not', &, . . ., in ordinary language. Supposing our interest still to be in the truth or falsity of formulae, rather than in shades of meaning, we now give rules for deciding the truth-values of $\neg p$, $p \,\&\, q$, $p \vee q$, $p \to q$, and $p \leftrightarrow q$ when those of p and q are given. (This is roughly the way we learn to *use*

† A variety of notations exists. For example, $\bar{p}$, $\sim p$, and p' for $\neg p$; $\wedge$ and $\cdot$ for &; and $\sim$ for $\leftrightarrow$; $\vee$ is used for 'or' because its usage resembles the *inclusive* Latin 'vel···vel' and not the *exclusive* 'aut···aut'.

the connectives in ordinary speech; we do not learn their 'meaning'.) The rules are specified by the following **truth-table.**

p	$\neg p$	p	q	$p \,\&\, q$	$p \vee q$	$p \rightarrow q$	$p \leftrightarrow q$
T	F	T	T	T	T	T	T
F	T	T	F	F	T	F	F
		F	T	F	T	T	F
		F	F	F	F	T	T

Fig. 8.4

Except perhaps for the column under '$p \rightarrow q$', this expresses quite well our usage of the connectives in ordinary conversation; and the column for '$p \rightarrow q$' will be justified in 8.5 below. As is customary, we use German letters $\mathfrak{A}$, $\mathfrak{B}$, $\mathfrak{C}$, . . ., etc., to denote formulae. Consider those formulae obtained by forming strings of sentences in S connected by connectives. Then we obtain the truth values of $\neg\mathfrak{A}$, $\mathfrak{A} \vee \mathfrak{B}$, etc., using the above table: we calculate the truth values of $\mathfrak{A}$, $\mathfrak{B}$ and substitute them in the p, q columns of the table.

8.3.2 EXAMPLE. Let $\mathfrak{A}$ denote the formula $\neg(p \,\&\, q)$. Its truth-table is then the first three columns of Fig. 8.5 in which the column for $p \,\&\, q$ is

p	q	$\mathfrak{A}$	$p \,\&\, q$
T	T	F	T
T	F	T	F
F	T	T	F
F	F	T	F

Fig. 8.5

inserted for convenience. Each entry in this column was inserted in the p column of Fig. 8.4, and the $\neg p$ column there then gave the $\mathfrak{A}$ column in Fig. 8.5. Notice, however, that this truth-table for $\mathfrak{A}$ is also that for the formula (see Fig. 8.6)

$$\mathfrak{B} = \neg p \vee \neg q$$

where we have used an obvious notation to indicate the calculation of the final column.

p	$\neg p$	q	$\neg q$	$\neg p \vee \neg q$
T	F	T	F	F ∨ F = F
T	F	F	T	F ∨ T = T
F	T	T	F	T ∨ F = T
F	T	F	T	T ∨ T = T

Fig. 8.6

Two formulae with identical truth-tables will be called **equivalent**, and we write

8.3.3 $$\mathfrak{A} \ .equ.\ \mathfrak{B}.$$

(We have in mind that $\mathfrak{A}$ and $\mathfrak{B}$ then correspond, in ordinary language, to sentences which 'mean' the same thing, even if they are not equal symbol by symbol.) Clearly, *equ* is a genuine equivalence relation. Figures 8.5 and 8.6 then show that

8.3.4 $$\neg(p \,\&\, q) \ .equ.\ \neg p \vee \neg q.$$

Exercise 7

Check the following equivalences, by inspecting truth-tables:

(i) $p \ .equ.\ \neg(\neg p)$;
(ii) $p \vee q \ .equ.\ q \vee p$; $p \,\&\, q \ .equ.\ q \,\&\, p$;
(iii) $p \vee (q \vee r) \ .equ.\ (p \vee q) \vee r$; $p \,\&\, (q \,\&\, r) \ .equ.\ (p \,\&\, q) \,\&\, r$;
(iv) $p \rightarrow q \ .equ.\ \neg p \vee q$;
(v) $p \leftrightarrow q \ .equ.\ (p \rightarrow q) \,\&\, (q \rightarrow p)$.

The formulae in (i)–(iii) of the last exercise show that the relation *equ* is a good model for the notion 'means the same as', since these formulae are those of 8.1.2–8.1.4$_p$ which we there agreed were normal usage. But to get wider scope we prepare to work with a system larger than our original system of straightforward sentences.

8.4 Extension to More General Formulae

The verbal description above can be clarified by using the notion of function. Our assumption that the elements of S can be assigned truth values means that there exists a function $\theta: S \rightarrow V$ where $V = \{0, 1\}$ and $\theta(p)$ is 0 or 1 according as p is false or true. (Different people might construct different functions θ, depending on the nature of the sentences in S and of the beliefs and knowledge of the people; but we assume that one particular θ has been chosen.) Now if we assign to V the operations $+$, $\cdot$, σ as in Exercise 5(ii), it becomes the Boolean algebra $\mathbb{B}_2$; and the truth-table in Fig. 8.4 shows that, for all $p, q \in S$, we obtain equations in $\mathbb{B}_2$:

8.4.1 $$\theta(\neg p) = \sigma(\theta(p)); \qquad \theta(p \,\&\, q) = \theta(p)\cdot\theta(q)$$
$$\theta(p \vee q) = \theta(p) + \theta(q).$$

Our earlier rule for finding the truth table of $\neg\mathfrak{A}$ or of $\mathfrak{A} \,\&\, \mathfrak{B}, \ldots$, etc., when those of $\mathfrak{A}, \mathfrak{B}, \ldots$, are known, is then to say that

8.4.2 $$\theta(\neg\mathfrak{A}) = \sigma(\theta(\mathfrak{A})), \qquad \theta(\mathfrak{A} \,\&\, \mathfrak{B}) = \theta(\mathfrak{A})\cdot\theta(\mathfrak{B})$$
$$\theta(\mathfrak{A} \vee \mathfrak{B}) = \theta(\mathfrak{A}) + \theta(\mathfrak{B}).$$

Thus, θ is beginning to look like a ***homomorphism*** of a B-algebra (whose elements are formulae $\mathfrak{A}, \mathfrak{B}, \ldots$) into $\mathbb{B}_2$, but more analysis is required.

We therefore now study the domain of θ, which we shall denote by S_∞; its elements will be 'formulae', but we have not yet said what a 'formula' is. To do this we use induction on n, and define the set S_n of all *n-formulae* as follows. Define S_0 to be the set S of simple sentences we started with prior to 8.3.1. Then S_{n+1} is to consist of S_n, together with $\neg\mathfrak{A}$, $\mathfrak{A} \vee \mathfrak{B}$, $\mathfrak{A} \,\&\, \mathfrak{B}$, whenever $\mathfrak{A}, \mathfrak{B} \in S_n$. This defines S_n for each $n \in \mathbb{N}$ by induction; and $S_0 \subseteq S_1 \subseteq \cdots \subseteq S_n \subseteq S_{n+1} \subseteq \cdots$. Now set

8.4.3
$$S_\infty = \bigcup_{n=0}^{\infty} S_n.$$

We then define a 'formula' to be any element of S_∞. The rules 8.4.2 are then used to extend the function θ, given prior to 8.4.1, from S to a function $\theta\colon S_\infty \to \mathbb{B}_2$. The extension proceeds by a stepwise procedure, first defining θ on S_0 $(= S)$, then extending θ from S_n to S_{n+1} by 8.4.2, using the inductive definition of S_{n+1}. Thus the extended $\theta\colon S_\infty \to \mathbb{B}_2$ still satisfies the equations of 8.4.2. Now S_∞ is not a B-algebra, since for example $\mathfrak{A} \,\&\, \mathfrak{B} \neq \mathfrak{B} \,\&\, \mathfrak{A}$ in general (in the sense of equality, letter by letter), so θ is not a homomorphism *of B-algebras.* It does, however, 'preserve the structure' in the general sense of 3.6.8, so we still call θ a 'homomorphism'.

Exercise 8

(i) Show that the extension of θ from S_n to S_{n+1} is the same as that obtained by applying the truth-tables of Fig. 8.4 to the set S_n.
(ii) Which other laws fail in S_∞, to prevent its being a B-algebra? In what sense is $\theta\colon S_\infty \to \mathbb{B}_2$ a homomorphism?

Our earlier definition (8.3.3) of equivalence between formulae becomes:

8.4.4 $\mathfrak{A}$ *.equ.* $\mathfrak{B}$ iff *for all homomorphisms* $\theta\colon S_\infty \to \mathbb{B}_2$
$$\theta(\mathfrak{A}) = \theta(\mathfrak{B});$$

and *equ* is an equivalence relation on the domain S_∞.

Warning: $\mathfrak{A}$ *.equ.* $\mathfrak{B}$ is *not* a formula belonging to S_∞!

We can then write the formulae in Exercise 7 above with general $\mathfrak{A}, \mathfrak{B} \in S_\infty$ in place of the p's and q's there; these follow automatically from the facts that in V, $\sigma^2 = \mathrm{id}$, while $+$, $\cdot$, are associative, commutative, idempotent and distributive with respect to each other. Hence we can list the following statements; since they hold in S_∞, and are the analogues of acceptable conventions about our usage of language, we have evidence that S_∞ is a good model.

8.4.5 *For all* $\mathfrak{A}, \mathfrak{B}, \mathfrak{C}, \mathfrak{D} \in S_\infty$, *the following statements hold.*

(i) $\neg(\neg\mathfrak{A})$ *.equ.* $\mathfrak{A}$;
(ii) $\mathfrak{A} \vee \mathfrak{B}$ *.equ.* $\mathfrak{B} \vee \mathfrak{A}$; $\mathfrak{A} \,\&\, \mathfrak{B}$ *.equ.* $\mathfrak{B} \,\&\, \mathfrak{A}$;
(iii) $\mathfrak{A} \vee (\mathfrak{B} \vee \mathfrak{C})$ *.equ.* $(\mathfrak{A} \vee \mathfrak{B}) \vee \mathfrak{C}$; $\mathfrak{A} \,\&\, (\mathfrak{B} \,\&\, \mathfrak{C})$ *.equ.* $(\mathfrak{A} \,\&\, \mathfrak{B}) \,\&\, \mathfrak{C}$;
(iv) $\mathfrak{A} \vee \mathfrak{A}$ *.equ.* $\mathfrak{A}$; $\mathfrak{A} \,\&\, \mathfrak{A}$ *.equ.* $\mathfrak{A}$;
(v) $\mathfrak{A} \vee (\mathfrak{B} \,\&\, \mathfrak{C})$ *.equ.* $(\mathfrak{A} \vee \mathfrak{B}) \,\&\, (\mathfrak{A} \vee \mathfrak{C})$;
(vi) $\mathfrak{A} \,\&\, (\mathfrak{B} \vee \mathfrak{C})$ *.equ.* $(\mathfrak{A} \,\&\, \mathfrak{B}) \vee (\mathfrak{A} \,\&\, \mathfrak{C})$.
(vii) *if* $\mathfrak{A}$ *.equ.* $\mathfrak{B}$ *and* $\mathfrak{C}$ *.equ.* $\mathfrak{D}$, *then*

$$\mathfrak{A} \vee \mathfrak{C} \text{ .equ. } \mathfrak{B} \vee \mathfrak{D}, \qquad \mathfrak{A} \,\&\, \mathfrak{C} \text{ .equ. } \mathfrak{B} \,\&\, \mathfrak{D},$$
$$\neg\mathfrak{A} \text{ .equ. } \neg\mathfrak{B}.$$

All these are analogues of 8.1.1, 8.1.2, etc., and some could be proved by the same methods as there; however, *all* are provable by the one method of evaluating θ for each formula. Similarly the laws of Pierce and De Morgan in the Boolean algebra $\mathbb{B}_2$ imply in S_∞:

8.4.6
$$\begin{cases} \text{(i)} & \neg(\mathfrak{A} \,\&\, \mathfrak{B}) \text{ .equ. } \neg\mathfrak{A} \vee \neg\mathfrak{B}, \\ \text{(ii)} & \neg(\mathfrak{A} \vee \mathfrak{B}) \text{ .equ. } \neg\mathfrak{A} \,\&\, \neg\mathfrak{B}. \end{cases}$$

As in 8.2.2, we can now form from S_∞ a B-algebra F whose elements are the cosets $[\mathfrak{A}]$ of S_∞ modulo the relation *equ*, and with $+$, $\cdot$, σ as there defined. However we can now prove what was foreshadowed in 8.2.7.

8.4.7 THEOREM. *F is a Boolean algebra.*

Proof. We have to find cosets $[\mathfrak{X}]$, $[\mathfrak{Y}]$, in F with the properties of 0 and 1 in Definition 8.2.5. We observe that for any $\mathfrak{A} \in S_\infty$, if $\mathfrak{A}_0$ denotes $\mathfrak{A} \,\&\, \neg\mathfrak{A}$, and $\mathfrak{A}_1$ denotes $\mathfrak{A} \vee \neg\mathfrak{A}$, then

$$\theta(\mathfrak{A}_0) = \theta(\mathfrak{A})\cdot\sigma(\theta(\mathfrak{A})), \text{ by 8.4.2, } = 0,$$

since $x\cdot\sigma(x) = 0$ in the Boolean algebra $\mathbb{B}_2$ of Exercise 5(ii). Thus if also $\mathfrak{B} \in S_\infty$, then $\theta(\mathfrak{B}_0) = 0$ also, whence $[\mathfrak{A}_0] = [\mathfrak{B}_0]$; we denote this element by $\mathbf{0}$. Also $\theta(\mathfrak{A}_1) = \theta(\mathfrak{A}) + \sigma(\theta(\mathfrak{A}))$, by 8.4.2, since $x + \sigma(x) = 1$ in $\mathbb{B}_2$. Thus for any $\mathfrak{B} \in S_\infty$, $[\mathfrak{A}_1] = [\mathfrak{B}_1]$, denoted by $\mathbf{1}$. We now check that $\mathbf{0}$ and $\mathbf{1}$ have the necessary properties. For any $[\mathfrak{B}] \in F$

$$\theta(\mathfrak{B} \vee \mathfrak{A}_0) = \theta(\mathfrak{B}) + \theta(\mathfrak{A}_0) = \theta(\mathfrak{B}) + 0 = \theta(\mathfrak{B}),$$

since $\mathfrak{B} \vee \mathfrak{A}_0$ *.equ.* $\mathfrak{B}$. In F, therefore,

$$[\mathfrak{B}] + \mathbf{0} = [\mathfrak{B}] + [\mathfrak{A}_0] = [\mathfrak{B} \vee \mathfrak{A}_0] = [\mathfrak{B}].$$

Similarly $[\mathfrak{B}] + \mathbf{1} = [\mathfrak{B}] + [\mathfrak{A}_1] = [\mathfrak{A}_1] = \mathbf{1}$,

$$\sigma(\mathbf{0}) = \sigma[\mathfrak{A}_0] = [\neg\mathfrak{A}_0] = [\mathfrak{A}_1] = \mathbf{1}.$$

Thus $[\mathfrak{A}_0]$, $[\mathfrak{A}_1]$ act as 0 and 1 in F, so F is a Boolean algebra. ∎

Finally, the homomorphism suspected after 8.4.2 is the function $\phi: F \to \mathbb{B}_2$ given by $\phi[\mathfrak{A}] = \theta(\mathfrak{A})$; ϕ is well-defined, since all elements $\mathfrak{B}'$ in one coset $[\mathfrak{B}]$ have the same value under θ, so $\theta(\mathfrak{B}') = \theta(\mathfrak{B})$, and $\phi[\mathfrak{B}]$ is independent of choice of representative $\mathfrak{B}$ from $[\mathfrak{B}]$.

Exercise 9

Verify in detail that $\phi: F \to \mathbb{B}_2$ is a homomorphism. When is ϕ onto?

8.5 Implication and Deduction

Let us now justify the table for $p \to q$ in Fig. 8.4. To reflect our usage in ordinary language (see 1.8.8), we feel that $\neg(p \to q)$ should correspond to 'p is true, yet q is false'; so we ought to have in our model $\neg(p \to q)$ *.equ.* $p \,\&\, \neg q$. Hence by 8.4.5(vii) we would then have

$$\neg(\neg(p \to q)) \textit{ .equ. } \neg(p \,\&\, \neg q)$$

so $p \to q$ *.equ.* $\neg(p \,\&\, \neg q)$ (by 8.4.5(i)), since *equ* is transitive,

.equ. $\neg p \vee \cdot \neg(\neg q)$ (by 8.4.6(i))

.equ. $\neg p \vee .q$ (by 8.4.5(i) and (vii)).

But already we have observed in Exercise 7(iv) that $p \to q$ and $\neg p \vee q$ have the same truth table; thus, if those for $\neg$ and $\vee$ are plausible, so is that for $\to$. Any apparent strangeness in it arises from associating 'if p, then q' in ordinary language with 'q can be deduced from p', (and we have not yet analysed formally the notion of a 'deductive proof': see 39.4, 39.5). However, the negation 'p true, yet q false' does not carry overtones of proof, and so *its* truth table seems plausible. To remain consistent, then, we define $\mathfrak{A} \to \mathfrak{B}$ to be the formula $\neg\mathfrak{A} \vee \mathfrak{B}$; since *equ* is reflexive, then even more:

8.5.1 $\mathfrak{A} \to \mathfrak{B}$ *.equ.* $\neg\mathfrak{A} \vee \mathfrak{B}$.

Using Exercise 7(v), we define similarly $\mathfrak{A} \leftrightarrow \mathfrak{B}$ by

$$\mathfrak{A} \leftrightarrow \mathfrak{B} = (\mathfrak{A} \to \mathfrak{B}) \,\&\, (\mathfrak{B} \to \mathfrak{A}).$$

It turns out that the definitions of $\to$, $\leftrightarrow$ give rise to the behaviour we would expect; thus the following results hold in our model S_∞, and correspond to working rules about methods of deduction, which mathematicians normally use in all mathematical work. Among such rules are:

8.5.2 (i) $\mathfrak{A} \to \mathfrak{B}$ *.equ.* $\neg\mathfrak{B} \to \neg\mathfrak{A}$;
(ii) If $(\mathfrak{A} \to \mathfrak{B}) \,\&\, (\mathfrak{B} \to \mathfrak{C})$, then $\mathfrak{A} \to \mathfrak{C}$;
(iii) If $\mathfrak{A}$ *.equ.* $\mathfrak{B}$ and $\mathfrak{C}$ *.equ.* $\mathfrak{D}$, then $\mathfrak{A} \to \mathfrak{C}$ *.equ.* $\mathfrak{B} \to \mathfrak{D}$;

for example, (i) expresses the convention that a deduction of $\neg\mathfrak{A}$ from $\neg\mathfrak{B}$ will suffice to demonstrate that $\mathfrak{B}$ follows from $\mathfrak{A}$. Observe that (iii) is a statement *about* S_∞; it is not a proposition *in* S_∞. This fact is recognized by the presence in (iii) of the relation *equ*.

Exercise 10

Prove the statements (i)–(iii) in 8.5.2.

A **tautology** is a formula $\mathfrak{A} \in S_\infty$ such that $\mathfrak{A}$ *.equ.* **1**, i.e., such that $\theta(\mathfrak{A}) = 1$, for *all* homomorphisms $\theta: S_\infty \to \mathbb{B}_2$. For example, $\mathfrak{A} \vee \neg\mathfrak{A}$ is a tautology, because (as we saw in the proof of 8.4.7), $\theta(\mathfrak{A} \vee \neg\mathfrak{A}) = 1$ for *any* θ. If we think of $\mathfrak{A}$ as a string of sentences $p_1, \ldots, p_n$ joined by logical connectives, then we are calling $\mathfrak{A}$ a tautology when $\mathfrak{A}$ is 'true' regardless of what truth-values are assigned to $p_1, \ldots, p_n$. As philosophers say, '$\mathfrak{A}$ is true in virtue of its form'. In fact, the statement '$\theta(\mathfrak{A}) = 1$ for any θ' corresponds in our model to the statement '$\mathfrak{A}$ is true' in ordinary language. Similarly, $\mathfrak{A}$ *.equ.* **0** corresponds to '$\mathfrak{A}$ is false for any truth-values of $p_1, \ldots, p_n$'. One objective of logic is to determine all tautologies. As an example, let us show

8.5.3 $\quad (\mathfrak{A} \mathbin{\&} (\mathfrak{A} \to \mathfrak{B})) \to \mathfrak{B}$ *is a tautology.*

(It expresses in our model the way we deduce in real life a new formula $\mathfrak{B}$ from an old one $\mathfrak{A}$: granted $\mathfrak{A}$, we are allowed $\mathfrak{B}$ if we can show that $\mathfrak{A}$ implies $\mathfrak{B}$. This method is called the **rule of Modus Ponens**.)

Proof of 8.5.3. Given any homomorphism $\theta: S_\infty \to \mathbb{B}_2$,

$$\begin{aligned}
&\theta[(\mathfrak{A} \mathbin{\&} (\mathfrak{A} \to \mathfrak{B})) \to \mathfrak{B}] \\
&= \theta[\neg(\mathfrak{A} \mathbin{\&} (\mathfrak{A} \to \mathfrak{B})) \vee \mathfrak{B}] && \text{by 8.5.1,} \\
&= \theta(\neg\mathfrak{A} \vee \neg(\neg\mathfrak{A} \vee \mathfrak{B})) + \theta(\mathfrak{B}), && \text{by 8.4.6 and 8.5.1}
\end{aligned}$$

—the sum in $\mathbb{B}_2$. We now omit the reasons for the remaining equalities; the last sum is now

$$\begin{aligned}
&= \theta(\neg\mathfrak{A} \vee (\mathfrak{A} \mathbin{\&} \neg\mathfrak{B})) + \theta(\mathfrak{B}) \\
&= \theta((\neg\mathfrak{A} \vee \mathfrak{A}) \mathbin{\&} (\neg\mathfrak{A} \vee \neg\mathfrak{B})) + \theta(\mathfrak{B}) \\
&= \theta(\neg\mathfrak{A} \vee \mathfrak{A}) \cdot \theta(\neg\mathfrak{A} \vee \neg\mathfrak{B}) + \theta(\mathfrak{B}) \\
&= 1 \cdot (\theta(\neg\mathfrak{A}) + \theta(\neg\mathfrak{B})) + \theta(\mathfrak{B}) \\
&= \theta(\neg\mathfrak{A}) + \sigma(\theta(\mathfrak{B})) + \theta(\mathfrak{B}) \\
&= \theta(\neg\mathfrak{A}) + 1 = 1,
\end{aligned}$$

since $x + 1 = 1$ for all x in a Boolean algebra. ■

Exercise 11

(i) Prove that the following formulae are tautologies:

$$\mathfrak{A} \to \mathfrak{A}, \qquad \mathfrak{A} \vee \mathfrak{A} \to \mathfrak{A}, \qquad \mathfrak{A} \to (\mathfrak{A} \vee \mathfrak{B}),$$

$$\mathfrak{A} \vee \mathfrak{B} \to \mathfrak{B} \vee \mathfrak{A}$$

$$(\mathfrak{A} \to \mathfrak{B}) \to [(\mathfrak{C} \to \mathfrak{A}) \to (\mathfrak{C} \to \mathfrak{B})].$$

(ii) Show that if $\mathfrak{A}$ *.equ.* $\mathfrak{B}$, then $\mathfrak{A} \leftrightarrow \mathfrak{B}$ is a tautology.
(iii) Verify the method of **reductio ad absurdum** in S_∞:

$$((\neg\mathfrak{B} \,\&\, \mathfrak{A}) \to \neg\mathfrak{A}) \to (\mathfrak{A} \to \mathfrak{B})$$

(i.e., to prove $\mathfrak{A}$ implies $\mathfrak{B}$, it suffices to show that $\mathfrak{A}$ and $\neg\mathfrak{B}$ together imply that $\mathfrak{A}$ is false).
(iv) Given θ, let $J_\theta = \{\mathfrak{A} \in S_\infty \,|\, \theta(\mathfrak{A}) = 1\}$. Prove that if $\mathfrak{A} \in J_\theta$, then $\mathfrak{A} \vee \mathfrak{B} \in J_\theta$ for all $\mathfrak{B} \in S_\infty$; and if $\mathfrak{B} \in J_\theta$ also, then $\mathfrak{A} \,\&\, \mathfrak{B} \in J_\theta$.

Prove also that the set of all tautologies in S_∞ is the intersection of all J_θ, for all possible homomorphisms $\theta: S_\infty \to \mathbb{B}_2$.
(v) Let P, Q be two switches, and for any circuit X let X_* denote the sentence 'X is on'. Prove that

$$(P + Q)_* \text{ .equ. } P_* \vee Q_*; \qquad (P \cdot Q)_* \text{ .equ. } P_* \,\&\, Q_*$$
$$(\sigma P)_* \text{ .equ. } \neg(P_*).$$

(vi) The members of a three-man committee vote secretly, each by pressing a button. A light glows to signify a majority vote. Design a switching circuit as follows. Let P, Q, R denote the three buttons, let p denote the sentence 'p is on', and let q, r denote similar sentences. Show that the light glows iff the formula $\mathfrak{A}$ is true, where $\mathfrak{A}$ is

$$(p \,\&\, q \,\&\, r) \vee (p \,\&\, q \,\&\, \neg r) \vee (p \,\&\, \neg q \,\&\, r) \vee (\neg p \,\&\, q \,\&\, r).$$

Show that $\mathfrak{A}$ *.equ.* $(p \,\&\, q) \vee (q \,\&\, r) \vee (r \,\&\, p)$. Hence design a suitable circuit.

8.5.4 We conclude this chapter by summarizing our position. In 8.4 we showed how to pass from the initial set S of propositions to the larger set S_∞; and we saw that a person's beliefs about the 'truth' or 'falsity' of the statements in S could be summarized by a function $\theta: S \to \mathbb{B}_2$. Such a person, if rational and curious, may wish to compute $\psi(\mathfrak{Q})$ for a given $\mathfrak{Q} \in S_\infty$ where $\psi: S_\infty \to \mathbb{B}_2$ is the extension of θ satisfying the rules 8.4.2. He will say that '$\mathfrak{Q}$ is true', or '$\mathfrak{Q}$ is false', according as he computes $\psi(\mathfrak{Q})$ to be 1 or 0; but in fact his primary interest will be in such conclusions of 'truth' and 'falsity' rather than in the direct computation of ψ. For that purpose short cuts are available, using tautologies. For example, from the tautology in 8.5.3 we know that for our person's ψ (as for that of any other person)

$$\psi((\mathfrak{A} \,\&\, (\mathfrak{A} \to \mathfrak{B})) \to \mathfrak{B}) = 1;$$

so that if $\psi(\mathfrak{A}) = 1$ and $\psi(\mathfrak{A} \to \mathfrak{B}) = 1$, then $\psi(\mathfrak{B})$ must be 1 also. Our person may express this conventionally by saying 'If $\mathfrak{A}$ is true, and if $\mathfrak{A} \to \mathfrak{B}$ is true, then $\mathfrak{B}$ is true', and he may possibly forget that the word 'true' here depends on his personal function θ. Similarly he might express the rule 8.5.2(ii) (say) as 'If $\mathfrak{A} \to \mathfrak{B}$ is true, and $\mathfrak{B} \to \mathfrak{C}$ is true, then $\mathfrak{A} \to \mathfrak{C}$ is true'. Once he has used these rules to conclude that a particular $\mathfrak{Q}$ in S_∞ is 'true', he will conventionally say that he has 'deduced' $\mathfrak{Q}$ from the original information contained in $\theta: S \to \mathbb{B}_2$. Now these rules, while extremely useful, are not sufficient for deducing most of the interesting propositions of mathematics; for the latter often require quantifiers and the substitution of some variables by more complicated ones. In Chapter 39, we shall extend these rules accordingly (see 39.4), and consider a more realistic and complicated notion of deducibility. Technically, we need to pass from the 'propositional calculus' of this chapter, to the 'predicate calculus'. For further details see Stoll [123].

Exercise 12

A harassed mother, resisting her seven-year-old son's demands, said, 'I cannot do what you ask, because I'm not God'. The son replied, 'But you don't believe in God, so how can you be not God?'. Show that the mother had (hastily) stated a tautology about her beliefs and that the implied doubt of the seven-year-old was in fact unfounded. [A complete analysis requires some of the theory from Chapter 39. The moral of this tale is that one should observe the logical processes of young children with the utmost respect.]

Part III
ARITHMETIC

Traditionally, the subject called 'arithmetic' in schools has meant a concern with the operations of addition and multiplication, with their inverses (subtraction and division), and with such problems as the factorization of integers and the construction of highest common factors (hcf's) and least common multiples (lcm's). These topics are all of great importance and potential interest. However, because of a preoccupation with applications to monetary transactions, (a field of considerable social concern), rapidity of calculation was grotesquely overemphasized, and the subject became identified with a dull rote-learning. Consequently, generations of schoolchildren have been bored and resentful at being turned into slow, inefficient computers, who think they loathe 'mathematics' when in fact they have never met a truly mathematical idea in all of the many hours spent in arithmetic classes.

In Chapters 9 to 13, which constitute this Part, we too deal with the same topics, neglecting the monetary ones. We look at the arithmetic operations in sets more general than the usual set $\mathbb{Z}^+$ of positive integers, thereby illuminating the various roles played by such properties as the commutative law of multiplication, the ordering of the elements of $\mathbb{Z}^+$ by size, and so on. In Chapter 9, after reviewing the basic arithmetic laws satisfied by the integers, we formulate the notion of a *ring*, and discuss the elementary algebraic structures called *abelian groups*, and *fields*, with their important subsets, the *sub-rings*, *subgroups*, and *sub-fields*. In Chapter 10, we consider in detail special examples of these structures, namely the sets $\mathbb{Z}_m$ of 'residues mod m'. These are the simplest such examples and have an interesting and elementary theory of their own; they are in a sense miniature approximations to $\mathbb{Z}$. We then pass to some general theory in Chapter 11, and study rings which have the extra property of possessing a 'norm', analogous to the absolute value $|n|$ of integers. Here we give some theory of hcf's and lcm's, with the accompanying Euclidean algorithm, in a form applicable to the theory of polynomials (in particular). This theory is continued in Chapter 12, where we consider factorization into primes, and its uniqueness, in various rings. Two applications are given in Chapter 13, to the problems of partial fractions and of continued fractions.

Chapter 9

COMMUTATIVE RINGS AND FIELDS

9.1 The Set of Integers as an Algebraic System

We shall assume given from the outset the set $\mathbb{Z}^+$ of *non-negative integers* $0, 1, 2, 3, \ldots$. This set is furnished with the order relation (by size) described in 4.4.12. Further, there is defined in $\mathbb{Z}^+$ the fundamental operation of addition of numbers, which we shall here presume well-known. We then define the operation of multiplication in $\mathbb{Z}^+$ by means of repeated addition, so that na denotes the result of adding a to itself n times; formally we use induction on n, and set:

9.1.1 $$0a = 0, \qquad (n + 1)a = na + a.$$

The operations of addition and multiplication satisfy certain basic laws of arithmetic, not all of which we will for the moment make explicit, but typical of such laws is the **commutative** law of multiplication:

9.1.2 $$ab = ba, \qquad (a, b \in \mathbb{Z}^+),$$

and the **distributive** law:

9.1.3 $$a(b + c) = ab + ac, \qquad (a, b, c \in \mathbb{Z}^+).$$

(Both follow from 9.1.1 by induction on a.) There are also certain laws connecting the addition and multiplication operations on the one hand with the order relation on the other; for example, for all $a, b, c \in \mathbb{Z}^+$, we have:

9.1.4 $$a \leqslant b \Leftrightarrow \exists n \in \mathbb{Z}^+ \text{ with } a + n = b;$$

9.1.5 $$a \leqslant b \Rightarrow a + c \leqslant b + c;$$

9.1.6 $$a \leqslant b \Rightarrow ac \leqslant bc.$$

Of course, the number n postulated in 9.1.4 is uniquely determined by a and b; it is customary to write $b - a$ for this number n and to describe n as obtained by subtracting a from b. However, if we confine attention to $\mathbb{Z}^+$ this number n exists only if $b \geqslant a$. In order to be able to subtract any number from any other it is necessary to pass from the set $\mathbb{Z}^+$ to the larger set $\mathbb{Z}$ of integers, $0, \pm 1, \pm 2, \ldots$, that is, to introduce (or *invent*) the negative numbers. There is a precise mathematical technique for doing this, but we defer it until Chapter 23. All we need to know, for the present, is that there is such a set $\mathbb{Z}$ containing $\mathbb{Z}^+$ in which the arithmetic operations of addition, subtraction, and multiplication are always possible according to the rules in 9.2 below. Also, if $a \in \mathbb{Z}$, then either $a \in \mathbb{Z}^+$ or $-a \in \mathbb{Z}^+$ but not both, unless $a = 0$.

We now propose to write down the important algebraic properties of the set $\mathbb{Z}$. However, since we wish to stress that the mathematics we shall do will depend on those properties alone and not on the particular nature of the set $\mathbb{Z}$, we shall write down the basic properties for a general set R and the reader will satisfy himself that each postulate asserted about the set R is indeed a true statement about the set $\mathbb{Z}$. Sets with these properties are called *commutative rings with unity element.*

9.2 Rings

Thus we consider a set R with two 'binary' operations written $+$ and $\cdot$ (the $\cdot$ symbol is usually omitted). These are functions $R \times R \to R$ (see Chapter 3) and we may require that, for all a, b, c in R, the following conditions shall be satisfied. [Conditions of this kind, which hold generally within a system, are usually called **axioms.** Some occur so often that they are given special names, in fact, the names indicated in brackets.]

$\mathfrak{A}_1$: $a + (b + c) = (a + b) + c$ (**associative law of addition**);
$\mathfrak{A}_2$: $a + b = b + a$ (**commutative law of addition**);
$\mathfrak{A}_3$: $\exists$ *in R an element* **zero** (*denoted by* 0) *such that for all* $a \in R$, $a + 0 = a$;
$\mathfrak{A}_4$: *to each* $a \in R$, $\exists$ *in R an element* (*denoted by* $-a$) *such that* $a + (-a) = 0$;
$\mathfrak{A}_5$: $a(bc) = (ab)c$ (**associative law of multiplication**);
$\mathfrak{A}_6$: $a(b + c) = ab + ac$, $(a + b)c = ac + bc$ (**distributive laws**);
$\mathfrak{A}_7$: $ab = ba$ (**commutative law of multiplication**);
$\mathfrak{A}_8$: $\exists$ *in R a* **unity** *element* (*denoted by* 1) *such that for all* $a \in R$, $a \cdot 1 = 1 \cdot a = a$; *moreover*, $1 \neq 0$.

For brevity, we often use a *triple* $(R, +, \cdot)$ to denote a set R with binary operations $+, \cdot$. Any triple $(R, +, \cdot)$ satisfying $\mathfrak{A}_1$–$\mathfrak{A}_6$ is called a **ring.** If it also satisfies $\mathfrak{A}_7$ it is called a **commutative ring.** If it satisfies $\mathfrak{A}_1$–$\mathfrak{A}_6$ and $\mathfrak{A}_8$ it is called a **ring with unity element.** If it satisfies all the axioms $\mathfrak{A}_1$–$\mathfrak{A}_8$ it is called a **commutative ring with unity element.** The reader will readily agree that $\mathbb{Z}$, the set of integers, with its usual addition and multiplication, is a commutative ring with unity element. Indeed it is naturally customary to refer to the two operations in R as addition and multiplication. If the operations $+$ and $\cdot$ in R are clear from the context, we permit ourselves to write R instead of the triple $(R, +, \cdot)$.

Axioms $\mathfrak{A}_1$–$\mathfrak{A}_8$ will figure very prominently in the sequel.

Not every triple satisfies the axioms, of course. For example, the set of all even integers, under ordinary addition and multiplication, is a commutative ring, but without unity element. The set $\mathbb{Z}^+$, under ordinary addition and multiplication, is not a ring since $\mathfrak{A}_4$ does not hold. Consider now some examples of systems which are rings. In these examples (as elsewhere!), we do not propose to give explicit proofs of all the statements made; and the reader should try to supply what we omit.

9.2.1 EXAMPLE. The set $\mathbb{Q}$ of rational numbers, the set $\mathbb{R}$ of real numbers, and the set $\mathbb{C}$ of complex numbers†, with their usual addition and multiplication, are all rings satisfying $\mathfrak{A}_1$–$\mathfrak{A}_8$.

9.2.2 EXAMPLE. The set $\mathbb{Z}$ with its usual addition, and with multiplication given by $a \cdot b = 0$ (for all $a, b \in \mathbb{Z}$), is a commutative ring without unity element. (This example indicates that different rings may have the same underlying set.)

9.2.3 EXAMPLE. Let $\mathbb{Z}[x]$ be the set of all polynomials‡ $f = a_0 + a_1x + \cdots + a_mx^m$, where $a_i \in \mathbb{Z}$ for each $i = 1, \ldots, m$. The a's are called the 'coefficients' of f, and, if all are zero, f is called the 'zero polynomial', denoted by 0. If all a's are zero except a_0, then f is just an integer, so $\mathbb{Z} \subseteq \mathbb{Z}[x]$. If also $g = b_0 + b_1x + \cdots + b_nx^n$, we define $f + g$ and fg to be the polynomials $\sum c_ix^i$, $\sum d_ix^i$, where $c_i = a_i + b_i$, $d_i = \sum_{j=0}^{i} a_{i-j}b_j$ and a_i is understood to be 0 if $i > m$, while $b_i = 0$ if $i > n$. Notice that $c_i = 0$ if $i > \max(m, n)$, while $d_i = 0$ if $i > m + n$, so that $f + g$, fg are indeed polynomials. Then $\mathbb{Z}[x]$, with this multiplication and addition, is a ring and satisfies $\mathfrak{A}_1$–$\mathfrak{A}_8$; its zero is the zero polynomial, and the unity element is the integer (= polynomial) 1.

9.2.4 EXAMPLE. The sets $\mathbb{Q}[x]$, $\mathbb{R}[x]$, $\mathbb{C}[x]$ of polynomials in x with rational, real, complex coefficients, respectively, with their usual addition and multiplication (defined as in Example 9.2.3) are all rings satisfying $\mathfrak{A}_1$–$\mathfrak{A}_8$.

9.2.5 EXAMPLE. Let m be a fixed positive integer and let $m\mathbb{Z}$ be the subset of $\mathbb{Z}$ consisting of integers divisible by m. Then we can add and multiply in $m\mathbb{Z}$, using the addition and multiplication in $\mathbb{Z}$, and we say that we have 'induced' addition and multiplication operations in $m\mathbb{Z}$. Under these operations, $m\mathbb{Z}$ becomes a ring satisfying $\mathfrak{A}_1$–$\mathfrak{A}_7$ (i.e., a commutative ring). But $m\mathbb{Z}$ fails to satisfy $\mathfrak{A}_8$ unless $m = 1$; for if $m \neq 1$, then $1 \notin m\mathbb{Z}$.

9.2.6 EXAMPLE. Let $\mathbb{Z}_m$ be the set of residue (or remainder) classes $[a]$ of $\mathbb{Z}$ modulo m (see 4.5.3 and Chapter 10). We define addition and multiplication in $\mathbb{Z}_m$ by the rules

$$[a] + [b] = [a + b],$$
$$[a][b] = [ab].$$

It may then be verified that $\mathbb{Z}_m$ satisfies $\mathfrak{A}_1$–$\mathfrak{A}_8$. (The reader will be familiar with the cases $m = 10$, $m = 12$ (hours, months, pence), $m = 7$ (days).)

† Precise definitions of $\mathbb{Q}$, $\mathbb{R}$, $\mathbb{C}$ will be found in Part VI.
‡ A precise definition of a polynomial will be found in Chapter 22.

9.2.7 EXAMPLE. Fix an integer $n > 0$, and let Γ_n be the subset of $\mathbb{C}$ consisting of all complex numbers of the form $a + ib\sqrt{n}$, such that a, b are integers (positive or negative). Then the rules of addition and multiplication in $\mathbb{C}$ induce in Γ_n (in the sense of 9.2.5) the structure of a commutative ring with unity element.

9.2.8 EXAMPLE. Let R be a ring and let S be any non-empty set. The set R^S of all functions $f: S \to R$ acquires the structure of a ring if we define $f + g$, fg, respectively, to be those functions which assign to each $s \in S$ the elements $f(s) + g(s)$, $f(s)g(s)$ in R. [E.g., the functions sin + cos, sin · cos are members of $\mathbb{R}^{\mathbb{R}}$.] The 'zero' is the constant function zero, and if R has unity element 1, then the constant function 1 is the unity element of R^S. If R is a commutative ring, so is R^S, and conversely. [This mode of construction is often referred to as 'adding (or multiplying) functions **pointwise**', i.e., by adding (or multiplying) their values.] Now consider the special case when S is the ring R. We define a second multiplication on R^R, by setting the 'product' of f, g to be the composite, $f \circ g: R \to R$. With this product, and the same sum as before, R^R becomes a ring in which the identity map is the unity element, and which is in general not commutative.

9.2.9 EXAMPLE. Let $M(n, R)$ be the set of $(n \times n)$-matrices with elements in the ring R. Then under the usual rules for adding and multiplying matrices, $M(n, R)$ is a ring with unity element, but it is commutative only in the trivial case when R is commutative and $n = 1$. (See Chapter 19.)

9.3 Consequences

We next detail some elementary consequences of the axioms, and in particular prove the famous proposition that 'minus times minus makes plus'.

9.3.1 PROPOSITION. *Let R be a ring. Then for all $a, b \in R$:*

(i) *the zero* 0 *is uniquely determined by the property that* $a + 0 = a$, i.e., *the only solution in R, of the equation $a + x = a$, is $x = 0$.*
(ii) *the element $-a$ is uniquely determined by the property $a + (-a) = 0$*; i.e., *the only solution in R, of the equation $a + x = 0$, is $x = -a$.*
(iii) $-(-a) = a$; $(-0) = 0$; $-(a + b) = (-a) + (-b)$;
(iv) $a0 = 0a = 0$;
(v) $a(-b) = (-a)b = -(ab)$;
(vi) $(-a)(-b) = ab$.

The statements (i)–(iii) are summaries, in additive notation (with $a + b$ for ab, $-a$ for a^{-1} and 0 for e) of the propositions of 3.6, since axioms $\mathfrak{A}_1$–$\mathfrak{A}_4$ guarantee that the hypotheses on A in 3.6 are here satisfied by R.

We will prove (iv) and (vi), leaving the remaining proofs to the reader.

Proof of (iv). We first prove that if $x + a = a$ for a particular a in R, then $x = 0$. For then

$$\begin{aligned} 0 &= a + (-a) && \text{by } \mathfrak{A}_4 \\ &= (x + a) + (-a) && \\ &= x + (a + (-a)) && \text{by } \mathfrak{A}_1 \\ &= x + 0 && \text{by } \mathfrak{A}_4 \\ &= x && \text{by } \mathfrak{A}_3. \end{aligned}$$

Now for any $a, b \in R$,

$$\begin{aligned} ab &= a(0 + b) && \text{by } \mathfrak{A}_3 \text{ (and } \mathfrak{A}_2) \\ &= a0 + ab && \text{by } \mathfrak{A}_6. \end{aligned}$$

It therefore follows from the argument above that $a0 = 0$; similarly $0a = 0$ (but note that we are *not* assuming $\mathfrak{A}_7$).

Proof of (vi). We have $0 = (-a)0$, by (iv), $= (-a)(b + (-b))$

$$\begin{aligned} &= (-a)b + (-a)(--b) \\ &= -ab + (-a)(-b) \qquad \text{by (v).} \end{aligned}$$

Thus $(-a)(-b) = -(-ab) = ab$ by (iii). ■

We will henceforth allow ourselves to write $b - a$ for $b + (-a)$, in any ring R.

9.4 Sub-rings

Let $(R, +, \cdot)$ be a ring and let S be a subset of R. Given $a, b \in S$, then we may calculate $a + b$ and ab in R, and it may happen that these elements lie in S also. We then say that S is **closed** under the addition and multiplication in R, and that these operations **induce** addition and multiplication (still denoted by $+, \cdot$) in S. We have met examples already in 9.2.5 and 9.2.7. The triple $(S, +, \cdot)$ now becomes a candidate for being itself a ring, and for this to be so $\mathfrak{A}_1$–$\mathfrak{A}_6$ must be satisfied in S. But $\mathfrak{A}_1$, $\mathfrak{A}_2$, $\mathfrak{A}_5$, $\mathfrak{A}_6$ hold already in R, so we need test only $\mathfrak{A}_3$, $\mathfrak{A}_4$. Thus:

9.4.1 PROPOSITION. *$(S, +, \cdot)$ is a ring provided that* (i) *S is closed under the induced operations,* (ii) *the zero of R lies in S, and* (iii) *for each a in S, the element $(-a)$ of R lies in S.* ■

The triple $(S, +, \cdot)$ is called a **sub-ring** of R. Since the operations are understood to be those induced from $(R, +, \cdot)$ we often say simply that 'S is a sub-ring of R'; recall that if the operations in R are fixed throughout a discussion, we speak of 'the ring R'. Plainly, any sub-ring $S \subseteq R$ is commutative if R is; and if R has a unity element which happens to belong to S, then this is the unity element of S. The converse need not hold. After this discussion, the following proposition is one of a kind which often occurs with 'substructures' such as sub-rings; e.g., the subgroups, sub-fields ··· we shall meet later.

9.4.2 PROPOSITION. *Let* $\{R_\alpha\}_{\alpha \in A}$ *be a family of sub-rings of* R. *Then* $S = \bigcap_{\alpha \in A} R_\alpha$ *is also a sub-ring of* R.

Proof. Given $a, b \in S$, then for each $\alpha \in A$, $a, b \in R_\alpha$; so $ab \in R_\alpha$ by 9.4.1(i). Hence ab lies in every R_α and so in S. Conditions (ii) and (iii) of 9.4.1 are verified similarly. ■

9.4.3 EXAMPLE. R is a sub-ring of itself, and so is the set $\{0\}$ consisting of the zero element of R. In the chain $\mathbb{Z} \subseteq \mathbb{Q} \subseteq \mathbb{R} \subseteq \mathbb{C}$, of 9.2.1, each set is a sub-ring of the next (observe also that the relation of being a sub-ring is transitive). In 9.2.3, $\mathbb{Z}$ is a sub-ring of $\mathbb{Z}[x]$. Similarly with 9.2.4. In 9.2.5, $m\mathbb{Z}$ is a sub-ring of $\mathbb{Z}$, but whereas $\mathbb{Z}$ has a unity element, $m\mathbb{Z}$ has not (unless $m = 1$). In 9.2.7, Γ_n is a sub-ring of $\mathbb{C}$ and it is commutative with unity element.

9.5 Commutative Groups

A set S, with *one* binary operation $\blacksquare$, is called a **commutative group** if it satisfies the following four axioms:

$\mathfrak{CG}_1$: $a \blacksquare (b \blacksquare c) = (a \blacksquare b) \blacksquare c$ (*associative law*);
$\mathfrak{CG}_2$: $a \blacksquare b = b \blacksquare a$ (*commutative law*);
$\mathfrak{CG}_3$: $\exists$ *in* S *an element* (*denoted by* e) *such that for all* $a \in S$, $a \blacksquare e = a$;
$\mathfrak{CG}_4$: *to each* $a \in S$, $\exists$ *in* S *an element* (*denoted by* $\bar{a}$) *such that* $a \blacksquare \bar{a} = e$.

Very often we use 'additive notation' to represent the binary operation in a commutative group. Thus, we often write $+$ for $\blacksquare$, and then we also write 0 for e and $-a$ for $\bar{a}$. When a commutative group is written additively we will usually call it an **abelian group**, after N. H. Abel (1802–29; see Bell [10], Ch. 17) who first studied commutative groups. We emphasize however, that the distinction between commutative groups in general and abelian groups is purely notational and not mathematical; the theories will be logically equivalent but expressed in slightly different notation and terminology.

We notice that, if we rewrite the axioms additively we get precisely axioms $\mathfrak{A}_1$–$\mathfrak{A}_4$. Thus we have our first, rather general, example of a commutative group, as observed already in the proof of Proposition 9.3.1.

9.5.1 EXAMPLE. Let $(R, +, \cdot)$ be a ring. Then if we ignore the multiplication, the pair $(R, +)$ is an abelian group said to **underlie** the ring $(R, +, \cdot)$.

We now give further examples of commutative groups.

9.5.2 EXAMPLE. Let X denote the unit circle $x^2 + y^2 = 1$ in the co-ordinate plane, and let $\mathbf{R}$ denote the set of all rotations of X about its centre; we count $\mathrm{id}_X\colon X \to X$ as a rotation. We take the binary operation to be composition;

for if T_1, T_2 are rotations through α_1, α_2 radians respectively, then $T_1 \circ T_2$ is a rotation through $\alpha_1 + \alpha_2$ radians, and lies in **R**. Hence $T_1 \circ T_2 = T_2 \circ T_1$. It is easily checked that $(\mathbf{R}, \circ)$ is a commutative group; the element $\overline{T}_1$ is rotation through α_1 radians in the direction opposite to that of T_1, and id_X plays the role† of the element e. Note that $\mathfrak{CG}_1$ follows from the associative law 2.8.6 for functions. There are many such geometrical examples, using also reflections and translations (see Chapter 18); however, $\mathfrak{CG}_2$ may be violated.

9.5.3 EXAMPLE. Consider the set $\mathbb{Q} - \{0\}$ of non-zero rationals, and let the binary operation $\bullet$ be ordinary multiplication. Then surely we obtain a commutative group with $e = 1$ and $\bar{a} = a^{-1}$. When multiplicative notation is used for a commutative group it is common to write 1 for e, and invariable to write a^{-1} for $\bar{a}$.

Before proceeding further we should explain that we have inserted this section on commutative groups here, apparently out of context, because of the relevance of abelian groups and fields (9.6) in arithmetic. Thus Examples 9.5.1, 9.5.3 are particularly relevant to our purpose. We did not feel it appropriate to discuss *groups* in detail in this chapter and have reserved a separate chapter for them (Chapter 18). But many readers will have realized that axioms $\mathfrak{CG}_1$, $\mathfrak{CG}_3$, $\mathfrak{CG}_4$ are just the standard group axioms, and axiom $\mathfrak{CG}_2$ asserts the group $(S, \bullet)$ to be commutative. A commutative group *is* therefore a group; thus much of the same language will be applied both to groups and to the more special commutative groups. [For example, we will talk of *subgroups* of commutative groups and the *group operation* in a commutative group.]

At the risk of underlining the obvious, we remark that any consequence of a subset E of the set of axioms $\mathfrak{A}_1$–$\mathfrak{A}_8$ must be a true statement in any system satisfying the axioms in E. In particular, only the axioms $\mathfrak{A}_1$–$\mathfrak{A}_4$ were needed to prove (i), (ii), and (iii) of 9.3.1. Hence we have at once:

9.5.4 PROPOSITION. *The statements* (i), (ii), (iii) *of* 9.3.1 *hold in any abelian group* $(S, +)$. ■ Of course, the corresponding statements (in 'multiplicative' notation) hold in any commutative group!

It is useful to have the concept of an abelian (or commutative) group in order to be able to state the next result—which is an important one—in appropriate generality.

9.5.5 THEOREM. *Let R be a non-empty set with a binary operation $+$. Then R is an abelian group if and only if it satisfies axioms $\mathfrak{A}_1$, $\mathfrak{A}_2$ and:*

$\mathfrak{A}_9$: *for each $a, b \in R$, the equation $a + x = b$ has a solution in R.*

In other words, axioms $\mathfrak{A}_1$, $\mathfrak{A}_2$, $\mathfrak{A}_3$, $\mathfrak{A}_4$ are equivalent to axioms $\mathfrak{A}_1$, $\mathfrak{A}_2$, $\mathfrak{A}_9$; of course this does *not* imply that axiom $\mathfrak{A}_9$ is equivalent to axioms $\mathfrak{A}_3$, $\mathfrak{A}_4$.

† A deliberate analogy with the casting of a play is intended. Here $(S, \cdot)$ corresponds to the script of a play, with its list of *dramatis personae*; and to let $\mathbb{R}$, $\circ$, id_X play the roles of S, $\bullet$, e in $\mathfrak{CG}_1$–$\mathfrak{CG}_4$ corresponds to an appropriate choice of real people from a drama company to act the parts.

Proof of 9.5.5. We easily infer $\mathfrak{A}_9$ from $\mathfrak{A}_1$–$\mathfrak{A}_4$; for $x = b - a$ is certainly a solution of $a + x = b$. Conversely, let R satisfy $\mathfrak{A}_1$, $\mathfrak{A}_2$, $\mathfrak{A}_9$. It follows from $\mathfrak{A}_9$ that, for each $a \in R$, there exists an element $0_a \in R$ with $a + 0_a = a$. Notice that we write 0_a rather than just 0 in order to stress that 0_a may depend on a. It is our next task to show that, in fact, 0_a is independent of a and does service as a zero throughout R. To this end let $b \in R$; we wish to show that $b + 0_a = b$. Now by $\mathfrak{A}_9$ there exists an element $x \in R$ with $a + x = b$. Thus

$$\begin{aligned} b + 0_a &= (a + x) + 0_a \\ &= (a + 0_a) + x \qquad \text{(by } \mathfrak{A}_1 \text{ and } \mathfrak{A}_2) \\ &= a + x \\ &= b. \end{aligned}$$

This shows that $0 = 0_a$ satisfies $b + 0 = b$ for all $b \in R$, so that $\mathfrak{A}_3$ is satisfied. Finally $\mathfrak{A}_4$ becomes just a special case of $\mathfrak{A}_9$ and so is also satisfied. ∎

Exercise 1

(i) Show that, in an abelian group A, if a, b are given in A, then the equation $a + x = b$ has a *unique* solution (for x), namely $x = (-a) + b$.
(ii) Show that, in a set R with two binary operations $+, \cdot$, axioms $\mathfrak{A}_1$, $\mathfrak{A}_3$, $\mathfrak{A}_4$, $\mathfrak{A}_6$, and $\mathfrak{A}_8$ imply $\mathfrak{A}_2$. (You may assume—or try to prove!—that $\mathfrak{A}_1$, $\mathfrak{A}_3$, $\mathfrak{A}_4$ imply that 0 is unique, $0 + a = a$, that $-a$ is uniquely determined by a, and $(-a) + a = 0$.)
(iii) Show that the set $\mathbb{Q}$ of rationals, under the binary operation $a * b = \frac{1}{2}(a + b)$ is not a commutative group. Which of the axioms $\mathfrak{CG}_1 - \mathfrak{CG}_4$ are satisfied by $(\mathbb{Q}, *)$? Do the same problem when $a * b$ is taken instead to be the ordinary difference, $a - b$.
(iv) Show that the set $\mathbb{R}$ of real numbers, under the binary operation $a * b = (a^3 + b^3)^{1/3}$, is a commutative group.
(v) Let $\mathfrak{p}X$ be the family of subsets of a set (see Chapter 1). Show that neither $(\mathfrak{p}X, \cap)$ nor $(\mathfrak{p}X, \cup)$ is a commutative group unless X is empty. Hence show that $(\mathfrak{p}X, \cup, \cap)$ is not a ring. Which of $\mathfrak{A}_1$–$\mathfrak{A}_8$ are satisfied?
(vi) Let $(S, +)$ be an abelian group, and let $U \subseteq S$ be such that U is closed under addition in S (compare 9.4). If conditions (i), (ii), (iii) of 9.4.1 hold, then U is called a **subgroup** of S. Prove that, if W is a sub-ring of a ring $(R, +, \cdot)$, then the underlying abelian group $(W, +)$ of W (see 9.5.1) is an abelian subgroup of $(R, +)$. Prove the analogue of 9.4.2 for abelian groups, that the intersection of subgroups of an abelian group is again a subgroup.
(vii) Given abelian groups† $(A, +)$, $(B, +)$, define their **direct sum** $A \oplus B$ as follows. As a set, $A \oplus B$ is the Cartesian product of A and B. Now define the sum $(a_1, b_1) + (a_2, b_2)$ to be $(a_1 + a_2, b_1 + b_2)$. Prove that $(A \oplus B, +)$ is an

† Of course, by writing $+$ for the operation both in A and in B we do not in any sense imply that the operations are the same. (Plainly they *cannot* be the same unless $A = B$, and even then they may not be the same.) We simply use $+$ as a generic symbol for the operation in an abelian group.

abelian group with $(0_A, 0_B)$ as zero, where 0_A, 0_B are the zeros in A, B. Prove that the sets

$$\{(a, 0_B)|a \in A\}, \qquad \{(0_A, b)|b \in B\},$$

with the induced addition, are subgroups of $A \oplus B$. A function $f: A \to B$ such that $f(a_1) + f(a_2) = f(a_1 + a_2)$ for all $a_1, a_2 \in A$, is called a **homomorphism.** Prove that the graph of a homomorphism is a subgroup of $A \oplus B$. The **kernel** K of f is the subset $f^{\flat}\{0_B\}$ of A; prove that K is a subgroup of A.
(viii) Let A be an abelian group and let S be a (non-empty) set. In the set A^S, define addition $(+)$ as in 9.2.8. Prove that $(A^S, +)$ is an abelian group. Suppose now that $(S, +)$ is an abelian group, and let $X \subseteq A^S$ denote the set of all homomorphisms $f: S \to A$. Prove that X is a subgroup of A^S.
(ix) When S in (viii) has only two elements, there is a bijection $f: A^S \to A \oplus A$ (see 3 Exercise 1(v)). Show that f is then a homomorphism between the group $(A^S, +)$ and the direct sum $A \oplus A$.

We shall study *groups* further in Chapter 18. As remarked, groups are pairs $(S, \cdot)$ where axioms $\mathfrak{CG}_1$, $\mathfrak{CG}_3$, $\mathfrak{CG}_4$ are satisfied, but not necessarily $\mathfrak{CG}_2$. Some results in the exercises still hold—which?

9.6 Fields

The axioms $\mathfrak{A}_1$–$\mathfrak{A}_8$ for a commutative ring with unity do not allow us necessarily to perform division; for example we cannot always solve for x the equation $ax = b$ in a ring R, even when $a \neq 0$ (e.g., $3x = 1$ has no solution in $\mathbb{Z}$). In some arithmetic systems, however, such an equation is always soluble. These systems are called *fields*, according to the following definition:

9.6.1 DEFINITION. A set F, with binary operations $+$, $\cdot$, is called a **field**, provided (i) $(F, +, \cdot)$ is a commutative ring with unity element 1, and (ii) to each non-zero x in F, $\exists$ in F an element, denoted by x^{-1}, such that $x \cdot x^{-1} = 1$.

For example, the sets $\mathbb{Q}$, $\mathbb{R}$, $\mathbb{C}$ in Example 9.2.1 are all fields with their usual addition and multiplication. So also is the ring $\mathbb{Z}_p$ of Example 9.2.6 when p is a prime number (see 10.2.2); and for such p, the set of pth roots of unity (see 33 Exercise 3(v)) with appropriate operations, forms a field.

Exercise 2

Let $s \in \mathbb{Q}$ be a fixed positive number, with $\sqrt{s} \in \mathbb{R} - \mathbb{Q}$. Show that the set X of all numbers $x + y\sqrt{s}$ in $\mathbb{R}$, with $x, y \in \mathbb{Q}$, is a field under the operations induced from $\mathbb{R}$. {Hint: If X were known to be a field, $(x + y\sqrt{s})^{-1}$ could be calculated by 'multiplying top and bottom by $x - y\sqrt{s}$'; thus we can (illegitimately) calculate what $(x + y\sqrt{s})$ should be, and then verify that it works. We must observe that if $x + y\sqrt{s} \neq 0$, then $x^2 - y^2 s \neq 0$; otherwise $\sqrt{s}$ would lie in $\mathbb{Q}$.}

Observe that in (ii) of 9.6.1, the phrase '$\exists$ in F' is vital; for example $3^{-1}(=1/3)$ exists in $\mathbb{Q}$ but not in $\mathbb{Z}$. Many students feel uncomfortable, at first, that x in (ii) is non-zero, and ask 'Why isn't 0^{-1} infinity?'. Two replies are: (a) 'infinity' does not belong to F; (b) if we arbitrarily declare that 0^{-1} is some fixed $z \in F$, then by definition $0 \cdot z = 1$, so $0 \cdot (0 \cdot z) = 0 \cdot 1$, whence by $\mathfrak{A}_5$ and 9.3.1(iv), we get $0 \cdot z = 0$, so that $0 = 1$ and then $0 = a$ for all $a \in F$! In other words, we cannot allow 0 to have an inverse in F, without throwing away some of the useful laws $\mathfrak{A}_1$–$\mathfrak{A}_8$ (or else restricting ourselves to the study of singletons!). Given the choice, we find it more interesting to retain the laws, so we do not assign an inverse to 0. Notice how explicit use of the axioms $\mathfrak{A}$ helps clarify the argument for (b) above.

A consequence of 9.6.1 is:

9.6.2 PROPOSITION. *Let F be a commutative ring with unity. Then F is a field iff the set $F - \{0\}$, of non-zero $x \in F$, is a commutative group with respect to the multiplication in F.*

Proof. Let F be a field. We must first show that $F - \{0\}$ is closed under multiplication, so that $(F - \{0\}, \cdot)$ is a genuine candidate to be a commutative group. Thus let $a, b \in F - \{0\}$; we must show that $ab \neq 0$. But if $b \neq 0$, then $\exists b^{-1}$ such that $bb^{-1} = 1$. Then, if ab were zero,

$$a = a1 = a(bb^{-1}) = (ab)b^{-1} = 0b^{-1} = 0 \qquad \text{(see Proposition 9.3.1(iv))},$$

contrary to hypothesis. So $ab \neq 0$ and $F - \{0\}$ is closed under multiplication.

Now we have an exercise in assigning roles (see footnote on p. 129); that is, we must check that if $(S, \cdot)$ is taken to be $(F - \{0\}, \cdot)$, then axioms $\mathfrak{CG}_1$–$\mathfrak{CG}_4$ are satisfied. Now $\mathfrak{CG}_1$, $\mathfrak{CG}_2$ hold because $\mathfrak{A}_5$, $\mathfrak{A}_7$ hold in F; the neutral element e in $F - \{0\}$ is the unity element 1 of F; while $\mathfrak{CG}_4$ holds in $F - \{0\}$ with a^{-1} playing the role of $\bar{a}$.

Conversely, if F is a commutative ring with unit element such that $(F - \{0\}, \cdot)$ is a group, then 9.6.1(ii) holds because inverse elements always exist in a commutative group, by $\mathfrak{CG}_4$. Therefore F is a field. ■

We now apply Proposition 9.5.4, Theorem 9.5.5 and the subsequent Exercise 1(i) to $F - \{0\}$, after translation into multiplicative notation. Thus 1 is unique; while $(x^{-1})^{-1} = x$, $1^{-1} = 1$, $(xy)^{-1} = x^{-1}y^{-1}$, and the solution of $ax = b$ is $x = a^{-1}b$, provided $a \neq 0$. We draw special attention to one of the facts established in proving Proposition 9.6.2.

9.6.3 PROPOSITION. *There are no 'divisors of zero' in F: that is, if $ab = 0$ in F, then either $a = 0$ or $b = 0$ (or both).* ■

By contrast, if the integer m is *not prime*, then the ring $\mathbb{Z}_m$ of Example 9.2.6 *cannot be a field.* For then m is a product, say $m = ab$, where a, b are integers > 1, whence taking residue classes gives $0 = [m] = [ab] = [a][b]$, yielding the non-zero elements $[a]$, $[b]$ as divisors of zero in $\mathbb{Z}_m$. See also 10.2.6.

9.6.4 SUB-FIELDS. The reader by now can probably guess what we shall mean by a sub-field of a field $(F, +, \cdot)$. Quite analogously to 9.4.1, we call a subset G of F a **sub-field** provided it is a sub-ring with more than one element, and if, given any non-zero element x of G, then the element x^{-1} (which exists in

F) lies in G. Hence $1 \in G$, since it is the product of the two elements x and x^{-1} from G. Observe that the sub-ring consisting only of 0 is not a sub-field (and it does not satisfy $\mathfrak{A}_8$). As with rings and abelian groups, we record:

9.6.5 PROPOSITION. *The intersection of any family of sub-fields of F is a sub-field.* ■

9.6.6 EQUATIONS. Fields arise particularly in connection with the theory of equations, since we cannot expect to be able to solve interesting equations unless we can divide. For example,

(i) $$a_0x^n + a_1x^{n-1} + \cdots + a_n = 0$$

is called a **polynomial equation in** F provided $a_0, \ldots, a_n \in F$; to **solve it in** F is to find the set Y of all $y \in F$ which make the left-hand side of (i) zero when x is replaced by y. Of course, Y may be empty. Its members (if any) are called **roots** of equation (i), and we shall show in Chapter 22 that Y never has more than n members, provided $a_0 \neq 0$. If $\#Y < n$, we ask: is there a larger field G containing F, in which equation (i) has its maximum number of roots $\alpha_1, \alpha_2, \ldots, \alpha_n \in G$? And what is the nature of the *smallest* such solution field? For example, if F is a subfield of $\mathbb{C}$ and z is an element of F, consider the equation $x^2 = z$, and suppose this has no solution in F. We know that it has exactly two (non-zero) roots in $\mathbb{C}$, and we denote one of these by $\sqrt{z}$. Then any field containing F and $\sqrt{z}$ must contain also all elements of $\mathbb{C}$ of the form $w = a + b\sqrt{z}$, with $a, b \in F$; and we now invite the reader to show that the set W of all such w is a sub-field of $\mathbb{C}$ (see 9.6.2). This then is the smallest solution field, since every $x \in F$ appears as $x + 0\sqrt{z}$ while $\sqrt{z} = 0 + 1\sqrt{z} \in W$. It is customary to denote W by $F[\sqrt{z}]$.

An interesting feature of $F[\sqrt{z}]$ is that the function $\theta: F[\sqrt{z}] \to F[\sqrt{z}]$, given by

9.6.7 $$\theta(x + y\sqrt{z}) = x - y\sqrt{z},$$

is an **automorphism**†; i.e., it is a bijection such that for all $u, v \in F[\sqrt{z}]$ we have:

$$\theta(u + v) = \theta(u) + \theta(v); \qquad \theta(uv) = \theta(u)\theta(v).$$

Proof. Consider the second equality. We can write u, v in the forms $u = a + b\sqrt{z}$, $v = c + d\sqrt{z}$ with $a, b, c, d \in F$; whence

$$\theta(uv) = \theta(ac + bdz + (ad + bc)\sqrt{z}) = ac + bdz - (ad + bc)\sqrt{z},$$
$$\theta(u)\theta(v) = (a - b\sqrt{z})(c - d\sqrt{z}) = ac + bdz - (ad + bc)\sqrt{z} = \theta(uv),$$

as required. The first equality above is proved similarly. ■
Observe that if $w \in F$, then $\theta(w) = w$.

† Etymological note: *auto-* and *morphism* are of Greek origin, referring respectively to *self* and *structure*; thus 'automorphism' carries the idea of a self-mapping, preserving structure (here carrying sums into sums, and products to products).

9.6.8 APPLICATIONS. The study of such automorphisms in the general case for equation (i) is the basis of 'Galois theory', and the interested reader may consult the relevant chapters of Birkhoff–MacLane [16] for details. This theory enabled the brilliant Galois, its creator [(1811–1832); see Bell [10], Ch. 20], to show that polynomial equations in $\mathbb{C}$ of degree $\geqslant 5$ cannot be solved by using only the algebraic operations of taking mth roots and the field operations of $\mathbb{C}$; he followed the work of N. H. Abel (1802–1829) who dealt with the case $n = 5$.

We can apply these remarks to the famous problem, formulated by the Greeks, of 'duplicating the cube'. This problem asks: given a line segment of unit length, construct a cube of volume 2 units. It is equivalent to giving a construction in the plane of a line segment of length $\sqrt[3]{2}$ units (given a unit length); and here 'construct' means 'use a straight-edge and compasses, a finite number of times'. Now, if line segments of lengths $l, m > 0$ have already been constructed, simple geometrical constructions of the sort allowed yield segments of lengths $l + m$, lm, l/m, and $\sqrt{l}$. Elaborating this idea, it can be shown (but, rest assured, we do not expect the reader to deduce for himself from this hurried account!) that if the required construction of $\sqrt[3]{2}$ were possible, then a finite diagram would result, among whose line segments would occur a finite set of lengths $q_1, \ldots, q_k = \sqrt[3]{2}$, with the following properties. Each q_i lies in some sub-field F_i of $\mathbb{R}$, where $\mathbb{Q} = F_0 \subseteq F_1 \subseteq \cdots \subseteq F_k$ and $q_i \notin F_{i-1}(1 \leqslant i \leqslant k)$; while each F_i is of the form $F_{i-1}[\sqrt{z_i}]$ (in the notation of 9.6.7) and $q_i = a_i + b_i\sqrt{z_i}$ for some $a_i, b_i, z_i \in F_{i-1}$. [We are here using the fact, known to the Greeks, that 2 has no rational cube root; see Theorem 24.1.2.] Then $q_k^3 = 2$; so applying an automorphism like θ in 9.6.7 we get

$$\begin{aligned} 2 = (a_k + b_k\sqrt{z_k})^3 = \theta(2) &= \theta((a_k + b_k\sqrt{z_k}))^3 \\ &= (\theta(a_k + b_k\sqrt{z_k}))^3 = (a_k - b_k\sqrt{z_k})^3; \end{aligned}$$

but 2 can have only one *real* cube root, whence $b_k = 0$. Thus $q_k = a_k \in F_{k-1}$, a contradiction. Hence no such construction of $\sqrt[3]{2}$ exists. For more details, see Courant–Robbins [25], where it is also shown, by a more complicated argument, that it is impossible to trisect an angle of 60° using ruler and compasses only. The same ideas as before can be used, since cos 20° satisfies the cubic equation $8c^3 - 6c - 1 = 0$. (Constructions *are* possible using ingenious evasions of the rules, or other geometrical implements.)

Of course, for all practical purposes, we can construct lengths in the allowed manner, as close as we like to $\sqrt[3]{2}$ or to cos 20°, but this is not the point. There is a well-known routine for bisecting an angle—in principle with complete accuracy—and *the routine works for all angles*; we do not require a special routine depending on which angle we are asked to bisect. The astonishing feature of the 'impossibility' proofs above is that they show that no amount of cleverness will ever lead us to the required routine. A fertile source of discussion is now opened up. These ideas have led mathematicians and others since 1800 to analyse such questions as: (i) What is a routine? (ii) Is there a routine which will provide the appropriate answer 'Yes' or 'No' to *every* mathematical question submitted? (iii) If not, which problems can be solved by routine methods? Although these questions appear to have only philosophical interest, they have led to the growth of the large electronic computer; for a 'routine' is essentially a finite chain of simple actions for which clear instructions can be given without ambiguity to be followed blindly, so it is something which a machine can do. This, on analysis, answers question (i); the answer to (ii) was given in the 1930's by Gödel, Turing, Church, and others, and is 'No' (see Nagel and Newman [92], and Chapter 39). Great interest therefore attaches to question (iii), whose ramifications extend beyond mathematics to the theory of computer programming, cybernetics, and automated processes generally.

Exercise 3

(i) Let $f: F \to F$ be an automorphism of a field (i.e., f is a bijection and satisfies 9.6.7). Prove that $f(0) = 0, f(1) = 1$, and for each non-zero $x \in F, f(-x) = -f(x)$, $f(x^{-1}) = (f(x))^{-1}$, and $f(x^n) = (f(x))^n$, $(n = \pm 1, \pm 2, \ldots)$. Let X denote the set of all $x \in F$ such that $f(x) = x$. Prove that X is a sub-field of F. Prove that $f^{-1}: F \to F$ is also an automorphism of F.
(ii) Let s be an element of a field F such that the equation $x^2 = s$ has no solution in F. Define binary operations $+, \cdot$ on $F \times F$ by the rules:

$$(x, y) + (u, v) = (x + u, y + v)$$
$$(x, y) \cdot (u, v) = (xu + yvs, xv + yu).$$

Prove that $(F \times F, +, \cdot)$ is a field, whose zero and unity element are respectively $(0, 0)$, $(1, 0)$; that $-(x, y) = (-x, -y)$ and, if $(x, y) \neq (0, 0)$, that

$$(x, y)^{-1} = (x/(x^2 - sy^2), -y/(x^2 - sy^2)).$$

Interpret all this when $F = \mathbb{R}$, $s = -1$.

Chapter 10

ARITHMETIC MOD m

10.1 Residue Classes, and the Ring $\mathbb{Z}_m$

One of the most fascinating rules which one learns—or used to learn—in elementary arithmetic is that of 'casting out 9's'. Suppose we wish to decide whether the number 758466 is divisible by 9. We may simply add the digits of the number, obtaining 36, and conclude that the number is divisible by 9 because 36 is divisible by 9. Indeed we may say more generally that if n is any number and if $\mu(n)$ is the sum of the digits of n (when n is written in the scale of 10) then $n - \mu(n)$ is always divisible by 9.

This rule is best described, and explained, by reference to the so-called *arithmetic mod 9*. We will describe arithmetic mod m as part of the theory of the integers; and are content to remark here that much of the theory of the integers generalizes to arbitrary integral domains (defined in 11.1.4).

Thus we consider the ring $\mathbb{Z}$ and let m be an arbitrary non-zero positive integer. Recall from 4.5 that we can set up an equivalence relation in $\mathbb{Z}$ by declaring that $a \sim b$ iff $m|(b - a)$. Using the hints given there, the reader should verify that this is indeed an equivalence relation. The equivalence classes are called *residue classes mod m* and are the elements of the set $\mathbb{Z}_m$ (see Example 9.2.6); we write $[a]$ for the residue class containing a. Thus if $m = 10$, there are 10 residue classes, namely $[0], [1], \ldots, [9]$. In general, $\mathbb{Z}_m$ is a *finite* set, containing m elements.

We now observe, and the reader should prove, that

10.1.1 *if* $a \sim a'$ *and* $b \sim b'$, *then* $a + b \sim a' + b'$ *and* $ab \sim a'b'$.

From 10.1.1 it follows that we may introduce operations of addition and multiplication into the set $\mathbb{Z}_m$ of residue classes mod m by setting (see Example 9.2.6):

10.1.2 $$[a] + [b] = [a + b], \qquad [a][b] = [ab].$$

Notice that 10.1.2 constitutes a *definition* of addition and multiplication in $\mathbb{Z}_m$ in terms of the (known) addition and multiplication in $\mathbb{Z}$. Observation 10.1.1 is relevant here because we need to know that the residue classes $[a + b]$, $[ab]$ are uniquely determined by the *residue classes* $[a]$ and $[b]$ and do not depend on the arbitrary choice of integers a, b from those residue classes. (Recall 4.5.9.)

There is an obvious and natural function $\rho: \mathbb{Z} \to \mathbb{Z}_m$, namely, that function which assigns to each integer a its residue class $\rho(a) = [a]$. In terms of the function ρ, the equations 10.1.2 assert that

10.1.3 $$\rho(a + b) = \rho(a) + \rho(b), \qquad \rho(ab) = \rho(a)\rho(b).$$

In other words, ρ carries sums into sums and products into products; such a function is called a **homomorphism**†. It is also plain that ρ is surjective; on the other hand it is plainly not injective, since $\rho(0) = \rho(m)$. (Compare 9.6.7.) Whenever there exists a surjective homomorphism ρ from one set with two binary operations $(+, \cdot)$ to another, it is not hard to see that axioms $\mathfrak{A}_1$–$\mathfrak{A}_8$ are inherited by the second set from the first, with $\rho(0)$ acting as the zero of the second set and $\rho(1)$ acting as the unity (see 3 Exercise 8(ii)). In this way, or in others, one may verify

10.1.4 THEOREM. *The set $\mathbb{Z}_m$, furnished with the operations of addition and multiplication* (10.1.2) *is a commutative ring with unity element. The zero of $\mathbb{Z}_m$ is* [0] *and the unity of $\mathbb{Z}_m$ is* [1]. ■

10.1.5 EXAMPLE. We may write down the complete addition and multiplication tables for $\mathbb{Z}_m$, for any given m. Thus the addition table for $\mathbb{Z}_5$ is

+	[0]	[1]	[2]	[3]	[4]
[0]	[0]	[1]	[2]	[3]	[4]
[1]	[1]	[2]	[3]	[4]	[0]
[2]	[2]	[3]	[4]	[0]	[1]
[3]	[3]	[4]	[0]	[1]	[2]
[4]	[4]	[0]	[1]	[2]	[3]

while the multiplication table is

·	[0]	[1]	[2]	[3]	[4]
[0]	[0]	[0]	[0]	[0]	[0]
[1]	[0]	[1]	[2]	[3]	[4]
[2]	[0]	[2]	[4]	[1]	[3]
[3]	[0]	[3]	[1]	[4]	[2]
[4]	[0]	[4]	[3]	[2]	[1]

† Etymological note: *homo-* = same; now compare footnote, Chapter 9, p. 133. The word 'homomorphism' is used also for mappings between systems with only one binary operation, when the appropriate equation in 10.1.3 is satisfied; e.g., in abelian groups, see 9 Exercise 1(vii).

Exercise 1

(i) Show that when $m = 10$, then the homomorphism ρ in 10.1.3 simply ignores everything but the 'units column', in arithmetic calculations performed in the scale of 10.
(ii) Write down the addition and multiplication tables for $\mathbb{Z}_2$, $\mathbb{Z}_6$, $\mathbb{Z}_7$, $\mathbb{Z}_{10}$. Find all elements in these sets which have square roots (e.g., in $\mathbb{Z}_6$, $[3] = [3]^2$, but in $\mathbb{Z}_7$, $[3] \neq x^2$ for any $x \in \mathbb{Z}_7$).

It is now convenient to use the notation of Example 4.5.3 for the equivalence relation yielding the residue classes, especially if we want to consider more than one modulus m at the same time. Thus we may write, as before,

10.1.6 $$a \equiv b \pmod m$$

to indicate that a and b belong to the same residue class mod m, i.e., that $m|(b - a)$. We may also express this by saying that a and b are **congruent** mod m, and we describe 10.1.6 as a **congruence** mod m. We would therefore be entitled to omit the square brackets in writing down the addition and multiplication tables, provided it is understood that the entry in row a and column b is just an integer congruent to ab. Alternatively, we may regard the tables of $\mathbb{Z}_m$ as giving binary operations $+$, $\cdot$ on the set $\{0, 1, \ldots, m - 1\}$, which then takes on the same structure as $(\mathbb{Z}_m, +, \cdot)$.

The reader should note that, in the addition table for $\mathbb{Z}_5$, each row consists of just a permutation of the elements of $\mathbb{Z}_5$. A similar statement holds, indeed, for $\mathbb{Z}_m$ for any modulus m. This fact has interesting applications in statistics.

Exercise 2

(i) Construct a **magic square** of order m; i.e., using m integers only, enter them on a grid with m rows and m columns, such that the sum of the integers in each row or column is a constant, k. (Harder: arrange for the sum along each diagonal to be k, as well).
(ii) k is an odd integer and is the sum of the squares of two integers. Show that $k \equiv 1 \bmod 4$.
(iii) Use the notion of an isomorphism to make precise the statement above, that the set in question "takes on the same structure as $(\mathbb{Z}_m, +, \cdot)$".

On the other hand, the pattern of the multiplication table is rather more subtle. Let us write down the multiplication table for $\mathbb{Z}_4$, namely (omitting the square brackets)

$\cdot$	0	1	2	3
0	0	0	0	0
1	0	1	2	3
2	0	2	0	2
3	0	3	2	1

In the multiplication table for $\mathbb{Z}_m$ we have, just as in the addition table, symmetry about the leading diagonal—this simply reflects the commutativity of multiplication (as of addition) in $\mathbb{Z}_m$. Equally obviously, the leading row and column of the multiplication table consist entirely of zeros. If we throw away the leading row and column, so that we concern ourselves only with the multiplication of non-zero residue classes, we notice a substantial difference between $\mathbb{Z}_4$ and $\mathbb{Z}_5$. In the latter case each non-zero class occurs once and once only in each row, i.e., each row is a permutation of the non-zero classes. In the case of $\mathbb{Z}_4$, however, the situation is different. For, whereas rows 1 and 3 consist of permutations of the non-zero residue classes (as in the case of $\mathbb{Z}_5$), row 2 consists of two occurrences of 2 and one occurrence of 0; thus it is not only *not* a permutation of the non-zero classes—it is not even confined to the non-zero classes. The reader will find that the multiplication table for $\mathbb{Z}_7$ resembles in essence that of $\mathbb{Z}_5$, whereas those of $\mathbb{Z}_6$ and $\mathbb{Z}_{10}$ reproduce the curious features of that of $\mathbb{Z}_4$. We now proceed to explain these phenomena—and other features of the addition and multiplication tables for $\mathbb{Z}_m$.

10.2 Theory of $\mathbb{Z}_m$

Recall the essence of Theorem 10.1.4; it is that:

10.2.1 *$\mathbb{Z}_m$ is a commutative ring with unity.*

One naturally then asks for what values of m, if any, is $\mathbb{Z}_m$ a field. The answer is given by Theorem 10.2.3 below, but we need some preparation. We shall assume here a fact to be proved later. Recall from Example 1.6.6 the notation $a|b$ ('a divides b') and the notion of the highest common factor, $\mathrm{hcf}(u, v)$, of the integers u, v. The Euclidean algorithm (11.6) shows how to find integers k, s, t such that $k = \mathrm{hcf}(u, v)$ and

10.2.2 $$k = su + tv.$$

Recall too that a positive integer p is said to be *prime* if $p > 1$ and the only positive integers which divide p exactly are p and 1. The first few primes are 2, 3, 5, . . . ; 1 is excluded for technical reasons (see Chapter 12).

10.2.3 LEMMA. *If p is a prime number such that $p|uv$, then $p|u$ or $p|v$.*

Proof. Suppose that $p\nmid u$. Then $\mathrm{hcf}(p, u) = 1$, since p is prime. Hence by 10.2.2, there exist integers s, t such that $1 = sp + tu$. Therefore $v = spv + tuv$, so p divides the right-hand side, since $p|uv$. That is, $p|v$. ■

10.2.4 COROLLARY. *Let a, b, c be integers such that* $\mathrm{hcf}(a, c) = 1 = \mathrm{hcf}(b, c)$. *Then* $\mathrm{hcf}(ab, c) = 1$.

Proof. If not, then some prime number p divides both ab and c. Therefore $p|a$ or $p|b$, so either $p|\mathrm{hcf}(a, c)$ or $p|\mathrm{hcf}(b, c)$. Since a prime number is by definition greater than 1, we have a contradiction. Hence $\mathrm{hcf}(ab, c) = 1$. ■

We use these facts to produce a new example of a commutative group.

10.2.5 THEOREM. *Let $\Phi_m = \{[a_1], \ldots, [a_k]\}$ denote the set of those residue classes mod m which are prime to m* (i.e., $\text{hcf}(a_i, m) = 1$ *for each* $i = 1, \ldots, k$). *Then Φ_m forms a commutative group under the multiplication induced by that in the ring $\mathbb{Z}_m$.*

Proof. Let us show first that Φ_m is closed under multiplication, i.e., that $[a][b] \in \Phi_m$ when $[a], [b] \in \Phi_m$. Now $[a][b] = [ab]$ by Definition 10.1.2, so we must show that $[ab]$ is prime to m, i.e., that $\text{hcf}(ab, m) = 1$. But $\text{hcf}(a, m) = 1$ since $[a] \in \Phi_m$, and similarly for b. Hence $\text{hcf}(ab, m) = 1$ by 10.2.4, whence Φ_m is closed. Just as in the proof of Proposition 9.6.2, the axioms $\mathfrak{CG}_1$ (associativity), $\mathfrak{CG}_2$ (commutativity), and $\mathfrak{CG}_3$ (existence of unity) are satisfied. Hence, it remains only to verify axiom $\mathfrak{CG}_4$ (existence of a multiplicative inverse).

Now the class [1] is the unity element (and we could take $[a_1] = [1]$). Suppose $[a]$ is a residue class prime to m, i.e., $\text{hcf}(a, m) = 1$. Then, by 10.2.2, there are integers b, n such that

$$ab + mn = 1,$$

and plainly $\text{hcf}(b, m) = 1$. But then $ab \equiv 1 \bmod m$, or $[a][b] = 1$. This shows that $[b]$ is the inverse of $[a]$ and completes the proof of Theorem 10.2.5. ■

Exercise 3

(In these exercises, as elsewhere, we write $[a]_m$ for elements of Φ_m to distinguish different m's.)

(i) Show that $\Phi_1 = \{[1]_1\}$, $\Phi_2 = \{[1]_2\}$, $\Phi_3 = \{[1]_3, [2]_3\}$, $\Phi_4 = \{[1]_4, [3]_4\}$, and describe Φ_{10}, Φ_{11}. Write out the multiplication tables of these commutative groups.

(ii) Since $\text{hcf}(2, 5) = 1$, and $3 \cdot 2 + (-1) \cdot 5 = 1$, we have $[3]_5[2]_5 = [1]_5$ in Φ_5 so $[2]^{-1} = [3]$. Calculate similarly the inverse of each element in Φ_5, Φ_{10}, Φ_{11}.

(iii) In Φ_4, the solution of the congruence $3x \equiv 1$ is $x = [3]_4^{-1}[1]_4 = [3]_4$ (using (i) above), but in Φ_2 the equation becomes $x = 1$ with solution $[1]_2$, since $3x = x + x + x \equiv x \bmod 2$. Solve the congruence in Φ_{10}, Φ_{11}, Φ_{12}. [We may refer to the congruence above as the equation $3x = 1$, where it is understood that 3, 1 stand for the residue classes containing them.]

(iv) Show that the equation $2x = 1$ has exactly one solution in $\mathbb{Z}_m$ if m is odd and no solution if m is even. How about the equation $2x = 2$ (which will not in general be the same as the equation $x = 1$)?

Before making a further study of the commutative group Φ_m, let us answer the question, raised above, about $\mathbb{Z}_m$.

10.2.6 THEOREM. *$\mathbb{Z}_m$ is a field $.\Leftrightarrow.$ m is prime.*

Proof. We have already observed (after 9.6.3) that if m is not a prime there are non-zero elements of $\mathbb{Z}_m$ whose product is zero. Thus in that case $\mathbb{Z}_m$ is certainly not a field. We now show that $\mathbb{Z}_m$ is a field if m is prime. By 9.6.2, this amounts to showing that the set N of non-zero residue classes mod m forms a commutative group under the multiplication of $\mathbb{Z}_m$. But since m is prime, $N = \Phi_m$ since every element $[a]$ in N is prime to m. Therefore N is indeed a group, by Theorem 10.2.5, and is of course commutative. ■

The last theorem explains, then, the strange features of the multiplication table for $\mathbb{Z}_m$, m not prime, to which attention has been drawn.

Exercise 4

(i) Give a formula for the number of occurrences of b in row a of the multiplication table for $\mathbb{Z}_m$.
(ii) Justify the rule of casting out 9's.
(iii) Let the integer n be written out as $a_r a_{r-1} \cdots a_0$ in the decimal scale. Define the function $v: \mathbb{Z}^+ \to \mathbb{Z}$ by $v(n) = \sum_{s=0}^{r} (-1)^s a_s$. Show that $n \equiv v(n) \bmod 11$. Hence test 69580379 for divisibility by 11.
(iv) Let $n = a_r a_{r-1} \cdots a_0$ as above. Let $m = b_{r-1} \cdots b_0$ be an integer such that

$$b_s = a_{s+1}, \qquad s > 0,$$
$$b_0 \equiv a_1 - 2a_0 \bmod 7.$$

Show that $m \equiv 0 \pmod 7$ if and only if $n \equiv 0 \pmod 7$. Can we infer that $m \equiv n \pmod 7$?
(v) Show that the simultaneous equations $3x + y = 1$, $5x + 4y = 2$ have a unique solution in all fields $\mathbb{Z}_p$, where p is prime and not 7; but in $\mathbb{Z}_7$ they have no solution.

10.3 Euler's Totient Function

We can draw some interesting deductions from Theorems 10.2.5 and 10.2.6. Let $\phi(m)$ be the number of positive integers prime to m and not greater than m. Thus $\phi(1) = 1$, $\phi(2) = 1$, $\phi(3) = 2$, $\phi(4) = 2, \ldots$. The function $\phi: \mathbb{N} \to \mathbb{N}$ is called Euler's **totient** function, and it is the number of elements in the multiplicative group Φ_m of Theorem 10.2.5; that is, $\phi(m)$ is the **order** of $\Phi(m)$. Plainly

$$\phi(m) = m - 1 \quad \text{if and only if } m \text{ is prime.}$$

10.3.1 THEOREM. (Euler's Theorem.) *Let a be prime to m. Then $a^{\phi(m)} \equiv 1 \bmod m$.*

Proof. This follows, in fact, from Theorem 10.2.5 and a general theorem of group theory [see 18.9.7], but we will give here a special argument. Since row a of the multiplication table for Φ_m consists of just a permutation of the entries in row 1, the product of the entries in the two rows is the same. On the other hand, the rows may be written

$$[a_1] \cdots [a_k], \quad \text{and}$$
$$[aa_1] \cdots [aa_k], \quad \text{respectively,}$$

where $k = \phi(m)$. Thus

$$[a_1 \cdots a_k] = [a^k a_1 \cdots a_k] = [a^k][a_1 \cdots a_k].$$

Now we may cancel $[a_1 \cdots a_k]$ to obtain $[a^k] = 1$ or $a^{\phi(m)} \equiv 1 \bmod m$. ■

The special case of this theorem when m is a prime number is due to Fermat [(1601–25; see Bell [10], Ch. 4], one of the founders of the theory of numbers. The reader should note that a deeper understanding of arithmetic has enabled us to deduce a result whose special cases would be far beyond the scope of even the most modern electronic computer.

Exercise 5

(i) Show that, if p is a prime number, then $a^p \equiv a \bmod p$.
(ii) Show, without a case-by-case examination, that we do not change the figure in the 'units' column when we raise a number to the fifth power.
(iii) Give a formal proof of the statement (in the proof of 10.3.1) that row a of the multiplication table for Φ_m consists of just a permutation of the entries in row 1. {Let $f: \Phi_m \to \Phi_m$ be the function such that $f([a_i)] = [aa_i]$. Prove that f is an equivalence $\Phi_m \to \Phi_m$.}
(iv) Given integers $m, n > 0$, define a function $g: \Phi_{mn} \to \Phi_m \times \Phi_n$ by the rule that $g([a]_{mn}) = ([a]_m, [a]_n)$, where $[\cdot]_q$ denotes residue classes in Φ_q. Prove that if $\Phi_m \times \Phi_n$ is given the multiplication defined in 9 Exercise 1(vii) (so as to become the *direct product*), then g is a homomorphism. Prove also that, if $\text{hcf}(m, n) = 1$, then g is an equivalence, and thus that $\phi(mn) = \phi(m)\phi(n)$. {You should use a special case of the *Chinese remainder theorem* (see p. 144).}

Deduce that if m factorizes into primes as $m = p_1^{n_1}p_2^{n_2}\cdots p_r^{n_r}$, then $\phi(m) = (p_1^{n_1} - p_1^{n_1-1})\cdots(p_r^{n_r} - p_r^{n_r-1})$. Hence compute $\phi(60)$ and check by crude enumeration of the elements of Φ_{60}.

10.3.2 THEOREM. **(Wilson's Theorem.)** *If p is prime, then*

$$(p-1)! \equiv -1 \ (mod\ p).$$

Proof. To each a, not divisible by p, there exists b such that $ab \equiv 1 \pmod p$. In this way we pair off the non-zero residue classes mod p into pairs $[a]$, $[b]$ with $[a][b] = [1]$. The only classes to get paired with themselves are $[1]$ and $[-1]$. (Prove this! The case $p = 2$ is a little special, since then $[1] = [-1]$, but it is utterly trivial.) Thus the product of the $(p-1)$ classes $[1], \ldots, [p-1]$ is just $[1][-1] = [-1]$, and so $[(p-1)!] = [-1]$ or

$$(p-1)! \equiv -1 \pmod p. \quad ■$$

Exercise 6

Let $(A, \cdot)$ be a finite commutative group. Prove that the product, x, of all the (distinct) elements in A, has the property that $x^2 = 1$. (This generalizes part of the proof of Theorem 10.3.2.)

10.4 Solution of Congruences

We close this chapter with a brief discussion of the solution of congruences. A congruence mod m, we recall, is just an equation in $\mathbb{Z}_m$. Thus the simplest type of congruence is the linear congruence

10.4.1 $$ax \equiv b \bmod m.$$

This can be solved in principle by simply consulting row a of the multiplication table for $\mathbb{Z}_m$ and observing in which columns the entry b occurs. The following criterion is important.

10.4.2 PROPOSITION. *The congruence* 10.4.1 *has a solution if and only if* $\text{hcf}(a, m)$ *divides* b.

We leave the details of the following proof to the reader. Let $\text{hcf}(a, m) = d$, $a = da'$, $m = dm'$. If a solution x of 10.4.1 exists, then $b = mk - ax$ for some k, so that $d|(mk - ax)$ or $d|b$. Conversely, suppose $d|b$, $b = db'$. Then x satisfies 10.4.1 if and only if it satisfies

10.4.3 $$a'x \equiv b' \bmod m'.$$

This congruence has a unique solution, in the sense that there is a unique residue class $[x_0]$ in $\mathbb{Z}_{m'}$ such that $a'x_0 \equiv b' \bmod m'$. Then the solutions of 10.4.1 are the residue classes $[x_0 + km']$ *in* $\mathbb{Z}_m$, where k runs from 0 to $d - 1$ inclusive. We further observe that the solution of 10.4.3 may be obtained by first solving the congruence $a'x \equiv 1 \bmod m'$ (e.g., by the methods of Exercise 3(ii)) and then multiplying the solution by b'. Alternatively, we may of course consult the multiplication table for $\mathbb{Z}_{m'}$. The existence and uniqueness of a solution of 10.4.3 follows from Theorem 9.5.5, since $\Phi_{m'}$ is a commutative group. ■

Let us take the problem just one stage further and consider simultaneous congruences

10.4.4 $$a_1x \equiv b_1 \bmod m_1,$$
$$a_2x \equiv b_2 \bmod m_2.$$

We proceed as before to solve the first congruence, obtaining the solution $x = x_0 + km_1'$ (if a solution exists). We now have to solve for k from the second congruence; thus

10.4.5 $$a_2x_0 + a_2km_1' \equiv b_2 \bmod m_2, \quad \text{or}$$
$$a_2m_1'k \equiv b_2 - a_2x_0 \bmod m_2.$$

This is now a congruence for the unknown k, all the other terms being known, and we proceed to solve it as above, assuming it has a solution. Notice that the linear congruence 10.4.1 certainly has a solution if $\text{hcf}(a, m) = 1$ (compare Exercise 3(ii)), and the solution is then unique (as a residue class mod m).

Thus if $hcf(a_1, m_1) = 1$, $hcf(a_2, m_2) = 1$ *and* $hcf(m_1, m_2) = 1$, *then the simultaneous congruences* (10.4.4) *have a unique solution as a residue class mod* m_1m_2. This is sometimes known as the **Chinese remainder theorem**; it plainly generalizes to any finite number of simultaneous congruences.

10.4.6 EXAMPLE. Solve the simultaneous congruences

$$7x \equiv 1 \bmod 4,$$
$$4x \equiv 2 \bmod 6.$$

The first congruence is obviously equivalent to $-x \equiv 1 \bmod 4$ or $x \equiv 3 \bmod 4$. Thus if $x = 3 + 4k$ we have, replacing the second congruence by the equivalent $2x \equiv 1 \bmod 3$,

$$6 + 8k \equiv 1 \bmod 3, \quad \text{or}$$
$$8k \equiv 1 \bmod 3.$$

This is plainly equivalent to $-k \equiv 1 \bmod 3$ or $k \equiv 2 \bmod 3$. Thus the solution is $x \equiv 11 \bmod 12$.

Exercise 7

Solve the following congruences

(i) $2x \equiv 3 \bmod 7$;
(ii) $4x \equiv 4 \bmod 7$;
(iii) $2(x + 1) \equiv 4(x - 1) \bmod 8$;
(iv) $6x \equiv 3 \bmod 3$;
(v) $13x \equiv 5 \bmod 9$;
(vi) $10x \equiv 5 \bmod 6$;
(vii) $3x \equiv 8 \bmod 11$, $5x \equiv 9 \bmod 3$;
viii) $6x \equiv 1 \bmod 7$, $7x \equiv 1 \bmod 8$.

(ix) Find all integers x, y satisfying the 'Diophantine equation' $lx + my = n$, where l, m, n are integers. {Reduce the problem to the case $\text{hcf}(l, m) = 1$, and solve congruences mod l and mod m.} Hence find the number of ways of spending 98 pennies on apples and oranges at threepence and fivepence each, respectively.
(x) (The monkey problem.) On a desert island, n sailors and a monkey pick a heap of N coconuts. During the night, one sailor divides the heap into n equal piles and finds that there is one coconut left, so he gives it to the monkey and takes his share. Ignorant of these happenings, a second sailor repeats the performance with the depleted pile; he is followed by a third sailor, and so on until all the sailors have taken what they think is their share. If m coconuts are left in the morning, and $n|m$, find all possible values of N. [See Gardner [39].]
(xi) Let R, S be commutative rings with unity element, and let $f: R \to S$ be a homomorphism (in the sense of 10.1.3). Let $I = \{r|r \in R \;.\&.\; f(r) = 0\}$. Then I is called the **kernel** of f; it is $f^\flat(0)$. Prove that I is a sub-ring of R with the property that, for all $c \in R$ and $r \in I$, $rc \in I$. Such a sub-ring is called an **ideal** of R. Show that, in 10.1.3, the kernel of ρ is $I = m\mathbb{Z}$. Prove that if an ideal J contains the unity element of R, then $J = R$.

(xii) Let I be an ideal in a commutative ring R with unit. On R define a relation T by: $aTb \;.\!\Leftrightarrow\!.\; a - b \in I$. Prove that T is an equivalence relation satisfying 10.1.1 in R. The set R/T of equivalence classes $[a]$ is customarily denoted by R/I; prove that it becomes a commutative ring when $+$ and $\cdot$ are defined by

$$[a] + [b] = [a + b], \qquad [a] \cdot [b] = [ab],$$

and that $[1]$ is a unity element. Prove that the canonical map $g: R \to R/I$ is a homomorphism and that its kernel $g^{\flat}\{[0]\} = I$. Describe the special case $R = \mathbb{Z}$.

There are many more problems involving congruences; for example, there is the fascinating question of quadratic congruences and, in particular, of *quadratic residues* mod p, where p is a prime. Here we call a a quadratic residue mod p if the congruence $x^2 \equiv a \bmod p$ has a solution. The reader is referred to Davenport [28] for a discussion of such questions.

Chapter 11

RINGS WITH INTEGRAL NORM

11.1 Integral Norms

An important property of the set $\mathbb{Z}$ of integers is that each integer $m \in \mathbb{Z}$ has an 'absolute value', usually denoted by $|m|$, which lies in $\mathbb{Z}^+ = \{0, 1, 2, \ldots\}$. More precisely, there exists a function $abs: \mathbb{Z} \to \mathbb{Z}^+$ defined by

11.1.1
$$abs(0) = 0, \qquad abs(m) = m \quad \text{if} \quad m > 0,$$
$$abs(m) = -m \quad \text{if} \quad m < 0.$$

It is customary to call $abs(m)$ the **modulus** of m, but we avoid the term in this chapter because of confusion with congruences mod m. The function abs is an example of an 'integral norm' on a ring, which we now explain.

Our purpose in this chapter is to separate out the roles played in the theory of factorization by (i) the ring-structure of $\mathbb{Z}$ and by (ii) the properties of the function abs. We therefore proceed more generally and study the theory of factorization in rings having an 'integral norm' to enrich their structure. The definition of a ring with integral norm is then as follows.

11.1.2 DEFINITION. Let R be a ring, and let $\phi: R \to \mathbb{Z}^+$ be a function. We call ϕ an **integral norm** on R, and (R, ϕ) a **ring with integral norm** ϕ, provided that the following axioms are satisfied:

$\mathfrak{N}_1$: $\phi(a) = 0 \;.\Leftrightarrow.\; a = 0$;

$\mathfrak{N}_2$: $\phi(ab) = \phi(a)\phi(b)$ for all $a, b \in R$.

R is then said to **admit** an integral norm. Before giving examples, we show that not all rings do admit integral norms, by giving the following necessary and sufficient condition.

11.1.3 PROPOSITION. *The ring R admits an integral norm ϕ iff it has no divisors of zero.*

For assume ϕ is a norm and let $ab = 0$. Then, applying ϕ, we get from $\mathfrak{N}_1$, $\mathfrak{N}_2$ that $\phi(a)\phi(b) = 0$. It thus follows (since $\phi(a)$, $\phi(b)$ are in $\mathbb{Z}^+$, and so ordinary integers) that $\phi(a) = 0$ or $\phi(b) = 0$. Axiom $\mathfrak{N}_1$ then implies that $a = 0$ or $b = 0$. Conversely, if R has no divisors of zero, then a norm ϕ is given by $\phi(0) = 0$, $\phi(a) = 1$, $a \neq 0$. ■

11.1.4 A commutative ring R with unity element and without divisors of zero is called an **integral domain** (since it has so much in common with the 'domain' of integers). Thus, Proposition 11.1.3 may be interpreted as saying that: a commutative ring with unity element, admitting an integral norm, has to be an integral domain. We saw after Proposition 9.6.3 that $\mathbb{Z}_m$ is not an integral domain when m is not prime; for such m, therefore, $\mathbb{Z}_m$ *admits no norm.* Other rings admitting no norm are rings of functions and of matrices (see Examples 9.2.8, 9.2.9). If the conditions on a norm are modified by permitting a range other than $\mathbb{Z}^+$, or by weakening axioms $\mathfrak{N}_1$, $\mathfrak{N}_2$, then very significant norms can be defined on rings of *continuous* functions; but then we are in the branch of calculus called 'normed rings'—see Gelfand [42]. Since we concern ourselves only with integral norms, that is, norms to the non-negative integers, we henceforth suppress the word 'integral' and speak simply of **norm.**

11.1.5 The most interesting norms ϕ on a ring R are those satisfying the **Euclidean axiom** and called **Euclidean**, after Euclid (330–275 B.C.).

$\mathfrak{N}_3$: *Given $a, b \in R$ with $b \neq 0$, there exist $q, r \in R$ such that $a = bq + r$ and* $\phi(r) < \phi(b)$.

We shall be principally concerned in this chapter with rings satisfying $\mathfrak{A}_1$–$\mathfrak{A}_8$ and admitting norms satisfying $\mathfrak{N}_1$–$\mathfrak{N}_3$, but since many of our concepts apply to arbitrary integral domains it will be useful to have this concept available.

11.1.6 We should next point out that $\mathbb{Z}$, with the norm function $abs: \mathbb{Z} \to \mathbb{Z}^+$ given by 11.1.1 is indeed an integral domain with Euclidean norm. Property $\mathfrak{N}_1$ is obvious; so, too, is property $\mathfrak{N}_2$. As to $\mathfrak{N}_3$, it simply asserts that, in $\mathbb{Z}$, we may divide any integer by any non-zero integer, obtaining a remainder smaller in absolute value than the divisor. This is familiar, of course, for positive integers and should be clear, after a moment's thought, for arbitrary integers.

11.2 Examples

We now return to the examples given in Chapters 9 and 10, and discuss whether they admit norms—and, if so, what norms. We confine attention to rings satisfying $\mathfrak{A}_1$–$\mathfrak{A}_8$.

11.2.1 (EXAMPLE 9.2.3 CONTINUED.) If f is not the zero polynomial, it has a **degree** d, which is the largest integer r such that $a_r \neq 0$; here, as before, $f = a_0 + a_1x + \cdots + a_mx^m$. We define $\phi: \mathbb{Z}[x] \to \mathbb{Z}^+$ by

$$\phi(0) = 0,$$
$$\phi(f) = 2^{\deg f}, \qquad f \neq 0.$$

In particular, if f is a constant (i.e., an element of $\mathbb{Z}$), and $f \neq 0$ then $\phi(f) = 2^0 = 1$.

The reader may readily verify $\mathfrak{N}_1$, $\mathfrak{N}_2$; on the other hand $\mathfrak{N}_3$ *does not hold for this* ϕ. To show this, it suffices to take a, b in $\mathfrak{N}_3$ to be x and $2x$, observing that we cannot find elements q, r in $\mathbb{Z}[x]$ with

$$x = q \cdot 2x + r$$

and $\phi(r) < \phi(2x)$, i.e., r is a constant. Indeed, $x = \frac{1}{2} \cdot 2x + 0$, but $\frac{1}{2}$ is *not* in $\mathbb{Z}[x]$. We will show later that $\mathbb{Z}[x]$ admits *no* Euclidean norm.

11.2.2 (EXAMPLE 9.2.1 CONTINUED.) The ring $\mathbb{Q}$ of rational numbers acquires a norm by defining

$$\phi(0) = 0, \qquad \phi(a) = 1 \quad \text{if} \quad a \neq 0.$$

This is a Euclidean norm, as the reader may verify; we will see later (in Exercise 5(ii)) that it is the only norm on $\mathbb{Q}$. Similar remarks apply to $\mathbb{R}$ and $\mathbb{C}$, to $\mathbb{Z}_p$ when p is a prime, and indeed to any field.

11.2.3 (EXAMPLE 9.2.4 CONTINUED.) The ring $\mathbb{Q}[x]$ acquires a norm just as in Example 9.2.3 above; that is to say, we define

$$\phi(0) = 0, \qquad \phi(f) = 2^{\deg f}, \qquad f \neq 0.$$

This is a Euclidean norm. The verification of this amounts to showing that, given two polynomials f, g in $\mathbb{Q}[x]$, with $g \neq 0$, we may divide f by g, getting a remainder r (a polynomial) which is zero or of lower degree than g. This division algorithm is taught early in the student's acquaintance with algebra. But unfortunately the vital distinction concerning $\mathfrak{N}_3$ between $\mathbb{Z}[x]$ and $\mathbb{Q}[x]$ receives no attention in most traditional courses on elementary algebra. The rings $\mathbb{R}[x]$, $\mathbb{C}[x]$ may be normed by the same formula as that given for $\mathbb{Q}[x]$; and these norms are also Euclidean. Of course there is not a *unique* Euclidean norm on $\mathbb{Q}[x]$; the norm $3^{\deg f}(f \neq 0)$ would evidently do as well.

11.2.4 (EXAMPLE 9.2.7 CONTINUED.) Here the behaviour of

$$\Gamma_n = \{a + ib\sqrt{n}, a, b \in \mathbb{Z}\}$$

is much influenced by the value of n. Certainly, for any n, the function $\phi: \Gamma_n \to \mathbb{Z}^+$, given by $\phi(a + ib\sqrt{n}) = a^2 + nb^2$, is a norm. (That it satisfies $\mathfrak{N}_2$ reflects the fact that the modulus of the product of two complex numbers is the product of their moduli.) However, this function ϕ is not Euclidean for all n; and, indeed, for some n there is no Euclidean norm on Γ_n. The case $n = 1$ is of special interest; here the ring Γ_1 is called the ring of **Gaussian integers.** We now prove that

$$\textit{when} \quad n = 1, \qquad \textit{the norm} \quad \phi: \Gamma_1 \to \mathbb{Z}^+ \quad \textit{is Euclidean.}$$

We will write Γ for Γ_1. The proof is, of course, only suitable for those familiar with the algebraic manipulation of complex numbers.

Thus let $\alpha = a + ib$, $\beta = c + id$ be two elements of $\Gamma = \Gamma_1$ with $\beta \neq 0$. To show that ϕ satisfies $\mathfrak{N}_3$ we must find $\gamma, \delta \in \Gamma$ such that $\alpha = \beta\gamma + \delta$ and $\phi(\delta) < \phi(\beta)$. Consider the complex number α/β which we may express as $\alpha/\beta = u + iv$. Of course, $u + iv$ is not necessarily an element of Γ, since there is no reason to suppose that u and v are integers. If, however, we look at α/β as a point in the Argand diagram, then there will be an element of Γ which is as close as possible to it (see Fig. 11.1). Let $p + iq$ be such an element of Γ; then α/β can be written

$$\alpha/\beta = u + iv = p + iq + r + is, \qquad \text{where } |r| \leqslant \tfrac{1}{2}, \qquad |s| \leqslant \tfrac{1}{2},$$

and $|\cdot|$ stands for the ordinary modulus of a complex number. Writing γ for $p + iq$, we have $\alpha = \beta\gamma + \beta(r + is) = \beta\gamma + \delta$, say. Since α, β, γ all belong to Γ, so does δ (since Γ is a ring; or the reader may verify this directly).

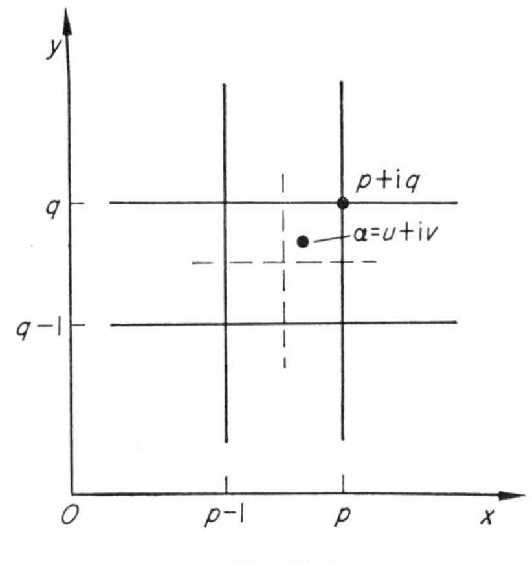

Fig. 11.1

Now any element $\xi \in \Gamma$ is, in particular, a complex number, and the square of its ordinary modulus is $\phi(\xi)$. Thus

$$\phi(\delta) = |\delta|^2 = |\beta|^2|r + is|^2 = \phi(\beta)(r^2 + s^2).$$

But $r^2 + s^2 \leqslant \frac{1}{2}$, so that $\phi(\delta) \leqslant \frac{1}{2}\phi(\beta) < \phi(\beta)$, since $\phi(\beta) \neq 0$. We have thus found γ, δ such that $\alpha = \beta\gamma + \delta$ and $\phi(\delta) < \phi(\beta)$, thus proving $\mathfrak{N}_3$ for the norm ϕ. ■

Exercise 1

(i) Why is the ordinary modulus not a norm on the ring $\mathbb{C}$ of complex numbers? Is it a norm on Γ?

(ii) Let R^S denote the ring of functions in Example 9.2.8. Suppose that $S = A \cup B$, where $S \neq A \neq \varnothing \neq B \neq S$. Choose $x \neq 0$ in R. Let $f, g \in R^S$ be

defined by: $f(s) = x$ if $s \in A$, $f(s) = 0$ if $s \notin A$; $g(s) = x$ if $s \in B$, $g(s) = 0$ if $s \notin B$. Prove that f and g are divisors of zero, and hence that R^S is not an integral domain. By drawing suitable smooth graphs, construct a similar example, but with f, g differentiable and $S = R = \mathbb{R}$.
(iii) In Example 9.2.9, construct a pair of non-zero 2×2 matrices with zero matrix product, to show that $M(2, \mathbb{R})$ admits no norm $\phi: M(2, \mathbb{R}) \to \mathbb{Z}^+$.
(iv) Is Euler's totient function (10.3) a norm on $\mathbb{Z}$ if we set $\phi(0) = 0$ and $\phi(-n) = \phi(n)$?

11.3 Factorization in Euclidean Domains

Let us call an integral domain with a norm ϕ a **normed domain**, and a **Euclidean domain** if the norm ϕ is Euclidean. We now discuss the arithmetic theory of the set $\mathbb{Z}$ as an example of a Euclidean domain; thus, we will be concerned with hcf's and lcm's in a Euclidean domain R.

Traditionally, factorization theory is done in the set $\mathbb{Z}^+$ (indeed, in $\mathbb{N} = \mathbb{Z}^+ - \{0\}$), rather than in $\mathbb{Z}$. Traditionally also, questions of the *uniqueness* of hcf's and lcm's and of factorization into primes are either ignored or, perhaps, dismissed as 'obvious'. Even so, the student is occasionally aware of some small feeling of discomfort in dealing with the number 1. If he knows a definition of a prime number, it seems that 1 fits that definition. Yet there is obviously something special about 1, because we cannot allow 1 to enter into an expression for a number as a product of primes without destroying uniqueness, and 1 itself cannot be factorized into primes.

In our more general approach, the special role of 1 in factorization theory becomes clear; even when passing to the Euclidean domain $\mathbb{Z}$ it becomes clear that the traditional theory carries over provided that a special role is also assigned to the integer -1. As we shall see, this special role springs from the fact that ± 1 are the only two *units* in $\mathbb{Z}$, according to the following definition.

Let R be a commutative ring with unity element. Then an element $u \in R$ is a **unit** if there exists $v \in R$ with $uv = 1$. [*Warning:* do not confuse 'unit' and 'unity element'. The unity element 1 is certainly a unit, but there are in general many units in R different from 1.] Thus, for example, in $\mathbb{Z}$ the units are ± 1, in $\mathbb{Q}$ every non-zero element is a unit, in $\mathbb{Q}[x]$ every non-zero constant polynomial is a unit (and only these polynomials are units).

Exercise 2

(i) Identify the units in $\mathbb{Z}[x]$ (compare $\mathbb{Q}[x]$!) and in $\Gamma(=\Gamma_1)$.
(ii) Prove that if every non-zero element of R is a unit, then R is a field, and conversely.
(iii) Prove that in the residue class ring $\mathbb{Z}_m$, the set of units is Φ_m (defined in Theorem 10.2.5).
(iv) Generalize (iii) by showing that the set of all units, in a commutative ring with unity element, is a group under multiplication induced from R. Prove that the sum of units is not necessarily a unit.

11.3.1 Let $a, b \in R$ where R is an integral domain. We say that a and b are **associates** if there is a unit $u \in R$ such that $b = au$. We write $a \sim b$ if a and b are associates. We leave the reader to prove the following result.

11.3.2 PROPOSITION. *$a \sim b$ is an equivalence relation.* ∎

Exercise 3

(i) Show that any two units of the integral domain R are associates. Hence show that the set U of all units of R constitutes a single equivalence class under the relation $(\sim)$ of association. (In particular, if $a \in U$, then for any $b \in R$,

$$a \sim b \;.\Leftrightarrow.\; b \in U.)$$

(ii) Let W denote the set of equivalence classes of $\mathbb{Z}$ under $(\sim)$. Prove that there is a bijection $W \to \mathbb{Z}^+$.

From now on we suppose without further mention that R is an integral domain. An important fact about a unit u is that its norm is (the integer) 1, as we now see.

11.3.3 THEOREM. *Let R be a normed domain with norm ϕ. If $u \in R$ is a unit, then $\phi(u) = 1$. Conversely if R is a Euclidean domain and $\phi(u) = 1$, then u is a unit.*

Proof. We first show that $\phi(1) = 1$. For, by $\mathfrak{N}_2$, $\phi(1) = \phi(1^2) = (\phi(1))^2$. Thus $\phi(1) = 1$ or 0; but, by $\mathfrak{N}_1$, $\phi(1) = 0$ implies $1 = 0$, contrary to $\mathfrak{A}_8$. Thus $\phi(1) = 1$.

Now let u be a unit. Then $\exists v \in R$ with $uv = 1$ so that, by $\mathfrak{N}_2$, $1 = \phi(1) = \phi(uv) = \phi(u)\phi(v)$. Now $\phi(u), \phi(v)$ are in $\mathbb{Z}^+$ with product 1; hence each must be 1. Thus $\phi(u) = 1$, as was to be proved.

The converse is more subtle and rests on the fact that the norm ϕ satisfies axiom $\mathfrak{N}_3$. We apply $\mathfrak{N}_3$ with $a = 1$, $b = u$. This is legitimate since, by $\mathfrak{N}_1$, $u \neq 0$ since $\phi(u) \neq 0$. Then $\exists q, r \in R$ with $1 = qu + r$ and $\phi(r) < \phi(u)$. But $\phi(u) = 1$ and the only element in $\mathbb{Z}^+$ less than 1 is 0. Thus $\phi(r) = 0$ whence, by $\mathfrak{N}_1$, $r = 0$, so that $1 = qu$ and u is a unit. ∎

Exercise 4

Show with the example of $\mathbb{Z}[x]$ that there are normed domains R and elements $u \in R$ such that $\phi(u) = 1$, yet u is not a unit.

We may derive an interesting corollary of Theorem 11.3.3 which is important in the theory of factorization into primes.

11.3.4 COROLLARY. *Let R be a Euclidean domain, $a, b, c \in R$ and $a = bc$, where none of a, b, c is zero and c is not a unit. Then $\phi(a) > \phi(b)$.*

For $\phi(a) = \phi(b)\phi(c) > \phi(b)$, since $\phi(c) > 1$ (notice that $\phi(c)$ is a positive integer equal to neither 0 nor 1). ∎

Exercise 5

(i) Use Theorem 11.3.3 to deduce that the units of Γ are precisely ± 1, $\pm \mathrm{i}$.
(ii) Use Theorem 11.3.3 to prove that if F is a field, the only norm on F is given by

$$\phi(0) = 0, \qquad \phi(a) = 1, \qquad a \neq 0 \in F,$$

and show that ϕ is a Euclidean norm.

11.4 Ideals

We now begin to develop the theory of hcf's. This theory is usually presented as a development of the theory of factorization into primes. We will give a presentation which is quite independent of the theory of primes but rests on the basic notion of an *ideal* in an integral domain. [This curious name arises because the 19th-century algebraists wished to enlarge a ring R by adding 'ideal' elements to it. See 10 Exercise 7 (xi).]

11.4.1 As far as the definition is concerned we may as well consider the more general setting of a commutative ring R. Then an **ideal** in R is a non-empty subset of R satisfying the two conditions:

$\mathfrak{I}_1$: *If* $a, b \in I$, *so does* $a - b$;
$\mathfrak{I}_2$: *If* $a \in I$, $r \in R$, *then* $ra \in I$.

The reader may be surprised that we do not ask that $a + b \in I$ if $a, b \in I$. We certainly require this, but it turns out to be guaranteed by $\mathfrak{I}_1$. For since I is non-empty, there is some $a \in I$. Thus, by $\mathfrak{I}_1$, $0 = a - a \in I$. Then if $b \in I$, again by $\mathfrak{I}_1$ we find $-b = 0 - b \in I$. Finally, if $a, b \in I$, then a, $-b \in I$, so $a + b = a - (-b) \in I$. On the other hand, the example $R = \mathbb{Z}$, $I = \mathbb{Z}^+$ shows that we may not replace $\mathfrak{I}_1$ by requiring that the sum of any two elements of I belongs to I. Notice also that we may restate $\mathfrak{I}_1$ by asserting that I is an abelian group (under addition). Also $\mathfrak{I}_1$ and $\mathfrak{I}_2$ together imply that I is itself a *sub-ring* of R. (See 9.4.) On the other hand, not all sub-rings are ideals. For if $R = \mathbb{Q}$ and $I = \mathbb{Z}$, then I is certainly a sub-ring of R but it fails to be an ideal, since the product of a rational number and an integer is not necessarily an integer. Some examples, in addition to those of 10 Exercise 7(xi)(xii), are as follows.

Example 9.2.5 (continued). In the ring $\mathbb{Z}$, the subset $m\mathbb{Z}$ consisting of integers divisible by m is an ideal in $\mathbb{Z}$. For, the difference of any two integers each divisible by m is again divisible by m; and if we take an integer divisible by m and multiply it by an arbitrary integer the product is divisible by m.

Example 9.2.3 (continued). In the ring $\mathbb{Z}[x]$ there are many ideals. We describe two of them. First consider the polynomials with zero constant term. It is easy to see that these form an ideal. Second consider the polynomials with even constant term. Again, it is not difficult to see that these polynomials form an ideal. The reader may well construct further ideals in $\mathbb{Z}[x]$.

Exercise 6

(i) Construct some ideals in $\mathbb{Q}[x]$.
(ii) Prove that the set Φ_m (see 10.2.5) is *not* an ideal (or sub-ring) of $\mathbb{Z}_m$.
(iii) Consider the ring R^S of all functions $S \to R$ (where R is a ring; see Example 9.2.8), and choose $s \in S$. Prove that $\{f \in R^S | f(s) = 0\}$ is an ideal.
(iv) Given a family of ideals in a ring, prove that their intersection is an ideal.

11.4.2 Now in any commutative ring R there are always two ideals, namely R itself and the ideal consisting of the zero element alone. Any other ideal in R is called **proper.**

11.4.3 PROPOSITION. *There are no proper ideals in a field F.*

We leave the proof to the reader, who should also notice that if I is an ideal of R and R possesses a unity element 1, then $I = R$ if and only if $1 \in I$. ■

Exercise 7

Show that the assertion of Exercise 6(iv) becomes false if we replace 'ideal' by 'proper ideal'.

11.4.4 Now let R be a commutative ring with unity element, let a be a fixed element of R and let I consist of all elements of R of the form ra as r ranges over R. Then the reader will observe that I is an ideal; it is called the **principal ideal generated by** a; of course $a \in I$, since $a = 1a$. We write $I = (a)$.

Exercise 8

(i) Prove that, in R, the ideal (a) is the intersection of all ideals $K \subseteq R$ such that $a \in K$. (Use Exercise 6(iv).)
(ii) Let $K_s \in R^S$ be the ideal in Exercise 6(iii). Define a function $\eta: R^S/K_s \to R$ by the rule: $\eta[f] = f(s)$, checking independence of representatives, and prove that η is an isomorphism. Hence prove that, when R is a field, K_s is **maximal**, in the sense that the only ideals containing K_s are K_s and R^S.

11.4.5 In the examples given prior to 11.4.3 the first two ideals are principal. [Of course, the improper ideals R and (0) are principal, being generated by 1 and 0 respectively.] Thus $m\mathbb{Z}$ is generated by m, and, in $\mathbb{Z}[x]$, the ideal of polynomials with zero constant term is generated by the polynomial x. On the other hand the ideal J in $\mathbb{Z}[x]$, of polynomials with even constant term, is *not* principal. This is seen by observing (i) that if a polynomial f did generate J, then f must be a factor of 2, since 2 (regarded as a polynomial) lies in J; (ii) the only factors of 2 are ± 1, ± 2; (iii) ± 1 do not belong to J; (iv) ± 2 do not generate the ideal since, for example, $x + 2$ is in J and is not divisible by ± 2 in $\mathbb{Z}[x]$. We have given here the steps of a rigorous proof of our assertion;

and the reader may remark that each step is obvious, indeed trivial. So, however, appear the steps of most proofs when they are properly understood! On the other hand we will soon see that in $\mathbb{Q}[x]$, as distinct from $\mathbb{Z}[x]$, *every* ideal is principal. This follows from the main theoretical result of this section, namely:

11.4.6 THEOREM. *Every ideal in a Euclidean domain is principal.*

Proof. Let I be an ideal in the Euclidean domain R. We wish to show that there exists an element $a \in I$ such that I simply consists of multiples of a; that is, we show that $I = (a)$.

Now if I is the zero ideal, then certainly I is principal, since then $I = (0)$. Thus we may suppose that $I \neq (0)$ and we consider the non-empty subset $S = \phi(I - (0))$ of $\mathbb{N}$. By the well-ordering (Chapters 4, 5) of $\mathbb{N}$ it follows that the set S has a smallest element, s, say. Let $a \in I - (0)$ be such that $\phi(a) = s$. We prove that $I = (a)$. Since $a \in I$ it follows that $(a) \subseteq I$; thus it only remains to show that if $b \in I$ then $b = qa$ for some $q \in R$. Now $a \neq 0$; thus by the Euclidean property $\mathfrak{N}_3$ (with the roles of a, b interchanged!) it follows that $\exists q, r \in R$ with $b = qa + r$ and $\phi(r) < \phi(a)$. Now $r = b - qa$. Since $a \in I$, then by $\mathfrak{I}_2$, $qa \in I$; and since $b \in I$ then by $\mathfrak{I}_1$, $r = b - qa \in I$. Suppose $r \neq 0$. Then $r \in I - (0)$ and $\phi(r) \in S$, $\phi(r) < s$. This contradicts the minimality of s, so the hypothesis $r \neq 0$ is false. It follows that $r = 0$, $b = qa$ and all is proved. ■ The reader is strongly advised to follow through this argument in the special case when R is the Euclidean domain $\mathbb{Z}$.

Knowing now that $I = (a)$, we still have the problem of trying to find a, given I. This problem is discussed in 11.6. We first deduce some properties of the generator of an ideal in a Euclidean domain.

11.5 HCF's

Let R be an integral domain. We are now going to consider problems of factoring elements in R. Thus, if $u, v \in R$, then we call u a **factor** of v, and write

$$u|v \quad (\textit{read: } `u \text{ divides } v')$$

whenever there exists $w \in R$ such that $uw = v$; if w is not a unit, then u is a **proper** factor. By $\mathfrak{A}_8$, $u|u$, while if $u|v$ and $v|r$, then $u|r$. [*Proof.* $\exists w, x$ such that $uw = v$, $vx = r$, whence $r = (uw)x = u(wx)$ by $\mathfrak{A}_5$. ■] Thus, the relation of being a factor is reflexive and transitive. (Is it an order relation?)

Exercise 9

(i) Use $\mathfrak{A}_8$ and 9.3.1(iv) to prove that for all $r \in R$, $1|r$ and $r|0$.

(ii) Prove that, if $u|a$ and $u|b$ in R, then $u|(a - b)$ and $u|ac$ for any $c \in R$. Hence prove that the set K of all $x \in R$ of which u is a factor, is an ideal in R, and that in fact $K = (u)$.

(iii) Give examples to show that the relation of being a factor is not symmetric. Prove that if $a|b$ and $b|a$, then a and b are associates (11.3.1); and conversely, if $a \sim b$, then $a|b$ and $b|a$.

In $\mathbb{Z}^+$, the highest common factor of a given pair of integers m, n, is frequently defined as the *biggest* integer k such that $k|m$ and $k|n$. This definition uses the fact that $\mathbb{Z}^+$ is ordered by size, a fact which is largely irrelevant to the problem of factorization, since many interesting rings are not ordered. We adopt a different definition, applicable in other situations.

11.5.1 DEFINITION. *We will call an element $d \in R$ a* **highest common factor (hcf)** *of the elements $a_1, \ldots, a_n$ of R if*

(i) *for each $i = 1, \ldots, n$, we have $d|a_i$, and*
(ii) *if $q \in R$ and $q|a_i$, $i = 1, \ldots, n$, then $q|d$.*

We have not yet shown (and it is not true!) that in every integral domain a finite set of elements has an hcf. But we will straight away show that *if* it has an hcf, then that hcf is unique. This assertion rests on the following lemma given above as an exercise. (Recall from 11.3.1 the relation of being associated.)

11.5.2 LEMMA. *Let R be an integral domain and let $a, b \in R$. Then a and b are associates if and only if $a|b$ and $b|a$.*

Proof. If a and b are associates then $b = au$, $a = bv$, so that $a|b$ and $b|a$. Conversely, let $a|b$ and $b|a$. If $a = 0$, then $b = 0$, and, of course, if $b = 0$, then $a = 0$. Then either $a = b = 0$—in which case a and b are associates —or, as we will henceforth assume, a and b are both non-zero. Then since $a|b$ and $b|a$, there are elements q, r in R such that $b = aq$, $a = br$. Hence (by associativity!) $b = brq$, so $b(1 - rq) = 0$ (by distributivity). Now since R is an integral domain, a product can only be zero if one of the factors is zero. Since $b \neq 0$, we conclude that $rq = 1$, so that r and q are units and a and b are associates. ∎

As a corollary we prove the *uniqueness theorem for hcf's.*

11.5.3 COROLLARY. *Let $a_1, a_2, \ldots, a_n$ be elements of R admitting a non-empty set H of hcf's and let $d \in H$. Then H is precisely the set of associates of d in R.*

Proof. Let K denote the set of associates of d in R. We must prove that $K = H$. The reader will quickly convince himself that, if d' is an associate of d, then d' certainly satisfies conditions (i) and (ii) of 11.5.1, postulated for d. Hence $K \subseteq H$. Conversely, if d' does satisfy conditions (i) and (ii), then from condition (i) for d and condition (ii) for d' we infer that $d|d'$. Similarly $d'|d$, so that d and d' are associates by Lemma 11.5.2. Hence $H \subseteq K$, so $H = K$. ∎

The following important theorem attests the *existence* of hcf's in Euclidean domains. We first observe that if R is an integral domain and $a_1, a_2, \ldots, a_n$

is a set of elements of R, we may consider the subset of R consisting of all elements $r_1a_1 + r_2a_2 + \cdots + r_na_n$ with $r_i \in R$. The reader can quickly verify that this subset is an ideal I in R. By an obvious extension of the terminology and notation for the case $n = 1$, we say that I is the ideal **generated** by the elements $a_1, a_2, \ldots, a_n$ and write $I = (a_1, a_2, \ldots, a_n)$. [We may even consider infinite subsets $\{a_i\}$ of R, but we will not enter into this generalization here.] This notion is a key to the theory of hcf's in Euclidean domains.

11.5.4 THEOREM. *Let $a_1, a_2, \ldots, a_n$ be elements of the Euclidean domain R. Then they possess an hcf, d, and, moreover, d is expressible in the form*

11.5.5 $$d = s_1a_1 + \cdots + s_na_n, \qquad s_i \in R.$$

Proof. The second assertion of the theorem gives the clue to its proof. We consider the ideal I generated by $a_1, a_2, \ldots, a_n$, $I = (a_1, a_2, \ldots, a_n)$. By Theorem 11.4.6, I is principal, say $I = (d)$. Of course d is of the form 11.5.5, so it remains to show that d satisfies conditions (i), (ii) of 11.5.1, for any hcf. But (i) follows immediately, since $a_i \in (d)$, and (ii) is an immediate consequence of 11.5.5. ■

We note that by this theorem we have proved more than is traditionally asserted for hcf's. For it is rare for students to be acquainted with the crucial fact that d is expressible in the form 11.5.5. We will give a decisive application to the theory of partial fractions in Chapter 13. Now let us justify the term 'highest', at least in $\mathbb{Z}$. We know from 11.5.4 that a non-empty set $a_1, \ldots, a_n$ of integers has a non-empty set H of hcf's; and if $d \in H$, then by 11.5.3 all other members d' of H are associates of d. Now, the only units in $\mathbb{Z}$ are 1 and -1, so d' must be $-d$. Thus, H has exactly the two elements d, $-d$ of which we suppose $d > 0$. All other common divisors of $a_1, \ldots, a_n$ divide d, so d is the biggest such (by 9.1.1 and 9.1.5). It is customary, when talking about $\mathbb{Z}$, to refer to d (and not $-d$) as 'the' hcf.

Exercise 10

(i) Find the hcf of (i) 9, 15; (ii) 28, 36; (iii) 120, 504, 700; and in each case, express the hcf as a linear combination of the numbers concerned, with integer coefficients.
(ii) Let R be a Euclidean domain and let $a, b \in R$ with hcf $(a, b) = 1$. Show that if $a|bc$, then $a|c$.
(iii) Prove that, in any ring R, the ideal generated by $a_1, \ldots, a_n$ is the intersection of all ideals $I \subseteq R$ such that $a_1, \ldots, a_n \in I$.
(iv) Using Lemma 11.5.2, prove that the ideals (a), (b) coincide (in an integral domain) if and only if a and b are associates. [This result suggests that, in a sense, *the* hcf should really be thought of as an ideal and not an element. This point of view is strongly recommended to those interested in extending the notion of hcf (and other arithmetical notions) to more general situations. It is reinforced by the next exercises.]
(v) In the integral domain R, let $[x]$ denote the coset of all associates of x, and let S denote the set of all such cosets. Show that S is closed under the multiplication given by $[x][y] = [xy]$, by checking independence of representatives.

(vi) Let J denote the family of all ideals of R. Define a function $f: S \to J$ by $f[x] = (x)$. Check independence of representatives, and show that f is one-one.

11.6 Euclid's Algorithm

As far as the theory has gone there is still one serious gap; namely, we have given no rule for calculating hcf's. We proceed now to describe a procedure (or routine) which yields the hcf in a finite number of steps. This fascinating procedure, essentially due to Euclid in the case of the integers, is called the **Euclidean algorithm**; but in its general application to Euclidean domains it is not (without further refinement) strictly algorithmic. However, the reader will easily be able to refine it in any particular Euclidean domain to yield a precise procedure capable of being programmed for a computer to yield the hcf in a finite number of operations. He should try writing out a 'flow diagram' for such an algorithm (i.e., a list of instructions, in the proper order).

Thus we suppose given a Euclidean domain R and we will describe how to obtain the hcf of two elements of R, which we will call a_0 and a_1 and which we may suppose to satisfy $\phi(a_0) \geqslant \phi(a_1) > 0$. Applying the Euclidean property $\mathfrak{R}_3$, we deduce that there are elements q_1, a_2 in R such that

$$a_0 = q_1 a_1 + a_2 \quad \text{and} \quad \phi(a_2) < \phi(a_1).$$

If $a_2 = 0$, the algorithm† terminates; otherwise we repeat the process to obtain elements q_2, a_3 such that

$$a_1 = q_2 a_2 + a_3 \quad \text{and} \quad \phi(a_3) < \phi(a_2).$$

We continue in this way, obtaining a sequence of elements $a_0, a_1, a_2, \ldots$ such that $\phi(a_i) > \phi(a_{i+1})$, $i = 1, 2, \ldots$. *Eventually the process must terminate*; for every strictly decreasing sequence of positive integers—and in particular the sequence $\phi(a_1), \phi(a_2), \ldots,$—must arrive at zero in a finite number of steps. Hence, we must eventually arrive at the equality

$$a_{r-1} = q_r a_r.$$

We have thus constructed the sequence of relations

$$\text{11.6.1} \qquad \begin{cases} a_0 = q_1 a_1 + a_2, & \phi(a_2) < \phi(a_1) \\ a_1 = q_2 a_2 + a_3, & \phi(a_3) < \phi(a_2), \\ \vdots & \\ a_{r-2} = q_{r-1} a_{r-1} + a_r, & \phi(a_r) < \phi(a_{r-1}), \\ a_{r-1} = q_r a_r. & \end{cases}$$

† I.e., the process of dividing and obtaining a remainder. The word 'algorithm' appears to derive from Al-Khwarizmi, a celebrated Arabic mathematician.

From this display, we can sketch the proof of:

11.6.2 THEOREM. *The element a_r is the hcf of a_0 and a_1.*

Proof. We must first verify that a_r divides a_0 and a_1. This we prove by climbing *up* the sequence 11.6.1. By the last equation a_r divides a_{r-1} and a_r. By the previous one, it therefore divides a_{r-2} and a_{r-1} and so on up, till we finally infer that a_r divides a_0 and a_1.

Secondly, we must show that if h divides a_0 and a_1, then h divides a_r. This we prove by climbing *down* the sequence 11.6.1. By the first equation h divides a_1 and a_2; by the next, it divides a_2 and a_3; and so on down until the penultimate equation shows that h divides a_r. ■ Observe that each of our two uses of 'and so on' covers an (easy) inductive proof.

Exercise 11

(i) Find the hcf of (i) 732, 20; (ii) -28, 36 by the Euclidean algorithm.
(ii) Adapt the Euclidean algorithm to deal with the hcf of three elements of R, and hence find the hcf of 120, 504, and 700.
(iii) Use the Euclidean algorithm to show that the hcf of a and b is a linear combination of a and b. [See also 9 Exercise 7(ix)].

The Euclidean algorithm, as described, involves possible choices at each stage of the process, since $\mathfrak{N}_3$ does not assert that q and r are *uniquely* determined by a, b and the relations $a = bq + r$, $\phi(r) < \phi(b)$. For example, in $\mathbb{Z}$, if $a = 5$, $b = 10$ we may take $q = 0$, $r = 5$ or $q = 1$, $r = -5$. In the case of the ring $\mathbb{Z}$, a rule may be given for choosing q and r uniquely (for example, that r should be non-negative). The reader is invited to formulate such a rule in the ring $\mathbb{Q}[x]$.

11.6.3 EXAMPLE. We now carry out the Euclidean algorithm in the ring $\Gamma = \Gamma_1$ of Gaussian integers (see 11.2.4). Let us consider the Gaussian integers $38 - 8\mathrm{i}$ and $11 - 3\mathrm{i}$. We carry out the process described in proving that Γ is a Euclidean domain. Thus

$$\frac{38-8\mathrm{i}}{11-3\mathrm{i}} = \frac{(38-8\mathrm{i})(11+3\mathrm{i})}{130} = \frac{442+26\mathrm{i}}{130} = \frac{17+\mathrm{i}}{5} = 3 + \frac{2+\mathrm{i}}{5}$$

(notice that $|\frac{2}{5}| \leqslant \frac{1}{2}$, $|\frac{1}{5}| \leqslant \frac{1}{2}$). Thus as the first stage we have

$$38 - 8\mathrm{i} = 3(11 - 3\mathrm{i}) + 5 + \mathrm{i}.$$

Then
$$\frac{11-3\mathrm{i}}{5+\mathrm{i}} = \frac{(11-3\mathrm{i})(5-\mathrm{i})}{26} = \frac{52-26\mathrm{i}}{26} = 2 - \mathrm{i}.$$

Therefore the algorithm is complete and $5 + \mathrm{i}$ is the hcf. However, as the reader may well imagine, it may be a long and tedious business to carry out the Euclidean algorithm in any particular instance. It is often possible to arrive at the answer more rapidly by a method of guesswork which presupposes familiarity with the process of factorizing ordinary integers. The essential observation which guides the guesswork we now state as a proposition.

11.6.4 PROPOSITION. *Let ϕ be a norm on the integral domain R. If $a, b \in R$ and $a|b$, then $\phi(a)|\phi(b)$.*

We simply apply $\mathfrak{N}_2$. ■

It follows, of course, from Proposition 11.6.4 that if d is the hcf of $a_1, a_2, \ldots, a_r$ in R, and if h is the hcf of $\phi(a_1), \phi(a_2), \ldots, \phi(a_r)$, regarded as elements of $\mathbb{Z}$, then $\phi(d)|h$.

Example 11.6.3 (continued). Now let us again consider the hcf of $38 - 8\text{i}$ and $11 - 3\text{i}$. Then $\phi(38 - 8\text{i}) = 1508$, $\phi(11 - 3\text{i}) = 130$ and the hcf of 1508 and 130 is 26. Thus if d is the hcf of $38 - 8\text{i}$ and $11 - 3\text{i}$, $\phi(d)|26$, so $\phi(d) = 26, 13, 2$, or 1. This gives us a (finite) set of Gaussian integers to test experimentally as potential hcf of $38 - 8\text{i}$ and $11 - 3\text{i}$. We recall that the units of Γ are $\pm 1, \pm \text{i}$, so that there are really four hcf's of any set of elements of Γ or, to put it another way, the number of possible choices for d is reduced by a factor of 4 from what it first appears to be. In fact we need only test as possible values of d the elements $5 \pm \text{i}$, $3 \pm 2\text{i}$, $1 \pm \text{i}$, and 1. We try these out (in this order) as possible factors of $11 - 3\text{i}$ (this is a process quite like that of using intelligent guesswork to look for factors of quadratic polynomials in $\mathbb{Z}[x]$, a process familiar to most readers). We find that, in fact, $5 + \text{i}$ is a factor. We finally show that $(5 + \text{i})|(38 - 8\text{i})$, and this *proves* that $5 + \text{i}$ is the hcf.

Exercise 12

Use the 'guessing' method to find the hcf of $38 - 8\text{i}$ and $11 - 13\text{i}$; and verify your answer by applying the Euclidean algorithm.

11.6.5 We remark that we may use Theorem 11.5.4 to show that certain domains are *not* Euclidean. Consider, for example, Γ_3, the ring of complex numbers $a + b\sqrt{-3}$, where $a, b \in \mathbb{Z}$; and consider the elements 2 and $1 + \sqrt{-3}$ in Γ_3. Since $\phi(2) = \phi(1 + \sqrt{-3}) = 4$ we see that if d was an hcf of 2 and $1 + \sqrt{-3}$, then $\phi(d)|4$, so $\phi(d) = 4, 2$, or 1. This shows that, essentially, $d = 2, 1 \pm \sqrt{-3}$, or 1 (note that the units of Γ_3 are ± 1, and that the equation $\phi(x) = 2$ has no solution in Γ_3). Now 2 is *not* a factor of $1 + \sqrt{-3}$, since the complex number $(1 + \sqrt{-3})/2$ does not belong to Γ_3. Similarly $1 \pm \sqrt{-3}$ is not a factor of 2. Thus if 2 and $1 + \sqrt{-3}$ have an hcf, it is 1. In fact this argument has shown that if *any* element of Γ_3 divides both 2 and $1 + \sqrt{-3}$ then it is ± 1, so that 1 is indeed the hcf.

Now if Γ_3 were Euclidean we know from 11.5.5 that we would be able to express 1 in the form $1 = 2\alpha + (1 + \sqrt{-3})\beta$, with $\alpha, \beta \in \Gamma_3$. Then $2 = 4\alpha + 2(1 + \sqrt{-3})\beta$. But $1 + \sqrt{-3}$ *is* a factor of 4, so this equation shows that $1 + \sqrt{-3}$ is a factor of 2, which we have already observed to be false. We have arrived at a contradiction; hence Γ_3 *is not Euclidean.* Note that we showed that 1 is the hcf of 2 and $1 + \sqrt{-3}$ by using the particular norm ϕ of Example 11.2.4; but the conclusion is that *no* norm on Γ_3 could be Euclidean. The reader may produce a similar argument to show that Γ_5 is not Euclidean.

11.7 LCM's

We now turn briefly to the question of lcm's. We will confine attention almost entirely to the case of the lcm of two elements in a Euclidean domain. We first prove two propositions.

11.7.1 PROPOSITION. *Let a, b be elements of an integral domain R with d as hcf. Then if $a = da'$, $b = db'$, the elements a' and b' have* 1 *as hcf.*

Proof. If $x|a'$, $x|b'$, then $dx|a$, $dx|b$, so $dx|d$ and x is a unit. ∎

11.7.2 PROPOSITION. *Let a, b be elements of a Euclidean domain R and let their hcf be* 1. *Then if $a|m$ and $b|m$, it follows that $ab|m$.*

Proof. By 11.5.5 we may find elements u, v in R such that $1 = au + bv$. Then $m = mau + mbv$. Since $b|m$, $ab|am$; and since $a|m$, $ab|mb$. Thus $ab|(mau + mbv)$, so $ab|m$. ∎ The reader should note that Proposition 11.7.2 is not true for arbitrary integral domains; for example, it fails in Γ_3 or Γ_5. [Why?]

11.7.3 We are now in a position to prove the main theorem on lcm's in a Euclidean domain. We will call an element $l \in R$ a **least common multiple (lcm)** of the elements $a_1, a_2, \ldots, a_n$ of R if

(i) $a_i|l$, $i = 1, \ldots, n$, *and*
(ii) if $a_i|m$, $i = 1, \ldots, n$, *then* $l|m$.

Just as for hcf's, we immediately prove from the definition that the lcm, if it exists, is unique up to multiplication by a unit. The main theorem we prove is:

11.7.4 THEOREM. *Let a, b be two non-zero elements of the Euclidean domain R. Then they possess an lcm l. Moreover $ab = dl$, where d is an hcf of a and b.*

[We refer to 'an' hcf since, for an arbitrary choice d of hcf, dl might differ from ab by a unit.]

Proof. Let us write $a = a'd$, $b = b'd$. Then the theorem effectively asserts that the element $ab'(=a'b)$ is an lcm of a and b. Plainly $a|ab'$ and $b|a'b$, so (i) is satisfied. Now let $a|m$, $b|m$. Then $m = ag = a'dg$, and $m = bh = b'dh$, for some elements g, h of R. Since R is an integral domain, it follows that $a'g = b'h$. Call this element m'. By Proposition 11.7.1, a', b' have hcf 1 and by Proposition 11.7.2 $a'b'|m'$. Thus $a'b'd|m$, verifying condition (ii). ∎

We can now prove that every finite set of elements of a Euclidean domain possesses an lcm, by induction on the number of elements in the set, using:

11.7.5 PROPOSITION. *Let $a_1, a_2, \ldots, a_n, b$ be elements of the integral domain R, let l' be the lcm of $a_1, a_2, \ldots, a_n$ and let l be the lcm of l' and b. Then l is the lcm of $a_1, a_2, \ldots, a_n, b$.*

We leave the proof as an exercise. ■ But we remark, emphatically, that a much neater approach to the question of the existence of lcm's in a Euclidean domain is to use Theorem 11.4.6, and we invite the reader to prove their existence that way.

Exercise 13

(i) Find the lcm of $38 - 8i$ and $11 - 13i$.
(ii) Find the lcm of $x^4 - 11x^3 + 45x^2 - 81x + 54$ and $x^3 - 2x^2 - 9x + 18$ in $\mathbb{Q}[x]$ and in $\mathbb{R}[x]$.
(iii) Let R be an integral domain and let R_0 be a subset of R which is also an integral domain with respect to the laws of addition and multiplication in R. Let R_0 be Euclidean. Show that if $a_1, \ldots, a_n$ belong to R_0, then their hcf as elements of R exists, and coincides with their hcf as elements of R_0. Prove the corresponding statement for lcm's, confining attention to the lcm of *two* elements of R_0.

Chapter 12

FACTORIZATION INTO PRIMES

Traditionally, the theory of hcf's and lcm's is developed—if so imposing a phrase may appropriately be applied—from the technique of factorizing whole numbers as a product of prime numbers. We have given, in the preceding chapter, a treatment which is quite independent of such factorization, and we hope the reader will have satisfied himself that the Euclidean algorithm, Theorem 11.5.4 and formula 11.5.5 give a clearer insight into the nature of hcf's than is provided by factorization into primes. However, it is, of course, a matter of profound importance that certain numbers are prime, that all whole numbers may be expressed as a product of prime numbers, and that such expressions are essentially unique. We proceed in this chapter to give precise meanings to these notions, and to prove them in a Euclidean domain. The final section of the chapter is devoted to transferring the theory from the Euclidean domain $\mathbb{Q}[x]$ to the non-Euclidean domain $\mathbb{Z}[x]$—a delicate operation often neglected in elementary and traditional treatments.

12.1 Prime Numbers

The reader will surely agree that, for the positive integers, the characteristic property of a prime number is that it cannot be *properly* factorized (see the first paragraph of 11.5). That is, a **prime number** is habitually defined to be any integer > 1 whose only factors are itself and 1. First, we shall prove a theorem of Euclid, which guarantees the existence of an infinity of prime numbers. We will lean heavily on the property that $\mathbb{Z}$ is ordered by size, a fact which makes the theorem very unusual among the rest in this chapter; the latter require different methods, since most rings are not ordered.

12.1.1 THEOREM. (Euclid.) *In $\mathbb{Z}$, the set of primes is infinite.*

Proof. We know that 2 is a prime number, so let us suppose that we have written down all the first n prime numbers as $2 = p_1$, $3 = p_2$, $5 = p_3, \ldots, p_n$ $(n > 2)$ and use induction on n. We first suppose that *there are no more primes*, to get a contradiction. Let $N = p_n! + 1$. If the italicized supposition were correct, then in particular N would not be prime; so N has a *proper* factor N_1. Now in $\mathbb{Z}^+$, if a is a *proper* factor of b, then $a < b$; whence $N_1 < N$. If N_1 is not a prime number, then it has a proper factor $N_2 < N_1$ (so $N_2 | N$). Since such a decreasing sequence of integers in $\mathbb{Z}^+$ must terminate [proof by induction!], we find a proper factor N_r of N which is a prime number. Hence, N_r is one of the finite list of primes above (by the italicized supposition), say $N_r = p_k$. Thus $p_k|N$, while of course $p_k|p_n!$, so $p_k|(N - p_n!)$ or $p_k|1$. This contradicts $p_k \geqslant 2$, so our supposition above was incorrect. Therefore there is a least prime number q between $p_n + 1$ and N; and q is the required $(n + 1)$st prime number p_{n+1}. The proof is now complete, by induction. ■ The reader may notice that we have been particularly scrupulous in this proof, not wishing to *assume* that if an integer has a proper factor, then it has a proper prime factor.

12.2 Irreducibility and Primes

Now let us consider what we should mean by a prime in a general integral domain R where no order relation may be available to help in proofs. The property of having no proper factors is what we prefer to call *irreducibility*; and we make the following definition (recall from 11.3 the definition of a *unit*).

12.2.1 DEFINITION. A non-zero element a of the integral domain R is called **irreducible** if it is not a unit and if the equation $a = bc$ implies that b or c is a unit.

That is to say, any factorization of a is trivial in that it simply expresses a as the product of an associate (see 11.3.1) of a by the appropriate unit. Plainly the irreducible elements of $\mathbb{Z}$ are just the numbers $\pm p$, where p is an arbitrary prime number in the habitual sense described in 12.1. Notice that the definition excludes units from the set of irreducible elements; this proves to be a very useful convention.

Now there is another characteristic property of prime numbers. Namely, suppose p is a prime number and is a factor of the product gh. Then (see 10.2.3) p must be a factor of g or h. We will, in fact, regard *this* property of prime numbers as the fundamental property and accordingly make the following definition.

12.2.2 DEFINITION. A non-zero element p of the integral domain R is **prime** if it is not a unit and if the relation $p|gh$ implies that $p|g$ or $p|h$.

As we have said, the notions 'prime' and 'irreducible' coincide in the ring $\mathbb{Z}$. We will shortly (see 12.2.4) give an example of an integral domain in which they do not coincide. However we first prove

12.2.3 THEOREM. *In an integral domain every prime element is irreducible. In a Euclidean domain every irreducible element is prime.*

Since $\mathbb{Z}$ is a Euclidean domain this theorem includes the remark that the two notions 'prime' and 'irreducible' coincide in $\mathbb{Z}$; it also includes the probably familiar fact that they coincide in $\mathbb{Q}[x]$, $\mathbb{R}[x]$, and $\mathbb{C}[x]$. It does not include the fact that they coincide in $\mathbb{Z}[x]$. We now prove the theorem.

Proof of 12.2.3. Let R be an integral domain, let p be prime and suppose that $p = bc$; we must show that b or c is a unit (so that the other is associated with p). Now if $p = bc$, certainly $p|bc$. Thus, p being prime, $p|b$ or $p|c$. Suppose $p|b$. But we have also $b|p$. By Lemma 11.5.2, p is associated with b, so c is a unit.

We now turn to the converse. Here we take R to be a Euclidean domain and we suppose that a is irreducible, that $a|bc$, but that $a\nmid b$; we must show that $a|c$. The argument is considerably more subtle than that above and the reader should follow it carefully with a familiar special case in mind.

We first observe that since R is a Euclidean domain, the elements a and b have an hcf. In fact we prove that their hcf is a unit. For if d is the hcf, then $a = da'$, $b = db'$, say. Since a is irreducible, d or a' is a unit. But a' is not a unit; for, if it were, d would be associated with a and so we would infer $a|b$, contrary to hypothesis. Thus a' is not a unit, and so d is a unit; that is, the hcf of a and b is a unit as claimed.

Now by 11.5.5 it follows that there are elements k, l in R such that $1 = ka + lb$. Then $c = kac + lbc$. But $a|ac$ and $a|bc$, so that $a|c$ and the theorem is proved. ■

12.2.4 (EXAMPLE 9.2.7 CONTINUED.) Let us consider the familiar example of Γ_3 which we already know not to be Euclidean (see 11.6.5). In Γ_3 we have the relation $2\cdot 2 = (1 + \sqrt{-3})(1 - \sqrt{-3})$. We prove that all the elements $2, 1 \pm \sqrt{-3}$ are irreducible but *none is prime*. Plainly 2 is not prime; for $2|(1 + \sqrt{-3})(1 - \sqrt{-3})$ but $2\nmid(1 \pm \sqrt{-3})$. Similarly $1 \pm \sqrt{-3}$ is not prime. On the other hand, suppose that $2 = bc$, where $b, c \in \Gamma_3$. Then $4 = \phi(2) = \phi(b)\phi(c)$. Since the equation $\phi(x) = 2$ has no solutions in Γ_3, then $\phi(b) = 1$ or $\phi(c) = 1$. If $\phi(b) = 1$, then $b = \pm 1$, so b is a unit. Thus 2 is irreducible; and similarly $1 \pm \sqrt{-3}$ is irreducible. Notice that the question whether an element is prime (or irreducible) depends on the integral domain to which it is regarded as belonging. Thus 2 is surely prime in $\mathbb{Z}$ but not in Γ_3. Indeed 2 is prime (and irreducible) in $\mathbb{Z}$, but is not prime (and not irreducible) in $\Gamma(=\Gamma_1)$, since $2 = (1 + i)(1 - i)$. We will return later to this important remark, but record now our experience that it is virtually never appreciated by students who claim to be familiar with the technique of factorizing polynomials. Indeed, *the question* 'Has the polynomial $x^2 + 1$ any factors?' *is strictly meaningless until one specifies the domain of coefficients.*

Exercise 1

(i) Let R be an integral domain. Let p be prime in R, and let $p|a_1a_2\cdots a_n$. Show that p divides some a_i, $1 \leqslant i \leqslant n$.

§(ii) Prove that, in $\mathbb{Z}^+$, every prime of the form $4k + 1$ can be expressed as the sum of two squares. The steps of the proof are as follows. (1) Show that if $p = 4k + 1$ the equation $x^{p-1} - 1 = 0$ has exactly $4k$ roots in the field $\mathbb{Z}_p$ (use 10.3.1); thus since $x^{2k} - 1 = 0$ has at *most* $2k$ roots (see 22.3.5), the equation $x^{2k} + 1 = 0$ has at *least* $2k$ roots in $\mathbb{Z}_p$. (2) There is then an integer a, $(0 < a < p)$, such that $p|(a^2 + 1)$. Consider this relation in the integral domain Γ_1 of 11.2.4. If p were prime in Γ_1, then $p|(a + i)$ or $p|(a - i)$, by the previous exercise; but this is clearly absurd. (Why?) Hence p is not prime in Γ_1, so $p = (x + iy)(u + iv)$ in Γ_1 where, by 12.2.3, neither factor is a unit of Γ_1. (3) Taking norms, we have $p^2 = (x^2 + y^2)(u^2 + v^2)$, so we have $p = x^2 + y^2 = u^2 + v^2$, since $x^2 + y^2 \neq \pm 1$, $u^2 + v^2 \neq \pm 1$.

(iii) Show that the expression of p as a sum of two squares given in (ii) above is essentially unique. [A similar proof, based on the skew-field of quaternions, shows that if p is of the form $4k + 3$, then p is expressible (essentially uniquely), as a sum of four squares (see Herstein [56]). And since every integer is a product of primes, it follows that *every positive integer is the sum of four squares,* since the product of two such sums is a third. This is **Lagrange's theorem.** Another, more elementary but less conceptual proof can be found in Davenport [28]. All this is part of **Waring's problem**: given n, to find a k such that every positive integer is the sum of k nth powers. For example, if $n = 3$, $k = 57$. A related problem is that of deciding in how many ways a given integer can be expressed as a sum of squares, and the solution uses an analogy with the theory of electric circuits. See Gardner [39].]

12.3 Existence and Uniqueness of Prime Factorization

We now prove the main result of this chapter, which we divide into two parts. We say that the non-zero element $a \in R$ **admits a factorization into primes** if there exists an equality $a = up_1 \cdots p_n$ where u is a unit and $p_1, \ldots, p_n$ are primes; the equality itself, once known, is called a **factorization of a into primes.** For convenience we regard a unit u as 'factorized into primes' by $u = u$ (with $n = 0$).

12.3.1 THEOREM. *In a Euclidean domain R every non-zero element admits a factorization into primes.*

Proof. We have a Euclidean norm $\phi: R \to \mathbb{Z}^+$ and we argue by induction on $\phi(a)$. Thus if $\phi(a) = 1$, a is a unit (Theorem 11.3.3) and so $a = a$ is a factorization into primes. Now suppose the assertion proved for all elements $r \in R$ with $\phi(r) < n$ and let $a \in R$ be such that $\phi(a) = n$. Then either a is prime, when $a = 1a$ is a factorization, or a is not prime and so not irreducible (Theorem 12.2.3). In the latter case $a = bc$, with neither b nor c a unit. Then $\phi(b) < n$, $\phi(c) < n$ (Corollary 11.3.4) so that, by the inductive hypothesis, we may find expressions

$$b = up_1 \cdots p_l, \qquad c = vq_1 \cdots q_m,$$

where u, v are units and $p_1, \ldots, p_l, q_1, \ldots, q_m$ are primes. Then

$$a = bc = uvp_1 \cdots p_lq_1 \cdots q_m,$$

and this is a factorization of a into primes (since uv, as a product of units, is itself a unit). This establishes the inductive hypothesis and proves the theorem. ■ This theorem, whose proof the reader should check by considering the familiar cases $R = \mathbb{Z}$, $R = \mathbb{Q}[x]$, asserts the *existence* of factorizations. Their *uniqueness* (in so far as uniqueness is possible) is attested by the following theorem.

12.3.2 THEOREM. *Let R be an integral domain and let*

$$up_1 \cdots p_m = vq_1 \cdots q_n$$

in R, where u, v are units of R and $p_1, \ldots, p_m, q_1, \ldots, q_n$ are primes in R. Then $m = n$ and, perhaps after renumbering the q's, p_i and q_i are associates for each i, $1 \leqslant i \leqslant m$.

Before proceeding to the proof, we explain the statement of the theorem by an example. In $\mathbb{Z}$ we may consider two factorizations of 12, namely

$$12 = 1 \cdot 2 \cdot (-3) \cdot (-2) = (-1) \cdot (3) \cdot (-2) \cdot (2).$$

These factorizations do not, however, differ essentially, since we may rearrange the factors in the second factorization in such a way that the factorizations differ, term by term, from each other at most by multiplication by -1. Furthermore it is clear that we could not possibly ask for more than this in the way of uniqueness, since we may obviously change the order of the primes in a factorization $a = up_1 \cdots p_n$ and then multiply each one by an arbitrary unit provided only that we adjust the 'unit factor' u accordingly. (In the case of $\mathbb{Z}$, the units are ± 1, so Theorem 12.3.2 asserts the maximum which we could hope to be true.)

Proof of 12.3.2. Let us write $\min(m, n)$ for the smaller of m and n, and argue by induction on $\min(m, n)$. Indeed there is no real loss of generality in supposing that $m \leqslant n$, so that we argue by induction on m—this is convenient just for notational simplicity. If $m = 0$, then $u = vq_1 \cdots q_n$; but if a product of elements is a unit, each element is a unit (proof as exercise!), so $n = 0$ and the theorem is proved in this case.

We deal with larger m by the method of induction, and must therefore make an 'inductive hypothesis'. The one we choose to make is somewhat involved (though completely natural) and we label it $P(m - 1)$ for reference.

Inductive Hypothesis $P(m - 1)$. Suppose we have an equation $up_1 \cdots p_{m-1} = v'q_1 \cdots q_{n'}$, where $m - 1 \leqslant n'$, while u, v are units and $p_1, \ldots, p_{m-1}, q_1, \ldots q_{n'}$ primes. Then $m - 1 = n'$ and, after renumbering, p_i and q_i are associates, $1 \leqslant i \leqslant m - 1$.

$P(0)$ has just been established. We proceed to deduce $P(m)$ from $P(m-1)$, and suppose given the equation $up_1 \cdots p_m = vq_1 \cdots q_n$ with $m \leqslant n$. Since $p_m | vq_1 \cdots q_n$ and p_m is prime, it follows (Exercise 1(i)) that $p_m | v$ or $p_m |$some q_i. But $p_m \nmid v$, since p_m is not a unit, so $p_m |$some q_i. Renumbering if necessary, we may suppose that $p_m | q_n$. But q_n is prime and hence irreducible. Thus p_m and q_n are associated, say $q_n = wp_m$ with w a unit. But then

$up_1 \cdots p_m = vwq_1 \cdots q_{n-1}p_m$ and, dividing by p_m (allowed since R is an integral domain), we conclude that

$$up_1 \cdots p_{m-1} = vwq_1 \cdots q_{n-1}.$$

We now bring the inductive hypothesis into play to deduce from $P(m-1)$ that $m - 1 = n - 1$, whence $m = n$, and, after renumbering, p_i and q_i are associates, $i = 1, \ldots, m - 1$. This establishes the induction. ■ Again, the reader is advised to follow through, on his own, the argument for the rings $\mathbb{Z}$, $\mathbb{Q}[x]$—or, indeed, $\mathbb{Z}[x]$, since here we have not assumed that R is Euclidean. [Of course, there are primes in $\mathbb{Z}[x]$ although $\mathbb{Z}[x]$ is not Euclidean. But Theorem 12.3.2 does not assert that there *are* any primes in R!]

Exercise 2

(i) Let R be an integral domain in which every non-zero element admits a factorization into primes. Show that every irreducible element in R is prime.
(ii) For each prime $p \in \mathbb{N}$ and integer $n \in \mathbb{N}$, define $h_p(n)$ to be the highest power of p dividing n. Prove that $h_p(nm) = h_p(n) + h_p(m)$; and hence use h_2 to prove the non-existence of a rational square root of 2.

12.4 Factorization in $\mathbb{Z}[x]$

In this section we sketch a logical development of a factorization theory in $\mathbb{Z}[x]$. We have already observed that $\mathbb{Z}[x]$ is certainly not a Euclidean domain; for the conclusion of Theorem 11.4.6 is false for $\mathbb{Z}[x]$, because in 11.4.5 we showed that the ideal of polynomials with even constant term is not principal. Indeed, the conclusion 11.5.5 is also false for $\mathbb{Z}[x]$, since the hcf of 2 and x in $\mathbb{Z}[x]$ is 1 and this is not expressible as a linear combination of 2 and x with coefficients in $\mathbb{Z}[x]$. Nevertheless there surely are hcf's and lcm's in $\mathbb{Z}[x]$, while polynomials in $\mathbb{Z}[x]$ may be factorized into primes. We show below how a theory may be established using the factorizations already available in $\mathbb{Z}$ and $\mathbb{Q}[x]$. The treatment is, in fact, capable of very substantial further generalization (to the case of polynomials in x with coefficients in an integral domain R in which a factorization theory has already been established), and thus enables us, in principle, to consider polynomials in several indeterminates; but our main aim here is not generality, but an explanation of the difficulties and subtleties of a theory which, in standard presentations of the subject, tends to be treated too lightheartedly as a piece of amusing technique.

The trick, then, is to regard $\mathbb{Z}[x]$ as a sub-ring of $\mathbb{Q}[x]$. However, we must be a little careful; for, suppose we take two polynomials in $\mathbb{Z}[x]$, regard them as belonging to $\mathbb{Q}[x]$ and then take their hcf in $\mathbb{Q}[x]$. There is no reason to suppose that this hcf lies in $\mathbb{Z}[x]$, and so it may, on such elementary grounds, fail to qualify as hcf in $\mathbb{Z}[x]$. Thus we must proceed a little more cautiously.

Recall first that the units of $\mathbb{Z}[x]$ are just ± 1, while the units of $\mathbb{Q}[x]$ are the non-zero constants (i.e., the non-zero rational numbers, regarded as constant polynomials). Now given any non-zero polynomial $f \in \mathbb{Z}[x]$, we may associate with it that *positive* integer $\delta(f)$ which is the hcf of its coefficients. Thus $\delta(f) \in \mathbb{N}$, so δ is a function $\mathbb{Z}[x] - (0) \to \mathbb{N}$. If $\delta(f) = 1$ we say that f is **primitive**, and the primitive polynomials play an important role in the theory. Thus $3x^2 + 5x$ is primitive, $3x^2 + 6$ is not. We now prove a very fundamental lemma.

12.4.1 LEMMA. *If f and g are primitive polynomials in $\mathbb{Z}[x]$, so is fg.*

Proof. For suppose $f = \sum_{i=0}^{m} a_i x^i$, $g = \sum_{j=0}^{n} b_j x^j$, and suppose f, g are primitive, but fg not primitive; we will obtain a contradiction. In fact, let p be a prime number dividing $\delta(fg)$, where δ is the function defined above. Since f, g are primitive, there are numbers k, l such that

$$p|a_0, p|a_1, \ldots, p|a_{k-1}, \qquad \text{while } p\nmid a_k,$$
$$q|b_0, q|b_1, \ldots, q|b_{l-1}, \qquad \text{while } q\nmid b_l.$$

That is, a_k is the first coefficient of f not divisible by p, and similarly for b_l. Now the coefficient of x^{k+l} in fg is $a_0 b_{k+l} + \cdots + a_k b_l + \cdots + a_{k+l} b_0$. [If $i > m$, we naturally take $a_i = 0$; and if $j > n$, we take $b_j = 0$.] But

$$p|(a_0 b_{k+l} + \cdots + a_{k-1} b_{l+1} + a_{k+1} b_{l-1} + \cdots + a_{k+l} b_0)$$

because, for each term, p divides the a or the b; while $p|\delta(fg)$, so that

$$p|(a_0 b_{k+l} + \cdots + a_{k-1} b_{l+1} + a_k b_l + a_{k+1} b_{l-1} + \cdots + a_{k+l} b_0)$$

by hypothesis. Thus $p|a_k b_l$. This is impossible, since $p\nmid a_k$, $p\nmid b_l$, and p is prime (Definition 12.2.2). ■

Now let f be any non-zero polynomial in $\mathbb{Q}[x]$. Then it is plain that we may write $f = (r/s)f'$, where r/s is a fraction in its 'lowest terms' and f' is primitive in $\mathbb{Z}[x]$. Plainly also, if $f \in \mathbb{Z}[x]$, then $s = 1$ and $r = \delta(f)$. Now r/s is a unit of $\mathbb{Q}[x]$; hence every non-zero polynomial in $\mathbb{Q}[x]$ is associated in $\mathbb{Q}[x]$ (in the technical sense) with a primitive polynomial in $\mathbb{Z}[x]$. We will prove the following result, in which δ_f means $\delta(f)$.

12.4.2 THEOREM. *Let f and g be non-zero polynomials in $\mathbb{Z}[x]$. Choose, as their hcf in $\mathbb{Q}[x]$, a primitive polynomial d' in $\mathbb{Z}[x]$. Let $c = hcf(\delta_f, \delta_g)$ in $\mathbb{Z}^+$. Then cd' is their hcf in $\mathbb{Z}[x]$.*

First, we state a proposition which follows from Lemma 12.4.1.

12.4.3 PROPOSITION. *Let $g, f \in \mathbb{Z}[x] - (0)$. Then*

$$g|f \quad \text{in } \mathbb{Z}[x]$$

if and only if

$$g|f \quad \text{in } \mathbb{Q}[x] \quad \text{and} \quad \delta(g)|\delta(f) \quad \text{in } \mathbb{Z}.$$

Proof. First observe that, as a consequence of Lemma 12.4.1, we infer (and the reader should!)

12.4.4 $\delta_{gh} = \delta_g \delta_h$.

Thus if $g|f$ in $\mathbb{Z}[x]$, then $\delta_g|\delta_f$ in $\mathbb{Z}^+$. Also, $g|f$ in $\mathbb{Q}[x]$, of course. Conversely, let $g|f$ in $\mathbb{Q}[x]$ and $\delta_g|\delta_f$ in $\mathbb{Z}^+$. We can write $g = \delta_g g'$, where g' is primitive; and similarly $f = \delta_f f'$. Then $g'|f'$ in $\mathbb{Q}[x]$, say $f' = g'h$ with $h \in \mathbb{Q}[x]$. We can also write $h = (r/s)h'$ with $r/s \in \mathbb{Q}$ and h' primitive in $\mathbb{Z}[x]$, so that $sf' = sg'h = g'sh = rg'h'$. But $g'h'$ is primitive, whence $s|r$, so r/s is an *integer* (indeed, the integer 1). This implies that $h \in \mathbb{Z}[x]$, and $f = \delta_f f' = \delta_f g'h = \delta_f(g/\delta_g)h = qgh$ with $q = \delta_f/\delta_g$. Thus $g|f$ in $\mathbb{Z}[x]$. ■
The reader should get a feeling for this proposition by considering special cases before proceeding to the proof of Theorem 12.4.2.

Proof of 12.4.2. With c, d' as in the statement of 12.4.2, we set $d = cd'$ so that $\delta_d = c$. To prove the theorem we must establish properties (i) and (ii) of 11.5.1 for d, to show it to be the hcf in $\mathbb{Z}[x]$ of f and g. Firstly $d|f$ in $\mathbb{Q}[x]$, since $d'|f$ in $\mathbb{Q}[x]$; secondly $\delta_d|\delta_f$ by construction. Thus by Proposition 12.4.3, $d|f$ in $\mathbb{Z}[x]$. Similarly $d|g$ in $\mathbb{Z}[x]$, so property (i) is verified. Now suppose that $h|f$ and $h|g$ in $\mathbb{Z}[x]$. Then $\delta_h|\delta_f$ and $\delta_h|\delta_g$, so $\delta_h|\delta_d$; and $h|f$, $h|g$ in $\mathbb{Q}[x]$, so $h|d$ in $\mathbb{Q}[x]$. Hence, again using Proposition 12.4.3, $h|d$ in $\mathbb{Z}[x]$, and property (ii) is thus verified also. ■

The reader should now be in a position to see how to extend the theory of hcf's in $\mathbb{Z}[x]$ to sets of more than two polynomials; and how to develop a satisfactory theory of lcm's. We are content to set down the facts relating to factorization into primes in $\mathbb{Z}[x]$.

12.4.5 PROPOSITION. *A polynomial p in $\mathbb{Z}[x]$, of positive degree, is prime if and only if it is primitive and prime in $\mathbb{Q}[x]$.*

We leave the proof as an exercise to the reader. ■

Now consider a polynomial $f \neq 0$ in $\mathbb{Z}[x]$; let us show how to factorize it into primes in $\mathbb{Z}[x]$. We set $f = qf'$ where $q = \delta_f$ and f' is primitive. Now we may factorize f' in $\mathbb{Q}[x]$ as a product of polynomials prime in $\mathbb{Q}[x]$; and there is no loss of generality in supposing these polynomials to be in $\mathbb{Z}[x]$ and primitive. Thus let $f' = (r/s)p_1 \cdots p_n$. By an easy extension of Lemma 12.4.1, $p_1 \cdots p_n$ is primitive; and since f' is primitive it readily follows that $r/s = \pm 1$. Now $f = qf'$, so we factorize q in $\mathbb{Z}$ as $q = k_1 \cdots k_m$, where $k_1, \cdots, k_m$ are primes in $\mathbb{Z}$. It is obvious that $k_1, \ldots, k_m$ are also prime in $\mathbb{Z}[x]$, so that, finally,

$$f = \pm k_1 \cdots k_m p_1 \cdots p_n,$$

and this is a factorization of f into primes in $\mathbb{Z}[x]$. We now have the full theory, since Theorem 12.3.2 is valid in $\mathbb{Z}[x]$, and Exercise 2(i) shows that the notions of prime polynomial and irreducible polynomial coincide in $\mathbb{Z}[x]$.

Exercise 3

Factorize into primes in $\mathbb{Z}[x]$ the following polynomials:

(i) $x^2 - 4$;
(ii) $x^2 + 4$;
(iii) 28;
(iv) $6x^3 - 12x^2 + 18x - 36$.

The domain $\mathbb{Z}[x]$ has now been exhibited as belonging to the important class of domains in which every non-zero element is expressible, in essentially only one way, as a product of primes. This class includes the Euclidean domains but is larger than it; it is called the class of **unique factorization domains** (UFD's). Then the whole theory of this section extends to UFD's in the following way. We consider a UFD, R, and prove that then $R[x]$ *is also a UFD*. The one additional piece of technique we require is the construction of a field F containing R which plays the role vis-à-vis R which $\mathbb{Q}$ played vis-à-vis $\mathbb{Z}$. This field F is called the *quotient field* of R and is defined for any integral domain R. The reader who studies the precise construction of $\mathbb{Q}$ from $\mathbb{Z}$ in 23.3 will easily imitate it to construct the quotient field of R; indeed, this is asserted in 23.3.8!. [See also 13.1.] The confident reader should be able to provide all details of the proof that $R[x]$ is a UFD if R is a UFD. He should also be able to provide a theory of hcf's and lcm's in $\mathbb{N}$ when they are encountered in a traditional elementary course of arithmetic. We believe that our presentation of the theory, for the class of Euclidean domains, is simpler and brings into prominence the crucial property (11.5.5) of an hcf in a Euclidean domain.

Chapter 13

APPLICATIONS OF THE THEORY OF HCF'S

In this chapter we discuss two applications of the theory of hcf's to interesting parts of mathematics. The first application, to the theory of partial fractions, will take the reader into familiar territory (if by an unfamiliar route!); the second, to the theory of continued fractions, may well be much less familiar to a 'modern' student.

13.1 Partial Fractions

The reader surely already appreciates the importance of the technique of expressing a rational function of x in partial fractions; for example, it is an essential tool in calculus, for the integration of differential equations. We take up here the question of the existence and uniqueness of such expressions.

The formal process of passing from the set of integers to the set of rational numbers will be described in Chapter 23. It is a special case of embedding an integral domain R in its *quotient field* $F(R)$ and may be applied to any integral domain. It permits us to divide an element of R by a non-zero element of R in such a way that the usual rules of arithmetic are preserved; indeed the elements of $F(R)$ may be regarded as such 'quotients' a/b and we agree, as usual, that

$$\frac{a}{b} = \frac{c}{d} \quad \text{if and only if} \quad ad = bc,$$

$$\frac{a}{b} + \frac{c}{d} = \frac{ad + bc}{bd},$$

$$\frac{a}{b} \cdot \frac{c}{d} = \frac{ac}{bd}.$$

If $R = \mathbb{Z}$, then $F(R) = \mathbb{Q}$; and if $R = \mathbb{Q}[x]$, then $F(R)$ is the field of rational functions in x with coefficients in $\mathbb{Q}$. These are the two examples we have principally in mind, but the theory applies to any Euclidean domain R, and so in particular to $R = F[x]$, where F is *any* field and $F[x]$ is the ring of polynomials in x with coefficients in F.

In the general case, observe that F is a function $X \to Y$, where X is the set of all integral domains, and Y is the set of all fields. The use of the capital

R should not mislead the reader into thinking that $F(R)$ is here the *image* of F; it is the *value* of F at $R \in X$.

Now let R be any Euclidean domain and $F(R)$ its quotient field. Let a, b be non-zero elements of R with hcf $(a, b) = 1$ and let $c \in R$. By 11.5.5 we have $1 = k'b + l'a$ for some $k', l' \in R$, so that $c = kb + la$ for some $k, l \in R$. Passing to $F(R)$ to allow division by $ab(\neq 0)$ we infer

13.1.1 $$\frac{c}{ab} = \frac{k}{a} + \frac{l}{b}.$$

We may clearly generalize 13.1.1 to the case of more than two factors in the denominator on the left. In particular, let g factorize (in R) into primes as $g = p_1^{n_1} \cdots p_r^{n_r}$. Then we may deduce:

13.1.2 THEOREM. *Let $f, g \in R$ with $g = p_1^{n_1} \cdots p_r^{n_r}$. Then, in $F(R)$,*

$$\frac{f}{g} = \frac{a_1}{p_1^{n_1}} + \cdots + \frac{a_r}{p_r^{n_r}}$$

for suitable elements $a_1, \ldots, a_r \in R$. ■

Exercise 1

(i) In 13.1.2, let $q_i = g/p_i^{n_i}$. Prove that the hcf of $q_1, \ldots, q_r$ is 1. Hence deduce Theorem 13.1.2.
(ii) Apply Theorem 13.1.2 to the fraction $\frac{17}{24}$ and to the rational function

$$\frac{\frac{3}{2}x + 7}{x^2 + \frac{1}{2}x - 5}.$$

Of course, Theorem 13.1.2 does not express the full force of partial fraction expansion in $\mathbb{Q}(x)$. In that case we infer after a few experiments, that if $\deg f < \deg g$, then we may choose $\deg a_i < n_i \deg p_i$ and the polynomials a_i are then *uniquely* determined. This last assertion generalizes in an obvious way to arbitrary Euclidean domains (replacing '$\deg f$' by $\phi(f)$ and '$n_i \deg p_i$' by $\phi(p_i)^{n_i}$); however, it is *false* in general. Thus, for example, we have, in $\mathbb{Q}$,

$$\tfrac{1}{6} = \tfrac{1}{2} - \tfrac{1}{3} = -\tfrac{1}{2} + \tfrac{2}{3},$$

giving *two* expressions for $\frac{1}{6}$ satisfying the required inequalities relative to the norm ϕ. *We must therefore look for an extra property of the norm ϕ in $\mathbb{Q}[x]$* [or, indeed, in $F[x]$ for any field F] *which accounts for the uniqueness of the partial fraction expansion.*

Let us say that the norm ϕ is **special** if it is Euclidean and satisfies

$$\mathfrak{N}_4: \qquad \phi(a + b) = \phi(a) \quad \text{if} \quad \phi(a) > \phi(b).$$

Plainly the norm $\phi: \mathbb{Q}[x] \to \mathbb{Z}^+$, given by $\phi(f) = 2^{\deg f}(f \neq 0)$, is special, since the degree of $(f + g)$ is the degree of f if $\deg f > \deg g$. On the other hand the absolute value (11.1.1) is not a special norm on $\mathbb{Z}$.

Exercise 2

(i) Let $\phi: R \to \mathbb{Z}^+$ be a special norm. Show that $\phi(a + b) \leqslant \phi(a)$ if $\phi(a) = \phi(b)$; and that $\phi(a + b) \leqslant \max(\phi(a), \phi(b))$, for any $a, b \in R$.
(ii) Let R be a Euclidean domain with a special norm ϕ. Show that, for given $a, b(b \neq 0) \in R$, the relation $a = qb + r$, $\phi(r) < \phi(b)$, guaranteed by $\mathfrak{N}_3$, uniquely determines q and r.

13.1.3 THEOREM. *Let R be an integral domain admitting a special norm ϕ. Let $f, g \in R$ with $\phi(f) < \phi(g)$ and let g factorize into primes as $g = p_1^{n_1} \cdots p_r^{n_r}$. Then there exist unique elements $a_1, \ldots, a_r \in R$ with $\phi(a_i) < \phi(p_i)^{n_i}$ and*

$$\frac{f}{g} = \frac{a_1}{p_1^{n_1}} + \cdots + \frac{a_r}{p_r^{n_r}}.$$

Proof. We will be content to show that if $\phi(f) < \phi(g)$ and $g = bc$ with hcf $(b, c) = 1$, then

13.1.4 $$\frac{f}{g} = \frac{k}{b} + \frac{l}{c}, \qquad \phi(k) < \phi(b) \quad \text{and} \quad \phi(l) < \phi(c)$$

where k, l are uniquely determined. It should be clear to the reader how to pass from 13.1.4 to the full statement of the theorem, using induction on r. To prove 13.1.4 we begin with the relation

$$f = k_1c + l_1b$$

which holds for some k_1, l_1 in R. By $\mathfrak{N}_3$ we may find q, k with $k_1 = qb + k$ and $\phi(k) < \phi(b)$. Thus we have expressed f as

13.1.5 $$f = kc + lb,$$

where $\phi(k) < \phi(b)$ (and $l = l_1 + qc$). We now show that $\phi(l) < \phi(c)$. For $\phi(kc) < \phi(bc) = \phi(g)$ and $\phi(f) < \phi(g)$, so $\phi(f - kc) < \phi(g)$, since ϕ is special (see Exercise 2(i)). But $f - kc = lb$, by 13.1.5, so $\phi(lb) < \phi(bc)$, whence $\phi(l) < \phi(c)$.

We now prove the uniqueness of 13.1.5, given that $\phi(k) < \phi(b)$, $\phi(l) < \phi(c)$. Suppose then that

$$kc + lb = k'c + l'b$$

with $\phi(k') < \phi(b)$, $\phi(l') < \phi(c)$ also. Now

$$(k - k')c = (l' - l)b.$$

Thus $c|(l' - l)b$; but hcf $(b, c) = 1$, so that (by 11 Exercise 10(ii)) $c|(l - l')$. But since ϕ is special, $\phi(l - l') \leqslant \max(\phi(l), \phi(l')) < \phi(c)$. Thus

$$\phi(c)|\phi(l - l') \quad \text{and} \quad \phi(c) > \phi(l - l')).$$

This is possible only if $\phi(l - l') = 0$, i.e., $l = l'$. Similarly $k = k'$ and so uniqueness is proved. ■

We have nothing to say at this stage about the standard techniques of finding the appropriate polynomials a_i in Theorem 13.1.3, although we will devote a little space in Chapter 22 to discussing substitution in polynomials. However we would wish to say a word about an alternative form of the partial fraction expansion in the case when $R = \mathbb{Q}[x]$, which is often more suitable. Namely, one replaces each a/p^n by a sum $\sum_{r=1}^{n} d_i/p^{n_i}$ where each d_i is a rational number. Such a replacement is possible, and the d_i are uniquely determined. We leave the proof of these statements to the reader, with the hint that the conclusion of Exercise 2(ii) is the key to the uniqueness of the d_i.

Finally we take up again the question of partial fraction expansion in $\mathbb{Q}$. We have seen that we cannot expect as clear-cut a statement to hold for $\mathbb{Z}$ as for a specially normed domain; nevertheless we can get close. We are content to state the analogue of 13.1.4 in the form:

13.1.6 THEOREM. *Let $f/g \in \mathbb{Q}$ be a (positive) fraction in its lowest terms with $f < g$ in $\mathbb{N}$; and let $g = bc$ with $b, c > 1$ in $\mathbb{Z}$ and hcf $(b, c) = 1$. Then we may express f/g as*

$$\frac{f}{g} = \frac{k}{b} + \frac{l}{c}, \qquad (k, l \in \mathbb{Z})$$

where $k > 0$, $k < b$, while $|l| < c$; and such an expression is unique.

Notice that we have insisted on taking $k > 0$ in the theorem. If we follow through the proof of Theorem 13.1.3, we observe that to obtain k we divide k_1 by b, and we may choose it positive (k is not zero, since f/g is in its lowest terms and $b > 1$). Then f and kc are both positive, so that $|f - kc| < \max(f, kc)$, and the proof that $|l| < c$ goes through as before. In proving uniqueness we infer, as before, that $b|(k - k')$, replacing the roles of l, l', c by k, k', b [this, of course, makes no difference to the proof of 13.1.4]. But $|k - k'| < \max(k, k') < b$, so $k = k'$ and, hence, $l = l'$. ■

Exercise 3

Generalize Theorem 13.1.6 to a fraction f/g with g expressed as a product of more than two factors.

13.2 Continued Fractions

We have described, in 11.6, the Euclidean algorithm for obtaining the hcf of two elements a_0 and a_1 in a Euclidean domain R. This algorithm gives us a rule for obtaining the coefficients k and l in an equation

$$d = ka_0 + la_1,$$

where d is the hcf of a_0 and a_1. As we have seen, k and l are not determined by a_0 and a_1, and indeed we may alter k by adding an arbitrary multiple of a_1, say ma_1, provided we alter l by subtracting ma_0. Now we may view the Euclidean algorithm in $\mathbb{Z}$ as a process of expanding the fraction a_0/a_1 in a certain way; namely, we suppose that all the remainders are to be taken *positive*, so that the a_i are determined and

$$\frac{a_0}{a_1} = q_1 + \frac{a_2}{a_1},$$

$$\frac{a_1}{a_2} = q_2 + \frac{a_3}{a_2},$$

$$\vdots$$

$$\frac{a_{r-2}}{a_{r-1}} = q_{r-1} + \frac{a_r}{a_{r-1}},$$

$$\frac{a_{r-1}}{a_r} = q_r.$$

Thus,

$$\frac{a_0}{a_1} = q_1 + \cfrac{1}{q_2 + \cfrac{a_3}{a_2}} = q_1 + \cfrac{1}{q_2 + \cfrac{1}{q_3 + \cfrac{a_4}{a_3}}}, \quad \text{etc.}$$

We write this as

$$\frac{a_0}{a_1} = q_1 + \frac{1}{q_2+}\,\frac{1}{q_3+}\cdots\frac{1}{q_r},$$

to avoid the unpleasantly slanting fraction, and we call

$$q_1,\ q_1 + \frac{1}{q_2},\ q_1 + \frac{1}{q_2+}\,\frac{1}{q_3}, \ldots,$$

the successive **convergents** to a_0/a_1. If we write these convergents $c_1, c_2, \ldots, c_r$ as ordinary fractions, then

$$c_1 = q_1,$$

$$c_2 = \frac{q_1q_2 + 1}{q_2}$$

$$c_3 = q_1 + 1\Big/\left(q_2 + \frac{1}{q_3}\right) = \frac{q_1q_2q_3 + q_1 + q_3}{q_2q_3 + 1}, \quad \text{etc.}$$

We now state some vital properties of the convergents c_i.

13.2.1 THEOREM. *Let c_i be expressed as a fraction u_i/v_i simply by straightforward expansion. Then u_i, v_i are given inductively by*

$$u_1 = q_1, \qquad u_2 = q_1q_2 + 1, \qquad u_i = q_iu_{i-1} + u_{i-2}, \qquad i > 2,$$

$$v_1 = 1, \qquad v_2 = q_2, \qquad v_i = q_iv_{i-1} + v_{i-2}, \qquad i > 2.$$

Proof. We use induction on i. The case $i = 3$ is plain; so we assume the formulae true for u_i and v_i, and we expand c_{i+1} by first replacing q_i by $q_i + 1/q_{i+1}$. The inductive hypothesis assures us (since the u_i, v_i are obtained by straightforward expansion) that

$$\frac{u_{i+1}}{v_{i+1}} = \frac{\left(q_i + \dfrac{1}{q_{i+1}}\right)u_{i-1} + u_{i-2}}{\left(q_i + \dfrac{1}{q_{i+1}}\right)v_{i-1} + v_{i-2}} = \frac{u_i + \dfrac{1}{q_{i+1}}u_{i-1}}{v_i + \dfrac{1}{q_{i+1}}v_{i-1}} = \frac{q_{i+1}u_i + u_{i-1}}{q_{i+1}v_i + v_{i-1}}.$$

Thus straightforward expansion of c_{i+1} yields $u_{i+1} = q_{i+1}u_i + u_{i-1}$, $v_{i+1} = q_{i+1}v_i + v_{i-1}$, and the proof by induction is complete. ■ [The proof is subtler than it may look at first sight. The phrase 'by straightforward expansion' is vital to the argument and would have to be made quite precise in a completely systematic description of the argument.]

13.2.2 THEOREM. $u_iv_{i-1} - u_{i-1}v_i = (-1)^i$, for each $i \geqslant 2$.

Proof. We first deduce from Theorem 13.2.1 that $u_iv_{i-1} - u_{i-1}v_i = -(u_{i-1}v_{i-2} - u_{i-2}v_{i-1})$, provided $i > 2$. Using induction on i, we continue in this way (provided $i > 3$) until we eventually reach

$$u_iv_{i-1} - u_{i-1}v_i = (-1)^{i-2}(u_2v_1 - u_1v_2).$$

But, by inspection, $u_2v_1 - u_1v_2 = 1$, so that $u_iv_{i-1} - u_{i-1}v_i = (-1)^i$. ■

13.2.3 COROLLARY. *hcf* $(u_i, v_i) = 1$ *for each* $i \geqslant 1$. ■

This corollary shows that, although in obtaining u_i and v_i we only employed 'straightforward expansion' and did not permit any cancellation of factors in numerator and denominator, in fact, no such cancellations would have been possible. Now

$$c_r = \frac{u_r}{v_r} = \frac{a_0}{a_1}.$$

Thus if hcf $(a_0, a_1) = 1$, then $u_r = a_0$, $v_r = a_1$. More generally, if hcf $(a_0, a_1) = d$, then $a_0 = du_r$, $a_1 = dv_r$. Thus we infer from Theorem 13.2.2:

13.2.4 COROLLARY. *If hcf* $(a_0, a_1) = d$ *and if* $c_i = u_i/v_i$ *are the successive convergents of the continued fraction expansion of* a_0/a_1, *with* $c_r = a_0/a_1$, *then*

$$a_0v_{r-1} - a_1u_{r-1} = (-1)^r d. \quad ■$$

13.2.5 EXAMPLE. Let us consider the case $a_0 = 910$, $a_1 = 624$. The continued-fraction expansion yields

$$\frac{910}{624} = 1 + \frac{1}{2+}\frac{1}{5+}\frac{1}{2}; \qquad q_1 = 1, \qquad q_2 = 2, \qquad q_3 = 5, \qquad q_4 = 2.$$

Thus

$$c_1 = 1, \qquad c_2 = 1 + \frac{1}{2} = \frac{3}{2}, \qquad c_3 = 1 + \cfrac{1}{2 + \cfrac{1}{5}} = 1 + \frac{5}{11} = \frac{16}{11},$$

and

$$c_4 = 1 + \cfrac{1}{2 + \cfrac{1}{5 + \cfrac{1}{2}}} = 1 + \cfrac{1}{2 + \cfrac{2}{11}} = 1 + \frac{11}{24} = \frac{35}{24}.$$

So

$$\begin{aligned} u_1 &= 1, & u_2 &= 3, & u_3 &= 16, & u_4 &= 35; \\ v_1 &= 1, & v_2 &= 2, & v_3 &= 11, & v_4 &= 24, \end{aligned}$$

and we verify that the formulae of Theorem 13.2.1 hold; in fact, of course, it is simplest to compute the u_i and the v_i from those formulae. If we compute $910 \cdot 11 - 624 \cdot 16$ we get 26; applying (and confirming) Corollary 13.2.4 (with $r = 4$) we see that hcf (910, 624) = 26 and we have expressed 26 as a linear combination of 910 and 624.

Exercise 4

Use Theorem 13.2.4 to compute the convergents to the continued fraction

$$2 + \frac{1}{2+}\,\frac{1}{3+}\,\frac{1}{3+}\,\frac{1}{2};$$

check the conclusion of Theorem 13.2.5 in this case; and draw a conclusion using Corollary 13.2.4.

In this chapter we have barely scratched the surface of the fascinating theory of continued fractions, which plays an important role in real-number theory, and provides approximations which are valuable in the use of numerical methods. The reader is referred to Davenport's delightful book [28] for further details of the theory, or to Olds [97] and the references there given.

Part IV

GEOMETRY OF $\mathbb{R}^3$

As in the case of the study of numbers, the twin drives of practical need and aesthetic attraction have interested mathematicians in spatial relationships. These relationships, when perceived in the space around us, have formed the basis of such arts as building and surveying, and they have been organized into the branch of mathematics we call geometry. The first such organization was that of Euclidean geometry, which was later subsumed by Descartes' invention of co-ordinate geometry. From then on geometrical notions have pervaded and inspired all mathematics; and in this Part we indicate such inspiration as we survey those parts of geometry which are concerned with three-dimensional space.

Beginning with Chapter 14 we assume an elementary knowledge of the Euclidean geometry of such things as lines, perpendiculars and planes as often presented at school; and we derive their properties using the notation of 'vectors'. Essentially we are here following in the footsteps of Descartes in the way he copied Euclidean geometry into co-ordinate geometry. The geometry then forces us to look at the algebra of vectors with the scalar and vector products, and we meet the vector-space structure of three-dimensional space and its subspaces. We also link up the vector spaces of pure mathematics with the vectors of the applied mathematician.

Corresponding to the geometrical problem of describing the intersection of three planes in space or more simply of two lines in a plane, the algebraic approach leads us in Chapter 14 to consider sets of simultaneous equations and their associated matrices. We must then pass to some elementary matrix algebra and the consideration of the associated functions—the linear transformations. In two dimensions each such transformation distorts areas in a fixed ratio called the 'determinant' of T, and similarly with volumes in three dimensions. Such determinants and the transformations are then studied in Chapter 15. In an appendix to the chapter we provide a more rigorous discussion of the notion of area and we have to look at the related notions of length of a continuous path and the question 'What is a polygon?'.

In Chapter 16 we reverse the procedure of Chapter 14, in that we use the algebra as a means of avoiding the difficult problem of building geometry rigorously from the axioms of Euclid. Technically, we make an algebraic model of geometry and as an example we take a non-trivial theorem of solid geometry and show how to prove it rigorously within our model By doing this we hope to throw light on some points of pedagogy and of axiomatics.

In Chapter 17 we pass to projective geometry. After discussing some axioms we look at some algebraic projective geometries, in particular the real and complex projective planes $\mathbb{RP}^2$ and $\mathbb{CP}^2$. We show how the Cartesian plane $\mathbb{R}^2$ can be embedded in $\mathbb{RP}^2$, and $\mathbb{RP}^2$ in $\mathbb{CP}^2$, so as to preserve many geometrical properties once these are formulated in an appropriate language. Then theorems about $\mathbb{R}^2$ appear as interpretations of more general theorems of $\mathbb{CP}^2$; and a considerable unification is effected. Finally we indicate how the ideas lead to Klein's notion of 'a geometry' as formulated in his Erlanger program. These ideas have been a most powerful unifying agent within mathematics.

Throughout the part we shall be making algebra the language of geometry in the sense that we construct and investigate algebraic *models* of geometries, and it will be clear that the geometry raises many pregnant questions within algebra. Such questions, among others, will be pursued later in Part V.

Chapter 14

VECTOR GEOMETRY OF $\mathbb{R}^3$

14.1 The Vector Space $\mathbb{R}^3$

It has been found by experience that a simple way of referring to a point P, in the three-dimensional space $\mathbf{S}$ in which we live, is to assign three 'co-ordinates' (x, y, z) to P. These are the projections of P onto three mutually perpendicular co-ordinate axes 'Ox, Oy, Oz' issuing from a fixed origin O. Now the co-ordinates x, y, z are numbers, and they are found in practice by a relatively complicated process of measurement and approximation. We cut out all the difficulties concerning the description of this process by taking as a 'model' of the space $\mathbf{S}$ the set $\mathbb{R}^3 = \mathbb{R} \times \mathbb{R} \times \mathbb{R}$; that is to say, *the correspondence* $P \to (x, y, z)$ *is assumed to be an equivalence* $\mathbf{S} \approx \mathbb{R}^3$. According to the theories of relativity and quantum mechanics, $\mathbb{R}^3$ fails to be a good model of $\mathbf{S}$ at very large and very small distances. Nevertheless, $\mathbb{R}^3$ is of practical interest as the *simplest* prototype of $\mathbf{S}$. For similar reasons, the set $\mathbb{R}^2 = \mathbb{R} \times \mathbb{R}$ is taken as a model of a plane, and $\mathbb{R}$ as a model of an (infinite) line. Since $\mathbb{R}$ has its algebraic structure, some of this algebra is inherited by $\mathbb{R}^2$ and $\mathbb{R}^3$, to make them examples of 'vector spaces'. We can use this structure to study the geometry of $\mathbb{R}^3$; and we first use the elements of Euclidean geometry freely in this discussion. Thus, we shall assume such things as Pythagoras's theorem and some simple facts about angles. Later we show how to *replace* Euclidean geometry by this algebraic study.

In order to obtain certain results for $\mathbb{R}^2$ and $\mathbb{R}^3$ together, we identify $\mathbb{R}^2$ with the subset $\Pi_z \subseteq \mathbb{R}^3$, where

14.1.1 $$\Pi_z = \{(x, y, z) \in \mathbb{R}^3 | z = 0\},$$

(the plane '$z = 0$'). Of course, the assignment $(x, y) \to (x, y, 0)$ is a function $f\colon \mathbb{R}^2 \to \mathbb{R}^3$, which is an injection (embedding) with image Π_z.

For typographical reasons, we shall denote points of $\mathbb{R}^3$ either by capital letters or by lower-case Clarendon type; thus $P = \mathbf{p}$. The co-ordinates of $\mathbf{p}$ will be (p_1, p_2, p_3), but occasionally we use $\mathbf{r} = (x, y, z)$. The **origin** O is the point $(0, 0, 0) = \mathbf{0}$. If A, B are distinct points in $\mathbb{R}^3$, then AB will denote the segment joining A to B, of length $|AB|$.

By definition, the co-ordinates p_i of $\mathbf{p} = (p_1, p_2, p_3)$ are the lengths of the projections of OP onto the axes. Thus we have three **projection** functions $\pi^i\colon \mathbb{R}^3 \to \mathbb{R}$, given by

14.1.2 $$\pi^i(\mathbf{p}) = p_i, \qquad (i = 1, 2, 3).$$

[The superscript i in π^i should not be confused with a square or a cube. There are technical reasons for placing it above rather than below.]

We now introduce, into $\mathbb{R}^3$, algebraic operations of 'multiplication by scalars' and of 'vector addition' as follows. Given $\lambda \in \mathbb{R}$ and $\mathbf{p} \in \mathbb{R}^3$, we define a new point $\lambda \cdot \mathbf{p} \in \mathbb{R}^3$ by

14.1.3 $$\lambda \cdot \mathbf{p} = (\lambda p_1, \lambda p_2, \lambda p_3)$$

where the products λp_i on the right are each in $\mathbb{R}$, of course. As usual with products, we often omit the dot, and write $\lambda \mathbf{p}$ for $\lambda \cdot \mathbf{p}$; for example

$$\tfrac{2}{3}(1, 5, 2) = (\tfrac{2}{3}, 3\tfrac{1}{3}, 1\tfrac{1}{3}).$$

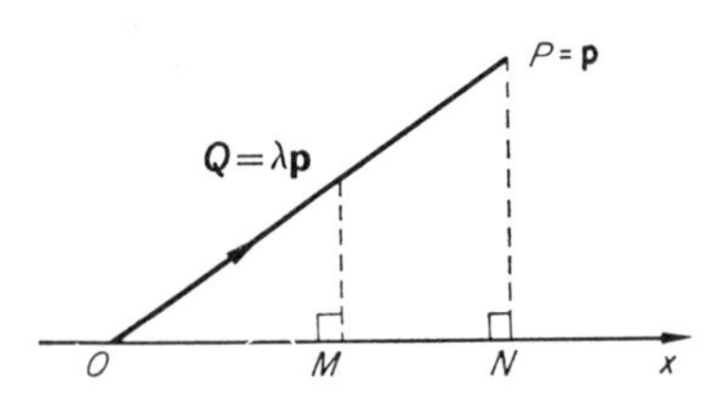

Fig. 14.1

Geometrically, the definition means the following. If we refer to Fig. 14.1, we see that $\lambda \mathbf{p}$ is the point Q, on the line through O and P, such that

$$|OQ|/|OP| = \lambda;$$

for $|ON| = p_1$, and by similarity of the right-angled triangles OMQ, ONP, we have

$$|OQ|/|OP| = |OM|/|ON| = \lambda p_1/p_1 = \lambda.$$

[As drawn, Q lies on the segment OP, but the same argument holds when P lies on OQ, or O on PQ. Also in Fig. 14.1, λ is assumed positive; for negative values of λ, $\lambda \mathbf{p}$ designates points on the line through O and P lying on the opposite side of O to P.]

Given also $\mathbf{q} \in \mathbb{R}^3$ we construct a new point, denoted by $\mathbf{p} + \mathbf{q}$, and defined by

14.1.4 $$\mathbf{p} + \mathbf{q} = (p_1 + q_1, p_2 + q_2, p_3 + q_3),$$

where the sums $p_i + q_i$ on the right are each in $\mathbb{R}$. For example

$$(1, 5, 2) + (-1, -1, 6) = (0, 4, 8).$$

Geometrically, 14.1.4 means the following. The three points OPQ (see Fig. 14.2) are coplanar if they are distinct, so we can complete the parallelogram $OPRQ$; *we then see that* $\mathbf{p} + \mathbf{q}$ *is* R. For,

$$\pi^1(R) = |OU| = |OS + SU| = |OT + OS|$$

since the congruent and parallel segments OP, QR must have congruent projections OS, TU on Ox. Hence $\pi^1(R) = \pi^1(\mathbf{q}) + \pi^1(\mathbf{p}) = p_1 + q_1$, and similarly for the other co-ordinates. Thus $R = \mathbf{p} + \mathbf{q}$, and this law of addition is often called the **parallelogram law.**

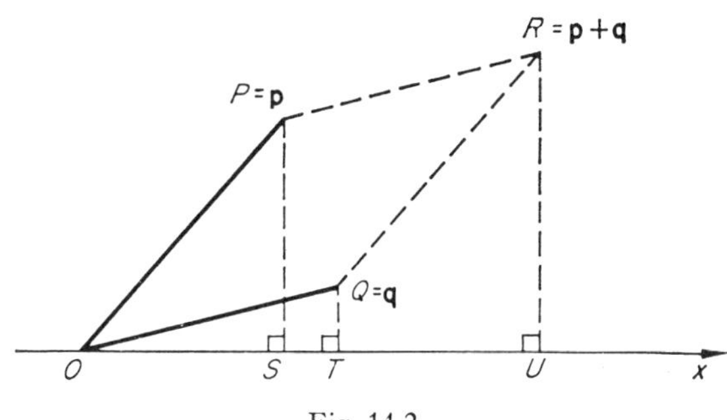

Fig. 14.2

14.1.5 The laws of addition and multiplication in $\mathbb{R}$ imply that $\mathbb{R}^3$ is an abelian group (see Chapter 9, $\mathfrak{A}_1$–$\mathfrak{A}_4$) with respect to the addition in 14.1.4, with $\mathbf{0}$ as zero, and $-1\cdot\mathbf{p}$ as the negative of $\mathbf{p}$. We therefore *define* $-\mathbf{p}$ to be $-1\cdot\mathbf{p}$. Moreover, multiplication by scalars satisfies, for any $\lambda, \mu \in \mathbb{R}$ and $\mathbf{p}, \mathbf{q} \in \mathbb{R}^3$, the following rules:

$\mathfrak{MS}_1$: $\lambda\cdot(\mu\mathbf{p}) = (\lambda\mu)\cdot\mathbf{p}$ (*associativity*);

$\mathfrak{MS}_2$: $\lambda\cdot(\mathbf{p} + \mathbf{q}) = \lambda\mathbf{p} + \lambda\mathbf{q}$ (*distributivity*);

$\mathfrak{MS}_3$: $0\cdot\mathbf{p} = \mathbf{0}$. $1\cdot\mathbf{p} = \mathbf{p}$.

Observe that the product $\lambda\mu$ on the right of $\mathfrak{MS}_1$ is in $\mathbb{R}$; and that the first zero and 1 in $\mathfrak{MS}_3$ are also in $\mathbb{R}$. To verify $\mathfrak{MS}_1$, $\mathfrak{MS}_2$, $\mathfrak{MS}_3$ we appeal to 14.1.3 and 14.1.4; let us check $\mathfrak{MS}_2$ for example. We have

$$\begin{aligned}\lambda\cdot(\mathbf{p} + \mathbf{q}) &= \lambda\cdot(p_1 + q_1, p_2 + q_2, p_3 + q_3) && \text{by 14.1.4,}\\ &= (\lambda(p_1 + q_1), \lambda(p_2 + q_2), \lambda(p_3 + q_3)) && \text{by 14.1.3,}\\ &= (\lambda p_1, \lambda p_2, \lambda p_3) + (\lambda q_1, \lambda q_2, \lambda q_3) && \text{by 14.1.4,}\end{aligned}$$

using the distributive law in $\mathbb{R}$,

$$= \lambda\mathbf{p} + \lambda\mathbf{q} \qquad \text{by 14.1.3;}$$

thus $\mathfrak{MS}_2$ is verified. We leave the other verifications to the reader. It is usual to call $\mathbb{R}^2$ and $\mathbb{R}^3$ 'real vector spaces', because this name is given to any abelian group which also admits multiplication by scalars satisfying $\mathfrak{MS}_1$, $\mathfrak{MS}_2$, $\mathfrak{MS}_3$. The general theory of vector spaces will be discussed in Chapter 19, and the reader will grasp that theory much better if he first understands what happens in $\mathbb{R}^2$ and $\mathbb{R}^3$.

Exercise 1

(i) Check that $\mathbb{R}^3$ is an abelian group with respect to $+$ in 14.1.4, by verifying the axioms $\mathfrak{A}_1$–$\mathfrak{A}_4$ in Chapter 9.
(ii) A **vector subspace** of $\mathbb{R}^3$ is any subset $X \subseteq \mathbb{R}^3$ which is a vector space relative to the operations of addition and scalar multiplication in $\mathbb{R}^3$. Thus it is necessary and sufficient that X be only 'closed' under these operations (i.e., $\mathbf{x} + \mathbf{y} \in X$ and $\lambda\mathbf{x} \in X$ whenever $\mathbf{x}, \mathbf{y} \in X$ and $\lambda \in \mathbb{R}$), since, for example, $\mathfrak{MS}_1$ holds in X because it already holds in $\mathbb{R}^3$. Prove that $\mathbf{0}$ lies in each subspace, and that if $\mathbf{x}$ lies in a subspace X, then $-\mathbf{x} \in X$. Prove also that $\mathbb{R}^2$ ($= \Pi_z$ in 14.1.1) and $\{0\}$ are subspaces of $\mathbb{R}^3$.
(iii) If $P \neq O$, show that all points on the entire line (OP) are of the form $\lambda\mathbf{p}$, $\lambda \in \mathbb{R}$. Prove also that (OP) is a subspace of $\mathbb{R}^3$.
(iv) Let $\mathbf{a}, \mathbf{b}, \mathbf{c}$ be three distinct non-zero points in $\mathbb{R}^3$, coplanar with $\mathbf{0}$, and such that O, A, B are not collinear. Show that $\exists$ points P, Q on the lines $(OA), (OB)$, such that $OPCQ$ is a parallelogram. Hence prove: $\exists\lambda, \mu \in \mathbb{R}$, such that $\mathbf{c} = \lambda\mathbf{a} + \mu\mathbf{b}$. State and prove the converse.
(v) In (iv) let D be a point in $\mathbb{R}^3$, which is not in the plane Π containing O, A, B. Given $P \in \mathbb{R}^3$, let the line through P parallel to (OD) meet Π in C. Prove that $\exists\alpha, \beta \in \mathbb{R}$ such that $\mathbf{p} = \alpha\mathbf{d} + \beta\mathbf{c}$, and hence (using (iv) again) that $\exists\rho, \sigma, \tau \in \mathbb{R}$ such that $\mathbf{p} = \rho\mathbf{a} + \sigma\mathbf{b} + \tau\mathbf{d}$. Pay special attention to the case when $P \in \Pi$.
(vi) Using the projections π^i in 14.1.2, prove that, for all $\lambda, \mu \in \mathbb{R}$ and $\mathbf{p}, \mathbf{q} \in \mathbb{R}^3$,

$$\pi^i(\lambda\mathbf{p} + \mu\mathbf{q}) = \lambda\pi^i(\mathbf{p}) + \mu\pi^i(\mathbf{q}).$$

(Observe that the sum and products on the right-hand side are in $\mathbb{R}$; those on the left-hand side are in $\mathbb{R}^3$.)
(vii) Given $\mathbf{r} \in \mathbb{R}^3$ and an integer $n > 0$, prove that $n\mathbf{r} = \mathbf{r} + \cdots + \mathbf{r}$, with n terms on the right-hand side.

14.2 Linear Dependence; Bases

Certain points in $\mathbb{R}^3$ which are of importance are the 'unit points' $\mathbf{i}, \mathbf{j}, \mathbf{k}$, given by

14.2.1 $$\mathbf{i} = (1, 0, 0), \quad \mathbf{j} = (0, 1, 0), \quad \mathbf{k} = (0, 0, 1).$$

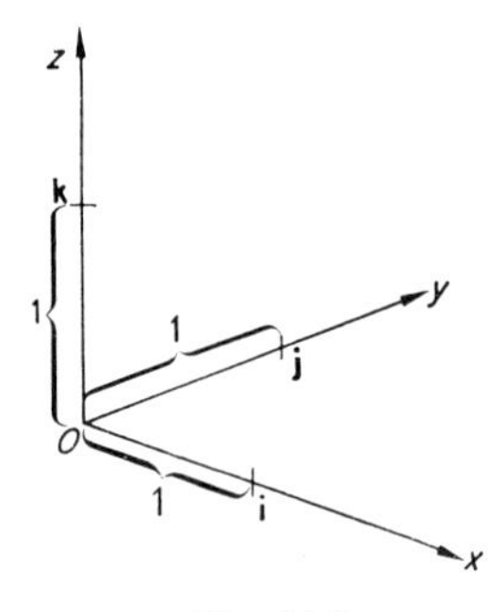

Fig. 14.3

Each point $\mathbf{r} = (x, y, z)$ in $\mathbb{R}^3$ can then be expressed as a sum

14.2.2
$$\mathbf{r} = x\mathbf{i} + y\mathbf{j} + z\mathbf{k}.$$

For, the first co-ordinate on the left is x, while the first on the right is

$$\begin{aligned}\pi^1(x\mathbf{i} + y\mathbf{j} + z\mathbf{k}) &= x\pi^1(\mathbf{i}) + y\pi^1(\mathbf{j}) + z\pi^1(\mathbf{k}) \qquad \text{(by Exercise 1(vi))}\\ &= x1 \quad + y0 \;\; + z0 = x;\end{aligned}$$

similarly for y and z. Given $\mathbf{r}$, the right-hand side of equation 14.2.2 is *uniquely determined*; for if also $\mathbf{r} = x'\mathbf{i} + y'\mathbf{j} + z'\mathbf{k}$, then $\mathbf{0} = u\mathbf{i} + v\mathbf{j} + w\mathbf{k}$ where $u = x - x'$, etc. The same argument as for 14.2.2 then gives $u = v = w = 0$, so $x = x'$, $y = y'$, $z = z'$.

14.2.3 DEFINITION. The n points $\mathbf{r}_1, \mathbf{r}_2, \ldots, \mathbf{r}_n \in \mathbb{R}^3$ are said to be **linearly independent** iff there is no equation $\lambda_1\mathbf{r}_1 + \cdots + \lambda_n\mathbf{r}_n = \mathbf{0}$ such that each $\lambda_i \in \mathbb{R}$ and some λ_j is non-zero. Otherwise the points are **linearly dependent,** and each is **linearly dependent on the others.**

14.2.4 EXAMPLE. (a) If $\mathbf{r}_1 = \mathbf{0}$ in 14.2.3, then the n points are linearly dependent, since $1\cdot\mathbf{r}_1 + 0\cdot\mathbf{r}_2 + \cdots + 0\cdot\mathbf{r}_n = \mathbf{0}$, yet $\lambda_1 = 1 \neq 0$. Every point is linearly dependent on itself (more precisely each $\mathbf{r} \in \mathbb{R}^3$ is linearly dependent on the singleton $\{\mathbf{r}\}$).

(b) Every $\mathbf{r} \in \mathbb{R}^3$ is linearly dependent on $\mathbf{i}$, $\mathbf{j}$, $\mathbf{k}$; for by 14.2.2, $1\mathbf{r} - x\mathbf{i} - y\mathbf{j} - z\mathbf{k} = \mathbf{0}$ and $\lambda_1 = 1 \neq 0$.

(c) The three points $\mathbf{i}, \mathbf{j}, \mathbf{k}$ are linearly independent, by the same argument as for equation 14.2.2.

(d) By Exercise 1(iv), any three points in $\mathbb{R}^3$ which are coplanar with $\mathbf{0}$ are linearly dependent.

(e) If $\mathbf{r}_1, \ldots, \mathbf{r}_n$ in 14.2.3 are linearly independent, so is any subset of them. If often happens in proofs that an equation $\lambda_1\mathbf{r}_1 + \cdots + \lambda_n\mathbf{r}_n = \mathbf{0}$ is obtained; thus, since $\mathbf{r}_1, \ldots, \mathbf{r}_n$ are linearly independent, then we may assert that each λ_i is zero, $i = 1, \ldots, n$.

We now prove a theorem which should be compared with the statements (c) and (d) of Example 14.2.4.

14.2.5 THEOREM. *Any four points* $\mathbf{u}, \mathbf{v}, \mathbf{w}, \mathbf{p} \in \mathbb{R}^3$ *are linearly dependent.*

Proof. If $\mathbf{u}, \mathbf{v}, \mathbf{w}$ are collinear we are through, by Exercise 1(iii). Otherwise, by Exercise 1(v) there exist $\lambda, \mu, \nu \in \mathbb{R}$ such that

$$\mathbf{p} = \lambda\mathbf{u} + \mu\mathbf{v} + \nu\mathbf{w},$$

so $\mathbf{u}, \mathbf{v}, \mathbf{w}, \mathbf{p}$ are linearly dependent. ∎

This theorem is, of course, susceptible of a purely algebraic proof. See Chapter 19.

By a **basis** for $\mathbb{R}^3$, we mean any finite non-empty set $\{\mathbf{r}_1, \ldots, \mathbf{r}_n\}$ of linearly independent points, such that every other point in $\mathbb{R}^3$ is linearly dependent on $\mathbf{r}_1, \ldots, \mathbf{r}_n$. Note that no basis contains $\mathbf{0}$. By Examples 14.2.4(b) and (c), $\mathbb{R}^3$ has at least one basis, namely $\{\mathbf{i}, \mathbf{j}, \mathbf{k}\}$. This leads us to prove:

14.2.6 THEOREM. *Every basis of* $\mathbb{R}^3$ *has exactly three elements.*

Proof. Let $\{\mathbf{r}_1, \ldots, \mathbf{r}_n\}$ be a basis. By Theorem 14.2.5, if $\{\mathbf{r}_1, \ldots, \mathbf{r}_n\}$ is a basis of $\mathbb{R}^3$, then $n < 4$, so $n = 1, 2$, or 3. If $n = 1$, then, for *any* $P \in \mathbb{R}^3$, $\mathbf{p} = \lambda \mathbf{r}_1$, so that P is collinear with O and R_1, an evident contradiction. If $n = 2$, then, for *any* $P \in \mathbb{R}^3$, $\mathbf{p} = \lambda \mathbf{r}_1 + \mu \mathbf{r}_2$, so that P is coplanar with O, R_1, R_2 (by Exercise 1(iv)), another evident contradiction. These contradictions eliminate the cases $n = 1, 2$, so $n = 3$; and this case is not contradictory, by the above remark that $\{\mathbf{i}, \mathbf{j}, \mathbf{k}\}$ is a basis. ∎

Exercise 2

(i) Prove that every basis for $\mathbb{R}^i$ ($i = 1, 2$) has exactly i elements.
(ii) If X is a vector subspace of $\mathbb{R}^3$, we define a basis for X just as for $\mathbb{R}^3$. Prove that all bases for X have the same number (1, 2, or 3) of elements. [This number is called the **dimension** of X.]
(iii) Prove that the dimension of X is 3, iff $X = \mathbb{R}^3$.

14.3 The Equation of a Line

Given distinct points $\mathbf{p}$, $\mathbf{q}$, let us find the form of a typical point $\mathbf{r}$ on the line $l = (PQ)$ through $\mathbf{p}$ and $\mathbf{q}$ (see Exercise 1(iii)). There is a line m through O, parallel to l, and there are points S, T on m so that (see Fig. 14.4) $OPRS$, $OPQT$ are parallelograms. Then $\mathbf{p} + \mathbf{t} = \mathbf{q}$ by the parallelogram law, so $\mathbf{t} = \mathbf{q} - \mathbf{p}$ (since $\mathbb{R}^3$ is a group). Hence $\mathbf{s} = \lambda \cdot (\mathbf{q} - \mathbf{p})$, by 14.1.3, where $\lambda = |OS|/|OT|$. But then, again by the parallelogram law, $\mathbf{r} = \mathbf{p} + \mathbf{s}$, so

14.3.1 $$\mathbf{r} = \mathbf{p} + \lambda \cdot (\mathbf{q} - \mathbf{p}).$$

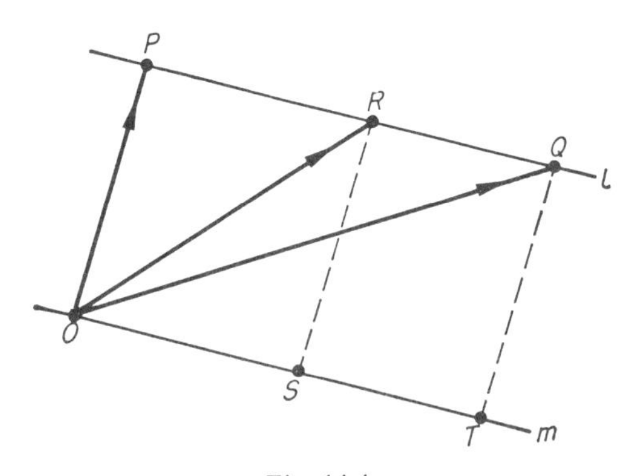

Fig. 14.4

Thus for example, the line joining (1, 2, 3) to (5, 7, 1) is the set of all $\mathbf{r}$ of the form $\mathbf{r} = (1, 2, 3) + \lambda\cdot(4, 5, -2) = (1 + 4\lambda, 2 + 5\lambda, 3 - 2\lambda)$. Starting from 14.3.1, the equation of l can be written in two ways; first as

14.3.2 $$\mathbf{r} = \mu\mathbf{p} + \lambda\mathbf{q}, \quad \text{with} \quad \lambda + \mu = 1$$

(where we have interchanged the roles of λ and μ compared with Exercise 1(iv)), and second as

14.3.3 $$\mathbf{r} = \mathbf{p} + \lambda\mathbf{t},$$

where $\mathbf{t}$ is fixed (here equal to $\mathbf{q} - \mathbf{p}$). The first form (14.3.2) is called the **vector equation** of the line (PQ), i.e.,

$$(PQ) = \{\mathbf{r}|\mathbf{r} \text{ is of the form 14.3.2}\};$$

and it tells us also that $\mathbf{r}$ *divides the segment* PQ *in the ratio* $\lambda:\mu$ (since $|PR|/|PQ| = |OS|/|OT| = \lambda$ and $\lambda + \mu = 1$). The second form (14.3.3) is also a vector equation of (PQ), but it gives (PQ) as the line through $\mathbf{p}$, parallel to a given direction OT. It is often written in the 'Cartesian' form (for $\mathbb{R}^3$):

14.3.4 $$\frac{x - p_1}{t_1} = \frac{y - p_2}{t_2} = \frac{z - p_3}{t_3}$$

(each ratio being equal to λ in 14.3.3), but this form ignores the possibility that a co-ordinate of $\mathbf{t}$ may be zero, and special cases must then be considered as well. The numbers t_1, t_2, t_3 in 14.3.4 are called the **direction ratios** of the line. When 14.3.3 is an equation concerning points in $\mathbb{R}^2$, then we have only two co-ordinates, and 14.3.3 becomes

$$t_2\cdot(x - p_1) = t_1\cdot(y - p_2)$$

—the familiar equation of a line in plane analytic geometry. It is clear from Fig. 14.4 and 14.3.2 that the *segment* PQ is the set of all $\mathbf{r}$ of the form

14.3.5 $$\mathbf{r} = \mu\mathbf{p} + \lambda\mathbf{q}, \qquad (0 \leqslant \mu \leqslant 1, \lambda + \mu = 1),$$

provided $P \neq Q$. If $P = Q$, then $\mathbf{r}$ in 14.3.5 is just $\mathbf{p}$; here then, PQ is simply the singleton $\{P\}$, and is called **degenerate**.

Exercise 3

(i) Show that the mid-point of AB is $\frac{1}{2}(\mathbf{a} + \mathbf{b})$. Show that the medians of a triangle ABC intersect in $\frac{1}{3}(\mathbf{a} + \mathbf{b} + \mathbf{c})$.
(ii) Show that the assignment $\mathbf{r} \to \lambda$ in 14.3.3 is an equivalence between l (regarded as the set of its points) and $\mathbb{R}$. Similarly show that this assignment, restricted to the segment PQ, is an equivalence $PQ \to \langle\!\langle 0, 1\rangle\!\rangle$ when $P \neq Q$.
(iii) Show that the points on a line l in $\mathbb{R}^3$ form a vector subspace, iff $\mathbf{0} \in l$.

(iv) Show that the set of all points inside and on a triangle ABC is

$$\{\mathbf{r}|\mathbf{r} = \alpha\mathbf{a} + \beta\mathbf{b} + \gamma\mathbf{c},\ \alpha + \beta + \gamma = 1,\ \alpha, \beta, \gamma \geqslant 0\}.$$

{Let the line from A to $\mathbf{r}$ join BC in M; use the fact that $\mathbf{r}$ and M are on the segments AM, BC respectively.}
Similarly, find an expression for the points inside and on a tetrahedron.
(v) A set $G \subseteq \mathbb{R}^3$ is said to be **convex** iff, given $P, Q \in G$, the whole segment PQ lies in G. Prove that the following sets are convex: a segment; a line; Π_z (in 14.1.1); $\mathbb{R}^3$; $\mathbb{R}^2$; the inside of a triangle in $\mathbb{R}^2$; the set $\{(x, y) \in \mathbb{R}^2|x^2/a^2 + y^2/b^2 \leqslant 1\}$ Show that the circle $x^2 + y^2 = 1$ in $\mathbb{R}^2$ is not convex. If G_1, G_2 are convex sets, prove that $G_1 \cap G_2$ is convex. Need $G_1 \cup G_2$ be convex?
(vi) A function $f: \mathbb{R}^3 \to \mathbb{R}$ is said to be **linear** iff it is of the form $f(x, y, z) = ax + by + cz + d$, $(a, b, c, d \in \mathbb{R})$. Prove that, on a segment PQ, a linear function f either is constant, or else takes its maximum and minimum values at the ends P, Q.
(vii) If $\mathbf{r} = \mathbf{q} + \mu\mathbf{s}$ is also an equation of the line 14.3.3, show $\exists a \in \mathbb{R}: \mathbf{t} = a\mathbf{s}$.

14.4 Lengths

Let us calculate the length of OP. If $O = P$ this length is clearly zero, so assume $O \neq P$. Drop a perpendicular PM onto Π_z (see Fig. 14.5) and drop a perpendicular MN from M onto the axis Ox. Then $|ON|$, $|NM|$ are the x and y co-ordinates of M in Π_z, so $|ON| = p_1$, $|MN| = p_2$, and $|PM| = p_3$.

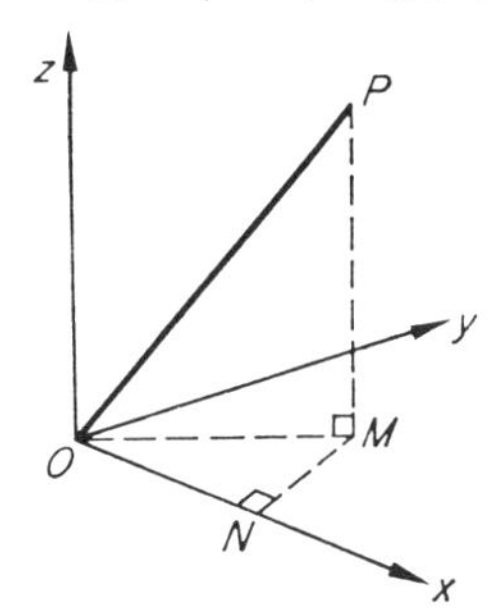

Fig. 14.5

By Pythagoras' theorem, $OP^2 = OM^2 + PM^2$ and $OM^2 = ON^2 + NM^2$; hence

$$OP^2 = p_1^2 + p_2^2 + p_3^2.$$

We therefore *define* the **norm** or **modulus** of $\mathbf{p}$ (and denote it by $|\mathbf{p}|$) to be

14.4.1 $$|\mathbf{p}| = \sqrt[+]{(p_1^2 + p_2^2 + p_3^2)}.$$

Thus $|\mathbf{i}| = |\mathbf{j}| = |\mathbf{k}| = 1$, and from Fig. 14.4 we see that

14.4.2 $$|\mathbf{p} - \mathbf{q}| = |OT| = |PQ|,$$

so $|\mathbf{p} - \mathbf{q}|$ *is the length of the segment* PQ.

Exercise 4

Let $\mathbf{r} = \mathbf{p} + \lambda\mathbf{t}$ be a point on the line through $\mathbf{p} = (1, 2, 3)$ parallel to (OT) where $\mathbf{t} = (-1, 5, 7)$. Let $\mathbf{q} = (2, 3, -1)$. Then $|\mathbf{r} - \mathbf{q}|^2$ is a function f of λ; show that it has exactly one minimum, and find the corresponding point $\mathbf{r}$.

14.5 Spheres

A **sphere** in $\mathbb{R}^3$ is defined to be any set of the form

14.5.1
$$\Sigma = \{\mathbf{r} \,|\, |\mathbf{r} - \mathbf{a}| = k\},$$

where $\mathbf{a} \in \mathbb{R}^3$ and $k \in \mathbb{R}$ are constants, called respectively the **centre** and **radius** of the sphere. Using 14.4.2, we get as the **equation** of the sphere

$$|\mathbf{r} - \mathbf{a}|^2 = k^2$$

i.e., $(x - a_1)^2 + (y - a_2)^2 + (z - a_3)^2 = k^2$; which simplifies to the form

14.5.2
$$x^2 + y^2 + z^2 + 2gx + 2fy + 2hz + c = 0.$$

If we work instead in the plane $\mathbb{R}^2$, the set 14.5.1 is called a **circle** or circumference, and equation 14.5.2 simplifies, because $z = 0$.

Exercise 5

(i) Show that any equation of the form 14.5.2 represents a sphere in $\mathbb{R}^3$ with centre $(-g, -f, -h)$, provided $f^2 + g^2 + h^2 - c \geqslant 0$; and similarly for circles in $\mathbb{R}^2$.
(ii) Prove that if l is the line with equation $\mathbf{r} = \mathbf{p} + \lambda\mathbf{q}$, and Σ is the sphere in 14.5.1, then $l \cap \Sigma$ consists of 0, 1 or 2 points, corresponding to certain values of $\mathbf{p}$ and $\mathbf{q}$.

14.6 Projections

Next let us calculate the length of the perpendicular projection of OP on a given line (OQ) when $\mathbf{p}$, $\mathbf{q}$ are distinct non-zero points in $\mathbb{R}^3$.

Let PM be the perpendicular from P onto (OQ), so that we wish to calculate $\mu = |OM|$ (see Fig. 14.6). Then

$$M = \frac{\mu}{|\mathbf{q}|} \cdot \mathbf{q} = \mathbf{m},$$

say. Therefore, since $OP^2 = OM^2 + MP^2$ and $MP^2 = |\mathbf{p} - \mathbf{m}|^2$ (by 14.4.2), then

$$\begin{aligned} OM^2 &= |\mathbf{p}|^2 - |\mathbf{p} - \mathbf{m}|^2 \\ &= (p_1^2 + p_2^2 + p_3^2) - \{(p_1 - m_1)^2 + (p_2 - m_2)^2 + (p_3 - m_3)^2\} \\ &= 2(p_1m_1 + p_2m_2 + p_3m_3) - (m_1^2 + m_2^2 + m_3^2). \end{aligned}$$

But $OM^2 = m_1^2 + m_2^2 + m_3^2$, so

$$OM^2 = (|OP| \cos \theta)^2 = p_1 m_1 + p_2 m_2 + p_3 m_3$$

and substituting $\mathbf{m} = \dfrac{\mu}{|\mathbf{q}|} \cdot \mathbf{q}$, $\mu = |OM|$, we get

14.6.1 $$|OP| \cos \theta = |OM| = (p_1 q_1 + p_2 q_2 + p_3 q_3)/|\mathbf{q}|.$$

The sum $p_1 q_1 + p_2 q_2 + p_3 q_3$ occurs so often that we have a special notation for it, calling it the **scalar product** (or **inner product**) of $\mathbf{p}$ and $\mathbf{q}$, and writing it $\mathbf{p} \cdot \mathbf{q}$.

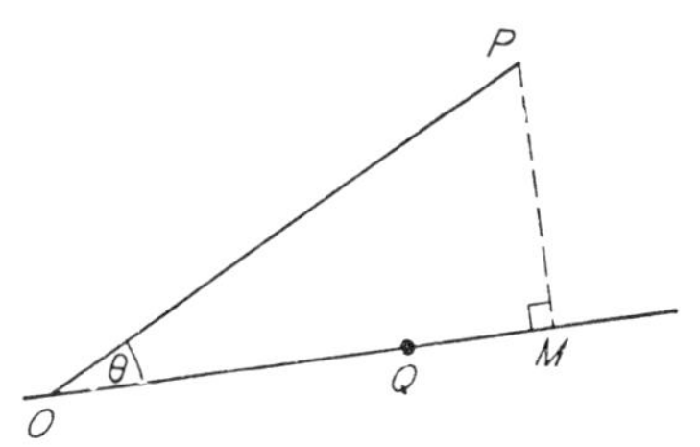

Fig. 14.6

14.6.2 DEFINITION. $\mathbf{p} \cdot \mathbf{q} = p_1 q_1 + p_2 q_2 + p_3 q_3$.
(Clearly $\mathbf{p} \cdot \mathbf{q} = \mathbf{q} \cdot \mathbf{p}$, since $p_i q_i = q_i p_i$ in $\mathbb{R}$.)

Exercise 6

(i) Evaluate $\mathbf{p} \cdot \mathbf{q}$ when $\mathbf{p} = (1, 2, 3)$, $\mathbf{q} = (-4, 6, 1)$.
(ii) For the unit points $\mathbf{i}, \mathbf{j}, \mathbf{k}$ of 14.2.1, evaluate $\mathbf{i} \cdot \mathbf{i}$, $\mathbf{i} \cdot \mathbf{j}$, etc.
(iii) Show that $(x, y, z) \cdot \mathbf{i} = x$, and similarly for y, z.

Returning to 14.6.1, in which $|\mathbf{q}| = |OQ|$, we get

14.6.3 $$|OP| \cdot |OQ| \cos \theta = \mathbf{p} \cdot \mathbf{q} = |\mathbf{p}| \cdot |\mathbf{q}| \cos \theta.$$

In particular, if $\mathbf{p} = \mathbf{q} \neq \mathbf{0}$, then $\theta = 0$; so $\cos \theta = 1$, and

14.6.4 $$\mathbf{p} \cdot \mathbf{p} = |\mathbf{p}|^2,$$

which is clear from 14.4.1 and 14.6.2, whether or not $\mathbf{p} = \mathbf{0}$. On the other hand, if $\theta = 90^\circ$, then $\cos \theta = 0$; so $\mathbf{p} \cdot \mathbf{q} = 0$. Conversely, if $\mathbf{p} \cdot \mathbf{q} = 0$, and $\mathbf{p} \neq 0 \neq \mathbf{q}$, then $\cos \theta = 0$; so $OP \perp OQ$, (briefly: $\mathbf{p} \perp \mathbf{q}$). Hence we may state:

14.6.5 THEOREM. *Suppose* $\mathbf{p} \neq 0 \neq \mathbf{q}$. *Then* $\mathbf{p} \perp \mathbf{q}$ *if and only if* $\mathbf{p} \cdot \mathbf{q} = 0$. ∎

From 14.6.4 and 14.6.5 follows:

14.6.6 $$\mathbf{i} \cdot \mathbf{i} = \mathbf{j} \cdot \mathbf{j} = \mathbf{k} \cdot \mathbf{k} = 1, \qquad \mathbf{i} \cdot \mathbf{j} = \mathbf{j} \cdot \mathbf{k} = \mathbf{k} \cdot \mathbf{i} = 0,$$

where $\mathbf{i}, \mathbf{j}, \mathbf{k}$ are given by 14.2.1.

Exercise 7

(i) In Exercise 4(i) show that the point $\mathbf{r} = \mathbf{p} + \lambda\mathbf{t}$, for which $f(\lambda)$ is least, is such that $(\mathbf{q} - \mathbf{r})\cdot\mathbf{t} = 0$. Now give a proof in general that the shortest segment from a point Q to a line l, is perpendicular to l (when Q is not on l).

(ii) A **translation** of $\mathbb{R}^3$ is any function $f: \mathbb{R}^3 \to \mathbb{R}^3$ of the form $f(\mathbf{r}) = \mathbf{r} + \mathbf{a}$ for some fixed $\mathbf{a}$. Prove that a translation preserves lengths and angles (i.e., $|\mathbf{p}-\mathbf{q}| = |f(\mathbf{p}) - f(\mathbf{q})|$, and the angle between the segments OP, OQ is that between the segments AP', AQ' where $P' = f(\mathbf{p})$, $Q' = f(\mathbf{q})$. Show that f does not preserve scalar products, i.e., in general $\mathbf{p}\cdot\mathbf{q} \neq f(\mathbf{p})\cdot f(\mathbf{q})$. Does f preserve linear independence?

(iii) In (ii) write $T(\mathbf{a})$ instead of f, to denote its dependence on $\mathbf{a}$. Prove that $T(\mathbf{a} + \mathbf{b}) = T(\mathbf{a}) \circ T(\mathbf{b})$, $T(\mathbf{0})$ = identity function on $\mathbb{R}^3$, $T(-\mathbf{a}) = (T(\mathbf{a}))^{-1}$. If $\lambda \in \mathbb{R}$, define $\lambda T(\mathbf{a})$ to be $T(\lambda\mathbf{a})$. Prove that with this definition of multiplication by scalars, the set of all translations of $\mathbb{R}^3$ is a real vector space, with addition taken to be composition of functions. [It turns out that this vector space is isomorphic to $\mathbb{R}^3$; see 15 Exercise 7(xiii)].

14.7 Vectors

We remarked above that $\mathbb{R}^2$ and $\mathbb{R}^3$ are vector spaces. Applied mathematicians, particularly, refer to objects called 'vectors' which we now explain. Let S denote the set of all (possibly degenerate†) segments $\underset{\sim}{PQ}$ in $\mathbb{R}^3$, and define an equivalence relation σ on S as follows, using the idea of a translation $f: \mathbb{R}^3 \to \mathbb{R}^3$ (see Exercise 7(ii)). Say that $\underset{\sim}{PQ}\sigma\underset{\sim}{UV}$ iff for some translation f of $\mathbb{R}^3$ (depending on $\underset{\sim}{PQ}$ and $\underset{\sim}{UV}$) $f(P) = U, f(Q) = V$. (Note that $\underset{\sim}{PQ}$ is not $\underset{\sim}{QP}$.) It then follows from 14.3.5 that f maps the segment $\underset{\sim}{PQ}$ onto $\underset{\sim}{UV}$. [Why?] We leave the reader to verify that σ is indeed an equivalence relation on S; moreover, any two degenerate segments are equivalent. The set $\mathbb{V} = S/\sigma$ is called the set of 'Euclidean vectors'; a **vector** is a σ-coset, and therefore consists of all segments $\underset{\sim}{UV}$ which are obtainable from a given one, $\underset{\sim}{PQ}$, by translations. Let $[\underset{\sim}{PQ}]$ denote the σ-coset of $\underset{\sim}{PQ}$; it contains one and only one special segment $\underset{\sim}{OT}$, starting from the origin. Define a function $\rho: \mathbb{V} \to \mathbb{R}^3$ by

$$\rho[\underset{\sim}{PQ}] = T.$$

There is also a function $\theta: \mathbb{R}^3 \to \mathbb{V}$, defined by

14.7.1 $$\theta(\mathbf{p}) = [\underset{\sim}{OP}].$$

Then $\theta \circ \rho$, and $\rho \circ \theta$, are, respectively, the identity functions on $\mathbb{V}$ and on $\mathbb{R}^3$; so by the inversion theorem (2.9.8)

14.7.2 $$\rho: \mathbb{V} \approx \mathbb{R}^3,$$

and $\theta = \rho^{-1}$. We can also introduce an algebraic structure in $\mathbb{V}$ (to make it a vector space) by the rules

14.7.3 $$[\underset{\sim}{PQ}] + [\underset{\sim}{UV}] = \theta(\rho[\underset{\sim}{PQ}] + \rho[\underset{\sim}{UV}]);$$

14.7.4 $$\lambda\cdot[\underset{\sim}{PQ}] = \theta(\lambda\cdot\rho[\underset{\sim}{PQ}]).$$

† $\underset{\sim}{PQ}$ is degenerate if $P = Q$; see the remark after 14.3.5.

On reflection, the reader will see that we are merely transferring the algebraic structure of $\mathbb{R}^3$ into $\mathbb{V}$ by means of the bijection θ—a common and useful device in algebra. Thus, in 14.7.3, we use ρ to map $[PQ]$ and $[UV]$ into $\mathbb{R}^3$, work out the sum of their images, and map it back into $\mathbb{V}$ by θ; the result is the right-hand side of 14.7.3. Similarly in 14.7.4.

It is left to the reader to verify that (a) $\mathbb{V}$, with this addition, is an abelian group, in which zero is the coset containing the degenerate segment OO, and where $-[PQ] = [QP]$; and that (b) the scalar multiplication given by 14.7.4 satisfies $\mathfrak{MS}_1$–$\mathfrak{MS}_3$ in 14.1.5 above.

Other definitions of vectors are to be found in books on applied mathematics (see, for example, Jeffreys–Jeffreys [68]). A common, but useless, definition (which is not in the reference above) is that a vector is 'anything having both magnitude and direction'. But then a ten-bedroomed house, facing North, thereby becomes a candidate for being a vector! Of course, it is necessary to be more precise: what is 'magnitude', and what is 'direction'? (See Exercise 8(i).)

Most books are not explicit about the equivalence relation σ, and at best say that 'a vector is a directed segment: two such segments PQ, UV are regarded as equal iff PQ can be mapped onto UV by a rigid translation'. The introduction of the equivalence relation σ is a mathematical way of summarizing the human activity of 'regarding as equal'. The same books sometimes define addition of vectors as follows. Referring back to Fig. 14.4 —for example—then such books define $OP + ST$ to be the third side of the triangle obtained by translating ST so that S is at P (thus $OP + ST = OR$, if in Fig. 14.4 $|PR| = |ST|$, since PR is parallel to ST). The same mode of definition gives $ST + OP = SQ$, so that to have commutative addition, OR has to be 'regarded' as equal to SQ. All this is looked after in the formulae 14.7.3, 14.7.4. Of course, in mechanics, it is sometimes necessary to work with the segment PQ rather than with the vector $[PQ]$, but then one must be careful with the algebraic laws. This is why, in pure mathematics, we prefer to work with the more basic space $\mathbb{R}^3$, rather than with $\mathbb{V}$; but the association often leads us to refer to the points of $\mathbb{R}^3$ as 'vectors'.

Exercise 8

(i) If $PQ\sigma UV$, show that PQ and UV are equal in length, and parallel. Hence invent suitable definitions of the 'magnitude' $m[PQ]$ and 'direction' of $[PQ]$. [If $m[PQ] = \lambda \cdot m[AB]$, where $\lambda \geqslant 0$ and AB represents a standard length like the metre bar in Paris, then λ is called the 'length of $[PQ]$ in metres'. This is the basis of a theory of units of measurement, and the reader should consider its relationship to such things as angle-measurement in degrees and in radians.]

(ii) In 14.7.3, 14.7.4 prove that

$$\rho([PQ] + [UV]) = \rho[PQ] + \rho[UV],$$
$$\rho(\lambda[PQ]) = \lambda \cdot \rho[PQ].$$

14.8 The Scalar Product

The product $\mathbf{p}\cdot\mathbf{q}$ given in 14.6.2 defines a function: $\mathbb{R}^3 \times \mathbb{R}^3 \to \mathbb{R}$ called the **scalar** (or **inner**) **product**; its value is always a 'scalar' (i.e., in $\mathbb{R}$, not in $\mathbb{R}^3$). To save going back to the definition each time in the future, we record some algebraic properties of the scalar product, as follows. For every $\lambda \in \mathbb{R}$, and $\mathbf{p}, \mathbf{q}, \mathbf{r} \in \mathbb{R}^3$, we have

$\mathfrak{S}\mathfrak{P}_1$: $\mathbf{p}\cdot\mathbf{q} = \mathbf{q}\cdot\mathbf{p}$ (*commutativity*);

$\mathfrak{S}\mathfrak{P}_2$: $\mathbf{p}\cdot(\mathbf{q} + \mathbf{r}) = \mathbf{p}\cdot\mathbf{q} + \mathbf{p}\cdot\mathbf{r}$ (*distributivity*);

$\mathfrak{S}\mathfrak{P}_3$: $\mathbf{p}\cdot(\lambda\mathbf{q}) = \lambda(\mathbf{p}\cdot\mathbf{q})$ (*scalar associativity*);

$\mathfrak{S}\mathfrak{P}_4$: $\mathbf{p}\cdot\mathbf{p} = 0 \;.\Leftrightarrow.\; \mathbf{p} = 0$.

These properties are immediate consequences of the definition 14.6.2 and the algebraic properties of $\mathbb{R}$. There is no question of associativity, since the formula $(\mathbf{a}\cdot\mathbf{b})\cdot\mathbf{c}$ is meaningless if both dots denote scalar products, and makes sense only if the second dot denotes scalar multiplication in the sense of 14.1.5. The student should resist any temptation to 'divide' by points of $\mathbb{R}^3$, for example to cancel $\mathbf{a}$ throughout an equation like $\mathbf{a}\cdot\mathbf{b} = \mathbf{a}\cdot\mathbf{c}$.

An important property of distance in $\mathbb{R}^3$ (hence in $\mathbb{R}^2$ and $\mathbb{R}$) is the so-called **triangle inequality.**

14.8.1 THEOREM. *Given* $\mathbf{p}, \mathbf{q}, \mathbf{r} \in \mathbb{R}^3$, *then*

$$|\mathbf{p} - \mathbf{q}| \leqslant |\mathbf{p} - \mathbf{r}| + |\mathbf{r} - \mathbf{q}|.$$

Proof. Geometrically, the inequality states that the length of the side PQ of triangle PQR does not exceed the sum of the lengths of the other two. An algebraic verification runs as follows, where we write p^2 for $\mathbf{p}\cdot\mathbf{p}$. Since a translation of R to O preserves lengths (see Exercise 7(ii)), we need only show that $|\mathbf{p} + \mathbf{q}| \leqslant |\mathbf{p}| + |\mathbf{q}|$ for all $\mathbf{p}, \mathbf{q} \in \mathbb{R}^3$. But

$$\begin{aligned}|\mathbf{p} + \mathbf{q}|^2 = (\mathbf{p} + \mathbf{q})\cdot(\mathbf{p} + \mathbf{q}) &= p^2 + q^2 + 2\mathbf{p}\cdot\mathbf{q} \qquad (\text{by } \mathfrak{S}\mathfrak{P}_1, \mathfrak{S}\mathfrak{P}_2)\\ &\leqslant p^2 + q^2 + 2|\mathbf{p}|\cdot|\mathbf{q}|\end{aligned}$$

by 14.6.3, since $|\cos\theta| \leqslant 1$. Thus $|\mathbf{p} + \mathbf{q}|^2 \leqslant (|\mathbf{p}| + |\mathbf{q}|)^2$ and we may take square roots since the lengths are positive, to get $|\mathbf{p} + \mathbf{q}| \leqslant |\mathbf{p}| + |\mathbf{q}|$ as required. ■

A proof without appeal to the cosine, and in all dimensions, will be given in 20.1.3.

Exercise 9

(i) Prove that equality occurs in 14.8.1 when, and only when, R is collinear with P and Q, and between them. Prove the inequality in one dimension without appeal to angles.

(ii) Verify $\mathfrak{S}\mathfrak{P}_1$–$\mathfrak{S}\mathfrak{P}_4$ in detail. [Some books define $\mathbf{p}\cdot\mathbf{q}$ in 14.6.2 as $|\mathbf{p}|\cdot|\mathbf{q}|\cos\theta$, but then the proof of $\mathfrak{S}\mathfrak{P}_2$ is not so easy.]

(iii) Show that if $\mathbf{p} * \mathbf{q}$ is another product, with values in $\mathbb{R}$, which satisfies the multiplicative rules 14.6.6 and $\mathfrak{SP}_1$, $\mathfrak{SP}_2$, $\mathfrak{SP}_3$, then $\mathbf{p} * \mathbf{q}$ is in fact the scalar product $\mathbf{p}\cdot\mathbf{q}$. {Substitute linear sums of the type 14.2.2 for $\mathbf{p}$ and $\mathbf{q}$ in $\mathbf{p} * \mathbf{q}$; the rules given then show that $\mathbf{p} * \mathbf{q} = p_1q_1 + p_2q_2 + p_3q_3$ by direct computation.}
(iv) Show that the equation of the sphere in 14.5.2 can be written in the form

$$|\mathbf{r}|^2 - 2\mathbf{a}\cdot\mathbf{r} + |\mathbf{a}|^2 = k^2.$$

(v) Introduce a product of vectors in $\mathbb{V}$ (see 14.7), by defining $[PQ]\cdot[UV]$ to be $\rho[PQ]\cdot\rho[UV]$. Show that this product satisfies $\mathfrak{SP}_1$, $\mathfrak{SP}_2$, $\mathfrak{SP}_3$.
(vi) Prove that, if the altitudes from the vertices A, B of a triangle ABC meet in H, then $CH \perp AB$. {In $\mathbb{V}$, $[AB] = [OA - OB]$ and $[AH]\cdot[BC] = 0$.}
(vii) Prove the rectangle property $|OP||OQ| = |OP'||OQ'|$ of a circle, for two chords PQ, $P'Q'$ meeting in O.

14.9 Planes

We can use the scalar product to derive the equation of a plane Π in $\mathbb{R}^3$, as follows. There is a unique line l through O perpendicular to Π; let P lie on this line with $|\mathbf{p}| = 1$, as in Fig. 14.7. If Π passes through $\mathbf{0}$, then since

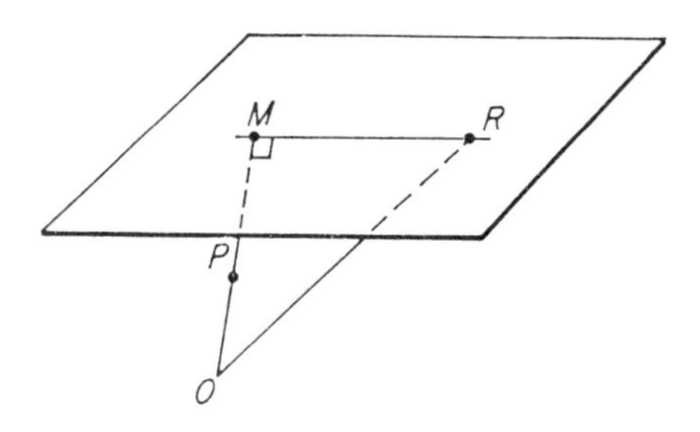

Fig. 14.7

l is perpendicular to every line in Π, we get $\mathbf{p}\cdot\mathbf{r} = 0$ for every $\mathbf{r} \in \Pi$, by 14.6.5. Conversely, if R is such that $\mathbf{p}\cdot\mathbf{r} = 0$, then $OR \perp l$, so R lies in Π. If $\mathbf{0} \notin \Pi$, let l meet Π in M. If R is any point on Π, then $l \perp MR$, so

$$|OM| = |OR|\cos MOR = |\mathbf{r}||\mathbf{p}|\cos MOR = \mathbf{r}\cdot\mathbf{p}$$

since $|\mathbf{p}| = 1$. Conversely, if S is any point such that $\mathbf{p}\cdot\mathbf{s} = |OM|$, then $|OS|\cos MOS = |OM|$; so OMR is a right angle, whence S lies in Π.

In either case, we can write

14.9.1 $$\Pi = \{\mathbf{r} | \mathbf{p}\cdot\mathbf{r} = |OM|\},$$

or equivalently, $\Pi = \{\mathbf{r} | \mathbf{q}\cdot\mathbf{r} = |\mathbf{q}||OM|\}$ for any number $|\mathbf{q}|$, with $\mathbf{q} = |\mathbf{q}|\mathbf{p}$. Now $|OM|$ is constant for Π, so the **equation** of Π is

14.9.2 $$\mathbf{q}\cdot\mathbf{r} = k, \qquad (k \text{ constant}).$$

or in co-ordinate form,

14.9.3 $$Ax + By + Cz = k, \qquad (A, B, C \in \mathbb{R}).$$

By arguing similarly in $\mathbb{R}^2$, the equation 14.9.3 takes the form $Ax + By = k$, which is again the familiar equation of a line in $\mathbb{R}^2$.

Conversely, any linear equation of the form 14.9.3 is the equation of some plane; for we can write 14.9.3 in the form 14.9.2 as $\mathbf{p}\cdot\mathbf{r} = k/|\mathbf{q}|$ where $\mathbf{p} = \mathbf{q}/|\mathbf{q}|$, and this represents a plane, by the 'converse' remark preceding 14.9.1. Hence, *a plane is the set of zeros of a linear function* $f\colon \mathbb{R}^3 \to \mathbb{R}$; where, in 14.9.3, f is given by

14.9.4 $$f(x, y, z) = Ax + By + Cz - k.$$

Of course, the zeros of f and of λf ($\lambda \in \mathbb{R}$, $\lambda \neq 0$) are identical.

Exercise 10

(i) Express the equation $x + 2y - 3z = 5$ in the form 14.9.2, to show that it represents a plane in $\mathbb{R}^3$.
(ii) If $\mathbf{a} = \lambda\mathbf{b} + \mu\mathbf{c}$, prove that in 14.9.4, $f(\mathbf{a}) = \lambda f(\mathbf{b}) + \mu f(\mathbf{c})$ iff $\lambda + \mu = 1$. Hence prove that if $\mathbf{b}, \mathbf{c} \in \Pi$, so does the entire line BC.
(iii) Prove that a plane forms a vector subspace of $\mathbb{R}^3$ (see Exercise 1(ii)) iff it passes through the origin (or, equivalently, iff $k = 0$ in 14.9.4).
(iv) Let $\mathbf{p}$ be a point in $\mathbb{R}^3$ and Π the plane in 14.9.2. Prove that the line through $\mathbf{p}$ and perpendicular to Π is $\mathbf{r} = \mathbf{p} + \lambda\mathbf{q}$. Hence find the foot of the perpendicular from $\mathbf{p}$ to Π, and show that the distance from $\mathbf{p}$ to Π is $f(\mathbf{p})/|\mathbf{q}|$.
(v) Show algebraically that the equation 14.9.4 cannot represent a line. (Geometrically this is obvious!)

INTERSECTIONS OF PLANES. Suppose Π_1, Π_2 are the zeros of linear functions f, g respectively; then $\forall\lambda, \mu \in \mathbb{R}$, $\lambda f + \mu g$ is still linear (i.e., of the form 14.9.3), so its set of zeros is a plane, which it is convenient to denote by $\lambda\Pi_1 + \mu\Pi_2$. (Exceptionally, $\lambda\Pi_1 + \mu\Pi_2$ might represent the whole of $\mathbb{R}^3$.)

We shall see in Exercise 12(vii) that if $\Pi_1 \neq \Pi_2$, then $\Pi_1 \cap \Pi_2$ is either empty (in which case Π_1 and Π_2 are said to be **parallel**) or a line in $\mathbb{R}^3$. Hence if in particular $\Pi_1 \cap \Pi_2$ contains at least three non-collinear points, then $\Pi_1 = \Pi_2$. This proves

14.9.5 PROPOSITION. *There is a unique plane through three non-collinear points of* $\mathbb{R}^3$. ■

The following proposition is important.

14.9.6 PROPOSITION. $\Pi_1 \cap \Pi_2 \subseteq \lambda\Pi_1 + \mu\Pi_2$.

Moreover, if $\Pi_1 \cap \Pi_2 \neq \varnothing$, *then every plane through* $\Pi_1 \cap \Pi_2$ *is of the form* $\lambda\Pi_1 + \mu\Pi_2$.

Proof. Let $\mathbf{p}$ be a point in the left-hand side. Then, simultaneously, $f(\mathbf{p}) = 0 = g(\mathbf{p})$, so $\lambda f(\mathbf{p}) + \mu g(\mathbf{p}) = 0$. Hence $\mathbf{p}$ is a zero of $\lambda f + \mu g$, so $\mathbf{p} \in \lambda\Pi_1 + \mu\Pi_2$; thus the required inclusion holds (whether or not $\Pi_1 \cap \Pi_2$ is empty).

As to the second assertion, it is obvious when $\Pi_1 = \Pi_2$, so suppose $\Pi_1 \neq \Pi_2$. We remarked above that then, since $\Pi_1 \cap \Pi_2$ is not empty, it is a line l in $\mathbb{R}^3$. Let Π denote a plane through l. We choose a point $\mathbf{p} \in \Pi - l$ (see Exercise 10(v)). Then the plane $\Pi' = \alpha\Pi_1 + \beta\Pi_2$ contains l (by the first part of the proof), where α, β respectively are the numbers $g(\mathbf{p})$, $-f(\mathbf{p})$; and $\mathbf{p} \in \Pi'$. Hence, since l contains at least two points $\mathbf{u}$, $\mathbf{v}$, the two planes Π, Π' each pass through the *three* non-collinear points $\mathbf{u}$, $\mathbf{v}$, $\mathbf{p}$. Therefore, by Proposition 14.9.5, Π cannot be distinct from Π', so Π is of the required form. ∎

The argument of the last proof applies to other loci besides planes; for example to lines and circles in $\mathbb{R}^2$. As an application, let us find the plane through (1, 1, 1) and the line l through (1, 2, 3) and (5, 7, 1).

14.9.7 SOLUTION. Write l in the Cartesian form 14.3.4 as

$$(x - 1)/4 = (y - 2)/5 = (z - 3)/(-2);$$

thus if $\mathbf{r} \in l$, then $\mathbf{r}$ satisfies simultaneously the linear equations $5(x - 1) = 4(y - 2)$, $-2(y - 2) = 5(z - 3)$, so $l = \Pi_1 \cap \Pi_2$ where Π_1, Π_2 are the planes given by the last two equations respectively. (The first equation is $5x - 4y + 0z = -3$, so Π_1 is parallel to the z-axis, being the locus swept out when l moves parallel to Oz; similarly Π_2 is parallel to Ox.) By 14.9.6, the plane Π we require is of the form $\lambda\Pi_1 + \mu\Pi_2$, so its equation is then

$$\lambda(5x - 4y + 3) + \mu(5z + 2y - 19) = 0.$$

Now choose λ, μ, to make Π pass through (1, 1, 1); so

$$\lambda(5 - 4 + 3) + \mu(5 + 2 - 19) = 0,$$

whence neither λ nor μ is zero, so $\lambda/\mu = 3$. Thus Π is $3\Pi_1 + \Pi_2$, with equation $3x - 2y + z = 2$.

The same technique would work with any other planes Π_1', Π_2' we might find, so long as $l = \Pi_1' \cap \Pi_2'$. It also requires only that we solve for the single number λ/μ, in contrast to some methods which solve for three unknowns or evaluate a determinant.

14.9.8 SEPARATION OF $\mathbb{R}^3$. A plane Π in $\mathbb{R}^3$, like that of 14.9.3, has the property that it 'separates' $\mathbb{R}^3$ in the following sense. Let f be a linear function as in 14.9.4, and define the sets

$$U = \{\mathbf{p} | f(\mathbf{p}) \leqslant 0\}, \qquad V = \{\mathbf{p} | f(\mathbf{p}) \geqslant 0\}$$

whose intersection is $U \cap V = \{\mathbf{p} | f(\mathbf{p}) = 0\}$, namely Π. Clearly $\mathbb{R}^3 = U \cup V$; it is usual to call U, V the two 'sides' of Π. Moreover (see the next Exercise) any two points in U can be joined by an unbroken path lying wholly in U, and similarly for V. On the other hand, if $\mathbf{p} \in U - \Pi$, $\mathbf{q} \in V - \Pi$, then there is no such path from $\mathbf{p}$ to $\mathbf{q}$ which does not cut Π. A line in $\mathbb{R}^2$ separates $\mathbb{R}^2$ (but not $\mathbb{R}^3$) similarly; and a point separates $\mathbb{R}$.

These features of the geometry of planes are the basis of the 'theory of linear inequalities' with its associated technique of 'linear programming'. An excellent detailed introduction to this is to be found in Kemeny–Snell–Thompson [72], to which for lack of space we refer the reader. However, we make the pedagogical remark that if schoolchildren met linear inequalities at school, to lessen the traditional emphasis on deriving formal *equalities*, then they might find other parts of mathematics (like inequalities in analysis, and the estimation of remainders and errors) easier than they now do. Further, the real-life problems to which the theory applies (of costing, ordering, dietary control, etc.) are enough to fire immediate enthusiasm.

Exercise 11

(i) Mark on graph paper the set of points (x, y) satisfying the simultaneous inequalities

$$x - y \leqslant 1, \qquad 3x + 5y \leqslant 2, \qquad x + 7y \leqslant -5.$$

(ii) In 14.9.8, show that the sets U, V, $U - \Pi$, $V - \Pi$ are all convex in the sense of Exercise 3(v). If $\mathbf{p} \in U - \Pi$, $\mathbf{q} \in V - \Pi$, show that any segment joining $\mathbf{p}$ to $\mathbf{q}$ cuts Π. More generally, show that any continuous path from $\mathbf{p}$ to $\mathbf{q}$ (in the sense of 15.4.1 below) cuts Π.
(iii) In the plane $\mathbb{R}^2$, let ABC be a triangle, and let U_A denote the side of the line (BC) which contains A. Let V_B, W_C be defined similarly. Prove that $U_A \cap V_B \cap W_C$ is the entire triangular disc bounded by ABC. [Compare Exercise 3(iv).]
(iv) Let $\mathbf{s}$ be a point on the surface of the sphere Σ, in 14.5.1. Find the equation of a plane through $\mathbf{s}$ and perpendicular to the radius through $\mathbf{s}$. Show that there is only one such plane, and that this plane meets Σ only in $\mathbf{s}$.
(v) Show that the line $\mathbf{r} = \mathbf{a} + \lambda\mathbf{b}$ and the plane $\mathbf{c}\cdot\mathbf{r} = k$ intersect in the single point $(k - \mathbf{c}\cdot\mathbf{a})/\mathbf{b}\cdot\mathbf{c}$, provided $\mathbf{b}\cdot\mathbf{c} \neq 0$; that if $\mathbf{b}\cdot\mathbf{c} = 0$ the line either lies in the plane, if $\mathbf{a}\cdot\mathbf{c} = k$, or the line and plane have no intersection if $\mathbf{a}\cdot\mathbf{c} \neq k$. [In this last case, the line and plane are said to be '**parallel**'; there is no $\mathbf{r} \in \mathbb{R}^3$ belonging to both, and to say (as is sometimes traditionally said) that they intersect 'at infinity' is meaningless, since 'infinity' is not a point in $\mathbb{R}^3$. See, however, the discussion of 17.6.]
(vi) If $\mathbf{a} \in \mathbb{R}^3$ is not zero, show that the set of all $\mathbf{b} \in \mathbb{R}^3$, such that $\mathbf{a} \perp \mathbf{b}$, forms a plane.
(vii) Prove that the circle $x^2 + y^2 = a^2$ separates $\mathbb{R}^2$ if $a \neq 0$, but it does not separate $\mathbb{R}^3$.
(viii) A **conic** is defined to be any subset of $\mathbb{R}^2$ with equation of the form $ax^2 + 2hxy + by^2 + 2gx + 2fy + c = 0$; it is called a **rectangular hyperbola** iff $a + b = 0$. Prove that the set of points formed by the union of a pair of perpendicular lines in $\mathbb{R}^2$ is a rectangular hyperbola. It can be shown (see Chapter 17) that given five points in $\mathbb{R}^2$, no four collinear, then there exists exactly one (possibly degenerate) conic through all of them. Hence use an argument like that for 14.9.6, to prove that any conic through the four points of intersection of two rectangular hyperbolae is also a rectangular hyperbola. Hence prove that the three altitudes of a triangle meet in a point.

(ix) In $\mathbb{R}^2$, prove that a disc S, bounded by a regular polygon, is convex. Prove that any real-valued linear function f on S takes its maximum at some vertex of S. If f is zero at three distinct non-collinear points of S, prove that f is zero everywhere on S.

(x) Find the shortest distance ρ between the lines l_1, l_2, whose equations are respectively

$$x = 1 = y - 1 = z - 1,$$

and

$$\frac{x-1}{2} = \frac{y-2}{3} = z - 3;$$

and find points P on l_1, Q on l_2 such that $|PQ| = \rho$. Find the equation of the plane through l_1 and Q.

14.10 The Vector Product

Given the non-zero points $\mathbf{a}, \mathbf{b} \in \mathbb{R}^3$, we know from Exercise 11(vi) that there is a plane of points $\mathbf{c}$ such that $\mathbf{a} \perp \mathbf{c}$. Let us ask then for the set C of all $\mathbf{c}$ such that $\mathbf{a} \perp \mathbf{c}$ and $\mathbf{b} \perp \mathbf{c}$. We shall prove:

14.10.1 PROPOSITION. *C is a line through O unless O, A, B are collinear; and then C is a plane.*

Proof. By 14.9.4, $\mathbf{c}$ must simultaneously satisfy the equations

$$\text{(i)} \qquad \mathbf{a}\cdot\mathbf{c} = 0 = \mathbf{b}\cdot\mathbf{c},$$

that is,

$$\begin{aligned} a_1c_1 + a_2c_2 + a_3c_3 &= 0 \\ b_1c_1 + b_2c_2 + b_3c_3 &= 0. \end{aligned}$$

Suppose for the moment that $a_1b_2 - a_2b_1 \neq 0$; then we can solve for c_1, c_2 in terms of c_3 to get

$$\mathbf{c} = \lambda(a_2b_3 - a_3b_2, a_3b_1 - a_1b_3, a_1b_2 - a_2b_1)$$

where $\lambda = c_3/(a_1b_2 - a_2b_1)$. Thus the set C of all possible points $\mathbf{c}$ satisfying the equations (i) is the whole of the line (OP), where P is the point (*denoted by* $\mathbf{a} \wedge \mathbf{b}$) whose co-ordinates are

$$\text{14.10.2} \qquad \mathbf{a} \wedge \mathbf{b} = (a_2b_3 - a_3b_2, a_3b_1 - a_1b_3, a_1b_2 - a_2b_1).$$

The same conclusion is reached if *any* one, rather than the third, of the co-ordinates of $\mathbf{a} \wedge \mathbf{b}$ is non-zero. If *all* three co-ordinates of $\mathbf{a} \wedge \mathbf{b}$ are zero, then we could not solve the equation (i) uniquely for $\mathbf{c}$; instead, since $\mathbf{a} \neq 0 \neq \mathbf{b}$, there exists $k \neq 0$ such that $\mathbf{a} = k\mathbf{b}$, whence O, A, B are *collinear*. Thus the equations (i) reduce to one equation saying that C is a plane through the origin.

The point $\mathbf{a} \wedge \mathbf{b}$ in 14.10.2 is called the **vector product** of $\mathbf{a}$ and $\mathbf{b}$ (denoted in some books also by $[\mathbf{a}, \mathbf{b}]$ or $\mathbf{a} \times \mathbf{b}$). It is a function $\mathbb{R}^3 \times \mathbb{R}^3 \to \mathbb{R}^3$, and by construction

14.10.3 $$\mathbf{a}\cdot(\mathbf{a} \wedge \mathbf{b}) = 0 = \mathbf{b}\cdot(\mathbf{a} \wedge \mathbf{b}).$$

The rather complicated formula 14.10.2 forces us for practical reasons to find simple characteristic algebraic properties of the vector product. We warn the reader first that it is neither commutative nor associative. It is not hard to verify that the vector product satisfies:

$\mathfrak{VP}_1$: $\mathbf{a} \wedge (\mathbf{b} + \mathbf{c}) = \mathbf{a} \wedge \mathbf{b} + \mathbf{a} \wedge \mathbf{c}$ (*distributivity*);
$\mathfrak{VP}_2$: $\mathbf{a} \wedge (\lambda\mathbf{b}) = \lambda(\mathbf{a} \wedge \mathbf{b})$ (*scalar associativity*);
(hence $\mathbf{a} \wedge \mathbf{0} = \mathbf{0}$)
$\mathfrak{VP}_3$: $\mathbf{a} \wedge \mathbf{b} = -(\mathbf{b} \wedge \mathbf{a})$ (*anti-commutativity*);
(hence $\mathbf{a} \wedge \mathbf{a} = \mathbf{0}$)
$\mathfrak{VP}_4$: the unit points $\mathbf{i}, \mathbf{j}, \mathbf{k}$ in 14.6 satisfy $\mathbf{i} \wedge \mathbf{j} = \mathbf{k}, \mathbf{j} \wedge \mathbf{k} = \mathbf{i}, \mathbf{k} \wedge \mathbf{i} = \mathbf{j}$.

The verification of these facts is left to the reader; they depend on the appropriate algebraic properties of $\mathbb{R}$. The properties $\mathfrak{VP}_1$–$\mathfrak{VP}_4$ determine the vector product completely, as we see in Exercise 12(i) below (so that if we forget the formula 14.10.2, we can always work it out from $\mathfrak{VP}_1$–$\mathfrak{VP}_4$). These properties will be useful throughout this Part.

Exercise 12

(i) Calculate $(1, 2, 3) \wedge (-1, 5, 7)$, $\mathbf{i} \wedge \mathbf{i}$, $\mathbf{i} \wedge \mathbf{j}$.
(ii) Verify the properties $\mathfrak{VP}_1$–$\mathfrak{VP}_4$.
(iii) Prove that, if $\mathbf{a} \times \mathbf{b}$ is any other product $\mathbb{R}^3 \times \mathbb{R}^3 \to \mathbb{R}^3$ satisfying $\mathfrak{VP}_1$, $\mathfrak{VP}_2$, $\mathfrak{VP}_3$, $\mathfrak{VP}_4$, then $\mathbf{a} \times \mathbf{b} = \mathbf{a} \wedge \mathbf{b}$.
(iv) Prove that $\mathbf{i} \wedge (\mathbf{i} \wedge \mathbf{j}) = -\mathbf{j}$, $(\mathbf{i} \wedge \mathbf{i}) \wedge \mathbf{j} = \mathbf{0}$, to see that the product is not associative. In fact, verify (by checking co-ordinates) the formula

$$\mathbf{a} \wedge (\mathbf{b} \wedge \mathbf{c}) = (\mathbf{a}\cdot\mathbf{c})\mathbf{b} - (\mathbf{a}\cdot\mathbf{b})\mathbf{c}.$$

(v) Prove that, if $\mathbf{a} \wedge \mathbf{b} = \mathbf{0}$ and $\mathbf{a} \neq \mathbf{0} \neq \mathbf{b}$, then $\exists k \neq 0$ such that $\mathbf{a} = k\mathbf{b}$. Hence show that we cannot also have $\mathbf{a}\cdot\mathbf{b} = 0$.
(vi) Two lines l, m in $\mathbb{R}^3$ intersect in a point N. Show that there is just one plane in $\mathbb{R}^3$ containing l and m, and its equation is

$$(\mathbf{a} \wedge \mathbf{b})\cdot(\mathbf{r} - \mathbf{n}) = 0$$

where l and m are parallel to OA, OB respectively.
(vii) Show that the planes $\mathbf{a}\cdot\mathbf{r} = p$, $\mathbf{b}\cdot\mathbf{r} = q$ intersect in a line parallel to (OC) where $C = \mathbf{a} \wedge \mathbf{b}$—provided $\mathbf{a} \wedge \mathbf{b} \neq \mathbf{0}$. What can you say if $\mathbf{a} \wedge \mathbf{b} = \mathbf{0}$? Find the equation of the line in the first case.
(viii) In the set $\mathbb{V}$ of vectors (14.7) introduce a product by defining $[PQ] \wedge [UV]$ to be $\theta(\rho[PQ] \wedge \rho[UV])$. Verify that $\mathfrak{VP}_1$–$\mathfrak{VP}_3$ are satisfied by this product. What about $\mathfrak{VP}_4$?

(ix) For any $\mathbf{a} \in \mathbb{R}^3$ prove that

$$2\mathbf{a} = \mathbf{i} \wedge (\mathbf{a} \wedge \mathbf{i}) + \mathbf{j} \wedge (\mathbf{a} \wedge \mathbf{j}) + \mathbf{k} \wedge (\mathbf{a} \wedge \mathbf{k}).$$

(x) Make up mnemonics to help you remember the formulae of 14.10.2 and that of (iv) above.

It is useful to consider the modulus of $\mathbf{a} \wedge \mathbf{b}$. Suppose that neither $\mathbf{a}$ nor $\mathbf{b}$ is zero, and let θ denote the angle between OA and OB.

14.10.4 PROPOSITION. $|\mathbf{a} \wedge \mathbf{b}| = |\mathbf{a}||\mathbf{b}| \sin \theta$.

Proof. By 14.9.2, we have $\cos \theta = \mathbf{a}\cdot\mathbf{b}/|\mathbf{a}||\mathbf{b}|$, so a simple algebraic manipulation, using $\sin^2 = 1 - \cos^2$, gives the required equation. ■

The equation in Proposition 14.10.4 gives an immediate explanation of the result of Exercise 12(v). Now the right-hand side of 14.10.4 is, of course, the area of the parallelogram with vertices $\mathbf{0}$, $\mathbf{a}$, $\mathbf{b}$ and $\mathbf{a} + \mathbf{b}$ (we denote it by $P(\mathbf{a}, \mathbf{b})$). If one of $\mathbf{a}, \mathbf{b}$ is zero, then $\mathbf{a} \wedge \mathbf{b} = \mathbf{0}$ by $\mathfrak{VP}_2$, so 14.10.4 still holds. The case when $\mathbf{a}, \mathbf{b}$ both lie in the plane Π_z of 14.1.1 is important, for then:

14.10.5 PROPOSITION. *The area of the parallelogram* P $(\mathbf{a}, \mathbf{b})$ (taken positive) is $|a_1 b_2 - a_2 b_1|$.

Proof. $|\mathbf{a} \wedge \mathbf{b}| = |(a_1, a_2, 0) \wedge (b_1, b_2, 0)| = |(0, 0, a_1 b_2 - a_2 b_1)|$. ■

For some purposes, it is convenient to think of the area together with an 'orientation' or normal direction. We may then just as well take the **oriented area** of $P(\mathbf{a}, \mathbf{b})$ to be $\mathbf{a} \wedge \mathbf{b}$, since the important information—size and normal direction—are here combined. [See also the Appendix to Chapter 15.] The **signed area** of $P(\mathbf{a}, \mathbf{b})$ will mean the third co-ordinate of $\mathbf{a} \wedge \mathbf{b}$.

14.11 Volumes

The remark immediately following 14.10.5 can be used to evaluate the volume V of the solid parallelepiped $P(\mathbf{a}, \mathbf{b}, \mathbf{c})$ determined as in Fig. 14.8 by three non-zero points $\mathbf{a}, \mathbf{b}, \mathbf{c} \in \mathbb{R}^3$.

14.11.1 PROPOSITION. $V = |\mathbf{a}\cdot(\mathbf{b} \wedge \mathbf{c})|$.

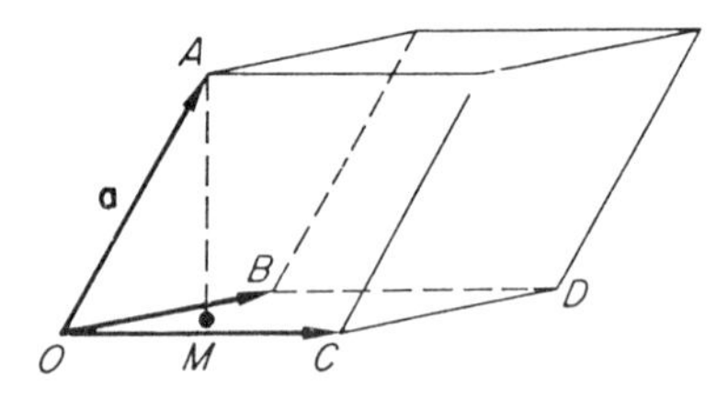

Fig. 14.8

Proof. Regarding $OBDC$ as the base, let AM denote the perpendicular from A to it, of length h. Then V is the product of the positive number $|AM|$, with the (positive) area of $OBDC$, so $V = |AM| \cdot |\mathbf{b} \wedge \mathbf{c}|$ by Proposition 14.10.4. Now AM is parallel to the direction of $\mathbf{b} \wedge \mathbf{c}$, since the latter is by definition perpendicular to both $\mathbf{b}$ and $\mathbf{c}$ and hence to $OBDC$. Thus $|AM|$ is equal to the projection of OA onto the direction of the unit vector $\mathbf{n}$ along $\mathbf{b} \wedge \mathbf{c}$; so (taking all lengths positive)

$$|AM| = |\mathbf{a} \cdot \mathbf{n}| = |\mathbf{a} \cdot ((\mathbf{b} \wedge \mathbf{c})/|\mathbf{b} \wedge \mathbf{c}|)|.$$

Therefore since $|\mathbf{b} \wedge \mathbf{c}|$ is a non-zero number, it cancels out of the earlier formula for V, so $V = |\mathbf{a} \cdot (\mathbf{b} \wedge \mathbf{c})|$ as required. ■

Observe that no brackets are really needed for the product in 14.11.1 since $(\mathbf{a} \cdot \mathbf{b}) \wedge \mathbf{c}$ is meaningless. This formula would still be correct if $\mathbf{b}$ and $\mathbf{c}$ were collinear with O (for then $OBDC$ in Fig. 14.8 has zero area, so $V = 0$); or when $\mathbf{a}$ is in the plane $OBDC$ (for then $|AM| = 0$).

If in 14.11.1 we take $\mathbf{b}$ in the role of $\mathbf{a}$, we obtain $V = |\mathbf{b} \cdot \mathbf{c} \wedge \mathbf{a}|$; so $\mathbf{a} \cdot \mathbf{b} \wedge \mathbf{c} = \pm(\mathbf{b} \cdot \mathbf{c} \wedge \mathbf{a})$. But by checking the coefficients of the term $a_1 b_2 c_3$ on each side we obtain

14.11.2 $$\mathbf{a} \cdot \mathbf{b} \wedge \mathbf{c} = \mathbf{b} \cdot \mathbf{c} \wedge \mathbf{a} = \mathbf{c} \cdot \mathbf{a} \wedge \mathbf{b}.$$

For algebraic reasons, it is convenient to consider volumes with a sign; so the **signed volume** of P $(\mathbf{a}, \mathbf{b}, \mathbf{c})$ is defined to be

14.11.3 $$\operatorname{Vol} P(\mathbf{a}, \mathbf{b}, \mathbf{c}) = \mathbf{a} \cdot \mathbf{b} \wedge \mathbf{c}.$$

By 14.11.2, $\operatorname{Vol} P(\mathbf{a}, \mathbf{b}, \mathbf{c}) = \operatorname{Vol} P(\mathbf{b}, \mathbf{c}, \mathbf{a})$; and by the property $\mathfrak{VP}_3$ it is also $-\operatorname{Vol} P(\mathbf{a}, \mathbf{c}, \mathbf{b})$.

Exercise 13

(i) Each point (x, y, z) is transformed into a new point (x', y', z') according to the rule

$$x' = 3x + 15y - 7z$$
$$y' = 5x + 2y - 3z$$
$$z' = -13x + 7y + 5z.$$

If the tetrahedron T with vertices $(0, 0, 0)$, (x_i, y_i, z_i), $i = 1, 2, 3$, transforms into a new tetrahedron T', show that the centroid of T transforms into that of T'. In the special case when $(x_1, y_1, z_1) = (1, 0, 0)$; $(x_2, y_2, z_2) = (0, 1, 0)$; $(x_3, y_3, z_3) = (0, 0, 1)$, find the new vertices and volume of T'.

(ii) Let $T(\mathbf{r})$ denote a translation (see Exercise 7(ii), (iii)). Give reasons for *defining* $\operatorname{Vol} T(P(\mathbf{a}, \mathbf{b}, \mathbf{c}))$ to be $\operatorname{Vol} P(\mathbf{a}, \mathbf{b}, \mathbf{c})$, where $T = T(\mathbf{r})$.

(iii) Prove that $P(\mathbf{a}, \mathbf{b}, \mathbf{c})$ is the set of all $\mathbf{r} \in \mathbb{R}^3$ of the form $\lambda\mathbf{a} + \mu\mathbf{b} + \nu\mathbf{c}$, where $0 \leqslant \lambda \leqslant 1$, $0 \leqslant \mu \leqslant 1$, $0 \leqslant \nu \leqslant 1$ and $\lambda + \mu + \nu = 1$, $(\lambda, \mu, \nu \in \mathbb{R})$.

(iv) Show by an example that $P(\mathbf{a}, \mathbf{b}, \mathbf{c} + \mathbf{d})$ need not contain $P(\mathbf{a}, \mathbf{b}, \mathbf{c})$. Show however that $\operatorname{Vol} P(\mathbf{a}, \mathbf{b}, \mathbf{c} + \mathbf{d}) = \operatorname{Vol} P(\mathbf{a}, \mathbf{b}, \mathbf{c}) + \operatorname{Vol} P(\mathbf{a}, \mathbf{b}, \mathbf{d})$, using the algebraic properties of the scalar and vector products.

(v) Prove that three points $\mathbf{a}, \mathbf{b}, \mathbf{c} \in \mathbb{R}^3$ are linearly independent (see 14.2) iff $\mathbf{a} \cdot \mathbf{b} \wedge \mathbf{c} \neq \mathbf{0}$. [Hence $\mathbf{a}$, $\mathbf{b}$, $\mathbf{c}$ form a basis iff $P(\mathbf{a}, \mathbf{b}, \mathbf{c})$ has non-zero volume.]
(vi) Let $\mathbf{a}, \mathbf{b}, \mathbf{c} \in \mathbb{R}^3$ be three linearly independent points. The lattice $\Omega = \Omega(\mathbf{a}, \mathbf{b}, \mathbf{c})$ is defined to be the set of all $\mathbf{r} \in \mathbb{R}^3$ of the form

$$\mathbf{r} = l\mathbf{a} + m\mathbf{b} + n\mathbf{c}$$

where l, m, n are arbitrary *integers*.
Show that Ω is an abelian group under addition. Let $T(\mathbf{r})$ denote translation by $\mathbf{r}$, as in (ii) above; prove that the image of $P(\mathbf{a}, \mathbf{b}, \mathbf{c})$ under $T(\mathbf{r})$ is a parallelepiped with all its vertices in Ω if $\mathbf{r} \in \Omega$. Such a parallelepiped is called a *cell* of Ω, and by (ii) its signed volume equals that of $P(\mathbf{a}, \mathbf{b}, \mathbf{c})$. Prove that $\mathbb{R}^3$ is the union of all the cells of Ω.
(vii) Do the problem analogous to (vi), in $\mathbb{R}^2$ and in $\mathbb{R}^1$.

Remark. In some books on linear algebra, the reader will find the product $\mathbf{a} \wedge \mathbf{b}$ used in a kind of algebra, called 'exterior algebra', where a vector space $V \wedge V$ is constructed from each vector space V. It turns out that V is isomorphic to $V \wedge V$ iff $\dim V = 3$ or 0. This is one of the many reasons why $\mathbb{R}^3$ is so important among the vector spaces (and why we may use the notation $\mathbf{a} \wedge \mathbf{b}$ in $\mathbb{R}^3$ itself).

Chapter 15

LINEAR ALGEBRA AND MEASURE IN $\mathbb{R}^3$

15.1 Matrices and Determinants

We often need to consider arrays of real numbers of the form

15.1.1
$$M = \begin{pmatrix} a & b \\ c & d \end{pmatrix}.$$

Here, M is called a 2×2 **matrix**, with real 'entries' a, b, c, d; or briefly, M is a real (2×2) matrix. These arise for example when solving the linear equations

$$ax + by = p,$$
$$cx + dy = q$$

to remind us of the positions of the coefficients; and the positions are so important† that we only say that M is also the matrix $\begin{pmatrix} u & v \\ w & t \end{pmatrix}$ when $a = u$, $b = v$, $c = w$, and $d = t$. Matrices serve a display function, then; for example, M can display a 'shopping list' on a certain date, giving the instruction to buy butter and sugar at two shops X, Y in the quantities a, b units of butter at X, Y respectively and c, d units of sugar similarly. This second application suggests certain algebraic operations with matrices. For suppose $N = \begin{pmatrix} e & f \\ g & h \end{pmatrix}$ is a shopping list at a later date for the two products at the two shops. Then the total purchase for the two occasions is given by a matrix naturally denoted by $M + N$, where

$$M + N = \begin{pmatrix} a + e, & b + f \\ c + g, & d + h \end{pmatrix}.$$

If the demand for butter and sugar were halved, then the shopping list becomes $\begin{pmatrix} \frac{1}{2}a, & \frac{1}{2}b \\ \frac{1}{2}c, & \frac{1}{2}d \end{pmatrix}$, naturally denoted by $\frac{1}{2}M$; more generally we define tM for each $t \in \mathbb{R}$ by

$$tM = \begin{pmatrix} ta & tb \\ tc & td \end{pmatrix}.$$

† Hence the use of the word 'matrix' (a term used by printers for a mould used in making type). The plural is 'matrices', usually pronounced 'maytrisseez', not 'mattresses'.

If nothing at all is bought, the shopping list becomes the 'zero' matrix

$$O = \begin{pmatrix} 0 & 0 \\ 0 & 0 \end{pmatrix};$$

Clearly, $0M = O$ for every M. And if instead of buying the items in M we sell them back, the resulting transaction can be recorded as $-M$, where

$$-M = \begin{pmatrix} -a & -b \\ -c & -d \end{pmatrix};$$

clearly $-1M = -M$. We now leave the reader to check that the set $\mathscr{M}_2$ of all 2×2 matrices with real entries satisfies the axioms $\mathfrak{A}_1$–$\mathfrak{A}_4$ for an abelian group under addition, and also the axioms $\mathfrak{MS}_1$–$\mathfrak{MS}_3$ for multiplication by scalars. Thus $\mathscr{M}_2$ forms a real vector space (see 14.1.5) which can be shown to have dimension 4. It has additional algebraic structure also; see Exercise 7(iv).

We define a function, det: $\mathscr{M}_2 \to \mathbb{R}$ by

15.1.2 $$\det M = ad - bc$$

with M as in 15.1.1. Although it is without significance on shopping lists, it is important geometrically. For, by Proposition 14.10.5, the number $\det M$ is the signed area of each of the parallelograms $P(\mathbf{u}, \mathbf{v})$, $P(\mathbf{r}, \mathbf{s})$ where

$$\mathbf{u} = (a, b), \qquad \mathbf{v} = (c, d); \qquad \mathbf{r} = (a, c), \qquad \mathbf{s} = (b, d) \in \mathbb{R}^2.$$

Thus also

15.1.3 $$\det M = \det \begin{pmatrix} a & c \\ b & d \end{pmatrix}.$$

A common notation for the right-hand side is $\begin{vmatrix} a & c \\ b & d \end{vmatrix}$, but it has the disadvantage of muddling the matrix $\begin{pmatrix} a & c \\ b & d \end{pmatrix}$ with the number $ad - bc$. The symbol $\begin{vmatrix} a & c \\ b & d \end{vmatrix}$ is then often called a '2×2 determinant'. In the formulation above, however, the determinant is a *function* with domain $\mathscr{M}_2$ and range $\mathbb{R}$.

Exercise 1

(i) Prove that $\det(tM) = t^2 \det M$.
(ii) Find examples to show that $\det(M + N)$ need not equal $\det M + \det N$.
(iii) If $\det M = 0$, prove that one row of M is a multiple of the other; and similarly for columns.

15.1.4 Next, we consider 3×3 matrices, and the reader may like to keep in mind the analogy of a 'shopping list' like the one used above (which obviously extends to '$m \times n$' matrices, for m products and n shops). This analogy will not be pursued here, since it is treated in detail in Kemeny–Snell–Thompson [72] and books on mathematical economics. A typical 3×3 matrix is a formula of the form

$$\begin{pmatrix} a & b & c \\ d & e & f \\ g & h & i \end{pmatrix}.$$

Two such matrices are called **equal** iff entries in corresponding positions are equal. Let M denote the 3×3 matrix shown above. Somewhat arbitrarily, it may seem, we define its determinant, det M, to be the number

15.1.5 $$\det M = a \det \begin{pmatrix} e & f \\ h & i \end{pmatrix} - b \det \begin{pmatrix} d & f \\ g & i \end{pmatrix} + c \det \begin{pmatrix} d & e \\ g & h \end{pmatrix}.$$

Here, then, det is a function with range $\mathbb{R}$ and domain the set $\mathscr{M}_3$ of all 3×3 matrices; the notation

$$\begin{vmatrix} a & b & c \\ d & e & f \\ g & h & i \end{vmatrix}$$

is often used for det M, but this can be confusing, since it muddles the *number* det M with the *matrix* M.

From 15.1.2 and 15.1.5, the definitions of scalar and vector products in Chapter 14 show at once that

15.1.6 $$\mathbf{a} \cdot \mathbf{b} \wedge \mathbf{c} = \det \begin{pmatrix} a_1 & a_2 & a_3 \\ b_1 & b_2 & b_3 \\ c_1 & c_2 & c_3 \end{pmatrix}$$

by expanding each side in full, and comparing terms. Direct calculation shows also that

15.1.7 $$\mathbf{a} \cdot \mathbf{b} \wedge \mathbf{c} = \mathbf{u} \cdot \mathbf{v} \wedge \mathbf{w}$$

where

$$\mathbf{u} = (a_1, b_1, c_1), \qquad \mathbf{v} = (a_2, b_2, c_2), \qquad \mathbf{w} = (a_3, b_3, c_3).$$

If M now denotes the matrix on the right-hand side of 15.1.6, we denote by M^T the matrix

15.1.8 $$\begin{pmatrix} a_1 & b_1 & c_1 \\ a_2 & b_2 & c_2 \\ a_3 & b_3 & c_3 \end{pmatrix}$$

which is called the **transpose** of M, and is obtained from M by interchanging the rows and the columns (or by reflecting M in its 'diagonal'). Thus, like 15.1.3, 15.1.7 says:

$\mathfrak{M}_1$: $$\det M = \det M^T.$$

To minimize the amount of computation, we formulate some rules concerning 3×3 determinants. By $\mathfrak{M}_1$ a rule about rows or columns of M is also a rule about the columns or rows, respectively, of M^T. Thus, the following rules, stated for columns, have direct analogues for rows which the reader is left to supply. Observe that the rules are valid also for the determinants of 2×2 matrices.

$\mathfrak{M}_2$: *If two columns of M are interchanged to make a new matrix N, then*

$$\det M = -\det N.$$

Proof. Suppose the interchanged columns are $\mathbf{a}$ and $\mathbf{b}$ in 15.1.8. Then by 15.1.6, $\det M = \mathbf{c}\cdot\mathbf{a} \wedge \mathbf{b}$, $\det N = \mathbf{c}\cdot\mathbf{b} \wedge \mathbf{a}$, and since $\mathbf{a} \wedge \mathbf{b} = -\mathbf{b} \wedge \mathbf{a}$, then $\det M = -\det N$. Similarly if $\mathbf{c}$ is one of the interchanged columns. ■

$\mathfrak{M}_3$: *If two columns of M are equal, then $\det M = 0$.*

Proof. If we interchange the two equal columns, M is unaltered, while its determinant changes sign (by $\mathfrak{M}_2$). Hence $\det M = -\det M$, so $\det M = 0$. ■

$\mathfrak{M}_4$: *If $\lambda \in \mathbb{R}$, and N is the matrix obtained from M by multiplying each entry in one column of M by λ, then* $\det N = \lambda\cdot\det M$.

Proof. Immediate from 15.1.6, using $\mathfrak{SP}_3$. ■

$\mathfrak{M}_5$: *If a multiple of one column of M is added to another, to form a new matrix N, then* $\det M = \det N$.

Proof. With an obvious notation, using 15.1.6,

$$\det N = (\mathbf{a} + \lambda\mathbf{b})\cdot(\mathbf{b} \wedge \mathbf{c}) = \mathbf{a}\cdot\mathbf{b} \wedge \mathbf{c} = \det M,$$

since $\mathbf{b}\cdot\mathbf{b} \wedge \mathbf{c} = 0$ by 14.10.3. ■

Matrices and determinants of higher order n will be defined in Chapter 19, and rules $\mathfrak{M}_1$–$\mathfrak{M}_5$ still hold for them. The reader should compare the proofs given here with those for general n.

Exercise 2

(i) Construct some 3×3 matrices with specific numbers as entries, and evaluate their determinants using the rules. Can you devise checks on the correctness of your computations?

(ii) By analogy with $\mathcal{M}_2$ in 15.1.2, introduce addition and scalar multiplication into the set $\mathcal{M}_3$ of all real 3×3 matrices, and verify that with your definitions, $\mathcal{M}_3$ is a real vector space (i.e., check that $\mathcal{M}_3$ is an additive group, satisfying conditions $\mathfrak{MS}_1$–$\mathfrak{MS}_3$ in 14.1.5).
(iii) Show that a 3×3 matrix may be regarded as a function $f\colon \mathbb{N}_3 \times \mathbb{N}_3 \to \mathbb{R}$, and that the definition of equality of functions then forces the definition of equality of matrices given prior to 15.1.3. [It is then customary to write f_{ij} instead of $f(i, j)$ and to denote f by (f_{ij}), meaning the matrix whose entry in the ith row and jth column is f_{ij}.]
(iv) Let $\{\mathbf{e}_1, \mathbf{e}_2, \mathbf{e}_3\}$, $\{\mathbf{b}_1, \mathbf{b}_2, \mathbf{b}_3\}$ be two bases for $\mathbb{R}^3$ (see 14.2.6). Then we can express the $\mathbf{b}$'s in terms of the $\mathbf{e}$'s by means of equations

$$\mathbf{e}_i = \sum_j p_{ij}\mathbf{b}_j, \qquad \mathbf{b}_i = \sum_j q_{ij}\mathbf{e}_j \qquad (i = 1, 2, 3),$$

(which incidentally display the economy of the 'f_{ij}' notation). Substitute the second equation in the first, to deduce that for each i, j

$$\sum_r p_{ir}q_{rj} = \delta_{ij},$$

where the 'Kronecker delta' δ_{ij} is defined to be 1 if $i = j$, and 0 otherwise. [It now follows (see Exercise 7(iv) below) that both matrices (p_{ij}), (q_{ij}) have non-zero determinant.]

15.2 Three Linear Equations

Let Π_1, Π_2, Π_3 be three planes in $\mathbb{R}^3$, with equations, respectively

15.2.1 $$\mathbf{a}\cdot\mathbf{r} = h_1, \qquad \mathbf{b}\cdot\mathbf{r} = h_2, \qquad \mathbf{c}\cdot\mathbf{r} = h_3.$$

Then it is often necessary to be able to describe the set $X = \Pi_1 \cap \Pi_2 \cap \Pi_3$. Before doing any calculations, let us describe X qualitatively. There are several possibilities, some being shown in Fig. 15.1.

In (i) the three planes coincide and each is X; in (ii) and (iii) they are parallel, and $X = \varnothing$. In the rest, all of the planes are distinct and no two are parallel; hence Π_1, Π_2 intersect in a line l (14 Exercise 12(vii)) which either is parallel to Π_3 (in (iv), so $X = \varnothing$), or lies in Π_3 (in (v), so $X = l$), or cuts Π_3 in just one point (in (vi), so X is a singleton). Thus forewarned of the complexity of the problem, let us now deal with it algebraically.

Any point $\mathbf{r} \in X$ must lie in Π_1, Π_2, and Π_3 and so must satisfy the three linear 'simultaneous' equations

15.2.2 $$\begin{aligned} a_1x + a_2y + a_3z &= h_1 \\ b_1x + b_2y + b_3z &= h_2 \\ c_1x + c_2y + c_3z &= h_3; \end{aligned}$$

conversely, any (x, y, z) satisfying these equations lies in X. *The problem of 'solving' these equations is precisely that of describing X fully.* Once this is

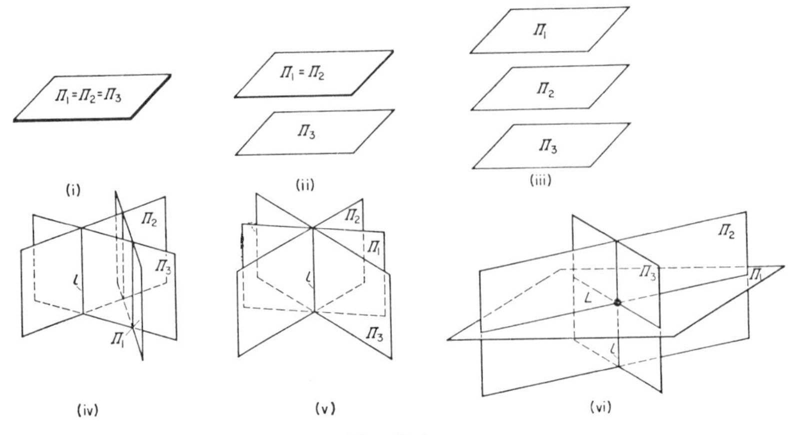

Fig. 15.1

understood, the various possible forms for X should cause no surprise: nevertheless, schoolboys who have dealt only with case (vi) and without the underlying geometry, are particularly aghast at the idea that X can be empty.

Let us now proceed to describe X algebraically, starting from the equations 15.2.2. They can be written as one single equation:

15.2.3 $$x\mathbf{u} + y\mathbf{v} + z\mathbf{w} = \mathbf{h}$$

where $\mathbf{u} = (a_1, b_1, c_1)$, $\mathbf{v} = (a_2, b_2, c_2)$, $\mathbf{w} = (a_3, b_3, c_3)$ and $\mathbf{h} = (h_1, h_2, h_3)$. Assuming that 15.2.3 is satisfied by some $\mathbf{r} \in X$, we can take the scalar product with $\mathbf{v} \wedge \mathbf{w}$; then using $\mathfrak{SP}_2$ with the facts (14.10.3) that $\mathbf{v}\cdot\mathbf{v} \wedge \mathbf{w} = \mathbf{w}\cdot\mathbf{v} \wedge \mathbf{w} = \mathbf{0}$, we have

15.2.4 $$x(\mathbf{u}\cdot\mathbf{v} \wedge \mathbf{w}) = \mathbf{h}\cdot\mathbf{v} \wedge \mathbf{w};$$

thus *if* there exists an $\mathbf{r} \in X$, and if $\mathbf{u}\cdot\mathbf{v} \wedge \mathbf{w} \neq 0$, then

15.2.5 $$x = \mathbf{h}\cdot(\mathbf{v} \wedge \mathbf{w})/\Delta, \qquad y = \mathbf{h}\cdot(\mathbf{w} \wedge \mathbf{u})/\Delta, \qquad z = \mathbf{h}\cdot(\mathbf{u} \wedge \mathbf{v})/\Delta,$$

where (using 14.11.2)

15.2.6 $$\Delta = \mathbf{u}\cdot\mathbf{v} \wedge \mathbf{w} = \mathbf{v}\cdot\mathbf{w} \wedge \mathbf{u} = \mathbf{w}\cdot\mathbf{u} \wedge \mathbf{v}.$$

But the triple (x, y, z), given by 15.2.5 *does* satisfy 15.2.3, as can be seen by writing out the products in full and collecting up terms. Hence, providing $\Delta \neq 0$, 15.2.3 has exactly one solution, given by 15.2.5. This is called **Cramer's rule**; it is not very efficient for practical computation, but it is useful for theoretical work.

If $\Delta = 0$, then we can obain 15.2.4 from 15.2.3, but 15.2.5 is now meaningless. We therefore ask ourselves just how Δ happens to be zero. First, we use 15.1.7 to assert that $\Delta = \mathbf{a}\cdot\mathbf{b} \wedge \mathbf{c}$. Then one reason for Δ being zero, might be that $\mathbf{a}$, $\mathbf{b}$, and $\mathbf{c}$ are *all* zero, a possibility which must be faced in a purely algebraic treatment of the equations 15.2.2, although we would then have no planes Π_i in 15.2.1. [Such 'strange' cases arise when considering the stationary points of functions of several variables.] In this case, the whole of $\mathbb{R}^3$ forms the set of solutions of 15.2.2, provided $\mathbf{h} = \mathbf{0}$; otherwise there are no solutions. Next, assume that $\mathbf{a}$, $\mathbf{b}$, $\mathbf{c}$ are not all zero, say $\mathbf{a} \neq \mathbf{0}$. Then think of Δ as a volume, as in 14.11.1; it might happen that the parallelepiped $P(\mathbf{a}, \mathbf{b}, \mathbf{c})$ has collapsed to a segment, so that $\mathbf{a}$, $\mathbf{b}$, $\mathbf{c}$ are all collinear with $\mathbf{0}$. Hence there exist numbers $\beta, \gamma \in \mathbb{R}$ such that

15.2.7 $$\mathbf{b} = \beta\mathbf{a}, \qquad \mathbf{c} = \gamma\mathbf{a}$$

and then the planes Π_i would look as in (i), (ii), or (iii) of Fig. 15.1. If there were an $\mathbf{r}$ satisfying 15.2.2, then

15.2.8 $$h_2 = \mathbf{b}\cdot\mathbf{r} = (\beta\mathbf{a})\cdot\mathbf{r} = \beta h_1, \qquad h_3 = \gamma h_1;$$

and conversely, if $h_2 = \beta h_1$ and $h_3 = \gamma h_1$, then $\Pi_1 = \Pi_2 = \Pi_3$ and the set X of solutions is Π_1 itself. Now we assumed above that $\mathbf{a} \neq 0$; thus some co-ordinate (say a_1) is non-zero. Hence

15.2.9 $$X = \Pi_1 = \{(x, y, z) | x = (h_1 - (a_2 y + a_3 z))/a_1\}.$$

If 15.2.7 is false, then *no* $\mathbf{r} \in \mathbb{R}^3$ satisfies 15.2.2, so the solution set X is empty. (In a numerical problem, all this can be seen at a glance.)

One remaining reason why Δ should be zero is that the parallelepiped $P(\mathbf{a}, \mathbf{b}, \mathbf{c})$ has a base but zero height; that is (contrary to 15.2.7) one of the products

$$\mathbf{a} \wedge \mathbf{b}, \qquad \mathbf{b} \wedge \mathbf{c}, \qquad \mathbf{c} \wedge \mathbf{a}$$

is non-zero, say $\mathbf{b} \wedge \mathbf{c} \neq \mathbf{0}$ (implying that $b_1c_2 - b_2c_1$, say, is not zero). Thus O, B, C are points in the base of P, which forms a non-zero plane area, in which $\mathbf{a}$ lies. Hence by 14 Exercise 1(iii), numbers $\lambda, \mu \in \mathbb{R}$ exist so that

15.2.10 $$\mathbf{a} = \lambda\mathbf{b} + \mu\mathbf{c}.$$

[That is, λ, μ can be found by solving the equations $a_1 = \lambda b_1 + \mu c_1$, $a_2 = \lambda b_2 + \mu c_2$ (since $b_1c_2 - b_2c_1 \neq 0$); the solution automatically satisfies $a_3 = \lambda b_3 + \mu c_3$.] Going back to 15.2.2, suppose that there is a solution $\mathbf{r} \in X$. Then (using $\mathfrak{SP}_2$, $\mathfrak{SP}_3$):

15.2.11 $$h_1 = \mathbf{a}\cdot\mathbf{r} = (\lambda\mathbf{b} + \mu\mathbf{c})\cdot\mathbf{r} = \lambda(\mathbf{b}\cdot\mathbf{r}) + \mu(\mathbf{c}\cdot\mathbf{r}) = \lambda h_2 + \mu h_3.$$

Hence by 14.9.6, Π_1 is a plane through $\Pi_2 \cap \Pi_3$; and $\Pi_2 \cap \Pi_3$ is a line, since $\mathbf{b} \wedge \mathbf{c} \neq \mathbf{0}$. Hence, as in Fig. 15.1(v), the solution-set X is the entire line $\Pi_2 \cap \Pi_3$, which is given by

15.2.12 $$X = \{\mathbf{r} | \mathbf{r} = \frac{1}{d}(h_2c_2 - h_3b_2, h_3b_1 - h_2c_1, 0) + t\mathbf{b} \wedge \mathbf{c}\},$$

where $d = b_1c_2 - b_2c_1 \neq 0$. If $h_1 \neq \lambda h_2 + \mu h_3$ (with this same λ, μ), then Π_1 is parallel to $\Pi_2 \cup \Pi_3$, so $X = \varnothing$, as in Fig. 15.1(iv).

This completes the analysis of all the possibilities for Δ in 15.2.6.

15.2.13 A FLOW CHART. When performing a numerical solution of a set of equations like 15.2.2, it is best to imagine the work arranged as in the following abbreviated 'flow chart', to cope with the various dilemmas as they occur.

Step 1. Is $\Delta \neq 0$?
Yes: apply Cramer's rule 15.2.6 (or any other, more efficient method to find the unique solution); *stop.*
No: proceed to:

Step 2. Is $\mathbf{a} = \mathbf{b} = \mathbf{c} = \mathbf{0}$?

Yes: $\left\{\begin{matrix}\text{If } \mathbf{h} = \mathbf{0}, \text{ then } X = \mathbb{R}^3 \\ \text{If } \mathbf{h} \neq \mathbf{0}, \text{ then } X = \varnothing\end{matrix}\right\}$; *stop.*

No: proceed to:

Step 3. Is $\mathbf{a} \wedge \mathbf{b} = \mathbf{b} \wedge \mathbf{c} = \mathbf{c} \wedge \mathbf{a} = \mathbf{0}$?

Yes: Find $\mathbf{a} \neq \mathbf{0}$, and find β, γ as in 15.2.7;

$$\left\{\begin{matrix}\text{If 15.2.8 holds, } X = \Pi_1 \text{ given by 15.2.9} \\ \text{If 15.2.8 false, } X = \varnothing\end{matrix}\right\}; \textit{stop.}$$

No: proceed to:

Step 4. Some product, say $\mathbf{b} \wedge \mathbf{c}$, $\neq \mathbf{0}$. Find λ, μ as in 15.2.10;

$$\left\{\begin{matrix}\text{If 15.2.11 holds, } X = \Pi_2 \cap \Pi_3 \text{ given by 15.2.12} \\ \text{If 15.2.11 false, } X = \varnothing\end{matrix}\right\}; \textit{stop.}$$

Our discussion of 15.2.2 was motivated by geometry, although the final solution makes sense independently of geometry. Hence, we can apply the method to sets of equations 15.2.2 in which the a's and x's *lie in any field*; and more general applications are possible: see Exercises 3(iii), (v) below.

Exercise 3

(i) If, in equations 15.2.2, $\mathbf{h} = \mathbf{0}$, then a solution $(x, y, z) \neq \mathbf{0}$ is called **non-trivial.** (Obviously, $\mathbf{0}$ is a solution—called the **trivial** solution.) Prove that a non-trivial solution exists iff $\Delta = 0$. {Use 15.2.7 and 15.2.10.} Hence show that $\mathbf{a}, \mathbf{b}, \mathbf{c}$ are linearly dependent iff $\mathbf{p}\cdot\mathbf{q} \wedge \mathbf{r} = 0$ where $\mathbf{p} = (\mathbf{a}\cdot\mathbf{a}, \mathbf{a}\cdot\mathbf{b}, \mathbf{a}\cdot\mathbf{c})$, $\mathbf{q} = (\mathbf{a}\cdot\mathbf{b}, \mathbf{b}\cdot\mathbf{b}, \mathbf{b}\cdot\mathbf{c})$, $\mathbf{r} = (\mathbf{a}\cdot\mathbf{c}, \mathbf{b}\cdot\mathbf{c}, \mathbf{c}\cdot\mathbf{c})$.
(ii) Hence prove that $\mathbf{a}, \mathbf{b}, \mathbf{c}$ are coplanar iff $\mathbf{a}\cdot\mathbf{b} \wedge \mathbf{c} = 0$.
(iii) Find all solutions of the equations

$$x - 2y + 3z = 1, \qquad -3x + y + 2z = 2, \qquad 5x + 4y + z = 5$$

when x, y, z lie in the ring $\mathbb{Z}_n$, and $n = 2, 3, 4, 5, 7$. [Follow through the procedure of the flow chart 15.2.13.]

(iv) Compose another exercise of your own, like (iii), taking care to adjust the coefficients to your satisfaction.

(v) Find all solutions of the equations
$x - 2y + 3z = (1, 1, 1), \quad -3x + y + 2z = (2, 2, 3), \quad 5x + 4y + 7 = (5, 1, 0)$
when x, y, z lie in $\mathbb{R}^3$. Will the method of 15.2.12 work when the third equation is replaced by

$$9x - 8y + 5z = -(1, 1, 1)?$$

§(vi) Find a solution of the differential equations (see Chapter 32):

$$Dx + 4y + 2z = 2, \qquad 3x + Dy + z = -1, \qquad y + z = 1 - Dz$$

when x, y, z are to be functions: $\mathbb{R} \to \mathbb{R}$, D denotes the differential operator d/dt. [Observe that equation 15.2.4 is still valid when the coefficients lie in a ring of operators on the variables x, y, z. Here, we obtain third-order differential equations $(D^3 + D^2 - 13D - 6)z + 6 = 0$, etc.]

15.3 Linear Transformations

There are two points of view which can be taken about the linear equations 15.2.2. The first is to regard them as equations to find x, y, z, given $\mathbf{a}$, $\mathbf{b}$, $\mathbf{c}$, and $\mathbf{h}$; and this gave rise to the theory of 15.2. The second point of view is to regard them as a means of producing $\mathbf{h}$ when (x, y, z) is given, keeping $\mathbf{a}$, $\mathbf{b}$, and $\mathbf{c}$ fixed. More precisely, the equations 15.2.2 define a *function* $T: \mathbb{R}^3 \to \mathbb{R}^3$ where for each $(x, y, z) \in \mathbb{R}^3$, $T(x, y, z) = (h_1, h_2, h_3)$ given by 15.2.2. The function T has two simple properties, which follow immediately from the linear nature of equations 15.2.2, namely, that for any $\mathbf{u}, \mathbf{v} \in \mathbb{R}^3$, and $\lambda \in \mathbb{R}$ we have:

$\mathfrak{LT}_1$: $T(\mathbf{u} + \mathbf{v}) = T(\mathbf{u}) + T(\mathbf{v})$;

$\mathfrak{LT}_2$: $T(\lambda\mathbf{u}) = \lambda T(\mathbf{u})$.

In particular, taking $\lambda = 0$ in $\mathfrak{LT}_2$ gives $T(\mathbf{0}) = \mathbf{0}$. Hence, $\mathfrak{LT}_1$ says that T maps the vertices of the parallelogram $P(\mathbf{u}, \mathbf{v})$ formed by O, $\mathbf{u}$, $\mathbf{v}$, and $\mathbf{u} + \mathbf{v}$, into the vertices of the quadrilateral $\underset{\sim}{Q}$ formed by O, $T(\mathbf{u})$, $T(\mathbf{v})$, $T(\mathbf{u} + \mathbf{v})$; and also that $\underset{\sim}{Q}$ is the parallelogram $P(T\mathbf{u}, T\mathbf{v})$ (by the parallelogram law of addition) since $T(\mathbf{u} + \mathbf{v}) = T(\mathbf{u}) + T(\mathbf{v})$. And $\mathfrak{LT}_2$ says that T maps lines through O into lines through O. Thus, although T may distort angles and lengths, it still respects the 'vector space' structure of $\mathbb{R}^3$. These properties may also be discussed in $\mathbb{R}^2$ and $\mathbb{R}^1$ $(=\mathbb{R})$; so we make a definition.

15.3.1 DEFINITION. Any function $f: \mathbb{R}^i \to \mathbb{R}^i$ $(i = 1, 2, 3)$ which satisfies $\mathfrak{LT}_1$ and $\mathfrak{LT}_2$ is called a **linear transformation.**

The simplicity of f in 15.3.1 is shown by:

15.3.2 THEOREM. *A linear transformation f is known fully as soon as we know the three values $f(\mathbf{u})$, $f(\mathbf{v})$, $f(\mathbf{w})$ for some basis $\{\mathbf{u}, \mathbf{v}, \mathbf{w}\}$ of $\mathbb{R}^3$.*

Proof. Since $\{\mathbf{u}, \mathbf{v}, \mathbf{w}\}$ forms a basis, then given $\mathbf{r} \in \mathbb{R}^3$, there exist numbers $l, m, n \in \mathbb{R}$ such that $\mathbf{r} = l\mathbf{u} + m\mathbf{v} + n\mathbf{w}$. Moreover l, m, n are uniquely determined by $\mathbf{r}$, since $\mathbf{u}$, $\mathbf{v}$, $\mathbf{w}$ are linearly independent. Hence

$$\begin{aligned} f(\mathbf{r}) &= f(l\mathbf{u} + m\mathbf{v} + n\mathbf{w}) \\ &= f(l\mathbf{u}) + f(m\mathbf{v}) + f(n\mathbf{w}) \qquad (\text{by } \mathfrak{LT}_1) \\ &= lf(\mathbf{u}) + mf(\mathbf{v}) + nf(\mathbf{w}) \qquad (\text{by } \mathfrak{LT}_2). \end{aligned}$$

Hence, to compute $f(\mathbf{r})$, we need only know $f(\mathbf{u})$, $f(\mathbf{v})$, $f(\mathbf{w})$ (since l, m, n are computable when we know $\mathbf{r}$, $\mathbf{u}$, $\mathbf{v}$, $\mathbf{w}$), because then we know the right-hand side of the last equation completely. ∎

Similar results hold in $\mathbb{R}$ and $\mathbb{R}^2$ (as always in the theory), and here as elsewhere the reader is advised to work out the analogues himself, as a check on his understanding of the work in $\mathbb{R}^3$. Almost all functions occurring in calculus are approximately linear, so that a study of linear transformations should come before calculus as a first approximation to considering differentiable functions (see Chapter 29).

Exercise 4

(i) Prove that a translation $T(\mathbf{a})\colon \mathbb{R}^3 \to \mathbb{R}^3$ (see 14 Exercise 7(ii)) is not a linear transformation unless $\mathbf{a} = \mathbf{0}$.
(ii) In $\mathbb{R}^3$, prove that the identity function and the functions $\mathbf{r} \to \mathbf{r} \wedge \mathbf{a}$, $\mathbf{r} \to \mathbf{0}$ are linear transformations ($\mathbf{a}$ constant).
(iii) In $\mathbb{R}^1$, prove that every linear transformation T is of the form $x \to ax$ where the number a depends only on T, not on x.
(iv) If S, $T\colon \mathbb{R}^3 \to \mathbb{R}^3$ are linear transformations, show that $S \circ T$ is one also.
(v) Prove that the image of a linear transformation $T\colon \mathbb{R}^3 \to \mathbb{R}^3$ is a linear subspace of $\mathbb{R}^3$. The **kernel** of T is defined to be $T^\flat(0)$. Prove that it is a linear subspace of $\mathbb{R}^3$, of dimension equal to 3 minus the dimension of $T^\flat(\mathbb{R}^3)$. By considering a difference of the form $\mathbf{x} - \mathbf{y}$, prove that T is one-one iff $T^\flat(0)$ is the singleton $\{0\}$. In general, show that for each $\mathbf{x} \in T(\mathbb{R}^3)$, there is a translation $f\colon \mathbb{R}^3 \to \mathbb{R}^3$ such that $f(T^\flat(0)) = T^\flat(x)$.

A linear transformation $f\colon \mathbb{R}^3 \to \mathbb{R}^3$ is called **non-singular** (in this theory) iff it is invertible.

In 15.3.6 below we shall give a simple criterion for deciding whether or not f is non-singular. If f is invertible, then by the inversion theorem (2.9.1), $f^{-1}\colon \mathbb{R}^3 \to \mathbb{R}^3$ exists and we now prove:

15.3.3 PROPOSITION. *f^{-1} is linear.*

Proof. We must check $\mathfrak{LT}_1$ and $\mathfrak{LT}_2$. For the first we have for any $\mathbf{u}, \mathbf{v} \in \mathbb{R}^3$, that $\mathbf{u} = f(\mathbf{a})$, $\mathbf{v} = f(\mathbf{b})$, say (since f is onto), so

$$\begin{aligned} f^{-1}(\mathbf{u} + \mathbf{v}) &= f^{-1}(f(\mathbf{a}) + f(\mathbf{b})) \\ &= f^{-1}(f(\mathbf{a} + \mathbf{b})) \qquad (f \text{ satisfies } \mathfrak{LT}_1), \\ &= \mathbf{a} + \mathbf{b} \qquad (f^{-1} \circ f = \text{identity}), \\ &= f^{-1}(\mathbf{u}) + f^{-1}(\mathbf{v}) \qquad (\mathbf{u} = f(\mathbf{a}) \;.\!\Rightarrow\!.\; \mathbf{a} = f^{-1}(\mathbf{u})), \end{aligned}$$

so $\mathfrak{LT}_1$ is satisfied. The checking of $\mathfrak{LT}_2$ proceeds similarly (see the next exercise). ■

Exercise 5

(i) Verify that f^{-1} satisfies $\mathfrak{LT}_2$ in 15.3.3.
(ii) Prove that the set of all non-singular linear transformations of $\mathbb{R}^3$ forms a group with respect to composition (see Chapter 18).
(iii) Let S, T: $\mathbb{R}^3 \to \mathbb{R}^3$ be linear transformations. Define functions

$$S + T: \mathbb{R}^3 \to \mathbb{R}^3, \; aT: \mathbb{R}^3 \to \mathbb{R}^3$$

(where $a \in \mathbb{R}$) by the equations

$$(S + T)(\mathbf{p}) = S(\mathbf{p}) + T(\mathbf{p}); \qquad (aT)(\mathbf{p}) = a \cdot T(\mathbf{p})$$

where each right-hand side is read in $\mathbb{R}^3$. Prove that $S + T$ and aT are linear transformations. If V: $\mathbb{R}^3 \to \mathbb{R}^3$ is linear, prove that

$$V \circ (S + T) = V \circ S + V \circ T; \qquad V \circ (aT) = a(V \circ T).$$

(iv) Let T: $\mathbb{R}^2 \to \mathbb{R}^2$ be a non-singular linear transformation. Prove that T preserves parallels and ratios but not necessarily distances (i.e., if l, m are lines with equal gradient, so are $T(l)$, $T(m)$; and if A, B, C are three collinear points, then $|T(AB)|/|T(BC)| = |AB|/|BC|$. An **affine equivalence** g· $\mathbb{R}^2 \to \mathbb{R}^2$ is a bijection of the form $k \circ T$, where k is a translation; show that g also preserves parallels and ratios. Two subsets X, $Y \subseteq \mathbb{R}^2$ are called **affine equivalent** iff there exists an affine equivalence g: $\mathbb{R}^2 \to \mathbb{R}^2$ such that $g(X) = Y$. Prove that this is an equivalence relation. Show that any two ellipses are affine equivalent, but that an ellipse and hyperbola never are. How about two parabolae? Prove that the affine equivalences of $\mathbb{R}^2$ form a group with respect to composition.
§(v) Let L denote the set of all linear transformations $\mathbb{R}^3 \to \mathbb{R}^3$ (L is often denoted by Hom($\mathbb{R}^3$, $\mathbb{R}^3$).) Show that L, with the addition and scalar multiplication of (iii), is a vector space over $\mathbb{R}$, and that its dimension is 9. If composition is included as a 'product', show that L is a linear algebra over $\mathbb{R}$.

A linear transformation f: $\mathbb{R}^3 \to \mathbb{R}^3$ has a very simple effect on volumes, because it distorts them all in the same ratio. We show this more precisely as follows.

First let $\mathbb{I}^3$ denote the 'unit cube'

$$\{(x, y, z) | 0 \leqslant x \leqslant 1, 0 \leqslant y \leqslant 1, 0 \leqslant z \leqslant 1\};$$

in the notation of 14.11, $\mathbb{I}^3 - P(\mathbf{i}, \mathbf{j}, \mathbf{k})$. Hence, if $\mathbf{r} \in \mathbb{I}^3$, then $\mathbf{r} = x\mathbf{i} + y\mathbf{j} + z\mathbf{k}$, so by the last equation in the proof of 15.3.2,

$$f(\mathbf{r}) = xf(\mathbf{i}) + yf(\mathbf{j}) + zf(\mathbf{k}).$$

Thus, by 14 Exercise 13(iii), f maps $\mathbb{I}^3$ onto the parallelepiped $P(\mathbf{a}, \mathbf{b}, \mathbf{c})$ where

15.3.4 $$\mathbf{a} = f(\mathbf{i}), \qquad \mathbf{b} = f(\mathbf{j}), \qquad \mathbf{c} = f(\mathbf{k}).$$

For reasons to appear, we define a function det: $L \to \mathbb{R}$ on the set L of all linear transformations of $\mathbb{R}^3$ to itself, by the rule:

15.3.5 $$\det f = \operatorname{Vol} P(\mathbf{a}, \mathbf{b}, \mathbf{c}).$$

For example, if $f = \text{id}$, then det (id) $= 1$, while det $(k \cdot \text{id}) = k^3$.

15.3.6 LEMMA. *f is non-singular iff $\det f \neq 0$.*

Proof. If $\det f \neq 0$ then by 15.3.5 and 14 Exercise 13(v), $\{\mathbf{a}, \mathbf{b}, \mathbf{c}\}$ is a basis for $\mathbb{R}^3$. Hence, given $\mathbf{r} \in \mathbb{R}^3$, there exist $l, m, n \in \mathbb{R}$ such that $\mathbf{r} = l\mathbf{a} + m\mathbf{b} + n\mathbf{c}$, so $\mathbf{r} = lf(\mathbf{i}) + mf(\mathbf{j}) + nf(\mathbf{k}) = f(l\mathbf{i} + m\mathbf{j} + n\mathbf{k})$, using 15.3.4 and the linearity of f. Therefore f is onto. To prove that f is one-one, suppose $f(\mathbf{x}) = f(\mathbf{y})$; then $f(\mathbf{x} - \mathbf{y}) = \mathbf{0}$ since f is linear, so

$$\begin{aligned} \mathbf{0} = f(\mathbf{x} - \mathbf{y}) &= f(r\mathbf{i} + s\mathbf{j} + t\mathbf{k}) \qquad \text{(say)} \\ &= r\mathbf{a} + s\mathbf{b} + t\mathbf{c} \end{aligned}$$

by the linearity of f. But $\mathbf{a}, \mathbf{b}, \mathbf{c}$ are linearly independent (being a basis), so $r = s = t = 0$. Hence $\mathbf{x} - \mathbf{y} = \mathbf{0}$, so $\mathbf{x} = \mathbf{y}$ and f is one-one. Therefore f is an equivalence (i.e., non-singular).

Conversely, suppose f is non-singular, and let us prove $\det f \neq 0$. By 14 Exercise 13(v), it suffices to prove $\mathbf{a}, \mathbf{b}, \mathbf{c}$ linearly independent. But if there were an equation $l\mathbf{a} + m\mathbf{b} + n\mathbf{c} = \mathbf{0}$, then as in the argument above for $\mathbf{r}$, $\mathbf{0} = f(l\mathbf{i} + m\mathbf{j} + n\mathbf{k})$ whence $l\mathbf{i} + m\mathbf{j} + n\mathbf{k} = \mathbf{0}$, since f is one-one. Thus, $l = m = n = 0$, since $\mathbf{i}, \mathbf{j}, \mathbf{k}$ are linearly independent; hence so are $\mathbf{a}, \mathbf{b}, \mathbf{c}$. ■ We can now state the main result of this section. [In fact, we prove rather more than is stated; see 15.3.8 and the Remark at the end of the proof.]

15.3.7 THEOREM. *Let $G \subseteq \mathbb{R}^3$ be a set with* (positive) *volume V. Then the set $f(G)$ has volume*

$$V|\det f|.$$

(Similarly for areas in $\mathbb{R}^2$.)

Proof. First consider a rectangular block $B = P(\alpha\mathbf{i}, \beta\mathbf{j}, \gamma\mathbf{k})$ whose edges are parallel to those of $\mathbb{I}^3$, and of lengths α, β, γ units. Then by 14.11.3,

$$\text{Vol } B = \alpha\mathbf{i} \cdot \beta\mathbf{j} \wedge \gamma\mathbf{k} = \alpha\beta\gamma(\mathbf{i} \cdot \mathbf{j} \wedge \mathbf{k}) = \alpha\beta\gamma$$

and the image of B by f is

$$\{f(\mathbf{r}) | \mathbf{r} = x\mathbf{i} + y\mathbf{j} + z\mathbf{k} \text{ and } 0 \leqslant x \leqslant \alpha,\ 0 \leqslant y \leqslant \beta,\ 0 \leqslant z \leqslant \gamma\}$$

which, by 14 Exercise 13(iii), is $P(\alpha\mathbf{u}, \beta\mathbf{v}, \gamma\mathbf{w})$, of (signed) volume

$$\begin{aligned} \alpha\mathbf{u} \cdot (\beta\mathbf{v} \wedge \gamma\mathbf{w}) &= \alpha\beta\gamma(\mathbf{u} \cdot \mathbf{v} \wedge \mathbf{w}) \\ &= \text{Vol } B|\det f|, \end{aligned}$$

by definition of $\det f$ in 15.3.5. Thus, f has transformed the block B into the parallelepiped $P(\alpha\mathbf{u}, \beta\mathbf{v}, \gamma\mathbf{w})$, and has changed its volume in the ratio $\det f$:1.

More generally, we can now see that f changes *all* volumes in the ratio $\det f$:1. For let G be a region in $\mathbb{R}^3$ possessing a volume. Briefly this means the following (for details see particularly 15.11 below). Let $\Lambda(\mathbf{a}, \mathbf{b}, \mathbf{c})$ denote the lattice of 14 Exercise 13(vi) and for each $t > 0$ let $s(t)$ and $S(t)$ denote respectively the sets of all cells σ of $\Lambda(t\mathbf{a}, t\mathbf{b}, t\mathbf{c})$ such that $\sigma \subseteq G$ and $\sigma \cap G \neq \varnothing$ respectively. Clearly $S(t) \supseteq s(t)$, and

(i) $$\bigcup_{\sigma \in s(t)} \sigma \subseteq G \subseteq \bigcup_{\sigma \in S(t)} \sigma.$$

Figure 15.2 shows the analogue in $\mathbb{R}^2$, with the lattice $\Lambda(\frac{1}{4}\mathbf{i}, \frac{1}{4}\mathbf{j})$, the inner unshaded cells being those of $s(\frac{1}{4})$ while the shaded ones are those of $S(\frac{1}{4}) - s(\frac{1}{4})$.

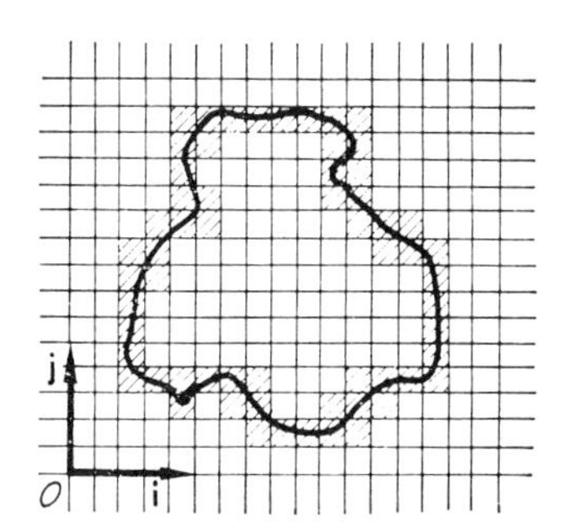

Fig. 15.2

Now each cell σ in $\Lambda(t\mathbf{a}, t\mathbf{b}, t\mathbf{c})$ has volume $t^3\Delta$ with $\Delta = \mathbf{a}\cdot\mathbf{b} \wedge \mathbf{c}$, so by (i) we have

(ii) $$t^3\Delta\cdot\#s(t) \leqslant \text{vol } G \leqslant t^3\Delta\cdot\#S(t),$$

where we take Δ and vol G *to be positive for the moment.* Then, as t becomes arbitrarily small, the two extremes become closer and closer to vol G (because this is what is meant in a precise treatment of the assertion 'G possesses a volume'!). If $\Delta \leqslant 0$, we proceed similarly but reverse all relevant inequalities; thus, *we allow* vol G *to be negative if Δ is*, and so use the *signed* volume Vol G. A similar construction applies to $f(G)$.

In particular then, take $(\mathbf{a}, \mathbf{b}, \mathbf{c})$ to be $(\mathbf{i}, \mathbf{j}, \mathbf{k})$ and write $u = f(\mathbf{i})$, $v = f(\mathbf{j})$, $w = f(\mathbf{k})$. Let $s(t)$, $s'(t)$, etc., refer respectively to $\Lambda(\mathbf{i}, \mathbf{j}, \mathbf{k})$, $\Lambda(\mathbf{u}, \mathbf{v}, \mathbf{w})$. Now each cell $\sigma \in s(t)$ maps by f onto a cell of $\Lambda(t\mathbf{u}, t\mathbf{v}, t\mathbf{w})$ with signed volume given by:

(iii) $$\text{Vol} f(\sigma) = t^3\cdot\det f\cdot\Delta.$$

Also $\sigma \subseteq G$, so $f(\sigma) \subseteq f(G)$, whence $f(\sigma) \in s'(t)$. Similarly, if $\sigma \in S(t)$, then $f(\sigma) \in S'(t)$; and by (i) above

(iv) $$\bigcup_{\sigma \in s(t)} f(\sigma) \subseteq f(G) \subseteq \bigcup_{\sigma \in S(t)} f(\sigma).$$

Suppose first that f is non-singular. Then since f^{-1} exists and is linear, it follows that $s(t) \approx s'(t)$ and $S(t) \approx S'(t)$, and the elements $f(\sigma)$ are all distinct members of $S'(t)$. Hence by (iv), we have the inequalities (when $\Delta' = \Delta \cdot \det f$ is positive):

$$\text{Vol}\, f(\sigma) \cdot \#s(t) \leqslant \text{Vol}\, f(G) \leqslant \text{Vol}\, f(\sigma) \cdot \#S(t),$$

so

$$t^3 \Delta' \cdot \#s'(t) \leqslant \text{Vol}\, f(G) \leqslant t^3 \Delta' \cdot \#S'(t)$$

(using (iii) and the above-remarked equivalences). If Δ' is negative, the inequalities are reversed. In either case the inequalities are just those (by analogy with (ii)) saying that the extremities each converge to $\text{Vol}\, f(G)$ as t converges to 0. On the other hand, by (ii) itself, these extremities converge to $\det f \cdot \text{Vol}\, G$. Hence

15.3.8 $$\text{Vol}\, f(G) = \det f \cdot \text{Vol}\, G,$$

which more than proves 15.3.7, being a statement about *signed* volumes.

If, however, f is singular, then $\det f = 0$, by 15.3.6. By (ii)

$$|\text{Vol}\, f(G)| \leqslant \sum |\text{Vol}\, f(\sigma)|,$$

and the right-hand side is zero by (iii), so 15.3.8 (and hence 15.3.7) still holds. ■

Remark. Observe that the proof has established 15.3.8, which is stronger than 15.3.7, since it refers to *signed* volumes. This strengthening is a dividend arising from our having to analyse the notion of volume, since the intuitive notion that suffices for *stating* 15.3.7 is not good enough for the proof. A consequence of 15.3.8 which 15.3.7 cannot yield, is:

15.3.9 THEOREM. *Let $f, g: \mathbb{R}^3 \to \mathbb{R}^3$ be linear transformations. Then*

$$\det (f \circ g) = \det f \cdot \det g.$$

Proof. Let G be a region in $\mathbb{R}^3$ possessing a non-zero signed volume v (e.g., a cube). Then by 15.3.8,

$$\text{Vol}\, (f \circ g)(G) = \det (f \circ g) \cdot v;$$

but the left-hand side is $\text{Vol}\, f(g(G))$ which by 15.3.8 is

$$\det f \cdot \text{Vol}\, g(G) = \det f \cdot \det g \cdot \text{Vol}\, G.$$

Comparing the two results gives 15.3.9, since $v \neq 0$. ■

Exercise 6

(i) If $f: \mathbb{R}^3 \to \mathbb{R}^3$ is a non-singular linear transformation, prove that $\det f^{-1} = (\det f)^{-1}$. {Use Theorem 15.3.9.}

(ii) If f is not non-singular, prove that f maps all sets possessing volume into sets of zero volume. [Thus f 'crushes' some sets, as would be expected if f is not one-one. Consider, for example, the function $(x, y, z) \to (x, y, 0)$.]
(iii) Prove in detail the remark in the proof of 15.3.7, that when f is non-singular, $s(t) \approx s'(t)$, $S(t) \approx S'(t)$. {Assign to each $\sigma \in S(t)$ its image $f(\sigma) \in S'(t)$.}
(iv) If $f(x, y, z) = (ax, by, cz)$ (with $abc \neq 0$), prove that the image under f of the spherical ball $B: x^2 + y^2 + z^2 \leqslant 1$ is the solid ellipsoid

$$E: x^2/a^2 + y^2/b^2 + z^2/c^2 \leqslant 1.$$

Hence prove that $\text{Vol } E = abc \cdot \text{Vol } B$.
§(v) Two straight cylindrical pipes intersect, their axes making an angle θ. Find the volume of their common part.
§(vi) In (iv), above, let $B' = B - \{0\}$. What is the volume of B'? Let B'' denote the complement in B of all points with rational co-ordinates. Has B'' a volume?

We have already seen how a triple of linear equations 15.2.2 gives rise to a linear transformation. Conversely:

15.3.10 PROPOSITION. *Every linear transformation $f: \mathbb{R}^3 \to \mathbb{R}^3$ is given by equations of the form* 15.2.2.

Proof. Given $\mathbf{r} \in \mathbb{R}^3$, then $\mathbf{r} = x\mathbf{i} + y\mathbf{j} + z\mathbf{k}$, so applying the last equation of the proof of 15.3.2 we have

$$f(\mathbf{r}) = xf(\mathbf{i}) + yf(\mathbf{j}) + zf(\mathbf{k}) = x\mathbf{a} + y\mathbf{b} + z\mathbf{c} \qquad \text{(say)}.$$

Thus, if $f(\mathbf{r}) = (h_1, h_2, h_3)$, then checking co-ordinates gives $h_i = xa_i + yb_i + zc_i$ $(i = 1, 2, 3)$, and we have equations of the form 15.2.2.

Exercise 7

(i) Let $\{\mathbf{e}_i\}$, $\{\mathbf{b}_i\}$ be bases of $\mathbb{R}^3$ as in Exercise 2(iv). Show that there is a linear transformation $M: \mathbb{R}^3 \to \mathbb{R}^3$ which maps the $\mathbf{e}$'s to the $\mathbf{b}$'s (i.e., such that $M(\mathbf{e}_i) = \mathbf{b}_i$ $(i = 1, 2, 3)$).
(ii) The matrix whose columns are $\mathbf{a}$, $\mathbf{b}$, $\mathbf{c}$ in 15.3.10 is called the 'matrix of f relative to the ordered basis $\mathbf{i}, \mathbf{j}, \mathbf{k}$'. Denote it by $m(f)$, and let $m_{rs}(f)$ denote the entry in its rth row and sth column. Write out $m(f)$ when f is each of the three functions in Exercise 4(ii), showing that $m(\text{id}) = (\delta_{rs})$, the matrix described in Exercise 2(iv). Do the same with $(\mathbf{j}, \mathbf{i}, \mathbf{k})$ replacing $(\mathbf{i}, \mathbf{j}, \mathbf{k})$. Similarly in $\mathbb{R}^2$, if $m(f) = \begin{pmatrix} a & b \\ c & d \end{pmatrix}$, show that $f(x, y) = (x', y')$ where $x' = ax + by$, $y' = cx + dy$.
(iii) With S, T as in Exercise 5(iii), prove that $m_{rs}(aS) = am_{rs}(S)$, $m_{rs}(S + T) = m_{rs}(S) + m_{rs}(T)$, and that

$$m_{rs}(S \circ T) = \sum_{t=1}^{3} m_{rt}(S) m_{ts}(T).$$

Show that $m_{rs}(S \circ T)$ is the scalar product of the rth row of $m(S)$ with the sth column of $m(T)$.
(iv) The last equation above is the definition of the product of any two matrices $M, N \in \mathscr{M}_i$ $(i = 2, 3)$; MN is that matrix whose (r, s)th entry is $\sum_{t=1}^{i} m_{rt} n_{ts}$ (the

scalar product of the rth row and sth column). Figure 15.3 may be found helpful as a way of remembering the rule. Explicitly,

$$\begin{pmatrix} a & b \\ c & d \end{pmatrix}\begin{pmatrix} p & q \\ r & s \end{pmatrix} = \begin{pmatrix} ap + br, & aq + bs \\ cp + dr, & cq + ds \end{pmatrix}.$$

The reader should practise by writing down matrices M, N, with integer entries, and multiplying out MN. The *reason* for the definition is that composition of linear transformations forces it upon us. Use (iii) to show that the multiplication is associative, distributive with respect to addition, and has $I = (\delta_{ij})$ as unit element. (I is therefore called the **unit matrix.**) Hence prove that the sets $\mathscr{M}_2$, $\mathscr{M}_3$ of 2×2 and 3×3 matrices are each rings with unit. Find matrices M, N such that $MN \neq NM$. Prove that $(MN)^T = N^T M^T$, where T denotes the transpose (see 15.1.8).

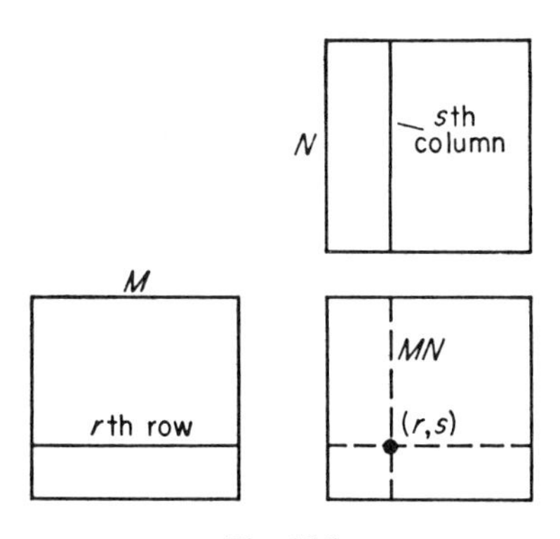

Fig. 15.3

(v) Two matrices M, N in $\mathscr{M}_3$ (or in $\mathscr{M}_2$) are said to be *inverses* if $MN = I = NM$. We then write $M = N^{-1}$. Prove that $(N^{-1})^{-1} = N$; and if M, N are the matrices $m(f)$, $m(g)$ of linear transformations $f, g\colon \mathbb{R}^3 \to \mathbb{R}^3$, then f, g are equivalences and $f^{-1} = g$. Compute M^{-1} for the 2×2 matrix M in (ii).
(vi) If $A, B \in \mathscr{M}_3$, prove that $\det AB = \det A \cdot \det B$ {use 15.3.9}. Similarly for $\mathscr{M}_2$.
(vii) In Exercise 2(iv), prove that the matrices $P = (p_{ij})$, $Q = (q_{ij})$ are multiplicative inverses, with non-zero determinants.
(viii) All the theory of problem (ii) holds when we replace $(\mathbf{i}, \mathbf{j}, \mathbf{k})$ by any basis of $\mathbb{R}^3$. Let $m_e(f)$, $m_b(f)$ denote the corresponding matrices relative to the bases $(\mathbf{e}_1, \mathbf{e}_2, \mathbf{e}_3)$, $(\mathbf{b}_1, \mathbf{b}_2, \mathbf{b}_3)$ of Exercise 2(iv). Thus $f(\mathbf{e}_i) = \sum_j n_{ij}\mathbf{e}_j$, where $(n_{ij}) = m_e(f)$. Prove that $m_e(f) = (P^T)^{-1} \cdot m_b(f) \cdot P^T$, with P as in problem (vii). Hence use problem (vi) to show that

$$\det m_e(f) = \det m_b(f) = \det f.$$

(ix) Work out the matrix $m(f)$ (in the sense of problem (ii)) when f is a reflection in a plane through the origin, and when f is a rotation through an angle θ about a

line through the origin. Do the same when the basis $(\mathbf{i}, \mathbf{j}, \mathbf{k})$ is replaced by $(\mathbf{j}, \mathbf{i}, \mathbf{k})$. {Work directly, or use problems (i) and (viii).}
(x) In (viii), every $\mathbf{r} \in \mathbb{R}^3$ can be expressed in the form $\mathbf{r} = \Sigma x_i \mathbf{e}_i$ and $f(r) = \Sigma y_j \mathbf{e}_j$. Show that $y_j = \sum_s k_{js} x_s$, where the matrix (k_{js}) is the transpose (see 15.1.8) of the matrix $(n_{ij}) = m_e(f)$.

The result of (viii) gives an important practical criterion for deciding whether or not a given linear transformation $f: \mathbb{R}^3 \to \mathbb{R}^3$ is invertible. For by choosing a suitable ordered basis $(\mathbf{e}_1, \mathbf{e}_2, \mathbf{e}_3)$, we compute $m_e(f)$, then work out its determinant. By (viii), this gives $\det f$, to which we now apply 15.3.6. The quickest way of working out f^{-1} is to solve the equations $\mathbf{r} = f(\mathbf{x})$ given by 15.3.10, to get $\mathbf{x} = f^{-1}(\mathbf{r})$. If f is known to be invertible but $\det f$ is not known, it is best to solve the equation using, say, the 'echelon rule' (see Birkhoff–MacLane [16]), since the required solution is known to exist and is unique. Cramer's rule is not efficient here. Then the coefficients on the right-hand side form $m_e(f^{-1})$.
(xi) The conditions $\mathfrak{LT}_1$, $\mathfrak{LT}_2$ for a function to be a linear transformation make sense for functions of the form $f: V \to W$, where V, W are lines or planes, through the origin in $\mathbb{R}^3$, or $\mathbb{R}^3$ itself. Such a function f is still called a linear transformation. Decide whether or not the following six functions are linear transformations:

$$V = \mathbb{R}^3, \quad W = \mathbb{R}, \quad f(x, y, z) = x; \quad x + y + z; \quad x^2;$$

$$V = \mathbb{R}^3; \quad W = \mathbb{R}^2, \quad f(x, y, z) = 2(x, y) + 3(y, z); \quad (0, xyz); \quad (0,0).$$

With the second and fourth functions, show that each is onto, while their sets of zeros form a plane P and a line L, respectively. Find linear transformations $p: P \to \mathbb{R}^2$, $q: L \to \mathbb{R}$, each of which is an equivalence.
(xii) Let L, M each denote sets of four lines through the origin in $\mathbb{R}^3$, such that no three lines of L are coplanar, and similarly for M. Show that there exists a linear transformation $T: \mathbb{R}^3 \to \mathbb{R}^3$ which maps each line $l \in L$ onto a prescribed line $f(l) \in M$. Show also that T is an equivalence iff $f: L \to M$ is an equivalence.
(xiii) Let V be the real vector space of all translations $\mathbb{R}^3 \to \mathbb{R}^3$ in 14 Exercise 7(ii). Show that the assignment $\mathbf{a} \to T(\mathbf{a})$ is an isomorphism $\mathbb{R}^3 \approx V$.
(xiv) A linear transformation $f: \mathbb{R}^2 \to \mathbb{R}^2$ is called **orthogonal** or an **isometry** if it preserves distances, i.e., if $|f(\mathbf{u}) - f(\mathbf{v})| = |\mathbf{u} - \mathbf{v}|$ for all $\mathbf{u}, \mathbf{v} \in \mathbb{R}^2$. Show that this occurs iff $x^2 + y^2 = x'^2 + y'^2$ where $\mathbf{u} = (x, y)$, $f(\mathbf{u}) = (x', y')$. Hence show that the matrix of f relative to the bases $(\mathbf{i}, \mathbf{j})$, $(\mathbf{i}, \mathbf{j})$ is of the form $P = \begin{pmatrix} l & -m \\ m & l \end{pmatrix}$ or $Q = \begin{pmatrix} l & m \\ m & -l \end{pmatrix}$ where $l^2 + m^2 = 1$, according as $\det f$ is 1 or -1. Show that $P = QR$ where $R = \begin{pmatrix} 1 & 0 \\ 0 & -1 \end{pmatrix}$, and show that P represents a rotation of $\mathbb{R}^2$ about O through an angle $\tan^{-1}(m/l)$, while R represents a reflection in the x-axis.
(xv) In Exercise 5(iv), suppose that T there is an isometry of $\mathbb{R}^2$. We then call g a **rigid motion**, while X and Y there are called **rigidly equivalent**. Show that we again have an equivalence relation, and that the rigid motions of $\mathbb{R}^2$ form a group under composition. Give an example to show that two ellipses in $\mathbb{R}^2$ need not be rigidly equivalent. When are two parabolae rigidly equivalent? §Show that any bijection of $\mathbb{R}^2$ which preserves distance is a rigid motion.§

(xvi) In (xiv) prove that $P \cdot P^T$ is the unit matrix, where P^T is the transpose of P (see 15.1.8 and $\mathfrak{M}_1$). Conversely, prove that if Q is a matrix satisfying $Q \cdot Q^T = I$, then Q is the matrix of an isometry of $\mathbb{R}^2$.
(xvii) Show that the set of all f with $\det f = 1$ forms a commutative group under composition, the group **R** of Example 9.5.2. {Multiply corresponding matrices.} Show by an example that this result is false for linear transformations $\mathbb{R}^3 \to \mathbb{R}^3$.
(xviii) Formulate and solve the problem in $\mathbb{R}^3$ analogous to problem (xiv).
(xix) Show that the function $j: \mathbb{Q} \to \mathscr{M}_2$ which assigns to each rational $u \in \mathbb{Q}$ the matrix $j(u) = u \cdot I$, is a linear transformation while also $j(uv) = j(u)j(v)$; and j is one-one but not onto. Thus j *embeds* $\mathbb{Q}$ in $\mathscr{M}_2$. Now find all solutions of the equation $x^2 = 2$ in $\mathscr{M}_2$ by regarding '2' here as $j(2) = 2I \in \mathscr{M}_2$. Why are there more than two solutions in $\mathscr{M}_2$?

Appendix: Length and Area

The rest of this chapter is really an appendix, following the ideas used in the proof of Theorem 15.3.7. In fact, we shall be less concerned with the notion of 'volume' than with the simpler (but still subtle) notions of 'length' and 'area'. We mainly discuss the problems of trying to assign a *length* to a continuous curve in space, and an *area* to a set of points in $\mathbb{R}^2$. At certain points in the discussion, we have to use limiting processes as used in Chapters 24 and 27; but even if the reader is not yet technically prepared to follow the details here, it is hoped that he will nevertheless appreciate the important general ideas and see the need for the critical approach, and for the large branch of mathematics to which these studies lead, which is called *measure theory*.

15.4 Paths

Consider, then, the notion of *length* of a continuous curve or path.

First, what *is* a continuous path? This question has given a great impulse to the subject of topology (see, for example the article by Hahn in Newman [93]), but we content ourselves with the following definition.

15.4.1 DEFINITION. By a **continuous path** from **a** to **b** in $\mathbb{R}^3$ we mean a continuous function $f: \langle\!\langle 0, p \rangle\!\rangle \to \mathbb{R}^3$ where $\langle\!\langle 0, p \rangle\!\rangle \subseteq \mathbb{R}$ and $f(0) = \mathbf{a}, f(p) = \mathbf{b}$.

Remarks. We do *not* say that the path is the image of f, and of course several paths may have the same image. To say that f is continuous, is to mean that the co-ordinates

15.4.2 $$x(t),\ y(t),\ z(t)$$

of $f(t)$ are continuous functions of t on the closed interval $\langle\!\langle 0, p \rangle\!\rangle$ in $\mathbb{R}$. They need not be differentiable; if for example f represents the Brownian motion of a particle at time t, the path is so jagged that there will not be a tangent at many points.

The **reverse**, $\bar{f}$, of the path f is given by $\bar{f}(t) = f(p - t)$. Then $\bar{f}$ is a path from **b** to **a**.

15.4.3 EXAMPLE. The simplest paths f are those whose image is the segment $\langle\!\langle \mathbf{a}, \mathbf{b} \rangle\!\rangle$ joining **a** to **b** in $\mathbb{R}^3$ and we call them **straight**; here we can give such a straight path, g, by means of a formula suggested by equation 14.3.5:

$$g(t) = t \cdot \mathbf{b} + (1 - t) \cdot \mathbf{a} \qquad t \in \langle\!\langle 0, 1 \rangle\!\rangle.$$

We define the length $L(f)$ of a straight path f to be $\|\mathbf{a} - \mathbf{b}\|$. Clearly, $\bar{f}$ is straight if f is and $L(\bar{f}) = L(f)$; further, for any other straight path h from **a** to **b**, $L(f) = L(h)$; for example, if $h(t) = f(t^2)$, for all $t \in \langle\!\langle 0, 1 \rangle\!\rangle$.

Returning to 15.4.1, suppose $g: \langle\!\langle 0, q \rangle\!\rangle \to \mathbb{R}^3$ is a path from **b** to **c**. Then we can form a path from **a** to **c**, denoted by $f * g: \langle\!\langle 0, p + q \rangle\!\rangle \to \mathbb{R}^3$, by joining f to g at **b** (see Fig. 15.4); in a formula

15.4.4
$$f * g(t) = \begin{cases} f(t), & \text{if } 0 \leqslant t \leqslant p, \\ g(t - p), & \text{if } p \leqslant t \leqslant p + q. \end{cases}$$

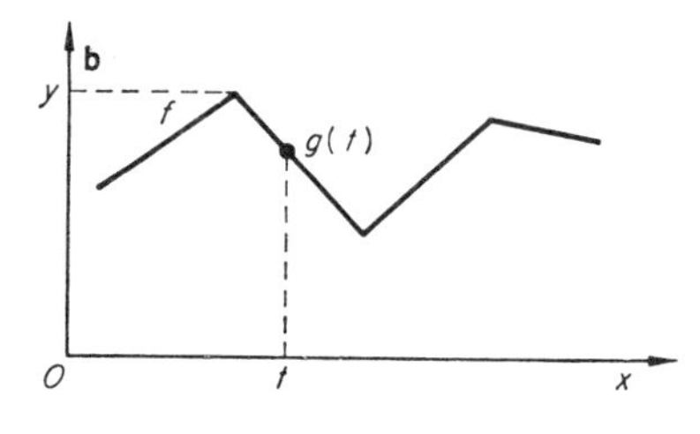

Fig. 15.4

Notice that the star operation $f * g$ is associative. Now, in particular, if for each $i = 1, 2, \ldots, n$ we have a straight path $f_i: \langle\!\langle 0, p_i \rangle\!\rangle \to \mathbb{R}^3$ from $\mathbf{a}_i$ to $\mathbf{b}_i$, where $\mathbf{a}_{i+1} = \mathbf{b}_i$ $(i = 1, \ldots, n - 1)$, then the product path

15.4.5
$$h = f_1 * f_2 * \cdots * f_n: \langle\!\langle 0, \Sigma p_i \rangle\!\rangle \to \mathbb{R}^3$$

is called a 'segmental path from $\mathbf{a}_1$ to $\mathbf{b}_n$'. If X is any subset of $\mathbb{R}^3$, we call it '**segmentally connected**' iff given any two points $\mathbf{a}, \mathbf{b} \in X$, there is in X a segmental path from **a** to **b**. For example $\mathbb{R}$, $\mathbb{R}^2$, $\mathbb{R}^3$ and the interior of a circle in $\mathbb{R}^2$ are segmentally connected, being *convex* (see 14 Exercise 3(v)).

Exercise 8

(i) Let E denote an annulus in $\mathbb{R}^2$ (i.e., E is of the form

$$\{(x, y) | u^2 \leqslant x^2 + y^2 \leqslant v^2\}$$

for some non-zero numbers u, v). Prove that E is segmentally connected, but that its two boundary circles are not.

(ii) Let $T: \mathbb{R}^3 \to \mathbb{R}^3$ be a linear transformation, and let X be a segmentally connected subset of $\mathbb{R}^3$. Show that $T(X)$ is also segmentally connected.
(iii) Show that a star-domain:

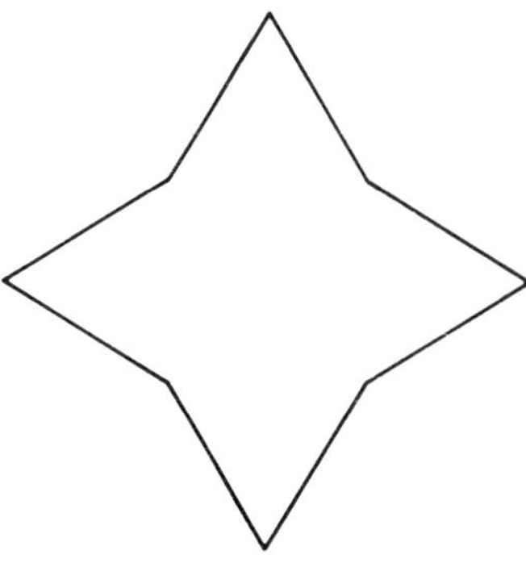

Fig. 15.5

is segmentally connected but not convex.

Consider again the segmental path h in 15.4.5. We define its length to be the sum

15.4.6
$$L(h) = \sum_{i=1}^{n} L(f_i),$$

which manifestly depends only on the points $\mathbf{a}_i$, $\mathbf{b}_i$, and not on the functions f_i. Moreover, it agrees with our intuitive notions of length which are based on piecing together such things as pieces of string. These notions lead us also to feel that we might be able to give a sensible definition of the length of a more general, curved path using approximations by segmental paths.

15.5 Rectifiability

Thus, let $f: \langle\!\langle 0, p \rangle\!\rangle \to \mathbb{R}^3$ be any path from $\mathbf{a}$ to $\mathbf{b}$, and let $0 = t_0 < t_1 < \cdots < t_n = p$ be a 'dissection' $\mathscr{D}$ of the interval $\langle\!\langle 0, p \rangle\!\rangle$ by the $n + 1$ points $t_0, \ldots, t_n$. Then we can add together the lengths of the segments $\langle\!\langle f(t_i), f(t_{i+1}) \rangle\!\rangle$ to form $\sum_{i=0}^{n-1} |f(t_{i+1}) - f(t_i)|$; we denote this number by $L(\mathscr{D})$. Now fix i, and let $\mathscr{D}'$ denote the dissection obtained by inserting one new point s between t_i and t_{i+1}. Then we see from Fig. 15.6 that

$$L(\mathscr{D}') - L(\mathscr{D}) = |f(t_i) - f(s)| + |f(t_{i+1}) - f(s)| - |f(t_{i+1}) - f(t_i)|,$$

which is $\geqslant 0$ by the triangle inequality (14.8.1). Thus *if we refine the dissection $\mathscr{D}$ by adding new points, we increase $L(\mathscr{D})$.* Now consider the set S of

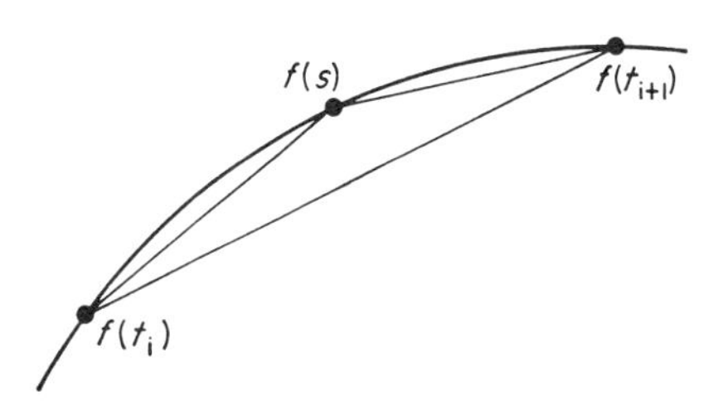

Fig. 15.6

all numbers $L(\mathscr{D})$ for all possible dissections $\mathscr{D}$ of $\langle\!\langle 0, p \rangle\!\rangle$; it is clearly non-empty. Hence, if S is bounded above, then it has a least upper bound (Theorem 24.6.9), which we *define* to be the length $L(f)$ of f:

15.5.1 DEFINITION. $$L(f) = \text{lub}_S\,(L(\mathscr{D})),$$

writing $L(f) = \infty$ when S is unbounded. If $L(f)$ is finite, then f is said to be *rectifiable*. We invite the reader to prove the easy consequences in the following exercises.

Exercise 9

(i) If f is a segmental path, prove that f is rectifiable and the two formulae 15.4.6 and 15.5.1 agree.
(ii) If f, g are rectifiable in 15.4.4, prove that $f * g$ is also, and $L(f * g) = L(f) + L(g)$.
(iii) By considering a dissection of $\langle\!\langle 0, 1 \rangle\!\rangle$ of the form $2/(2n - 1)\pi$ $(n = 1, \ldots, m$, for suitable $m)$ prove that the path $f\colon \langle\!\langle 0, 1 \rangle\!\rangle \to \mathbb{R}$ given by $f(x) = x \sin(1/x)$ $(x \neq 0)$, $f(0) = 0$, is not rectifiable. {Hint: $\sum_{n=1}^{\infty} (1/n)$ is divergent.}
(iv) Prove that if the continuous paths $f, g\colon \langle\!\langle 0, 1 \rangle\!\rangle \to \mathbb{R}^3$ from $\mathbf{a}$ to $\mathbf{b}$ are both one-one and rectifiable, with the same image, then $L(f) = L(g)$. (f and g are said to be 'differently parametrized'; they represent the same set of points, but traversed at different 'times' t, with different velocities when f, g are differentiable.)

A simple criterion for rectifiability (which is not the most general: see Olmsted [98]) is:

15.5.2 THEOREM. *Suppose that the co-ordinates $x(t)$, $y(t)$, $z(t)$ of $f(t)$ in* 15.1.2 *are differentiable functions of t on $\langle\!\langle 0, p \rangle\!\rangle$. Then f is rectifiable and*

$$L(f) = \int_0^p \sqrt{(\dot{x}^2(t) + \dot{y}^2(t) + \dot{z}^2(t))}\, dt$$

(where $\dot{x}(t) = dx/dt$, etc.).

A plausible argument is this: let $s(u) = L(f \mid \langle\!\langle 0, u \rangle\!\rangle)$, so that, in infinitesimal notation:

$$\begin{aligned}(\delta s)^2 &= (\delta x)^2 + (\delta y)^2 + (\delta z)^2 && \text{approximately}\\ &= (\dot{x}\,\delta t)^2 + (\dot{y}\,\delta t)^2 + (\dot{z}\,\delta t)^2 && \text{approximately;}\end{aligned}$$

so
$$ds/dt = \lim_{\delta t \to 0} (\delta s/\delta t) = \sqrt{(\dot{x}^2 + \dot{y}^2 + \dot{z}^2)},$$
whence
$$s(p) = \int_0^p \sqrt{(\dot{x}^2 + \dot{y}^2 + \dot{z}^2)}\, dt.$$

This argument assumes that $s(u)$ exists and is differentiable, and gives no estimate of how close we can make the two approximations used in the proof. These points can be cleared up in a rigorous proof, at the cost of some detail.

15.5.3 EXAMPLE. (*The circle and* π.)

Let C_r denote the circle $x^2 + y^2 = r^2$ in $\mathbb{R}^2$; we can express it as the image of the path $u * v$, where $u, v\colon \langle 0, 2r \rangle \to \mathbb{R}^3$ are paths given by

$$u(t) = (t - r, (r^2 - (t - r)^2)^{1/2}), \qquad v(t) = (t - r, -(r^2 - (t - r)^2)^{1/2}).$$

We can then use 15.5.2 directly to show that $u * v$ is rectifiable and calculate the perimeter $L(C_r)$ of C_r as $L(u) + L(v)$. However, if we go back to the definition 15.5.1, we can avoid the calculus and incidentally obtain approximations for π, as follows. Let P_{rn}, Q_{rn} ($n = 2, 3, \ldots$) denote respectively the regular polygons of 2^n sides, inscribed and escribed in C_r (see Fig. 15.7).

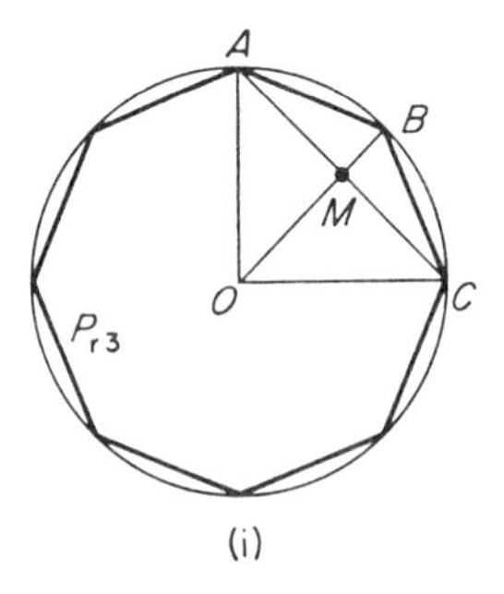

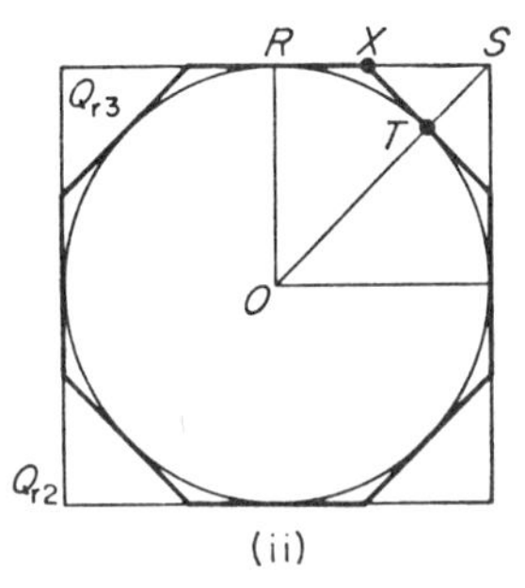

Fig. 15.7 $RX = XT = b_{n+1}/2$. $XS = \frac{1}{2}(b_n - b_{n+1})$.

Then for each n, we have

15.5.4 $$L(P_{rn}) \leqslant L(P_{r,n+1}) \leqslant L(C_r) \leqslant L(Q_{r,n+1}) \leqslant L(Q_{rn})$$

[We state this as 'obvious', but a rigorous proof is slightly complicated]; so the sequences $\{L(P_{rn})\}$, $\{L(Q_{rn})\}$ of numbers increase and decrease respectively to one and the same limit $L(C_r)$. Now let λP_{rn} denote the polygon with vertices $\lambda \mathbf{v}_i$, where $\mathbf{v}_i$ runs through the vertices of P_{rn}; and similarly for λQ_{rn}. Clearly $r^{-1}P_{rn} = P_{1n}$, $r^{-1}Q_{rn} = Q_{1n}$. Then dividing the above inequalities by r, we get

$$L(P_{1n}) \leqslant L(P_{1,n+1}) \leqslant L(C_r)/r \leqslant L(Q_{1,n+1}) \leqslant L(Q_{1,n}),$$

whence $L(C_r)/r = L(C_1)$. It is customary to denote $\frac{1}{2}L(C_1)$ by the symbol π (=Greek 'p', for 'perimeter'), whence

15.5.5 $$L(C_r) = 2\pi r.$$

We can also use the 'limit' notation and write

15.5.6 $$2\pi = \lim_{n\to\infty} L(P_{1n}).$$

15.5.7 To obtain approximations for π, let a_n, b_n denote respectively the length of a side of P_{1n}, Q_{1n}. Then from Fig. 15.7(i) we obtain, on applying Pythagoras's theorem, that $AB^2 = AM^2 + MB^2$, i.e.,

$$a_{n+1}^2 = a_n^2/4 + (1 - \sqrt{(1^2 - a_n^2/4)})^2 = 2(1 - \sqrt{(1 - a_n^2/4)}).$$

From Fig. 15.7(ii), we obtain, similarly, $b_{n+1} = 4(\sqrt{(1 + b_n^2/4)} - 1)/b_n$.

Now $L(P_{1n}) = 2^n a_n$, $L(Q_{1n}) = 2^n b_n$ and $a_2 = \sqrt{2}$, $b_2 = 2$, so $2\sqrt{2} \leqslant \pi \leqslant 4$ while $a_3^2 = 2 - \sqrt{2}$, $b_3 = 2(\sqrt{2} - 1)$; so we get a better estimate

$$4(\sqrt{2}(\sqrt{2} - 1))^{1/2} \leqslant \pi \leqslant 8(\sqrt{2} - 1),$$

and so on. Note that $a_n^2 = 2 - \sqrt{(2 + \sqrt{(2 + \sqrt{(2 + \cdots + \sqrt{2}))\cdots))}}}$ with $n - 2$ square roots.

Exercise 10

(i) Show that $3 < \pi < 4$.
(ii) Is π greater or less than 22/7?
(iii) Write a flow diagram for an iterative process to compute π to an accuracy of n decimal places.
(iv) Use the inequalities of 15.5.4 to prove that $L(C_r)$ exists.

15.6 Jordan Arcs and Curves

Let $f: \langle 0, 1\rangle \to \mathbb{R}^2$ be a path from $\mathbf{a}$ to $\mathbf{b}$. If f is one-one, we call f a **Jordan arc**, after the French mathematician C. Jordan (1838–1922). If also $\mathbf{a} = \mathbf{b}$, then f is called a **Jordan curve**, or a **simple closed curve**. In either case, the image of f cannot cross itself, since f is one-one on $\langle 0, 1\rangle$. Jordan seems to have been the first to point out that a proof was required to establish the seemingly obvious fact that a Jordan curve divides the plane into an 'inside' and an 'outside'. The fact is 'obvious' only for unsophisticated curves; it is not so obvious even for such a relatively simple one as that in Fig. 15.8. A full proof is fairly difficult and is called the 'Jordan curve theorem'; see for example Chapter 25 and Newman [94], Ch. V. We need to discuss here the special case when the Jordan curve is 'polygonal'. Thus by a 'polygonal curve $J = J(\mathbf{v}_1, \mathbf{v}_2, \ldots, \mathbf{v}_n)$' we mean a segmental path of the form 15.4.5 where the points $\mathbf{v}_1, \ldots, \mathbf{v}_n$ are in $\mathbb{R}^2$ and

15.6.1 $$J = f_1 * f_2 * \cdots * f_n, \qquad (n \geqslant 2)$$

where each f_i is the straight path from $\mathbf{v}_i$ to $\mathbf{v}_{i+1}$ and $\mathbf{v}_{n+1} = \mathbf{v}_1$. If this path J is also a Jordan curve, we call it a 'Jordan polygon'; thus no two edges intersect except that

$$\langle\!\langle \mathbf{v}_i, \mathbf{v}_{i+1} \rangle\!\rangle \cap \langle\!\langle \mathbf{v}_{i+1}, \mathbf{v}_{i+2} \rangle\!\rangle = \mathbf{v}_{i+1}, \qquad (i = 1, 2, \ldots, n).$$

The image of the function J is called the 'locus' of J, and denoted by $|J|$.

Fig. 15.8

The **Jordan curve theorem** for J states:

15.6.2 THEOREM. $\mathbb{R}^2 - J = A \cup B$, *where* A, B *are sets such that* $A \cap B = \varnothing$ *and*

(a) *A and B are each segmentally connected* (defined prior to Exercise 8);
(b) *if* $\mathbf{a} \in A$, $\mathbf{b} \in B$ *and f is any path in* $\mathbb{R}^2$ *from* $\mathbf{a}$ *to* $\mathbf{b}$, *then f cuts* $|J|$ *somewhere.*

This last statement is valid for any Jordan curve J, but we here consider a polygonal Jordan curve P. A good proof can be found in Courant–Robbins [25], Chapter V, so we shall omit it here. (See also Chapter 25.) However, the proof depends on the following observation, which we shall use later. Let $\mathbf{d}$ be a fixed direction in $\mathbb{R}^2$, not parallel to an edge of P. Then, for each point $\mathbf{z}$ of $\mathbb{R}^2$ not on an edge of P, let r be the ray through $\mathbf{z}$ parallel to $\mathbf{d}$; i.e.,

$$r = \{\mathbf{v} | \mathbf{v} = \mathbf{z} + t\mathbf{d}, \ t \geqslant 0\}.$$

Define $W(\mathbf{z})$, the **order** of $\mathbf{z}$ with respect to P, to be the number of intersections of $|P|$ with r, if r does not pass through a vertex of P. If r passes through the vertex V, count that intersection twice if both edges of P at V are on the same side of r, otherwise count it once. Thus, in Fig. 15.9, $W(\mathbf{z}) = 2$, $W(\mathbf{z}') = 4$, $W(\mathbf{z}'') = 1$, $W(\mathbf{z}''') = 0$. In the general case, $W(\mathbf{z})$ is finite, and is either odd or even. We define the 'interior' and 'exterior' of P, respectively, to be the sets

15.6.3 $$\operatorname{Int} P = \{\mathbf{z} | W(\mathbf{z}) \text{ is odd}\}, \qquad \operatorname{Ext} P = \{\mathbf{z} | W(\mathbf{z}) \text{ is even}\}.$$

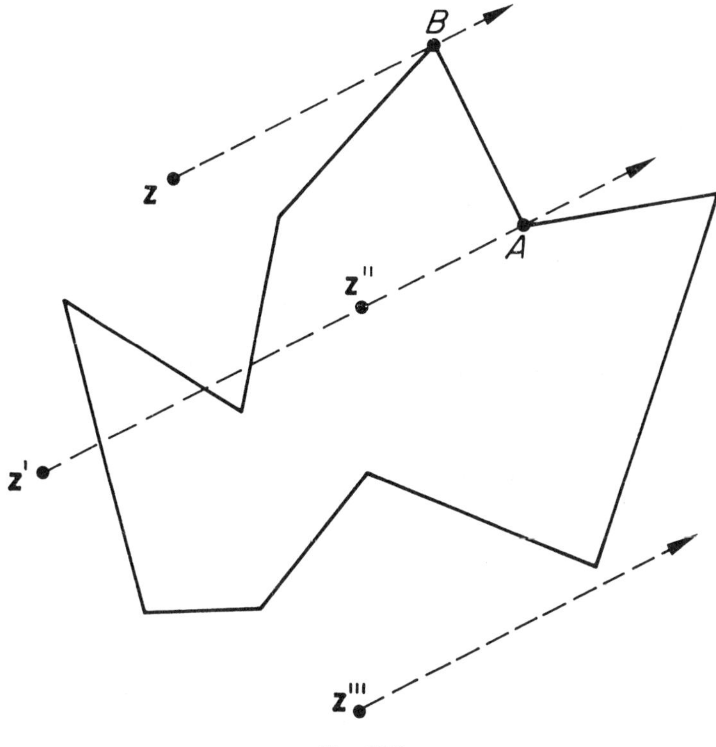

Fig. 15.9

Figure 15.9 shows that these are reasonable definitions to make, and the proof consists in showing that Int P, Ext P have the properties (a), (b) required of A, B in the statement of 15.6.2. We naturally use the obvious terminology and call Int P, Ext P the 'inside' and 'outside' of P; observe that we can distinguish them since Ext P contains points at arbitrarily large distances from P while Int P is 'bounded'. It can be shown that both these sets are open in $\mathbb{R}^2$ (in the sense of 25.6.5).

Exercise 11

(i) Prove that Int P, Ext P in 15.6.3 satisfy the conditions on A, B in 15.6.2.
(ii) In the statement 15.6.2, suppose $|J|$ is the circle $x^2 + y^2 = 1$. Show that A, B are the regions $x^2 + y^2 < 1$, $x^2 + y^2 > 1$, in either order.
(iii) Prove the above remark that Ext P contains points at arbitrarily large distances from P. {Does every line meet P? Zero is even!}

Topological questions of the kind considered above, are important when we study the notion of area, to which we now turn.

15.7 Area

The notion of the area of a plane region is so ingrained in us, that a *theory* of area is quite subtle; ingrained notions are the hardest to analyse. However, we are going to look for a way of assigning to each member G of a (preferably) wide class $\mathscr{M}$ of subsets of the plane, a number $\mathscr{A}(G)$, with at least the properties:

$\mathfrak{PA}_1$: *$\mathscr{A}$ is 'translation-invariant'*, i.e., *$\mathscr{A}(G) = \mathscr{A}(G + v)$ for each $v \in \mathbb{R}^2$;*
$\mathfrak{PA}_2$: *$\mathscr{A}$ is additive*, i.e., *if A, B and $A \cap B$ are in $\mathscr{M}$, and $\mathscr{A}(A \cap B) = 0$, then $A \cup B \in \mathscr{M}$ and*

$$\mathscr{A}(A \cup B) = \mathscr{A}(A) + \mathscr{A}(B);$$

$\mathfrak{PA}_3$: *If $A \subseteq B$ in $\mathscr{M}$, then $\mathscr{A}(A) \leqslant \mathscr{A}(B)$;*
$\mathfrak{PA}_4$: *The empty set, finite sets, and all images of Jordan polygons lie in $\mathscr{M}$, all having $\mathscr{A}$ equal to zero.*

From the practical point of view of measuring carpets or similar floor-coverings, $\mathfrak{PA}_1$ ensures that a carpet bought in a shop is the same size when we bring it home; while $\mathfrak{PA}_2$ ensures that the total carpet required for a house is the sum of the requirements for each room; $\mathfrak{PA}_3$ requires smaller carpets for smaller rooms, while $\mathfrak{PA}_4$ requires that the cracks between the floorboards should need no extra carpet.

15.7.1 RECTANGLES. Consider first the case when G is a rectangle R. Are we stating a theorem or a definition, when we say that the area of R is its length times its breadth? If we want to cover R with squares of side 1, and if R is l units long and b units broad, then an inductive *proof* (using the distributive law) tells us that $l \times b$ squares are required. But if R has dimensions $\sqrt{2}$, π, say, what do we do then? Of course, carpet-layers do not meet this problem, since they have to measure all lengths in integer multiples of some unit like a centimetre, but we still need an idealized theory partly for simplicity, and partly also to cover any potential refinement of measurement.

We begin, then, by *defining* the area, $\mathscr{A}(R)$, of a rectangle R to be its length times its breadth in *all* cases. A carpet-cutting argument then shows that the area of a parallelogram $ABCD$ ought to be defined as its base times its height (equal to that of the rectangle $CDEE'$ in Fig. 15.10). By elementary trigonometry, this becomes

15.7.2 $$\mathscr{A}(ABCD) = AD \cdot AB \sin DAB,$$

which we used in 14.10. The same kind of argument (see the figure) leads to the definition of the area of a triangle as one-half its base times height (equal to half the area of parallelogram $ABCD$ in Fig. 15.10). This is consistent with $\mathfrak{PA}_1$, $\mathfrak{PA}_2$, and $\mathfrak{PA}_4$, but we consider $\mathfrak{PA}_3$ later (15.9 below). By 15.7.2,

15.7.3 $$\mathscr{A}(ABC) = \tfrac{1}{2}AB \cdot AC \sin BAC,$$

whence 'triangles on the same base and between the same parallels (i.e., of equal height and base-lengths) have the same area'. To extend these definitions to polygons with more sides, two methods spring to mind:

Method A: cut up such a polygon P into triangles, and add their areas together.

Method B: use the classical geometrical proof (based on the remark following 15.7.3) that there exists an algorithm for constructing† a square of the same area as P.

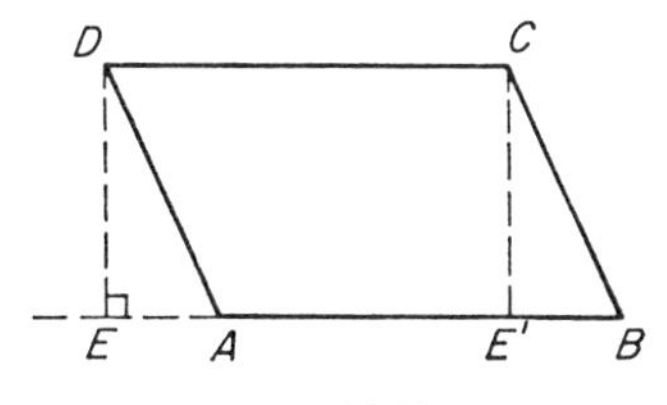

Fig. 15.10

The main objection to each method is that we would have to show that the result of each calculation is independent of the particular dissections of P that we happened to use. This is possible; Method A is treated in detail in Hilbert [57]. We, however, will proceed rather differently, as we wish to use analytic geometry and eventually to be able to assign an area to plane regions with curved boundaries.

Exercise 12

Does the classical algorithm (Method B) work for a non-convex polygon?

15.8 Polygons

Consider polygons first. By a **polygon** $P(\mathbf{v}_1, \mathbf{v}_2, \ldots, \mathbf{v}_n)$ with **vertices** $\mathbf{v}_1, \mathbf{v}_2, \ldots, \mathbf{v}_n$, we mean simply a choice of the distinct points $\mathbf{v}_1, \mathbf{v}_2, \ldots, \mathbf{v}_n \in \mathbb{R}^2$, *in that order*. [This notation will not be confused with the notations $P(\mathbf{a}, \mathbf{b})$, $P(\mathbf{a}, \mathbf{b}, \mathbf{c})$ of 14.10 and 14.11.] We distinguish it from the polygonal curve $J(\mathbf{v}_1, \ldots, \mathbf{v}_n)$ of 15.6.1 (to which it is closely related). For each $r = 1, 2, \ldots, n$ we also regard as *equal* the polygons

15.8.1 $$P(\mathbf{v}_1, \mathbf{v}_2, \ldots, \mathbf{v}_n), \qquad P(\mathbf{v}_{r+1}, \mathbf{v}_{r+2}, \ldots, \mathbf{v}_n, \mathbf{v}_1, \mathbf{v}_2, \ldots, \mathbf{v}_r),$$

(whereas the associated curves J are not equal). Thus for example, when $n = 3$,

15.8.2 $$P(ABC) = P(BCA) = P(CAB) \neq P(BAC),$$

† With ruler and compasses only. The problem of extending this to an algorithm for finding a square with the same area as a given circle, was suggested by the Greeks under the name 'squaring the circle', and was proved impossible in the 19th century.

and when $n = 4$,

15.8.3 $$P(ABCD) = P(BCDA), \text{ etc.}, \neq P(BACD).$$

We make the convention that $\mathbf{v}_{n+1} = \mathbf{v}_1$. The equation in the following Lemma is vital:

15.8.4 LEMMA. *For any* $\mathbf{p} \in \mathbb{R}^2$, $\sum_{i=1}^{n} \mathbf{v}_i \wedge \mathbf{v}_{i+1} = \sum_{i=1}^{n} (\mathbf{v}_i - \mathbf{p}) \wedge (\mathbf{v}_{i+1} - \mathbf{p})$, *where* $\wedge$ *represents vector product.*

Proof. Since $\mathbf{p} \wedge \mathbf{p} = \mathbf{0}$, the right-hand side is

$$\sum \mathbf{v}_i \wedge \mathbf{v}_{i+1} - \mathbf{p} \wedge \sum \mathbf{v}_{i+1} - \left(\sum \mathbf{v}_i\right) \wedge \mathbf{p}$$

and since $\mathbf{v}_{n+1} = \mathbf{v}_1$, $\sum \mathbf{v}_{i+1} = \sum \mathbf{v}_i$. Now use property $\mathfrak{VP}_3$ of 14.10. ■

A plane region in space has a direction normal to it, and it is often convenient to incorporate this direction with the area. As suggested by 14.10.5, we therefore associate with P a *vector* $\boldsymbol{\alpha} = \boldsymbol{\alpha}P$ given by:

15.8.5 DEFINITION. $\boldsymbol{\alpha}P(\mathbf{v}_1, \mathbf{v}_2, \ldots, \mathbf{v}_n) = \frac{1}{2} \sum_{i=1}^{n} \mathbf{v}_i \wedge \mathbf{v}_{i+1}$. We note that this is consistent with 15.8.1. Since we regard $\mathbb{R}^2$ as the plane $z = 0$ in $\mathbb{R}^3$, we are therefore considering the vector $\boldsymbol{\alpha}P$ as having a direction parallel to Oz.

15.8.6 EXAMPLE. Let $n = 3$, $\mathbf{v}_1 = \mathbf{0}$; then for the triangle $\mathbf{v}_1\mathbf{v}_2\mathbf{v}_3$, $\boldsymbol{\alpha}$ is $\frac{1}{2}(\mathbf{v}_2 \wedge \mathbf{v}_3)$, whose magnitude agrees with the earlier formula for the area of $\mathbf{v}_1\mathbf{v}_2\mathbf{v}_3$. On the other hand, the triangle $\mathbf{v}_2\mathbf{v}_1\mathbf{v}_3$ has $\boldsymbol{\alpha}$ equal to $\frac{1}{2}(\mathbf{v}_3 \wedge \mathbf{v}_2)$. This is $-\frac{1}{2}(\mathbf{v}_2 \wedge \mathbf{v}_3)$, consistently with the inequality in 15.8.2.

15.8.7 EXAMPLE. Let $n = 4$, $\mathbf{v}_1 = \mathbf{0}$, $\mathbf{v}_2 = \mathbf{i}$, $\mathbf{v}_3 = \mathbf{i} + \mathbf{j}$, $\mathbf{v}_4 = \mathbf{j}$. Then

$$\boldsymbol{\alpha}P(\mathbf{v}_1, \mathbf{v}_2, \mathbf{v}_3, \mathbf{v}_4) = \tfrac{1}{2}((\mathbf{0} \wedge \mathbf{i}) + (\mathbf{i} \wedge (\mathbf{i} + \mathbf{j})) + ((\mathbf{i} + \mathbf{j}) \wedge \mathbf{j}) + (\mathbf{j} \wedge \mathbf{0}))$$

which is $\mathbf{k}$, of modulus 1, using the rules $\mathfrak{VP}$ in 14.10. On the other hand (see Fig. 15.11),

$$\boldsymbol{\alpha}P(\mathbf{v}_1, \mathbf{v}_3, \mathbf{v}_4, \mathbf{v}_2) = \tfrac{1}{2}(\mathbf{0} \wedge (\mathbf{i} + \mathbf{j}) + ((\mathbf{i} + \mathbf{j}) \wedge \mathbf{j}) + \mathbf{j} \wedge \mathbf{i} + \mathbf{i} \wedge \mathbf{0}) = \mathbf{0}.$$

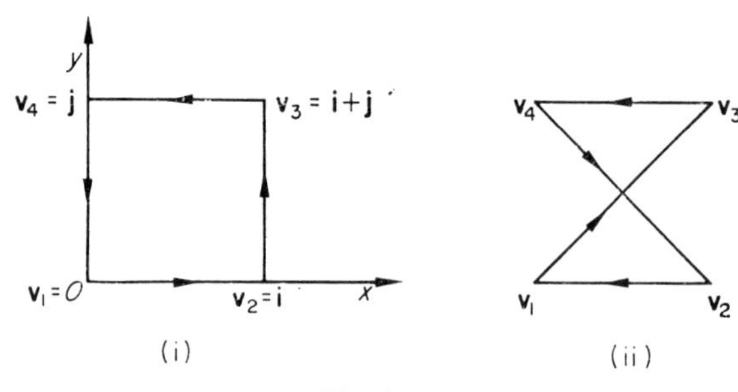

Fig. 15.11

Similarly, if $\mathbf{v}_1 = \mathbf{0}$ and $P(\mathbf{v}_1\mathbf{v}_2\mathbf{v}_3\mathbf{v}_4)$ is a parallelogram, then $\mathbf{v}_3 = \mathbf{v}_2 + \mathbf{v}_4$, so $\boldsymbol{\alpha}P(\mathbf{v}_1\mathbf{v}_2\mathbf{v}_3\mathbf{v}_4) = \mathbf{v}_2 \wedge \mathbf{v}_4$ whose modulus is in accordance with 15.7.2. We have, therefore, a true extension of the 'area' of a parallelogram as defined in 15.7.2.

15.9 Properties of $\boldsymbol{\alpha}$

Let us prove some simple properties of the function $\boldsymbol{\alpha}$.

15.9.1 LEMMA. *$\boldsymbol{\alpha}$ is translation-invariant in the sense of $\mathfrak{PA}_1$.*

Proof. If P is translated through $\mathbf{a}$ to $P + \mathbf{a}$, then $\boldsymbol{\alpha}(P + \mathbf{a})$ is $\frac{1}{2}\sum (\mathbf{v}_i + \mathbf{a}) \wedge (\mathbf{v}_{i+1} + \mathbf{a})$ which is $\boldsymbol{\alpha}(P)$ by Lemma 15.8.4. ■

For the second property we must say what we mean by a 'subdivision' of P. If we insert a point $\mathbf{w}$ on one of the edges of P, say on $\langle\!\langle \mathbf{v}_i, \mathbf{v}_{i+1} \rangle\!\rangle$, then $P(\mathbf{v}_1, \ldots, \mathbf{v}_i, \mathbf{w}, \mathbf{v}_{i+1}, \ldots, \mathbf{v}_n)$ is called a **simple** subdivision of P; and a polygon Q is a **subdivision** of P iff $\exists$ polygons $P = P_1, P_2, \ldots, P_k = Q$, each a simple subdivision of the preceding one.

15.9.2 LEMMA. *$\boldsymbol{\alpha}$ is 'invariant under subdivision', in the sense that $\boldsymbol{\alpha}(P) = \boldsymbol{\alpha}(Q)$.*

Proof. It suffices to assume that Q is a simple subdivision of P, say by the insertion of $\mathbf{w}$ on the edge $\langle\!\langle \mathbf{v}_i, \mathbf{v}_{i+1} \rangle\!\rangle$. But then $\boldsymbol{\alpha}(Q)$ contains the term $\mathbf{v}_i \wedge \mathbf{w} + \mathbf{w} \wedge \mathbf{v}_{i+1} = \mathbf{v}_i \wedge \mathbf{v}_{i+1}$ since $\mathbf{w} = t\mathbf{v}_i + (1 - t)\mathbf{v}_{i+1}$ for some t with $0 \leqslant t \leqslant 1$; and all other terms are the same as those in $\boldsymbol{\alpha}(P)$. ■

By the 'locus' $|P|$ of P, we mean the set of all points on the edges of P. Clearly (using 15.9.2):

15.9.3 LEMMA. *For any subdivision Q of P, the loci $|P|$, $|Q|$ are equal.* ■

Corresponding to $\mathfrak{PA}_2$, we have:

15.9.4 LEMMA. *Let $P(\mathbf{v}_1, \mathbf{v}_2, \ldots, \mathbf{v}_n)$, $Q(\mathbf{w}_1, \mathbf{w}_2, \ldots, \mathbf{w}_m)$ be polygons such that there exist $i, p, j \in \mathbb{N}$, for which $\mathbf{v}_{i+t} = \mathbf{w}_{j+p-t}$, $t = 0, 1, \ldots, p$ (see Fig. 15.12). Then $P(\mathbf{v}_1, \mathbf{v}_2, \ldots, \mathbf{v}_i, \mathbf{w}_{j+p+1}, \ldots, \mathbf{w}_m, \ldots, \mathbf{w}_j, \mathbf{v}_{i+p+1}, \ldots, \mathbf{v}_n)$ is called the **sum** $P + Q$ of P and Q, and $\boldsymbol{\alpha}(P) + \boldsymbol{\alpha}(Q) = \boldsymbol{\alpha}(P + Q)$.*

Proof. The sum of P and Q is so defined that each term $\mathbf{v}_{i+t} \wedge \mathbf{v}_{i+t+1}$, $0 \leqslant t \leqslant p - 1$, in $\boldsymbol{\alpha}(P)$ cancels with the term $\mathbf{w}_{j+p-(t+1)} \wedge \mathbf{w}_{j+p-t}$ in $\boldsymbol{\alpha}(Q)$; and the remaining terms add exactly to $\boldsymbol{\alpha}(P + Q)$. ■

15.9.5 The polygons $P = P(\mathbf{v}_1, \ldots, \mathbf{v}_n)$, $P' = P(\mathbf{v}_n, \mathbf{v}_{n-1}, \ldots, \mathbf{v}_1)$ are distinct since $n \geqslant 1$, and $\boldsymbol{\alpha}(P) = -\boldsymbol{\alpha}(P')$ in view of the anticommutativity of the vector product. Now $\boldsymbol{\alpha}(P)$ is a vector of the form $(0, 0, z)$; and if $z \geqslant 0$, we say that P is 'positively oriented', and P' is 'negatively oriented'. Thus, for example, a sum of positively oriented polygons is positively oriented, by Lemma 15.9.4.

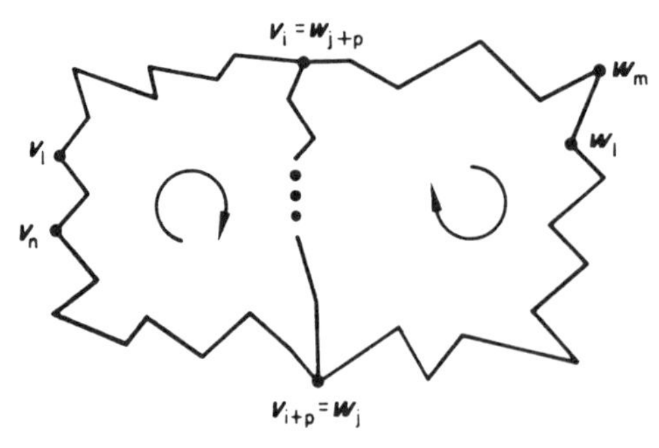

Fig. 15.12

Exercise 13

(i) For each real number λ, let λP denote $P(\lambda\mathbf{v}_1, \ldots, \lambda\mathbf{v}_n)$. Prove that $\boldsymbol{\alpha}(\lambda P) = \lambda^2\boldsymbol{\alpha}(P)$.
(ii) With P, Q as in 15.9.4, prove that $\lambda(P + Q) = \lambda P + \lambda Q$.
(iii) If P in (i) is positively oriented, show that λP is also, provided $\lambda \neq 0$.
(iv) Prove that if P is a Jordan polygon in (i), so is λP, provided $\lambda \neq 0$.
(v) For each $\mathbf{a} \in \mathbb{R}^2$, let $\mathbf{a} + P(\mathbf{v}_1, \ldots, \mathbf{v}_n)$ denote $P(\mathbf{a} + \mathbf{v}_1, \ldots, \mathbf{a} + \mathbf{v}_n)$. Prove that $\boldsymbol{\alpha}(\mathbf{a} + P) = \boldsymbol{\alpha}(P)$, and P is positively oriented or a Jordan polygon iff the same holds for $\mathbf{a} + P$.

Recall that $\boldsymbol{\alpha}(P)$ is of the form $(0, 0, z)$. To take account of the sign of z, denote this z by $\boldsymbol{\alpha}_3(P)$.

Corresponding to property $\mathfrak{PA}_3$ above, we can now prove:

15.9.6 LEMMA. *Let P be a Jordan polygon whose locus lies entirely inside a Jordan polygon J. Then if P and J are positively oriented, $\boldsymbol{\alpha}_3(P) \leqslant \boldsymbol{\alpha}_3(J)$.*

Proof. The basic idea of the proof is best seen by an example (Fig. 15.13). We have $|P| \subseteq \text{Int } J$. Choose a point $\mathbf{z} \in \text{Int } P$ so that no segment from $\mathbf{z}$

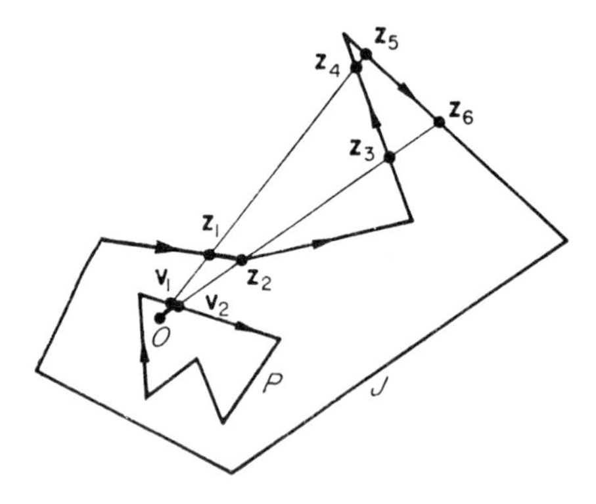

Fig. 15.13

to a vertex of P or J contains an edge of P or J; such a $\mathbf{z}$ exists, since we need avoid only a finite number of lines (along these edges). By property $\mathfrak{PA}_1$ we may then assume $\mathbf{z} = \mathbf{0}$. Next subdivide both P and J so that if $\mathbf{v}$ is a vertex of P or J then the ray $\langle\langle 0, \mathbf{v}\rangle\rangle$ meets J or P only in (possibly new) vertices. By 15.9.3 $\boldsymbol{\alpha}(P)$ and $\boldsymbol{\alpha}(J)$ are left unaltered. Now refer to Fig. 15.13. The sum giving $\boldsymbol{\alpha}(P)$ is made up of terms like $\mathbf{v}_i \wedge \mathbf{v}_{i+1}$; and we compare it with the terms

$$\text{(i)}\ \mathbf{z}_1 \wedge \mathbf{z}_2, \qquad \mathbf{z}_3 \wedge \mathbf{z}_4, \qquad \mathbf{z}_5 \wedge \mathbf{z}_6,$$

occurring in the sum for $\boldsymbol{\alpha}(J)$, where the $\mathbf{z}$'s are vertices of J on the segments $\langle\langle 0, \mathbf{v}_i\rangle\rangle$, $\langle\langle 0, \mathbf{v}_{i+1}\rangle\rangle$. By the definition of Int P in 15.6.3, the number $w(0)$ is odd, so there is an odd number of terms in (i) above. Now, since $\mathbf{z}_1$ (for example) lies on $\langle\langle \mathbf{z}, \mathbf{v}_1\rangle\rangle$ then there exists $r_1 \in \mathbb{R}$, such that $r_1 \geqslant 1$ and $\mathbf{z}_1 = r_1\mathbf{v}_1$. Similarly, $\mathbf{z}_4 = s_1\mathbf{v}_1$, $\mathbf{z}_5 = t_1\mathbf{v}_1$, $\mathbf{z}_2 = r_2\mathbf{v}_2$, etc., with $1 \leqslant r_1 < s_1 < t_1$ and similarly for r_2, etc. Thus the term $T = \mathbf{z}_1 \wedge \mathbf{z}_2 + \mathbf{z}_3 \wedge \mathbf{z}_4 + \mathbf{z}_5 \wedge \mathbf{z}_6$ in $\boldsymbol{\alpha}(J)$ is—using the anticommutativity of the vector product—

$$T = (r_1r_2 - s_1s_2 + t_1t_2)(\mathbf{v}_1 \wedge \mathbf{v}_2) = u\cdot(\mathbf{v}_1 \wedge \mathbf{v}_2), \quad \text{say},$$

where $u \geqslant 1$, since $1 \leqslant r_1r_2 < s_1s_2 < t_1t_2$. Hence the modulus of $\mathbf{v}_1 \wedge \mathbf{v}_2$ does not exceed that of T. Arguing similarly for each term $\mathbf{v}_i \wedge \mathbf{v}_{i+1}$ in the sum for $\boldsymbol{\alpha}(P)$, we therefore obtain $\boldsymbol{\alpha}_3(J) \geqslant \boldsymbol{\alpha}_3(P)$, justifying the order in which we took the terms in T above.

A full proof differs from this only in that inductive definitions must be given of the term corresponding to T above, to show that its odd number of terms always cancel so conveniently. ■

We can now define an 'area' for the set $\mathscr{M}_p$ of all 'polygonal regions' in $\mathbb{R}^2$, where we mean by such a region a set of the form Int P, where P is a polygon. Thus we define the area of Int P by

15.9.7 $$\mathscr{A}(\text{Int } P) = |\boldsymbol{\alpha}_3(P)|,$$

which depends only on $|P|$, by Lemmas 15.9.2 and 15.9.3. The function $\mathscr{A}: \mathscr{M}_p \to \mathbb{R}^+$ satisfies the conditions $\mathfrak{PA}_1$–$\mathfrak{PA}_3$ above, by Lemmas 15.9.1, 15.9.4, and 15.9.6. We can satisfy $\mathfrak{PA}_4$ by simply throwing into $\mathscr{M}_p$ the sets there mentioned and setting their area to be zero.

15.10 Curved Boundaries

We next seek to enlarge the family $\mathscr{M}_p$ to include bounded regions $G \subseteq \mathbb{R}^2$, whose frontiers are not necessarily polygons but more general curves. Our method is that of 'exhaustion', which goes back to Archimedes, and we look first at the problem of defining an area for a circular disc.

15.10.1 EXAMPLE. Consider first the circular region D_r: $x^2 + y^2 \leqslant r^2$ in $\mathbb{R}^2$, and the (positively oriented) inscribed and escribed polygons P_{rn}, Q_{rn} of

Example 15.5.3. Clearly, the sequences $\{\mathscr{A}P_{rn}\}$, $\{\mathscr{A}Q_{rn}\}$ are increasing and decreasing respectively, the first bounded above (by $\mathscr{A}Q_{r1}$) the second bounded below (by $\mathscr{A}P_{r1}$). Hence each sequence tends to a limit, which in this case is the same for both sequences; and this common limit is what we *define* to be the area $\mathscr{A}D_r$ of D_r. Thus, for each n, we have

$$\mathscr{A}P_{rn} \leqslant \mathscr{A}P_{r,n+1} \leqslant \mathscr{A}D_r \leqslant \mathscr{A}Q_{r,n+1} \leqslant \mathscr{A}Q_{rn}$$

and division throughout by r^2 gives

$$\mathscr{A}P_{1n} \leqslant \mathscr{A}P_{1,n+1} \leqslant \mathscr{A}D_r/r^2 \leqslant \mathscr{A}Q_{1,n+1} \leqslant \mathscr{A}Q_{(n)}$$

whence, letting $n \to \infty$, we obtain

$$\mathscr{A}D_r = r^2\mathscr{A}D_1.$$

Now in Fig. 15.7(i) the area of triangle OAB is (by 15.7.3) $\frac{1}{2}\text{AC}\cdot\text{OB} = \frac{1}{4}a_n$, so $\mathscr{A}P_{1,n+1} = 2^{n+1}a_n/4 = \frac{1}{2}L(P_{1,n+1})$. Therefore

$$\mathscr{A}D_1 = \lim_{n\to\infty} \mathscr{A}P_{1n} = \tfrac{1}{2} \lim_{n\to\infty} L(P_{1n}),$$

and by 15.5.6 the last limit is $\frac{1}{2}L(C_1) = \pi$, so

$$\mathscr{A}D_r = \pi r^2.$$

15.10.2 We now copy this procedure for a more general bounded set $G \subseteq \mathbb{R}^2$. Suppose that there exist two sequences $\{P_n\}$, $\{Q_n\}$ of positively oriented Jordan polygons, such that†

(i) $|P_n| \subseteq \text{Int}\, P_{n+1} \subseteq G \subseteq \text{Int}\, Q_{n+1} \subseteq \text{Int}\, Q_n$,

(ii) $G = \bigcup \text{Int}\, P_n$,

(iii) Given a Jordan Polygon J with $G \subseteq \text{Int}\, J$, some Q_n lies in J.

By (ii) G can be shown to be an open connected subset of $\mathbb{R}^2$.

Such a region G will be called 'almost polygonal'. Then by Lemma 15.9.6 the sequence $\{\mathscr{A}P_n\}$ increases with n and is bounded above by $\mathscr{A}Q_1$, so it tends to a limit p. On the other hand $\{\mathscr{A}Q_n\}$ decreases as n increases and is bounded below by $\mathscr{A}P_1$, so it tends to a limit $q \geqslant p$. If it happens that $p = q$, then we say that 'G has area p', and we (temporarily) denote p by $\mathscr{A}'G$. We now show:

15.10.3 LEMMA. *$\mathscr{A}'G$ does not depend on the particular sequences $\{P_n\}$, $\{Q_n\}$ chosen in* 15.10.2.

Proof. Let $\{P'_n\}$ be another sequence of positively oriented Jordan polygons with

$$|P'_n| \subseteq \text{Int}\, P'_{n+1} \subseteq G, \quad (\text{so Int}\, P'_n \subseteq \text{Int}\, P'_{n+1})$$

† These conditions are chosen for technical convenience; they are not satisfied by the polygons of Fig. 15.7, although it is easy to construct sequences that close down on the circular disc as required by conditions (i)–(iii).

Let $B_n = \text{Int } P_n \cup |P_n|$, the interior of P_n plus the boundary. We have to borrow a topological result to assert that there exists an integer m such that $B_n \subseteq \text{Int } P'_m$. The technical reason (see 25.8, and 25 Exercise 9(i)) is that B_n is compact and hence by 15.10.2 lies in the union of a *finite* number of the open sets Int P'_n, of which there is then a biggest (by 15.10.2(i)), say Int $P'_{m(n)}$. Therefore by Lemma 15.9.6, $\mathscr{A}P_n \leqslant \mathscr{A}P'_m$; so $p \leqslant p'$. Similarly $p' \leqslant p$, whence $p = p'$. Using 15.10.2(iii) instead of compactness, we may show in the same way that the limit q does not depend on the choice of sequence $\{Q_n\}$. This proves the lemma—and, indeed, proves more, as the reader will see, since the argument applies even if $p \neq q$. ■

15.10.4 COROLLARY. *If J is a Jordan polygon, then* $\mathscr{A}'(\text{Int } J) = \mathscr{A}(\text{Int } J)$. (Take $J = Q_n$ for each n, and for P_n shrink J inwards.) ■

We can therefore, without ambiguity, write $\mathscr{A}G$ for $\mathscr{A}'G$, and we have now widened the class $\mathscr{M}_p$ to the class $\mathscr{M}$ of all almost polygonal regions G (in the sense of 15.10.2).

This definition ($\mathscr{A}'$) of area is clearly translation-invariant (by 15.9.1), so property $\mathfrak{PA}_1$ holds. By Lemma 15.9.6, $\mathfrak{PA}_3$ holds. However, $\mathfrak{PA}_2$ is harder to prove, partly because of the inconvenience of the requirement that the polygons P_n, Q_n be Jordan polygons.

Exercise 14

Show that if (iii) is omitted from 15.10.2 we need not have $p = q$, but that if G is bounded, then there exists a sequence $\{Q'_n\}$ which satisfies (iii). (The required construction is difficult!)

15.11 Lattices

We therefore make yet another (technically more convenient) definition of area, which nevertheless turns out to agree with the above function $\mathscr{A}: \mathscr{M} \to \mathbb{R}$. (Compare the proof of Theorem 15.3.7 above.) Let $\mathbf{a}$, $\mathbf{b}$ be two points in $\mathbb{R}^2$ such that $\mathbf{a} \wedge \mathbf{b}$ has positive third co-ordinate. For each positive real number t, the set of all points of the form

$$nt\mathbf{a} + mt\mathbf{b}, \quad m, n \in \mathbb{Z}$$

is called a **lattice** and we denote it by $t \cdot \Lambda(\mathbf{a}, \mathbf{b})$, or $t\Lambda$ for short. (Thus, if $t = 1$, $1 \cdot \Lambda = \Lambda$.) The 'fundamental region' of $t\Lambda$ is the set σ_0 of all $\mathbf{z} \in \mathbb{R}^2$ of the form

15.11.1 $$ut\mathbf{a} + vt\mathbf{b}, \qquad (0 \leqslant u \leqslant 1, 0 \leqslant v \leqslant 1),$$

and the 'cells' of $t\Lambda$ are the sets $\mathbf{z} + \sigma_0$, $\mathbf{z} \in t\Lambda$.

We note that

15.11.2 $$\mathscr{A}(\mathbf{z} + \sigma_0) = \mathscr{A}(\sigma_0) = t^2 \cdot \mathbf{a} \wedge \mathbf{b} = t^2\Delta, \quad \text{say.}$$

Let G be a bounded set in $\mathbb{R}^2$; we approximate it by cells of $t\Lambda$ as follows. Consider the sets of cells:

$$s(t) = \{\sigma \in t\Lambda | \sigma \subseteq G\},$$
$$S(t) = \{\sigma \in t\Lambda | \text{Int } \sigma \cap G \neq \varnothing\}$$

where $\text{Int } \sigma = \{\mathbf{z} \in \sigma | 0 < u < 1, 0 < v < 1\}$ in 15.11.2 (see Fig. 15.2).

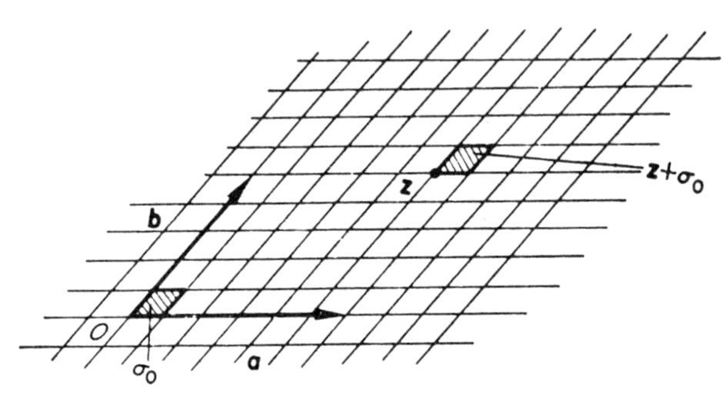

Fig. 15.14

We associate with the cell $\sigma = \mathbf{z} + \sigma_0$ the polygon

$$P(\sigma) = \mathbf{z} + tP(\mathbf{0}, \mathbf{a}, \mathbf{a} + \mathbf{b}, \mathbf{b})$$

in the notation of Exercise 13(v). Then by forming sums as in 15.3.7 we associate polygons $P(s(t))$, $P(S(t))$, with $s(t)$, $S(t)$, given with an obvious notation by:

15.11.3 $$P(s(t)) = \sum_{\sigma \in s(t)} P(\sigma); \qquad P(S(t)) = \sum_{\sigma \in S(t)} P(\sigma).$$

Since Δ in 15.11.2 is positive, these polygons are positively oriented. Then, using 15.9.4 to compute the $\mathscr{A}$'s, we have

15.11.4 $$\mathscr{A}P(s(t)) = \#s(t) \cdot t^2\Delta \leqslant \#S(t) \cdot t^2\Delta = \mathscr{A}P(S(t)).$$

Now for each positive integer n, the sets s, S satisfy:

$$\bar{s}(2^{-n}) \subseteq \bar{s}(2^{-(n+1)}) \subseteq \bar{S}(2^{-(n+1)}) \subseteq \bar{S}(2^{-n}),$$

where $\bar{s}(t)$ denotes the set of all points in the cells of $s(t)$, so

$$\mathscr{A}P(s(2^{-n})) \leqslant \mathscr{A}P(s(2^{-(n+1)})) \leqslant \mathscr{A}P(S(2^{-(n+1)})) \leqslant \mathscr{A}P(S(2^{-n})).$$

Hence the sequence $\{\mathscr{A}P(s(2^{-n}))\}$ increases to a limit s (say) and the sequence

$\{\mathscr{A}P(S(2^{-n}))\}$ decreases to a limit S (say); and $s \leqslant S$. If it should happen that $s = S$, then we take this number to be the 'area' of G and write

15.11.5 $$\mathscr{A}_\Lambda(G) = s.$$

It is not difficult to verify that $\mathscr{A}_\Lambda$ satisfies the conditions $\mathfrak{PA}_1$–$\mathfrak{PA}_3$. Also, analogues of conditions (i)–(iii) of 15.10.2 hold when the P's, Q's, and J are polygons of the forms $P(S(t))$, $P(S(u))$, $P'(S(v))$—possibly from different lattices Λ, Λ'; here G has to be *open*. A compactness argument like that in the proof of 15.10.3 shows that different lattices yield the same area. Hence we have (provided G is open, bounded):

15.11.6 THEOREM. *If $\mathscr{A}_\Lambda(G) = s$, and if Λ' is a second lattice, then $\mathscr{A}_{\Lambda'}(G)$ exists and is also equal to s.* ■

This result was used as being intuitively evident in the proof of Theorem 15.3.7.

15.12 $\mathscr{A}_\Lambda$ Related to $\mathscr{A}$

The basic theorem is now the following.

15.12.1 THEOREM. *Let G be a plane region, whose boundaries consist of a finite number of rectifiable Jordan curves $C_1, \ldots, C_k$. Then $\mathscr{A}_\Lambda G$ exists and $\mathscr{A}_\Lambda(G) = \mathscr{A}_\Lambda(\bar{G})$ where $\bar{G} = G \cup |C_1| \cup \cdots \cup |C_k|$. If $k = 1$, then also $\mathscr{A}G$ exists and $\mathscr{A}G = \mathscr{A}_\Lambda G$.* (Of course, by Theorem 15.11.6, $\mathscr{A}_\Lambda(G)$ is independent of which lattice Λ is used.)

We shall not prove this theorem in full here (see Olmsted [98]), but we give a proof when $k = 1$ and C_1 is a Jordan polygon. By Theorem 15.11.6 we may, and shall, use the lattice $\Lambda = \Lambda(\mathbf{i}, \mathbf{j})$.

15.12.2 LEMMA. *Let s be a segment $\langle\mathbf{u}, \mathbf{v}\rangle$ in $\mathbb{R}^2$. Then the number of cells of $t\Lambda(t > 0)$ whose interior meets s, is at most* $2(1 + t^{-1}|\mathbf{u} - \mathbf{v}|)$.

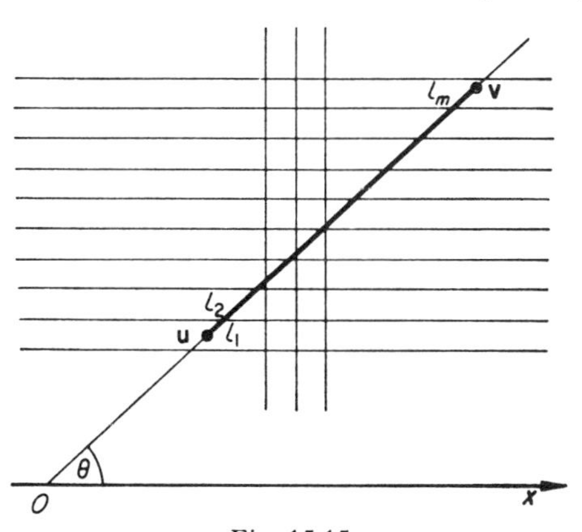

Fig. 15.15

Proof. For brevity, we call lines parallel to the x-axis and y-axis which pass through points of $t\Lambda$ 'horizontal' and 'vertical' lines of $t\Lambda$, respectively. Let s make an angle θ with the x-axis, and suppose $0 \leqslant \sin\theta \leqslant \cos\theta$ (the cases for other angles θ are similar). Then the number of intersections of s with vertical lines of $t\Lambda$ is $m \leqslant 1 + t^{-1}|\mathbf{u} - \mathbf{v}|$, and these divide s into segments $l_1, l_2, \ldots, l_m$. None of these can contain two intersections (except at end-points) with horizontal lines of $t\Lambda$, because $\sin\theta \leqslant \cos\theta$. If l_p (say) contains one such intersection at an interior point z, then z meets exactly two cells of $t\Lambda$ whose interiors meet s. Hence s meets at most $2m$ cells of $t\Lambda$. ■

15.12.3 COROLLARY. *If J is a Jordan polygon, then $|J|$ meets at most $4t^{-1}L(J)$ cells of $t\Lambda$.* ■

Hence, 15.11.4 above tells us that

$$0 \leqslant \mathscr{A}P(S(t)) - \mathscr{A}P(s(t)) \leqslant 4t^{-1}L(J)\cdot t^2\Delta = 4t\Delta\cdot L(J)$$

and the right-hand side tends to 0 with t since $2\Delta L(J)$ is constant. Hence the limits s, S are equal, so by 15.11.5 we may assert:

15.12.4 $$\mathscr{A}_\Lambda(\text{Int } J) \quad \textit{exists.}$$

The last inequalities also tell us that $\mathscr{A}_\Lambda(|P|) = 0$ for any Jordan polygonal arc P, since $s(t) = \varnothing$ here. Hence $\mathfrak{PA}_4$ holds for $\mathscr{A}_\Lambda$, and since $\mathscr{A}_\Lambda$ satisfies $\mathfrak{PA}_2$, then for any Jordan polygonal curve J,

$$\mathscr{A}_\Lambda(\text{Int } J) = \mathscr{A}_\Lambda(\text{Int } J \cup |J|)$$

where the left-hand side is given by 15.10.4; and we may as well prove the more general:

15.12.5 THEOREM. *If G is almost polygonal, then $\mathscr{A}_\Lambda(G)$ exists iff $\mathscr{A}(G)$ exists and then $\mathscr{A}(G) = \mathscr{A}_\Lambda(G)$.*

*Proof.** Since G is almost polygonal, there exist sequences $\{P_n\}$, $\{Q_n\}$ as in 15.10.2. A compactness argument, as in the proof of Lemma 15.10.3, shows that for each $n \in \mathbb{N}$ there exists $m \in \mathbb{N}$, such that

$$P(s(2^{-n}) \subseteq \text{Int } P_m, \qquad P(S(2^{-m})) \subseteq \text{Int } Q_n.$$

Taking limits as $n \to \infty$, we have in the notations of 15.11.5 and 15.10.2:

$$s = \lim_{n\to\infty} \mathscr{A}P(s(2^{-n})) \leqslant \lim \mathscr{A}P_n = p;$$

similarly, $S = \lim \mathscr{A}P(S(2^{-n})) \leqslant \lim \mathscr{A}Q_n = q$. Similarly again, $p \leqslant s$ and $q \leqslant S$, so $p = s$ and $q = S$. Hence $p = q$ iff $s = S$, so $\mathscr{A}G$ exists iff $\mathscr{A}_\Lambda(G)$ exists; and then $\mathscr{A}G = \mathscr{A}_\Lambda(G)$. ■

* Our intention is that the reader understand the structure of the argument, even if certain of the details elude him.

To summarize, we have made three apparently different definitions of the area of a Jordan polygon, all of which have turned out to agree with each other. The $\mathscr{A}_\Lambda$ definition is the most general, and satisfies the conditions $\mathfrak{PA}_1$–$\mathfrak{PA}_4$; it is also the basis for the theory of double integrals in calculus. Moreover, since we can always define a lattice in space $\mathbb{R}^n$, the $\mathscr{A}_\Lambda$ definition generalizes to volumes in $\mathbb{R}^3$ as in the proof of Theorem 15.3.7 and to the so-called 'Jordan content' in $\mathbb{R}^n$. The other definitions are not satisfactory in higher dimensions, as the following two instances show.

Analogously to 15.8, define a *polyhedron* $R(\sigma_1, \ldots, \sigma_n)$ in $\mathbb{R}^3$ to be a set of triangles $\sigma_1, \ldots, \sigma_n$ such that any two meet at most in edges or vertices, and exactly two meet along each edge. Then using 15.8.5 as a guide, we could try to define the 'volume' μR of R as

$$\mu R = \tfrac{1}{6} \sum_{i=1}^{n} \mathbf{a}_i \cdot \mathbf{b}_i \wedge \mathbf{c}_i$$

where $\mathbf{a}_i$, $\mathbf{b}_i$, $\mathbf{c}_i$ are the vertices of σ_i. This then presupposes a choice of orientation of σ_i in the order $\mathbf{a}_i$, $\mathbf{b}_i$, $\mathbf{c}_i$. In 'good' cases it happens that we get the 'right' volume. But suppose we try to prove the analogue of Lemma 15.8.4, that for each $\mathbf{p} \in \mathbb{R}^3$

$$\mu R = \tfrac{1}{6} \sum (\mathbf{a}_i - \mathbf{p}) \cdot (\mathbf{b}_i - \mathbf{p}) \wedge (\mathbf{c}_i - \mathbf{p});$$

then the right-hand side is

$$\tfrac{1}{6}\left(\sum \mathbf{a}_i \cdot \mathbf{b}_i \wedge \mathbf{c}_i - \sum \mathbf{p} \cdot \mathbf{b}_i \wedge \mathbf{c}_i - \sum \mathbf{a}_i \cdot \mathbf{p} \wedge \mathbf{c}_i - \sum \mathbf{a}_i \cdot \mathbf{b}_i \wedge \mathbf{p}\right)$$

and the last three sums will add to zero only if we can prove that, if $\langle\!\langle \mathbf{b}_i, \mathbf{c}_i \rangle\!\rangle$ is an edge of σ_i in the chosen orientation, then $\langle\!\langle \mathbf{c}_i, \mathbf{b}_i \rangle\!\rangle$ is an edge of some other σ_j in *its* chosen orientation. Thus, we need to know whether we can 'orient' $R(\sigma_1, \ldots, \sigma_n)$, in the sense of choosing an orientation for each σ_i such that the required cancellations occur. This can be done, but it is complicated; and if R is also allowed to have a boundary, so that some edges lie in only one triangle, then orientation is not always possible, e.g., if R is a Möbius band (see 25.12).

The way in which the method of 15.5.3 can fail is shown by a famous example of the nineteenth-century mathematician Schwarz, in which we attempt to calculate the surface area A of a right circular cylinder of radius r, height h. Just as segmental paths were used in 15.5.1 to approximate the length of a curve, we now use polyhedral approximations to the area A. We divide the surface by n circles parallel to the base and distant h/n apart. Then we take m points on each circle equally spaced; and with those on any circle midway between those on the circles beneath or above, as in Fig. 15.16. Now form a set of $2nm$ plane triangles with these points as vertices, $2m$ triangles in each row. Each triangle has base $2r \sin \pi/m$ and height

$$(r^2(1 - \cos \pi/m)^2 + h^2/n^2)^{1/2},$$

so using the approximations π/m, $\pi^2/2m^2$ for $\sin \pi/m$, $1 - \cos \pi/m$ respectively, we find that the set of all triangles has total area

$$2nmr\frac{\pi}{m}\left(\frac{h^2}{n^2} + r^2\frac{\pi^4}{4m^4}\right)^{1/2}$$

approximately. If now $n = m$, this tends (as n tends to ∞) to $2\pi rh$, as we would expect. But if $n = m^2$ it tends to $2\pi r\sqrt{(h^2 + (r^2\pi^4)/4)}$ and if $n = m^3$ it tends to ∞. Thus this mode of approximation does not lead to a unique area, and other methods must be used. For other 'paradoxes' concerning areas and volumes, see Hadwiger–Debrunner–Klee [46], p. 50.

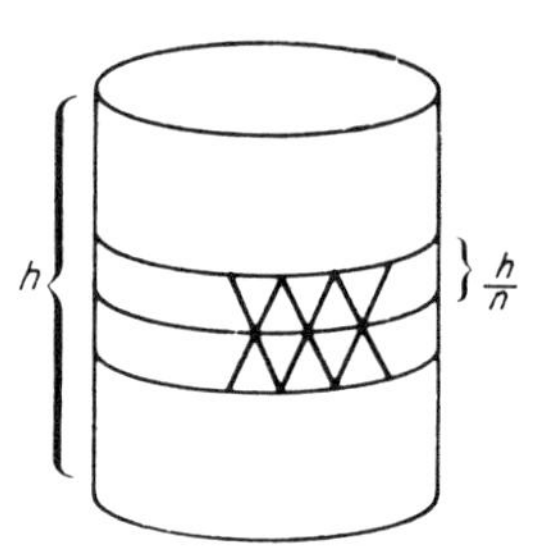

Fig. 15.16

Other approaches to area are those of Lebesgue and of Hausdorff, and the reader is now referred to books on the subject called 'measure theory' and on various notions of 'integral' in calculus. [See, e.g., Halmos [49a], Apostol [4], and Courant [24], Vol. 2.] Relatively general theories are found to be necessary, particularly for the theory of probability, where it is necessary to assign 'volumes' to very random portions of high-dimensional space. [See, e.g., Kingman–Taylor [74a].]

Chapter 16
THE LOGIC OF GEOMETRY

16.1 Philosophies of the Greeks and Others

The preceding theory of $\mathbb{R}^3$ (in Chapters 14, 15) was based on Euclidean geometry, since for example we used Pythagoras' theorem, together with results about similar triangles and existence of perpendiculars. Now, Euclidean geometry was primarily conceived by Euclid and his school to be a theory about the space **S** in which we live. The object was to find out all about the geometry of **S**, i.e., to formulate statements about the properties of **S** and to decide which of these were true. Different individuals formulate such statements by different methods, like guessing or dreaming; and then comes the problem of convincing others (and themselves) of the truth of their hunch. Apart from inspirational and personal methods such as artists and mystics use, three possible methods are known.

The first is barbaric, but in common use by mankind. It amounts to saying 'Statement K is true because the Boss (or custom or Government, etc.) says so', and disbelievers are 'convinced' by force of some kind. It has a long-term disadvantage also, because a huge project based on principles imposed by a Boss may very well turn out to be wrong, and waste considerable time, effort, and resources. Consider for example the effects on Germany of the Nazi treatment of Einstein and other scientists; or the Tizard–Cherwell controversy (Snow [120]).

The other two methods available belong to a different level of civilization from the first, and their discovery marked great strides in human progress†. One of them is the inductive method of natural science, discussed in Chapter 5. Applied to the statement K: 'For a triangle ABC, right-angled at B,

(i) $AC^2 = AB^2 + BC^2$',

it would argue something like this: 'K has been verified within the limits of experimental error, by examining a large sample of right-angled triangles. We shall therefore believe that K is true until a counter-example is produced, in which event we shall modify our beliefs accordingly'. Unfortunately, we cannot then be sure that the theorem should not say either

(ii) $AC^2 = AB^2 + BC^2 + (t - t_1)(t - t_2)\cdots(t - t_n)$

† Notice we do not claim that all change is progress!

where t is the time now and the t_i's are the times of former experimental verifications; or

(iii) $AC^2 = AB^2 + BC^2 + d$

where $d \neq 0$ is a constant, but too small to detect experimentally, especially if AC is of cosmic magnitude. A physicist would probably then choose to believe (i) because it is simpler than (ii) or (iii), on the ground that 'Nature is simple'. But this introduces an unproved (though useful) hypothesis, and mathematical history affords many examples where what is 'complex' to one generation is 'simple' to the next. Moreover, the uncharted seas of mathematical imagination are potentially vaster than the physical world and we need a more reliable method of navigation.

The discovery of the third method was apparently an achievement peculiar to the civilization of the Greeks. Recall that the problem is to find a way by which one individual A, can make some claim K, and get it accepted by another individual B, who is possibly disbelieving at the start. (Perhaps $A = B$.) The method found by the Greeks consists in formulating an impartial procedure of rules, by which B listens† to A's argument in support of K, and agrees to let A proceed at each stage if certain conditions, laid down beforehand, are met. If A meets these conditions each time until his argument is complete, then B *agrees* to be convinced. The procedure need not be confined to mathematics, and is nowadays of course used by decision-makers in many spheres of rational activity, particularly in certain courts of law, and to decide certain contests.

The Greeks applied this method in geometry as follows. First they made certain *definitions* to say what lines, points, etc., are, and then they laid down certain *axioms* or *postulates* which are of two kinds. The first kind are rules of logic (e.g., 'equals added to equals give equals') and are necessary for mathematics generally. The second kind of axiom is geometrical in nature, and asserts that the geometrical objects, given by the definitions, have certain properties. The point of making such definitions and axioms was to limit argument; so that in establishing any proposition, a start could be made at the axioms (which were thought to state truths which no sane person could deny). Thus, an argument in support of K was called a *proof*, and K was then agreed to be true, if it proceeded using only the allowed logical rules, and appealed only to the axioms or else propositions previously established by the same method.

This highly civilized procedure has long been regarded as an ideal, which some philosophers (e.g., Spinoza) and some politicians (e.g., the writers of the American Constitution) have sought to emulate: while all mathematicians

† The assumption is made that B is able and willing to listen. Thus this method presupposes good will on B's part. Compare also Disraeli's dictum that 'It is the duty of Her Majesty's loyal Opposition to oppose'.

have tried to build the various branches of mathematics according to the Euclidean model. Nowadays, however, there is the exception that mathematicians no longer insist that the axioms be self-evident. Indeed, the entities of an axiomatic system are logically quite free of physical meaning, so that no question of self-evidence arises. This is *not* to say that mathematics has lost its potential usefulness for studying the physical world.

Exercise 1

(i) Which of the above three methods of establishing a claim are used by a missionary, a poet, a painter, a musician, Freud, Marx?
(ii) Suppose an individual holds a set of cherished beliefs which imply logically that he should perform a certain action. What happens if (a) he disapproves of the action and you do not, (b) you disapprove of the action and he does not?

16.2 Hilbert

Curiously, however, it was not until the publication in 1902 of a fundamental book [57] by Hilbert (1862–1943), that Euclidean geometry itself was freed from blemishes. The latter were of two kinds. First, the definitions were not satisfactory, in that, e.g., lines were defined in terms of 'length' and 'breadth', which notions, when analysed, presupposed the knowledge of what a line is. Secondly, the proofs used concepts not given in the axioms; for example, the fact that the diagonals of a parallelogram P intersect, and do so *inside* P, was not proved; nor was the term 'inside P' even defined, but merely inferred from sketches (see Example 16.5.3 below). Thus the 'proofs' of Euclid did not live up to the standards envisaged in the rules.

Hilbert realized that geometry, like all other deductive systems, must be allowed a set T of some undefined terms. These are to be regarded as basic, and all other terms are to be defined by using only the elements of T. The axioms then tell us what we are allowed to do with the undefined terms. He chose a particular set T, and built up, with the aid only of logic, a system of theorems which encompassed all those of Euclidean geometry. (Other writers have since obtained the same system of theorems with other choices of T; if T' is such a choice, then of course, the objects of T have to be constructed from those in T'.) Besides putting Euclidean geometry on a proper footing, Hilbert's study also threw new light on the interrelationships between the various theorems. This aspect has its own aesthetic appeal apart from the *meanings* of the theorems; and the same kind of 'theorem about theorems' is now to be found in many other branches of mathematics. Moreover, his work led to the construction of systems which satisfied some axioms of T and not others, and introduced an awareness of the notion of mathematical 'structure', so that known structures could be used to make 'models' of others. A glimpse of this idea will be given below; we first consider geometrical examples, and later take up the idea of proof more generally.

16.3 Pedagogy

Now, the resulting discipline, of stating and proving Euclid's theorems properly, is long and arduous, and probably unsuitable for beginners. The facts of the theorems can be first understood and demonstrated as in physics, as part of the physics of the natural world; but sometime in school we must begin a study of the theorems from the point of view of pure mathematics. How then are we to begin?

One solution is for gifted teachers to take Hilbert's book and teach from it, reading the Preface of Veblen–Young [130] for encouragement, and the recent book by Moise [90] as a guide. This process will take time (and energy!); and while we await its results we can adopt another approach which at the same time imparts skills needed urgently by the student in other branches of the subject.

Thus, having obtained the theory of Chapters 14 and 15 by a mixture of physics and algebra, we consolidate it by following the procedure surreptitiously introduced into Chapter 14. That is, we presuppose only a knowledge of the properties of $\mathbb{R}$ together with some facts of the physical geometry of $\boldsymbol{S}$ (for guidance). Suppose also that the idea of a real three-dimensional vector space has been acquired as in Chapter 14, with the notions of inner product and (hence) of vector product whose properties are summarized by $\mathfrak{S}\mathfrak{P}_1$–$\mathfrak{S}\mathfrak{P}_4$ in 14.8 and $\mathfrak{V}\mathfrak{P}_1$–$\mathfrak{V}\mathfrak{P}_4$ in 14.10. Then we shall say that we shall study $\boldsymbol{S}$ by means of a 'model', described in 16.4 below, which has the virtue of being easily described with precision. Since we incidentally want to know how good a model of $\boldsymbol{S}$ we have got (for the benefit of applied mathematics), we must deduce the mathematical properties of the model without appeal to $\boldsymbol{S}$ except for inspiration. Hence, a rough guide as to what arguments to allow is to imagine we are trying to convey to a blind man (who knows algebra and set theory) what $\boldsymbol{S}$ is like; any use of his eyes is barred. In [115], Sawyer uses the analogous discipline of telephoning an account of geometry to an angel in Heaven. The reader might invent other situations, or replace the blind man by a computing machine.

16.4 An Algebraic Model of $\mathbb{R}^3$

To describe the model, we must first say which items will correspond to the words 'point', 'line', 'plane', 'space', 'on', 'passes through'; so 'space' now means $\mathbb{R}^3$, and 'point' means an element of $\mathbb{R}^3$. A 'line' is any set of the form $\{\mathbf{r} \in \mathbb{R}^3 | \mathbf{r} = \mathbf{a} + \lambda\mathbf{b}, \lambda \in \mathbb{R}\}$ with $\mathbf{a}$, $\mathbf{b}$ constant and $\mathbf{b} \neq 0$; as before, we say for short that this line has equation $\mathbf{r} = \mathbf{a} + \lambda\mathbf{b}$. Similarly, a plane is any set with equation of the form $\mathbf{p}\cdot\mathbf{r} = k$ where $\mathbf{p}$ is constant and non-zero. An item X is 'on' Y if X forms a subset of Y, when also Y 'passes through' X. The 'distance' between points $\mathbf{u}$, $\mathbf{v}$ is defined to be $\|\mathbf{u} - \mathbf{v}\|$, and 'angle' may be introduced as follows. For brevity, we work in $\mathbb{R}^2$, and in essence we merely reverse the flow of ideas behind 14 Exercise 7(ii) and 15 Exercise 7(xiv), observing that the transformations defined there are purely algebraic notions. We saw there that if we knew already about angles in geometry, then to each angle α corresponds an orthogonal transformation T rotating $\mathbb{R}^2$ through α about O, and *conversely*. Since we can describe T algebraically, we can give an algebraic description of α, as we now show. We want to define angles which remain invariant under translation, and hence we merely need to define angles between segments OP, OQ, where P, Q lie on the unit circle $\mathbb{S}^1$ (now *defined* as the set of all $\mathbf{a} \in \mathbb{R}^2$ with $\|\mathbf{a}\| = 1$) with centre at the origin O. Let $\mathbf{R}$ denote the set of all orthogonal transformations T of $\mathbb{R}^2$ with determinant 1; the elements of $\mathbf{R}$ are called **rotations**, for brevity, and as we saw earlier, they form a commutative group under composition. We shall associate with the ordered triple (P, O, Q) a rotation in $\mathbf{R}$ denoted by $P\hat{O}Q$ and called the **angle** between OP and OQ, as follows.

Recall from 15 Exercise 7(xiv) that relative to the basis $(\mathbf{i}, \mathbf{j})$ of $\mathbb{R}^2$, every rotation $T \in \mathbf{R}$ has matrix $\mu(T)$ of the form $\begin{pmatrix} l & -m \\ m & l \end{pmatrix}$ with $l, m \in \mathbb{R}$ and $l^2 + m^2 = 1$. Thus $T(\mathbb{S}^1) = \mathbb{S}^1$ and

$$\text{(i)}\ \ T(x, y) = (lx - my,\ mx + ly), \qquad \mu(T^{-1}) = \begin{pmatrix} l & m \\ -m & l \end{pmatrix},$$

whence $T(\mathbf{i}) = (l, m)$. Therefore $\mu(T)$ and hence T are determined completely by stipulating where T is to map $\mathbf{i}$.

We may therefore define a function $f\colon \mathbb{S}^1 \to \mathbf{R}$ by assigning to each $p \in \mathbb{S}^1$ that rotation $f(\mathbf{p})$ in $\mathbf{R}$ which sends $\mathbf{i}$ to $\mathbf{p}$, and the above remark about $T(\mathbf{i})$ shows that f is a bijection. Thus we may transfer the commutative group structure of $\mathbf{R}$ to $\mathbb{S}^1$, by setting† $\mathbf{p} \centerdot \mathbf{q} = f^{-1}(\tilde{\mathbf{p}} \circ \tilde{\mathbf{q}})$ (where $\tilde{\mathbf{p}} = f(\mathbf{p})$, etc.); hence the neutral element of $\mathbb{S}^1$ is $\mathbf{i}$, since $\tilde{\mathbf{i}}$ keeps $\mathbf{i}$ fixed and is therefore the identity mapping $\mathscr{I} \in \mathbf{R}$. Thus f *is now an isomorphism of the group* $(\mathbb{S}^1, \centerdot)$ *onto* $(\mathbf{R}, \circ)$.

Now observe that, for any $\mathbf{p} \in \mathbb{S}^1$, $T \in \mathbf{R}$,

$$\text{(ii)}\ \ f(T(\mathbf{p})) = T \circ \tilde{\mathbf{p}};$$

for, the effect of the right-hand side U on $\mathbf{i}$ is to send it via $\mathbf{p}$ to $T(\mathbf{p})$, so U is $f(T(\mathbf{p}))$ by definition of f. There follows

$$\text{(iii)}\ \ \text{for each } \mathbf{t} \in \mathbb{S}^1, \qquad \tilde{\mathbf{t}}(\mathbf{p}) = \mathbf{t} \centerdot \mathbf{p}$$

because we may replace T by $\tilde{\mathbf{t}}$ in (ii) to get

$$\tilde{\mathbf{t}}(\mathbf{p}) = f^{-1}(\tilde{\mathbf{t}} \circ \tilde{\mathbf{p}}) = \mathbf{t} \centerdot \mathbf{p}$$

by definition of multiplication in $\mathbb{S}^1$.

† The dot here is a temporary notation not to be confused with the scalar product.

Given $P, Q \in \mathbb{S}^1$, we know by definition that $\tilde{\mathbf{p}}$ maps $\mathbf{i}$ to $\mathbf{p}$, so $\tilde{\mathbf{p}}^{-1} = f(\mathbf{p})^{-1}$ maps $\mathbf{p}$ to $\mathbf{i}$. We therefore define the angle $P\hat{O}Q$ by

16.4.1 $$P\hat{O}Q = \mathbf{p}^{-1} \cdot \mathbf{q} \in \mathbb{S}^1.$$

In our model $\mathbb{R}^2$, this definition corresponds to the physical process (in **S**) of rotating I to Q (using $\tilde{\mathbf{q}}$), then Q back to Q_* (using $\tilde{\mathbf{p}}^{-1}$: see Fig. 16.1) and taking $P\hat{O}Q$ to be the total angle through which I must be rotated to get to Q_*; but this does not *prove* equation 16.4.1—it simply makes the definition seem a plausible candidate, and we need now to show that this definition has the properties we feel 'angle' should have.

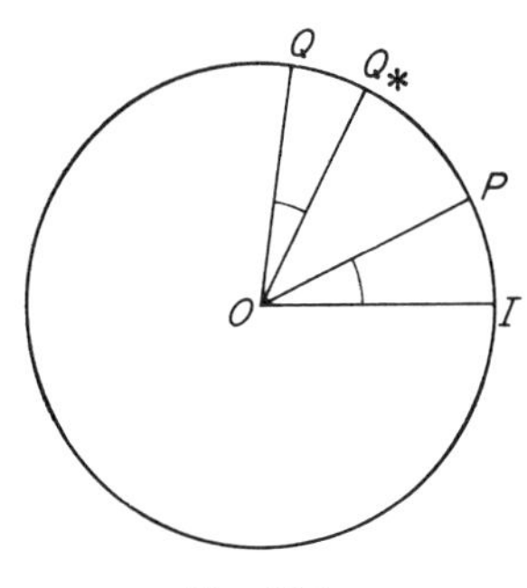

Fig. 16.1

If we rotate P to P' and Q to Q' by $T \in \mathbf{R}$, then $P'\hat{O}Q'$ 'ought' to be $P\hat{O}Q$; but this follows from (iii) since $\mathbf{p}' = \mathbf{t} \cdot \mathbf{p}$, $\mathbf{q}' = \mathbf{t} \cdot \mathbf{q}$ (where $\tilde{\mathbf{t}} = T$) whence by 16.4.1,

16.4.2 $$P'\hat{O}Q' = (\mathbf{t} \cdot \mathbf{p})^{-1}(\mathbf{t} \cdot \mathbf{q}) = (\mathbf{p}^{-1} \cdot \mathbf{t}^{-1})(\mathbf{t} \cdot \mathbf{q}) = \mathbf{p}^{-1} \cdot \mathbf{q} = P\hat{O}Q.$$

In particular, taking P', Q' to be diametrically opposite to P, Q (i.e., with $\mu(T) = \begin{pmatrix} -1 & 0 \\ 0 & 1 \end{pmatrix}$), we see that '*vertically opposite*' *angles are equal.* Now, as remarked above, $\mathbf{i}$ is the neutral element of $\mathbb{S}^1$. Hence if $\mathbf{q}_* = \mathbf{p}^{-1} \cdot \mathbf{q}$ in 16.4.1, then

16.4.3 $$P\hat{O}Q = \mathbf{q}_* = \mathbf{i}^{-1} \cdot \mathbf{q}_* = I\hat{O}Q_*,$$

the last equality by definition of $I\hat{O}Q_*$ using 16.4.1. Thus $P\hat{O}Q$ and $I\hat{O}Q_*$ are the same element of $\mathbb{S}^1$. Further, in any multiplicative group, $(x^{-1}y)^{-1} = y^{-1}x$; hence in 16.4.1 we have

16.4.4 $$P\hat{O}Q = (Q\hat{O}P)^{-1}.$$

We saw from (i) that $T(\mathbf{i}) = (l, m)$, and this leads us (with hindsight!) to define the sine and cosine of T to be m and l respectively. Since $l^2 + m^2 = 1$ these are functions

16.4.5 $\sin: \mathbf{R} \to \langle\!\langle -1, 1 \rangle\!\rangle, \qquad \cos: \mathbf{R} \to \langle\!\langle -1, 1 \rangle\!\rangle, \qquad \sin^2 + \cos^2 = 1.$

To obtain 'addition' formulae, let $S \in \mathbf{R}$ have matrix $\begin{pmatrix} u & -v \\ v & u \end{pmatrix}$, so that $\sin S = v$, $\cos S = u$ and matrix multiplication gives

$$\mu(ST) = \mu(T)\mu(S) = \begin{pmatrix} l & -m \\ m & l \end{pmatrix}\begin{pmatrix} u & -v \\ v & u \end{pmatrix} = \begin{pmatrix} lu - mv, & -lv - mu \\ mu + lv, & -mv + lu \end{pmatrix};$$

therefore, reading off the first column (in a more familiar order!) we get

16.4.6
$$\sin(ST) = vl + um = \sin S \cos T + \cos S \sin T$$
$$\cos(ST) = ul - mv = \cos S \cos T - \sin S \sin T.$$

All this carries over to angles; we use 16.4.3, to define the sine and cosine of $P\hat{O}Q$ by

$$\sin P\hat{O}Q = \sin I\hat{O}Q_* = \sin f(\mathbf{q}_*); \qquad \cos P\hat{O}Q = \cos f(\mathbf{q}_*)$$

(so $\sin^2 + \cos^2 = 1$ as before). Of course, we now write the operation in $\mathbb{S}^1$ additively, so that in 16.4.1, $P\hat{O}Q$ becomes $\mathbf{q} - \mathbf{p}$, while the equation 16.4.4 then translates 'additively' to $P\hat{O}Q = -Q\hat{O}P$; also $\mathbf{i}$ is now zero, $= I\hat{O}I = P\hat{O}P$, and it has the expected property that, for all $Q \in \mathbb{S}^1$,

16.4.7
$$I\hat{O}I + I\hat{O}Q_* = \mathscr{I}f(\mathbf{q}_*) = f(\mathbf{q}_*) = I\hat{O}Q_*.$$

The equations 16.4.6 now translate at once to the familiar form (with $\alpha = P\hat{O}Q$, $\beta = Q\hat{O}R$):

16.4.8 $\sin(\alpha + \beta) = \sin(f(\mathbf{q}_*)f(\mathbf{r}_*)) = \sin\alpha\cos\beta + \cos\alpha\sin\beta$,
and similarly $\cos(\alpha + \beta) = \cos\alpha\cos\beta - \sin\alpha\sin\beta$.

Exercise 2

(i) Let I' denote the other end of the diameter of $\mathbb{S}^1$ through I. Given $P \in \mathbb{S}^1$, prove that $I\hat{O}P + P\hat{O}I' = R$, where $\mu(R) = \begin{pmatrix} -1 & 0 \\ 0 & -1 \end{pmatrix}$. [Thus R is the *angle* 180°.]

(ii) Using translations to define angles between arbitrary pairs of lines in $\mathbb{R}^2$ (by translating their intersection to O), prove the equality of 'corresponding' angles and of 'alternate' angles when a line cuts a pair of parallel lines. {Use equality of vertically opposite angles.}

(iii) Prove that the angles of a triangle in $\mathbb{R}^2$ add (in the sense of 16.4.7) to R (in (i)).

Having got so far, it is worth digressing a little further to say how we now define sines and cosines of *real numbers*. For this we use the fact that $\mathbb{S}^1$ has length 2π (see 15.5.3) and define a function $\eta: \mathbb{R} \to \mathbb{S}^1$ by saying that $\eta(x) = y$, where y is the point on $\mathbb{S}^1$ whose arc-length from $\mathbf{i}$ (in the direction from $\mathbf{i}$ to $-\mathbf{i}$ via $\mathbf{j}$) is z, where z satisfies the inequalities

$$x = 2k\pi + z, \qquad 0 \leqslant z < 2\pi, \qquad (k \in \mathbb{Z}).$$

[In the terminology of group theory, we identify $\mathbb{S}^1$ with the quotient group $\mathbb{R}/M$ where M consists of all integral multiples of 2π; then η is the natural epimorphism and it can be shown that $\eta(x) = e^{iz}$.] Finally we define the function sin: $\mathbb{R} \to \langle\!\langle -1, 1 \rangle\!\rangle$ to be the composite

16.4.9 $$\mathbb{R} \xrightarrow{\eta} \mathbb{S}^1 \xrightarrow{f} \mathbf{R} \xrightarrow{\sin} \langle\!\langle -1, 1 \rangle\!\rangle,$$

where the last function is that given by 16.4.5; and similarly for the function cos: $\mathbb{R} \to \langle\!\langle -1, 1 \rangle\!\rangle$. The study of these functions on $\mathbb{R}$, from the point of view of calculus, is now a matter of investigating f, η and the functions in 16.4.5, and is no longer relevant to our purposes here (see 29.8). We return now to the question of setting up a geometry of $\mathbb{R}^3$ which will satisfy the blind man postulated at the end of 16.3.

So far so good: our blind man can certainly understand these definitions, even if he does not appreciate why we made them. And in order to apply the model to the space **S**, we can check by plotting graphs and using protractors that these 'points', 'planes', 'angles', etc., are good representations of the corresponding items we draw on paper or build in engineering. For any application of the mathematics to the world, the definitions are reasonable ones to make.

Next, it is necessary to verify that the geometrical axioms of Euclid hold in this model, if only to make sure that we have given our blind man the right impression about **S**. This task consists of a long sequence of (easy) deductions, each of the magnitude of an exercise. For lack of space we give only some sample verifications, but urge the reader to complete for himself the verification of others on the list as given in, say, Hilbert [57]. Observe that each verification of an axiom constitutes a *theorem* about $\mathbb{R}^3$. In these verifications, we sometimes take short cuts, suggested by our later knowledge that $\mathbb{R}^3$ is a good model of **S**: to our blind man, they will merely look artificial (as with the definition of angle) or like cunning tricks. They will still be valid mathematics.

This procedure is not too bad for giving a model of the plane geometry of **S** (where the model is really $\mathbb{R}^2$). But we shall shortly give a portion of the theory for $\mathbb{R}^3$, partly to give an example of the kind of sustained—even tedious—argument that serious mathematics involves, particularly in a piece of research. This should help to correct the widespread impression that short, 'clever', arguments are the stuff of mathematics. Intellectual *stamina* is more useful than cleverness, and can best be acquired by practice on long arguments of the following kind (some of which can be applied in $\mathbb{R}^2$).

As our next sample of the theory, then (and we do not propose in this chapter to do more than give samples), we shall verify enough axioms to prove *in our model* (i.e., for our blind man) the following theorem of Euclid.

16.4.10 THEOREM. *Any two distinct non-parallel lines in space have one and only one common perpendicular.*

Since the words 'parallel' and 'perpendicular' have not yet been defined in our model, we cannot even *state* the analogue of 16.4.10 in terms of the model. This is one reason for the length of the treatment, in 16.6 below, and it may be omitted on a first reading. We postpone it for a moment, in order to add the following remarks.

16.5 The Pay-off

A person might object to all this procedure on the grounds that the definitions given are artificial without a prior knowledge of Euclidean geometry; to which we can only reply that *once* they are given, a logically watertight system of geometry can be developed, and it matters not a whit how the definitions happened to be suggested. This is typical of the way in which a great deal of mathematics has grown. First a terminology and body of results is obtained by logically unsatisfactory (but often aesthetically pleasing) methods. Then the theory is seen to have flaws which prevent further progress. Finally the theory is reset in a possibly different language, and derived in a correct logical manner, usually gaining greater power. For example, we have just seen how to model Euclidean geometry in $\mathbb{R}^3$, whose study is included in the general study of $\mathbb{R}$ to which we can bring all the machinery of calculus. The new associations also raise new questions which enrich the older theory. For example, the usual theory of conic sections in $\mathbb{R}^2$ leads to the study of the general equation of the second degree with real coefficients:

16.5.1 $$ax^2 + by^2 + 2hxy + 2gx + 2fy + c = 0;$$

the *notation* then suggests the study of equations of higher degree, which then leads to the theory of higher plane curves and to differential geometry. These disciplines could scarcely be imagined, let alone described, within the original geometry of Euclid. (See also Exercise 3(v) below.)

The equation at once raises the questions: what happens if we allow all numbers to be complex instead of real? This question is forced on us when we consider problems about intersections of curves. For example, if l is the line $ux + vy + w = 0$ in $\mathbb{R}^2$, and C is the curve whose equation is 16.5.1, then $l \cap C$ is either empty, a singleton, or a pair of points. But if we change $\mathbb{R}$ to the set $\mathbb{C}$ of complex numbers, then the equation of l now represents the set

16.5.2 $$L: \{(x', y') \in \mathbb{C} \times \mathbb{C} | ux' + vy' + w = 0\}$$

and 16.5.1 represents some set $C' \subseteq \mathbb{C} \times \mathbb{C}$; and then $L \cap C'$ is always either a singleton or a pair of points. Once we have become accustomed to the change from $\mathbb{R}$ to $\mathbb{C}$, we may as well consider what happens when $\mathbb{R}$ is replaced by *any* field $\mathbf{F}$.

16.5.3 EXAMPLE. Suppose $\mathbf{F}$ is $\mathbb{Z}_2$, the field of residues mod 2 (see Chapter 10). The plane $\mathbb{R}^2$ is replaced by $\mathbb{Z}_2 \times \mathbb{Z}_2$, and consists of only four points and six lines (see Fig. 16.2). In particular, the four points form a parallelogram, since the sides have equations $y = 0$ (consisting of $(0, 0)$ and $(1, 0)$), $y = 1$, $x = 0$, $x = 1$; and the first two are parallel (having no common point), as are the second two. The diagonals have equations $y = x$, $y = 1 - x$, and these *do not intersect* (as can be verified by inspection). The 'fact' of the intersection of diagonals is therefore not so 'obvious' as Euclid thought, and in fact fails to occur when we use any field (like $\mathbb{Z}_2$) in which $2 = 0$; for the equations $y = x$, $y = 1 - x$ are soluble if and only if $\exists x$ such that $2x = 1$ *in the field*; and this is impossible when $2 = 0$.

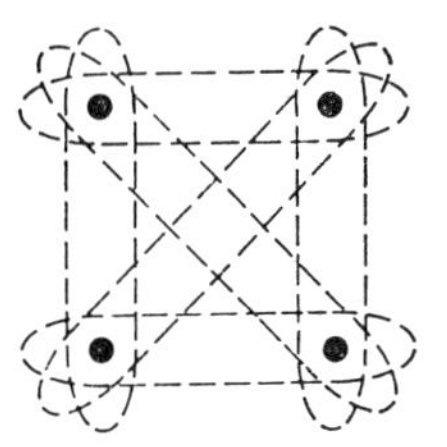

Fig. 16.2

Exercise 3

(i) Show that the parallelogram in the example above has neither an 'inside' nor an 'outside' in any reasonable sense, since it is the whole space. Show also that $\mathfrak{S}\mathfrak{P}_4$ in 14.8 fails.
(ii) In $\mathbb{Z}_2 \times \mathbb{Z}_2 \times \mathbb{Z}_2$, show that $\mathbf{a} = \mathbf{a} \wedge \mathbf{b}$ when $\mathbf{a} = (0, 1, 1)$, $\mathbf{b} = (1, 1, 1)$; and that the plane $\mathbf{b}\cdot\mathbf{r} = 0$ consists of the line $\mathbf{r} = (1, 1, 0) + t\mathbf{a}$ together with $\mathbf{0}$.
(iii) Show that when $\mathbb{R}$ is replaced by $\mathbb{Z}_3$, then the plane $\mathbb{Z}_3 \times \mathbb{Z}_3$ has 9 points, and 12 lines. How many sets of parallel lines are there? Do the same problem replacing $\mathbb{R}$ by $\mathbb{Z}_n$.
(iv) Show that a copy of the geometry of $\mathbb{Z}_3 \times \mathbb{Z}_3$ is given by the following array, in which the letters represent the 'points' and the columns the 'lines':

$$\begin{array}{cccccccccccc} A & D & G & A & B & C & A & B & C & A & C & B \\ B & E & H & D & E & F & F & D & G & E & D & F \\ C & F & I & G & H & I & H & I & E & I & H & G. \end{array}$$

(v) Show that, in $\mathbb{R}^2$, the line through $(-1, 0)$ with gradient t cuts the circle $C\colon x^2 + y^2 = 1$ in the point $((1 - t^2)/(1 + t^2), 2t/(1 + t^2))$. Hence show that every point on C with *rational* co-ordinates is of this form, with t rational, and conversely. Hence prove that the triples of integers (x, y, z), not all zero, such that $x^2 + y^2 = z^2$, are given by

$$(b^2 - a^2, 2ab, b^2 + a^2), \qquad \text{as } a, b \text{ run through } \mathbb{Z}.$$

[The analogous problem, of solving $x^n + y^n = z^n$ when n is an integer >2 was conjectured by Fermat to have no solutions other than trivial ones like $x = z$, $y = 0$. For the history of this unsolved problem (called 'Fermat's last theorem') see Bell [11].]

16.6 Plan for a Proof

We shall now give the treatment, promised above, of the theory necessary to prove Theorem 16.4.10. Let us first consider our plan of campaign. We are in the situation envisaged in 16.1, because we have a hunch that 16.4.10 can be stated and proved in our model, and we must convince our blind man that our hunch is correct. Before presenting our case to him, let us think of a 'visual' proof to satisfy ourselves, after the manner of Euclidean geometry proper. Such a proof has the following steps, and is suggested by attempts to draw the required perpendicular from the data (see Fig. 16.3). For brevity

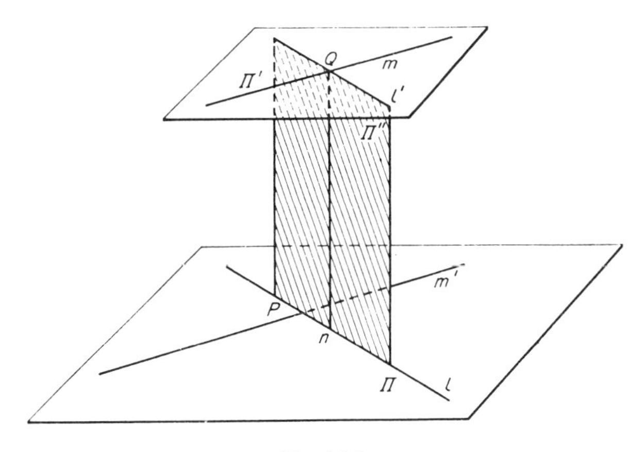

Fig. 16.3

$\Pi, \Pi', \ldots$ denote planes, while capital and small italic letters denote points and lines respectively; in particular, l and m will stand for the particular lines between which we wish to establish the existence and uniqueness of a common perpendicular. The notation '$\exists!\,\alpha$:' means 'there is one and only one α such that'. Notice that we are proving an 'existence and uniqueness' theorem, because we must show first that a perpendicular exists and then that there is only one such perpendicular. The proof of existence runs:

(i) l is a non-empty set of points: thus we can choose $P \in l$ such that
(ii) $\exists!\, m'$: $P \in m'$ and $m' \| m$, and $m' \neq l$ (since $l \nparallel m$, given);
(iii) $\exists!\, \Pi$: $l \subseteq \Pi$ and $m' \subseteq \Pi$;
(iv) $\exists!\, \Pi'$: $m \subseteq \Pi'$ and $\Pi \| \Pi'$;
(v) $\exists!\, \Pi''$: $l \subseteq \Pi''$ and $\Pi \perp \Pi''$;
(vi) $\exists!\, l'$: $\Pi'' \cap \Pi' = l'$;
(vii) $\exists!\, Q$: $l' \cap m = Q$;
(viii) $\exists!\, n$: $Q \in n$ and $n \perp \Pi'$;
(ix) $n \subseteq \Pi''$;
(x) in Π'', $n \perp l'$, $l' \| l$, so $n \perp l$.

Then n is the required mutual perpendicular. Its uniqueness follows from the fact that, if n' were a second perpendicular, then n, m, n' would lie in a plane containing l, contrary to the assumption that l and m are not coplanar.

To translate this for our blind man, we must assign meanings within our model to the symbols $\|$, $\perp$ (since he already has meanings for all the other symbols and words in (i)–(x)). Then we must prove algebraically each of the assertions (i)–(x), which are all statements about the way the items in our geometry behave, and are either axioms in Euclidean geometry, or short deductions from them. Some details must still be left to the reader, as exercises, for lack of space, but they will be such that our blind man (a competent algebraist) can also complete them for himself. The completed proof (see 16.8.22) is then a mixture of algebraic and logical methods. [Another proof, at least of the existence of the perpendicular, can be given using calculus (see 37 Exercise 2(v)); but the logical structure is *technically* not so simple as the one here.] We now consider the verifications of the steps (i)–(x).

16.7 Verifications

To our sight, a point is clearly not a line. But to convince our blind man, we remark that a point is a singleton, whereas a line l is not, being equivalent, as a set, to $\mathbb{R}$ (hence $l \neq \varnothing$). Now let us prove for him that:

16.7.1 *The line* $\mathbf{r} = \mathbf{a} + t\mathbf{b}$ *cannot be a plane.*

Proof. If the line l were also the plane $\mathbf{p}\cdot\mathbf{r} = k$, then $\mathbf{p}\cdot\mathbf{a} = k$ (taking $t = 0$) and $\mathbf{p}\cdot\mathbf{b} = 0$. Hence the line m: $\mathbf{r} = \mathbf{a} + t(\mathbf{b} \wedge \mathbf{p})$ also lies in the plane which by hypothesis is l. In particular then, $\exists s \in \mathbb{R}$ such that $\mathbf{a} + \mathbf{b} \wedge \mathbf{p} = \mathbf{a} + s\mathbf{b}$. Hence $s \neq 0$, otherwise $\mathbf{b} \wedge \mathbf{p} = \mathbf{0}$; and $\mathbf{p}\cdot\mathbf{b} = 0$, which is impossible by 14 Exercise 12(v), since $\mathbf{b} \neq \mathbf{0} \neq \mathbf{p}$. Thus $\mathbf{b} \wedge \mathbf{p} = s\mathbf{b}$, $s \neq 0$; so taking the scalar product with $\mathbf{b}$ gives $0 = s\mathbf{b}^2$, contrary to the facts that $s \neq 0$ and $\mathbf{b} \neq \mathbf{0}$. ∎

16.7.2 COROLLARY (of Proof). *If a line lies in a plane, then it is a proper subset of that plane.* ∎

We leave the reader to complete the proof of:

16.7.3 *A plane is not empty; a line is equivalent to* $\mathbb{R}$. ∎

In our earlier treatment (in Chapter 14), it was geometrically obvious when two equations

16.7.4 $$\mathbf{p}\cdot\mathbf{r} = c, \qquad \mathbf{q}\cdot\mathbf{r} = d \qquad (\mathbf{p} \neq \mathbf{0} \neq \mathbf{q})$$

represented the same plane, simply by looking at the perpendiculars from the origin. But for our blind man we must prove algebraically:

16.7.5 *The equations* 16.7.4 *represent the same plane iff* $\exists s \neq 0$ *such that* $\mathbf{p} = s\mathbf{q}$, $c = sd$.

Proof. If $s \neq 0$ exists, then each equation implies the other. To prove the converse, suppose $\Pi_1 = \Pi_2$. By 16.7.3, there is then at least one $\mathbf{r}_0 \in \Pi_1$, so $\mathbf{r}_0$ satisfies each equation in 16.7.4. There are two alternatives:

(i) $\mathbf{p} \wedge \mathbf{q} = \mathbf{0}$. Then by 14 Exercise 12(v), $\exists s \neq 0$ such that $\mathbf{p} = s\mathbf{q}$, so $c = \mathbf{p}\cdot\mathbf{r}_0 = s\mathbf{q}\cdot\mathbf{r}_0 = sd$, as required.

(ii) $\mathbf{p} \wedge \mathbf{q} \neq \mathbf{0}$. Then one co-ordinate is not zero, say $p_1q_2 - p_2q_1 \neq 0$. Hence, we can solve equations 16.7.4 for x_0, y_0 in terms of z_0, to get

$$x_0 = [z_0(p_2q_3 - q_2p_3) + q_2c - p_2d]/\Delta, \qquad (\Delta = p_1q_2 - p_2q_1)$$
$$y_0 = [z_0(p_3q_1 - q_3p_1) - q_1c + p_1d]/\Delta, \qquad z_0 = z_0,$$

that is, $\mathbf{r}_0 = \mathbf{a} + \lambda\mathbf{b}$, where $\lambda = z_0/\Delta$, $\mathbf{b} = \mathbf{p} \wedge \mathbf{q}$, and

$$\mathbf{a} = (q_2c - p_2d, p_1d - q_1c, 0)/\Delta.$$

Reversing the calculation shows that *every* point of the form $\mathbf{a} + t\mathbf{b}$ satisfies both equations 16.7.4. Hence $\Pi_1 \cap \Pi_2$ is the line $\mathbf{r} = \mathbf{a} + t\mathbf{b}$, which is a proper subset of Π_1 by 16.7.2. Therefore $\Pi_1 \neq \Pi_2$, otherwise the intersection would be all Π_1.

Since $\Pi_1 = \Pi_2$, alternative (ii) is eliminated. ■

16.7.6 COROLLARY (of Proof). *If equations* 16.7.4 *represent distinct planes, then either* $\exists s \neq 0$ *such that* $\mathbf{p} = s\mathbf{q}$, $c \neq sd$, *and the planes do not intersect; or* $\mathbf{p} \wedge \mathbf{q} \neq \mathbf{0}$ *and the planes intersect in a line.* ■

Analogously for lines: suppose

16.7.7 $$\mathbf{r} = \mathbf{a} + t\mathbf{b}, \qquad \mathbf{r} = \mathbf{c} + s\mathbf{d} \qquad (\mathbf{b} \neq \mathbf{0} \neq \mathbf{d})$$

are equations of lines l, m respectively.

16.7.8 $l = m$ iff $\mathbf{a} \in l \cap m$ *and* $\exists k \neq 0$ *such that* $\mathbf{b} = k\mathbf{d}$.

Proof. If $l = m$, then $l \cap m = l$, so $\mathbf{a} \in l \cap m$. Also $\exists u$ such that $\mathbf{a} = \mathbf{c} + u\mathbf{d}$, so each $\mathbf{r} \in l$ is simultaneously of the forms $\mathbf{c} + u\mathbf{d} + t\mathbf{b}$, $\mathbf{c} + s\mathbf{d}$, whence $t\mathbf{b} = (s - u)\mathbf{d}$. Taking $\mathbf{r} = \mathbf{a} + \mathbf{b}$ gives $\mathbf{b} = (s - u)\mathbf{d}$; and $s - u \neq 0$ since $\mathbf{b} \neq \mathbf{0}$. Hence $s - u$ is the required k. Similar arguments prove the converse. ■

Similar arguments also show

16.7.9 PROPOSITION. *The line* l: $\mathbf{r} = \mathbf{a} + t\mathbf{b}$ *can also be written in the forms*

$$\{\mathbf{r} | \mathbf{r} = u\mathbf{c} + v\mathbf{d}, u + v = 1\}, \qquad \mathbf{r} = \mathbf{c} + t(\mathbf{c} - \mathbf{d})$$

for any two points $\mathbf{c}, \mathbf{d} \in l$. ■ [Here, as elsewhere in this section, 'two' means 'two distinct'.]

16.7.10 COROLLARY. *There is exactly one line through two points.*

Proof. Let the two points be $\mathbf{c}$, $\mathbf{d}$. Then one line, l, through $\mathbf{c}$ and $\mathbf{d}$ is $\mathbf{r} = s\mathbf{c} + t\mathbf{d}$, $s + t = 1$ (with $\mathbf{c}, \mathbf{d}$ corresponding respectively to $s = 1$, $s = 0$). If a second line m: $\mathbf{r} = u\mathbf{a} + v\mathbf{b}$, $u + v = 1$, passed through $\mathbf{c}$ and $\mathbf{d}$, then by the second form in 16.7.9 m is also $\mathbf{r} = p\mathbf{c} + q\mathbf{d}$, $p + q = 1$, so $l = m$. ■

This result enables us to speak of 'the' line (CD).

16.7.11 COROLLARY. *If two points* $\mathbf{c}$, $\mathbf{d}$ *lie in a plane, then so does the whole line* (CD). (By the linearity of $\mathbf{p}\cdot\mathbf{r}$, with $\mathbf{r}$ as in 16.7.9.) ■

16.7.12 COROLLARY. *Two lines intersect in at most a single point.* ■

16.7.13 PROPOSITION. *Exactly one plane passes through three non-collinear points.*

Proof. Let the points be A, B, C. Then one plane Π passes through these points, as is shown by the method of Solution 14.9.7. If a second plane Π' passed through them also, then $\Pi \cap \Pi' \neq \varnothing$. Hence by 16.7.6 Π must be Π', otherwise $\Pi \cap \Pi'$ is a line containing the three non-collinear points A, B, C. ■

16.7.14 PROPOSITION. *If two lines intersect, there is exactly one plane through both of them.*

Proof. The lines l, m meet in exactly one point A, by 16.7.12. By 16.7.3, $\exists B \in l$, $C \in m$ with $B \neq A \neq C$. Now apply 16.7.13 followed by 16.7.11. ■

16.8 Parallels and Perpendiculars

A line through O will be called a **direction**: it is of the form $\mathbf{r} = t\mathbf{b}$. Two directions, d: $\mathbf{r} = t\mathbf{b}$ and d': $\mathbf{r} = s\mathbf{b}'$ will be called **perpendicular** iff $\mathbf{b}\cdot\mathbf{b}' = 0$, and we then write $d \perp d'$. Since $\mathbf{b}\cdot\mathbf{b}' = \mathbf{b}'\cdot\mathbf{b}$, the relation of perpendicularity is symmetric: it is neither reflexive nor transitive [why?]. To the plane Π: $\mathbf{b}\cdot\mathbf{r} = k$, and the line l: $\mathbf{r} = \mathbf{a} + t\mathbf{b}$, we assign the direction $\mathbf{r} = t\mathbf{b}$, denoted respectively by $\operatorname{dir}\Pi$, $\operatorname{dir} l$. By 16.7.5 and 16.7.8 this direction is independent of our choice of writing the equations of Π and l. Two results we need are:

16.8.1 *If* C, D *are two points, then the line* CD *has direction* $\mathbf{r} = t(\mathbf{c} - \mathbf{d})$. (Apply 16.7.9.) ■

16.8.2 *Given a point* $\mathbf{q}$ *and a direction* d, *there exist exactly one line* l *and one plane* Π *through* $\mathbf{q}$ *with direction* d.

Proof. If d is $\mathbf{r} = t\mathbf{b}$, then l is $\mathbf{r} = \mathbf{q} + t\mathbf{b}$ and Π is $\mathbf{b}\cdot(\mathbf{r} - \mathbf{q}) = 0$. Uniqueness of Π and l follows from 16.7.5 and 16.7.8 respectively. ■

The notion of being parallel is now extended:

16.8.3 DEFINITION. Two lines l, l' are **parallel** (written $l \| l'$) iff their directions are equal sets. Two planes Π, Π' are **parallel** ($\Pi \| \Pi'$) iff their directions are equal. Finally l, Π are **parallel** ($l \| \Pi$ and $\Pi \| l$) iff their directions are perpendicular and $l \nsubseteq \Pi$.

Clearly, the 'parallel' relation is symmetric and transitive. The definition says nothing about reflexivity, and it is convenient here to agree that the relation is *never* reflexive. The following results can now be proved from the definitions.

16.8.4 *Parallel lines do not intersect.* (Apply 16.8.2.) ■

16.8.5 *Two parallel lines lie in exactly one plane.* (If the lines are as in 16.7.7, they lie in the plane $(\mathbf{c} - \mathbf{a}) \wedge \mathbf{b} \cdot (\mathbf{r} - \mathbf{a}) = 0$. It is unique by 16.7.13.) ■

16.8.6 *Two planes are parallel iff they do not intersect.* (Apply 16.7.6.) ■

16.8.7 *A line and plane are parallel iff they do not intersect.* (Apply 14 Exercise 11(v).) ■

16.8.8 *If $l \| \Pi$, there is exactly one plane Π' through l and parallel to Π.* (Use the method of 14.9.7 to find Π'; uniqueness follows from 16.8.7.) ■

We next consider perpendicularity, already defined for directions.

16.8.9 DEFINITION. Two lines l, l' are **perpendicular** (written $l \perp l'$) iff their directions are perpendicular. Two planes Π, Π' are **perpendicular** ($\Pi \perp \Pi'$) iff $\operatorname{dir} \Pi \perp \operatorname{dir} \Pi'$. And l and Π are **perpendicular** ($l \perp \Pi$ or $\Pi \perp l$) iff their directions are equal.

Some consequences are:

16.8.10 *If two lines are perpendicular to the same plane, then they are parallel.* (All three directions are the same.) ■

16.8.11 *Given a point Q and a plane Π, there is exactly one line l such that $Q \in l$ and $l \perp \Pi$. Also, $l \cap \Pi$ is a single point* (called the 'foot of the perpendicular from Q to Π').

Proof. If $\operatorname{dir} \Pi$ is $\mathbf{r} = t\mathbf{b}$, then l is $\mathbf{r} = \mathbf{q} + t\mathbf{b}$; its uniqueness follows from 16.8.10 and 16.8.4. ■

16.8.12 *As* 16.8.11 *but with Π replaced throughout by a line m, not through Q.*

Proof. If m has equation $\mathbf{r} = \mathbf{c} + s\mathbf{d}$, then $\operatorname{dir}(QB)$ is $\mathbf{r} = t(\mathbf{q} - \mathbf{b})$, by 16.8.1. Hence, if the result were true with $B = \mathbf{c} + s'\mathbf{d}$ the foot of the perpendicular, then $\operatorname{dir}(QB) \perp \operatorname{dir} m$, so $(\mathbf{q} - \mathbf{b}) \cdot \mathbf{d} = 0$, whence $s' = \mathbf{d} \cdot (\mathbf{q} - \mathbf{c})/d^2$. This gives uniqueness of QB; and existence follows by direct verification that this B $(= \mathbf{c} + \mathbf{d} \cdot (\mathbf{q} - \mathbf{c})\mathbf{d}/\mathbf{d}^2)$ gives $QB \perp m$. ■

16.8.13 COROLLARY. *A line and perpendicular plane intersect in exactly one point.* ■

16.8.14 *If $l \subseteq \Pi$, then* dir $l \perp$ dir Π. (If l is $\mathbf{r} = \mathbf{a} + t\mathbf{b}$ and Π is $\mathbf{p}\cdot\mathbf{r} = c$, then taking $t = 0$ gives $\mathbf{p}\cdot\mathbf{a} = c$, so $\mathbf{p}\cdot\mathbf{b} = 0$.) ■

16.8.15 *If $m \perp \Pi$ then m is perpendicular to every line in Π.* (For $m \perp \Pi$ iff dir m = dir Π; now use 16.8.14.) ■

16.8.16 *If $l \subseteq \Pi$, $m \nsubseteq \Pi$, and $m \| l$, then $m \| \Pi$.* (Use 16.8.14.) ■

16.8.17 *Let l, m be non-parallel coplanar lines. Then $l \cap m$ is a single point.*

Proof. Suppose the lines lie in a plane Π, and have equations

$$l\colon \mathbf{r} = \mathbf{a} + t\mathbf{b}, \qquad m\colon \mathbf{r} = \mathbf{c} + s\mathbf{d}.$$

By 16.8.14, dir $l \perp$ dir Π and dir $m \perp$ dir Π, so we can take the equation of Π to be $(\mathbf{b} \wedge \mathbf{d})\cdot\mathbf{r} = k$. Hence, since $l \subseteq \Pi$ and $m \subseteq \Pi$,

$$(\mathbf{a} - \mathbf{c})\cdot(\mathbf{b} \wedge \mathbf{d}) = k - k = 0.$$

Therefore the equations $u(\mathbf{a} - \mathbf{c}) + t\mathbf{b} - s\mathbf{d} = \mathbf{0}$ have a non-trivial solution (by 15.2.12) with $u \neq 0$ (since $\mathbf{b} \wedge \mathbf{d} \neq 0$ because dir $l \neq$ dir m. Hence there exist $v = t/u$, $w = s/u$ such that $\mathbf{a} + v\mathbf{b} = \mathbf{c} + w\mathbf{d}$; thus $l \cap m \neq \varnothing$, and 16.8.17 now follows from 16.7.12. ■

16.8.18 *Two perpendicular planes intersect in a line.* (Apply $\mathfrak{SP}_4$ and 16.7.6.) ■

16.8.19 *Given l, Π, there exists exactly one plane Π' through l and perpendicular to Π.*

Proof. Use the method of Solution 14.9.7, arguing for existence and uniqueness as in 16.8.12. ■

By 16.8.18, the planes Π, Π' in 16.8.19 intersect in a line, called the **perpendicular projection** of l on Π.

16.8.20 *If $l \| \Pi$, then l is parallel to its perpendicular projection on Π.*

Proof. If the projection is p, then p and l are coplanar in Π' in 16.8.19. If they were not parallel, then by 16.8.17 they would intersect in $\underset{\sim}{Q}$, say. Thus $\underset{\sim}{Q} \in l$ and $\underset{\sim}{Q} \in p \subseteq \Pi$, so $\underset{\sim}{Q} \in l \cap \Pi$, contrary to $l \| \Pi$. ■

16.8.21 *Let two perpendicular planes Π_1, Π_2 intersect in l, let $Q \in l$, and let m be the perpendicular to Π_2 through Q. Then $m \subseteq \Pi_1$.*

Proof. Let Π_i have equation $\mathbf{p}_i\cdot\mathbf{r} = c_i$ $(i = 1, 2)$. Since dir m = dir Π_2, we can take the equation of m to be $\mathbf{r} = \mathbf{q} + t\mathbf{p}_2$ by 16.7.8. Hence $\mathbf{p}_1\cdot\mathbf{r} = \mathbf{p}_1\cdot\mathbf{q}$ since $\mathbf{p}_1\cdot\mathbf{p}_2 = 0$; but $\mathbf{p}_1\cdot\mathbf{q} = c_1$, since $\mathbf{q} \in l \subseteq \Pi_1$. Thus $m \subseteq \Pi_1$. ■

16.8.22 We have now done enough to prove each of the statements (i)–(x) in 16.6. Thus 16.7.3 implies (i), while 16.8.2 implies (ii); (iii) follows from 16.7.14, and (iv) from 16.8.8 and 16.8.16 (since $m \| m' \subseteq \Pi$ by (ii)). From 16.8.19 follows (v), and (vi) from 16.8.18. If (vii) were false, then $l' \| m$ by 16.8.17, so $l \| l' \| m$ using 16.8.20, whence $l \| m$ by transitivity, contrary to the fact. From 16.8.11 follows (viii), and (ix) from 16.8.21. Finally, (x) is a special case of 16.8.17. This completes the proof within our model of the 'existence' part of Theorem 16.4.10. The reader is now invited to show that n, m, n' are coplanar, as required by the commentary following (x) in 16.6. This task completes the proof, within our model, of Theorem 16.4.10. ∎

Exercise 4

(i) Find an alternative proof of 16.4.10, by using the method of Solution 14.9.7 to construct a plane Π_1 through m perpendicular to l. {Since $\Pi_1 \cap l$ is a point Q, there is a line q in Π_1 such that $Q \in q \perp m$ (a rigorous proof of this is required!). Then also $q \perp l$.}

(ii) Make a flow-diagram, in the form of a linear graph, to which of the theorems of 16.7 and 16.8 are needed to prove the later ones. [This same technique is helpful in most parts of mathematics, for understanding the way a theory hangs together.]

Chapter 17
PROJECTIVE GEOMETRY

17.1 A Commercial

A beautiful example of the growth of a theory is the subject called projective geometry, at least as far as the theory of conics. With this restriction, it is not particularly difficult, and is packed with enough influential ideas to be a miniature mathematics itself. This latter aspect is not apparent in the traditional, sterile, English Scholarship approach to the subject, where the prime motivation is to beat Scholarship examiners (an activity independent of mathematics). Since this sterility is beginning to lead to the abolition of projective geometry in certain institutions, we here 'plug' the subject, *because of its excellence for showing what genuine mathematics is.*

17.2 Perspective

First let us see how Euclidean geometry grew into projective geometry. In the geometry of $\mathbb{R}^3$, exceptions often have to be made about parallel lines and planes. For instance, consider the theorem of Desargues (1593–1662) stated in the following 'Euclidean' form. We first need a definition: Triangles ABC, $A'B'C'$ are 'in central perspective' when the joins AA', BB', CC' of corresponding vertices are concurrent or parallel. They are 'in strong axial perspective' when each pair of corresponding sides has an intersection and the three intersections are collinear. [See Fig. 17.1.] A form of Desargues' theorem then says:

17.2.1 *Two triangles are in central perspective if they are in strong axial perspective.*

We shall prove this theorem in 17.6.2 below, although the reader is invited to prove it using vector algebra. For the moment observe how the statement of the theorem—when the terms 'central perspective' and 'strong axial perspective' are unravelled—involves irritating provisos: '. . . if intersections exist'. The irritation becomes worse if we try to formulate a converse of 17.2.1, especially since we feel that the provisos are not inherent in the problem, but accidents of some kind. Mathematicians naturally feel an urge to unify the theory so that the accidents are taken care of, and led by Monge, Poncelet, and others in the 19th century (see Bell [10], Chs. 12, 13), they built a theory in which pairs of parallel lines are no harder to handle than any other pairs. This theory is called *projective* geometry, essentially because it arose

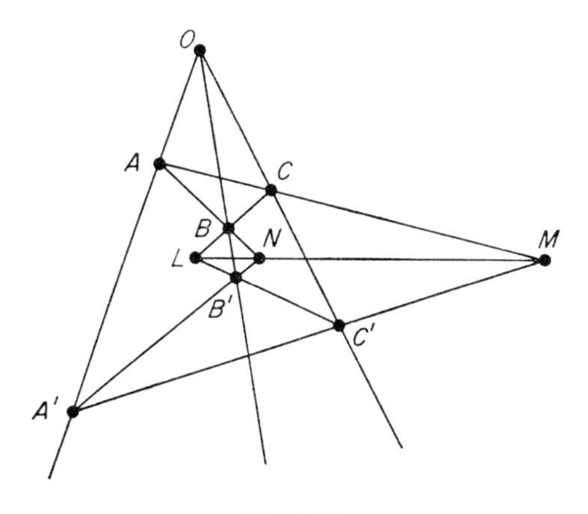

Fig. 17.1

from the practical requirements of draughtsmen in the Renaissance, who wanted to know how to make correct perspective drawings (hence the appearance of the word 'perspective' in 17.2.1). For this aspect of the theory, see the article 'Perspective Drawing' in *Encyclopædia Britannica* (1946); it is unfortunate that standard courses on projective geometry do not include some of this material, since a helpful question when doing a problem is to ask: could I draw the figure accurately from the data?

17.3 Plane Projective Geometries

The first attempts to deal with parallel lines in the plane were based on the idea of allowing them to meet, but 'at infinity', although no such place exists in the plane $\mathbb{R}^2$, of course. The next step was to enlarge the lines to *make* an intersection point exist; and a good description of this process is given in Courant–Robbins [25], Ch. V. After a long history, the abstract notion of a 'plane projective geometry' was evolved, and it seems at first sight remote from its origins. However, it conveniently summarizes the idea that we are studying a system containing 'points' and 'lines', and that points can be on lines, and lines can pass through points.

17.3.1 DEFINITION. A **plane projective geometry** $\mathscr{P}$ consists of three disjoint sets, $\mathbf{P}$, $\mathbf{L}$, $\mathbf{I} \subseteq \mathbf{P} \times \mathbf{L}$, called respectively the **set of points**, **set of lines** and **incidence relation** of $\mathscr{P}$. [If $P \in \mathbf{P}$, $l \in \mathbf{L}$, and $(P, l) \in \mathbf{I}$, we write $P\mathbf{I}l$ or $l\mathbf{I}P$.] Two conditions must be satisfied:

$\mathfrak{PP}_1$: *Given two points P, $Q \in \mathbf{P}$, there exists exactly one $l \in \mathbf{L}$ such that $P\mathbf{I}l$ and $Q\mathbf{I}l$;*

$\mathfrak{PP}_2$: *Given two lines l, $m \in \mathbf{L}$, there exists exactly one $P \in \mathbf{P}$ such that $l\mathbf{I}P$ and $m\mathbf{I}P$.*

17.3.2 To help us to remember what we have in mind, we say P is 'on' l (and l is 'on' P) when $P\mathbf{I}l$. Thus $\mathfrak{PP}_1$ says that:

17.3.3 *given two points, they lie on a unique line;*

while $\mathfrak{PP}_2$ says that:

17.3.4 *given two lines, they lie on a unique point* (in brief: they *intersect*—but this expression is merely a reminder, or conventional form of words). Observe that Definition 17.3.1 is about as complicated as that of a function (cf. 3.4.1); and that $\mathfrak{PP}_1$ determines a function $G \to \mathbf{L}$, where

$$G = \{(P, Q) \in \mathbf{P} \times \mathbf{P} | P \neq Q\},$$

whose value at (P, Q) is the line l on P, Q; similarly with $\mathfrak{PP}_2$. We shall abbreviate 'a plane projective geometry' to 'a PPG', to distinguish it from 'projective geometry'—the subject we are studying.

17.3.5 EXAMPLES. (i) The most trivial PPG is that in which $\mathbf{P}$, $\mathbf{L}$, $\mathbf{I}$ are all empty. Here $\mathfrak{PP}_1$, $\mathfrak{PP}_2$ are vacuously satisfied.

(ii) The system $\mathscr{E}(\mathbb{R})$ of points and lines of $\mathbb{R}^2$, where $P\mathbf{I}l$ means '$P \in l$', is *not* a PPG, since $\mathfrak{PP}_2$ is not always true. [Why not?]

(iii) Let $\mathbf{P}$, $\mathbf{L}$ consist respectively of all lines and all planes through the origin in $\mathbb{R}^3$; let $P\mathbf{I}l$ mean that the line P is contained in the plane l. Then $\mathfrak{PP}_1$ and $\mathfrak{PP}_2$ hold, by 14 Exercise 12(vi), (vii). The triple $(\mathbf{P}, \mathbf{L}, \mathbf{I})$ is therefore a PPG according to the definition; it is called **the $\mathbb{R}^3$-model**, below.

(iv) Let $\mathbf{F}$ be any field, $\mathbf{F}^3 = \mathbf{F} \times \mathbf{F} \times \mathbf{F}$. On $\mathbf{F}^3 - \mathbf{0}$, define an equivalence relation R by: $\mathbf{a}R\mathbf{b}$ iff $\exists t \in \mathbf{F}$ such that $t \neq 0$ and $\mathbf{a} = t\mathbf{b}$ (so $\mathbf{0}$, $\mathbf{a}$, $\mathbf{b}$ are collinear). Let X denote the set of all linear functions f: $\mathbf{F}^3 \to \mathbf{F}$, of the form $f(\mathbf{r}) = \mathbf{a} \cdot \mathbf{r}$, where $\mathbf{a} \in \mathbf{F}^3 - \mathbf{0}$. On X define an equivalence relation S by: fSg iff $\exists t \in \mathbf{F}$ such that $t \neq 0$ and $f = tg$. Define $\mathbf{P}$, $\mathbf{L}$ respectively to be the families $(\mathbf{F}^3 - \mathbf{0})/R$, X/S, of cosets; and define a relation $\mathbf{I}$ by setting $(\mathbf{a}R)\mathbf{I}(fS)$ whenever $f(\mathbf{a}) = 0$. (Check for independence of representatives of the cosets $\mathbf{a}R$, fS.) Then $\mathfrak{PP}_1$ holds; for, given distinct points $\mathbf{a}R$, $\mathbf{b}R$, they lie on the 'line' fS, where $f(\mathbf{r}) = (\mathbf{a} \wedge \mathbf{b}) \cdot \mathbf{r}$; and $\mathfrak{PP}_2$ holds, because given distinct 'lines' fS, gS with $f(\mathbf{r}) = \mathbf{a} \cdot \mathbf{r}$, $g(\mathbf{r}) = \mathbf{b} \cdot \mathbf{r}$, then they lie on the 'point' $(\mathbf{a} \wedge \mathbf{b})R$. (To verify these statements use 14 Exercise 12(vi),(vii) with $\mathbb{R}$ replaced by $\mathbf{F}$.) Thus $(\mathbf{P}, \mathbf{L}, \mathbf{I})$ forms a PPG denoted by $\mathscr{P}(\mathbf{F})$. When $\mathbf{F} = \mathbb{R}$, the PPG here is isomorphic to that in (iii) above. (See Exercise 2(ix) below.) The set $\mathbf{P}$ of $\mathscr{P}(\mathbb{R})$ is then called the **real projective plane**, denoted by $\mathbb{RP}^2$. $\mathscr{P}(\mathbb{R})$ is customarily described by saying that a 'line' is an equation $lx + my + nz = 0$ where (l, m, n) is a non-zero point of $\mathbb{R}^3$ [and two such equations are 'the same' if one is a multiple of the others], while a 'point' is a non-zero point of $\mathbb{R}^3$, where we regard $\mathbf{a}$ as being the same as $\mathbf{b}$ iff $\exists t \neq 0$ in $\mathbb{R}$ such that $\mathbf{a} = t\mathbf{b}$ (!!). The 'point' $\mathbf{a}$ is on the 'line' $lx + my + nz = 0$ iff $(l, m, n) \cdot \mathbf{a} = 0$. At the very least, it is safer to denote the point **a** by [**a**], to make quite clear that a 'point' is an equivalence class (viz., **a**R in our terminology above).

Exercise 1

(i) Verify in detail that the relations R, S in (iv) of the last example are equivalence relations.
(ii) Show that when $\mathbf{F} = \mathbb{Z}_2$ in Example 17.3.5(iv), the sets $\mathbf{P}$, $\mathbf{L}$ each have just seven elements, with just three points on each line, and three lines on each point. [If the seven points are represented by $A, B, \ldots, G$ as in Fig. 17.2, then six of the

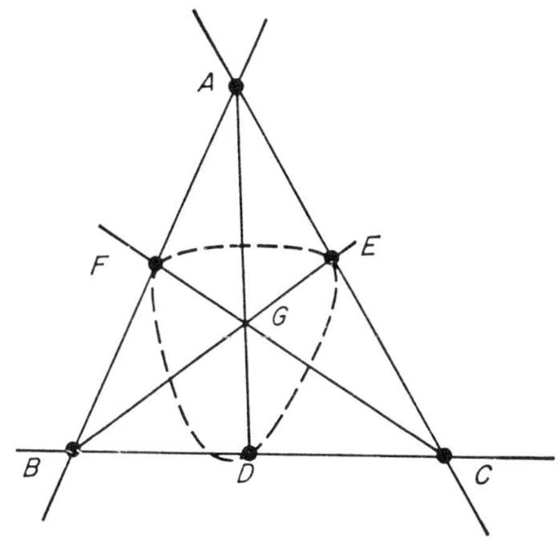

Fig. 17.2

lines are represented by the six lines drawn in the figure, and the seventh line is given by the set E, F, D. This is not 'straight', but Definition 17.3.1 says nothing about straightness.]
(iii) Reconcile the traditional description of $\mathscr{P}(\mathbb{R})$ as given at the end of Example 17.3.5(iv) with its 'coset' description at the beginning.

17.4 Duality

If we interchange the words 'line' and 'point' in the literary versions 17.3.3, 17.3.4 of $\mathfrak{PP}_1$, $\mathfrak{PP}_2$, then we obtain 17.3.4, 17.3.3 respectively. It is customary to call such statements 'duals' of each other. A notable feature of projective geometry is that each theorem occurs with a dual, and if we proceed properly, one proof will yield two dual theorems. (This phenomenon, of duality, is now recognized in other branches of mathematics, e.g., in Boolean algebra and topology; see also Chapter 38 on categories.) Since we have not yet proved any *theorems* of projective geometry, it will help to look at some before we formulate the notion of duality more precisely. First, however, we formulate the 'dual' of a PPG. Let $\mathscr{P}$ be as in 17.3.2; we assign to it a new PPG denoted by $\mathscr{P}^* = (\mathbf{P}^*, \mathbf{L}^*, \mathbf{I}^*)$; where $\mathbf{P}^* = \mathbf{L}$, $\mathbf{L}^* = \mathbf{P}$, and, for each $P \in \mathbf{P}^*$, $l \in \mathbf{L}^*$, $P\mathbf{I}^*l$ means $l\mathbf{I}P$ in $\mathscr{P}$. The above remark about interchanging 17.3.3 and 17.3.4 shows that $\mathscr{P}^*$ is a PPG; it is called the **dual** of $\mathscr{P}$. Clearly $(\mathscr{P}^*)^* = \mathscr{P}$.

Roughly speaking, we are going to take theorems which are true of 'most' PPG's; and interpret each theorem in $\mathscr{P}$ and $\mathscr{P}^*$ to get two theorems for the price of one. But now let us consider some theorems.

As we have defined it, a projective plane is a very general kind of object, and we cannot expect an interesting theory without adding more hypotheses. For example, it is natural to assume that the following 'axioms of extension' hold in $\mathscr{P}$, thus excluding uninteresting examples like 17.3.5(i).

$\mathfrak{E}_1$: *There is at least one line* (i.e., $\mathbf{L} \neq \varnothing$);
$\mathfrak{E}_2$: *Not all points are on the same line;*
$\mathfrak{E}_3$: *There are at least three points on each line.*

We shall prove the

17.4.1 THEOREM. *If $\mathscr{P}$ satisfies $\mathfrak{E}_1$, $\mathfrak{E}_2$, $\mathfrak{E}_3$, so does $\mathscr{P}^*$.*

Before starting the proof proper, we introduce a notation, and let $\mathfrak{E}_i^*$ denote the dual of $\mathfrak{E}_i$, obtained by interchanging the words 'point (of $\mathscr{P}$)' and 'line (of $\mathscr{P}$)'; thus $\mathfrak{E}_1^*$ is: 'There is at least one point (of $\mathscr{P}$) (i.e., $\mathbf{P} \neq \varnothing$)', and similarly for $\mathfrak{E}_2^*$, $\mathfrak{E}_3^*$. But then $\mathfrak{E}_1^*$ also means: 'There is at least one line (of $\mathscr{P}^*$) (i.e., $\mathbf{L}^* \neq \varnothing$)'. Hence, to prove the theorem, it suffices to prove $\mathfrak{E}_i^*$, $i = 1, 2, 3$, *for $\mathscr{P}$ itself*. That is, we shall prove instead of 17.4.1 the equivalent:

17.4.2 THEOREM. *If $\mathscr{P}$ satisfies $\mathfrak{E}_i$ $(1 \leqslant i \leqslant 3)$, then $\mathscr{P}$ satisfies $\mathfrak{E}_i^*$ $(1 \leqslant i \leqslant 3)$.*

Proof of 17.4.2. To prove $\mathfrak{E}_1^*$: there is at least one line l (of $\mathscr{P}$), and at least three points (of l), so there is at least one point (of $\mathscr{P}$); hence $\mathfrak{E}_1^*$ holds. To prove $\mathfrak{E}_2^*$: we must show that, for any point P, there is a line not on P. Now certainly there is a line l (by $\mathfrak{E}_1$). If l is *not* on P, then not all lines are on P, which is $\mathfrak{E}_2^*$. If l *is* on P there exists $Q \neq P$ on l (by $\mathfrak{E}_3$) and R not on l (by $\mathfrak{E}_2$). Hence by $\mathfrak{P}\mathfrak{P}_1$, there exists a line m which we can denote by RQ, and $m \neq l$, since R is on m and not on l. Thus P is not on m, otherwise $l = PQ = m$ (by $\mathfrak{P}\mathfrak{P}_1$ again). Hence m is a line not on P, so $\mathfrak{E}_2^*$ holds. To prove $\mathfrak{E}_3^*$: let P be a point; we must find at least three lines on P. Now, there is a line l not on P by $\mathfrak{E}_2^*$, and by $\mathfrak{E}_3$ there are at least three points A, B, C on l. Hence by $\mathfrak{P}\mathfrak{P}_1$, the lines PA, PB, PC exist and are on P. It remains to prove them distinct. But if $PA = PB = m$, say, the distinct points A, B are on l and on m, which is contrary to $\mathfrak{P}\mathfrak{P}_2$ unless $l = m$. But then P is on $m = l$, contrary to the fact; so $\mathfrak{E}_3^*$ holds. ■

We now infer immediately

17.4.3 THEOREM. *The converse of Theorem 17.4.2 holds.*

Proof. Theorem 17.4.1 is true for any plane projective geometry and hence for $\mathscr{P}^*$. But replacing $\mathscr{P}$ by $\mathscr{P}^*$ in 17.4.1 simply yields the converse! ■

These examples should suffice to illustrate the

17.4.4 **principle of duality.** *If* $s(\mathscr{P})$ *is a statement about* $\mathscr{P}$, *then* (as prior to 17.4.2) *its* **dual** $s^*(\mathscr{P})$, *is the same statement* $s(\mathscr{P}^*)$ *about* $\mathscr{P}^*$; *the statement* s *holds for* $\mathscr{P}$ *iff the statement* s^* *holds for* $\mathscr{P}^*$. [We ought to define the word 'statement' more precisely, but its meaning is obvious in the applications. A correct general statement is somewhat subtle to make!] Notice that, more generally, the duality principle asserts that if S is a set of plane projective geometries and S^* the set of dual geometries and s any statement about geometries, then s holds for the members of S if and only if s^* holds for the members of S^*.

Exercise 2

(i) If $\mathscr{P}$ satisfies $\mathfrak{E}_1$, $\mathfrak{E}_2$ but the words 'at least three' in $\mathfrak{E}_3$ are changed to 'exactly two', show that $\mathscr{P}$ consists of just three points and three lines; investigate what happens if we replace 'exactly two' by 'one'.
(ii) Prove that the PPG in Exercise 1(ii) satisfies $\mathfrak{E}_1$, $\mathfrak{E}_2$, and $\mathfrak{E}_3$ with 'at least' replaced by 'exactly' in $\mathfrak{E}_3$.
(iii) Apply (ii) to the 'problem of the lunches' in O'Hara and Ward [96], p. 17.
(iv) Let $\mathscr{P}$ be a PPG satisfying $\mathfrak{E}_i$ $(1 \leqslant i \leqslant 3)$ and such that there is a line with exactly $n + 1$ $(\geqslant 3)$ points on it. Prove that there are exactly $n + 1$ points on *each* line, and exactly $n + 1$ lines on each point; with exactly $n^2 + n + 1$ points and $n^2 + n + 1$ lines in all. {Use duality.}
(v) Prove that no PPG satisfying $\mathfrak{E}_i$ $(1 \leqslant i \leqslant 3)$ can have just 23 points.
(vi) By taking a field of residues mod p, with p prime, prove that there exists a PPG satisfying the conditions of (iv), with $n = p$. {Use Example 17.3.5(iv).} [It can be shown that there is such a PPG if $n = p^\alpha$, where p is prime. On the other hand, it is unknown whether such a PPG exists when $n = 10$. The theorem of Bruck and Ryser states that no PPG exists when (a) $n \equiv 1$ or $2 \bmod 4$; and (b) if $n = n'd$, where n' is the largest *squared* integer with d an integer, then d is divisible by at least one prime $\equiv 3 \bmod 4$. For example, take $n = 2p$, $p \equiv 3 \bmod 4$. See Ryser [113].]
(vii) Let $\mathscr{P} = (\mathbf{P}, \mathbf{L}, \mathbf{I})$ be a PPG as in 17.3.1. For each $l \in \mathbf{L}$, define

$$S_l = \{P \in \mathbf{P} | P\mathbf{I}l\}.$$

Now form a new PPG, denoted by $\mathscr{P}' = (\mathbf{P}', \mathbf{L}', \mathbf{I}')$ where $\mathbf{P}' = \mathbf{P}$, $\mathbf{L}'$ is the set of all S_l, $l \in \mathbf{L}$, and $P\mathbf{I}'S_l$ iff $P \in S_l$. Verify that $\mathscr{P}'$ is a PPG. Prove that the function $g: \mathbf{L} \to \mathbf{L}'$ given by $g(l) = S_l$, is an equivalence. {Consider the cases $\#l \geqslant 2, \#l < 2$.} Show that $\mathscr{P}'$ satisfies $\mathfrak{E}_i$ if $\mathscr{P}$ does $(1 \leqslant i \leqslant 3)$. [Thus, we have found a copy, $\mathscr{P}'$ of $\mathscr{P}$, in which 'P on l' means $P \in l$. The word 'copy' is analysed in the next problem.]
(viii) Let $\mathscr{P} = (\mathbf{P}, \mathbf{L}, \mathbf{I})$, $\mathscr{Q} = (\mathbf{Q}, \mathbf{M}, \mathbf{J})$ be two PPG's (in which $\mathbf{Q}$ denotes the set of points of $\mathscr{Q}$, etc.). A **map** $f: \mathscr{P} \to \mathscr{Q}$ is defined to be a pair of functions $f: \mathbf{P} \to \mathbf{Q}$, $g: \mathbf{L} \to \mathbf{M}$ such that, if $P\mathbf{I}l$ in $\mathscr{P}$, then $f(P)\mathbf{J}g(l)$ in $\mathscr{Q}$. Use 3 Exercise 4(ii) to show that the function $f \times g: \mathbf{P} \times \mathbf{Q} \to \mathbf{L} \times \mathbf{M}$ maps $\mathbf{I}$ into $\mathbf{J}$. Writing

$\not{f} = (f, g)$ for brevity, prove that if $\not{g} = (h, k)\colon \mathcal{Q} \to \mathcal{R}$ is a map of projective planes, then the composite $(h \circ f, k \circ g)\colon \mathcal{P} \to \mathcal{R}$ is a map. Prove also that $(\mathrm{id}, \mathrm{id})\colon \mathcal{P} \to \mathcal{P}$ is a map, and that if f, g are both equivalences, then $(f^{-1}, g^{-1})\colon \mathcal{Q} \to \mathcal{P}$ is a map. In this case, if also $(f \times g)(\mathbf{I}) = \mathbf{J}$, then $\not{f}\colon \mathcal{P} \to \mathcal{Q}$ is called an **isomorphism** ($\mathcal{Q}$ is a 'copy' of $\mathcal{P}$). In (vii) above, prove that $(\mathrm{id}, g)\colon \mathcal{P} \to \mathcal{P}'$ is an isomorphism. Construct two PPG's with *bijections* f, g, but such that (f, g) is not an isomorphism.
(ix) Prove the statement in Example 17.3.5(iv) that the real PPG, $\mathcal{P}(\mathbb{R})$, is isomorphic to the PPG of Example 17.3.5(iii).

17.5 The Geometry $\mathcal{P}(\mathbb{R})$

We shall next investigate the system $\mathcal{P}(\mathbb{R})$ more closely, and show how to obtain from it some results about the geometry of $\mathbb{R}^2$. This feature is akin to the added power we have when one number system can be embedded in a richer one, but it is often not mentioned in traditional textoooks. The latter will start with a bare, unmotivated, description of $\mathcal{P}(\mathbb{R})$ (or even of $\mathcal{P}(\mathbb{C})$), and may mystify the reader by supporting the text with sketches of 'genuine' Euclidean lines, points, and conics whose relationship with $\mathcal{P}(\mathbb{R})$ is not made clear. This comes of the desire to train the reader to beat a parochial examination system, rather than to satisfy his curiosity. We certainly do not deny the important role of such sketches in developing intuition, and we shall show in 17.6 below why their presence is legitimate.

Let us recapitulate the description of $\mathcal{P}(\mathbb{R})$ in more traditional terms. As remarked in Example 17.3.5(iv), the set of points of $\mathcal{P}(\mathbb{R})$ is denoted by $\mathbb{RP}^2$ and a point $[\mathbf{a}]$ is an equivalence class of non-zero points of $\mathbb{R}^3$, where $\mathbf{b} \in [\mathbf{a}]$ iff $\exists t \neq 0$ such that $\mathbf{a} = t\mathbf{b}$. The 'line $l\colon \mathbf{p}\cdot\mathbf{r} = 0$' ($\mathbf{p} \neq \mathbf{0}$) is defined to be the set of all $[\mathbf{a}]$ such that $\mathbf{p}\cdot\mathbf{a} = 0$. And $[\mathbf{a}]$ is 'on' l iff $\mathbf{p}\cdot\mathbf{a} = 0$. [If $\mathbf{a} = t\mathbf{b}$, then also $\mathbf{p}\cdot\mathbf{b} = 0$, so these definitions are independent of representatives.] The two points $A = [\mathbf{a}]$, $B = [\mathbf{b}]$ lie on the (unique) line $(\mathbf{a} \wedge \mathbf{b})\cdot\mathbf{r} = 0$, denoted by† AB; since $A \neq B$, $\mathbf{a} \wedge \mathbf{b} \neq \mathbf{0}$. The lines $\mathbf{p}\cdot\mathbf{r} = 0$, $\mathbf{q}\cdot\mathbf{r} = 0$ intersect in the (unique) point $[\mathbf{p} \wedge \mathbf{q}]$. It will be helpful always to keep in mind the fact that $\mathcal{P}(\mathbb{R})$ is isomorphic to the PPG of Example 17.3.5(iii).

17.5.1 PROPOSITION. *The line AB is the set of points of the form $P = [\lambda\mathbf{a}+\mu\mathbf{b}]$, $\lambda, \mu \in \mathbb{R}$, for a fixed choice of* $\mathbf{a}, \mathbf{b}$ *from the sets* $[\mathbf{a}'] = A$, $[\mathbf{b}'] = B$.

Proof. Since $(\mathbf{a} \wedge \mathbf{b})\cdot(\lambda\mathbf{a} + \mu\mathbf{b}) = 0$, then $P \in AB$. Conversely, if $[\mathbf{r}]$ satisfies $(\mathbf{a} \wedge \mathbf{b})\cdot\mathbf{r} = 0$, then the numbers λ, μ exist by the theory of 15.2.10 since $\mathbf{a} \wedge \mathbf{b} \neq \mathbf{0}$. ∎

17.5.2 COROLLARY. *Three points,* $A = [\mathbf{a}]$, $B = [\mathbf{b}]$, $C = [\mathbf{c}]$ *are collinear, iff* $\mathbf{a}\cdot\mathbf{b} \wedge \mathbf{c} = 0$.

† Rather than by (AB) as in Chapter 14, since we do not need to consider segments in the present chapter.

17.5.3 REMARK. Had we chosen to represent A, B by $[s\mathbf{a}]$, $[t\mathbf{b}]$ respectively, then, for a given point P on AB, λ, μ would change accordingly. [How?] Some books write P in 17.5.1 as $\lambda A + \mu B$, but this can be confusing, as they usually say that $\lambda A = A$ (although they mean $[\lambda\mathbf{a}] = [\mathbf{a}]$).

The three points $X = [\mathbf{i}]$, $Y = [\mathbf{j}]$, $Z = [\mathbf{k}]$ constitute the **triangle of reference**; its sides are the lines $x = 0$, $y = 0$, $z = 0$ of which the first is YZ and does not pass through X, and similarly for the rest. The point $R = [\mathbf{r}]$ satisfies $[\mathbf{r}] = [x\mathbf{i} + y\mathbf{j} + z\mathbf{k}]$ (by 14.2.2), and the numbers x, y, z are called the **homogeneous co-ordinates** of R—because if we had chosen to write $[\mathbf{s}]$ for R, then the ratios $x:y:z$ and $s_1:s_2:s_3$ would be the same. Because of the simplicity of the homogeneous co-ordinates of X, Y, Z and some related points, the algebra of any configuration containing these points is simplified. For this reason, it is useful to be able to take *any* triangle as XYZ, by which is meant the following (indeed stronger) statement:

17.5.4 THEOREM. *Let $W = [(1, 1, 1)]$, and let A, B, C, D be any four points of $\mathbb{RP}^2$, no three of which are collinear. Then $\exists$ a function $f: \mathbb{RP}^2 \to \mathbb{RP}^2$, which is an equivalence, maps lines onto lines, preserves incidence, and is such that $f(A) = X$, $f(B) = Y$, $f(C) = Z$, $f(D) = W$.* [Recall that $\mathbb{RP}^2$ is the set of all points in $\mathscr{P}(\mathbb{R})$, and that 'f preserves incidence' means that if the point P is on the line l, then $f(P)$ is on $f(l)$.]

Proof. By 15 Exercise 7(xii), there exists a non-singular linear transformation $T: \mathbb{R}^3 \to \mathbb{R}^3$ which maps the lines $\mathbf{r} = t\mathbf{a}, \ldots, \mathbf{r} = t\mathbf{d}$ onto the lines $\mathbf{r} = t\mathbf{i}, \ldots, \mathbf{r} = t\cdot(1, 1, 1)$ (where $A = [\mathbf{a}]$, etc.), because no three of the first four lines are coplanar, by 17.5.2. Define the required $f: \mathbb{RP}^2 \to \mathbb{RP}^2$ by: $f[\mathbf{r}] = [T\mathbf{r}]$. By 17.5.1, any point P on the line UV is $[\lambda\mathbf{u} + \mu\mathbf{v}]$, so $f(P) = [T\mathbf{p}] = [\lambda T\mathbf{u} + \mu T\mathbf{v}]$, whence f maps the whole line UV onto the whole line joining fU to fV (since T is nonsingular); moreover, f thereby preserves incidence. ■

Remark. A function $g: \mathbb{RP}^2 \to \mathbb{RP}^2$ which (like f in 17.5.4) is a bijection, preserves incidence and maps lines onto lines, is called a **collineation**. We leave the reader to verify that the set of all collineations of $\mathbb{RP}^2$ forms a group under composition (and in particular, the inverse f^{-1} of f in 17.5.4 is a collineation).

As an application, let us prove Desargues' theorem in $\mathscr{P}(\mathbb{R})$. We first reformulate the definitions given for 17.2.1, the irritating provisos being no longer necessary because of the axiom $\mathfrak{P}\mathfrak{P}_2$.

17.5.5 DEFINITION. Two triangles ABC, $A'B'C'$ are **in central perspective** iff the joins AA', BB', CC' are all concurrent. They are **in axial perspective** iff the three intersections of pairs of corresponding sides are all collinear. [We must of course assume $A \neq A'$, etc.]

17.5.6 The **theorem of Desargues** in $\mathscr{P}(\mathbb{R})$ states: *Two triangles are in central perspective iff they are in axial perspective.*

Proof. Central perspective $\Rightarrow$ axial perspective. For, suppose AA', BB', CC' all meet in D. By 17.5.4, there is a collineation f which maps A, B, C, D respectively onto X, Y, Z, W. Let $f(A') = X'$, etc. Then X' lies on XW, and so by 17.5.1, X' is of the form $[\mathbf{w} + p\mathbf{i}] = [(1 + p, 1, 1)]$; similarly $Y' = [(1, 1 + q, 1)]$, $Z' = [(1, 1, 1 + r)]$. The equation of XY is $z = 0$, so XY meets $X'Y'$ in $R = [(p, -q, 0)]$, so by symmetry of the notation, YZ meets $Y'Z$ in $P = [(0, q, -r)]$ and ZX meets $Z'X'$ in $Q = [(-p, 0, r)]$. But these three points are collinear, say on l, by 17.5.2. Now apply f^{-1}; then $f^{-1}P, f^{-1}Q, f^{-1}R$ lie on $f^{-1}l$, since f^{-1} preserves incidence. But $f^{-1}P$ is the intersection of $f^{-1}(XY)$ and $f^{-1}(X'Y')$, i.e., of AB, $A'B'$, and similarly for $f^{-1}Q, f^{-1}R$. Thus triangles ABC, $A'B'C'$ are in axial perspective. The converse can be proved by similar calculations and use of f, and we leave such a proof to the reader. ∎ It is, however, instructive to prove the converse using the notion of **reciprocal duality**, as follows. $\mathscr{P}(\mathbb{R})$ has the property that it is isomorphic to its dual, $\mathscr{P}^*(\mathbb{R})$:

17.5.7 *There is an isomorphism* $f: \mathscr{P}(\mathbb{R}) \to \mathscr{P}^*(\mathbb{R})$.

Proof. Recall from Exercise 2(viii) that f has to be a pair (d_1, d_2) of equivalences $d_1: \mathbf{P} \to \mathbf{L}$, $d_2: \mathbf{L} \to \mathbf{P}$ (where $\mathbf{P}$ here is $\mathbb{RP}^2$, and $\mathbf{L}$ is the set of lines), such that incidence is preserved by f and f^{-1}. To define d_1, d_2 we let $d_1[\mathbf{a}]$ be the line $l: \mathbf{a}\cdot\mathbf{r} = 0$, and let $d_2(l)$ be $[\mathbf{a}]$. If $[\mathbf{b}]$ is on l, then obviously $d_2(l)$ is on $d_1[\mathbf{a}]$, so incidence is preserved by f; similarly f^{-1} preserves incidence. Hence 17.5.7 holds. ∎ [This kind of duality is called 'reciprocal', to distinguish it from that in 17.4, since here each point P and line l have specific duals, $d_1(P)$, $d_2(l)$. In the general duality of 17.4.2, individual points and lines *do not have their own duals*; only the *concepts* 'point' and 'line' are dual (or $\mathbf{P}$ and $\mathbf{L}$ are dual).]

We now apply this to prove that

axial perspective implies central perspective.

Since central perspective implies axial perspective in $\mathscr{P}(\mathbb{R})$, it follows from the *principle of duality* (17.4.4) that axial perspective implies central perspective in $\mathscr{P}^*(\mathbb{R})$; but $\mathscr{P}^*(\mathbb{R})$ is isomorphic to $\mathscr{P}(\mathbb{R})$, so axial perspective implies central perspective in $\mathscr{P}(\mathbb{R})$. ∎

Caution. We cannot replace this argument by the general duality of 17.4.2 without invoking 17.5.7, since the co-ordinate methods of proof in 17.5.6 do not dualize in the general sense.

17.5.8 REMARK. Everything we have said so far about $\mathscr{P}(\mathbb{R})$ is valid for $\mathscr{P}(\mathbf{F})$, where $\mathbf{F}$ is any field.

Exercise 3

(i) Check the truth of 17.5.8, in particular as it concerns the proof of 17.5.4.

(ii) Let A, B, C, D be four points in $\mathscr{P}(\mathbf{F})$, no three collinear. Let AB, CD; AD, BC; AC, DB meet in F, G, H respectively. Use 17.5.4 to prove that F, G, H are not collinear unless $2 = 0$ in $\mathbf{F}$.

(iii) Prove the **theorem of Pappus** in $\mathscr{P}(\mathbf{F})$: ABC, DEF are two triples of points on distinct lines l, m respectively. Let AB, DE; BC, EF; CD, FA meet in U, V, W respectively. Then U, V, W are collinear. Formulate and prove the dual, using 17.5.7 (see 17.7 below).
(iv) Let $C = [\mathbf{c}]$, $D = [\mathbf{d}]$ be two points on the line AB in Proposition 17.5.1. Show that $\exists \alpha, \beta, \gamma, \delta \in \mathbb{R}$ such that P in 17.5.1 is $[\lambda'\mathbf{c} + \mu'\mathbf{d}]$, where $\lambda' = \lambda\alpha + \mu\beta$, $\mu' = \lambda\gamma + \mu\delta$.

17.6 Relevance to $\mathbb{R}^2$

Let us now see how $\mathbb{R}^2$ appears in projective geometry. Its system of points and lines of $\mathbb{R}^2$ with the incidence relation 'on' will be denoted by $\mathscr{E}(\mathbb{R}^2)$; although $\mathscr{E}(\mathbb{R}^2)$ is not a PPG, we can map it into $\mathscr{P}(\mathbb{R})$ as follows. To each $(u, v) \in \mathbb{R}^2$, we assign the point $e_1(u, v) = [(u, v, 1)] \in \mathbb{RP}^2$, and to each line l: $ax + by + c = 0$ we assign the line $e_2(l)$: $ax + by + cz = 0$ in $\mathscr{P}(\mathbb{R})$. Clearly, if $(u, v) \in l$ in $\mathscr{E}(\mathbb{R}^2)$, then $e_1(u, v) \in e_2(l)$ in $\mathscr{P}(\mathbb{R})$, so incidence relations are preserved. Moreover, e_1 and e_2 are each one-one functions, so we say they combine to form an **embedding**

17.6.1 $$e\colon \mathscr{E}(\mathbb{R}^2) \to \mathscr{P}(\mathbb{R}).$$

Observe that the only points of $\mathbb{RP}^2$ which are not in the image $e_1(\mathbb{R}^2)$ are those of the form $[(x, y, 0)]$, and these form the line $z = 0$. Moreover, if m is a line in $\mathbb{R}^2$, parallel to l (given above), then it has equation $ax + by + c' = 0$ for some $c' \neq c$, so $e_1(l)$, $e_2(m)$ intersect in $[(bc' - bc, ca - c'a, 0)]$ (which is $[b, -a, 0]$ and lies in $\mathbb{RP}^2$ since $c \neq c'$). Thus, if $l \| m$ in $\mathbb{R}^2$, then $e_2(l)$ meets $e_2(m)$ on $z = 0$. For this reason, $z = 0$ is often called the 'line at infinity', denoted by l_∞; but this name is used only for historical reasons; *there is no 'infinity'*. Also, l_∞ is the only line not in the image of e_2.

We can now *deduce the Euclidean theorem* 17.2.1 *from Desargues' in* $\mathscr{P}(\mathbb{R})$ as follows.

17.6.2 PROOF. Let ABC, $A'B'C'$ be triangles in $\mathbb{R}^2$, in strong axial perspective (in the sense of 17.2.1). The system K, consisting of the sides of the triangles and their various intersections, is then a configuration in $\mathscr{E}(\mathbb{R})$ which is mapped into $\mathscr{P}(\mathbb{R})$ by e of 17.6.1 to form the Desargues configuration of 17.5.6. By 17.5.6 then, the triangles $A_1B_1C_1$, $A_1'B_1'C_1'$ (where $A_1 = e_1(A)$, etc.) are in axial perspective, hence in central perspective. Thus A_1A_1', B_1B_1', C_1C_1' meet in a point $V \in \mathbb{RP}^2$, and either $V \in l_\infty$ or $V = e_1(U)$ for some $U \in \mathbb{R}^2$. Therefore in the first case AA', BB', CC' were parallel, and in the second case they met in U. Hence triangles ABC, $A'B'C'$ were in central perspective, in $\mathbb{R}^2$. ■

Exercise 4

(i) What happens to the x- and y-axes of $\mathbb{R}^2$ under the embedding 17.6.1?
(ii) Use the method of 17.6.2 to formulate and prove some kind of converse of 17.2.1.

(iii) Use 17.6.1 to explain why a sketch on paper can have relevance to an argument in $\mathscr{P}(\mathbb{R})$.

17.7 Conics

The next most simple loci in $\mathbb{R}^2$ after lines are the 'conics'—the circle, parabola, ellipse, and hyperbola, all known to the Greeks as plane sections of a cone. Each required separate study by special methods until in the 19th century a unified treatment was made possible through projective geometry. We shall give a brief account of this attractive study. Analytic geometry in $\mathbb{R}^2$ showed that the conics were all given by a second-degree equation with real coefficients of the form

17.7.1 $$ax^2 + by^2 + 2hxy + 2gx + 2fy + c = 0.$$

[See Sommerville [121] for a proof that all plane sections of a cone have equations of this form; and for some historical remarks.] Its **discriminant** is the determinant $D = \det A$, where A is the matrix

$$A = \begin{pmatrix} a & h & g \\ h & b & f \\ g & f & c \end{pmatrix}.$$

The following statements can be proved (see, for example, Sommerville [121]). Suppose first that $D \neq 0$. Then 17.7.1 represents a **parabola** if $h^2 = ab$, an **ellipse** if $h^2 < ab$, and a **hyperbola** if $h^2 > ab$. If an ellipse has $h = 0$ and $a = b$, it is a circle; if a hyperbola has $a + b = 0$ it is a rectangular hyperbola. If $D = 0$, then 17.7.1 factorizes into a pair of real linear factors, and represents a pair of lines (see Example 1.6.4) which are parallel or equal if $h^2 = ab$, and perpendicular if $a + b = 0$. Thus, it is convenient to regard a pair of lines as a conic, which is a parabola if they are parallel or equal, and a rectangular hyperbola if they are perpendicular. Knowing all this, the originators of projective geometry studied the 'homogeneous' equation,

17.7.2 $$ax^2 + by^2 + cz^2 + 2fyz + 2gzx + 2hxy = 0,$$

(where the alphabetical order of coefficients is responsible for the apparently curious choice in 17.7.1). In the $\mathbb{R}^3$-model of Example 17.3.5(iii), 17.7.2 represents a cone; in $\mathscr{P}(\mathbb{R})$, the set of all points of $\mathbb{RP}^2$ satisfying 17.7.2 is called a **conic**. Let us give an application to $\mathbb{R}^2$. We have stated the theorem of Pappus in Exercise 3(iii); it holds also when the pair of lines is replaced by a circle (and both can be proved within Euclidean geometry, but not easily, using the 'theorems of Ceva and Menelaus'). Since a pair of lines and a circle both have equations of the form 17.7.1, we might suspect that the theorem would still hold with the pair of lines replaced by *any* conic. The resulting general theorem is proved fairly easily in books on projective geometry, under the name of **Pascal's theorem.** Let us now deduce from it the first two forms in $\mathbb{R}^2$.

We shall use the embedding e of 17.6.1. First, we extend e to the conics of $\mathbb{R}^2$ by assigning to each conic S of the form 17.7.1 the conic $e_3(S)$ in $\mathscr{P}(\mathbb{R})$ given by 17.7.2. [In traditional textbooks, equation 17.7.2 is 17.7.1 'made homogeneous'.] If $P \in S$, then it is easily verified that $e_1(P) \in e_3(S)$, so incidences are preserved; and e_3 is one-one. Thus, if for example, S is a

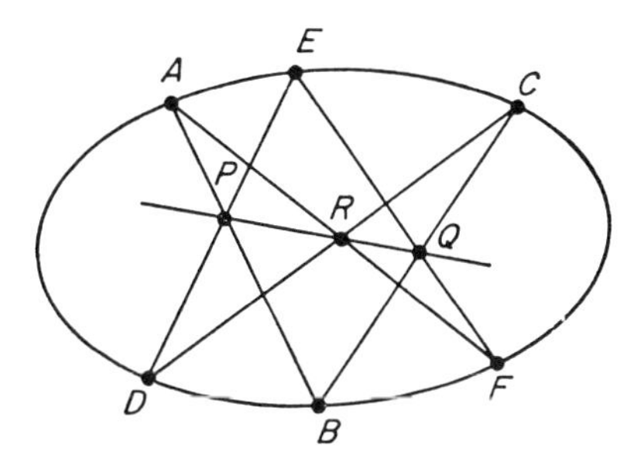

Fig. 17.3

circle, and $ABCDEF$ are six points on S, then $A_1 = e_1(A)$, B_1, C_1, D_1, E_1, F_1 are six points on $e_3(S)$ in $\mathscr{P}(\mathbb{R})$, so by Pascal's theorem, the intersections P_1 of A_1B_1, D_1E_1; Q_1 of B_1C_1, E_1F_1; R_1 of C_1D_1, F_1A_1 are collinear on some line l_1. If l_1 is not l_∞, then P_1, Q_1, R_1 are the images by e_1 of the points P, Q, R in $\mathbb{R}^2$, where P is the intersection of AB, DE, etc. Since $l_1 \neq l_\infty$, $l_1 = e_2(l)$ for some line l in $\mathbb{R}^2$. Hence P, Q, R are collinear on l, because e preserves incidence. If $l_1 = l_\infty$, then P, Q, R do not exist, so $AB \| DE$, etc. If P, say, does not exist but Q, R do, then l exists and is parallel to AB and DE. The theorem of Pappus is proved similarly. ■

If we try to obtain a dual of Pascal's theorem by applying the reciprocal duality of 17.5.7, we meet some difficulties. To resolve them, we first say that a line l is a **tangent** to a conic S at P if $l \cap S$ is the singleton $\{P\}$. [If S were not a conic, this definition would be unsuitable; for the reasons behind it, see Exercise 5(vii) below.] We can now state the following theorem, which explains what happens when we apply the duality of 17.5.7 to the conic S of 17.7.2. For a proof, see any book on projective geometry (e.g., O'Hara and Ward [96]), and Exercise 5(iv) below.

17.7.3 THEOREM. *The set of all points P on S is mapped by d_1 onto the set T of all tangents to a conic S' in $\mathscr{P}(\mathbb{R})$.* [*The equation of S' need not concern us here.*] *Moreover, $d_2(T) = S$.* ■

We can now use this to apply reciprocal duality to Pascal's theorem, thereby obtaining the 'theorem of Brianchon' (which was first proved directly, about 100 years before it was seen to be the dual of Pascal's theorem):

17.7.4 (**Brianchon's theorem.**) *Let a, b, c, d, e, f, be six tangents to a conic in $\mathscr{P}(\mathbb{R})$. Let ab, etc., denote the intersection of a and b, and let l, m, n respectively denote the joins of ab, de; bc, ef; cd, fa. Then l, m, n are concurrent.* [Cf. Exercise 3(iii), last sentence.] ■

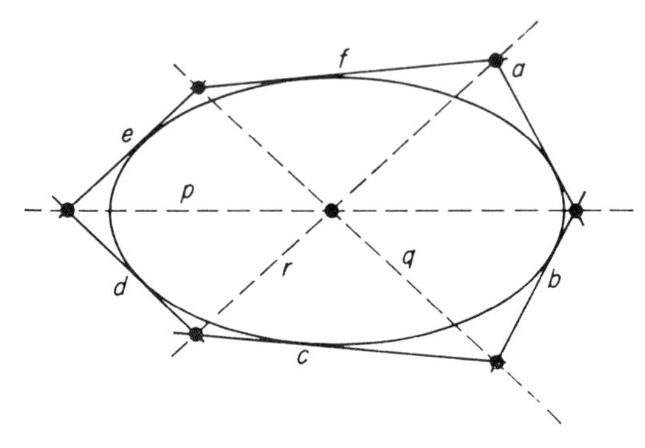

Fig. 17.4

Exercise 5

(i) Prove for yourself the summary of results immediately following 17.7.1.
(ii) Prove that a plane section of a cone in $\mathbb{R}^3$ is one of the curves mentioned after 17.7.1. [See Sommerville [121].]
(iii) Prove that, if the conic 17.7.2 passes through the vertices of the triangle of reference and $[(1, 1, 1)]$, then its equation is of the form $fy \cdot (z - x) + gx \cdot (z - y) = 0$. Hence prove that there is exactly one such conic which passes through a given fifth point not collinear with the previous four. Use this to prove that, in any $\mathscr{P}(\mathbf{F})$, there is exactly one conic through a given five points (no three of which are collinear). Now use 17.7.3 to prove that, given five lines in $\mathscr{P}(\mathbf{F})$, no three of which are concurrent, there exists a unique conic to which the lines are tangent.
(iv) Show that the conic $S: y^2 = xz$ in $\mathscr{P}(\mathbf{F})$ passes through the vertices X, Z of the triangle of reference, and tangents at X and Z meet in Y. Every point on S is of the form $[(1, t, t^2)]$ except Z; show that the tangent at such a point is $xt^2 - 2yt + z = 0$, and hence that the line $lx + my + nz = 0$ is a tangent to S iff $m^2 = 4ln$. Thus the set of tangents to S constitutes the 'line conic' $m^2 = 4ln$. Now use a collineation of the sort 17.5.4 to prove Theorem 17.7.3 (provided $\#\mathbf{F} \geqslant 4$), proving that such a collineation maps conics and tangents to conics and tangents.
(v) Hence show that two conics S, T in $\mathscr{P}(\mathbf{F})$ intersect in at most four points. If $S \cap T$ has just four points, and U is any conic through them such that $U \neq T$, prove that there is exactly one $t \in \mathbf{F}$ such that U has equation $A + tB = 0$, where the equations of S, T are $A(x, y) = 0$, $B(x, y) = 0$, respectively.

(vi) Given the matrix A as in 17.7.1, define a function $v: \mathbf{F}^3 \times \mathbf{F}^3 \to \mathbf{F}$ by

$$v(r, r') = axx' + byy' + czz' + f\cdot(yz' + y'z) + g\cdot(zx' + xz') + h\cdot(xy' + yx').$$

To remind us of A, it is convenient to write rAr' for $v(r, r')$. Prove that this 'multiplication' of r by r' is commutative and distributive, and associative with respect to scalars; i.e., that $rAr' = r'Ar$ and $rA\cdot(sp + tg) = s\cdot rAp + t\cdot rAq$. Show that the conic 17.7.2 can be written: $rAr = 0$. (Matrix multiplication may be used instead.)

(vii) Show that the line PQ cuts S (in 17.7.2) in $[\lambda p + \mu q]$ iff

$$\lambda^2 pAp + 2\lambda\mu pAq + \mu^2 qAq = 0.$$

Hence prove that a line cuts a conic (in $\mathscr{P}(\mathbf{F})$) in two, one, or no points, and that the single point occurs iff the roots of the last equation coincide, i.e., iff

$$(pAq)^2 = (pAp)(qAq).$$

[In $\mathscr{P}(\mathbb{R})$, we can think of this case as the limit, when the two distinct roots ultimately coincide; the line would then be called the 'tangent', and this name is taken over for a general field $\mathbf{F}$, since the case of coincident roots makes sense without any ideas of limits.]

(viii) Given a line l and four points, none on l and no three collinear, prove by using (v) and (vii) that there are at most two conics through the four points and tangent to l; and that when the field $\mathbf{F}$ is that of complex numbers, two conics will occur. [This example shows a fallacy in one 'solution' to (iii). For, it is sometimes argued that as the conic S in 17.7.2 is determined by the five ratios $a:b:c:d:e:f$, then S 'has five degrees of freedom, so it is uniquely determined by five conditions'. The trouble lies in not defining 'degree of freedom' or 'condition'.]

(ix) Suppose P, Q are two points on the conic S. Let the tangents to S at P, Q intersect in V. Prove that the equation of PQ is $vAr = 0$. Prove that, if a line through V cuts S in L and M, then the tangents at L, M intersect on PQ. For any point $[\mathbf{w}]$, the line with equation $wAr = 0$ is called the **polar** of $[w]$ with respect to S; $[\mathbf{w}]$ is called its **pole**. Show that every line has a unique pole. [N.B. In Fig. 17.5, V need not be 'outside' S as shown; for S need have no 'outside'.] In (iv), show that $y = 0$ is the polar of Y.

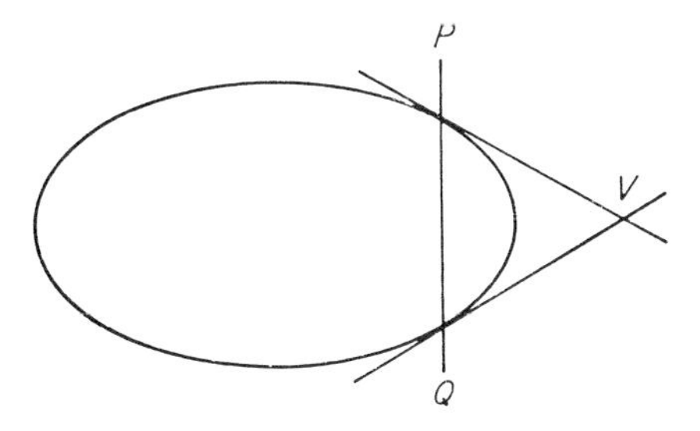

Fig. 17.5

(x) Define a new kind of duality $\not{p}: \mathscr{P}(\mathbf{F}) \to \mathscr{P}^*(\mathbf{F})$ by assigning to each point P its polar $p_1(P)$ with respect to S, and to each line l its pole $p_2(l)$. Prove that incidence is preserved, that p_1, p_2 are one-one, and that $\not{p}$ is an isomorphism. Hence show that f in 17.5.7 is '$\not{p}$' when S is taken to be the conic $x^2 + y^2 + z^2 = 0$ (which may be empty—in $\mathscr{P}(\mathbb{R})$, for example).

17.8 Models of $\mathbb{RP}^2$

By this time, the reader will probably have been surprised that the discussion so far has given no idea what $\mathbb{RP}^2$ looks like. No such idea is necessary from the logical point of view of doing strict projective geometry, since $\mathbb{RP}^2$ is simply what we have defined it to be. However, to ask what $\mathbb{RP}^2$ 'looks like' is to ask for a model: we show a person a plastic or visual model when he asks what a dinosaur looks like, since we cannot show him the real thing. The first model of $\mathbb{RP}^2$ is the $\mathbb{R}^3$-model of Example 17.3.5(iii); each point P of $\mathbb{RP}^2$ corresponds to a line l through $\mathbf{0}$ in $\mathbb{R}^3$. This is still hard to 'see'.

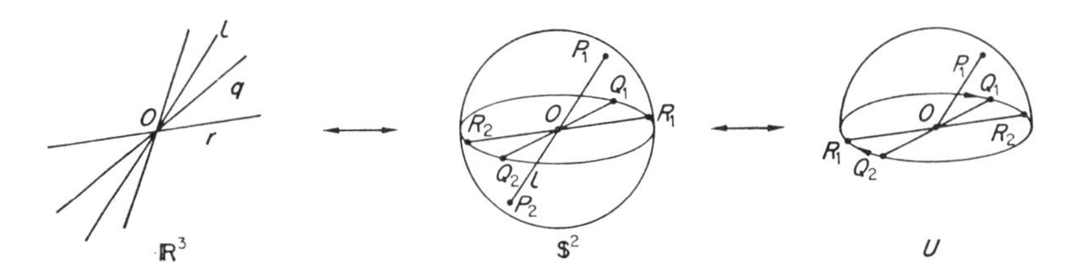

Fig. 17.6

For a second model, let $\mathbb{S}^2$ denote the unit sphere: $x^2 + y^2 + z^2 = 1$ in $\mathbb{R}^3$; then l intersects $\mathbb{S}^2$ in exactly two points (said to be 'antipodal'), so our second model of $\mathbb{RP}^2$ is the set of all antipodal pairs $\{P_1, P_2\}$ on $\mathbb{S}^2$. This is perhaps easier to imagine than the first model, but the reader probably wants to ask: why not choose just one point from each pair $\{P_1, P_2\}$? All right, but which

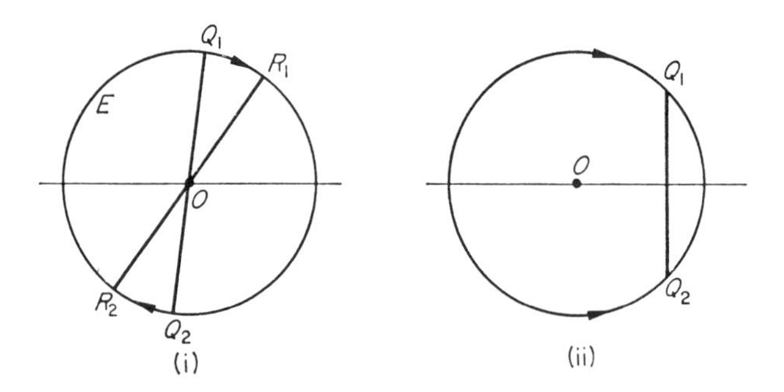

Fig. 17.7

one shall we choose, since the set $\{P_1, P_2\}$ is unordered? As an attempt, let U denote the upper hemisphere of $\mathbb{S}^2$ (i.e., $z \geqslant 0$) and choose from $\{P_1, P_2\}$ that point which lies in U. We then get a unique point except when P_1 and P_2 both lie on the equator $z = 0$ of $\mathbb{S}^2$. However, the resulting model is not too bad: at first sight it looks as though we could make it from a circular disc of paper by putting glue on the boundary and sticking together diametrically opposed points (see Fig. 17.7(i)); in a similar manner to the way we could make a ball or purse by bringing together vertically opposite boundary points like Q_1, Q_2 in Fig. 17.7(ii). Unfortunately the gluing is not possible in our three-dimensional space, as the interior of the disc keeps getting in the way. A further clarification of the model is shown, however, in Fig. 17.8:

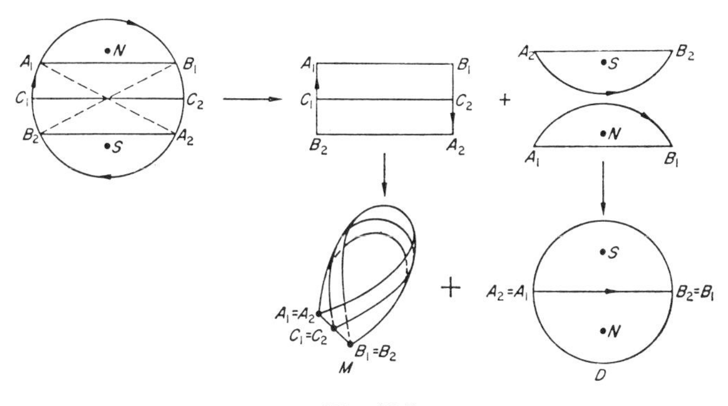

Fig. 17.8

where we have bent curved edges a little. In (iii) the two curved slices are glued along A_1B_1 and B_2A_2 (in that order) to leave a circular disc D; to glue the short edges of the rectangular strip $A_1B_1A_2B_2$ as indicated by the arrows, we must give the strip a half-twist and get what is called a **Möbius band** M, which has only one edge. Thus *our model of* $\mathbb{RP}^2$ *now consists of the union of* M *and* D, *these being united along their boundaries.* (It is this final union which cannot be performed in $\mathbb{R}^3$; see 25.12.11.)

One thing is shown clearly by the model, E, in Fig. 17.7(i). For, a line m of $\mathbb{RP}^2$ corresponds to a plane Π through $\mathbf{0}$ in the $\mathbb{R}^3$-model; and Π cuts U in a semicircle whose end-points $\{Q_1, Q_2\}$ lie on the equator, thereby becoming glued together to form a closed loop λ in E. The points of λ in E correspond to points of m in $\mathbb{RP}^2$, *so lines in* $\mathbb{RP}^2$ *correspond to loops in E.*

A further representation of $\mathbb{RP}^2$ is given in the following figure, which also shows how to cut $\mathbb{RP}^2$ neatly into triangles T_i like a polyhedron. In the construction of E the triangle T_6, for example, is glued together from the two positions shown.

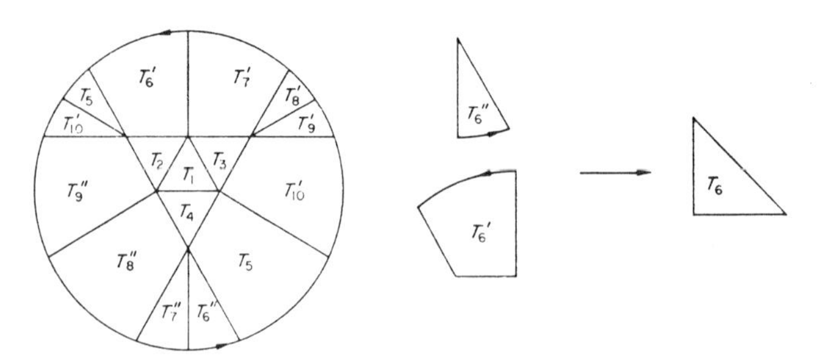

Fig. 17.9

Exercise 6

(i) Let a_0, a_1, a_2 denote the numbers of vertices, edges and faces of the 10 triangles on E (using Fig. 17.9). Prove that $a_0 - a_1 + a_2 = 1$.
(ii) To which loop in E does l_∞ correspond? Show that not all loops on E correspond to lines in $\mathbb{RP}^2$.
(iii) If S is a conic in $\mathbb{RP}^2$, to what does it correspond in E?
(iv) Prove that the centre of the disc D in Fig. 17.8 can be taken to correspond to any point $[\mathbf{a}] \in \mathbb{RP}^2$; e.g., if $\mathbf{a} = (x, y, 1)$, then D can be taken to correspond to all points $[(x', y', 1)]$ such that $(x - x')^2 + (y - y')^2 < d^2$, where d can be prescribed in advance. Consider similarly a point $[\mathbf{a}]$ of the form $[(x, y, 0)]$.
(v) Show that the triangles $T_i (i > 4)$ in Fig. 17.9 together form a Möbius band.
(vi) Show that, if we worked in four-dimensional space $\mathbb{R}^4$ instead of $\mathbb{R}^3$, then we could glue E together in the prescribed way. [Erect a cone in $\mathbb{R}^4$ on the boundary of M, analogously to the way we erect a cone in $\mathbb{R}^3$ on a circle in $\mathbb{R}^2$.]

17.9 Embedding $\mathscr{P}(\mathbb{R})$ in $\mathscr{P}(\mathbb{C})$

When we ask for intersections of curves in $\mathscr{P}(\mathbb{R})$ as in Exercise 5(v),(vii),(viii), we constantly have to consider exceptions according as the relevant equations have real roots or not. This trouble consequently disappears in $\mathscr{P}(\mathbb{C})$, because of the 'fundamental theorem of algebra' (22.5.2); and so we now embed $\mathscr{P}(\mathbb{R})$ in $\mathscr{P}(\mathbb{C})$ as follows, where the embedding simply makes explicit the notion of 'regarding every real number as complex'. Thus, to each point $[\mathbf{a}]$ of $\mathscr{P}(\mathbb{R})$ we assign the point $c_1[\mathbf{a}]$ (denoted by $[\mathbf{a}]'$ for clarity) of $\mathscr{P}(\mathbb{C})$ which is the equivalence class of all $\mathbf{b} \in \mathbb{C}^3 - \mathbf{0}$ such that: $\mathbf{b} \in [\mathbf{a}]'$ iff $\exists z \in \mathbb{C}$ such that $\mathbf{b} = z\mathbf{a}$. Assign to the line $l\colon \mathbf{p}\cdot\mathbf{r} = 0$ in $\mathscr{P}(\mathbb{R})$ the line $c_2(l)$ with the same equation, in $\mathscr{P}(\mathbb{C})$. Clearly, incidence is preserved, and c_1, c_2 are one-one functions; so we have an *embedding*

17.9.1 $$c\colon \mathscr{P}(\mathbb{R}) \to \mathscr{P}(\mathbb{C}).$$

Observe that the line $c_2(l)$ is a set containing the image $c_1(l)$, as well as many more points which, however, are not in $c_1(\mathbb{RP}^2)$. Similarly, if to each conic

S in $\mathscr{P}(\mathbb{R})$ we assign the conic $c_3(S)$ in $\mathscr{P}(\mathbb{C})$ with the same equation, then incidence is preserved, and $c_3(S)$ is a set containing $c_1(S)$ as well as other points not in $c_1(\mathbb{RP}^2)$. The line $z = 0$ in $\mathscr{P}(\mathbb{C})$ will now be denoted by l_∞; it contains the 'line at infinity' of $\mathscr{P}(\mathbb{R})$.

It is of particular interest to compose c with the embedding e of 17.6.1 and to consider the embedding

17.9.2 $$c \circ e = s: \mathscr{E}(\mathbb{R}^2) \to \mathscr{P}(\mathbb{C})$$

(including also the conics). For example, let K be the circle $x^2 + y^2 + 2gx + 2fy + c = 0$ in $\mathbb{R}^2$, so that $s_3(K) = c_3 \circ e_3(K)$ is the conic $x^2 + y^2 + 2gxz + 2fyz + cz^2 = 0$ in $\mathscr{P}(\mathbb{C})$. The latter cuts $z = 0$ where $x^2 + y^2 = 0$, i.e., in the points

$$I = [(1, \mathrm{i}, 0)]', \qquad J = [(1, -\mathrm{i}, 0)]' \qquad (\mathrm{i}^2 = -1);$$

these are called the '**circular points at infinity**', and are independent of K. [This statement is often expressed as 'all circles intersect in I, J'; but this is meaningless, since a circle is in $\mathbb{R}^2$ and I, J are not. Stated properly, the true result is less surprising, since $s_3(K)$ is a much bigger set than K.] More generally, if S is the conic of 17.7.1, in $\mathbb{R}^2$, then $s_3(S)$ cuts $l_\infty: z = 0$ where

17.9.3 $$ax^2 + 2hxy + by^2 = 0.$$

Thus S is a circle iff $s_3(S) \cap l_\infty = \{I, J\}$ (for then $a = b$, $h = 0$). Moreover, by applying to 17.9.3 the criteria following 17.7.1, S is a parabola if $s_3(S) \cap l_\infty$ is a single point P (when also l_∞ is a tangent to $s_3(S)$ at P); similarly when S is an ellipse or hyperbola (see Exercise 7(iii),(iv)). Observe that a conic in $\mathscr{P}(\mathbb{C})$ may cut l_∞ in I, J and yet need not be $s_3(S)$ for any S in $\mathbb{R}^2$; e.g., $x^2 + y^2 + \mathrm{i}z^2 = 0$. This point is sometimes overlooked in traditional texts.

As an application let us prove:

17.9.4 THEOREM. *Let P be a parabola in $\mathbb{R}^2$, with focus S, and let ABC be the vertices of the triangle formed by three tangents to P. Then the circumcircle of ABC passes through S.*

Proof. By a change of co-ordinates in $\mathbb{R}^2$, if necessary, we can assume that the parabola has equation $y^2 = 4ax$, with focus S at $(a, 0)$. Now use the embedding s, of 17.9.2. It is easily verified that in $\mathscr{P}(\mathbb{C})$ the lines p, q joining $s_1(S)$ to I, J are tangents to $s_3(P)$; while l_∞ is a tangent to $s_3(P)$, by the remarks above. Hence, if t_1, t_2, t_3 are the given three tangents to P, we have six tangents—p, q, l_∞, and $s_2(t_i)$ $(1 \leqslant i \leqslant 3)$—to $s_3(P)$. A standard theorem in any $\mathscr{P}(\mathbf{F})$ states that if the sides of a hexagon touch a conic, then the vertices lie on another conic; so some conic T in $\mathscr{P}(\mathbb{C})$ passes through the circular points I, J, and $s_1(A)$, $s_1(B)$, $s_1(C)$, $s_1(S)$. But the first five points lie on $s_3(K)$, where K is the circumcircle of ABC. Hence T and $s_3(K)$ have five common points, so by Exercise 5(iii) they are equal. Thus $s_1(S) \in s_3(K)$, so $S \in K$, since s is an embedding. ■

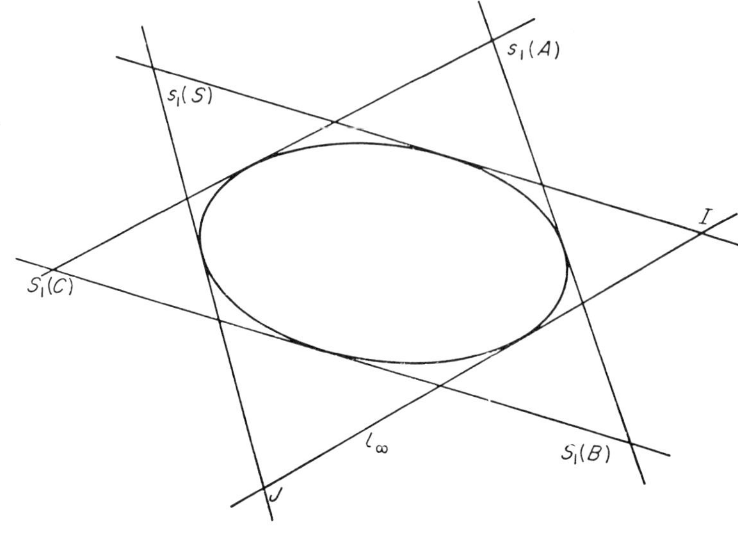

Fig. 17.10

For other applications, see Durell [34]; in particular, the notion of perpendicularity in $\mathbb{R}^2$ has to be translated into terms of $\mathscr{P}(\mathbb{C})$. Sometimes the process can be reversed, and a complicated configuration in $\mathscr{P}(\mathbb{C})$ can be analysed by showing it to be the image of a simple configuration in $\mathbb{R}^2$. But care is necessary here since the maps s_i are not onto; however, a result might be suggested and then a correct proof can be given.

Exercise 7

(i) Try to prove the 'standard theorem' quoted in the last proof. Verify also the stated remark that the lines p, q are tangents to $s_3(P)$.
(ii) In $\mathbb{R}^2$, prove that the four circumcircles of the triangles formed by four lines taken in threes, meet in a point. (No three of the lines are to concur.) [Use Exercise 5(iii) to find a parabola touching the lines, and then use 17.9.4.]
(iii) If H is a hyperbola in $\mathbb{R}^2$, use 17.9.3 to prove that in $\mathscr{P}(\mathbb{C})$, $s_3(H) \cap l_\infty$ is a pair of distinct points $c_1(e_3(H) \cap m)$, where m is $z = 0$ in $\mathscr{P}(\mathbb{R})$.
(iv) If E is an ellipse in $\mathbb{R}^2$, use 17.9.3 to prove that in $\mathscr{P}(\mathbb{C})$, $s_3(E) \cap l_\infty$ is a pair of distinct points not in $c_1(\mathbb{RP}^2)$ and so not in $s_1(\mathbb{R}^2)$.
(v) Let C be the centre of a circle K in $\mathbb{R}^2$. Prove that, in $\mathscr{P}(\mathbb{C})$, l_∞ is the polar of C, so C is the pole of l_∞. If K' is a circle also with centre C, and l a line through C, describe the configuration in $\mathscr{P}(\mathbb{C})$ formed by embedding C, K, K', l and the tangents to K, K' at the intersections of l with K, K'. Use your description to guess a theorem about tangents at the ends of a chord, in $\mathscr{P}(\mathbb{C})$. Does the theorem follow directly from your knowledge of concentric circles? [This is one way in which examiners discover Scholarship questions; set your own paper by this means!]

17.10 Projection in $\mathbb{R}^3$

The collineation in 17.5.4 has its origin in a problem concerning a pair of planes Π, Π' in $\mathbb{R}^3$. Thus, suppose the planes intersect in a line l, and let N be a point not on the planes. If $P \in \Pi$, then NP meets Π' in a unique point, P', except when NP is parallel to Π'. This exception occurs when P lies on

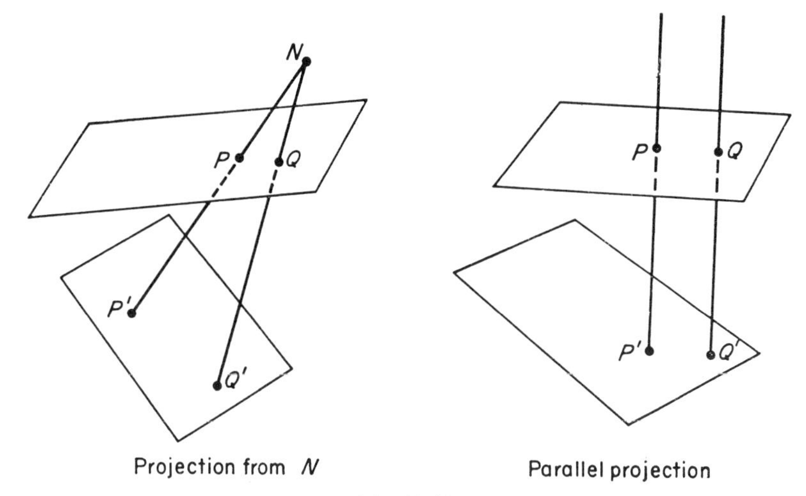

Fig. 17.11

the line $m = \Pi' \cap \Pi''$ where Π'' is the plane through N parallel to Π'. The correspondence $P \to P'$ is then a function, $p: \Pi - m \to \Pi'$, called a '**perspectivity, centre** N'; and it arises in practical problems of drawing, since lines in Π which intersect on m project into non-intersecting (hence parallel) lines in Π'. We therefore say that 'p sends m to infinity'. To simplify and clarify the construction, we embed $\mathbb{R}^3$ in the **real projective space** $\mathbb{RP}^3$ as follows. A point $[\mathbf{a}]$ of $\mathbb{RP}^3$ is defined to be an equivalence class of points $\mathbf{a} \in \mathbb{R}^4 - \mathbf{0}$, where $\mathbf{b} \in [\mathbf{a}]$ iff $\exists t \in \mathbb{R}$ such that $t \neq 0$ and $\mathbf{b} = \mathbf{a}t$. The function $q: \mathbb{R}^3 \to \mathbb{RP}^3$, defined by $q(x, y, z) = [(x, y, z, 1)]$ is one-one, and is the required embedding of points. A **line** AB in $\mathbb{RP}^3$ is defined to be the set of all points of the form $[t\mathbf{a} + s\mathbf{b}]$ where $s, t \in \mathbb{R}$; and a **plane** in $\mathbb{RP}^3$ is defined to be the set of all points $[(x, y, z, w)]$ satisfying an equation of the form $ax + by + cz + dw = 0$. We extend the embedding q to lines and planes of $\mathbb{R}^3$, by assigning to each line AB and plane $\mathbf{p} \cdot \mathbf{r} = c$ of $\mathbb{R}^3$, the line $q_1(l)$ in $\mathbb{RP}^3$ joining qA to qB, and the plane $q_2(\Pi): p_1x + p_2y + p_3z - cw = 0$, respectively. The functions q_1, q_2 are easily verified to be one-one, and incidence is preserved. The above projection process can now be carried out in $\mathbb{RP}^3$, and here there are no exceptions, because when NP is parallel to Π' in $\mathbb{R}^3$, $q_1(NP)$ meets $q_2(\Pi')$ on the 'plane at infinity' $\Pi_\infty: w = 0$ in $\mathbb{RP}^3$. Hence we obtain a function $g: q_2(\Pi) \to q_2(\Pi')$ which has all the properties of a collineation, and which agrees with p on the subset $q(\Pi - m) \subseteq q_2(\Pi)$ in the sense that $g(q(P)) = q(p(P))$ for each $P \in \Pi - m$. Moreover, $g(q_1(m))$ is the line $q_2(\Pi) \cap \Pi_\infty$. Now $q_1(NP)$ is the line joining qN to qP; and in $\mathbb{RP}^3$, we can perform the same construction with any point M replacing qN. In particular, take M on the plane at infinity, so that the lines joining M to qP are the images by q_1 of *parallel* lines in $\mathbb{R}^3$. This kind of correspondence is called a **parallel projection** $\Pi \to \Pi'$ in the **direction** common to all the parallel lines.

If in $\mathbb{R}^3$ we compose a finite sequence of parallel projections and/or perspectivities, with different centres or directions and starting with plane Π_0 and finishing with plane Π_n, then we obtain a function with range Π_n and domain a portion of Π_0; it is called a **projectivity** from Π_0 to Π_n. Analogously, we may define a projectivity: $\Pi \to \Pi'$ between planes of $\mathbb{RP}^3$, and it can be verified that the collineation f of 17.5.4 is a projectivity $\Pi_\infty \to \Pi_\infty$ when Π_∞ is $w = 0$, identified with $\mathbb{RP}^2$; see Exercise 8(iv) below.

Exercise 8

(i) In Exercise 3(iii), project UV to infinity, and then prove Pappus's theorem for the resulting simplified figure in $\mathbb{R}^2$. Hence deduce the theorem in $\mathbb{R}^2$ for the general case.
(ii) Use the method of (i) to prove the $\mathbb{R}^2$-form of Desargues' theorem (see also Courant–Robbins [25]).
(iii) Let Π be a plane in $\mathbb{RP}^3$. Prove that the set of all projectivities $\Pi \to \Pi$ forms a group, when 'projectivity' is defined by analogy with the definition for $\mathbb{R}^3$.
(iv) Show that the correspondence $[x, y, z] \to [x, y, z, 0]$ induces an incidence-preserving embedding of $\mathbb{RP}^2$ in $\mathbb{RP}^3$ with image the plane $w = 0$.

17.11 Invariants: The Erlanger Program

Granted the idea of a projectivity in $\mathbb{R}^3$, a new question arises: given sets $U, V \subseteq \Pi_0$, how can we tell whether or not there is in $\mathbb{R}^3$ a projectivity from

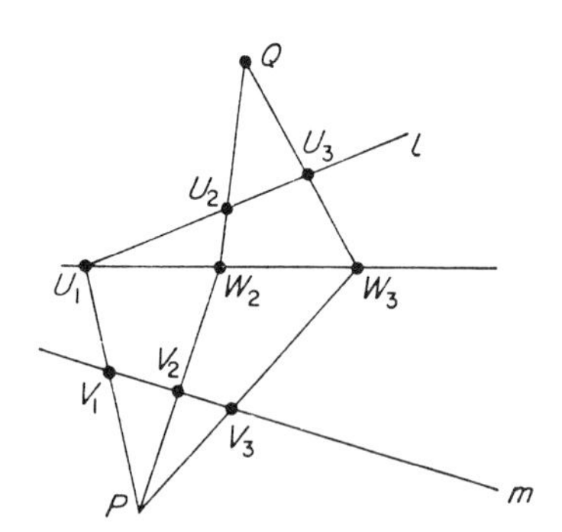

Fig. 17.12

Π_0 to Π_0 which maps U onto V? It is instructive to consider the problem one dimension lower, when U, V lie on lines l, m, respectively, of $\mathbb{R}^2$, and we look for a projectivity $f: l \to m$ constructed in $\mathbb{R}^2$ by analogy with the constructions stated above for $\mathbb{R}^3$. For example, if U, V each consist of one point or two points, a projection obviously always exists. It exists, less obviously, if each of U, V is a collinear set of three points in $\mathbb{R}^2$ (see Fig. 17.12; for, choose P in $\mathbb{R}^2 - (l \cup m)$ and project V_1, V_2, V_3 to U_1, W_2, W_3 on a third line, through U_1; then project W_2, W_3 onto U_2, U_3 as shown (perhaps W_2U_2 may be parallel to W_3U_3).

But, for collinear *quadruples*, the problem is more complicated. To cope with it, we need to define the **cross-ratio** of four collinear points in $\mathbb{R}^2$:

7.11.1 DEFINITION. $\mathrm{R}_x(P_1P_2P_3P_4) = |P_1P_2| \cdot |P_3P_4| / |P_1P_4| \cdot |P_3P_2|$ (where $|PQ|$ here is $-|QP|$). Thus, for example, $\mathrm{R}_x(1234) = 1 \cdot 1/3 \cdot (-1) = -\frac{1}{3}$ on the x-axis $\mathbb{R}$.

17.11.2 THEOREM. *Let each set U, V be a collinear quadruple of points in $\mathbb{R}^2$. There exists a projectivity in $\mathbb{R}^2$ carrying U onto V iff they have equal cross-ratios.*

Proof. Observe first that parallel projection of $P_1P_2P_3P_4$ to $Q_1Q_2Q_3Q_4$ clearly preserves cross-ratios: $\mathrm{R}_x(P_1P_2P_3P_4) = \mathrm{R}_x(Q_1Q_2Q_3Q_4)$. But so also

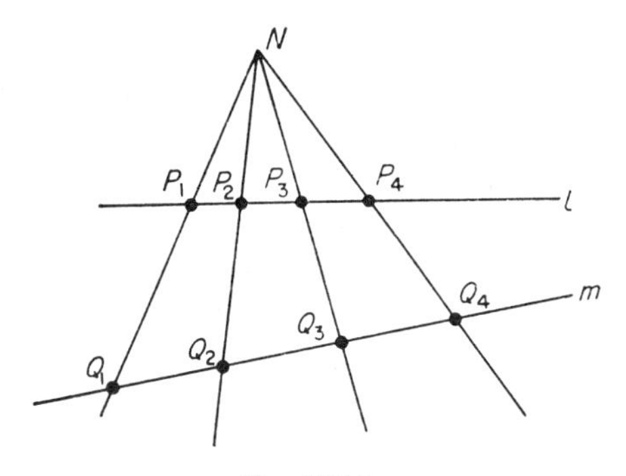

Fig. 17.13

does a perspectivity, centre N (see Fig. 17.13), since the fact that the triangles NP_1P_2, NP_3P_4, etc., all have the same height implies that

$$\mathrm{R}_x(P_1P_2P_3P_4) = \sin P_1NP_2 \cdot \sin P_3NP_4 / \sin P_1NP_4 \cdot \sin P_3NP_2$$

(using the formula: area $= \frac{1}{2}bc \sin A = \frac{1}{2}$ base $\times$ height). The right-hand side depends only on the angles at N, not on the points P_i, so the left-hand side is $\mathrm{R}_x(Q_1Q_2Q_3Q_4)$. Hence any finite sequence of parallel projections and perspectivities preserves cross-ratio; so if a projectivity exists carrying U onto V, then the cross-ratios are equal.

Conversely, suppose that $\text{R}(P_1P_2P_3P_4) = \text{R}(Q_1Q_2Q_3Q_4)$. By the argument for Fig. 17.12, there is in $\mathbb{R}^2$ a projectivity f from l to m, such that $f(P_i) = Q_i, i = 1, 2, 3$. Let $f(P_4) = X$. Then

$$\text{R}(P_1P_2P_3P_4) = \text{R}(Q_1Q_2Q_3X),$$

by the first part of the proof, so $\text{R}(Q_1Q_2Q_3X) = \text{R}(Q_1Q_2Q_3Q_4)$. Hence $X = Q_4$, by the definition of the cross-ratio. ■

The first half of the last proof shows that the cross-ratio is *invariant under projectivity*: if two quadruples U, V have different cross-ratios, then neither can be mapped onto the other by a projectivity. This suggests that we should look for other (not necessarily numerical) invariants, of more complicated figures so that at the very least we could say that if some invariant of K is different from the corresponding item for K', then K cannot be mapped onto K' by a projectivity. Ultimately we should like to work out a 'complete' set $I(K)$ of invariants of K such that, conversely, if $I(K) = I(K')$, then K can be mapped onto K' by a projectivity. (Also, to be useful, we require that $I(K)$ be computable in a finite number of steps from a knowledge of K.)

The same point of view applies in projective geometry also. Thus, to get an analogue of Theorem 17.11.2 in projective geometry, we need some analogue of the cross-ratio which does not use the lengths required in Definition 17.11.1. A clue is provided when we take P_i in 17.11.1 in the vector form $\mathbf{p} + \lambda_i\mathbf{t}$, where l is the line $\mathbf{r} = \mathbf{p} + \lambda\mathbf{t}$ of 14.3.3. For then we easily compute

17.11.3 $$\text{R}(P_1P_2P_3P_4) = (\lambda_2 - \lambda_1)(\lambda_4 - \lambda_3)/(\lambda_4 - \lambda_1)(\lambda_2 - \lambda_3);$$

it is assumed that the points are all distinct, so that the denominator is not zero. Recall now Proposition 17.5.1 in projective geometry; if l is a line in $\mathscr{P}(\mathbf{F})$ for some PPG with co-ordinates in a field $\mathbf{F}$, then each $P_i \in l$ is of the form $[\lambda_i\mathbf{a} + \mu_i\mathbf{b}]$ with $\mathbf{a}$, $\mathbf{b}$ fixed in $\mathbf{F}^3$ and $[\mathbf{a}], [\mathbf{b}] \in l$. Thus it is a small step from 17.11.3 to consider as a candidate for the cross-ratio

17.11.4 $$\text{R}(P_1P_2P_3P_4) = \frac{(\lambda_2\mu_1 - \lambda_1\mu_2)(\lambda_4\mu_3 - \lambda_3\mu_4)}{(\lambda_4\mu_1 - \lambda_1\mu_4)(\lambda_2\mu_3 - \lambda_3\mu_2)},$$

where the products $\lambda_i\mu_j$ occur rather than the ratios λ_i/μ_i to avoid zero denominators. We leave the reader to verify (using Exercise 3(iv)) that the right-hand side of 17.11.4 is independent of the choice of $\mathbf{a}$, $\mathbf{b}$ from $[\mathbf{a}], [\mathbf{b}]$, and also of the choice of $[\mathbf{a}], [\mathbf{b}]$ from l. Secondly we show:

17.11.5 PROPOSITION. *The cross-ratio* R *in* 17.11.4 *is invariant under a projectivity in* $\mathscr{P}(\mathbf{F})$.

Proof. In the notation of Fig. 17.13 let $P = [\mathbf{p}] \in l$, and let m be the line $\mathbf{m}\cdot\mathbf{r} = 0$. Then the line NP has equation $(\mathbf{n} \wedge \mathbf{p})\cdot\mathbf{r} = 0$ and intersects m in $Q = f(P)$ where $f: l \to m$ is a bijection and $Q = [\mathbf{m}\cdot\mathbf{n} \wedge \mathbf{p}]$. Now $P = [\lambda\mathbf{a} + \mu\mathbf{b}]$, so by the linearity of the products, $Q = [\lambda\mathbf{a}' + \mu\mathbf{b}']$ where $\mathbf{a}' = \mathbf{m}\cdot\mathbf{n} \wedge \mathbf{a}$ and $\mathbf{b}' = \mathbf{m}\cdot\mathbf{n} \wedge \mathbf{b}$. We may express this briefly by saying that 'f preserves the parameters (λ, μ)', in the sense that if we choose these special triples $\mathbf{a}'$, $\mathbf{b}'$ to represent $f(A), f(B)$ respectively, then $Q_i = f(P_i) = [\lambda_i \mathbf{a}' + \mu_i \mathbf{b}']$. Therefore we may use the same parameters (λ_i, μ_i) to compute $\text{R}\!\!\!/(Q_1Q_2Q_3Q_4)$ according to 17.11.4, and it equals $\text{R}\!\!\!/(P_1P_2P_3P_4)$. Hence a single perspectivity f preserves $\text{R}\!\!\!/$, and therefore so does a composition of perspectivities (possibly from different centres). A projectivity in $\mathscr{P}(\mathbf{F})$ is by definition such a composition, whence $\text{R}\!\!\!/$ is an invariant as asserted. ■

We now have what we need, to infer immediately

17.11.6 COROLLARY. *The necessary and sufficient condition in* 17.11.2 *holds also in* $\mathbb{RP}^2$. ■

Exercise 9

(i) Get the 'feel' of $\text{R}\!\!\!/$ by feeding some numbers into 17.11.3 and 17.11.4, and computing. Also prove that if in projective geometry, $\text{R}\!\!\!/(P_1P_2P_3A) = \text{R}\!\!\!/(P_1P_2P_3B)$, then $A = B$. [The latter fact is needed in the proof of 17.11.6, corresponding to the analogous step in proving Theorem 17.11.2.]
(ii) Using the embedding e of $\mathbb{R}^2$ in $\mathbb{RP}^2$ (see 17.6.1), let $Q_i = e_1 P_i$ where P_i $(1 \leqslant i \leqslant 4)$ are the points of $\mathbb{R}^2$ in 17.11.3. Show that e preserves cross-ratios in the sense that $\text{R}\!\!\!/(P_1P_2P_3P_4) = \text{R}\!\!\!/(Q_1Q_2Q_3Q_4)$. {Hint: use the form 14.3.2 of equation of a line, rather than 14.3.3.}
(iii) In (ii) let Q_∞ denote the intersection of $e_2(l)$ with l_∞. Show that P_2 is the mid-point of P_1P_3 iff $\text{R}\!\!\!/(Q_1Q_2Q_3Q_\infty) = -1$. Hence formulate and prove a theorem of projective geometry which (using e) implies that the medians of a triangle in $\mathbb{R}^2$ are concurrent.
(iv) Let $g: l \to m$ be a bijection between two projective lines (considered as sets of points) and suppose that g preserves parameters, in the sense used in the proof of Proposition 17.11.5. Show that g is a projectivity in $\mathscr{P}(\mathbf{F})$. State and prove the dual.
(v) Let P, Q be distinct points on the conic $S: xz = y^2$ in $\mathscr{P}(\mathbf{F})$. Define a bijection $h: L_P \to L_Q$ between the sets of all lines through P and through Q as follows: if $l \in L_P$, then l cuts S again in some unique point R, and $h(l)$ is defined to be the line QR. Show that h preserves parameters, and hence that h is a 'line projectivity' dual to that in (iv). What is $h(PQ)$?
(vi) Conversely to (v), suppose that there is a line projectivity $k: L_P \to L_Q$. Prove that the locus of all intersection points $l \cap k(l)$ is a conic S as l runs through L_P (where S is a single line iff k is a perspectivity). [This fact, and its dual, form the basis of a deeper treatment of conics, and it can be carried out 'synthetically', i.e., in projective geometries of the general kind 17.3.1, where co-ordinates may not be available. These facts, too, give short proofs of Pascal's theorem and the 'standard theorem' quoted in the proof of Theorem 17.9.4.]

17.11.7 The general problem, mentioned after Theorem 17.11.2, makes sense also when we replace 'projection' by some other type of transformation like 'rigid motion' (15 Exercise 7(xv)) or 'affine transformation' (15 Exercise 5(iv)), and if we replace $\mathbb{R}^2$ by $\mathbb{RP}^2$ or $\mathbb{RP}^3$ and 'projectivity' by 'collineation'.

For example, if U is a pair of points $\mathbf{a}, \mathbf{b} \in \mathbb{R}^2$, then the distance $|\mathbf{a} - \mathbf{b}|$ is an invariant under any rigid motion T, because $|\mathbf{a} - \mathbf{b}| = |T\mathbf{a} - T\mathbf{b}|$. Moreover, this one number constitutes a complete set of invariants of U under rigid motions. Again if U is a triangle, then the congruence theorems of Euclid effectively assert that the set of lengths of the sides of U is a complete set of invariants (for rigid motions). But even with rigid motions, U need not be very complicated before the problem of describing a complete set of invariants of U becomes difficult or impossible.

It was F. Klein (1849–1925) who first realized that it was the generality and depth of this problem, which was *the* characteristic of geometry. In his 'Erlanger Program' he pointed out that the different branches of geometry were distinguished by the kinds of bijective transformations they allowed—rigid motions in Euclidean geometry, collineations in projective geometry, and so on—but the fundamental question in them all was the same: find complete sets of invariants. This search was Klein's 'programme'. Moreover, the allowed bijective transformations usually form a group [the projections in $\mathbb{R}^3$ may not, but extended to collineations as in 17.10 they can be embedded in one]. Thus Klein's general formulation of a 'geometry' was the study of a problem of the form:

17.11.8 Given a set T, subset T_0, and a subgroup G of the group of all bijections $T \to T$, find a complete set S of properties ('G-invariants') such that

(i) *T_0 possesses each property in S, as also does $g(T_0)$ for every $g \in G$;*
(ii) *If T_1 is any subset of T possessing all the properties in S, then $\exists g \in G$ such that $g(T_1) = T_0$.*

Strictly, this defines the 'geometry with respect to G, of T_0 in T'; for example, when T, T_0, G are respectively $\mathbb{R}^2$, a triangle, and the group of translations of $\mathbb{R}^2$. More generally, one usually allows several sets $T_0, T_1, \ldots$ at once, as with triangles and rectangles in $\mathbb{R}^2$. The reader is invited to show that, in 17.11.8, we get the same geometry if T_0 is replaced by $g(T_0)$, where $g \in G$.

In practice, it is the possible 'properties' (vaguely defined here) which influence the choice of G, and much of the particular geometry studies relationships between properties, rather than the fundamental problem posed above. For example, Pythagoras's theorem gives a relation of this kind, and does not help with the fundamental problem. However, from Klein's point of view, this theorem is a pretty pebble gathered on the road to the final goal. Moreover, his view was so general that it helped to break down barriers between subjects, so that techniques successful in one were boldly applied to another—notably by Hilbert, who introduced powerful geometrical techniques into the calculus of variations and the theory of integral equations. And this point of view dominates modern mathematics, particularly through the 'geometry' called topology (see Chapter 25).

Exercise 10

Is there a projectivity (a) $\mathbb{R} \to \mathbb{R}$, (b) $\mathbb{R}^2 \to \mathbb{R}^2$, carrying the integers 0, 1, 2, 3 on the x-axis to the integers 3, 0, 1, -1?

(ii) Show that, given any two lines in $\mathscr{P}(\mathbf{F})$ there is a collineation carrying one onto the other, and similarly for two conics. Is there a collineation carrying a given line onto a given conic?

(iii) Prove that, given two ellipses in $\mathbb{R}^2$, there is an affine transformation, but not necessarily a rigid motion, carrying one onto the other, but no mapping of either kind will carry a given ellipse onto a given hyperbola. Consider similar problems for parabolae, circles, and pairs of straight lines.

(iv) In the notation of 17.11.8, introduce an equivalence relation R into the family $\mathfrak{p}T$ by: T_1RT_2 if $\exists g \in G$ such that $g(T_1) = T_2$. Prove that R is an equivalence relation (since G is a group). [T_1, T_2 are then said to be G-**equivalent**. A generalization of 17.11.8 is to determine the nature of T/R.] Prove that a circle and an ellipse in $\mathbb{R}^2$ are 'affine equivalent' (i.e., G-equivalent when G is the group of affine transformations).

[The last part of (iv) can be restated as: assign the various kinds of conic in 17.7.1 to their affine-equivalence classes.]

Part V
ALGEBRA

It is not easy, and should not be easy, to say where arithmetic ends and algebra begins. The content of this part draws heavily on the concepts and content of Part III, but the sort of questions we ask—and answer in most cases—are not so directly inspired by elementary arithmetic. Thus we were very largely concerned in Part III with divisibility questions in the integers and in systems similar to the integers, that is, integral domains.

In this part we again consider systems similar to the integers, but our interest changes. We consider various algebraic systems, each inspired by some familiar arithmetic object, but now we consider the members of a particular system, say, the system of groups, and consider how different members of this system may resemble or differ from each other. Thus, we are concerned with a study of the classification of different groups in terms of the structure which groups carry. We are also concerned with transformations of groups; these transformations will only be regarded as admissible (to group theory) if they transport group structure. (Chapter 18 is in fact devoted to group theory.)

Similar remarks apply to vector spaces and inner product spaces, the subject matter of Chapters 19 and 20, where the inspiration comes also from the geometry of $\mathbb{R}^3$. It is somewhat of an innovation of this book to include this material for students at the level of a first-year university course, but we believe that the material is not only interesting and important but furnishes an introduction for the student to the axiomatic method, perhaps the most powerful and certainly the most characteristic of modern mathematics. It is certainly not too soon for the student to get an understanding of what mathematics is as well as of what mathematics does.

Chapter 21 introduces a further element of structure, namely that of an order relation which is required to be compatible with the algebraic operations present within the system. A special feature of this chapter is a discussion of Dedekind's definition of the real numbers and a brief treatment of Boolean algebras. The reader may well prefer to postpone his study of the former; but we would not recommend him to delay coming to grips with the latter. For—again, apart from their intrinsic interest and importance—they carry the important message that *algebra is not necessarily about numbers.* Boolean algebra is important; moreover the axioms of a Boolean algebra are easily understood and have the valuable feature, from the teaching point of view, that they are violated by any system incorporating the arithmetic of the integers.

In the final chapter of this part we give a precise definition of a polynomial, a concept which has, of course, figured prominently in previous chapters and, indeed, in previous parts. We discuss solutions of polynomial equations without any attempt at completeness of exposition but, primarily, in order to show the importance of precise formulations and to set the main problems in evidence.

Chapter 18
GROUPS

18.1 Notion of a Group

We have already, in Section 9.5, met the notion of an *abelian* (or *commutative*) *group* with examples. Although this chapter is self-contained, the reader will find it helpful to read 9.5 again. The basic notion which we discuss in this chapter is that of a *group*, and an *abelian group* is then just a group in which the group operation is written as addition and is commutative (see $\mathfrak{G}_4$ below). In fact, the notion of a group plays a fundamental role in algebra and geometry and an important role in the calculus and mathematical analysis. For example, the theory of groups virtually arose (in the hands of Galois, Abel, and others) out of the attempt to understand why equations of degree $\leqslant 4$ could be solved by standard methods, whereas those methods failed for equations of degree $\geqslant 5$. For the history of this successful attack on a very natural and classical problem the reader is referred to one of E. T. Bell's books ([10] is probably the most attractive reference); for an account of the Galois theory of equations, the reader may consult Postnikov [103], the relevant chapter of Birkhoff–MacLane [16], or the more sophisticated text [6] by Artin.

As remarked in 17.11, the notion of a group was pointed out by Felix Klein to be central to geometry, since it enabled one to describe quite precisely what was meant by *a geometry*. Thus we may consider the co-ordinate plane $\mathbb{R}^2$ of all ordered pairs of real numbers (x, y). Now it may be shown that for any set X, the set Perm X of all bijective transformations $X \to X$ forms a group under the composition of transformations (see Example 18.2.2(xiii) below). Thus we get a geometry, according to Klein, by choosing a subgroup G of the group of self-equivalences of $\mathbb{R}^2$. If F is a geometrical figure, i.e., a subset of $\mathbb{R}^2$, then a property of F is an invariant of the geometry specified by G if it is also a property of every image $g(F)$ of F as g runs through the elements of G. Thus if we take G to be the group of 'Euclidean' motions generated by all translations, rotations and reflections, the resulting geometry is the Euclidean geometry of the plane. For a further discussion of this aspect of group theory, see Bell [12].

We will demonstrate the applicability of elementary group theory to the theory of linear differential equations in the course of Chapter 32. We hope that this brief preliminary propaganda will encourage the reader to study this section and that, after doing so, he will be convinced of the importance of the concept of a group in mathematics and ready and anxious to explore it further. If so, we recommend Ledermann [80]. If not, we recommend that he reconsider his relationship to mathematics.

18.2 Definition of a Group

We will first repeat the definition of a group, with a slight change of notation from that used in Section 9.5. We say that a set G together with a binary operation† $\circ$ forms a *group* if it satisfies the following axioms:

$\mathfrak{G}_1$: (associative law): $a \circ (b \circ c) = (a \circ b) \circ c$, *for all* $a, b, c \in G$;
$\mathfrak{G}_2$: (existence of neutral element): $\exists$ *an element* $e \in G$ *such that* $e \circ a = a \circ e = a$ *for all* $a \in G$;
$\mathfrak{G}_3$: (existence of inverses): *To each* $a \in G$, $\exists$ *an element* $\bar{a} \in G$ *such that* $a \circ \bar{a} = \bar{a} \circ a = e$.

18.2.1 REMARK. It is sufficient to demand in $\mathfrak{G}_2$ that $a \circ e = a$ and in $\mathfrak{G}_3$ that $a \circ \bar{a} = e$; one may then deduce, using $\mathfrak{G}_1$, that $e \circ a = a$ and $\bar{a} \circ a = e$. This is significant because we do *not* insist that $a \circ b = b \circ a$ for all $a, b \in G$.

Exercise 1

Show that e is uniquely determined; and that $\bar{a}$ is uniquely determined by a; and that $\bar{e} = e$, $\bar{\bar{a}} = a$, $\overline{a \circ b} = \bar{b} \circ \bar{a}$. Show also that if $ag = a$ for some element a, then $g = e$. {Compare 3.6}.

We now give a list of examples of the group concept. The reader should verify that each of the given sets, together with the prescribed operation, does form a group.

18.2.2 EXAMPLE.

(i) $G = \mathbb{Z}$, $\circ = +$ (then $e = 0$, $\bar{a} = -a$);
(ii) $G = m\mathbb{Z}$, $\circ = +$;
(iii) $G = \mathbb{Z}_m$, $\circ = +$ ($e = [0]$, $\overline{[a]} = [-a]$;
(iv) $G = \mathbb{Q}$, $\circ = +$;
(v) $G = \mathbb{Z}_p - [0]$, $\circ = \times$, p prime (here $e = [1]$, and we use Euler's Theorem (10.3.1) to take $\overline{[a]} = [a]^{p-2}$).
(vi) $G = \mathbb{Q} - \{0\}$, $\circ = \times$ $\left(e = 1, \text{ the inverse of } \frac{p}{q} \text{ is } \frac{q}{p}\right)$;
(vii) $G = \mathbb{R}$, $\circ = +$;
(viii) $G = \mathbb{R} - \{0\}$, $\circ = \times$;
(ix) $G = \mathbb{C}$, $\circ = +$;
(x) $G = \mathbb{C} - \{0\}$, $\circ = \times$;
(xi) $G = \mathbb{R}[x]$, $\circ = +$ (see Example 9.2.4);
(xii) $G =$ set of Euclidean transformations (i.e., rigid motions) in $\mathbb{R}^3$ transforming a given square into itself, $\circ =$ composition;

† I.e., a function $\circ: G \times G \to G$; but we write $g_1 \circ g_2$ instead of $\circ(g_1, g_2)$.

(xiii) $G = S_n$, the set of permutations of the set $(1, 2, \ldots, n)$, $\circ$ = composition (see 18.2.3 below). S_n is called the **symmetric group on n symbols.** More generally, we defined in Chapter 8 the set Perm A of all bijections $A \to A$, for any set A. Then Perm A is a group with respect to composition; for, if $f, g \in$ Perm A, then $f \circ g \in$ Perm A, so Perm A is *closed* under composition. The associativity of functions (2.7.6) implies that Perm A satisfies $\mathfrak{G}_1$; the neutral element required by $\mathfrak{G}_2$ is id: $A \to A$ (by 2.7.3); and the inversion theorem 2.9.1 shows that the group inverse of f is f^{-1}: $A \to A$, still in Perm A; so $\mathfrak{G}_3$ holds, whence (Perm A, $\circ$) is a group.

Examples (i)–(xi) above are distinguished from the last two inasmuch as they are examples of **commutative groups**; that is, groups $(G, \circ)$ such that $\mathfrak{G}_4$ holds.

$\mathfrak{G}_4$: *For all $a, b \in G$, $a \circ b = b \circ a$.*

As remarked in 9.5, such groups are also called **abelian**, after the great Norwegian mathematician Abel, but recall that this terminology is normally only employed if the group operation is written as addition. We will demonstrate below that example (xii) is non-commutative and example (xiii) is non-commutative, provided $n > 2$. Examples (i), (ii), (iv), (vi)–(xi) in 18.2.2 are distinguished from the others inasmuch as the sets G are in these cases infinite, whereas in the other cases they are finite. In general the number of elements in a group is called the **order** of the group. In the case of a finite group we may write down the **group table** which records all the values of the binary operation $\circ$ on elements of $G \times G$. Such a group table for the group $(\mathbb{Z}_5, +)$ appears in Example 10.1.5; others appear below.

If the group operation is written as $+$, we call the group *additive*; and we call the group *multiplicative* if it is written as $\times$ or $\cdot$ or simply by juxtaposition (that is, we omit the symbol for the operation and denote the effect of the operation on the ordered pair (a, b) simply by ab). We may also say that the group operation is *written additively* or *multiplicatively*; of course it is no intrinsic property of a group that the group operation is written additively or multiplicatively—it is simply a question of the choice of symbol to represent the group operation. It is naturally customary, if additive notation is used, to use 0 for the neutral element and $-a$ for $\bar{a}$; similarly, if multiplicative notation is used, it is customary to use 1 for the neutral element (though e is sometimes retained) and a^{-1} for $\bar{a}$. Recall the convention that 1 is used for Id_X, which is the neutral element of the group Perm X. The reader should not, of course, confuse 1 and e here with the numbers 1 and e. The group G of example (v) is usually called the multiplicative group of $\mathbb{Z}_p$; notice that it has $(p - 1)$ elements. The group table for the multiplicative group of $\mathbb{Z}_5$ is displayed in Example 10.1.5; strictly speaking the group table is obtained by suppressing the first row and column of the multiplication table given there.

Examples (v), (vi), (viii) and (x) of 18.2.2 are all subsets of fields, and each is the *multiplicative group* of the corresponding field; see Prop. 9.6.2.

18.2.3 EXAMPLE. We give the group table for S_3. (See Example 18.2.2(xiii).) Let us regard S_3 as the group of permutations of the numbers 1, 2, 3 and denote an element of S_3 by its effect on the ordered triple (1, 2, 3). Thus the six elements of S_3 are:

$$(1\ 2\ 3),\quad (1\ 3\ 2),\quad (2\ 1\ 3),\quad (2\ 3\ 1),\quad (3\ 1\ 2),\quad (3\ 2\ 1).$$

Write these elements as $e, s_1, s_2, s_3, s_4, s_5$. Then, for example, $s_1 = (1, 3, 2)$ means that $s_1(1) = 1$, $s_1(2) = 3$, $s_1(3) = 2$; and it is now plain that e is the neutral element of S_3. The group table is then:

	e	s_1	s_2	s_3	s_4	s_5
e	e	s_1	s_2	s_3	s_4	s_5
s_1	s_1	e	s_3	s_2	s_5	s_4
s_2	s_2	s_4	e	s_5	s_1	s_3
s_3	s_3	s_5	s_1	s_4	e	s_2
s_4	s_4	s_2	s_5	e	s_3	s_1
s_5	s_5	s_3	s_4	s_1	s_2	e

Here the appearance, for example, of s_1 in the row labelled by s_3 and the column labelled by s_2 indicates that if we carry out the permutation s_3 and *follow* it by the permutation s_2 we obtain the permutation $s_1 = s_2 \circ s_3$.

One may immediately recognize whether a finite group is commutative by looking at its table—the group is commutative if and only if the table is symmetric about its leading diagonal. Thus S_3 is *not* commutative ($s_1 s_2 \neq s_2 s_1$)

Exercise 2

(i) *Deduce* that S_n is not commutative for $n \geqslant 3$.
(ii) Prove that, in S_3, $s_3^3 = e$ and $s_3^2 = s_4$. What other elements have 'square roots' in S_3?
(iii) On the set S_3, define a new operation $*$ by the rule: $a * b = b \circ a$. Prove that $(S_3, *)$ is a group, whose multiplication table is obtained from that above by reflecting in the leading diagonal $e \cdots e$. [The *leading diagonal* goes from top left to bottom right.] If we reflect in the diagonal $s_5 \cdots s_5$, do we still get a group?

18.2.4 EXAMPLE. We give the table for the group of Example 18.2.2(xii). Let us denote the vertices of the square by A, B, C, D. Plainly any Euclidean (i.e., rigid) motion of $\mathbb{R}^3$ which maps the square onto itself effects a permutation of the vertices; and a given permutation can be achieved by at most one motion. [This is intuitively obvious; but its proof requires some linear algebra.] Thus it remains to list those permutations which are so achieved.

We can achieve any *cyclic* permutation of A, B, C, D; thus

$$(A\ B\ C\ D),\quad (B\ C\ D\ A),\quad (C\ D\ A\ B),\quad (D\ A\ B\ C).$$

We can then carry out a reflection in one of the diagonals (say top left to bottom right); thus we achieve

$$(C\ B\ A\ D),\quad (D\ C\ B\ A),\quad (A\ D\ C\ B),\quad (B\ A\ D\ C).$$

We certainly cannot achieve a permutation which simply interchanges two adjacent vertices; and from this it is an easy deduction that the list of eight

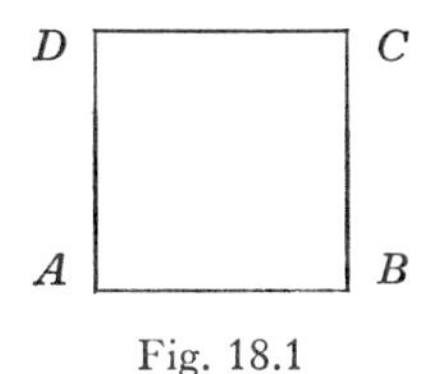

Fig. 18.1

elements above is the complete list of group elements. Thus the group is of order 8; denoting its elements (in the order above) by e, t_1, t_2, ..., t_7 we find that the group table is

	e	t_1	t_2	t_3	t_4	t_5	t_6	t_7
e	e	t_1	t_2	t_3	t_4	t_5	t_6	t_7
t_1	t_1	t_2	t_3	e	t_7	t_4	t_5	t_6
t_2	t_2	t_3	e	t_1	t_6	t_7	t_4	t_5
t_3	t_3	e	t_1	t_2	t_5	t_6	t_7	t_4
t_4	t_4	t_5	t_6	t_7	e	t_1	t_2	t_3
t_5	t_5	t_6	t_7	t_4	t_3	e	t_1	t_2
t_6	t_6	t_7	t_4	t_5	t_2	t_3	e	t_1
t_7	t_7	t_4	t_5	t_6	t_1	t_2	t_3	e

This group is non-commutative (since, e.g., $t_1t_4 \neq t_4t_1$). The reader is advised to study this example carefully and to check the group table experimentally.

Exercise 3

(i) Show that the elements e, t_1, t_2, t_3 form a commutative group and interpret this fact.

(ii) To find the inverse of, say, t_1, we look down the 't_1' column to find e; hence $t_1 = t_1^{-1}$. Find the inverses of all other elements.

11

We now give some of the basic facts and concepts of group theory. Naturally we cannot aim at completeness. [See Ledermann [80] and the texts there listed for further information.] Our first result gives an alternative characterization of a group and also introduces in a systematic way the idea of an *equation*. (Compare Proposition 9.5.5.)

18.2.5 THEOREM. *Let* $(G, \circ)$ *be a non-empty set with an associative binary operation* $\circ$. *Then* G *is a group if and only if the equations* $a \circ x = b$, $y \circ a = b$ *have a solution in* G *for each* $a, b \in G$.

(Notice that we say 'G is a group'; this is a standard abbreviation which will henceforth be adopted. We will also allow ourselves to write a group *multiplicatively* where convenient.)

Proof. Let G be a group. Then $a(a^{-1}b) = (aa^{-1})b = eb = b$, so that the equation $ax = b$ has a solution. Similarly, $y = ba^{-1}$ is a solution of the equation $ya = b$.

Conversely we suppose that the equations $ax = b$, $ya = b$ have a solution in G for each $a, b \in G$. Thus in particular the equation $ax = a$ has a solution, say $x = e_a$. We prove that e_a is a neutral element for G. For any $b \in G$ there exists $y \in G$ with $ya = b$. Thus $be_a = (ya)e_a = y(ae_a) = ya = b$. Hence $e = e_a$ is a right neutral element in G (i.e., $be = b$ for all $b \in G$). Similarly there exists a left neutral element e' in G. But then

$$\begin{aligned} e'e &= e' && \text{(because } e \text{ is a right neutral element)} \\ &= e && \text{(because } e' \text{ is a left neutral element).} \end{aligned}$$

Thus $e = e'$ is a neutral element in G and axiom $\mathfrak{G}_2$ is satisfied. Axiom $\mathfrak{G}_3$ is also satisfied, since there are elements a', a'' such that $a'a = e$, $aa'' = e$. But then $a' = a'e = a'(aa'') = (a'a)a'' = ea'' = a''$, so that $a' = a'' = a^{-1}$ is (the) inverse of a in G. ∎

In the light of Remark 18.2.1 the proof above is unnecessarily elaborate. For it is sufficient to show that the solvability of the equations $ax = b$, $ya = b$ guarantees the existence of a *right* neutral element and a *right* inverse.

Exercise 4

(i) Show that the equations $ax = b$, $ya = b$ have a *unique* solution in the group G. Solve the equation $axb = c$ in G.
(ii) Solve the equation $xt_1x = t_3$ in the group of Example 18.2.4. {Use the multiplication table.}
(iii) Solve the equation $x^2s_3 = s_4$ in the group S_3 of Example 18.2.3.

18.3 Indices; Subgroups

Consider an element a belonging to a group G. Using multiplicative notation it is natural to write a^2 for aa and, in general, a^n for the product of n

copies of the element a. Similarly one writes a^{-n} for $(a^{-1})^n$ and the usual **laws of indices** hold in G. That is, with the convention $a^0 = e$,

18.3.1 $$a^m a^n = a^{m+n}, \qquad (a^m)^n = a^{mn}, \qquad (a \in G;\ m, n \in \mathbb{Z}).$$

The proof of 18.3.1, as indeed the definition of a^n, $n \in \mathbb{N}$, is by induction, and the reader may care to provide it in detail to acquire practice. Observe that if we put $n = -1$ in the second equation of 18.3.1, then we get $(a^m)^{-1} = a^{-m}$.

Now let H be a non-empty subset of G with the following property.

$\mathfrak{P}$: *If $a, b \in H$, then $ab^{-1} \in H$.*

The reader will readily verify that H is then a group, and conversely, that if H is a subset of G which is a group with respect to the binary operation $\circ$ defining the group G, then H has property $\mathfrak{P}$. We call such a subset H a **subgroup** of G. For example, G and the singleton $\{e\}$ are subgroups of G. The following proposition is evident.

18.3.2 PROPOSITION. *Let G be a group and let $a \in G$. Then the collection of all powers of a forms a subgroup of G.* ■

We call this subgroup the subgroup **generated by** a. If this subgroup is the whole of G, then G is said to be a **cyclic** group generated by a. Of course if G is generated by a, it is also generated by a^{-1}. The reader should now consider these remarks in 'additive' notation. For example, in place of $a^2 = aa$ we have $2a = a + a$, and the second equation of 18.3.1 becomes $n(ma) = mna$. And then the subgroup generated by a is the set

$$\{0, \pm a, \pm 2a, \ldots\};$$

in particular, $\mathbb{Z}$ is cyclic, generated by $a = 1$ (or by $a = -1$).

Exercise 5

(i) State which of the groups of Example 18.2.2 are cyclic and indicate their generators.

(ii) Prove the remark above, that the singleton $\{e\}$ is a subgroup of G, where e is the neutral element.

Now let G be a cyclic group generated by a. Then G is finite or infinite. If G is finite, not all the positive powers of a can be distinct. Suppose then that $e, a, \ldots, a^{n-1}$ are all distinct but that a^n is equal to one of them, say $a^n = a^q$ for some q with $0 \leqslant q \leqslant n - 1$. Then $a^{n-q} = e$; but since $e, a, \ldots, a^{n-1}$ are all distinct we must have $q = 0$, $a^n = e$. We now observe that every element of G must be equal to one of the elements $e, a, \ldots, a^{n-1}$. For an arbitrary element of G has the form a^k, where k is an integer. Now by the Euclidean property (11.1.5) of $\mathbb{Z}$ we may write $k = qn + r$ where $0 \leqslant r \leqslant n - 1$. Then, using the laws of indices (18.3.1),

$$a^k = a^{qn+r} = (a^n)^q a^r = e \cdot a^r = a^r.$$

It therefore follows that G is *of order* n; we also say that the element a *has order* n. This latter terminology carries over to elements of arbitrary groups, so that the **order** of an element is just the order of the subgroup it generates; if this order is finite, the element is called **periodic.** We also note from the calculation above that $a^k = e$ if and only if $n|k$, where n is the order of a; this gives us an alternative characterization of the order. Finally, we observe that our argument also shows us that if G is cyclic infinite, generated by a, then all the positive powers of a must be distinct; indeed, *all* the powers of a are distinct, as the reader will readily prove.

Exercise 6

Let A be an *abelian* group. Prove that the set of all periodic elements of A forms a subgroup of A.

18.4 Generators of a Group

We now generalize the notion of the generator of a cyclic group. We first draw the reader's attention to the following lemma, whose proof he should supply (see Prop. 9.4.2).

18.4.1 LEMMA. *Let H_α, $\alpha \in A$, be a family of subsets of the group G, indexed by the set A; and let $H = \bigcap_{\alpha \in A} H_\alpha$. Then if each H_α is a subgroup of G, so is H.* ■

That is, the intersection of subgroups is a subgroup.

Now let S be a subset of G and let H_S be the intersection of the subgroups of G which contain S. Then H_S is a subgroup of G containing S and it is plainly the smallest subgroup of G containing S, in the sense that any other subgroup of G containing S also contains H_S. We call H_S the subgroup of G **generated** by S, and S is a set of **generators** of G if $H_S = G$.

Exercise 7

Verify that if $S = \{a\}$, then H_S is just the cyclic subgroup of G generated by a.

We now give a more constructive interpretation of H_S. For simplicity we confine attention to the case in which S is a finite set, say $S = \{a_1, a_2, \ldots, a_n\}$, $a_i \in G$. Now consider the set T of all elements which are products of the form

18.4.2 $$a_{i_1}^{r_1} \cdots a_{i_k}^{r_k}$$

in G where $1 \leqslant i_j \leqslant n$ (and, of course, the r_j are arbitrary integers). *Then T is a subgroup.* For it is certainly non-empty (and even if S were empty we would conventionally include e in T as the unique element 18.4.2 with $k = 0$). Moreover, if $g \in T$, then by Exercise 1 and induction on k, $g^{-1} = a_{i_k}^{-r_k} \cdots a_{i_1}^{-r_1}$, so $g^{-1} \in T$; hence if also $h \in T$, then hg^{-1} is still a product of the form 18.4.2, therefore in T. Thus T has property $\mathfrak{P}$ of 18.3, and so T is a subgroup of G.

Now let H be any subgroup of G containing S. Then it is easy to show that every element 18.4.2 belongs to H. Thus $T \subseteq H$, and this shows that $T = H_S$. Thus we have proved:

18.4.3 THEOREM. *Let S be a subset of G. Then the subgroup generated by S is the subgroup of G consisting of all elements of the form* 18.4.2. ■

Notice that if G is not commutative, then there may well be repetitions of the suffixes in the element 18.4.2. For example, the element $b^{-1}ab$ belongs to the subgroup of G generated by $\{a, b\}$. However, if G is commutative and $S = \{a_1, \ldots, a_n\}$, then an element of H_S may be written in the convenient form

18.4.4 $$a_1^{m_1} \cdots a_n^{m_n},$$

or, if additive notation is used,

18.4.5 $$m_1 a_1 + \cdots + m_n a_n.$$

The notion of a set of generators plays a fundamental role in group theory. It is not possible at this stage to indicate how fundamental (see Ledermann [80]), but we will give one simple application which will simultaneously indicate the broad scope of the theory.

18.4.6 EXAMPLE. (*Möbius Transformations.*) Consider the set M of all 'Möbius'† functions T of $z \in \mathbb{C}$, where

(i) $T(z) = (az + b)/(cz + d), \qquad (a, b, c, d \in \mathbb{C};\ ad - bc \neq 0).$

If $c = 0$, then $d \neq 0$, since the **discriminant**, $ad - bc$, is not zero; thus in this case T is a function $\mathbb{C} \to \mathbb{C}$. If on the other hand $c \neq 0$, then the denominator vanishes when $z = -d/c$, so T is a function $\mathbb{C} - \{d/c\} \to \mathbb{C}$. To avoid this distinction, we add a new point labelled '∞' and called 'infinity' to $\mathbb{C}$ and regard T as a function $\mathbb{C}_* \to \mathbb{C}_*$ (where $\mathbb{C}_* = \mathbb{C} \cup \{\infty\}$), such that

(ii) $T(\infty) = a/c, \qquad T(-d/c) = \infty, \qquad (T(\infty) = \infty \text{ if } c = 0).$

With certain evident conventions about ∞, the equations (ii) are deducible from (i), so we call (i) the 'equation of $T\colon \mathbb{C}_* \to \mathbb{C}_*$'. The set M is now taken to be the set of all such extended functions. In particular, when $c = 0 = b$, and $a = 1 = d$, then T is $\mathrm{id}\colon \mathbb{C}_* \to \mathbb{C}_*$. Also, in the general case, T is a bijection. For if we solve the equation $v = (au + b)/(cu + d)$ when $cu + d \neq 0$, we get $u = (-du + b)/(cv - a)$, which suggests that $T^{-1}(z) = (-dz + b)/(cz - a)$, $(T^{-1}(\infty) = -d/c$, $T^{-1}(a/c) = \infty)$ ought to be the equation of the inverse of T; plainly this is the case, and $T^{-1} \in M$ since $(-d)(-a) - bc \neq 0$.

† After A. F. Möbius (1790–1868), a German mathematician and astronomer.

We assert that M is a group under the operation of composition of functions. If S, $T \in M$, we must first prove that $S \circ T \in M$. Let T be as above and let S have equations

$$S(z) = (pz + q)/(rz + s), \qquad ps - qr \neq 0$$
$$\text{(so that } S(\infty) = p/r, \qquad S(-s/r) = \infty).$$

Then

$$\text{(iii)} \quad S \circ T(z) = \left(p\left(\frac{az + b}{cz + d}\right) + q\right)\Big/\left(r\left(\frac{az + b}{cz + d}\right) + s\right)$$
$$= ((ap + cq)z + bp + dq)/((ar + cs)z + br + ds)$$

and the discriminant of $S \circ T$ is

$$(ap + cq)(br + ds) - (ar + cs)(bp + dq) = (ad - bc)(ps - rq) \neq 0,$$

since S, T each have non-zero discriminant. We check that all is well 'at infinity', so that the function in M, whose equation is given by (iii), is verified to be $S \circ T$, and M is closed under composition. We now verify that $(M, \circ)$ satisfies the group axioms. $\mathfrak{G}_1$ (associativity) holds, since composition of functions is associative. We observed that $\text{id}: \mathbb{C}_* \to \mathbb{C}_*$ is in M, so $\mathfrak{G}_2$ holds; while the existence of inverses was shown above, whence $\mathfrak{G}_3$ holds, and $(M, \circ)$ is a group.

We now proceed to look for a set of generators of M. We observe that, provided $c \neq 0$, we may write T as

$$T = T_4 \circ T_3 \circ T_2 \circ T_1,$$

where

$$T_1(z) = z + \frac{d}{c},$$

$$T_2(z) = \frac{1}{z}$$

$$T_3(z) = \frac{bc - ad}{c^2} z,$$

$$T_4(z) = z + \frac{a}{c};$$

while, if $c = 0$,

$$T = U_2 \circ U_1,$$

where

$$U_1(z) = z + \frac{b}{a},$$

$$U_2(z) = \frac{a}{d} z.$$

(Notice that if $c = 0$, then neither a nor d is zero, since $ad - bc \neq 0$.) From this analysis it follows that M is generated by the set Σ consisting of

$$\text{18.4.7}\begin{cases} \textbf{translations} & T(z) = z + a, \quad a \in \mathbb{C}, \\ \text{the } \textbf{reflection-inversion} & T(z) = \dfrac{1}{z}, \quad \text{and} \\ \textbf{magnifications} \text{ and } \textbf{rotations} & T(z) = kz, \quad k \in \mathbb{C}, \quad k \neq 0. \end{cases}$$

[The given transformation is a magnification by $|k|$ and a rotation through $\arg k$ (see Chapter 24).]

Consider now an application of the result that Σ generates M.

There is an important theorem which asserts that the **cross-ratio**

$$\text{R}\!\!\!/(z_1, z_2, z_3, z_4) = \frac{(z_1 - z_2)(z_3 - z_4)}{(z_1 - z_3)(z_2 - z_4)},$$

of four not necessarily collinear points, is invariant under any transformation $T \in M$, i.e., that $\text{R}\!\!\!/(z_1, z_2, z_3, z_4) = \text{R}\!\!\!/(Tz_1, Tz_2, Tz_3, Tz_4)$. Thus, in Klein's sense mentioned in 18.1, T is an element of the 'geometry of $\mathbb{R}^2$ under Möbius transformations'. We may of course prove this by 'slogging it out'; that is, we may take an arbitrary $T \in M$ and show that

$$\text{18.4.8} \qquad \frac{(z_1 - z_2)(z_3 - z_4)}{(z_1 - z_3)(z_2 - z_4)} = \frac{(Tz_1 - Tz_2)(Tz_3 - Tz_4)}{(Tz_1 - Tz_3)(Tz_2 - Tz_4)}.$$

However, it is plain that, in verifying that some property is invariant under the operations of a group G, *it is sufficient to test the invariance for each generator of some set of generators of G.* Thus using 18.4.7 we need only test 18.4.8 for

$$T(z) = z + a, \qquad T(z) = kz \quad \text{and} \quad T(z) = \frac{1}{z}.$$

In these three cases the assertion is trivially verified, and so it is true for any Möbius transformation T.

Exercise 8

(i) With T as in equation (i) of Example 18.4.6, let $f(w, z) = czw - az + dw - b$. Prove that $f(w, z) = 0$ iff $T(z) = w$. (Here w, z are to lie in $\mathbb{C}$.) Prove also that $f(w_1 + w_2, z) = f(w_1, z) + f(w_2, z)$; $f(w, z_1 + z_2) = f(w, z_1) + f(w, z_2)$. Thus f is linear in each variable, and is called a 'bilinear' function $\mathbb{C} \times \mathbb{C} \to \mathbb{C}$.
(ii) The group M in Example 18.4.6 contains the subset N of all T with a, b, c, d real. Prove that N is a subgroup of M. Given $T \in N$, prove that T maps the sets $\{y = 0\}$, $\{y \neq 0\}$, into themselves. How does T map a circle?
(iii) If, further, $L = \{T \mid T \in M \text{ and } a, b, c, d \in \mathbb{Z}\}$ prove that L is a subgroup of M. Let L_0 consist of those $T \in L$ admitting equations $T(z) = (az + b)/(cz + d)$ with $ad - bc = \pm 1$. Show that L_0 is a subgroup of L.
(iv) Let $T(z) = (az + b)/(cz + d)$ with $a, b, c, d \in \mathbb{Z}$ and $ad - bc = \pm 1$. Let $T_1(z) = z + 1$, $T_2(z) = -z$, $T_3(z) = 1/z$, and let L_1 be the subgroup of L

generated by T_1, T_2, and T_3. Show that if $a = 0$ or $c = 0$, then $T \in L_1$. Now suppose $a \neq 0$, $c \neq 0$. Prove by induction on min (a, c) that $T \in L_1$, inferring that $L_1 = L_0$. {Use the Euclidean algorithm.}
(v) Show that $T_1^2 = T_3^2 = \text{id} = T_4^3$, where $T_4(z) = 1/(1 - z)$ and express T_4 in terms of T_1, T_2, T_3. $\{T^{n+1} = T^n \circ T$, by the rule 18.3.1.$\}$
(vi) Let $\mathbb{N}_n = \{1, 2, \ldots, n\}$ and let G denote the set of all functions $f\colon \mathbb{N}_n \to \mathbb{Z}$; with addition defined as in 9 Exercise 1(viii), G is an abelian group, called the **free abelian group of rank** n. Prove that G is generated by the n functions f_i, where $f_i(i) = 1$, $f_i(j) = 0$ when $i \neq j$. Prove also that $f_1, \ldots, f_n$ are *independent*, i.e., that there is no equation $\sum_{i=1}^{n} m_i f_i = 0$ with each $m_i \in \mathbb{Z}$, unless every m_i is zero.
(vii) Prove that a finitely generated abelian group is finite if each generator is periodic.

18.5 Subgroups

We now return to the question of subgroups. We have agreed that a subgroup H of a group G is a non-empty subset H of G with the property:

$\mathfrak{P}$: *If* $a, b \in H$, *then* $ab^{-1} \in H$.

This is a readily applicable definition; and it is rendered acceptable by the remark, which the reader has already been invited to verify, that a subgroup H is then no other than a subset of G which is a group with respect to the defining operation $\circ$ of G; that is, we require that $h \circ h' \in H$ if $h, h' \in H$, so that $\circ$ induces an operation on H and this shall be a group operation.

However, we can give the following modification of the criterion for a subgroup. Let $\mathfrak{Q}$ be the property:

$\mathfrak{Q}$: *If* $a, b \in H$, *then* ab *and* a^{-1} *belong to* H.

18.5.1 PROPOSITION. *A non-empty subset H of G is a subgroup if and only if it has property $\mathfrak{Q}$.*

Proof. Let H have property $\mathfrak{Q}$. Then if $b \in H$, $b^{-1} \in H$. Thus if $a, b \in H$, then $a, b^{-1} \in H$ and so $ab^{-1} \in H$. Thus H has property $\mathfrak{P}$.

Conversely, let H have property $\mathfrak{P}$. Since H is non-empty, $a \in H$ for some a in G. Thus $aa^{-1} = e \in H$. Again applying property $\mathfrak{P}$ we deduce that, if $a \in H$, then $ea^{-1} = a^{-1} \in H$. Finally if $a, b \in H$ then $a, b^{-1} \in H$ so that $a(b^{-1})^{-1} = ab \in H$, so H has property $\mathfrak{Q}$. ■

18.5.2 COROLLARY. *Let H be a finite, non-empty subset of G with the property that if $a, b \in H$, then $ab \in H$. Then H is a subgroup of G.*

Proof. We have to show that if $a \in H$, then $a^{-1} \in H$. Certainly if $a \in H$, then $a^k \in H$ for all *positive* powers of a. But since H is finite, so must the set of distinct positive powers of a be. That is, a is an element of finite order, say $a^n = e$ where we may assume $n > 1$. Then $a^{-1} = a^{n-1}$, so $a^{-1} \in H$ and so H is a subgroup. ■

Exercise 9

(i) Show by an example that we cannot dispense with the finiteness condition in Corollary 18.5.2.
(ii) We saw in Lemma 18.4.1 that the intersection of a family of subgroups is a subgroup. Show that, on the other hand, the union of a pair of subgroups is not necessarily a subgroup.
(iii) Prove that a subgroup of a subgroup B of a group C is itself a subgroup of C.
(iv) Show that the construction of the direct sum $A \oplus B$ of two abelian groups (see 9 Exercise 1(vii)) works for non-abelian groups also. We denote it then by $A \times B$. In $A \times B$, show that $xy = yx$ provided $x \in A$ and $y \in B$. {You must first show how A and B may be regarded as subgroups of $A \times B$.}
(v) Given a subset $X \subseteq \mathbb{Z}$ which is a group under the ordinary multiplication of integers, prove that the neutral element is the integer 1, and hence that X is either the singleton $\{1\}$ or the pair $\{-1, 1\}$.

18.6 Homomorphisms of Groups

It is natural that, in the theory of groups, we should wish to study functions ϕ from the group G to the group H, for any two groups G, H. The functions ϕ which are appropriate to group theory are those which 'preserve group structure' in the sets G and H; that is to say, we would surely require that

18.6.1 $$\phi(gg') = \phi(g)\phi(g'), \qquad (g, g' \in G)$$

(where the product on the right is in H), and we would also ask that ϕ map the neutral element of G to the neutral element of H and that $\phi(g^{-1}) = \phi(g)^{-1}$. We have already met such a function, viz., $\rho: \mathbb{Z} \to \mathbb{Z}_m$ in 10.1.3, which had other properties also. A function $\phi: G \to H$ satisfying 18.6.1 is called a **homomorphism**; and it is a fortunate fact that all homomorphisms behave well with respect to neutral elements and inverses: they *preserve* them:

18.6.2 PROPOSITION. *Let $\phi: G \to H$ be a homomorphism. Then*

(i) $\phi(e_G) = e_H$, *where e_G, e_H are the neutral elements of G, H;*
(ii) $\phi(g^{-1}) = (\phi(g))^{-1}$.

Proof. (i) We have, for any $g \in G$,

$$\phi(g) = \phi(ge_G) = \phi(g)\phi(e_G).$$

Thus (see Exercise 1) $\phi(e_G) = e_H$.

(ii) $e_H = \phi(e_G) = \phi(gg^{-1}) = \phi(g)\phi(g^{-1})$.

Thus

$$\phi(g^{-1}) = (\phi(g))^{-1}. \quad ■$$

Some important examples of homomorphisms are given in Exercise 10 below. The reader may be helped to appreciate this vitally important concept by seeing some very concrete examples. Thus let $\mathbb{R}_+$ be the set of strictly positive real numbers. Then

$$\log: (\mathbb{R}_+, \times) \to (\mathbb{R}, +)$$

and

$$\exp: (\mathbb{R}, +) \to (\mathbb{R}_+, \times)$$

are homomorphisms; that is, $\log ab = \log a + \log b$, $\exp(x + y) = \exp x \exp y$, for any $a, b \in \mathbb{R}_+$, $x, y \in \mathbb{R}$. (Each of these equations is of the form 18.6.1, with one side written multiplicatively, the other additively.) As a further example consider the differential operator

$$D: \mathscr{D}(I) \to \mathbb{R}^I$$

(see Chapter 29). Then D is a homomorphism of additive abelian groups. In fact, D preserves more than just group structure; each of $\mathscr{D}(I)$, $\mathbb{R}^I$ has a *vector space* structure over $\mathbb{R}$ and D is linear (see Chapters 29 and 19).

Exercise 10

(i) Let G be a group and H a subgroup of G. Let $i: H \to G$ be the inclusion map of H in G. Show that i is a homomorphism. Conversely, suppose that H is a subset of the group $(G, \cdot)$ with a binary operation $*$ such that $(H, *)$ is a group and $i: (H, *) \to (G, \cdot)$ is a homomorphism. Prove that $(H, \cdot)$ is a subgroup of G, and that $(H, \cdot) = (H, *)$.
(ii) Let $\phi: A \to B$ be a homomorphism of groups which is *constant*, i.e., $\phi(a) = \phi(a')$ for all $a, a' \in A$. Prove that $\phi(a)$ is the neutral element of B. [ϕ is then called the **trivial** homomorphism, and especially if the groups are abelian, we often write $\phi = 0$.]
(iii) A very important property of $(n \times n)$ matrices (with real coefficients, say) is that they can be multiplied, by the rule 19.4.8, and that

$$\det AB = \det A \cdot \det B,$$

where det stands for the determinant. Restricting attention to the case of non-singular matrices, express this equality in the form that a certain function of groups is a homomorphism. (A is **non-singular** if $\det A \neq 0$.)
(iv) Let n, m be positive integers such that $n|m$. Prove that in $\mathbb{Z}$, the coset $[a]_m$ lies in $[a]_n$. Show that the correspondence $[a]_m \to [a]_n$ is a function (independent of representatives) and a homomorphism $\mathbb{Z}_m \to \mathbb{Z}_n$.
(v) Let $\phi: G \to H$ be a homomorphism of groups. Prove that $\operatorname{Im}(\phi)$ is a subgroup of H. If $g \in G$, prove that $\phi(g^n) = (\phi(g))^n$, $n = 0, \pm 1, \ldots$. Hence prove that if g is periodic of order k, then $\phi(g)$ is periodic with order dividing k.
(vi) Prove that the function $g \to g^{-1}$ is a homomorphism of the group G into itself, if G is commutative. Also prove the converse.

18.7 Isomorphisms

We have met in Chapter 2 injective, surjective, and bijective functions. An injective homomorphism is called a **monomorphism** (see Exercise 10(i));

a surjective homomorphism is called an **epimorphism** (see Exercise 10(iii)); and a bijective homomorphism is called an **isomorphism.** Now given any bijective function $\phi: G \to H$ there is by the inversion theorem (2.9.1) a unique inverse $\psi: H \to G$. The reader will readily prove

18.7.1 PROPOSITION. *If $\phi: G \to H$ is an isomorphism, so is ψ.* ■

Thus, for example, $\exp: \mathbb{R} \to \mathbb{R}_+$ is the inverse of $\log: \mathbb{R}_+ \to \mathbb{R}$ and both are isomorphisms. The reader will also prove without difficulty:

18.7.2 PROPOSITION. *Let A, B, C be groups and let $\alpha: A \to B$, $\beta: B \to C$ be functions. Then $\beta \circ \alpha: A \to C$ is a homomorphism (monomorphism, epimorphism, isomorphism) if both α and β are homomorphisms (monomorphisms, epimorphisms, isomorphisms).* ■

Two groups A, B are called **isomorphic** whenever there is an isomorphism $A \to B$; and we write $A \approx B$, a relation on the class of all groups, which we see to be an equivalence relation from these last two propositions (together with the evident fact that the identity function $G \to G$ is an isomorphism). It follows that we may consider *equivalence classes of mutually isomorphic groups.* It is essentially true to say that the proper study of group theory is not so much the groups themselves as these equivalence classes. That is to say, we do not insist on distinguishing between a group G and a group H isomorphic to it, since G and H have the same algebraic structure (from the point of view of group theory) and simply differ in the actual nature of their elements. For example, $(\mathbb{R}_+, \times)$ and $(\mathbb{R}, +)$ are isomorphic groups and are thus entirely equivalent in group theory (though not with respect to other theories). Another (important!) example may help to bring this point home to the reader.

18.7.3 THEOREM. *Two cyclic groups are isomorphic* iff *they have equal orders.*

Proof. If $\phi: A \to B$ is an isomorphism, then in particular $A \approx B$, so $\#A = \#B$ (see 2.11.1), and the groups have equal orders. Conversely, suppose they have equal orders and suppose first that this order is finite, say n. Then $A = \{e_A, a, \ldots, a^{n-1}\}$, $B = \{e_B, b, \ldots, b^{n-1}\}$ and the function $f: A \to B$, given by $f(a^k) = b^k$ $(k = 0, \ldots, n-1)$ is plainly a bijection. It is also a homomorphism, since $f(a^j \cdot a^k) = f(a^{j+k}) = b^{j+k}, f(a^j) \cdot f(a^k) = b^j \cdot b^k = b^{j+k}$, so A and B are isomorphic. The proof when the order is infinite is similar (we have $A = \{e_A, a^{\pm 1}, a^{\pm 2}, \ldots\}$, $B = \{e_B, b^{\pm 1}, b^{\pm 2}, \ldots\}$ and again set $f(a^k) = b^k$, $k = 0, \pm 1, \pm 2, \ldots$). ■ Thus it is natural to talk of *the* infinite cyclic group, or *the* cyclic group of order n. The additive group of residue classes mod 4 and the multiplicative group of non-zero residue classes mod 5 are both *copies* of the cyclic group of order 4; and any algebraic statement about the one may be translated into a statement about the other; we have only to set up the necessary isomorphism, which may be done by mapping $[1]_4$ to $[2]_5$, as for f in the proof of 18.7.3 above.

Exercise 11

(i) Describe in detail the isomorphism referred to in the previous paragraph. Is this isomorphism unique?
(ii) Set up an isomorphism between the symmetric group S_3 (see Example 18.2.3) and the group of symmetries of an equilateral triangle.
§(iii) (J. W. Craggs.) Let A, B denote thin lenses, and let $C = A * B$ denote the lens formed by placing one face of A against a face of B. If C is regarded as thin, the signed focal lengths c, a, b of C, A, B satisfy $c^{-1} = a^{-1} + b^{-1}$, where $a^{-1} = 0$ if A is flat. Denote by L the set of all thin lenses and let ARB mean that A has the same focal length as B. Show that L/R forms a commutative group under the operation induced by $*$; and that this group is isomorphic to $\mathbb{R}_+$ under multiplication, using an isomorphism h, where $h(AR) = e^{1/a}$. Does $(L, *)$ form a group?
Can we work similarly with the set of *thick* lenses? How about the set of all resistances in parallel, connecting two terminals?
(iv) Let H be a subgroup of G and let $x \in G$. Show that the set of elements of the form $xhx^{-1} (h \in H)$, is a subgroup of G isomorphic to H. (We write xHx^{-1} for this subgroup and call it a **conjugate** of H in G.)
(v) If G is a group, then an isomorphism $\phi: G \to G$ is called an **automorphism** of G. Prove that the set of all automorphisms of G forms a group, Aut G, under composition; and that Aut G is a subgroup of Perm G. Show that Aut $\mathbb{Z} \approx \mathbb{Z}_2$.
(vi) Given x in the group G, define a function $h_x: G \to G$ by $h_x(g) = xgx^{-1}$ for each $g \in G$. Prove that h_x is an automorphism of G; it is called an **inner** automorphism. Now show that $h_x \circ h_y = h_{xy}$, $h_e =$ id, and $h_{x^{-1}} = (h_x)^{-1}$, and that the set of all inner automorphisms of G forms a subgroup of Aut G. Show that the function $\phi: G \to$ Aut G, given by $\phi(x) = h_x$, is a homomorphism.
(vii) Use the last part of 9 Exercise 1(viii) to prove that the free abelian group of rank n in Exercise 8(vi) is isomorphic to the direct sum $\mathbb{Z} \oplus \cdots \oplus \mathbb{Z}$ of n copies of $\mathbb{Z}$.
(viii) For any groups A, G let Hom (A, G) denote the set of homomorphisms from A to G. Thus Hom (A, G) is a subset of G^A. Show that Hom (A, G) is a sub*group* of G^A if G is commutative.

Show how a homomorphism $\phi: A \to B$ induces (by restricting ϕ^G, ϕ_G of 6.2)

$$\phi_*: \text{Hom}(G, A) \to \text{Hom}(G, B)$$
$$\phi^*: \text{Hom}(B, G) \to \text{Hom}(A, G).$$

Show that (a) ϕ_* is injective if ϕ is a monomorphism;
(b) ϕ^* is injective if ϕ is an epimorphism;
(c) ϕ_* may fail to be surjective even if ϕ is an epimorphism;
(d) ϕ^* may fail to be surjective even if ϕ is a monomorphism.

(Contrast the good behaviour of ϕ^G, ϕ_G!)
(ix) Elements of Hom (A, A) are called **endomorphisms** of A, provided A is commutative. Show that Hom (A, A) is then a ring with unity element, when multiplication is taken to be composition of functions.
(x) Prove Cayley's theorem, that every finite group G is isomorphic to a group of permutations, as follows. Given $g \in G$, define a function $\theta_g: G \to G$ by $\theta_g(x) = gx$.

Prove that θ_g is a bijection of G. Prove that $\theta_h \circ \theta_g = \theta_{hg}$, and hence that $\theta: G \to \operatorname{Perm} G$ given by $\theta(g) = \theta_g$ is a homomorphism. Finally show that θ is an injection, hence a monomorphism. Show that $\theta_g \in \operatorname{Aut} G$ iff $g = e$ and that $\theta(G) \neq \operatorname{Perm} G$ unless $G = \{e\}$.

We should not leave this matter of isomorphism of groups without repeating that we are here looking at an idea which is one of the most basic in the whole of mathematical thought. We start out with the concept of a set and we impose a pattern on that set. In the case of group theory the pattern is provided by the group operation, but there are other algebraic patterns (for example, those of rings and vector spaces) and there are non-algebraic patterns as well, very important among which is the pattern imposed by a *topology* or 'nearness' concept in the set. Thus for example the set $\mathbb{R}$ may be furnished with an addition operation, a product operation, and a *metric* or distance function. Each type of pattern automatically carries with it the notion of pattern-preserving transformation and hence of pattern-preserving equivalence. Much of mathematics consists of the search for similarity of pattern in different sets; and hence, on a more technical level, of the search for properties of 'patterned sets' which are preserved under the appropriate isomorphism. To us it does not seem fanciful to claim that this is one of the most essential ways in which men organize their experience of the external world, both in recognition and perception *and* in their art. See also Weyl [134].

18.8 Kernels and Images

Let $\phi: G \to H$ be a homomorphism. We pick out two sets K, I where $K \subseteq G$, $I \subseteq H$, as follows. The set I is just the image of ϕ, $I = \phi(G)$. The set K which is called the **kernel** of ϕ consists of all elements $g \in G$ such that $\phi(g) = e_H$, $K = \phi^\flat\{e_H\}$. We often write $I = \operatorname{Im}(\phi)$, $K = \ker(\phi)$. The reader may prove:

18.8.1 PROPOSITION. *K is a subgroup of G; I is a subgroup of H.* ■

Of course, the subset of any group G, consisting of just the neutral element alone, is a subgroup. We now draw the reader's attention to

18.8.2 THEOREM. *ϕ is a monomorphism if and only if $K = \{e_G\}$.*

Proof. If ϕ is injective and $\phi(g) = e_H$, then $\phi(g) = \phi(e_G)$ by 18.6.2(i), so $g = e_G$. Conversely, if $K = \{e_G\}$ and $\phi(g) = \phi(g')$, then by using 18.6.2(ii): $\phi(g'g^{-1}) = \phi(g')\phi(g^{-1}) = \phi(g')(\phi(g))^{-1} = e_H$, so $g'g^{-1} \in \phi^{-1}\{e_H\} = K$; and $g = g'$, since $K = \{e_G\}$. Thus ϕ is injective. ■ The importance of this theorem is that it very considerably simplifies the test whether a function ϕ is an injection in the case that ϕ is a homomorphism. It is sufficient to 'localize' the test at e_H.

We now point out a very important property of the subgroup K. We describe a subgroup N of the group G as **normal** if $g^{-1}ng \in N$ whenever $g \in G$, $n \in N$. [Current synonyms for 'normal' are 'self-conjugate' and 'invariant'.] Notice that *all subgroups of a commutative group are normal*, so that the notion of normality is only significant for non-commutative groups. In particular, $\{e\}$ and G are normal in G.

18.8.3 THEOREM. *The kernel of the homomorphism $\phi: G \to H$ is normal in G.*

Proof. Let $g \in G$, $k \in K$. Then

$$\phi(g^{-1}kg) = \phi(g^{-1})\phi(k)\phi(g) = (\phi(g))^{-1}e_H\phi(g) = (\phi(g))^{-1}\phi(g) = e_H. \quad \blacksquare$$

Notice on the other hand that $I = \phi(G)$ need not be normal in H.

Exercise 12

(i) The **centralizer** $C(x)$ of x in the group G is the set of all g such that $gx = xg$. Prove that $C(x)$ is a subgroup of G, and $x^n \in C(x)$ for all integers n. Show in Example 18.2.4 that $t_1t_4t_1^{-1} = t_5 \notin C(t_4)$, so that $C(t_4)$ is not a normal subgroup.
(ii) The **centre** C of a group G is the set of all $g \in G$ such that g commutes with *every* element of G. Prove that C is a normal subgroup of G, and that $C = \bigcap_{x \in G} C(x)$.
(iii) In Exercise 11(vi), show that the kernel of $\phi: G \to \text{Aut } G$ is the centre, C, of G. [Hence, by Theorem 18.8.3, we see again that C is normal in G.] By construction of ϕ, Im (ϕ) is the group of inner automorphisms of G. Show that this group is normal in Aut G.
(iv) Given a homomorphism $h: A \to B$, prove that Im (h) is abelian if A is; if A is generated by $a_1, \ldots, a_n$, (see 18.4.3), prove that Im (h) is generated by $h(a_1), \ldots, h(a_n)$, some of which may be equal. In particular, prove that Im (h) is cyclic if A is.
(v) A **commutator** in a group G is any element of the form $xyx^{-1}y^{-1}$, $(x, y \in G)$. In Examples 18.2.3, 18.2.4, decide which elements of the groups are commutators. Show that the set of commutators of G (which need not form a subgroup) generates a *normal* subgroup G', the **commutatorsabgroup** or **derived** group of G. Given a homomorphism $f: G \to H$, show that $f(G') = (\text{Im }(f))' \subseteq H'$. Hence prove that if H is abelian, then $G' \subseteq \ker(f)$.

Next, we proceed to explore the significance of the notion of normality. In the following discussion it may well help the reader to keep in mind the way in which the additive group $\mathbb{Z}_m$ of residue classes mod m was defined and the relation of this group to the groups $\mathbb{Z}$ and $m\mathbb{Z}$. It may also reassure the reader to know that the concepts of the coset space and the quotient group, to which he is about to be introduced, are one which university students frequently find to be extremely difficult (indeed, impossibly difficult in many cases!). We hope that the familiarity with set-theoretical ideas, which the reader now has, will have helped to eliminate the psychological blocks which stand in the path of the traditionally trained student. In particular, he should recall the content of Chapter 4.

18.9 Subgroups, Quotient Spaces, and Quotient Groups

Let G be a group and N a subgroup. We set up an equivalence relation R_N on G (said to be **induced** by N) by declaring

$$aR_Nb \quad \text{if and only if} \quad ab^{-1} \in N.$$

The reader should verify that R_N is indeed an equivalence relation (recall, in particular the case $G = \mathbb{Z}$, $N = m\mathbb{Z}$). The equivalence classes are called (left) **cosets** of N in G, and we will write $[a]_N$ for the coset containing a, perhaps abbreviating to $[a]$. [In Chapter 4, we would write aR_N.] The set of cosets is written G/N and called the (left) **coset space** of N in G [or (left) **quotient space** of G by N]. Thus if $G = \mathbb{Z}$, $N = m\mathbb{Z}$, G/N is nothing other than the set $\mathbb{Z}_m$. We point out that b belongs to the coset $[a]$ if and only if $b = na$ for some n in N. Among the cosets is $[e_G]$ and this is nothing other than N itself. One may write Na for the coset $[a]$ to emphasize that the members of $[a]$ are just the elements of G of the form na with $n \in N$. It is easily verified that the function $f: N \to Na$, given by $f(n) = na$, is an equivalence; it follows (see Chapter 7) that each coset of N in G has the same cardinal number of elements, *that number being the order of N*. Thus, the relation R_N partitions G into cosets $[a]$, the number $\#[a]$ being constant and equal to $\#N$. If G is a finite group, then obviously the number of cosets of N in G is also finite; this number is called the **index** of N in G. Hence, the argument above yields the fundamental relation expressed by:

18.9.1 THEOREM. *Let G be a finite group and N a subgroup. Then the order of G is just the product of the order of N and the index of N in G.* ■ This is known as **Lagrange's theorem**, after Lagrange (1736–1813); see Bell [10], Ch. 10.

18.9.2 COROLLARY. *The order of a subgroup exactly divides the order of the group. In particular, if G is a group and $a \in G$, then the order of a exactly divides the order of G.* ■

Up to now, no assumption has been made as to the normality of N. We now prove the crucial

18.9.3 LEMMA. *Let N be a subgroup of G and let $a_iR_Nb_i$, $a_i, b_i \in G$, $i = 1, 2$. Then if N is normal, $a_1a_2R_Nb_1b_2$.*

Proof. We know that $a_i = n_ib_i$ for some $n_i \in N$, $i = 1, 2$. Thus

$$a_1a_2 = n_1b_1n_2b_2 = n_1b_1n_2b_1^{-1}b_1b_2 = n_1(b_1n_2b_1^{-1})b_1b_2.$$

Now since N is normal in G, $b_1n_2b_1^{-1} \in N$. Thus $n_1(b_1n_2b_1^{-1}) \subset N$, so that $a_1a_2 = nb_1b_2$ for a certain $n \in N$ and hence $a_1a_2R_Nb_1b_2$. ■ The reader is advised, as an exercise, to consider the case $G = S_3$, $N =$ subgroup generated by the permutation $J = (1\ 3\ 2)$, in the notation of Example 18.2.3 [J is the permutation keeping 1 fixed and interchanging 2 and 3]. He should find that the conclusion of Lemma 18.9.3 is false in this case, so that the given subgroup is *not* normal in S_3.

It follows from Lemma 18.9.3 that if N is a *normal* subgroup of G, then we may introduce a binary operation $\circ$ into the coset space G/N by defining

18.9.4 $$[a_1] \circ [a_2] = [a_1 a_2];$$

Lemma 18.9.3 assures us that the coset on the right depends only on the *cosets* $[a_1]$ and $[a_2]$ and not on any particular choice of elements in G belonging to those cosets. As an example, this operation in the coset space $\mathbb{Z}_m = \mathbb{Z}/m\mathbb{Z}$ is just the addition given in 10.1.2 with respect to which the set $\mathbb{Z}_m$ is an additive group. More generally, we see from the next theorem that we get a group when N is normal in G, by breaking up G into 'chunks' (cosets of N in G) and multiplying the chunks according to the rule 18.9.4. For examples see Exercise 13.

18.9.5 THEOREM. *Let N be a normal subgroup of G. Then the coset space G/N becomes a group under the operation* 18.9.4. *Moreover the function $\nu: G \to G/N$, given by*

$$\nu(a) = [a], \qquad a \in G,$$

is then an epimorphism of G onto G/N with kernel N.

We leave the proof as an exercise. ■ The reader should note that we now have a converse of Theorem 18.8.3, in that every normal subgroup has been proved to be the kernel of some homomorphism. The group G/N which we have constructed is called the **quotient group** (or **factor group**) of G by N. If finite, its order is, of course, just the index of N in G.

We are now in a position to apply the diagram appearing in 4 Exercise 10(i). For, as the reader will already have observed in proving that the kernel of $\nu: G \to G/N$ is just N, the equivalence relation set up on G by ν partitions G into the cosets of N in G. Now let $\phi: G \to H$ be a homomorphism with kernel N. Then ν and ϕ induce the same partition of G, so that we have the diagram

18.9.6
$$\begin{array}{ccccc} G & \xrightarrow{\phi_0} & I & \xrightarrow{i} & H \\ & {\scriptstyle\nu}\searrow & \uparrow{\scriptstyle j} & & \\ & & G/N & & \end{array}$$

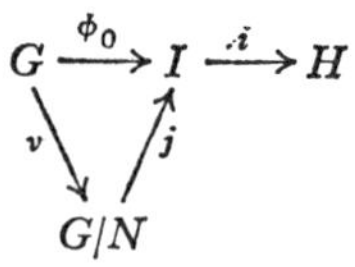

Fig. 18.2

It is plain that ϕ_0 and i are homomorphisms. We know that j is an equivalence of sets; to prove it an isomorphism we have only to show that it is a homomorphism. Now j is defined by

18.9.7 $$j[a] = \phi(a).$$

Thus the fact that j is a homomorphism is an immediate consequence of 18.9.4, together with the fact that ϕ is a homomorphism. Notice that 18.9.6 yields a 'factorization' of every homomorphism,

$$\phi = ij\nu,$$

into the composition of: a projection ν onto a quotient group, an isomorphism j, and an inclusion i. Thus $\nu\colon G \to G/N$, $j\colon G/N \approx \operatorname{Im}\phi$, $i\colon \operatorname{Im}\phi \subseteq H$, where $N = \ker\phi$.

We close this section by giving one simple application of Corollary 18.9.2. We gave (Theorem 10.3.1) a somewhat improvised proof of Euler's theorem. We now give a slick proof!

18.9.8 THEOREM. *Let a be prime to m and let ϕ be Euler's totient function. Then*

$$a^{\phi(m)} \equiv 1 \pmod m.$$

Proof. By 10.2.5 the residues prime to m form a group Φ_m of order $\phi(m)$ under multiplication. By Corollary 18.9.2 the order of $[a]$ divides $\phi(m)$ so that, using Exercise 10(v), we have

$$[a^{\phi(m)}] = [a]^{\phi(m)} = [1]. \quad \blacksquare$$

Of course, Fermat's famous theorem that $a^p \equiv a \pmod p$ for any prime p is just a special case of Euler's theorem, since $\phi(p) = p - 1$.

Exercise 13

(i) Prove that $G/\{e\} \approx G$.
(ii) Let $\phi\colon G \to H$ be a homomorphism. Show that the solution set of the equation $\phi(x) = h$ (h fixed in H, x in G) is empty or a coset of K in G where K is the kernel of ϕ.
(iii) Let $A \times B$ be the direct product of groups A, B (see Exercise 9(iv)). Prove that A is normal in $A \times B$ and that $(A \times B)/A \approx B$.
(iv) Let G be a group and $m > 0$ a fixed integer. Prove that the set X, of all mth powers of elements of G, generates a normal subgroup of G which coincides with X if G is commutative. If $G = \mathbb{Z} \times \mathbb{Z}$, prove that $G/X \approx \mathbb{Z}_m \times \mathbb{Z}_m$.
(v) Let $h\colon A \to B$ be a homomorphism. Prove that the graph H of h in $A \times B$ is a subgroup, but not in general normal. If A, B are abelian, prove that $(A \oplus B)/H \approx B$. Consider in particular the cases $h = 0$, $h = \mathrm{id}$.
(vi) Let G' denote the derived group of a group G (see Exercise 12(v)). Prove that G/G' is abelian (the **abelianizer** of G). Show that if $G' \subseteq H \subseteq G$, then H is normal and G/H is commutative. State a converse. Find the abelianizer of S_3.
(vii) In Exercise 12(iii), prove that the group $I(G)$ of inner automorphisms of G is isomorphic to G/C, where C is the centre. [Note that the centre of G/C need not consist of the neutral element alone, unlike the derived group of G/G' in (vi).]
(viii) In 15 Exercise 7(xiv), show that the set G of orthogonal linear transformations of $\mathbb{R}^2$ is a group; and that the subset H, of those $T \in G$ with $\det T = 1$, is a normal subgroup of G. Prove that $G/H \approx \mathbb{Z}_2$, and G/H is generated by the coset of the reflection in the x-axis.

(ix) Let A be an abelian group, and let P denote its subgroup of periodic elements. Prove that A/P has no non-zero periodic elements. Hence prove that if A is finitely generated, then A/P has a finite set of generators which are independent (in the sense of Exercise 8(vi)). {Consult Ledermann [80], Ch. VI.} Deduce from Exercise 11(vii) that A/P is isomorphic to the free abelian group A_n on n generators for some n, and hence that $A \approx P \oplus A/P$, a direct sum of a finite, and a free, abelian group. We call n the **rank** of A.

18.9.9 The result given in (ix) above is called the **decomposition** theorem for abelian groups. Now P can itself be decomposed into a direct sum $\mathbb{Z}_{d(1)} \oplus \mathbb{Z}_{d(2)} \oplus \cdots \oplus \mathbb{Z}_{d(k)}$ where each integer $d(i)$ divides $d(i + 1)$, $i = 1, \ldots, k - 1$. This solves the 'isomorphism problem' for finitely generated abelian groups, in the following sense. If A, B are such groups, of ranks n, m, with integers $(d(1), \ldots, d(k))$, $(e(1), \ldots, e(j))$, then $A \approx B$ if and only if $n = m$, $j = k$, $d(i) = e(i)$ $(1 \leqslant i \leqslant k)$. Moreover, it is a finite process to compute these integers, given A, B. No such solution is known for non-abelian groups, but a solution which is 'computable' is known to be impossible, in general. It is not even known how many different (non-isomorphic) groups there are, of a given order n (when n is larger than about 200). These are problems of the 'pure' theory of groups, and they generate much present-day research about the structure of special classes of groups. There is also an 'applied' theory. For example, the *crystallographic groups* yield information about the structure of crystals (see Weyl [134]) and the *knot groups* are studied in connection with the (unsolved) problem of enumerating all possible types of knot (see Crowell–Fox [26]). The latter theory is a flourishing branch of the subject called topology, and has fed ideas back into group theory; throughout topology, groups are used to express complicated geometrical problems in algebraic terms, when they often become easier to handle. And physicists study groups in connection with quantum theory and the study of fundamental particles. Lack of space prevents our giving detailed examples of these fascinating applications, since the passage from the initial problem (e.g., crystal, knot, particle) to a problem about groups needs a lengthy description.

Exercise 14

Use the decomposition theorem to prove that every finitely-generated commutative group of order >2 has a non-trivial automorphism (i.e., different from the identity automorphism). Hence prove that every finitely-generated group of order >2 has a non-trivial automorphism.

18.10 Rings

Since normal subgroups of a group G arise as kernels of homomorphisms of G, it is appropriate to compare this situation with that for rings. Thus, let $h\colon R \to S$ be a homomorphism of (not necessarily commutative) rings. If we regard R and S merely as abelian groups, then the kernel K of h is $h^{\flat}(0_S)$, an abelian subgroup of R, as we know. However, it is easily checked that K is not only a subring of R, but has the stronger property:

$$\forall r \in R, \quad \textit{and} \quad \forall k \in K, \quad \textit{both} \quad rk \quad \textit{and} \quad kr \in K.$$

Any subgroup H of R with this property is called a (two-sided) **ideal** of R. We now leave the reader to verify that the quotient R/H of R by a two-sided ideal H is a ring if multiplication of H-cosets is defined by

$$[r][s] = [rs]$$

and the function $r \to [r]$ analogous to ν in 18.9.5 is then a ring-homomorphism $R \to R/N$. Since any invertible ring-homomorphism has a ring-homomorphism as inverse [prove this!], we call it a ring-isomorphism. As an example, the reader may verify that the function analogous to j in 18.9.7, which sends $[r]$ to $h(r)$, is a ring-isomorphism $R/K \approx \operatorname{Im} h$.

Exercise 15

Let R be the ring of all 2×2 matrices with integer entries. Let H denote the subset of R consisting of all matrices whose entries are divisible by a fixed integer n. Show that H is an ideal in R, and show that R/H is isomorphic (as a ring) to the ring S of all matrices with entries in $\mathbb{Z}_n$. {Show that the homomorphism $\mathbb{Z} \to \mathbb{Z}_n$ induces an obvious homomorphism $R \to S$.}

Chapter 19

VECTOR SPACES AND LINEAR EQUATIONS

The reader will easily believe that much of the difficulty of group theory may be attributed to the fact that the group operation is not assumed to be commutative. Indeed we mentioned in 18.9.9 towards the end of Chapter 18 a result on finitely generated *abelian* groups, which gives a complete statement of the algebraic structure of such groups. No such statement is known at present for arbitrary finitely-generated groups, and the problems involved in analysing their structure are indeed formidable. In this chapter we discuss a simpler theory even than that of commutative groups, namely *vector spaces.* However, it is not just in the interests of simplicity that we introduce the notion of a vector space. It is rather because vector spaces play a decisive role in Euclidean geometry and the theory of linear equations that they figure so prominently in modern mathematics courses.

As the name 'vector space' implies, the notion is really familiar to anybody who has met 'vectors' in two- or three-dimensional Euclidean space; and the reader of Chapters 14 and 15 will find the present chapter a natural extension of the ideas introduced there. From the algebraic viewpoint there are two main features of such vectors: (a) any two may be added together to produce a third, and (b) any vector may be multiplied by a scalar, which is just a classy name for a real number, to produce a new vector. We leave on one side for the moment scalar and vector multiplication of vectors, which belong to the more sophisticated reaches of the theory.

We now introduce the abstract concept of a *vector space*: however, we hope that by following the definition with a list of examples we will give the reader the necessary confidence!

19.1 Preliminary Definitions

19.1.1 DEFINITION. Let F be a field. Then by a **vector space** V over F we mean an (additive) abelian group together with an operation† of F on V which satisfies the following axioms.

$\mathfrak{V}_1$: $x(u + v) = xu + xv$, *for all* $x \in F$, $u, v \in V$;
$\mathfrak{V}_2$: $(x + y)u = xu + yu$, *for all* $x, y \in F$, $u \in V$;
$\mathfrak{V}_3$: $(xy)u = x(yu)$, *for all* $x, y \in F$, $u \in V$;
$\mathfrak{V}_4$: $1u = u$, *for all* $u \in U$, *where* 1 *is the unity element of* F.

† That is, a function $\mu: F \times V \to V$; as usual, $\mu(x, v)$ is denoted by xv. We no longer use Clarendon type for elements of a general V.

Exercise 1

Distinguishing the zeros of F and V by 0 and $\mathbf{0}$, respectively, prove

(i) $x\mathbf{0} = \mathbf{0}$ for each $x \in F$ (put $v = \mathbf{0}$ in $\mathfrak{V}_1$);
(ii) $0u = \mathbf{0}$ for each $u \in V$ (put $y = 0$ in $\mathfrak{V}_2$);
(iii) $x(-u) = -(xu)$ (put $v = -u$ in $\mathfrak{V}_1$);
(iv) $(-x)u = -(xu)$ (put $y = -x$ in $\mathfrak{V}_2$, use (ii)).

It is necessary to use known facts about abelian groups (see 9.5.4).

19.1.2 EXAMPLE. (i) Let V be the trivial abelian group consisting of just the zero element, 0. V becomes the 'trivial vector space' under the rule: $x0 = 0$ for all $x \in F$.

(ii) Let $F = \mathbb{R}$, the field of real numbers, and let V be the set of ordered pairs (x_1, x_2) of real numbers. Then V is an abelian group under the rule

$$(x_1, x_2) + (x_1', x_2') = (x_1 + x_1', x_2 + x_2'),$$

and becomes† a vector space under the rule

$$x(x_1, x_2) = (xx_1, xx_2).$$

This is the familiar vector-space structure on $\mathbb{R}^2$. Plainly we may generalize this example (a) by replacing $\mathbb{R}$ by any other field F (for example, by the complex field $\mathbb{C}$), and (b) by replacing the set of ordered *pairs* by the set $F \times \cdots \times F$ (Cartesian product) of ordered n-tuples $(x_1, x_2, \ldots, x_n)$, $x_i \in F$. The resulting vector space is written F^n; notice that F^1 is just F, viewed as a vector space over itself. For example, $\mathbb{Z}_2^2$ is a vector space over $\mathbb{Z}_2$ and has just four elements.

(iii) Let $F = \mathbb{R}$ and let V be the set of real-valued functions defined on some set S. Then V is an abelian group under

$$(f_1 + f_2)(s) = f_1(s) + f_2(s), \qquad (s \in S, f_1, f_2 \in V),$$

and becomes a vector space under the rule

$$(xf)s = xf(s), \qquad (s \in S, f \in V, x \in \mathbb{R}).$$

By 6.1.4, if S is finite with n elements, then there is a bijection $V \approx \mathbb{R}^n$. A converse result is given by Corollary 19.4.11. In particular also the set of real-valued functions defined on some interval $I \subseteq \mathbb{R}$ is a vector space over $\mathbb{R}$. Certain subsets of this set are also vector spaces over $\mathbb{R}$, as the reader may verify (and/or recall); thus the set of continuous functions, the set of differentiable functions, the set of integrable functions (from I to $\mathbb{R}$) are all vector spaces with respect to the operations described above. (See Chapter 26.)

(iv) $\mathbb{R}$ may be regarded as a vector space over $\mathbb{Q}$, the field of rational numbers. For $\mathbb{R}$ certainly forms an abelian group under addition and $\mathbb{Q}$ operates on $\mathbb{R}$ by the ordinary multiplication in $\mathbb{R}$, since $\mathbb{Q}$ is a subset of $\mathbb{R}$.

† With dogmatic assertions of this kind, it is assumed that the reader will check for himself that the axioms $\mathfrak{V}_1$–$\mathfrak{V}_4$ are satisfied.

(v) Let $f(D)$ be any linear differential operator (see Chapter 29). Then the set of solutions of the differential equation $f(D)y = 0$ is a vector space over $\mathbb{R}$.

(vi) Let P_n denote the set of all polynomials of degree $\leqslant n$, with real coefficients. Then P_n is a vector space over $\mathbb{R}$ under the same operations as in Example 19.1.2(ii). [The constant polynomial, zero, is taken in P_n.]

19.2 Bases

Now let V be a vector space over the field F. A set of elements $u_1, u_2, \ldots, u_n$ of V is said to be **linearly independent** if the relation

$$x_1u_1 + x_2u_2 + \cdots + x_nu_n = 0,$$

$x_i \in F$, implies that each x_i is zero. In the contrary case the set $u_1, u_2, \ldots, u_n$ is said to be **linearly dependent**; and if there is, in particular, a relation $x_1u_1 + \cdots + x_nu_n = 0$ with $x_i \neq 0$, then u_i is said to be linearly dependent on the remaining u's. The set $u_1, u_2, \ldots, u_n$ is said to **generate** or **span** V if every element of V is expressible in the form $x_1u_1 + \cdots + x_nu_n$. A linearly independent generating set is called a **basis**. *We will henceforth in this Chapter only be concerned with vector spaces admitting a finite basis*, though the ambitious reader may well like to see for himself how the arguments need to be modified if vector spaces not admitting finite bases (e.g., Example 19.1.2(iii), (iv)) are under discussion. [Chapter 29 is rich in examples of 'big' vector spaces (e.g., $\mathscr{D}(I)$, $\mathbb{R}^I$) and the reader should extract them. It is not too difficult to see how to adapt the definitions of linear independence, generating set, and basis to the case when there are infinitely many u_i. It is then a *theorem* (granted Zorn's lemma) that every vector space has a basis; see 7 Exercise 8(i). Indeed much of modern analysis is concerned with the transfer to arbitrary vector spaces of notions which are susceptible to fairly elementary treatment in the case of the so-called *finite-dimensional* vector spaces.]

Exercise 2

Prove that $e_1, \ldots, e_n$ is a basis for F^n (F a field) where all co-ordinates of e_i are zero, except the ith, which is the unity element of F.

The basic lemma in the whole theory is the following **exchange lemma.**

19.2.1 LEMMA. *Let $u_1, u_2, \ldots, u_m$ be linearly independent in V and let $v_1, v_2, \ldots, v_n$ generate V. Then we may replace certain of the v's (m in number) by all the u's in such a way that the resulting set of elements still generates V.*

Proof. We describe a step-by-step method of exchanging u's for v's. Formally, then, we suppose that we have exchanged r of the u's ($0 \leqslant r < m$), so that (renumbering the v's if necessary) the vectors $u_1, \ldots, u_r, v_{r+1}, \ldots, v_n$ span V. Then there is an expression

$$u_{r+1} = x_1u_1 + \cdots + x_ru_r + y_1v_{r+1} + \cdots + y_{n-r}v_n, \qquad (x_i, y_j \in F).$$

Moreover, it is not possible that all y_j are zero, since $u_1, \ldots, u_r, u_{r+1}$ are linearly independent. Renumbering if necessary, we suppose $y_1 \neq 0$ so that

$y_1^{-1} \in F$ and hence v_{r+1} belongs to the space spanned by $u_1, \ldots, u_r, u_{r+1}, v_{r+2}, \ldots, v_n$. This implies that this last set of vectors spans F; so we have taken the process one step further by exchanging u_{r+1} for v_{r+1}. The process cannot end before all the u's have been exchanged. ■

It follows of course immediately from the lemma that $m \leqslant n$. Thus we infer

19.2.2 THEOREM. *Let V be a vector space admitting a finite basis. Then each basis of V contains the same number of elements.*

Proof. If $u_1, \ldots, u_m$ and $v_1, \ldots, v_n$ are bases, then $m \leqslant n$ and $n \leqslant m$, so $m = n$. ■

The number of elements in a basis for V is called the **dimension** (or, sometimes, **rank**) of V. We prove two further useful theorems about bases for a vector space before giving examples.

19.2.3 THEOREM. *Let $v_1, \ldots, v_n$ be a generating set for V. Then we may choose as a basis for V a subset of this set.*

Proof. We proceed through the list $v_1, \ldots, v_n$ and throw out every v_i which is linearly dependent on its predecessors. We claim that the remaining v's constitute a basis. Let us call these retained v's $v_{i_1}, \ldots, v_{i_m}$ and the throw-outs $v_{j_1}, \ldots, v_{j_l}$ (so that $l + m = n$). Certainly $v_{i_1}, \ldots, v_{i_m}$ are linearly independent. For, if not, there exists a relation

$$\sum_{r=1}^{m} x_r v_{i_r} = 0$$

with not all the x_r equal to 0. Let k be the largest suffix such that $x_r \neq 0$. Then v_{i_k} is linearly dependent on its predecessors, contrary to hypothesis.

It remains to show $v_{i_1}, \ldots, v_{i_m}$ generate V. It is plainly sufficient to show that each v_{j_s} is dependent on the elements $v_{i_1}, \ldots, v_{i_r}$, with $i_r < j_s$. [Why?] We show this for each v_j in turn. It is plainly true for v_{j_1}, since v_{j_1} was the first to be thrown out. Suppose it proved for $v_{j_1}, \ldots, v_{j_{s-1}}$. Since v_{j_s} was thrown out, we must have

$$v_{j_s} = \sum_{i_q < j_s} x_q v_{i_q} + \sum_{t=1}^{s-1} y_t v_{j_t}, \qquad (x_q, y_t \in F).$$

But, by hypothesis, v_{j_t} is expressible in terms of the v_{i_q} for $t < s$, and so therefore is v_{j_s}. This proves the theorem. ■ We will shortly give some exercises (Exercise 3(iii), (iv), (vi)) to help the reader to grasp the inwardness of this theorem and its proof. We give one immediate consequence, whose proof we leave to the reader.

19.2.4 COROLLARY. *Let V be n-dimensional and let $v_1, \ldots, v_n$ be a generating set for V. Then it is also a basis for V.* ■

Our second theorem asserts that every linearly independent set forms part of a basis. Precisely, we have

19.2.5 THEOREM. *Let V be a finite-dimensional vector space and let $u_1, \ldots, u_m$ be linearly independent in V. Then we may find elements $u_{m+1}, \ldots, u_n$ in V such that $u_1, \ldots, u_m, u_{m+1}, \ldots, u_n$ is a basis.*

Proof. Let $v_1, \ldots, v_n$ be any basis for V. By the exchange lemma we may replace m of the v's by the u's and we still have a generating set for V. But this set still has n elements and so, by Corollary 19.2.4, it is a basis for V. ∎

A consequence of this theorem, whose proof the reader may supply, is:

19.2.6 COROLLARY. *Let V be n-dimensional and let $u_1, \ldots, u_n$ be linearly independent. Then they form a basis.* ∎

Exercise 3

(i) Show that $\mathbb{R}^2$ is a two-dimensional vector space; and more generally that $\mathbb{R}^n$ is n-dimensional. (Compare the treatment of $\mathbb{R}^3$ in Chapter 14.)
(ii) Show that the set of vectors (x_1, x_2, x_3) of $\mathbb{R}^3$ satisfying the relation $x_1 + x_2 + x_3 = 0$ is a two-dimensional vector space and find a basis for it.
(iii) Show that the vectors $(1, 0, 1)$, $(1, 1, 1)$, $(1, 2, 1)$, $(2, 0, 2)$, $(2, 1, 2)$. $(2, 1, 1)$ generate $\mathbb{R}^3$ and extract a basis from this generating set (see Theorem 19.2.3). Express the vector $(-3, 7, \frac{1}{2})$ as a linear combination of the basis elements you have extracted.
(iv) Show that the vectors $(1, 1, 2)$, $(\frac{1}{2}, \frac{1}{2}, 1)$, $(0, 1, 1)$, $(-\frac{1}{2}, \frac{1}{2}, 0)$, $(\frac{1}{2}, \frac{1}{2}, \frac{1}{2})$ generate $\mathbb{R}^3$ and extract a basis from this generating set.
(v) Show that the polynomials $1, x, x^2, \ldots, x^n$ form a basis for the space P_n of Example 19.1.2(vi).
(vi) Show that the rule we gave in proving Theorem 19.2.3 to construct a basis for V would be unchanged if we replaced 'predecessors' by 'surviving predecessors'.

19.3 Subspaces

It is (or should be) perfectly clear what is meant by a (vector) subspace U of a given vector space V; U is a subgroup of the abelian group V and $xu \in U$ whenever $x \in F$ and $u \in U$. (Exercise 3(ii) above is concerned with a certain two-dimensional subspace of $\mathbb{R}^3$.) Theorem 19.2.5 enables us to infer the very important result that, if V is a vector space and U is a subspace of V, then we may choose a basis for V, say $v_1, \ldots, v_n$, such that $v_1, \ldots, v_m$ forms a basis for U. (Consequently, denoting, as is customary, the dimension of V by $\dim V$, we have $\dim U \leqslant \dim V$.) This result enormously simplifies vector space theory by comparison with group theory, or even the theory of abelian (commutative) groups. It means, to express the matter informally, that there is essentially only one way in which an m-dimensional subspace can sit in an n-dimensional vector space. More formally, given any n-dimensional vector space V and m-dimensional subspace U, there always exists an $(n - m)$-dimensional subspace W such that each vector $v \in V$ is *uniquely expressible* in the form

$$v = u + w, \qquad u \in U, \qquad w \in W.$$

For we may take W to be the subspace of V spanned by $v_{m+1}, \ldots, v_n$, where $v_1, \ldots, v_n$ is the basis referred to above. We write

19.3.1 $$V = U \oplus W$$

and call V the **direct sum** of the subspaces U, W. However, we particularly stress the fact that the vector space W is not itself uniquely determined by V and U, since there will in general be many bases $v_1, \ldots, v_n$ of V such that $v_1, \ldots, v_m$ span U. Suppose for example that $V = \mathbb{R}^2$ and that U is the one-dimensional subspace spanned by the vector $(1, 0)$. Then we may take for W the one-dimensional subspace W_a spanned by $(a, 1)$, where a is any element of $\mathbb{R}$; and W_a and W_b intersect only in the zero vector if $a \neq b$.

Exercise 4

(i) Find a suitable subspace W for the vector space V and subspace U of Exercise 3(ii).
(ii) Show that if V is a finite-dimensional vector space and if U is a subspace of V of the same dimension as V, then $U = V$.
(iii) In Example 19.1.2(vi), prove that P_{n-1} is a subspace of P_n if $n > 0$. Find a suitable W if $U = P_{n-1}$.
(iv) Let $U_0 \supseteq U_1 \supseteq \cdots \supseteq U_n \supseteq \cdots$ be a sequence of subspaces of the vector space U_0. If dim U_0 is finite, prove that $\exists n$ such that $U_n = U_{n+k}$, $k = 0, 1, 2, \ldots$.
(v) A **flat**, Q, in the vector space V is a subset of the form $\{q + u\}$ where q is fixed and u runs through a subspace U of V. Prove that Q is not a subspace unless $q = 0$. Prove that, in P_n, the set of all polynomials f such that $f(0) = 1$ is a flat.
(vi) Prove that the intersection of a family of subspaces of a vector space is a subspace (cf. 9.4.2).
(vii) If U is a subspace of the vector space V over F, we can form the quotient group V/U as in 18.9. Prove that V/U is a vector space over F. [It is called the **quotient space** of V by U.] Show that each coset of U in V is a flat. If dim V is finite, show that dim V/U = dim V − dim U. Describe $\mathbb{R}^3/\mathbb{R}^2$, and V/U when V, U are the spaces of polynomials and even polynomials over F, of dimension $\leqslant n$. (A polynomial $p(x)$ is called *even* if the coefficients of odd powers of x are all zero.)
(viii) Let V be a vector space over F. For each $v \in V$, define a function $T_v: V \to V$ (called a **translation** through v) by $T_v(w) = v + w$. Prove that, in (v), the flat Q is $T_q(U)$. Show that $T_u \circ T_v = T_{u+v}$, $T_0 = \text{id}$. For each $x \in F$ define xT_v by T_{xv}. Prove that the set of all translations, with these operations, forms a vector space T_V over F.
(ix) Let V, W be vector spaces over F. Show that $V \times W$ is a vector space over F if we use the rules of Example 19.1.2(i) with x_1, x_2 there taken in V, W, respectively. Find a basis for $V \times W$, given bases $(v_1, \ldots, v_n)$, $(w_1, \ldots, w_m)$ for V, W.

19.4 Homomorphisms: Matrices

We now consider *homomorphisms* (or **linear transformations**) of vector spaces. The concept is quite simple and obvious (recall 15.3): if U and V

are vector spaces over the same field F and if $\phi: U \to V$ is a function from U to V, then ϕ is a **homomorphism** if it is a homomorphism with respect to their abelian group structures, and if, also,

$$\phi(xu) = x\phi(u), \qquad (x \in F,\ u \in U)$$

Thus,

19.4.1 $$\phi(xu + yv) = x\phi(u) + y\phi(v), \qquad (x, y \in F,\ u, v \in U).$$

We need not repeat all the notions which are relevant here and are analogous to those of group theory. Thus we will speak of the **kernel** and **image** of ϕ, and we will use the terms monomorphism, epimorphism, isomorphism without more ado; the reader may find the term **null space** instead of kernel in some texts on linear algebra.

19.4.2 EXAMPLES. The operator $D: P_n \to P_{n-1}$ which assigns to each polynomial in P_n (see Example 19.1.2(vi)) its derivative, is a homomorphism, with image P_{n-1} and kernel the set of all constant polynomials. The inclusion map $P_{n-1} \to P_n$ is a homomorphism with image P_{n-1} and kernel $\{0\}$ (i.e., a monomorphism). The function which assigns to $f(x) = a_0 + \cdots + a_n x^n$ the $(n + 1)$-tuple $(a_0, \ldots, a_n)$ is an isomorphism $P_n \to \mathbb{R}^{n+1}$.

Exercise 5

(i) Prove the assertions in 19.4.2.
(ii) Given a homomorphism $\phi: U \to V$, prove that the kernel and image of ϕ are subspaces of U, V respectively.
(iii) When $k < n$, prove that the function $F^n \to F^k$ given by $(x_1, \ldots, x_n) \to (x_1, \ldots, x_k)$ is an epimorphism, and determine its kernel.
(iv) Let U be a subspace of the vector space V over F. Let $\nu: V \to V/U$ be the natural map (see 18.9). Prove that ν is a homomorphism of vector spaces. If $V = U \oplus W$ (see 19.3.1), prove that V/U is isomorphic to W.
(v) In the space $V \times W$ of Exercise 4(ix), prove that the functions which send $v \in V$ to $(v, 0)$, $w \in W$ to $(0, w)$, are monomorphisms $\alpha: V \to V \times W$, $\beta: W \to V \times W$, and that $\mathrm{Im}\,(\alpha) \cap \mathrm{Im}\,(\beta) = \{(0, 0)\}$. Then show that $V \times W$ is the direct sum (see 19.3.1) $\mathrm{Im}\,\alpha \oplus \mathrm{Im}\,\beta$. [Thus, $V \times W$ is spanned by the union of a copy each of V and of W, these copies intersecting only in $(0, 0)$.] Show that $V \times W \approx V \oplus W$.
(vi) Conversely, suppose a vector space U contains subspaces A, B such that $A \cap B = \{0\}$, while $A \cup B$ spans U (cf. 19.3.1). Prove that $U \approx A \oplus B$.

Now, a homomorphism $\phi: U \to V$ *is entirely specified by its effect on the elements of a basis for* U. For, suppose in fact that U has the basis $u_1, \ldots, u_m$ and V has the basis $v_1, \ldots, v_n$, while ϕ is prescribed by the equations

19.4.3 $$\phi(u_i) = \sum_{j=1}^{n} a_{ij} v_j, \qquad (a_{ij} \in F).$$

Certainly these equations do determine a homomorphism, since any element of U is uniquely expressible as $\sum_{i=1}^{m} b_i u_i$, $b_i \in F$, and by the linearity of ϕ (see 19.4.1)

$$\phi\left(\sum_{i=1}^{m} b_i u_i\right) = \sum_{i=1}^{m} b_i \phi(u_i). \tag{19.4.4}$$

The set of elements a_{ij} in 19.4.3, *with their relative positions* when the linear equations 19.4.3 are written out in full, is best displayed in the form

$$A = \begin{pmatrix} a_{11}, \ldots, a_{1n} \\ \vdots \qquad \vdots \\ a_{m1}, \ldots, a_{mn} \end{pmatrix}$$

and the rectangular array A is called an $m \times n$ **matrix** (with entries or 'coefficients' in F) sometimes abbreviated to $A = (a_{ij})$. By 19.4.3, we have assigned to each homomorphism $\phi: U \to V$ an $m \times n$ matrix $A = M(\phi)$. Thus we have a function

$$M: H(U, V) \to \mathscr{M}_{mn}$$

where $H(U, V)$ is the set of all homomorphisms $U \to V$ and $\mathscr{M}_{mn}$ is the set of all $m \times n$ matrices with coefficients in F. Observe that the construction of M is *in terms of the given ordered bases* (u_i) of U, (v_j) of V; that is, the base $(u_1, u_2, \ldots, u_n)$ is different from, say, $(u_2, u_1, u_3, \ldots, u_n)$, and the corresponding matrices will have their first two rows interchanged. By 19.4.3, M is one-one; while any set of mn elements a_{ij} in F determine a homomorphism ϕ with $(a_{ij}) = M(\phi)$, by the rule 19.4.3. Thus, *M is a bijection.*

We may give $H(U, V)$ the structure of a vector space over F by the rules

$$\begin{aligned} (\phi + \psi)(u) &= \phi(u) + \psi(u), \qquad && (\phi, \psi \in H(U, V), u \in U), \\ (x\phi)(u) &= x\phi(u), && (x \in F, \phi \in H(U, V), u \in U), \end{aligned} \tag{19.4.5}$$

analogously to the structure given to R^S in Example 19.1.2(iii). Also, we may regard the matrix A, above, as a function: $N_{mn} \to F$, where

$$N_{mn} = \{(i, j) | i = 1, \ldots, m; j = 1, \ldots, n\}.$$

Thus, the same analogy leads us to give a vector space structure to $\mathscr{M}_{mn}$ by the rules

$$\begin{aligned} (a_{ij}) + (b_{ij}) &= (a_{ij} + b_{ij}), \\ x(a_{ij}) &= (xa_{ij}); \end{aligned} \tag{19.4.6}$$

and it is easy to check that M is now a *homomorphism* of vector spaces.

Since we saw that M was a bijection, we conclude:

19.4.7 THEOREM. *$M: H(U, V) \to \mathcal{M}_{mn}$ is an isomorphism.* ■

We want to stress at this stage that the isomorphism $M: H(U, V) \to \mathcal{M}_{mn}$ depended on a choice of bases in U and V; a different choice of bases would have led to a different isomorphism. The linear transformations themselves should be regarded as the basic ('invariant') objects of study, because it is only through an arbitrary choice of bases that we may associate a matrix with a linear transformation T as a help in describing T. This viewpoint is in marked contrast with the classical viewpoint which represented matrices as things in themselves and perhaps never mentioned vector spaces or linear transformations at all! Also, this lack of mention makes the uninformed reader feel that the classical rule for multiplying matrices is quite capricious. The rule is as follows.

If $A = (a_{ij})$ is an $(m \times n)$–matrix and $B = (b_{jk})$ is an $(n \times p)$–matrix, then the **product matrix** $AB = C$ is defined as the $(m \times p)$ matrix (c_{ik}) such that

19.4.8 $$c_{ik} = \sum_j a_{ij} b_{jk}.$$

The reason for the definition is this: if A is the matrix of $\phi: U \to V$ with respect to (ordered) bases $(u_1, \ldots, u_m)$, $(v_1, \ldots, v_n)$, and if B is the matrix of $\psi: V \to W$ with respect to bases $(v_1, \ldots, v_n)$, $(w_1, \ldots, w_p)$, then $C = AB$ is the matrix of $\psi \circ \phi: U \to W$ with respect to the bases $(u_1, \ldots, u_m)$, $(w_1, \ldots, w_p)$. (There is still an awkward switch of order here, but we shall ask why in Exercise 6(vi). See also 20 Exercise 7(ii).) As written, however, we have $M''(\psi \circ \phi) = M(\phi)M'(\psi)$, where M is the function of 19.4.7 and M', M'' are the analogues with domains $H(V, W)$, $H(U, W)$, relative to the bases (u), (v), (w). But now we have a quick proof of the *associative law for the product* (19.4.8), since (with an obvious notation) we use the associativity of functions to give

19.4.9 $$(AB)D = (M(\phi)M(\psi))M(\chi) = M(\psi \circ \phi)M(\chi) = M(\chi \circ \psi \circ \phi)$$

and, similarly,

$$A(BD) = M(\chi \circ \psi \circ \phi). \quad ■$$

Exercise 6

(i) Verify in detail that $H(U, V)$ and $\mathcal{M}_{mn}$ are vector spaces with the structures given in 19.4.5, 19.4.6; and that M is a homomorphism.

(ii) Verify the above statement that $M''(\psi \circ \phi) = M(\phi)M'(\psi)$.

(iii) Use the method of proof for 19.4.9, to show that matrix multiplication 19.4.8 is distributive. Give a direct proof also, just by manipulating the $\sum$ sign.

(iv) Prove that, when $m = n > 1$, $\mathcal{M}_{nn}$ is a (non-commutative) ring with unity element (δ_{ij}) where δ_{ij} is the 'Kronecker delta' (after L. Kronecker (1823–91); see Bell [10], Ch. 25), i.e., $\delta_{ij} = 0$ if $i \neq j$ and $\delta_{ii} = 1$. Find two non-commuting elements of $\mathcal{M}_{nn}$. By considering the linear independence of powers of a given matrix $x \in \mathcal{M}_{nn}$, prove that x satisfies a polynomial equation, with coefficients in F, of degree $\leqslant n^2$. What if $n = 1$?

(v) Show that, when $U = V$, then the subset J of $H(V, V)$, consisting of all automorphisms of V, is a group. Show that J is non-commutative if $n > 1$. If $\dim V = n$, show that the group of units in the ring $\mathscr{M}_{nn}$ is $M(J)$, with $M: H(V, V) \to \mathscr{M}_{nn}$ as in 19.4.7. [An element of this group of units is called a **non-singular** $(n \times n)$ matrix. Thus, the non-singular $(n \times n)$ matrices over F form a group under multiplication; it is isomorphic to J.]
(vi) Suggest a way of avoiding the switch of order between linear transformations and their associated matrices! (Why should AB be the matrix of $\psi \circ \phi$ if A is the matrix of ϕ and B of ψ??)
(vii) Let G be a finite group of order n, and regard G as a group of permutations on $\{1, \ldots, n\}$ (see 18 Exercise 11(x)). To each $g \in G$, assign a homomorphism $\phi_g = \phi: F^n \to F^n$ by setting $\phi(u_i) = u_{g(i)}$, where u_i runs through a basis for F^n. Prove that the assignment $g \to \phi_g$ is a monomorphism θ of G in the group J of (v) above; then $M \circ \phi: G \to \mathscr{M}_{nn}$ is called a **representation** of G by $n \times n$ matrices. The theory of group representations is an extensive branch of group theory, with applications in theoretical physics (see Wigner [135a]).
(viii) In Exercise 4(viii), prove that the space T_V of translations is isomorphic with V. [The space T_V may be used for an approach to vectors in $\mathbb{R}^3$, different from our construction of $\mathbb{V}$ in 14.7. See Patterson [102].
(ix) Let $(v_1, \ldots, v_n)$, $(w_1, \ldots, w_n)$ be two bases of V in (v) above, and let A, B denote the matrices of a homomorphism $\phi: V \to V$ with respect to (v_i) and (w_i) respectively. Thus we have equations of the form

$$\phi(v_i) = \sum a_{ij} v_j, \qquad \phi(w_i) = \sum b_{ij} w_j \quad \text{and} \quad v_i = \sum p_{ij} w_j.$$

Show that $A = PBP^{-1}$ and hence (using 18 Exercise 10(iii)) that $\det A = \det B$. (This common value is *defined* to be $\det \phi$.)
§(x) Given a homomorphism $\phi: V \to V$, it is important in applications to know if there exist $\lambda \in \mathbb{R}$, $v \in V$ such that $\phi(v) = \lambda v$ with $v \neq 0$. Such a λ, v are called, respectively, an **eigenvalue** and **eigenvector** of ϕ. Show that ϕ maps the whole line through 0 and v into itself. Prove that a *necessary* condition for λ to be an eigenvalue is that, just as we have seen when $n = 2$ or 3, $\det(\phi - \lambda \cdot \mathrm{id}_V) = 0$. [This equation is a polynomial equation of degree $\dim V$, and of course some of its roots may not be real. It can be shown that the product of the roots is $\det \phi$.] Show that if λ is an eigenvalue of ϕ, then the set of all w, satisfying $\phi(w) = \lambda w$, is a subspace W of V. [There is a considerable theory concerning the dimension of W and methods of finding a basis for W. See Mirsky [88].]

Next, we show that every vector space V over F, of finite dimension n, is isomorphic to F^n. This is why classical texts often define a vector space as a 'space of n-tuples', but in so doing, they severely restrict the range of application of the theory. The result we want is:

19.4.10 THEOREM. *Two (finite-dimensional) vector spaces over the field F are isomorphic if and only if they have the same dimension.*

Proof. Let U and V be vector spaces of dimension n, with bases $u_1, \ldots, u_n$; $v_1, \ldots, v_n$ respectively. Then the homomorphism $\phi: U \to V$ given by

$$\phi(u_i) = v_i$$

is plainly invertible and hence an isomorphism. Conversely, let U have a basis $u_1, \ldots, u_n$ and let $\phi: U \to V$ be an isomorphism. Then we shall show that $\phi(u_1), \ldots, \phi(u_n)$ *is a basis for* V, and so V is n-dimensional. We must show, then, that $\phi(u_1), \ldots, \phi(u_n)$ span V and are linearly independent.

First, let $v \in V$. Since ϕ is surjective, $v = \phi(u)$ for some $u \in U$. Then u may be expressed as $u = \sum_{i=1}^{n} x_i u_i$, $(x_i \in F)$, so by 19.4.4,

$$v = \phi\left(\sum_{i=1}^{n} x_i u_i\right) = \sum_{i=1}^{n} x_i \phi(u_i);$$

therefore $\phi(u_1), \ldots, \phi(u_n)$ span V.

Second, suppose $\sum_{i=1}^{n} y_i \phi(u_i) = 0$, $y_i \in F$. Then $\phi\left(\sum_{i=1}^{n} y_i u_i\right) = 0$, again by 19.4.4. But ϕ is injective, so $\sum_{i=1}^{n} y_i u_i = 0$. Since $u_1, \ldots, u_n$ are linearly independent, each $y_i = 0$, so that $\phi(u_1), \ldots, \phi(u_n)$ are linearly independent. ■

19.4.11 COROLLARY. *If V is a finite-dimensional vector space over the field F, then V is isomorphic to F^n for some n (vis.: $n =$* dim *V).* ■

Theorem 19.4.10 and its Corollary have the expected analogues for vector spaces of arbitrary dimension.

19.5 Rank of a Linear Transformation

We now analyse an arbitrary homomorphism $\phi: U \to V$ in a manner similar to that of 18.9.6. However, the problem is easier because of the richer and more special structure of vector spaces. First we make an important definition.

19.5.1 DEFINITION. The **rank** of $\phi: U \to V$ is the dimension of $\phi(U)$.

Now let K be the kernel of ϕ. By Theorem 19.2.5 we may choose a basis $u_1, \ldots, u_m$ for U such that $u_{r+1}, \ldots, u_m$ is a basis for K. We prove:

19.5.2 PROPOSITION. $\phi(u_1), \ldots, \phi(u_r)$ is a basis for $\phi(U)$.

Proof. Let $v \in \phi(U)$. Then

$$\begin{aligned} v = \phi(u) &= \phi\left(\sum_{i=1}^{m} x_i u_i\right) && \text{(say)} \\ &= \sum_{i=1}^{m} x_i \phi(u_i) && \text{(by 19.4.1)} \\ &= \sum_{i=1}^{r} x_i \phi(u_i), \end{aligned}$$

since ϕ is zero on each of $u_{r+1}, \ldots, u_m \in K$. Thus $\phi(u_1), \ldots, \phi(u_r)$ span $\phi(U)$. To prove their linear independence, let $\sum_{i=1}^{r} y_i\phi(u_i) = 0$. Then $\phi\left(\sum_{i=1}^{r} y_i u_i\right) = 0$, so that $\sum_{i=1}^{r} y_i u_i \in K$. Since $u_{r+1}, \ldots, u_m$ is a basis for K,

$$\sum_{i=1}^{r} y_i u_i = \sum_{j=r+1}^{m} z_j u_j, \qquad \text{for some } z_j \in F.$$

But $u_1, \ldots, u_m$ are linearly independent, so that each $y_i = 0$ (and each $z_j = 0$). This proves that $\phi(u_1), \ldots, \phi(u_r)$ are linearly independent. ∎

We may now invoke Theorem 19.2.5 again to infer that there is a basis $v_1, \ldots, v_n$ for V with $v_i = \phi(u_i)$, $i = 1, \ldots, r$. Thus, given a homomorphism $\phi: U \to V$, then, *with respect to suitably chosen bases $u_1, \ldots, u_m$ in U and $v_1, \ldots, v_n$ in V, ϕ takes the form*

19.5.3 $$\begin{cases} \phi(u_i) = v_i, & i = 1, \ldots, r \\ \phi(u_i) = 0, & i = r + 1, \ldots, m, \end{cases}$$

where r is the rank of ϕ. This shows that the rank is the only intrinsic invariant of ϕ. [This phrase has a precise meaning; but here the reader's attention is simply drawn to the fact that the *form* of 19.5.3 is entirely specified by the number r.]

Exercise 7

(i) Show that if $\phi: U \to V$, then

$$\dim\,(\text{kernel of } \phi) + (\text{rank of } \phi) = \dim U.$$

[In classical terms, 'nullity + rank = dimension'.]
(ii) What is the matrix of ϕ with respect to the bases referred to in 19.5.3?
(iii) Find suitable bases for the homomorphism $\phi: \mathbb{R}^3 \to \mathbb{R}^3$ given by

$$\phi(x_1, x_2, x_3) = (2x_1 + x_2 + x_3, -x_2 + x_3, x_1 + x_3),$$

so that ϕ takes the form (19.5.3). What is the rank of ϕ?
(iv) Let U, V be n-dimensional and let $\phi: U \to V$ be a homomorphism. Show that ϕ is an isomorphism if it is a monomorphism or an epimorphism.
(v) If U is m-dimensional and V n-dimensional, show that the dimension of $H(U, V)$ is mn. How would you choose a basis for $H(U, V)$? How would you set up an isomorphism between $H(U, F)$ and U, where U is a vector space over F? Consider particularly also the case when $U = F$.
(vi) Let $0 \xrightarrow{f_0} U_0 \xrightarrow{f_1} U_1 \to \cdots \xrightarrow{f_n} U_n \xrightarrow{f_{n+1}} 0$ be a sequence of finite-dimensional vector spaces U over the same field, and homomorphisms $f_i: U_i \to U_{i+1}$ such that $f_{i+1} \circ f_i$ is the zero homomorphism, $i = 0, \ldots, n$. Prove that, for each i,

$\ker f_{i+1} \supseteq \operatorname{Im} f_i$. Hence we can form the quotient spaces (Exercise 4(vi)) $H_i = \ker f_{i+1}/\operatorname{Im} f_i$. Use Exercise 7(i) to prove

$$\sum_{i=0}^{n} (-1)^i \dim U_i = \sum_{i=0}^{n} (-1)^i \dim H_i.$$

[The left-hand side is called the **Euler characteristic** of the sequence $\{U_i, f_i\}$.]

19.6 Linear Equations

We consider two vector spaces U and V over the same field F, of dimensions m and n respectively, and a homomorphism or linear map $\phi: U \to V$. We fix a vector b in V and ask for (a description of) the set of all vectors $u \in U$ such that

19.6.1 $$\phi(u) = b.$$

Then 19.6.1 is called a **linear equation** in the unknown vector u; and the set $\phi^\flat\{b\}$ of all vectors u satisfying 19.6.1 is called the **solution set** of the equation. If we choose bases $u_1, \ldots, u_m$ for U, $v_1, \ldots, v_n$ for V, then we may write, uniquely,

$$b = \sum_{i=1}^{n} b_i v_i, \qquad \text{(for some } b_i \in F\text{)}$$

and

$$u = \sum_{i=1}^{m} x_i u_i,$$

where $x_1, \ldots, x_m$ are 'unknown' elements of F to be determined by the equation 19.6.1. Suppose that, with respect to the bases $(u_1, \ldots, u_m)$, $(v_1, \ldots, v_n)$, ϕ assumes the form

$$\phi u_i = \sum_{j=1}^{n} a_{ij} v_j, \qquad i = 1, \ldots, m.$$

Then the single equation 19.6.1 is equivalent to the set of n 'simultaneous' equations

19.6.2 $$\sum_{i=1}^{m} x_i a_{ij} = b_j, \qquad j = 1, \ldots, n.$$

The reader may well be more familiar with the form 19.6.2 than with 19.6.1; we regard it as very important that he understand the equivalence of the two forms. For it is plain that, conversely, given the set of n equations 19.6.2 in the m unknowns $x_1, \ldots, x_m$ in F, we may regard the matrix (a_{ij}) as specifying, for given bases, a linear transformation ϕ from U to V; and 19.6.2 may then be telescoped into the single vector equation 19.6.1. The latter is far more convenient for a theoretical study of linear equations, for it puts the problem in suitably 'invariant' form. That is, if we consider 19.6.1, we are at liberty to choose any bases we please for U and V, and in this way we can greatly simplify the study of the nature of the solution. In fact, any of the familiar procedures for effective solution of the set of equations 19.6.2 amounts to a systematic process of basis-changing, except for Cramer's rule for writing down the unique solution, where a unique solution exists; but this rule is almost useless in practice. We cannot add to the excellent account in P. M. Cohn's little book [23] of the *technique* of solving linear equations.

However, we do want to give the reader a clear but elementary account of the theory. First we stress the fact that *the equation* 19.6.1 *may have no solution*, i.e., the solution set may be empty. For we have not demanded that ϕ be an epimorphism, that is, that ϕ map U onto V; and, if ϕ is not epimorphic, then b may well not belong to the image, $\phi(U)$, of ϕ. Suppose, however, that 19.6.1 has a particular solution $u = a$ in U, so that $\phi a = b$. Since ϕ is linear, we conclude that the vector u_0 is also a solution if and only if $\phi(u_0 - a) = 0$, that is, if and only if $u_0 - a$ belongs to the kernel of ϕ. Thus we have proved:

19.6.3 PROPOSITION. *The solution set Σ of the equation $\phi u = b$ either is empty, or consists of all vectors of the form $k + a$, where k runs through the kernel of ϕ and a is an arbitrary solution of the equation.* ■

In the terminology of Exercise 4(v), we can say more briefly:

19.6.4 *Either $\Sigma = \varnothing$ or Σ is the flat $a + \ker \phi$, where a is any element of Σ.*

[Note that, so far as this proposition is concerned, no use has been made of the finite-dimensionality of U and V. Indeed, the assertion of this proposition is very familiar to anybody who has studied *linear differential equations* (see Chapter 29). For a linear differential equation,

19.6.5 $$Df = g,$$

where D is some linear differential operator (say

$$D = d^2/dx^2 + 6x(d/dx) + x^2 - 1),$$

may be regarded as belonging to the family of equations of type 19.6.1, where $F = \mathbb{R}$, and $U = \mathscr{C}^2(\mathbb{R})$, $V = \mathscr{C}^1(\mathbb{R})$ in the notation of 29.6.1; and Proposition 19.6.3 asserts that if 19.6.5 has a solution, then the general solution may be obtained by finding an arbitrary solution (the 'particular integral') and adding to it an arbitrary solution of the differential solution

$$Df = 0$$

(the 'complementary function'). In present-day scientific and mathematical research it is a very important problem to study *non-linear* differential equations [for example, $d^2y/dx^2 - (dy/dx)^2 + y^2 = e^x$], since so many of the subtle processes of nature seem to be most effectively modelled in mathematics by such equations. The differential operator $D: \mathscr{C}^2 \to \mathscr{C}^1$ is here no longer linear, i.e., it is not a homomorphism. In general the extreme difficulty of the study is largely attributable to the fact that we cannot apply Proposition 19.6.3 and obtain a particular integral and the complementary function.]

We know from Exercise 7(i) that

$$\dim(\text{kernel } \phi) = \dim U - \text{rank } \phi.$$

Thus we know that if the equation $\phi u = b$ has a solution, then the solution set Σ has dimension $m - r$ (where $\dim U = m$, rank $\phi = r$) in the following sense. Although Σ is not a vector subspace of U unless $b = 0$, it is a flat obtained by *translating* the subspace kernel ϕ by the vector a, and it is thus natural to attribute to Σ the same dimension as kernel ϕ. We augment 19.6.4 then, by:

19.6.6 *If* $\Sigma \neq \varnothing$, *then* $\dim \Sigma = \dim U - \text{rank } \phi$.

In particular, suppose that rank $\phi = m$. Then $\dim(\text{kernel } \phi) = 0$, so kernel $\phi = 0$ and ϕ is a monomorphism. We may then invoke Exercise 7(iv) to prove

19.6.7 PROPOSITION. *Let* $\dim U = \dim V = m$ *and let* $\phi: U \to V$ *be a linear transformation of rank* m. *Then the equation* $\phi u = b$ *has a unique solution in* U *for each* $b \in V$. ∎

This proposition shows the importance that attaches to the question of computing the rank of a linear transformation. In the case of homomorphisms from U to V where $\dim U = \dim V$, the associated matrix (relative to a given choice of bases in U and V) is *square* (i.e., it possesses an equal number of rows and columns) and we can associate with any square matrix an element of F called its *determinant*, just as in the case $n = 3$, discussed in Chapter 15. We will not enter into a description of the $(n \times n)$ determinant in this book, the reader being referred either to Hodge–Pedoe [66], Vol. 1, or to Halmos [49], for a systematic treatment, which in particular establishes the rules $\mathfrak{M}_1$–$\mathfrak{M}_5$ of 15.1 in the $n \times n$ case, and shows that det is multiplicative (see 18 Exercise 10(iii)). We are content to remark here that if $\dim U = \dim V$, then $\phi: U \to V$ has maximum rank ($= \dim U$) if and only if the determinant of the associated matrix relative to any choice of bases in U and V is non-zero. The texts quoted (as well as others more familiar to the reader) give methods of computing rank. The importance of deciding the rank of a linear transformation is brought out by the answer to the general question whether the equation $\phi u = b$ has a solution, when $\dim V$ is not necessarily equal to $\dim U$. So we return to 19.6.1.

We construct a new vector space $\overline{U} = U \times F$ over F; it consists of pairs (u, x), where $u \in U$, $x \in F$, and the vector-space structure on $\overline{U}$ is the natural one:

$$(u, x) + (u', x') = (u + u', x + x'), \qquad (u, u' \in U;\ x, x' \in F);$$
$$y(u, x) = (yu, yx), \qquad (u \in U,\ x, y \in F).$$

We now define a function $\bar{\phi}: \overline{U} \to V$ by the rule

19.6.8 $$\bar{\phi}(u, x) = \phi u + xb, \qquad (b \text{ in } 19.6.1).$$

and it is easily checked that $\bar{\phi}$ is a homomorphism. The main theorem then reads

19.6.9 THEOREM. *Let U, V be finite-dimensional vector spaces and let $\phi: U \to V$ be a linear map. Then the equation $\phi u = b$ has a solution if and only if*

$$\operatorname{rank} \phi = \operatorname{rank} \bar{\phi},$$

where $\bar{\phi}$ is given by 19.6.8.

Proof. Recall that $\operatorname{rank} \phi = \dim \phi(U)$, $\operatorname{rank} \bar{\phi} = \dim \bar{\phi}(\overline{U})$. Plainly $\bar{\phi}(\overline{U}) \supseteq \phi(U)$. Thus, if $\dim \bar{\phi}(\overline{U}) = \dim \phi(U)$, then (see Exercise 4(ii)) $\bar{\phi}(\overline{U}) = \phi(U)$; but since, clearly, $b \in \bar{\phi}(\overline{U})$, we infer that $b \in \phi(U)$. Conversely, if $b \in \phi(U)$, then, plainly, $\phi u + xb \in \phi(U)$ for all $u \in U$, $x \in F$. That is to say, $\bar{\phi}(\overline{U}) \subseteq \phi(U)$, whence $\bar{\phi}(\overline{U}) = \phi(U)$ and so $\operatorname{rank} \phi = \operatorname{rank} \bar{\phi}$. ■

19.6.10 REMARKS. (i) The reader may not feel convinced that this theorem advances the problem of deciding whether the equation $\phi u = b$ has a solution. However, suppose that we associate with ϕ the matrix A, having chosen bases $u_1, \ldots, u_m$ for U and $v_1, \ldots, v_n$ for V. Now we may regard U as a subspace of $\overline{U}$ by identifying u with $(u, 0)$ and we may then choose the basis $u_1, \ldots, u_m$, 1 for $\overline{U}$, where we have written 1, both for $(0, 1) \in \overline{U}$ and for the unity element of F. With respect to this basis (and the basis $v_1, \ldots, v_n$ for V) the matrix $\overline{A}$ associated with $\bar{\phi}$ is obtained by placing the row $b_1 \cdots b_n$ at the bottom of A. Then, in traditional language, we have to decide whether the rank of the 'augmented' matrix $\overline{A}$ exceeds that of A. This is a standard computational problem. The practical reader may still legitimately object that the best way to decide whether the equation has a solution is to set about solving it! However, to make sure *all* solutions are obtained, one needs 19.6.6. See Cohn [23] for further discussion.

(ii) The reader should note that, since $\operatorname{rank} \phi \leqslant \min(\dim U, \dim V)$ for any $\phi: U \to V$, Proposition 19.6.7 follows immediately from Theorem 19.6.9.

(iii) Of course, the equation $\phi u = 0$ always has a solution, namely $u = 0$. In this case, the equation itself or the associated set of equations (19.6.2) is described as **homogeneous**; and a complete solution of the problem is provided by Proposition 19.6.3, the solution set being a subspace of U of dimension dim U − rank ϕ. It is common to call the solution $u = 0$ of the equation $\phi u = 0$ the **trivial solution** and then to say that the equation possesses **non-trivial** solutions if and only if dim $U >$ rank ϕ. (See 15 Exercise 3(i).)

(iv) The more theoretically-inclined reader may note that in the sketch we have given of the theory of linear equations, the finite-dimensionality of V plays no role; only that of U is important.

We close this section on linear equations by advising the reader again to compare the general theory we have given with the theory of linear equations and Euclidean geometry in Euclidean three-dimensional space, as outlined in Chapters 14, 15.

Chapter 20

INNER PRODUCT SPACES AND DUALITY

20.1 Scalar Products; Distance

In Euclidean geometry we have more than the structure of a vector space in $\mathbb{R}^3$; we have notions of *length* and of *perpendicularity* (or *orthogonality*) of two vectors; and, more generally, a notion of the *angle* between two vectors. We describe now, very briefly, how we may, in a general abstract fashion, enrich the structure of a vector space so that these notions take on a meaning. Our purpose is not to be abstract, but to transfer ideas from $\mathbb{R}^3$ to other situations. However, we will confine attention to the case when the ground field F is the field $\mathbb{R}$ of real numbers.

Let, then, V be a vector space over the field $\mathbb{R}$. We generalize first the scalar product in $\mathbb{R}^3$ (see 14.8). By an **inner product** on V we mean a bilinear, symmetric function $V \times V \to \mathbb{R}$; that is, it associates with any pair of vectors $u, v \in V$ a number, denoted by $\prec u, v \succ$, in $\mathbb{R}$. The **symmetry** of the inner product means that

$$\mathfrak{IP}_1\colon \prec u, v \succ = \prec v, u \succ;$$

and the **bilinearity** means that, for all $\alpha_1, \alpha_2 \in \mathbb{R}$ and $u_1, u_2, v \in V$,

$$\mathfrak{IP}_2\colon \prec \alpha_1 u_1 + \alpha_2 u_2, v \succ = \alpha_1 \prec u_1, v \succ + \alpha_2 \prec u_2, v \succ.$$

(*Deduce* that $\prec u, \beta_1 v_1 + \beta_2 v_2 \succ = \beta_1 \prec u, v_1 \succ + \beta_2 \prec u, v_2 \succ$; it is not necessary to *assume* this.) In addition we require that the inner product have the following very strong property:

$$\mathfrak{IP}_3\colon \prec u, u \succ \geqslant 0; \qquad \prec u, u \succ = 0 \quad \textit{only if} \quad u = 0.$$

20.1.1 EXAMPLE. Let $V = \mathbb{R}^3$ and let $\prec \mathbf{u}, \mathbf{v} \succ$ be the usual *scalar product* $\mathbf{u} \cdot \mathbf{v}$ of the vectors $\mathbf{u}$ and $\mathbf{v}$ as in 14.8. Clearly $\mathfrak{IP}_1$–$\mathfrak{IP}_3$ hold, since they are just another way of expressing $\mathfrak{SP}_1$–$\mathfrak{SP}_4$ in 14.8. Now, with respect to a given Cartesian co-ordinate system, we have $\mathbf{u} = (a_1, a_2, a_3)$, $\mathbf{v} = (b_1, b_2, b_3)$, so

$$\prec \mathbf{u}, \mathbf{v} \succ = a_1 b_1 + a_2 b_2 + a_3 b_3, \qquad a_i, b_j \in \mathbb{R}.$$

This formula clearly generalizes to $\mathbb{R}^n$. Moreover, $\prec \mathbf{u}, \mathbf{u} \succ$ is the square of the length of $\mathbf{u}$; thus if the vector $\mathbf{u}$ is thought of geometrically as leading from the origin O to the point P, then $\prec \mathbf{u}, \mathbf{u} \succ$ is the square of the length of OP. We generalize this below.

We will call a vector space V furnished with an inner product an **inner product space.** Then if V is an inner product space we define the **length** of a vector $u \in V$ to be

$$\|u\| = \sqrt{\langle u, u \rangle};$$

and, more generally, the **distance** between two vectors $u, v \in V$ to be

20.1.2 $$\rho(u, v) = \|u - v\|.$$

To justify the term 'distance' we must show that ρ satisfies certain axioms which would be asked of any function deserving to be called a distance. The three properties required of a distance function or **metric** (that is, a function $\rho: V \times V \to \mathbb{R}^+$) are

$\mathfrak{D}_1$: $\rho(u, v) = 0$ *if and only if* $u = v$;

$\mathfrak{D}_2$: $\rho(u, v) = \rho(v, u)$,

$\mathfrak{D}_3$: **(triangle inequality)** $\rho(u, v) + \rho(v, w) \geqslant \rho(u, w)$.

These properties are satisfied by the ordinary distance between pairs of points in Euclidean geometry (see Theorem 14.8.1), but many other such metrics exist, as we shall see below.

Consider the particular ρ given by 20.1.2. Then $\mathfrak{D}_1$ follows immediately from $\mathfrak{IP}_3$, and $\mathfrak{D}_2$ from the relation $\langle u, v \rangle = \langle -u, -v \rangle$, itself a consequence of the bilinearity of the inner product. To prove $\mathfrak{D}_3$ we first observe on changing $u - v$, $v - w$ to u, v that it is equivalent to

20.1.3 $$\|u\| + \|v\| \geqslant \|u + v\|,$$

and we will prove this inequality as a consequence of the following important lemma:

20.1.4 LEMMA. **(Schwarz's inequality.)** $\|u\| \cdot \|v\| \geqslant |\langle u, v \rangle|$. *Equality occurs if and only if u and v are linearly dependent.*

Proof. We consider the vector $\alpha u + \beta v$ where $\alpha, \beta \in \mathbb{R}$. Then

$$0 \leqslant \langle \alpha u + \beta v, \alpha u + \beta v \rangle = \alpha^2 \langle u, u \rangle + 2\alpha\beta \langle u, v \rangle + \beta^2 \langle v, v \rangle.$$

We thus have a homogeneous quadratic polynomial (or quadratic form) in α, β and are told that it never takes negative values. It follows (see 22 Exercise 10(iii)) that $\langle u, u \rangle \langle v, v \rangle \geqslant \langle u, v \rangle^2$ or, if we take positive square roots,

$$\|u\| \cdot \|v\| \geqslant |\langle u, v \rangle|.$$

Plainly equality holds if u and v are linearly dependent. Conversely, if equality holds, then $\alpha^2 \langle u, u \rangle + 2\alpha\beta \langle u, v \rangle + \beta^2 \langle v, v \rangle$ takes on the value 0 for some pair of real numbers α, β, not both zero. This means that $\langle \alpha u + \beta v, \alpha u + \beta v \rangle = 0$, so that $\alpha u + \beta v = 0$ and u, v are linearly dependent. ■

20.1.5 We return to the proof of 20.1.3. We have

$$\begin{aligned}\|u + v\|^2 = \langle u + v, u + v\rangle &= \langle u, u\rangle + 2\langle u, v\rangle + \langle v, v\rangle \\ &\leqslant \|u\|^2 + 2\|u\|\cdot\|v\| + \|v\|^2, \\ &\qquad\qquad \text{by Schwarz's inequality} \\ &= (\|u\| + \|v\|)^2,\end{aligned}$$

and 20.1.3 results by taking positive square roots. ■ In fact, we see that we may improve 20.1.3 by adding that equality holds if and only if u and v are linearly dependent. Reverting to the triangle inequality ($\mathfrak{D}_3$) we infer that (for the ρ of 20.1.2) $\rho(u, v) + \rho(v, w) = \rho(u, w)$ if and only if there are real numbers α, β, γ not all zero such that

$$\alpha u + \beta v + \gamma w = 0, \quad \text{and} \quad \alpha + \beta + \gamma = 0.$$

Exercise 1

(i) Let V be an inner product space with a finite basis $v_1, v_2, \ldots, v_n$. For each $x = \sum x_i v_i$, $y = \sum y_i v_i$ in V (where $x_i, y_i \in \mathbb{R}$) show that $\langle x, y\rangle = \sum_{i,j} x_i y_j e_{ij}$, where $e_{ij} = \langle v_i, v_j\rangle$. Describe the form of the matrix $E = (e_{ij})$.
(ii) Conversely, given a vector space W over $\mathbb{R}$ with finite dimension n and a *suitable* matrix E, show how to define an inner product on W. Find an inner product on $\mathbb{R}^2$ with respect to which $\|u\| = s^2 + 12t^2 + 5st$, for each $u = (s, t) \in \mathbb{R}^2$. What if 12 is replaced here by 6?
(iii) A **homomorphism** $h: U \to V$ between inner product spaces is defined to be a linear transformation such that h preserves products, i.e., $\langle x, y\rangle_U = \langle hx, hy\rangle_V$ for all $x, y \in U$ (where $\langle\ \rangle_U$ denotes the product in U). Prove that h also preserves distances, i.e., $\rho_U(x, y) = \rho_V(hx, hy)$. Hence prove that h is a monomorphism. [It is usual to call h an **isometry**.]
(iv) Let V be the vector space of all continuous functions $J \to \mathbb{R}$, where J is the interval $[-1, 1] \subseteq \mathbb{R}$. For each $f, g \in V$ define $\langle f, g\rangle = \int_{-1}^{1} f(x)g(x)\,dx$. Prove that V, with this product, is an inner product space. Interpret Schwarz's inequality, using a graph. [Compare (i) above: remember that an integral arises from a sum!]

20.2 Geometry in V

We turn now to the question of orthogonality in V and, more generally, to the introduction of a notion of angle in V. Motivated by geometry (14.6.3), we declare the **angle** α between two non-zero vectors u and v to be the angle given by

20.2.1 $$\cos\alpha = \frac{\langle u, v\rangle}{\|u\|\cdot\|v\|},$$

and we declare two vectors u and v to be **orthogonal** if $\langle u, v\rangle = 0$.

This definition makes good sense in view of Schwarz's inequality and it means that non-zero orthogonal vectors are at right angles. Of course, it leaves an ambiguity of sign in the angle α, but this is to be expected and can be resolved by a convention. The definition may be taken as the starting point for the systematic development of plane trigonometry, where incidentally, $\cos \alpha$ occurs more often than the pure angle α. We now pursue this line briefly.

Let, then, V be an n-dimensional inner product space. We shall develop a 'geometry' in V (whatever its dimension) like Euclidean geometry, as follows. For example, define a 'triangle' PQR in V to be a set $\{p, q, r\}$ of three distinct points, and define its 'sides' to be the segments PQ, QR, RP, where a segment is any set of the form

20.2.2 $$\{r \in V \mid r = su + tv,\ s + t = 1,\ 0 \leqslant s \leqslant 1\}.$$

The **length** of PQ is denoted by $|PQ|$ and is defined to be $\rho(p, q)$. Then $\mathfrak{D}_3$ translates to $|PQ| + |QR| \geqslant |PR|$, or

20.2.3 '*the sum of the lengths of two sides of a triangle is not less than the length of the third*',

—a familiar theorem of Euclidean geometry which now holds in *any* inner-product space V (for example, in that of Exercise 1(iv)).

Next, we prove Pythagoras's theorem in our geometry. Motivated by 14.6.5, we say that in V, triangle PQR is **right-angled** at Q iff

$$\langle p - q, r - q \rangle = 0.$$

Then we have *Pythagoras's theorem for* V (and its converse):

20.2.4 $$|PR|^2 = |PQ|^2 + |QR|^2 \;:\Leftrightarrow:\; PQR \quad \textit{is right-angled at } Q.$$

Proof. For brevity write u^2 for $\langle u, u \rangle$. Then

$$\begin{aligned} |PQ|^2 + |QR|^2 - |PR|^2 &= (p - q)^2 + (q - r)^2 - (p - r)^2 \\ &= 2(\langle p, r \rangle - \langle p, q \rangle - \langle q, r \rangle + q^2) \\ &= 2(\langle p - q, r - q \rangle) \end{aligned}$$

so the left-hand side is zero iff the right-hand side is zero. ∎

20.2.5 We define a **plane** in V by analogy with 14.9.2 to be any set of the form $\{v \in V \mid \langle v, p \rangle = k\}$, where $p \in V$ and $k \in \mathbb{R}$ are fixed. Just as in 14.9.8, such a plane separates V into the regions $\langle v, p \rangle < k$, $\langle v, p \rangle > k$, and the notion of convexity (14 Exercise 3(v)) applies directly. See also 14 Exercise 11. Similarly, we define a **line** in V to be any set of the form $\{v \in V \mid v = at + b\}$ where a, b are fixed in V, $a \neq 0$, and $t \in \mathbb{R}$; and planes then intersect lines in a manner analogous to the theory of 14 Exercise 11(v). [These ideas are the basis for the theory of linear programming; see Kemeny *et al.* [73].] It follows also from the remarks in 20.1.5 that:

20.2.6 *if* $|PQ| + |QR| = |PR|$, *then* P, Q, R *lie on a line.*

We can go on to introduce 'spheres' in V, by analogy with the treatment in 14.5. Thus, we define a **sphere** in V with centre c and radius a to be the set of all $v \in V$ such that $\|c - v\| = a$. It separates V into the 'inside ball' $\{v \mid \|c - v\| \leqslant a\}$ and the 'exterior' $\{\|c - v\| \geqslant a\}$, the former of these subsets being convex. Moreover, the line $v = qt + b$ meets the sphere in at most two points; and if they coincide to the single point p, then $\prec p - c, q \succ = 0$ (and the line is called a 'tangent' to the sphere at p). Similarly for tangent planes to spheres in V; the *algebra* is exactly the same as that in 14 Exercise 11(iv), but of course we cannot appeal to pictures for proofs (although they are indispensable for the motivation of proofs).

Exercise 2

(i) Prove that a plane X or a line l is a vector subspace of V iff $0 \in X$ or $0 \in l$. Prove that if a line l intersects a plane X, then $l \cap X$ is a singleton or l: similarly when l intersects a line.
(ii) Let S be a sphere in V with centre c. A **chord** of S is defined to be any segment AB, where $A, B \in S$. If M is the mid-point of AB (i.e., $M \in AB$ and $|AM| = |MB|$), prove that $m = \frac{1}{2}(a + b)$, and that $\prec c - m, a - b \succ = 0$. Prove also the 'rectangle property': if lines l, m in V meet in J, and if they meet S in A, B, C, D, respectively, then $|JA| \cdot |JB| = |JC| \cdot |JD|$. (Cf. 14 Exercise 9(vii).)
(iii) Show that the triangle PQR cannot be right-angled at both P and Q.
(iv) Let l, m be lines in V, which intersect in Q. They are **perpendicular** iff for some $P \in l$, $R \in m$, PQR is a right-angled triangle at Q. Prove that, for any $P' \in l - Q$, $R' \in m - Q$, $P'QR'$ is right-angled at Q. For any triangle ABC in V, a segment AP is called an **altitude** (from A) iff $P \in BC$ and APB is right-angled at P. Prove that APC is right-angled at P and that there exists exactly one such P. Prove that the three altitudes of ABC meet in exactly one point.
(v) The lines $l: v = at + b$, $m: v = ct + d$ in V are called **parallel** (written $l \| m$) iff there exists an $s \in \mathbb{R} - \{0\}$ such that $a = sc$. Prove that this is an equivalence relation on the set of all lines of V. Prove that, given l and any $p \in V$ not on l, then there is exactly one line containing p and parallel to l. Suppose V has the property: given any two lines u, v, then either u meets v or $u \| v$. Prove that $\dim V = 2$, and conversely. [The previous statement, when $\dim V = 2$, is the so-called 'parallel postulate for Euclidean geometry'. There are geometries, called non-Euclidean, in which it is false. Their discovery was an epoch-making event in human thought; see Bell [10] on Bolyai and Lobachevsky.]
(vi) If X is a plane in V and $p \in V - X$, prove that there exists a unique point $q \in X$ such that $\rho(p, q)$ is least (i.e., $\rho(p, q) < \rho(p, r)$ for every $r \in X - q$). Do the same problem when X is replaced by a sphere with centre not at p.
(vii) In the space of Exercise 1(iv) prove that the functions $\sin 2\pi nx$, $\cos 2\pi mx$ are all mutually orthogonal if $n \neq m$ $(n, m \in \mathbb{Z})$.
(viii) Let U be a subspace of the inner product space V. Let W be the set of all $v \in V$ such that $\prec u, v \succ = 0$ for *every* $u \in U$. Show that W is a subspace of V Draw a picture when $V = \mathbb{R}^3$ and U is a plane.

20.3 Orthogonality

We study further the notion of orthogonality. Let V be an n-dimensional inner product space. A basis $(u_1, \ldots, u_n)$ for V is said to be **orthonormal** if

20.3.1
$$\langle u_i, u_j \rangle = 0 \quad \text{if} \quad i \neq j,$$
$$= 1 \quad \text{if} \quad i = j.$$

For example, in $\mathbb{R}^3$, the vectors **i**, **j**, **k** form an orthonormal basis. More generally:

20.3.2 THEOREM. *The inner product space V possesses an orthonormal basis.*

Proof. Let $v_1, \ldots, v_n$ be a basis for V. We proceed to construct an orthonormal basis $u_1, \ldots, u_n$ out of the basis $v_1, \ldots, v_n$. We begin by setting $u_1 = v_1/\|v_1\|$; this is legitimate, since $v_1 \neq 0$. Then $\langle u_1, u_1 \rangle = 1$ and $u_1, v_2, \ldots, v_n$ is a basis. We suppose, as an inductive hypothesis, that we have constructed a basis $u_1, \ldots, u_m, v_{m+1}, \ldots, v_n$ for V such that 20.3.1 holds for $i, j \leqslant m$; and we suppose $m < n$ (otherwise there is nothing more to do!). Set $w = v_{m+1} - \sum_{i=1}^{m} \langle v_{m+1}, u_i \rangle u_i$. It then follows from 20.3.1 and $\mathfrak{IP}_2$ that $\langle w, u_i \rangle = 0$, $i = 1, \ldots, m$. Also $w \neq 0$, since $u_1, \ldots, u_m, v_{m+1}$ are linearly independent. Thus we may set $u_{m+1} = w/\|w\|$. Then $u_1, \ldots, u_m, u_{m+1}, v_{m+2}, \ldots, v_n$ is again a basis for V and 20.3.1 holds for $i, j \leqslant m + 1$. Proceeding in this way we construct an orthonormal basis $u_1, \ldots, u_n$ for V. ■ This is called the **Gram–Schmidt** process.

Exercise 3

(i) In 19 Exercise 3(iv), extract an orthonormal basis from the generating set given there, using the process given in the proof of Theorem 20.3.2.
(ii) Let V, W be inner product spaces of finite dimension n, and let $(v_1, \ldots, v_n)$, $(w_1, \ldots, w_n)$ be orthonormal bases for V, W respectively. Prove that there is an isometry (see Exercise 1(iii)) $h: V \to W$ such that $h(v_i) = w_i$, $1 \leqslant i \leqslant n$. [Thus *all inner product spaces of the same finite dimension are isomorphic.*]
(iii) In the inner product space V with orthonormal basis $v_1, \ldots, v_n$, each element x is of the form $\sum x_i v_i$. Prove that $\|x\| = \sum x_i^2$. Hence prove that we cannot define an inner product on the vector space $\mathbb{Z}_p^n$ (for example).

We stressed in 19.3.1 that if U is a subspace of the vector space V, then the relation $V = U \oplus W$ does not uniquely determine W. However, if V is an inner product space, then there is a unique subspace W satisfying this relation which is orthogonal to U. This is called the *orthogonal complement* of U. Indeed, let us define the **orthogonal complement**, $U^\perp$, of U to be the set of vectors in V orthogonal to each vector in U. The bilinearity ($\mathfrak{IP}_2$) of the inner product ensures that $U^\perp$ is a subspace of V (see Exercise 2(viii)) and we prove

20.3.3 THEOREM. $V = U \oplus U^{\perp}$.

Proof. First, $U \cap U^{\perp} = 0$; for if $v \in U \cap U^{\perp}$, then $\langle v, v \rangle = 0$, so by $\mathfrak{IP}_3$ $v = 0$. Now let v be an arbitrary vector in V. Choose an orthonormal basis $(u_1, \ldots, u_m)$ for U and consider the vectors $u = \sum_{i=1}^{m} \langle v, u_i \rangle u_i$, $w = v - u$. Then $u \in U$ and $\langle w, u_i \rangle = 0$, $i = 1, \ldots, m$, so that $w \in U^{\perp}$. Thus every vector is expressible as $v = u + w$, where $u \in U$ and $w \in U^{\perp}$; and by 19.3.1 the theorem is proved. ■

This theorem has the consequence that, in an inner product space V, every subspace U is furnished with a function (called the **canonical projection**) $p: V \to U$. For since each v is uniquely expressible as $v = u + w$, ($u \in U$, $w \in U^{\perp}$), then the map $v \to u$ defines a projection $p: V \to U$, that is, a linear map $V \to U$ such that $p(u) = u$ for each $u \in U$. Of course, there is such a projection for a subspace U of any vector space V, but the projection depends on the choice of space W satisfying $V = U \oplus W$; the special feature in an inner product space is the existence of a preferred or *canonical* projection.

Exercise 4

(i) Prove that $\dim V = \dim U + \dim U^{\perp}$.
(ii) Show that a plane has dimension $n - 1$.
(iii) In (i), prove that $V/U \approx U^{\perp}$.

20.4 Duality

In 19 Exercise 7(v) we (surreptitiously) introduced the *dual vector space* V^* of a given vector space V. Namely, if V is a vector space over F, then its **dual** V^* is the set $H(V, F)$ of all linear functions $V \to F$, and the vector-space structure in V^* is given as in 19.4.5 by

20.4.1 $$(x\phi + y\psi)(u) = x\phi(u) + y\psi(u), \qquad \phi, \psi \in V^*, \quad x, y \in F, \quad u \in V.$$

When V is of finite dimension n, the same exercise showed that $\dim V^* = n \cdot \dim F = n$; so by 19.4.10, V is isomorphic to V^*. We now look further at this situation. First, besides defining the dual of a *vector space*, we must define the dual of a *homomorphism*. Thus, given a homomorphism $\delta: U \to V$, define its **dual**, $\delta^*: V^* \to U^*$ as follows. The value of δ^* at $\phi \in V^*$ is that function $\delta^*(\phi): U \to F$ whose value at $u \in U$ is $\phi(\delta u)$. [We warn the reader that this definition needs to be chewed over! However, though unfamiliar, it is essentially quite simple.] Observe that U, V here need not be finite-dimensional. Observe also that δ^* goes from V^* to U^*, *not* from U^* to V^*.

Exercise 5

(i) Verify that δ^* is a homomorphism.
(ii) If U is a subspace of V and δ is the inclusion map, prove that $\delta^*(\phi) = \phi | U$. If $U = V$ and $\delta = \text{id}$, prove that $\delta^* = \text{id}$.

(iii) If, in (i), δ is onto, prove that δ^* has kernel zero. If δ has kernel zero and V is finite-dimensional, prove that δ^* is onto.
(iv) Given linear transformations $U \xrightarrow{\delta} V \xrightarrow{\eta} W$, prove the 'contravariance' rule (see 6.2.9) $(\eta \circ \delta)^* = \delta^* \circ \eta^*: W^* \to U^*$. Hence (using (ii) and 2.9.8), prove that if δ is an isomorphism, so is δ^*, and $(\delta^*)^{-1} = (\delta^{-1})^*$.

We remarked that $V \approx V^*$, provided V is finite-dimensional. However, the isomorphism given by Theorem 19.4.10 is not 'canonical'; that is to say, to describe an isomorphism we had to pick (somewhat arbitrarily) a basis for V. So far all we have said has nothing to do with the inner product and could have been discussed in Chapter 19. However, we now show that if $F = \mathbb{R}$ and V is an inner product space, then an isomorphism between V and V^* may be defined which is independent of the choice of basis.

Thus, let V be a (finite-dimensional) inner product space over $\mathbb{R}$. Consider the map $\lambda: V \to V^*$, whose value at $u \in V$ is the function $V \to \mathbb{R}$ given by

20.4.2 $$(\lambda(u))(v) = \langle u, v \rangle, \qquad (u, v \in V).$$

Then the bilinearity of the inner product ensures that $\lambda(u) \in V^*$ and λ is a homomorphism. Property $\mathfrak{IP}_3$ shows that λ is a monomorphism. Hence, since we know that V and V^* are of the same dimension (being isomorphic vector spaces) we deduce from 19 Exercise 7(iv):

20.4.3 THEOREM. *The map* $\lambda: V \to V^*$ *given by* 20.4.2 *is a vector-space isomorphism.* ■ Observe that since λ is onto, every linear function $f: V \to F$ is given by a projection onto some direction (by 20.4.2).

We will use λ to induce an inner product structure in V^* in the obvious way. Thus if $\phi, \psi \in V^*$, then $\phi = \lambda(u)$, $\psi = \lambda(v)$ for some unique $u, v \in V$ and we set $\langle \phi, \psi \rangle = \langle u, v \rangle$.

Exercise 6

(i) Given any vector space V over F, define $e: V \to V^{**}$ by

$$e(v)(\phi) = \phi(v), \qquad v \in V, \qquad \phi \in V^*.$$

Show that e is a monomorphism. [You may assume that, even in an infinite-dimensional vector space, every subspace is a direct summand, i.e., $U \subseteq V \Rightarrow V = U \oplus W$ for some W.] Hence deduce that e is an isomorphism (indeed, a canonical isomorphism) if V is finite-dimensional.
(ii) Show that if $\delta: U \to V$, then the diagram

$$\begin{array}{ccc} U & \xrightarrow{\delta} & V \\ \downarrow e & & \downarrow e \\ U^{**} & \xrightarrow{\delta^{**}} & V^{**} \end{array}$$

is 'commutative', $\delta^{**}e = e\delta: U \to V^{**}$. Hence (using (i) above), deduce the second assertion of Exercise 5(iii) from the first.

Next, let $(u_1, u_2, \ldots, u_n)$ be an orthonormal basis for V. We find a basis for V^* as follows. Each $v \in V$ can be expressed uniquely as $v = \sum x_j u_j$ $(x_j \in \mathbb{R})$, whence by 20.4.2,

20.4.4
$$\lambda(u_i)(v) = \prec u_i, v \succ = \sum_j x_j \prec u_i, u_j \succ = x_i.$$

Thus $\lambda(u_i)$ is the 'ith co-ordinate function' u_i^*, which assigns to each v its ith co-ordinate relative to the basis $(u_1, \ldots, u_n)$. Since $\lambda: V \to V^*$ is an isomorphism, it follows that

20.4.5 *An orthonormal basis for V^* is given by the set of co-ordinate functions u_i^*.*

By 20.4.4 we see that $u_i^*(u_j)$ is 1 if $i = j$, and zero otherwise. We call $(u_1^*, \ldots, u_n^*)$ the basis for V^* *dual* to $(u_1, \ldots, u_n)$. Plainly a dual basis (to a given basis) may be defined in any vector space by using this characterization.

Now let $\delta: V \to V$ be a homomorphism, and let A be its matrix $M(\delta)$ (see 19.4.7) with respect to the given orthonormal basis $(u_1, \ldots, u_n)$. We look for the matrix of $\delta^*: V^* \to V^*$, with respect to the basis $(u_1^*, \ldots, u_n^*)$ of V^*. This is really a question about dual bases, and we answer it without reference to inner products.

Given any matrix A, we define the **transpose** A^T of A to be the matrix B given by

$$b_{ij} = a_{ji}.$$

That is, A^T is obtained from A by interchanging rows and columns. We prove

20.4.6 THEOREM. *If A is the matrix associated with $\delta: V \to V$ with respect to the basis $u_1, \ldots, u_n$, then A^T is the matrix associated with $\delta^*: V^* \to V^*$ with respect to the dual basis $u_1^*, \ldots, u_n^*$.*

Proof. Now

$$\begin{aligned} u_i^*(u_j) &= 0 \quad \text{if} \quad i \neq j \\ &= 1 \quad \text{if} \quad i = j. \end{aligned}$$

Thus by definition of δ^* following 20.4.1,

$$(\delta^* u_i^*)(u_k) = u_i^*(\delta u_k) = u_i^*\left(\sum_l a_{kl} u_l\right) = a_{ki},$$

using the definition of A. But a_{ki} is also $\left(\sum_j a_{ji} u_j^*\right)(u_k)$. Since this is true for each k, then by definition of equality of functions we have

$$\delta^* u_i^* = \sum_j a_{ji} u_j^*,$$

which establishes the theorem. ■

Exercise 7

(i) If A, B are two $n \times n$ matrices over $\mathbb{R}$, prove that $(A + B)^T = A^T + B^T$, $(\lambda A)^T = \lambda A^T (\lambda \in \mathbb{R})$ and $(AB)^T = B^T A^T$.
(ii) In 19.4.7, let $U = V$, $m = n$; and define $M^T: H(V, V) \to \mathscr{M}_{nn}$ by $M^T(\phi) = (M(\phi))^T$. Prove that M^T is still an isomorphism of vector spaces, but that now $M^T(\phi \circ \psi) = M^T(\phi) M^T(\psi)$, thus dealing with the reversal of order mentioned in 19 Exercise 6(vi).
(iii) In the notation of Theorem 20.4.6, let the ith co-ordinate of $x \in V$ be x_i $(= u_i^* x)$, and let x_i' be that of $\delta(x)$. Prove that $x_i' = \sum_j a_{ij} x_j$. [Thus, A tells us how δ transforms basis elements, while A^T tells us how to transform co-ordinates. Since the latter were taken to be of basic interest in the classical theory, the definition (19.4.8) of multiplication of matrices got formulated as in (ii) above, and not satisfying the 'reversed' form of 19 Exercise 6(ii).]
(iv) Given a non-zero $f \in V^*$, prove that the set $P = \{v \in V \mid f(x) = 0\}$ is a plane (see 20.2.5) of the form $\{v \in V \mid \prec v, p \succ = 0\}$; where $f = \lambda(p)$ in the isomorphism λ of 20.4.2. Show that the line (see 20.2.5) through f and the origin of V^*, is $\lambda(P^\perp)$ (orthogonal complement in V). [Thus V^* can be identified, geometrically speaking, with the set of all planes through the origin O of V; and for each $v \in V$, $\lambda(v)$ is then the plane orthogonal to v, through O.]

20.5 Orthogonal Transformations

Our last definition in this chapter introduces one of the most important notions of linear geometry (see also 15 Exercise 7(xiv)). A linear transformation $\delta: V \to V$ of an inner product space into itself is said to be **orthogonal** if it preserves the inner product; that is, if

$$\prec u, v \succ = \prec \delta u, \delta v \succ, \qquad u, v \in V,$$

(so δ is an *isometry* in the sense of Exercise 1(iii)). Plainly such a linear transformation is, in fact, an isomorphism; for if $\delta u = 0$, then $\prec u, u \succ = \prec \delta u, \delta u \succ = 0$, so that $u = 0$ and δ is a monomorphism, whence by 19 Exercise 7(iv), δ is an isomorphism.

Exercise 8

An automorphism of a vector space V is an isomorphism of V onto itself (as a vector space). Thus an isometry of V is an isomorphism of V onto itself (as an inner product space). Show that the orthogonal transformations form a subgroup of the group of all automorphisms $V \to V$. Is this subgroup normal?

Following the example of 15 Exercise 7(xiv), we take up the question of the general form of a *matrix* corresponding to an orthogonal transformation.

We say that a matrix A is **orthogonal** if it is the matrix of an orthogonal transformation with respect to an orthonormal basis.

20.5.1 THEOREM. *The matrix is orthogonal if and only if $AA^T = I$ (where I is the identity matrix (δ_{ij}) of 19 Exercise 6(iv)); equivalently, if and only if*

$$A^T = A^{-1}.$$

Proof. Let $\delta: V \to V$ be orthogonal. Now δ induces $\bar{\delta}: V^* \to V^*$ by the rule $\bar{\delta}\lambda = \lambda\delta$, with λ as in 20.4.2. That is, writing u^* for $\lambda(u)$, $u \in V$,

$$\bar{\delta}(u^*) = (\delta(u))^*.$$

Of course $\bar{\delta}$ *is not to be confused with* δ^*. Indeed, the matrix of $\bar{\delta}$ with respect to the basis $u_1^*, \ldots, u_n^*$ is plainly just the same matrix A as that of δ with respect to the basis $u_1, \ldots, u_n$. Now δ is orthogonal if and only if

$$\prec \delta u, \delta v \succ = \prec u, v \succ.$$

But $\prec u, v \succ = u^*(v)$, by 20.4.2, while

$$\prec \delta u, \delta v \succ = (\delta u)^*(\delta v) = (\delta^*(\delta u)^*)(v) = ((\delta^*\bar{\delta})(u^*))(v).$$

Thus δ is orthogonal if and only if $\delta^*\bar{\delta} = 1$, the identity transformation of V^*. Passing to the associated matrices and recalling Theorem 20.4.6 and the first fact about matrix multiplication, we find that δ is orthogonal if and only if $AA^T = I$, as asserted. ■

Exercise 9

Let $(u_1, \ldots, u_n)$, $(v_1, \ldots, v_n)$ be two orthonormal bases for V and let $v_i = \sum a_{ij}u_j$. Prove that the matrix $A = (a_{ij})$ is orthogonal. {Let $u_i = \sum b_{ij}v_j$ and compute $\prec u_i, v_j \succ$ two ways to prove $b_{ij} = a_{ji}$.} Conversely, show that if an orthonormal basis is transformed by an orthogonal transformation, the result is again an orthonormal basis.

Chapter 21

INEQUALITIES AND BOOLEAN ALGEBRA

21.1 Inequalities

It is a common experience that students who have studied algebra in a traditional school course are well able to handle equations but are quite unable to manipulate (and even, perhaps, to understand) inequalities. We have therefore included this chapter in order to exhibit the algebraic nature of inequalities.

The name 'inequality' is misleading; for we do not propose simply to study the relation $x \neq y$ on a given set S. Instead, we suppose that the set S is provided with an *order relation* $x \geqslant y$ (to be read† 'x is greater than or equal to y') and it is with such order relations that we will be concerned. An important example which will receive major attention here is the set $\mathbb{R}$ of real numbers in which the order relation $x \geqslant y$ simply means that $x - y \in \mathbb{R}^+$. However we will begin with a general definition; recall the definition of order relation in 4.4. We shall be concerned here with the case when the 'law of trichotomy' (4.4.6) holds.

21.1.1 DEFINITION. Let S be a set and let $\geqslant$ be a relation on S. Then $\geqslant$ is a **linear ordering** if it is

$\mathbb{O}_1$: *transitive*:	$(a \geqslant b$ and $b \geqslant c) \Rightarrow a \geqslant c$;
$\mathbb{O}_2$: *antisymmetric*:	$(a \geqslant b$ and $b \geqslant a) \Rightarrow a = b$;
$\mathbb{O}_3$: *total*:	*for all* $a, b \in S$, *either* $a \geqslant b$ *or* $b \geqslant a$.

A reflexive relation satisfying $\mathbb{O}_1$ and $\mathbb{O}_2$ is called an **ordering** or a **partial ordering**; this more general notion is of great importance in mathematics. Given an order relation $\geqslant$ on S, we write $a > b$ ('a is greater than b') if $a \geqslant b$ and $a \neq b$.

Exercise 1

(i) Let $\geqslant$ be a linear ordering on S. Define a new relation $\leqslant$ ('less than or equal to') by the rule: $a \leqslant b$ if and only if $b \geqslant a$. Show that $\leqslant$ is a linear ordering.

† In particular, $x \geqslant x$ for all x; for since $x = x$, it is a true proposition that 'x is greater than x or $x = x$'. This usage is a common cause of difficulty with students, who are often amazed that $2 \geqslant 2$. Warning: with a general order relation $\geqslant$, no question of 'bigness' (as between numbers) need be involved.

(ii) Is the relation 'softer than' ($<$), in music, a linear ordering?
(iii) Set up an order relation in $\mathbb{Z}[x]$.

If we were primarily concerned with set theory as such, we would now develop the theory of linear orderings (or partial orderings) on sets. However, we are here concerned with algebra and so we pass immediately to the study of linear orderings in integral domains. For such orderings to yield algebraic information it is, of course, essential that the order relation be intimately tied to the algebraic structure. We are thus led to the following definition.

21.1.2 DEFINITION. Let R be an integral domain. A subset R^+ of R is called an **order** or **ordering subset** if it is closed under addition and multiplication, $0 \in R^+$, and for all non-zero $a \in R$, either $a \in R^+$ or $-a \in R^+$, but not both.

It follows that, if we write R^- for the set of all $-a$, $(a \in R^+)$, then

21.1.3 $$R^+ \cup R^- = R, \qquad R^+ \cap R^- = \{0\}.$$

Plainly, $R^- \cdot R^- \subseteq R^+$, since R^+ is closed under multiplication and

$$(-a)(-b) = ab.$$

Let us write $a \geqslant b$, if $a, b \in R$ and $a - b \in R^+$. This obviously generalizes the familiar order $\geqslant$ on $\mathbb{R}$.

21.1.4 PROPOSITION. *The relation $\geqslant$ is an order relation on R with the following properties:*

(i) *if $a \geqslant b$, then $a + c \geqslant b + c$, all $c \in R$;*
(ii) *if $a \geqslant b$, then $ac \geqslant bc$, all $c \in R^+$.*

Proof. The relation is transitive, since if $a - b \in R^+$ and $b - c \in R^+$, then $a - c \in R^+$, since R^+ is closed under addition. The relation is antisymmetric, since if $a - b \in R^+$ and $b - a \in R^+$, then $a = b$. Thirdly, it is total, since either $a - b \in R^+$ or $b - a \in R^+$.

As to the algebraic properties (i), (ii) above, (i) follows immediately, since $a - b = (a + c) - (b + c)$, and (ii) is a consequence of the fact that R^+ is closed under multiplication. ■

21.1.5 PROPOSITION. *Let R^+ be an order on R. Then $a^2 > 0$ for all non-zero $a \in R$; in particular, $1 > 0$. Further, $a^2 + b^2 = 0$ if and only if $a = b = 0$.*

We leave the proof to the reader, who should generalize the final assertion to the case $a_1^2 + \cdots + a_n^2 = 0$. ■

Exercise 2

(i) Show that it is impossible to define an order on $\mathbb{C}$, or on the fields $\mathbb{Z}_p$, p prime.
(ii) Show that if $a, b \in R$ and $ab > 0$, $b > 0$, then $a > 0$. Hence show that if F is a field and $a > 0$, $a \in F$, then $1/a > 0$. Deduce that if $x, y \in F$, and $x > y > 0$, then $1/y > 1/x$.

Proposition 21.1.4 has a converse which we now enunciate; we leave the proof as an exercise.

21.1.6 PROPOSITION. *Let R be an integral domain and let $\geqslant$ be an order relation on R satisfying conditions* (i) *and* (ii) *of Proposition* 21.1.4.

Set $R^+ = \{a \in R | a \geqslant 0\}$. Then R^+ is an order in R such that the corresponding order-relation is just $\geqslant$. ■

In the light of this proposition, we will feel entitled to refer to either R^+ or the corresponding order relation, as an order on R.

Exercise 3

Show how an order on R may be used to induce an order on the quotient field of R [see 23.3.8, 23 Exercise 8(iii)].

If R and S are ordered sets and $f: R \to S$ is a function, then f is order-preserving if the relation $a \geqslant b$ implies $fa \geqslant fb$. If now R and S are ordered rings (i.e., integral domains with orders R^+, S^+ defined in them), then we define a homomorphism $\phi: R \to S$ to be **order-preserving** if $\phi(R^+) \subseteq S^+$. Plainly ϕ is then order-preserving as a function, $a \geqslant b \;.\Rightarrow.\; \phi a \geqslant \phi b$; and conversely, if ϕ is a ring-homomorphism and an order-preserving function, then $\phi(R^+) \subseteq S^+$.

21.1.7 EXAMPLE. If R is an ordered ring and $a \in R^+$, then the function $\phi_a: R \to R$ given by $\phi_a(b) = ab$ is an order-preserving function but not a ring-homomorphism, since it fails to preserve products. The embeddings $\mathbb{Z} \subseteq \mathbb{Q} \subseteq \mathbb{R}$ are all order-preserving homomorphisms.

We are mainly concerned here with the standard orders on $\mathbb{Z}$, $\mathbb{Q}$, and $\mathbb{R}$. These orders convert $\mathbb{Z}$ into an **ordered ring** and $\mathbb{Q}$, $\mathbb{R}$ into **ordered fields.** The embeddings $\mathbb{Z} \subseteq \mathbb{Q} \subseteq \mathbb{R}$, as stated above, are *order-preserving* in the sense that $\mathbb{Z}^+ \subseteq \mathbb{Q}^+ \subseteq \mathbb{R}^+$. Indeed, if we read in Part VI the way in which the number system is built up, then we find that, given the natural order on $\mathbb{Z}$, there is a unique order on $\mathbb{Q}$ making $\mathbb{Z}$ an ordered sub-ring of $\mathbb{Q}$, and the natural order on $\mathbb{R}$ is simply that which ensures that the limit of a sequence of elements of $\mathbb{Q}^+$ is in $\mathbb{R}^+$. Despite our main concern, we have nevertheless thought it worthwile to talk a little about general ordered rings in order to illustrate (a) that we could define order relations other than the familiar ones and (b) that the manipulations with inequalities (which we should be able to carry out with facility) are valid in any ordered ring or field (see Exercise 2(ii)).

Exercise 4

(i) Show that $2^{341} > 4^{321}$.

(ii) Use the ordering on $\mathbb{R}$ to show that the equation $x^n = 1$ has exactly two roots in $\mathbb{R}$ if n is even and exactly one root if n is odd. Generalize to an arbitrary ordered field.

21.2 Some Applications

We confine attention now to the ordered field $\mathbb{R}$; we have already described one important inequality in $\mathbb{R}$, namely Schwarz's inequality (Lemma 20.1.4). In fact, by applying it to $\mathbb{R}^n$ we obtain the classical form of the inequality:

21.2.1 $(a_1^2 + \cdots + a_n^2)(b_1^2 + \cdots + b_n^2) \geqslant (a_1b_1 + \cdots + a_nb_n)^2$,

where equality occurs if and only if $a_i = kb_i$, $i = 1, \ldots, n$. The reader may observe that this inequality holds in any ordered ring.

We give two further examples of important inequalities. First, we consider the quadratic polynomial

$$p(x) = ax^2 + 2bx + c \qquad (a \neq 0)$$

in $\mathbb{R}[x]$. Then we assert:

21.2.2 THEOREM. *If $p(\alpha) \neq 0$ for all α in $\mathbb{R}$, then $p(\alpha)$ has the same sign as a for all $\alpha \in \mathbb{R}$.*

Proof. This can be proved by a simple continuity argument. However, we can give a simple algebraic proof by observing that

21.2.3 $$a \cdot p(x) = (ax + b)^2 + ac - b^2.$$

Then p has a zero in $\mathbb{R}$ if and only if $b^2 \geqslant ac$. Now if $b^2 < ac$, then $a \cdot p(\alpha) > 0$ for all $\alpha \in \mathbb{R}$, which establishes the assertion. ■

Exercise 5

Does the conclusion of Theorem 21.2.2 hold for any ordered field?

We continue to study the polynomial p in $\mathbb{R}[x]$ and prove

21.2.4 THEOREM. *Let the quadratic polynomial $p \in \mathbb{R}[x]$ have zeros in $\mathbb{R}$. Then $p(\alpha)$ has the same sign as a except between the zeros of p.*

Proof. Let p have zeros β, γ in $\mathbb{R}$ (note that if p has one zero in $\mathbb{R}$, then by equation 21.2.3 it has two—possibly coincident). Thus

$$p(x) = a(x - \beta)(x - \gamma),$$

whence

$$p(\alpha) = a(\alpha - \beta)(\alpha - \gamma).$$

If $\beta = \gamma$, then $a \cdot p(\alpha) > 0$ unless $\alpha = \beta$, establishing the result. If $\beta \neq \gamma$, suppose without loss of generality that $\gamma < \beta$. Then $a \cdot p(\alpha) > 0$ if $\alpha < \gamma$ or if $\alpha > \beta$; and $a \cdot p(\alpha) < 0$ if $\gamma < \alpha < \beta$. This completes the proof. ■

Exercise 6

(i) Does the conclusion of Theorem 21.2.4 hold for any ordered field?
(ii) Illustrate Theorem 21.2.4 by drawing a graph of the function

$$y = -3x^2 + 9x - 6.$$

Our second example concerns elements of $\mathbb{R}^+$. Given any n elements $a_1, \ldots, a_n$ of $\mathbb{R}^+$, we may form their **arithmetic mean** $(a_1 + \cdots + a_n)/n$ and their **geometric mean** $(a_1 a_2 \cdots a_n)^{1/n}$; we refer to these as A.M.$(a_1, \ldots, a_n)$, G.M.$(a_1, \ldots, a_n)$ and prove the following result, which has applications particularly in probability theory.

21.2.5 THEOREM. *A.M. $\geqslant$ G.M., equality holding if and only if $a_1 = \cdots = a_n$.*

Before proving this, we note that if $a_1 = \cdots = a_n$, then $a_1 = \cdots = a_n =$ A.M. $=$ G.M. We now give the proof.

Proof. We first prove this for $n = 2$; it is, of course, trivial for $n = 1$. We must show that

$$\frac{a_1 + a_2}{2} \geqslant \sqrt{(a_1 a_2)},$$

equality holding if and only if $a_1 = a_2$. But

$$\left(\frac{a_1 + a_2}{2}\right)^2 - a_1 a_2 = \left(\frac{a_1 - a_2}{2}\right)^2 \geqslant 0,$$

equality holding iff $a_1 = a_2$. Since the a's are in $\mathbb{R}^+$, it is legitimate to take square roots; thus the assertion is established when $n = 2$.

We next prove the assertion if $n = 2^k$, by induction on k. The case $k = 1$ is already established. Write $S(n)$ for $\sum_{i=1}^{n} a_i$, $P(n)$ for $a_1 \cdots a_n$; we want to show that $S(n)/n \geqslant (P(n))^{1/n}$, equality implying that all a_i are equal. For typographical reasons, we shall write (temporarily) $p(k)$ for 2^k. Thus

$$\begin{aligned} S(2^k)/2^k &= S(p(k))/p(k) = \left(\sum_{i=1}^{p(k-1)} a_i + \sum_{i=p(k-1)+1}^{p(k)} a_i\right)\Big/p(k) \\ &\geqslant \tfrac{1}{2}((a_1 \cdots a_{p(k-1)})^{1/p(k-1)} + (a_{p(k-1)+1} \cdots a_{p(k)})^{1/p(k-1)}), \\ &= \tfrac{1}{2}(A + B), \quad \text{say.} \end{aligned}$$

This follows by the inductive hypothesis that, for *any* collection of 2^{k-1} numbers in $\mathbb{R}^+$, A.M. $\geqslant$ G.M. We apply it to the collection $a_1, \ldots, a_{p(k-1)}$, and to the collection $a_{p(k-1)+1}, \ldots, a_{p(k)}$. Moreover, the inductive hypothesis implies that equality occurs here if and only if $a_1 = \cdots = a_{p(k-1)} = A$ and $a_{p(k-1)+1} = \cdots = a_{p(k)} = B$. We now apply the theorem for $n = 2$ to infer that, if we do write the last expression as $\frac{1}{2}(A + B)$, then

$$\begin{aligned} \tfrac{1}{2}(A + B) &\geqslant \sqrt{(AB)}, \quad \text{equality occurring only if} \quad A = B \\ &= (P(2^k))^{1/p(k)}. \end{aligned}$$

Thus $S(2^k)/p(k) \geqslant (P(2^k))^{1/p(k)}$ and equality holds if and only if $a_1 = \cdots = a_{p(k-1)} = \cdots = a_{p(k)}$. This establishes the induction, and so proves the theorem when $n = 2^k$ $(= p(k))$.

Finally, we deal with the general case. Given any n, choose k so that $p(k) = 2^k \geqslant n$ [why is this possible?], and set $a_{n+1} = \cdots = a_{p(k)} = \text{A.M.}(a_1, \ldots, a_n)$. Then plainly

$$\text{A.M.}(a_1, \ldots, a_n) = \text{A.M.}(a_1, \ldots, a_{p(k)});$$

set this equal to α. Moreover, by what we have already proved,

$$\alpha^{p(k)} \geqslant a_1 \cdots a_n \alpha^{p(k)-n},$$

equality occurring only if $a_1 = \cdots = a_n \ (=\alpha)$. Thus

$$\alpha^n \geqslant a_1 \cdots a_n$$

$$\text{or} \quad \text{A.M.} \geqslant \text{G.M.},$$

equality occurring only if $a_1 = \cdots = a_n$, and the theorem is completely proved. ■

Exercise 7

(i) The **harmonic mean** of n positive numbers, H.M. $(a_1, \ldots, a_n)$, is defined by

$$\frac{1}{\text{H.M.}(a_1, \ldots, a_n)} = \text{A.M.}\left(\frac{1}{a_1}, \ldots, \frac{1}{a_n}\right).$$

Show that H.M. $\leqslant$ G.M., equality occurring only if $a_1 = \cdots = a_n$.
(ii) A man cycles from A to B at 10 mile/h, and back from B to A at 15 mile/h. What is his average speed? (Compare your answer with methods of computing land- and water-speed records!)
(iii) The equation $x^{10} - 20x^9 + a_2x^8 + a_3x^7 + a_4x^6 + a_5x^5 + a_6x^4 + a_7x^3 + a_8x^2 + a_9x + 1024 = 0$ has all its roots real and positive; find $a_2, \ldots, a_9$.
(iv) Show that Theorem 21.2.5, interpreted as relating to the inequality $S(n)^n \geqslant n^nP(n)$, holds in any ordered ring.
(v) If $x > 0$ in $\mathbb{R}$, and $n \in \mathbb{N}$, prove that $x < y \Rightarrow x^{1/n} < y^{1/n}$.
(vi) Find all $n \in \mathbb{N}$ such that $(\frac{3}{4})^n < \frac{1}{100}$.
(vii) Find all $\alpha \in \mathbb{R}$ such that $5\alpha^2 + 12\alpha < -4$.
(viii) Find all $n \in \mathbb{N}$ such that $n^{1/n} < (n+1)^{1/(n+1)}$. {*Hint:* $(1 + 1/n)^n \to$ e.} Hence show that $\lim_{n\to\infty} n^{1/n} = 1$.
(ix) Find a geometrical interpretation of 21.2.5. {When $n = 2$, consider the rectangle property of a semicircle of diameter $a_1 + a_2$. When $n > 2$, the problem becomes more difficult!}
See also the problems on linear inequalities in Chapter 16.

21.3 Dedekind's Completion of the Rationals

In this section, which may be regarded as an adjunct to Chapter 24, we show how the notions of 21.1 lead to a construction of the real numbers $\mathbb{R}$ starting from $\mathbb{Q}$. The construction follows the method of Richard Dedekind [(1831–1916); see Bell [10], Ch. 27], who in 1872 was the first to give a *proof*, awaited since the Greeks, of the existence of $\mathbb{R}$. His proof consisted in first prescribing *precisely* what requirements $\mathbb{R}$ had to satisfy, and then to make such a system, using $\mathbb{Q}$ as the building blocks. He decided that $\mathbb{R}$ had to be a linearly ordered field, containing a copy of $\mathbb{Q}$, and also satisfying the following condition, which we state before explaining:

21.3.1 (*The lub Condition.*) *If E is a non-empty subset of $\mathbb{R}$, and E is bounded above, then E has a least upper bound.*

Here, 'E is bounded above' means that $\exists M \in \mathbb{R}$ such that M exceeds every element of E; and M is then *an* 'upper bound' of E. Such an upper bound M_0 is **least** (briefly, 'M_0 is a **lub**') if it does not exceed any other upper bound of E; clearly there can be at most one lub, so M_0 is then *the* lub. Geometrically, 21.3.1 states that if E lies on the x-axis in $\mathbb{R}^2$, and does not go off to infinity as x increases to the right, then there is a point M_0 (the lub) on the axis, such that E lies to the left of M_0 and M_0 is the smallest such point. Just why this condition should be so important is discussed in 24.6; but observe that it is not satisfied by $\mathbb{Q}$, since the set $E = \{1, 1.4, 1.41, \ldots\}$ of decimal approximations increasing to $\sqrt{2}$ in $\mathbb{R}$ has $\sqrt{2}$ as lub in $\mathbb{R}$ but has no *least* upper bound in $\mathbb{Q}$, although $3/2$ (for example) is an upper bound of E in $\mathbb{Q}$.

Exercise 8

(i) Show that $\mathbb{N}$ is not bounded above.
(ii) Is the empty set bounded above? Has it a lub?
(iii) Find upper bounds for the sets $\{n^{1/n}\}_{n \in \mathbb{N}}$, and $\{(1 + 1/n)^n\}_{n \in \mathbb{N}}$.
(iv) Verify the **axiom of Eudoxus** (or of Archimedes), that if condition 21.3.1 holds, then, given any $x \in \mathbb{R}$, there is an integer greater than x. {*Hint:* let E be the set of integers $\leqslant x$, and add 1 to its lub.}

21.3.2 The important step in Dedekind's construction is to find a linearly ordered set T which (i) satisfies all the conditions on a linearly ordered field except that subtraction may not always be possible, (ii) satisfies condition 21.3.1, and (iii) contains a copy of the set $\mathbb{Q}^+$ of all rationals $\geqslant 0$ (i.e., we must find a one-one function $\eta\colon \mathbb{Q}^+ \to T$ which preserves addition, multiplication and order; $\mathbb{Q}^+$ is then *embedded* in T). It is then quite easy to embed T in a field F in which subtraction is possible, by the method for constructing $\mathbb{Z}$ out of $\mathbb{Z}^+$ explained in 23.2, and F is then the system $\mathbb{R}$ we require. We shall therefore construct T here; this separation avoids certain technical difficulties concerning negative signs.

Thus, given $q \in \mathbb{Q}^+$, let $S_q = \{x \in \mathbb{Q}^+ | 0 \leqslant x < q\}$, where we exclude q for technical reasons. Hence $0 \in S_q \neq \mathbb{Q}^+$, and if $x \in S_q$ then $S_x \subseteq S_q$; moreover, S_q has a lub (viz.: q), but $q \notin S_q$. This led Dedekind to define a **section** of $\mathbb{Q}^+$ to be any proper subset S, containing 0 but *not* containing its lub (if any), and such that

21.3.3 $$\text{if} \quad x \in S, \quad \text{then} \quad S_x \subseteq S.$$

Since $S \neq \mathbb{Q}^+$, $\exists z \in S - \mathbb{Q}^+$ and S is bounded above by z although no *least* upper bound of S may exist. It was Dedekind's idea to work with sections as substitutes for the lub's which are missing from $\mathbb{Q}^+$. Let, then, T denote the family of all sections of $\mathbb{Q}^+$. We embed $\mathbb{Q}^+$ in T via the function

21.3.4 $$\eta: \mathbb{Q}^+ \to T$$

defined by $\eta(q) = S_q$; η is clearly one-one. Next, we order T by writing

$$S \leqslant S' .\Rightarrow. S \subseteq S',$$

and assert

21.3.5 *T is linearly ordered by $\leqslant$.*

Proof. The properties of inclusion give conditions $\mathbb{O}_1, \mathbb{O}_2$: as to $\mathbb{O}_3$, if sections S, S' are given and $S' \nleqslant S$, then $\exists y, y \in S'$, and $y \notin S$; we claim that then $S \leqslant S'$, to satisfy $\mathbb{O}_3$. For, any s in S satisfies either $s < y$ or $y < s$ by the total ordering of $\mathbb{Q}^+$ (since $s \neq y$) and we cannot have $y < s$, since otherwise condition 21.3.3 would force $y \in S$; thus $s < y \in S'$, so $s \in S'$ by 21.3.3, whence $S \subseteq S'$, implying $S \leqslant S'$. ■

Next we observe that the embedding η in 21.3.4 is *order-preserving*; for if $q \leqslant r$ in $\mathbb{Q}^+$, then $S_q \subseteq S_r$, so $\eta(q) \leqslant \eta(r)$ in T.

Let us now verify that T satisfies the lub condition 21.3.1. We must show that every non-empty subset $E = \{S_\alpha\}_{\alpha \in A}$ of sections in T, for which there is a section S such that $(\forall\alpha)S_\alpha \leqslant S$, has a lub in T. In fact, we claim:

21.3.6 *The lub of E is*

$$S_* = \bigcup_{\alpha \in A} S_\alpha.$$

Proof. We first show that $S_* \in T$.

Plainly $S_* \subseteq S$, so $S_* \neq \mathbb{Q}^+$; $(\forall\alpha)0 \in S_\alpha$ so $0 \in S_*$; and if S_* has lub u in $\mathbb{Q}^+$, then $u \notin S_*$, otherwise $(\exists\alpha)u \in S_\alpha$ and u, being an upper bound of S_*, is an upper bound, and hence the lub of S_α, contrary to the condition that no section contains its lub. Finally, to verify 21.3.3 for S_*, let $x \in S_*$; then $(\exists\alpha)x \in S_\alpha$ whence by 21.3.3 for S_α, $S_x \subseteq S_\alpha \subseteq S_*$, so S_* satisfies 21.3.3. Thus $S_* \in T$. Next, S_* is *an* upper bound of E, since each $S_\alpha \subseteq S_*$, whence $(\forall\alpha)S_\alpha \leqslant S_*$. Finally, if M is any upper bound of E, then $S_* \leqslant M$; for, since each $S_\alpha \leqslant M$, $S_\alpha \subseteq M$, so $\bigcup_{\alpha \in A} S_\alpha \subseteq M$, i.e., $S_* \leqslant M$. ■

Our next job is to introduce into T the operations of addition and multiplication. Define, for $S, S' \in T$:

21.3.7 $$\begin{cases} S + S' = \{s + s' \mid s \in S \text{ and } s' \in S'\}, \\ SS' = \{ss' \mid s \in S \text{ and } s' \in S'\}. \end{cases}$$

Consider $S + S'$ first; does it lie in T? We need to observe that if $x, y \in \mathbb{Q}^+$, then $S_x + S_y = S_{x+y}$ [check this!], i.e., the embedding η in 21.3.4 satisfies

21.3.8 $$\eta(x) + \eta(y) = \eta(x + y),$$

so η is an *additive homomorphism.* Now $S + S'$ contains $0 = 0 + 0$; and it is not all $\mathbb{Q}^+$, since $\exists x, y \in \mathbb{Q}^+$ with $S \subseteq S_x$, $S' \subseteq S_y$ (prove this) and $S + S' \subseteq S_x + S_y = S_{x+y} \neq \mathbb{Q}^+$. Suppose that $S + S'$ has a lub $u \in \mathbb{Q}^+$; we must show that $u \notin S + S'$. If, instead, $u \in S + S'$, then $u = s + s'$ for some $s \in S$, $s' \in S'$, by definition of $S + S'$. Thus s is an upper bound of S, for if $s < s_1 \in S$, then $u < s_1 + s' \in S + S'$, contrary to the assumption that u is an upper bound of $S + S'$. Therefore s is the *least* upper bound of S, contrary to the requirement that no section contains its lub. To avoid this contradiction, we must admit that $u \notin S + S'$, as required. Finally, to verify 21.3.3 for $S + S'$, let $x \in S + S'$. Then $x = s + s'$ (say) where $s \in S$, $s' \in S'$; so $S_x = S_{s+s'} = S_s + S_{s'} \subseteq S + S'$, by using 21.3.5 again. Thus $S_x \leqslant S + S'$, and 21.3.3 is verified: so $S + S' \in T$.

Observe that addition in T is associative and commutative, since the same is true of addition in $\mathbb{Q}^+$. Also $S_0 + S = S$, so S_0 $(=\{0\})$ acts as a zero.

Products are dealt with similarly (see Exercise 9 below), and the distributive law can be verified. Moreover, the embedding η in 21.3.4 is a *multiplicative homomorphism*, since the definition of product in 21.3.7 implies:

21.3.9 $$\forall x, y \in \mathbb{Q}^+, \qquad S_x S_y = S_{xy} \quad \text{(i.e., } \eta(x)\eta(y) = \eta(xy)\text{)}.$$

Thus η preserves order, addition and multiplication. Hence the programme outlined in 21.3.2 is completed.

Exercise 9

(i) Let S be a section, and let $x \in S$. Prove that $\exists y \in S$ such that $x < y$.
(ii) Verify in detail the homomorphism properties 21.3.8 and 21.3.9. Use the latter to verify that SS' in 21.3.7 lies in T.
(iii) Verify that the product SS' in 21.3.7 is associative and commutative, that $S_1 S = S$ (so S_1 acts as unit element) and that if $S \neq S_0$ in T, then $SU = S_1$, where

$$U = \bigcap_{0 < x \in S} S_{x^{-1}};$$

and verify that $U \in T$. Thus reciprocals exist in T.
(iv) Verify the distributive law in T.
(v) Using the embedding η in 21.3.4, show that the image by η, of a section $S \in T$, has a least upper bound in T, namely S.
(vi) Show that if we define 'sections' in T analogously to those in $\mathbb{Q}^+$, and let W denote the resulting set of sections, then the analogue of η in 21.3.4 is a bijection $T \to W$. Thus *if we repeat Dedekind's construction on T instead of $\mathbb{Q}^+$ we get nothing essentially new.*
(vii) Prove that if $S + S' = S + S''$ in T, then $S' = S''$. Hence suggest how to introduce 'negatives'.

21.4 Boolean Algebra

Next we turn briefly to a type of algebra which is not concerned with numbers at all, but where an order relation is of interest. Our main purpose in including this section is to emphasize to the reader that algebra need not be about numbers. To make the treatment independent of the material in Chapter 8, we first collect together the defining axioms for a Boolean algebra.

We consider a set R with two binary operations $+$, $\cdot$, satisfying the following axioms.

$\mathfrak{B}_1$: $a + (b + c) = (a + b) + c$ *and* $a(bc) = (ab)c$;

$\mathfrak{B}_2$: $a + b = b + a$ *and* $ab = ba$;

$\mathfrak{B}_3$: $a(b + c) = ab + ac$ *and* $a + (bc) = (a + b)(a + c)$.

$\mathfrak{B}_4$: *there exist distinct elements* $0, 1 \in R$ *such that*
$a + 0 = a$ *and* $a \cdot 1 = a$;

$\mathfrak{B}_5$: *to each* $a \in R$, *there exists* $a' \in R$ *such that*

$a + a' = 1$ *and* $aa' = 0$.

Comparing these axioms with axioms $\mathfrak{A}_1$–$\mathfrak{A}_8$ of 9.2 (which characterize a commutative ring), the reader will notice that $\mathfrak{A}_4$ is missing, while, if he has read Chapter 8, he may wonder about the Pierce–De Morgan law; but see 21.4.5 below. Also, the second part of $\mathfrak{B}_3$, together with $\mathfrak{B}_5$, are new. Of course, each part of $\mathfrak{B}_5$ holds in any field, but no a' in a field will do both jobs (except in $\mathbb{Z}_2$), and the second part of $\mathfrak{B}_3$ looks at first sight very strange. However, we have so arranged the axioms above that each axiom comprises two statements, one of which can be obtained from the other by interchanging $+$ and $\cdot$. Thus the strange second part of $\mathfrak{B}_3$ is just the counterpart of the familiar distributive law. However, it is certainly false for $\mathbb{Z}$ (or any ring containing $\mathbb{Z}$).

What, then, is the relevance of this algebraic system? It is just that it is the algebraic system appropriate to the study of subsets of a given set. Thus let S be a set and let $\mathfrak{p}S$ be the set of subsets of S. Then $\mathfrak{p}S$ is turned into an algebraic system satisfying $\mathfrak{B}_1$–$\mathfrak{B}_5$ by taking $+$ to be set-union and $\cdot$ to be set-intersection; moreover 0 is then the empty set, 1 is S itself and a' is the complement of a. The reader will readily verify the axioms, using the rules for sets given in 1.7. This interpretation of the given algebraic system justifies its introduction; indeed, it was introduced by George Boole (1815–1864) to study set-theory and formal logic, and the system has come to be known as a **Boolean algebra.** We should admit that, in a very precise sense, it has no other justification! For it may be proved that *every Boolean algebra is isomorphic to an algebra of sets*; see Exercise 13(i) below.

The symmetry of the axioms with respect to $+$ and $\cdot$ enables us to deduce, from any given theorem on Boolean algebras, a new theorem called its **dual,** obtained by interchanging the roles of $+$ and $\cdot$, and 0 and 1.

The reader is advised to write down for himself the duals of the assertions which follow. (For a precise formulation of this duality, see 8 Exercise 6(v).) Compare also the results of 3.6.

21.4.1 THEOREM. *The element a' is uniquely determined by $\mathfrak{B}_5$, while $a'' = a$.*

Proof. Suppose also $a + \bar{a} = 1$, $a\bar{a} = 0$. Then

$$a' = a' \cdot 1 = a'(a + \bar{a}) = a'a + a'\bar{a} = 0 + a'\bar{a} = a'\bar{a},$$

by $\mathfrak{B}_4$, $\mathfrak{B}_5$, etc.

Similarly $\bar{a} = \bar{a}a'$; but $a'\bar{a} = \bar{a}a'$, so $a' = \bar{a}$. Since $a' + a = 1$ and $a'a = 0$, we have $a'' = a$. ■

We call the operation $a \to a'$ (of forming a' from a) **complementation** and call a' the **complement** of a.

21.4.2 PROPOSITION. *For all $b \in R$, $b^2 = b$ (i.e., multiplication in R is* **idempotent**).

Proof. Put $a' = \bar{a} = b$ in the second line of the proof of Theorem 21.4.1. ■

21.4.3 PROPOSITION. *For all $a \in R$, $a \cdot 0 = 0$.*

Proof. $$a \cdot 0 = a \cdot (aa') = a^2a' = aa' = 0. \quad ■$$

We observe that we would arrive at a contradiction if we endeavoured to incorporate axiom $\mathfrak{A}_4$ into our system. For suppose that there existed $(-a)$ with $a + (-a) = 0$. Then

$$\begin{aligned} 0 = a + (-a) &= (a + a) + (-a) \qquad \text{(by the } \textit{dual} \text{ of 21.4.2)} \\ &= a + (a + (-a)) \\ &= a + 0 \\ &= a. \end{aligned}$$

But since $1 \neq 0$, the equality $a = 0$ does *not* hold throughout R! ■

21.4.4 COROLLARY (by duality). $a + 1 = 1$. ■

Next, we prove the 'rules of Pierce and De Morgan' in R; we state one of them, since the other follows by duality.

21.4.5 PROPOSITION. *For all a, b in R,*

$$(a + b)' = a'b'.$$

Proof. By Theorem 21.4.1, it suffices to show that $a'b'$ is the complement of $a + b$, i.e., that $\mathfrak{B}_5$ holds in the form

$$a + b + a'b' = 1, \qquad (a + b)a'b' = 0.$$

Now by the second half of $\mathfrak{B}_3$,

$$a + b + a'b' = (a + b + a')(a + b + b') = (1 + b)(1 + a)$$

(using associativity and commutativity)

$$= 1 \cdot 1 = 1$$

by 21.4.4 and 21.4.2. The second equation is proved similarly, so $a'b'$ satisfies the requirements for being $(a + b)'$. ■

Exercise 10

(i) Show that $0' = 1$, $1' = 0$.
(ii) State and prove the dual of the rule 21.4.5.
(iii) Show that the equation $a + x = b$ has a solution if and only if $a + b = b$. Is the solution unique (look at a Venn diagram)? Dualize.

21.5 Ordering a Boolean Algebra

Let R be a Boolean algebra as before. We shall now introduce an order-relation into R, based on the following proposition.

21.5.1 PROPOSITION. *If $a + b = b$, then $ab = a$.*

Proof. Let $a + b = b$, then $ab' = ab' + 0 = ab' + bb' = (a + b)b' = bb' = 0$. Thus $a = a \cdot 1 = a(b + b') = ab + ab' = ab + 0 = ab$. ■

Since the converse of this proposition is the dual, we conclude that the two conditions $a + b = b$, $ab = a$ are equivalent. We use these conditions to introduce a partial ordering into R, namely,

21.5.2 DEFINITION. $a \leqslant b$ in R if $a + b = b$ (equivalently, if $ab = a$).

We must verify transitivity ($\mathbb{O}_1$) and antisymmetry ($\mathbb{O}_2$). To demonstrate the former, suppose $a + b = b$, $b + c = c$. Then $a + c = a + (b + c) = (a + b) + c = b + c = c$, so that $a \leqslant c$. Antisymmetry is quite obvious. Of course this partial ordering is not, in general, total ($\mathbb{O}_3$). This may be seen by observing that, in the case of a Boolean algebra of sets, then $a \leqslant b$ means that a is contained in b; and obviously, given two subsets of a given set, it can happen that neither is contained in the other.

The ordering of R, given by 21.5.2, is said to be **induced** by the Boolean algebra structure of R. Some algebraic properties are therefore to be expected:

21.5.3 THEOREM. *The partial order relation $a \leqslant b$ has the following algebraic properties: if $a \leqslant b$, then*

(i) $a + c \leqslant b + c$;
(ii) $ac \leqslant bc$;
(iii) $b' \leqslant a'$.

(Compare Proposition 21.1.4.)

Proof. Given $a + b = b$, then for (i) we have

$$a + c + b + c = a + b + c + c = a + b + c = b + c,$$

so

$$a + c \leqslant b + c;$$

(ii) we have $ac + bc = (a + b)c = bc$, so $ac \leqslant bc$; (iii) we invoke the rule (21.4.5) of Pierce and De Morgan: $(a + b)' = a'b'$, to deduce that $b' = a'b'$, or $b' \leqslant a'$. ■

Exercise 11

(i) Show that when R is the Boolean algebra $\mathfrak{p}S$, then $\leqslant$ is set-inclusion.
(ii) In a Boolean algebra R prove that $ab \leqslant a \leqslant a + b$ for all a, b. Hence show that $0 \leqslant a \leqslant 1$ for all a.
Show further that 0 is the unique element a_0 such that $a_0 \leqslant a$, all $a \in R$; and characterize 1 similarly.
(iii) What can be inferred from $x \leqslant x'$?

A Boolean algebra has the following very important property with respect to the partial ordering we have been discussing.

21.5.4 PROPOSITION. *If* $a_1 \leqslant a$ *and* $a_2 \leqslant a$, *then* $a_1 + a_2 \leqslant a$.

Proof. We have $a_1 + a = a$, $a_2 + a = a$; thus

$$a_1 + a_2 + a = a_1 + a_2 + a + a = (a_1 + a) + (a_2 + a) = a + a = a,$$

so $a_1 + a_2 \leqslant a$. ■ This, together with Exercise 11(ii), establishes that $a_1 + a_2$ is the **least upper bound** of a_1, a_2 (denoted by lub (a_1, a_2)); that is, it is an element c (and, therefore, the *unique* element c) such that

(i) $a_1 \leqslant c$, $a_2 \leqslant c$ and
(ii) if $a_1 \leqslant a$, $a_2 \leqslant a$, then $c \leqslant a$.

Similarly a_1a_2 is the **greater lower bound** of a_1, a_2 (denoted by glb (a_1, a_2)). A partially ordered set in which any two elements have a least upper bound and a greatest lower bound is called a **lattice**; thus a Boolean algebra is a special case of a lattice. Lattices have been extensively studied in mathematics, because of their relevance to many algebraic systems. (See Birkhoff [15].)

Exercise 12

(i) Let G be a group and consider the set of subgroups of G. If H_1, H_2 are two such subgroups, let lub (H_1, H_2) be the subgroup of G generated by H_1 and H_2, and let glb $(H_1, H_2) = H_1 \cap H_2$. Show that the set of subgroups is a lattice with lub (H_1, H_2) as least upper bound of H_1 and H_2, and glb (H_1, H_2) as greatest lower bound of H_1 and H_2.
(ii) Let R be the algebra of subgroups of G given by $H_1 + H_2 =$ lub (H_1, H_2), $H_1H_2 =$ glb (H_1, H_2). Show that axioms $\mathfrak{B}_1$, $\mathfrak{B}_2$, $\mathfrak{B}_4$ for a Boolean algebra are satisfied but that axiom $\mathfrak{B}_3$ is not.

(iii) In a Boolean algebra R, prove that if $x + z = 1$, then $x' \leqslant z$.
If $xz = 0$, prove that $z \leqslant x'$.
If $x \leqslant y$ and $x \neq y$, prove that $yx' \neq 0$.

(iv) An element $a \in R$ is called an **atom** iff $a \neq 0$ and if $x \leqslant a$ then $x = 0$ or $x = a$. Show that if a is an atom, then for any $y \in R$, either $a \leqslant y$ or $a \leqslant y'$ but not both. {Consider ay.} Which elements are atoms in $\mathfrak{p}S$? In the dual (see 8 Exercise 5(v)) of $\mathfrak{p}S$?
(v) Suppose S is a *finite* Boolean algebra. Show that, given $x \in S$ ($x \neq 0$), there exists some atom $\leqslant x$. Let $A(x)$ denote the set of *all* atoms $\leqslant x$, and let $\bar{x}$ denote the sum of all the atoms in $A(x)$. Prove that $\bar{x} = x$. {$\bar{x} \leqslant x$, and if $\bar{x} \neq x$ apply the third part of (iii).}
(vi) In (v), define E to be the set $A(1)$, and put $A(0) = \phi \subseteq E$. Show that, given $B \subseteq E$, there exists $x \in S$ such that $B = A(x)$. Prove that for all $x \in S$, $A(x') = E - A(x) = \complement_E A(x)$. If $x \leqslant y$, prove that $A(x) \subseteq A(y)$. Hence prove that $A(x + z) = A(x) \cup A(z)$, $A(xz) = A(x) \cap A(z)$. If $A(x) = A(y)$, prove that $x = y$.

21.6 Homomorphisms

If R_1, R_2 are Boolean algebras, then we call a function $\phi: R_1 \to R_2$ **additive** if

(i) $\phi(a + b) = \phi(a) + \phi(b)$;

multiplicative if

(ii) $\phi(ab) = \phi(a) \cdot \phi(b)$;

and **complementary** if

(iii) $\phi(a') = \phi(a)'$.

These equations state various ways in which the mapping ϕ can 'preserve structure'.

21.6.1 EXAMPLE. The prime example has already occurred in 2 Exercise 5(iii); it has all three properties. Thus let $f: S_1 \to S_2$ be a function inducing the function $f^\flat: \mathfrak{p}S_2 \to \mathfrak{p}S_1$ (where $f^\flat(X) = f^{-1}(X)$ for each $X \subseteq S_2$). Then $\mathfrak{p}S_1$ and $\mathfrak{p}S_2$ are Boolean algebras, and we saw in Chapter 2 that $f^\flat$ satisfies the above conditions (i)–(iii). On the other hand, the function $F: \mathfrak{p}S_1 \to \mathfrak{p}S_2$ given by $F(Y) = f(Y)$, $Y \subseteq S_1$ is in general only additive. {See 2 Exercise 5(i).}

Now the reader should notice: the rules of Pierce and De Morgan show that, in any Boolean algebra, multiplication is determined by addition and complementation, since the dual of 21.4.5 states that

21.6.2 $$ab = (a' + b')';$$

and similarly (or dually) addition is determined by multiplication and complementation, by the rule 21.4.5 and 21.4.1; for then

21.6.3 $$a + b = (a'b')'.$$

It readily follows that a complementary function $\phi: R_1 \to R_2$ is additive if and only if it is multiplicative. Thus we are led to call a function $\phi: R_1 \to R_2$ an **(algebra)-homomorphism** if it is complementary and additive or, equivalently, if it is complementary and multiplicative. The function $f^\flat: \flat S_2 \to \flat S_1$ in Example 21.6.1 is therefore an algebra homomorphism, while F there is not. Indeed one sees that $f^\flat$ is a far more 'natural' construction than F precisely because $f^\flat$ is a Boolean algebra homomorphism, unlike F. The following consideration immediately arises. Of the conditions (i), (ii), (iii) above, we have seen that (i) & (iii) $.\Rightarrow.$ (ii) and (ii) & (iii) $.\Rightarrow.$ (i). Is it true that (i) & (ii) $.\Rightarrow.$ (iii)? The answer is no; for the function $\phi: R_1 \to R_2$, given by $\phi(a) = 0$, all $a \in R_1$, plainly satisfies (i), (ii) but not (iii) (find another such function!). Thus it is natural to ask about the relation of conditions (i), (ii) to condition (iii). Our first result in this direction greatly simplifies the problem of verifying that a given ϕ satisfies condition (iii).

21.6.4 PROPOSITION. *Let $\phi: R_1 \to R_2$ be a map of Boolean algebras satisfying conditions* (i), (ii). *Then ϕ satisfies condition* (iii) *if and only if* $\phi(0) = 0$, $\phi(1) = 1$.

Proof. Suppose condition (iii) is satisfied. Then for all $a \in R$, $\phi(0) = \phi(aa') = \phi(a)\phi(a') = \phi(a)\phi(a)' = 0$; and similarly $\phi(1) = 1$. Conversely, suppose $\phi(0) = 0$, $\phi(1) = 1$. Then $\phi(a) + \phi(a') = 1$, $\phi(a)\phi(a') = 0$, so that $\phi(a') = \phi(a)'$ by 21.4.1. ■

We preface our next observation about homomorphisms of Boolean algebras with a remark about the order relation in a Boolean algebra.

21.6.5 LEMMA. *Let $\phi: R_1 \to R_2$ be an additive or multiplicative function of Boolean algebras. Then ϕ is an order-preserving function with respect to the partial orderings in R_1, R_2 induced by their Boolean algebra structure.*

Proof. Assume ϕ additive. Then

$$a \leqslant b \Leftrightarrow a + b = b \Rightarrow \phi(a) + \phi(b) = \phi(b) \Leftrightarrow \phi(a) \leqslant \phi(b).$$

Now apply the duality principle to deal with a multiplicative ϕ. ■

We may now prove

21.6.6 THEOREM. *Let $\phi: R_1 \to R_2$ be a surjection satisfying conditions* (i) *and* (ii). *Then ϕ also satisfies condition* (iii).

Proof. We have remarked that, in a Boolean algebra R, 0 is characterized by the property $0 \leqslant a$, for all $a \in R$ (see Exercise 11). Thus if $\phi: R_1 \to R_2$ is additive, it follows from Lemma 21.6.5 that $\phi(0) \leqslant \phi(a_1)$, all $a \in R_1$. But if ϕ is surjective, then $\phi(0) \leqslant a_2$, all $a_2 \in R_2$, so that $\phi(0) = 0$. Similarly $\phi(1) = 1$ and the theorem follows from Proposition 21.6.4. ■ For other examples of homomorphisms, see Exercises 6, 8 and 9 of Chapter 8.

Exercise 13

(i) In Exercise 12(vi), show that when S is a finite Boolean algebra, the assignment $x \to A(x)$ is an isomorphism (i.e., an invertible homomorphism) between S and $\mathfrak{p}E$, where E is the set $A(1)$ of all atoms of S.
[This proves that every finite Boolean algebra is (isomorphic to) a Boolean algebra of sets. If the algebra is not finite, it may not be isomorphic to all of $\mathfrak{p}E$ for any E; see Stoll [123].]
(ii) Show that every finite Boolean algebra has 2^n elements, for some n. [Hence not every set can be given the structure of a Boolean algebra.]
(iii) Let $\theta: S_1 \to S_2$ be an isomorphism of Boolean algebras. Show that $\theta^{-1}: S_2 \to S_1$ is also such an isomorphism.
(iv) In Example 21.6.1, which of the conditions (i)–(iii) does the function F satisfy? When is F complementary?
(v) Show that an intersection of Boolean sub-algebras of a Boolean algebra, is again a Boolean sub-algebra.
(vi) How many sub-algebras are there of $\mathfrak{p}U$, when U is a finite set?
§(vii) Let X be a topological space, and let $O(X)$, $C(X)$ denote respectively the families of open and of closed sets in X. Show that $(O(X), \cup, \cap)$ and $(C(X), \cap, \cup)$ are lattices but not in general Boolean algebras. Show that the function $c: O(X) \to C(X)$, given by $c(A) = X - A$, is additive, multiplicative, and order-preserving (with respect to the operations indicated). Would you call c an isomorphism?
§(viii) Show that the smallest Boolean sub-algebra $B(X)$ of $\mathfrak{p}X$ to contain $O(X)$ also contains the *set* $C(X)$. [B is called the family of **Borel sets** of X; it is of importance in probability theory.]
§(ix) Let X, Y be topological spaces, and $f: X \to Y$ a function. Show that f is continuous iff $f^\flat: \mathfrak{p}Y \to \mathfrak{p}X$ carries $O(Y)$ and $C(Y)$ to $O(X)$, $C(X)$ respectively. Show that, if f is continuous, then $f^\flat(B(Y) \subseteq f^\flat(B(X)$; and if f is a homeomorphism, then $f^\flat|O(Y)$ is an isomorphism $O(Y) \approx O(X)$. Is the converse true?
(For the requisite topological notions, see Chapter 25.)

Chapter 22

POLYNOMIALS AND EQUATIONS OF DEGREE n

22.1 Polynomial Forms

Having dealt with linear equations in Chapter 19, we take up in this chapter the question of the solution of equations of degree greater than 1.

The general problem is then the following. Let $p(x)$ be a polynomial of degree n over the field F, thus

22.1.1 $$p(x) = a_0 + a_1x + \cdots + a_nx^n, \qquad (a_i \in F,\ a_n \neq 0).$$

We seek the set of all elements $\alpha \in F$ such that $p(\alpha) = 0$. To understand the problem quite precisely we have to realize what is involved in the passage from the polynomial $p(x)$, which is an element of $F[x]$, to the element $p(\alpha)$ of F. This process of passing from $p(x)$ to $p(\alpha)$ is called **substitution** (of α for x); but it should be realized that x is a special polynomial, i.e., a special element of the polynomial ring $F[x]$, while α is an element of F. To stress this remark, we point out that, in the expression

22.1.2 $$p(\alpha) = a_0 + a_1\alpha + \cdots + a_n\alpha^n,$$

obtained by substituting α for x in 22.1.1, the '+' sign does not have, strictly speaking, the same meaning that it has in 22.1.1; for in 22.1.1 we 'add' in $F[x]$, while in 22.1.2 we 'add' in F. Naturally, it would be intolerable to be under the necessity constantly to keep this fine distinction in mind; but we need to see what is involved in overlooking it before doing so. To this end we first recall the nature of the polynomial ring $F[x]$. In our references to polynomials in previous chapters, it was not important to stress the distinction —as now we must—between a **polynomial function** and a **polynomial.** For example, consider the polynomials x, x^2 in $\mathbb{Z}_2[x]$. The associated *functions* $\mathbb{Z}_2 \to \mathbb{Z}_2$ are equal (to id: $\mathbb{Z}_2 \to \mathbb{Z}_2$) since each has value 0 or 1 when x is 0 or 1 respectively. On the other hand, the degrees of the polynomials are different, and they are also different polynomials because corresponding coefficients are different. Thus, strictly speaking, the values of the associated polynomial function are not relevant; it is the set of coefficients which distinguishes the polynomial, and we speak of a **polynomial form** to indicate that we are not speaking of polynomial functions. We therefore call each member† f of $F[x]$ a **polynomial form**, and regard it as an infinite sequence of elements of F, thus $f = (a_0, a_1, \ldots, a_n, \ldots)$, where *all but a finite number of the terms are zero.* The rules for adding and multiplying polynomial forms (see Example 9.2.3) then become:

22.1.3 $$(a_0, a_1, \ldots, a_n, \ldots) + (b_0, b_1, \ldots, b_n, \ldots) = (a_0 + b_0, a_1 + b_1, \ldots, a_n + b_n, \ldots),$$

22.1.4 $$(a_0, a_1, \ldots, a_n)(b_0, b_1, \ldots, b_n) = (c_0, c_1, \ldots, c_n, \ldots),$$

where

$$c_i = \sum_{j+k=i} a_j b_k.$$

The reader should notice how the convention of 'adding on zero terms' simplifies the statement of these rules.

Exercise 1

(i) Verify that the set $F[x]$ of polynomial forms is a commutative ring under the laws of composition 22.1.3, 22.1.4. (The only difficulty lies in verifying the associativity of multiplication.)
(ii) Multiply out the product of real polynomials $(1 - 2x + 3x^2 - 5x^3) \times (2 + x + 2x^2 + 7x^4)$ in the traditional way. Compare the result with the multiplication of $(1, -2, 3, -5, 0, 0, \ldots)$, $(2, 1, 2, 0, 7, 0, 0, \ldots)$ by the rule 22.1.4.

Now we may embed F in $F[x]$ by associating with each $a \in F$, the polynomial form $k(a) = (a, 0, \ldots, 0, \ldots)$; this is the familiar process of identifying a constant with the corresponding constant polynomial. This embedding, $k: F \to F[x]$, is one-one and preserves addition and multiplication, so *it enables us to regard F as a sub-ring* of $F[x]$.

So far we have kept the nature of 'x' somewhat mysterious. We clear up the mystery by saying that x is just the polynomial form $(0, 1, 0, \ldots, 0, \ldots)$. It will help to avert confusion if we use a special fount of type, and write

$$\mathfrak{x} = (0, 1, 0, \ldots).$$

For convenience, let $\mathbf{0}_n$ denote a row of n zeros, with a similar meaning for $\mathbf{0}_\infty$. Thus by 22.1.4, we have

$$\mathfrak{x} = (\mathbf{0}_1, 1, \mathbf{0}_\infty), \qquad \mathfrak{x}^2 = \mathfrak{x}\mathfrak{x} = (\mathbf{0}_2, 1, \mathbf{0}_\infty)$$

and, by induction on n,

22.1.5 $$\mathfrak{x}^n = (\mathbf{0}_n, 1, \mathbf{0}_\infty);$$

by convention, $\mathfrak{x}^0 = (1, \mathbf{0}_\infty) = 1$.

† We write f rather than $f(x)$ to stress that f is *not* to be thought of as a function.

It then follows immediately that, in $F[x]$,

22.1.6 $$a_0 + a_1\mathfrak{x} + \cdots + a_n\mathfrak{x}^n = (a_0, a_1, \ldots, a_n, \ldots),$$

22.1.7 $$\mathfrak{x}^n(b_0, b_1, \ldots, b_m, \ldots) = (\mathbf{0}_n, b_0, b_1, \ldots, b_m, \ldots).$$

Notice that 22.1.6 is *proved* by using the identification of constants a_m with constant polynomial forms $(a_m, \mathbf{0}_\infty) = k(a_m)$, the meaning of $\mathfrak{x}$, and the rules 22.1.3, 22.1.4. In this way we obtain once more the usual notation for polynomials in that very notation (but writing $\mathfrak{x}$ for x). We will naturally revert to the usual notation once we have given a complete description of the ring $F[x]$.

Exercise 2

(i) Verify that the embedding $k\colon F \to F[x]$ has the properties stated above. Is $k(F)$ an ideal in $F[x]$?
(ii) Prove 22.1.5, 22.1.6, and 22.1.7 in detail.

It remains to recall from 11.2.1 the notion of degree. The **zero polynomial form** is just the polynomial form $(0, 0, 0, \ldots, 0, \ldots)$, in which each term is zero. Given any *non-zero* polynomial form $f = (a_0, a_1, \ldots, a_m, \ldots)$, there is a largest integer n such that $a_n \neq 0$; this integer is then called the **degree** of f. If we can write f as $a_0 + a_1\mathfrak{x} + \cdots + a_n\mathfrak{x}^n$, with $a_n \neq 0$, then the degree is n. The non-zero constants are the polynomial forms of degree 0, **linear** polynomial forms $a_0 + a_1\mathfrak{x}$ have degree 1 ($a_1 \neq 0$), and **quadratic** polynomial forms $a_0 + a_1\mathfrak{x} + a_2\mathfrak{x}^2$ have degree 2 ($a_2 \neq 0$).

Exercise 3

(i) The degree defines a function deg: $F[x] - \{0\} \to \mathbb{Z}^+$. Show that

$$\deg fg = \deg f + \deg g,$$
$$\deg (f + g) = \deg f \text{ if } \deg f > \deg g,$$
$$\deg (f + g) \leqslant \deg f \text{ if } \deg f = \deg g, \text{ and } f \neq -g.$$

Can strict inequality hold in the last relation?
(ii) Show that, in $\mathbb{Z}_m[x]$ there are m^{k+1} polynomial forms of degree $\leqslant k$. How many are there of degree exactly k?

22.2 Substitution

We now take up the question of substitution. Given any element $\alpha \in F$, the rule whereby we pass from 22.1.1 to 22.1.2 determines a function

22.2.1 $$S_\alpha\colon F[x] \to F,$$

which we may call the **α-substitution function.** For example, $S_\alpha(\mathfrak{x}) = \alpha$, $S_\alpha(\mathfrak{x}^2 + 1) = \alpha^2 + 1$. We now make the crucial observation:

22.2.2 THEOREM. *The function S_α is a ring-homomorphism: $F[x] \to F$.*

Proof. We give the proof, not because it is difficult (it isn't!), but because the reader may not see clearly what precisely has to be proved. Believing that this can be seen most easily if we use the 'sequence' notation for polynomials, we temporarily adopt that notation again. Thus let $f = (a_0, \ldots, a_n, \ldots)$, $g = (b_0, \ldots, b_n, \ldots)$ be two polynomial forms, that is, elements of $F[x]$. Then, if N is chosen large enough [take $N = \max(\text{degree} f, \text{degree} g)$, with an appropriate modification if f or g is the zero polynomial],

22.2.3 $$S_\alpha(f) = a_0 + a_1\alpha + \cdots + a_N\alpha^N,$$

22.2.4 $$S_\alpha(g) = b_0 + b_1\alpha + \cdots + b_N\alpha^N,$$

22.2.5 $$f + g = (a_0 + b_0, \ldots, a_n + b_n, \ldots),$$

22.2.6 $$S_\alpha(f + g) = a_0 + b_0 + (a_1 + b_1)\alpha + \cdots + (a_N + b_N)\alpha^N;$$

It now follows from the axioms for a field (indeed, from the axioms for a *ring*) that

$$S_\alpha(f + g) = S_\alpha(f) + S_\alpha(g)$$

in F. This shows that S_α preserves addition.

Passing to multiplication, we recall from 22.1.4 that $fg = (c_0, c_1, \ldots, c_n, \ldots)$, where

22.2.7 $$c_n = \sum_{p+q=n} a_p b_q.$$

Thus, again supposing M taken large enough, say $M = 2N$,

22.2.8 $$S_\alpha(fg) = c_0 + c_1\alpha + \cdots + c_M\alpha^M,$$

and it remains to show that, in F,

$$(a_0 + a_1\alpha + \cdots + a_N\alpha^N)(b_0 + b_1\alpha + \cdots + b_N\alpha^N)$$
$$= c_0 + c_1\alpha + \cdots + c_{2N}\alpha^{2N},$$

where c_n is given by 22.2.7. This equality, however, follows from the axioms for a field (indeed, from the axioms for a *commutative ring*). ■ Notice that commutativity of multiplication is crucial to the last argument.

We have now established the validity of the usual substitution processes in $F[x]$. Indeed, we have established them in the polynomial ring $R[x]$, where R is any commutative ring; and *everything* we have said so far in this chapter (including the remarks on degree) applies provided R is an integral domain. This is important, since one plainly wishes to consider polynomials with integer coefficients.

Exercise 4

(i) Fixing $\alpha \in R$, where R is a commutative ring, show that S_α is onto, and that its kernel is an ideal in $R[x]$. Enumerate this kernel when $R = \mathbb{Z}_4$, $\mathbb{Z}_5$ and $\alpha = [2]_4$, $[2]_5$, respectively.
(ii) Find a statement made about degree which is false in $R[x]$ if R is not an integral domain.
(iii) Carrying on the tradition of 22.1.2, write $p(\alpha)$ for $S_\alpha(p)$. Show that Theorem 22.2.2 allows us to write

$$(f + g)(\alpha) = f(\alpha) + g(\alpha), \qquad (fg)(\alpha) = f(\alpha)g(\alpha).$$

We may consider then a polynomial ring $R[x]$, where R is an integral domain, and fix a non-zero polynomial p in $R[x]$. We may then ask for what elements $\alpha \in R$, is p in the kernel of the substitution function $S_\alpha: R[x] \to R$; that is, for what elements α do we have $p(\alpha) = 0$? (Such an element α is called a **zero** (or **root**) of p.) However, this is again not quite a perfect picture of what we understand by solving an equation. Consider for example the polynomial $2x - 1$ in $\mathbb{Z}[x]$. There is no $\alpha \in \mathbb{Z}$ such that $2\alpha - 1 = 0$, yet we would like to be permitted to say that $x = \frac{1}{2}$ is a solution of the equation $2x - 1 = 0$. Again, we would like to be permitted to say that $x = \sqrt{2}$ is a root of the equation $x^2 - 2 = 0$; and we would like to be permitted to say that the equation $x^2 + 1 = 0$ has no real roots but does have the complex roots $\pm i$.

All that is needed is that we should specify a set S in which the solution is to be sought and that there should exist an integral domain $\bar{R}$ containing R and S. Then we consider the functions $S_\alpha: \bar{R}[x] \to \bar{R}$; our polynomial p may be considered as a polynomial in $\bar{R}[x]$ and we look for those $\alpha \in S$ such that $p(\alpha) = 0$. It is plain that the subset of S consisting of solutions of the equation $p(x) = 0$ does not depend on the choice of 'enveloping domain' $\bar{R}$. We call S the **solution-domain.**

Let us continue further with this discussion of the *meaning* of 'solving equations' before we return to the more technical side of the question. We see that, in order to have a well-defined notion of solution, one must specify a set S within which the solutions are to be sought. Thus the question 'Does the equation $p(x) = 0$ have a solution?' is incomplete and so unanswerable. Much of the mystery surrounding the existence and nature of complex numbers is due to the bad design of the question 'Does the equation $x^2 + 1 = 0$ have a solution?'. The answer is that it has no solution in the set of real numbers but it does have solutions in the set of complex numbers. In this respect the question has the same status as the apparently less troublesome one 'Has the equation $x + 1 = 0$ a solution?', where the answer is that it has no solution in $\mathbb{Z}^+$ but does have a solution in $\mathbb{Z}$. Of course we are not discussing here the *method* of construction of complex numbers (or negative numbers) which is dealt with in Chapter 24. That method (see 24.7.2) is indeed part of the whole algebraic theory of equations, where one considers algebraic 'extensions' of the coefficient field within which a given polynomial may be factorized as a product of linear factors. This is a very beautiful part of algebra but would take us too far from our present objective. We hope, however, to have said enough to dissipate one of the traditional sources of obscurity in school mathematics.

22.3 The Remainder Theorem

We now return to the consideration of polynomials in $R[x]$ where R is an integral domain; we allow ourselves to drop the word 'form' henceforth. The reader will be familiar with the process of dividing one polynomial by another. We will insist here that the divisor g is **monic**, that is, a (non-zero) polynomial of the form

$$g = b_0 + b_1\mathfrak{x} + \cdots + b_n\mathfrak{x}^n, \qquad \text{with } b_n = 1.$$

Then, just as in the case $R = \mathbb{Z}$, we may divide an arbitrary polynomial f in $R[x]$ by g, obtaining a quotient q and a remainder r, both in $R[x]$, such that

22.3.1 $$f = qg + r,$$

where $r = 0$ or $\deg r < \deg q$ (see Chapter 11; since g is monic we may carry out the Euclidean algorithm). It is often convenient, as here, to assign to the polynomial 0 the degree $-\infty$, and thus remove its exceptional nature in certain statements.

In particular, let $\alpha \in R$ and let $g = \mathfrak{x} - \alpha$. Then either $r = 0$ or $\deg r = 0$, that is, r is a constant polynomial, $r \in R$. Thus

22.3.2 $$f = q\cdot(\mathfrak{x} - \alpha) + r, \qquad r \in R,$$

where we have inserted the dot to eliminate the belief that the right-hand side involves a function q with argument $\mathfrak{x} - \alpha$.

We immediately infer the **remainder theorem**:

22.3.3 THEOREM. *Given $f \in R[x]$, $\alpha \in R$, there exists a polynomial $q \in R[x]$ such that*

$$f = q\cdot(\mathfrak{x} - \alpha) + f(\alpha).$$

Proof. We substitute α into 22.3.2, that is, we apply S_α to each side. Since S_α is a ring homomorphism (Theorem 22.2.2) we infer that $f(\alpha) = q(\alpha)\cdot 0 + r$, so that $r = f(\alpha)$. ■

22.3.4 COROLLARY. *If also α is a zero of f, then $f = q\cdot(\mathfrak{x} - \alpha)$.* ■

We are now ready to prove:

22.3.5 THEOREM. *A polynomial f in $R[x]$ of degree m cannot have more than m zeros.*

Proof. Notice first that it is not necessary to specify the solution domain S in this theorem. Plainly the degree of a polynomial f does not change when we enlarge the domain of the coefficients, so it is sufficient in proving this theorem, to suppose the solution domain to be R itself. Of course, had we wanted to prove that a given polynomial f has *precisely* m zeros, it would have been necessary to specify S to give meaning to our assertion.

After this preamble, we proceed with the proof. We argue by induction on m. If $m = 0$, then f is just a non-zero constant polynomial and so it has no zeros. Thus we suppose the theorem proved for polynomials of degree $< m$ and consider the polynomial f of degree m. If f has no zeros we are home! If, on the other hand, α is a zero of f, then, by Corollary 22.3.4,

$$f = q\cdot(\mathfrak{x} - \alpha).$$

Computing degrees (see Exercise 3), we have $m = \deg q + 1$, whence $\deg q = m - 1$ (or the reader may legitimately say this is obvious!). Thus, by the inductive hypothesis, q has at most $(m - 1)$ zeros. Moreover, it is plain that any zero of f is α or a zero of q; for if β is neither, then by applying the substitution S_β, we find that $f(\beta) = q(\beta)(\beta - \alpha) \neq 0$ since $q(\beta) \neq 0$, $\beta - \alpha \neq 0$ and R does not have divisors of zero. Thus we have proved that f has at most m zeros and the inductive step is complete. ■ Notice that, in this theorem, zeros of f are counted with their appropriate **multiplicities**; thus if, in fact, $(\mathfrak{x} - \alpha)^k$ is a divisor of f, then α has multiplicity k and counts k times towards the tally of zeros of f.

Exercise 5

(i) Consider polynomials with coefficients in $\mathbb{Z}_{12}$. Show that [1], [5], [7], [11] are all zeros of $\mathfrak{x}^2 - [1]$. Why does this fact not violate Theorem 22.3.5?

(ii) Let p be prime, so that $\mathbb{Z}_p$ is a field. Show that the polynomial $\mathfrak{x}^{p-1} - 1$ in $\mathbb{Z}_p[x]$ has $(p - 1)$ zeros in $\mathbb{Z}_p$. {Use Euler's Theorem, 10.3.1.} Deduce from this and Theorem 22.3.5 that if $q|(p - 1)$, then the polynomial $\mathfrak{x}^q - 1$ has q zeros in $\mathbb{Z}_p$. {Cf. 12 Exercise 1(ii).}

(iii) Suppose the polynomial $p \in R[x]$ has no root in R, where R is an integral domain. Let I be the ideal generated by p, and let $E = R[x]/I$ (see 10 Exercise 7(xii) and 11 Exercise 10(iv)). Let $\nu: R[x] \to E$ be the natural map. Prove that $\nu|R$ is a monomorphism (where, as usual, we identify R with the set of constant polynomials in $R[x]$). Hence we may also identify R with $\nu(R)$ in E, so that E is an 'extension' of the smaller ring R. We can then regard each $f = a_0 + a_1\mathfrak{x} + \cdots + a_n\mathfrak{x}^n \in R[x]$ as a member also of $E[x]$. Show that $\nu(f) = a_0 + a_1\xi + \cdots + a_n\xi^n$ in E where $\xi = \nu(\mathfrak{x})$, $\xi \in E$; and hence show that, if $p = b_0 + \cdots + b_m\mathfrak{x}^m$, then $b_0 + \cdots + b_m\xi^m = 0 = \nu(p)$ in E. That is, $S_\xi(p) = 0$, where $S_\xi: E[x] \to E$ is the ξ-substitution function associated with $E[x]$. Thus $\xi \in E$ is a root of p, regarded as a polynomial in $E[x]$.

(iv) If $R[x]$ is a Euclidean domain and p is irreducible over R, show that E is a field. {Mimic the proof of the existence of inverses in Theorem 10.2.5.} Examine the special case when $R = \mathbb{R}$, and p is $\mathfrak{x}^2 + 1$. Show that, here, $E \approx \mathbb{C}$ and ξ plays the role of $\sqrt{-1}$.

(v) In 10 Exercise 7(xii) suppose that R is an integral domain, while the ideal I is principal, generated by $p \in R$. Show that R/I is an integral domain if and only if p is prime in R. Show further that if R is a Euclidean domain and p is prime in R, then R/I is a field. This generalizes Theorem 10.2.6. See also 18.10.

22.4 Polynomial Functions

We can now take up an important point of definition in the notion of a polynomial. We recall that a polynomial in $R[x]$ is, ultimately, nothing but a sequence of elements of R; the special feature of such a sequence being that all but a finite number of its terms are zero. However the process of substitution, which we have described in detail in 22.2, associates with every polynomial f in $R[x]$ a function $R \to R$ which transforms each $\alpha \in R$ into $S_\alpha(f) \in R$. In fact we have written in 22.1.2, according to custom, $p(\alpha)$ for $S_\alpha(p)$ and have thus failed to distinguish notationally between the polynomial p and the associated function $R \to R$. Are we justified in this notational device?

To answer this question, let us first introduce some further notation! We will write $\tilde{f}(\alpha) = S_\alpha(f)$ (instead of $f(\alpha)$ as in 22.1.2); thus $\tilde{f}$ is by definition a **polynomial function** $R \to R$. Then $\tilde{f} \in R^R$ and we denote by

22.4.1 $$\Phi\colon R[x] \to R^R$$

the mapping which transforms the polynomial f into the polynomial function $\tilde{f}$; that is, $\Phi(f) = \tilde{f}$. In particular, $\Phi(\mathfrak{x}) = \mathrm{id}_R$. We remind the reader that R is an integral domain; and that, for any set S, R^S acquires a ring-structure from that of R in the obvious way (see Example 9.2.8).

22.4.2 THEOREM. *The mapping Φ is a ring-homomorphism. Moreover, if R is infinite, then Φ is one-one.*

Proof. The first statement is an easy consequence of Theorem 22.2.2 (see Exercise 4(iii)); it depends then only on the fact that R is a commutative ring.

The second statement is an easy consequence of Theorem 22.3.5. For let $\tilde{f} = 0 \in R^R$. This means that, for any $\alpha \in R, \tilde{f}(\alpha) = 0$; that is, α is a zero of f. But if R is infinite and $f \neq 0$ there must by Theorem 22.3.5 be some $\alpha \in R$ such that α is *not* a zero of f. This shows that $f = 0$ if $\tilde{f} = 0$, and hence that Φ is one-one. ■

This theorem may be said to justify the failure to distinguish between f and $\tilde{f}$ in elementary texts. For there, R is always infinite ($R = \mathbb{Z}$ or $\mathbb{Q}$ or $\mathbb{R}$ or $\mathbb{C}$, though the precise coefficient domain is often only implicit!); and if Φ is a ring-monomorphism it allows us to identify $R[x]$ with the subring of R^R consisting of the polynomial functions (so we 'regard' f and $\Phi(f)$ as being the same thing).

We ourselves would argue that the distinction between these two concepts is worth making in any case. Divisibility properties of polynomials are much better handled in $R[x]$; for example, in many texts proofs of the remainder theorem seem to require the reader to divide the function $f(x)$ by the function $x - \alpha$ and then to put $x = \alpha$ and this understandably troubles the brighter students. Again it is often asserted, if not just assumed, in school texts that there is only one 'expression' for a polynomial function and that a polynomial function arises from the product of two given polynomial functions. (But recall that functions are multiplied by multiplying their values. It is therefore not a trivial fact that the product of two polynomial functions is a *polynomial* function.) So far as we can see, this is usually established by the celebrated method of 'proof by authority'.

Of course if R is finite then the distinction between polynomials and polynomial functions is crucial, as we saw above in the example of $\mathbb{Z}_2[x]$. See also Exercise 6(iii) below.

Exercise 6

(i) Let p be an odd prime. Find two distinct polynomials in $\mathbb{Z}_p[x]$ whose corresponding polynomial functions coincide.
(ii) Let R be a finite commutative ring with unity, and for each $a \in R$ let $\delta_a \in R^R$ denote that function which is 1 at a and zero elsewhere. Prove that the function $\Phi: R[x] \to R^R$ of 22.4.1 is onto, iff for each $a \in R$, δ_a is a polynomial function. Hence prove that Φ is onto when $R = \mathbb{Z}_p$, p prime. {Consider a product of factors $x - u$.}
(iii) In $\mathbb{Z}_4[x]$, prove that for any polynomial f, $\Phi(f) = \Phi(g)$ for some g of degree $\leqslant 3$. Find a non-zero polynomial in $\mathbb{Z}_4[x]$, of degree 3, but with 4 roots. Show that (in the notation of (ii)) $\delta_2 \notin \operatorname{Im} \Phi$, so that Φ here is not onto.
(iv) For any $f \in R[x]$, let $e_f: R[x] \to R[x]$ be given by

$$e_f(g) = a_0 + a_1 f + \cdots + a_n f^n,$$

where

$$g = a_0 + a_1 \mathfrak{x} + \cdots + a_n \mathfrak{x}^n.$$

Show that e_f is a ring-homomorphism and that $S_\alpha \circ e_f = S_{f(\alpha)}$. Investigate the dependence of e_f on f. Show that $x^n - 1$ is a factor of $x^{mn} - 1$.

22.5 Real and Complex Polynomials

Although the techniques of proof are not purely algebraic, the following theorems on polynomials with real or complex coefficients should be stated here.

22.5.1 THEOREM. *Every polynomial in $\mathbb{R}[\mathfrak{x}]$ of odd degree has a zero in $\mathbb{R}$.*

Proof. Let f have odd degree, say $f = a_0 + a_1\mathfrak{x} + \cdots + a_n\mathfrak{x}^n$, $a_i \in \mathbb{R}$, with $a_n \neq 0$ and n odd. Without real loss of generality for our problem we may suppose $a_n = 1$. Then plainly $f(\alpha)$ is positive if α is large enough and positive; and $f(\beta)$ is negative if β is large enough and negative. But $\gamma \to f(\gamma)$ may be viewed as a *continuous* function from $\mathbb{R}$ to $\mathbb{R}$. (It is the function $\tilde{f}$ in the sense of 22.4.1.) Thus, by a basic theorem on continuous functions (see 28.5.1), $f(x)$ has a zero somewhere between α and β. ■ Notice that $x^2 + 1$ is a polynomial in $\mathbb{R}[x]$ of *even* degree with no zero in R.

A crucial theorem in algebra which we will not prove here is the following, relating to polynomials with complex coefficients. It is often called the **fundamental theorem of algebra** and was first proved correctly by Gauss. For an attractive proof, see Courant–Robbins [25]; the fundamental theorem is:

22.5.2 THEOREM. *Every polynomial in* $\mathbb{C}[x]$ *of positive degree has a zero in* $\mathbb{C}$. □

Notice that if we combine this theorem with Corollary 22.3.4 we find that *every polynomial in* $\mathbb{C}[x]$ *of degree n has n zeros in* $\mathbb{C}$, *and factorizes in* $\mathbb{C}[x]$ *as a product of n linear factors.* Now let p be a polynomial in $\mathbb{C}[x]$, say, $p = (a_0, a_1, \ldots, a_n, \ldots)$, $a_n \in \mathbb{C}$. We may take complex conjugates in $\mathbb{C}$ (i.e., replace $b + ic$ by $b - ic$, where $b, c \in \mathbb{R}$) and thus obtain a new polynomial $\bar{p}$ in $\mathbb{C}[x]$ by the rule $\bar{p} = (\bar{a}_0, \bar{a}_1, \ldots, \bar{a}_n, \ldots)$, where $\bar{a}_n$ is the complex conjugate of a_n. Since the function $a \to \bar{a}$ is an *automorphism* of $\mathbb{C}$ (see 9.6.7 ff.), it follows easily (and the reader should verify) that

22.5.3 $$\overline{p(\alpha)} = \bar{p}(\bar{\alpha})$$

for all $\alpha \in \mathbb{C}$. From this we conclude

22.5.4 THEOREM. *α is a zero of $p \in \mathbb{C}[x]$ if and only if $\bar{\alpha}$ is a zero of $\bar{p}$. In particular, if $p \in \mathbb{R}[x]$, then $\alpha \in \mathbb{C}$ is a zero of p if and only if $\bar{\alpha}$ is a zero of p.* ■

By Theorem 22.5.2, a polynomial $p \in \mathbb{R}[x]$ of degree m has m zeros in $\mathbb{C}$. Some of these may actually be in $\mathbb{R}$. The rest will be outside $\mathbb{R}$ and must, according to Theorem 22.5.4, occur in complex conjugate pairs; thus in particular there must be an *even* number of such complex roots of the equation $p(x) = 0$.

Exercise 7

(i) Given that $2 - 3i$ is a root of the equation $3x^3 - 4x^2 + 7x + 104 = 0$, find the other roots.
(ii) Prove that the function $y = x^n \sin^2 x$ is not a polynomial function, where n is an integer.

22.6 Derivation

An important *algebraic* operation on a ring of polynomials is that of **derivation.** Given the polynomial $p = a_0 + a_1\mathfrak{x} + \cdots + a_N\mathfrak{x}^N$ in $R[x]$, where R is an integral domain, we define a new polynomial, p', in $R[x]$ by

22.6.1 $$p' = a_1 + 2a_2\mathfrak{x} + \cdots + Na_N\mathfrak{x}^{N-1},$$

and call p' the **derivative** of p. The reader familiar with the calculus will recognize the resemblance between p' and the derivative of the function $p(x)$ when $R = \mathbb{R}$, but he should notice that our definition is purely algebraic and involves no notions of the differential calculus (such as limits!). We consider the function $D: R[x] \to R[x]$, given by $D(p) = p'$; from the definition 22.6.1 we have at once that $D(\mathfrak{x}^n) = n\mathfrak{x}^{n-1}$ ($n = 1, 2, \ldots$), while if f is constant (i.e., of degree zero), then $Df = 0$ and $D(fp) = fDp$. Hence by 22.6.1,

22.6.2 $$p' = D\left(\sum_{r=0}^{N} a_r\mathfrak{x}^r\right) = \sum_{r=0}^{N} a_r D\mathfrak{x}^r.$$

We now prove the usual rules for calculating the derivatives of sums and products of polynomials:

22.6.3 PROPOSITION. (i) $D(f + g) = Df + Dg$;
(ii) $D(fg) = (Df)g + f(Dg)$.

Proof. The verification of (i) is immediate. To prove (ii) we revert to the sequence notation of 22.1.3–22.1.5. Thus let

$$f = (a_0, \ldots, a_n, \ldots), \qquad g = (b_0, \ldots, b_n, \ldots).$$

Now

$$\begin{aligned} \mathfrak{x}^n g &= (\mathbf{0}_n, 1, \mathbf{0}_\infty)(b_0, \ldots, b_m, \ldots) \\ &= (\mathbf{0}_n, b_0, b_1, \ldots, b_m, \ldots) \qquad \text{(by 22.1.7)} \end{aligned}$$

so $$D(\mathfrak{x}^n g) = (\mathbf{0}_{n-1}, nb_0, (n+1)b_1, \ldots, (n+m)b_m, \ldots).$$

Collecting terms then gives

$$\begin{aligned} D(\mathfrak{x}^n g) &= (\mathbf{0}_{n-1}, nb_0, nb_1, \ldots, nb_m, \ldots) + (\mathbf{0}_n, b_1, 2b_2, \ldots, mb_m, \ldots) \\ &= n\mathfrak{x}^{n-1}g + \mathfrak{x}^n g', \qquad \text{(by 22.1.7 again)} \end{aligned}$$

thus verifying (ii) when $f = \mathfrak{x}^n$. Also, (ii) holds when f is constant, as observed. Hence, if $f = (a_0, a_1, \ldots, a_n, \mathbf{0}_\infty)$, then by 22.1.6, $f = \sum_{r=0}^{n} a_r\mathfrak{x}^r$; and so, using the special cases just proved, we have

$$\begin{aligned} D(fg) &= D\left(\sum a_r\mathfrak{x}^r g\right) = \sum a_r D(\mathfrak{x}^r g) \qquad \text{(by (i))} \\ &= \left(\sum a_r(D\mathfrak{x}^r)\right)g + \left(\sum a_r\mathfrak{x}^r\right)(Dg) \\ &= (Df)g + f(Dg) \qquad \text{(by 22.6.2).} \end{aligned}$$

This completes the proof of (ii). ■

22.6.4 COROLLARY. *For any $n > 0$, $D(f^n) = nf^{n-1}D(f)$.*

The reader should supply the proof by induction on n, noting the dependence of the result on the commutativity of R.

The connection with the solution of polynomial equations is provided by the next theorem. We call α an ***m*-fold zero** of f if $(\mathfrak{x} - \alpha)^m$ is a factor of f but $(\mathfrak{x} - \alpha)^{m+1}$ is not. We then have

22.6.5 THEOREM. *Let α be a zero of the polynomial f in $\mathbb{C}[x]$. Then, for any $m \geqslant 1$, α is an m-fold zero of f if and only if it is an $(m-1)$-fold zero of f'.*

Proof. Let α be an m-fold zero of f. Then

$$f = (\mathfrak{x} - \alpha)^m q \quad \text{and} \quad q(\alpha) \neq 0.$$

By Prop. 22.6.3 and Corollary 22.6.4,

$$\begin{aligned} f' &= m(\mathfrak{x} - \alpha)^{m-1} q + (\mathfrak{x} - \alpha)^m q' \\ &= (\mathfrak{x} - \alpha)^{m-1}(mq + (\mathfrak{x} - \alpha)q'); \end{aligned}$$

and $mq(\alpha) + 0 \cdot q'(\alpha) \neq 0$. Thus α is an $(m-1)$-fold zero of f'.

Now, by Theorem 22.3.5, every zero of the polynomial f is an m-fold zero for some m, and if it is a k-fold zero of f', then we have just shown that $m - 1 = k$. Thus the converse also holds. [Conventionally, $m = \infty$, iff f is the zero polynomial.] ■ Observe that all we required about $\mathbb{C}$ in this theorem was that $ma \neq 0$ if $a \neq 0$. Thus the theorem holds if $\mathbb{C}$ is replaced by any integral domain with the property that 1 generates, additively, a cyclic infinite group. Such domains are said to have **zero characteristic**; notice that the field $\mathbb{Z}_p$, where p is prime, is *excluded* from the full force of the theorem, since $p1 = 0$ in $\mathbb{Z}_p$. On the other hand the argument remains valid even in this case provided $p \nmid m$. Indeed, the case $m = 1$ of Theorem 22.6.5 is important. A 1-fold zero is called a **simple** zero; and a simple zero is thus recognized by the fact that $f(\alpha) = 0, f'(\alpha) \neq 0$. Similarly a 2-fold zero is recognized by the fact that $f(\alpha) = f'(\alpha) = 0$, $f''(\alpha) \neq 0$ (provided R is not of characteristic 2 (see Exercise 8(iv) below)).

Exercise 8

(i) Show that 2 is a 3-fold zero of $\mathfrak{x}^4 - \mathfrak{x}^3 - 18\mathfrak{x}^2 + 52\mathfrak{x} - 40$ (a) by establishing that $(\mathfrak{x} - 2)^3$ is a factor but $(\mathfrak{x} - 2)^4$ is not, (b) by 'long division'.
(ii) A polynomial f in $\mathbb{Z}[x]$ has degree 5, while 0 and $-\frac{1}{2}$ are both 2-fold zeros of f'. If $f(0) = 1, f(1) = 65$, find f.
(iii) Suppose the polynomials f, g in $R[x]$ have all their roots in R. The sets of roots are F, G respectively. Which polynomials have $F \cap G$, $F \cup G$ as sets of roots?
(iv) Show that the 'line' $y - p(\alpha) = p'(\alpha)(x - \alpha)$ 'touches' the 'curve' $y = p(x)$. {Precisely, show that α is at least a 2-fold zero of the polynomial $p(x) - p(\alpha) - p'(\alpha)(x - \alpha)$.}
(v) Show that the binomial theorem and Leibniz theorem hold in $R[x]$ for any commutative ring R with unity (see 5 Exercise 1(xvi)).
(vi) Any function from a ring to itself, satisfying (i) and (ii) of 22.6.3, is called a **derivation**. Show that the set of all derivations of a ring form an abelian group G. Show further that if $d_1, d_2 \in G$, then $d_1 d_2 - d_2 d_1 \in G$. (Here multiplication is composition of functions.)
(vii) Let R be an integral domain. Define the **characteristic** of R to be the smallest positive n such that $n1 = 0$, if such integers exist. Show that the characteristic is prime.

If we take the special case $R = \mathbb{R}$, then f' is just the differential coefficient of f in the sense of the differential calculus—at least, provided we identify f and f' with $\Phi(f)$ and $\Phi(f')$ as allowed by Theorem 22.4.2. The mean value theorem (29.7.4) for the differential calculus then asserts that, if f is a differentiable function of x in the interval $a \leqslant x \leqslant b$, then the relation

22.6.6
$$f(b) = f(a) + (b - a)f'(\xi)$$

holds for some ξ in $a < x < b$. This leads immediately to the following conclusion (cf. Rolle's theorem, 29.7.1).

22.6.7 THEOREM. *Between any two zeros of the polynomial f in $\mathbb{R}[x]$ lies a zero of f'.* ■

Of course, this is not strictly a theorem about zeros of polynomials, since we may replace the polynomial f in its enunciation by any differentiable function; but it deserves to be stated here, since plainly it gives useful information on the location of zeros of polynomials in $\mathbb{R}[x]$.

Exercise 9

(i) Show that the polynomial $\frac{1}{5}x^5 + x^3 + 2x + 19$ has precisely one zero in $\mathbb{R}[x]$.
(ii) In the identification of f with Φf in the remarks preceding 22.6.6, show that more precisely $\Phi(Df) = (d/dx)(\Phi f)$. Interpret this equation in words.

22.7 Solution of Polynomial Equations

So far we have given little clue as to how zeros of polynomials are located. This problem can be understood in a precise and an approximate sense. That is, we may wish to obtain the solutions of the polynomial equation $p(x) = 0$ quite precisely, or we may be content with approximate solutions. It is, in our view, a defect of many standard school texts that they place so much emphasis on the former problem to the virtual exclusion of the latter. We regard the latter question as far more important for users of mathematics; particularly is it the case that modern computers can easily be programmed to obtain solutions to any conceivably desired degree of accuracy. See also Newton's method (Chapter 34).

However, if we take up the first problem, the deepest and most interesting result, due to Abel and Galois, is the following. If we are only allowed to use the ordinary operations of arithmetic, together with the extraction of roots, then *it is impossible to lay down a systematic procedure for finding the zeros of polynomials of degree greater than 4.* A discussion of this result will be found in the works of Bell [10, 11, 12]; and a proof of the result, in the works of Artin [6] and Postnikov [103].

Methods of solution of polynomial equations of degree 1, 2, 3, or 4 had been known long before Galois' time; and Galois' theorem put an end to centuries of effort to solve the quintic (the equation of degree 5). It should be understood that we are not saying that no quintic can be solved; Galois' result means that we cannot give a *general rule* for solving all quintics.

We will be content to describe the solution of the quadratic equation, and consider only the polynomial ring $\mathbb{C}[x]$; but the reader will observe that the solution is valid in $F[x]$, where F is any field in which we may take square roots. Our omission of any description of the solution of the cubic or quartic (degree 3 or 4) springs from our conviction that these solutions are of purely historic interest. [See, for example, Birkhoff–MacLane [16].] We omit the solution of the linear equation on the grounds of its simplicity (see also Chapter 19).

Thus, consider the general quadratic polynomial $a_0 + a_1x + a_2x^2$ in $\mathbb{C}[x]$. We will revert to more traditional notation and write this polynomial as $ax^2 + 2bx + c$. Thus the problem is to solve the equation $ax^2 + 2bx + c = 0$, where $a, b, c \in \mathbb{C}$ and $a \neq 0$. Of course if we can factorize the polynomial $ax^2 + 2bx + c$ as a product of linear factors, then we may immediately pass to the solution of the equation. If we are able to spot the factors, that is certainly the quickest method of solution. What we give is essentially a procedure for finding the factors.

Let us write $f(x) = 0 \Leftrightarrow g(x) = 0$ to mean:

$x = \alpha$ is a solution of the equation $f(x) = 0$ if and only if it is a solution of the equation $g(x) = 0$.

Then (extending slightly but in an obvious way the significance of the symbol $\Leftrightarrow$) we have

$$
\begin{aligned}
& ax^2 + 2bx + c = 0 \\
\Leftrightarrow\ & a^2x^2 + 2abx + ac = 0 \qquad \text{(since } a \neq 0\text{)} \\
\Leftrightarrow\ & (ax + b)^2 - b^2 + ac = 0 \\
\Leftrightarrow\ & (ax + b)^2 = b^2 - ac \\
\Leftrightarrow\ & ax + b = \pm\sqrt{(b^2 - ac)} \\
\Leftrightarrow\ & ax = -b \pm \sqrt{(b^2 - ac)} \\
\Leftrightarrow\ & x = \frac{-b \pm \sqrt{(b^2 - ac)}}{a}.
\end{aligned}
$$

Notice that every complex number has a square root, so the procedure is universally valid. Notice too that if $a, b, c \in \mathbb{R}$ then the solutions are in $\mathbb{R}$ if and only if $b^2 \geqslant ac$. We also point out that our solution provides a verification of Theorem 22.5.4, since if a, b, c are real and $b^2 < ac$ then the solutions of the given equation are complex conjugates.

Exercise 10

(i) Solve the following equations in $\mathbb{C}$:

(a) $2ix^2 - (1 + i)x + 1 + i = 0$;
(b) $3x^3 - 5x - 2 = 0$;
(c) $x^2 - x - 4 - 3\sqrt{2} = 0$.

(ii) For what elements $a \in \mathbb{Z}_7$ has the equation $3x^2 - 5x + a = 0$ roots in $\mathbb{Z}_7$?
(iii) Prove that, in $\mathbb{R}[x]$, $ax^2 + 2bx + c > 0$ iff $a > 0$ and $b^2 < ac$. Explain this fact by means of a graph. Similarly analyse the inequality $ax^2 + 2bx + c \geqslant 0$.
(iv) In $\mathbb{R}$ criticize the following 'solution' of the equation

$$\sqrt{(x + 2)} + \sqrt{(4x + 1)} = 5.$$

'Square each side to get $5x - 22 = -2\sqrt{(4x^2 + 9x + 2)}$, and square again to get $9x^2 - 256x + 476 = 0$. Hence $x = 2$ or $x = 26\frac{4}{9}$. The latter value does not satisfy the original equation, hence $x = 2$.' {Use the symbol $\Leftrightarrow$ as in the text.}

22.8 Application to Finite Fields

We have already (in Exercise 8(vii)) introduced the *characteristic* of an integral domain. If 1 is the unity element of the integral domain R, then the additive subgroup of R generated by 1 is infinite or of order p, where p is prime (see Exercise 8(vii)). In the former case the characteristic of R is said to be zero, in the latter case it is p. Plainly R is infinite if the characteristic of R is zero, so if R is finite its characteristic is p for some prime p.

Let then R be a finite integral domain of characteristic p. We first prove

22.8.1 THEOREM. *R contains p^n elements for some $n \geqslant 1$.*

Proof. Plainly $px = 0$ for any $x \in R$, since $px = p(1 \cdot x) = (p1)x = 0x = 0$. Thus $\mathbb{Z}_p$ operates on the additive group of R by the rule

$$[m]x = mx \qquad ([m] \in \mathbb{Z}_p).$$

It follows that R admits a vector-space structure over $\mathbb{Z}_p$. [Check the axioms $\mathfrak{V}_1$–$\mathfrak{V}_4$ in 19.1.] It thus has a basis containing, say, $n > 0$ elements; therefore R contains p^n elements by Corollary 19.4.11. ■

We now specialize to the case when R is a finite field F, of characteristic p; and we prove

22.8.2 THEOREM. *There exists, up to isomorphism, precisely one field with p^n elements, $(n = 0, 1, 2, \ldots)$.*

Proof. Suppose we have a field Γ of characteristic p in which the equation $x^{p^n} - x = 0$ has p^n solutions. Plainly the argument in Theorem 22.6.5 still works in Γ to guarantee that all these solutions are distinct; for the derivative of $x^{p^n} - x$ is $p^n x^{p^n - 1} - 1 = -1$ since Γ is of characteristic p, so there are no multiple roots of $x^{p^n} - x$ in Γ.

We now show that the solutions form a sub-field of Γ which will then be a field with p^n elements. First we observe that if $(a + b)^{p^n}$ is expanded by the binomial theorem, then all binomial coefficients except the first and the last are divisible by p. Therefore

$$(a - b)^{p^n} = a^{p^n} + (-1)^{p^n} b^{p^n} \quad \text{in } \Gamma.$$

Thus if a, b are roots of $x^{p^n} - x$, then

22.8.3 $$(a - b)^{p^n} = a^{p^n} + (-1)^{p^n} b^{p^n} = a + (-1)^p\, b = a - b.$$

The last equality is obvious if p is odd and follows if $p = 2$ from the fact that then $2b = 0$, so $b = -b$. Thus 22.8.3 shows that $a - b$ is also a root of $x^{p^n} - x$. Plainly ab is also a root, and so too is a^{-1} provided $a \neq 0$. Thus the set of solutions of $x^{p^n} - x = 0$ forms a field F of p^n elements; whence, to establish the existence of such a field F, it suffices to construct a suitable field Γ of the kind postulated above.

Such a Γ is a **splitting** field for the equation $x^{p^n} - x = 0$, that is, a field in which $x^{p^n} - x$ breaks up into linear factors. We construct it by adjoining solutions of this equation to the field $\mathbb{Z}_p$. The process generalizes that of Exercise 5(iii) and we will not go into details here, beyond pointing out that in fact we consider the equation $g(x) = 0$ where

$$x^{p^n} - x = (x^p - x)g(x),$$

since the elements of $\mathbb{Z}_p$ are already zeros of $x^p - x$. The polynomial $g(x)$ will have no zeros in $\mathbb{Z}_p$, so we can steadily enlarge $\mathbb{Z}_p$, as in Exercise 5(iii) to enable us to factorize $g(x)$ into linear factors. Thus we have demonstrated the existence of a splitting field Γ as required, and hence the existence of a field F of p^n elements.

We are also content merely to sketch the proof of uniqueness. Any such field F is a splitting field for $g(x)$ over $\mathbb{Z}_p$; and indeed it is the minimum splitting field since it contains the minimum number of elements. Now the minimum splitting field of a polynomial is unique up to isomorphism. The reader wishing more details should consult Birkhoff–MacLane [16]. ■

Our real concern in this section is to demonstrate a remarkable property of the field F of p^n elements. We state it as:

22.8.4 THEOREM. *The multiplicative group of F is cyclic.*

The key observation is that of Exercise 5(ii), slightly generalized. We write $F^* = F - \{0\}$, $q = p^n - 1$, and prove:

22.8.5 PROPOSITION. *Let $r|q$. Then $x^r - 1$ has exactly r zeros in F.*

Proof of 22.8.5. $x^q - 1$ has q zeros in F, since F^* is a group of order q under multiplication. Now

$$x^q - 1 = (x^r - 1)h(x),$$

where $h(x)$ is of degree $q - r$. Thus by Theorem 22.3.5 $x^r - 1$ has at most r zeros and $h(x)$ has at most $q - r$ zeros in F. But every one of the q zeros of $x^q - 1$ is a zero of $x^r - 1$ or of $h(x)$. Thus $x^r - 1$ has r zeros in F. ■

The proof of Theorem 22.8.4 is now pure group-theory. We need:

22.8.6 THEOREM. *Let G be a commutative group of order q. Suppose that for every $r|q$, there are exactly r elements of G satisfying $x^r = e$. Then G is cyclic.*

Proof. Let q factorize into primes as $q = p_1^{m_1} \cdots p_k^{m_k}$. The hypothesis implies that for each $i = 1, 2, \ldots, k$, there is an element $g_i \in G$ such that

22.8.7 $$g_i^{p_i^{m_i}} = e, \qquad g_i^{p_i^{m_i - 1}} \neq e.$$

Plainly such an element is of order $p_i^{m_i}$. We claim that if $g = g_1 g_2 \cdots g_k$, then g is of order q. For suppose g is of order $s < q$. Then $s|q$ (Corollary 18.9.2), so that

$$s = p_1^{n_1} \cdots p_k^{n_k}, \qquad 0 \leqslant n_i \leqslant m_i, \qquad (i = 1, 2, \ldots, k)$$

and for some i, $n_i < m_i$. Suppose for definiteness that $n_1 < m_1$, and set

$$t = p_1^{n_1} p_2^{m_2} \cdots p_k^{m_k}.$$

Then $g^t = e$; but $g^t = g_1^t$, so the order of g_1 (which is $p_1^{m_1}$ by 22.8.7) divides t. This is impossible since $n_1 < m_1$, which contradiction shows that g is of order q. Hence G is cyclic (generated by g). ■

In the light of Proposition 22.8.5, Theorem 22.8.4 is an immediate corollary of Theorem 22.8.6. ■

Exercise 11

Find generators of the multiplicative group of $\mathbb{Z}_p$ for $p = 7, 11, 13, 17, 19, 23$.

Part VI

NUMBER SYSTEMS AND TOPOLOGY

In the preceding parts we permitted outselves to refer freely to natural numbers, integers, rational numbers, real numbers and complex numbers, in order to describe the framework for the arithmetic, algebra and geometry we were doing and in order to provide examples illustrating our results. It is our belief that it is a sound pedagogical principle to introduce concepts first in a somewhat intuitive way and then, once interest in them has been aroused (we hope!), to reintroduce them somewhat more formally. This process of reacquaintance should not necessarily be completed at the second time of meeting; indeed, it is probably never desirable and rarely possible to complete it so soon. For with increasing sophistication the student sees more and more of what lies behind the deceptively simple—and the deceptively obscure—fundamental concepts of mathematics. If the student has gained a significant grasp of elementary arithmetic and algebra he should now be ready to take a closer look at the process whereby our number system is built up. We should explain that, here, by 'process' we understand 'logical process' and not 'psychological process' or 'historical process'; these latter 'processes' are of great interest and importance, but we will not be concerned with them here. We do hope that the reader will be fired to read up for himself the history of the development of the number system, because the intuitive interest of this history is enormous and it is also very instructive to compare the historical development with the logical development. The latter is what we present in this Part, but we simply now state some of the historical *motivation.*

Briefly, the Greeks knew about the set $\mathbb{N}$ of natural numbers, 1, 2, . . ., and about the positive rationals p/q where $p, q \in \mathbb{N}$. After the invention of algebra by the Arabs (influenced from India), the notation itself suggested solving equations like $3x + 2 = 0$, thus forcing a widening of the concept of number to include what we now call 'negative numbers'. Thus, the system $\mathbb{Q}$ of all rationals (including the ring $\mathbb{Z}$ of all integers) was used by the Arabs, even though they did not prove its existence; negative numbers were called 'imaginary'. By the 14th century, Italian mathematicians had introduced the new 'imaginary number' $i = \sqrt{-1}$, since they found that a number system even larger than $\mathbb{Q}$ was necessary for their (successful) solution of the quadratic, cubic and quartic equations. Thus was evolved the system $\mathbb{C}$ of complex numbers, which Gauss showed, with his 'fundamental theorem of algebra' (see 22.5.2), in the 19th century, to suffice for providing solutions of *all* polynomial equations.

Moreover, Gauss showed how to build $\mathbb{C}$, given the system $\mathbb{R}$ of real numbers—a system known intuitively to earlier mathematicians and discussed profoundly by the classical Greek Eudoxus. With $\mathbb{R}$ everybody felt he knew what a number was, but nobody could describe it satisfactorily. Did real numbers *exist*? What does this question mean? Thus began the great logical progress of the 19th century, ending in the creation of $\mathbb{R}$ from $\mathbb{Q}$, of $\mathbb{Q}$ from $\mathbb{Z}$, and $\mathbb{Z}$ from $\mathbb{N}$, together with the discovery that all the properties of $\mathbb{N}$ could be deduced from the two famous axioms of Peano. A result of this creative process was the invention of the language of set theory, since traditional language was too clumsy to express the necessary subtleties.

The reader may find this Part, and particularly Chapter 24, the most difficult in the book; but this should not be surprising, considering the vast effort which mathematicians needed to think out the work. Fortunately, also, the reader can delay digesting this material, provided he knows the *rules* for working with $\mathbb{Z}$, $\mathbb{Q}$, $\mathbb{R}$, and $\mathbb{C}$ as embodied in the appropriate sets of axioms; once these are stated properly, then we are proving, in this Part, one big *existence theorem*, viz., that, granted the existence of $\mathbb{N}$, systems exist which satisfy the requirements of working mathematicians. We do *not* answer the question: 'What is a number?' since mathematicians do not need to ask this question—and it would never occur to anybody else!

Chapter 23
THE RATIONAL NUMBERS

23.1 The Peano Axioms

One must start somewhere in the logical construction of the number system. It is possible to found the system on set theory, and to build up the natural numbers from the basic operations of set theory. We ourselves would be quite content if the reader took the properties of the natural numbers, 0, 1, 2, . . ., and the arithmetical operations associated with them for granted. Such a reader may omit from here as far as 23.2, except that he should acquaint himself with the principle of induction.

However, for the benefit of those who prefer to go further back, we list below the **Peano axioms** for the natural numbers. Thus we have a set S; this set has a **distinguished element** 0; and it is furnished with a function suc: $S \to S$, to be read 'successor of'. The Peano axioms are then the following:

$\mathfrak{P}_1$: suc *is injective* (i.e., $\operatorname{suc} x = \operatorname{suc} y \;.\Rightarrow.\; x = y$) *and* $\operatorname{suc} S = S - \{0\}$;
$\mathfrak{P}_2$: **(principle of induction)** *let E be a subset of S; if* $0 \in E$ *and* $\operatorname{suc} E \subseteq E$, *then* $E = S$.

We may write 1 for suc (0), 2 for suc (1), etc. Further, it follows from $\mathfrak{P}_2$ that every $x \in S$ is reached from 0 after a finite chain suc (0), suc (suc (0)), . . ., of iterated applications of the function suc; this may be proved by using the proper set-theoretical definition of finitude. [No circularity is in fact involved here.] It follows from $\mathfrak{P}_1$ and $\mathfrak{P}_2$ that the equation $\operatorname{suc} \cdots \operatorname{suc} x = x$ has no solutions and that we may identify x with the iterated application of suc required to obtain x from 0. Briefly ,we 'count' to x by applying suc x times; and a copy of every whole number in our experience (say 1,253,431) appears in S as $\operatorname{suc}^{1,253,431}(0)$.

We now explicitly assume that there does exist a set satisfying the Peano axioms. Choosing one particular such set, we denote it by $\mathbb{Z}^+$.

In terms of the Peano axioms we may introduce addition and multiplication into $\mathbb{Z}^+$. We first make the crucial remark that, to define a function f from $\mathbb{Z}^+$ to some set B, it is sufficient to give the value of $f(0)$ and the value of $f(\operatorname{suc} n)$ in terms of $f(n)$. For then the set of elements of $\mathbb{Z}^+$ for which f is defined will be the whole of $\mathbb{Z}^+$, by $\mathfrak{P}_2$. Thus we define the function '$+m$': $\mathbb{Z}^+ \to \mathbb{Z}^+$, $m \in \mathbb{Z}^+$, by

$$(+m)(0) = m$$
$$(+m)(\operatorname{suc} n) = \operatorname{suc}\,(+m)(n);$$

we write, according to custom, $m + n$ for $(+m)(n)$. This last definition is an example of the 'inductive' or 'recursive' method of definition, which we mentioned in 5.3.

Exercise 1

(For the diligent reader! A detailed account is given in Thurston [125].)
(i) Deduce that addition, so defined, is associative and commutative.
{*Hint:* Given $a \in \mathbb{Z}^+$, let $E = \{m | a + m = m + a\}$. By definition of + above, $0 \in E$; while $m \in E \;.\Rightarrow.\; \operatorname{suc} m \in E$. Hence $E = \mathbb{Z}^+$, by $\mathfrak{P}_2$, so $a + m = m + a$ for all $m \in \mathbb{Z}^+$, and therefore for all a also. This pattern of proof applies in the other exercises of this section.}
(ii) Show that if $m + n = m' + n$ in $\mathbb{Z}^+$, then $m = m'$.

The set $\mathbb{Z}^+$ has a natural order relation (see 4.4) built into it; for it is clear that if x, y are distinct elements of $\mathbb{Z}^+$, then either y may be obtained from x by iterated application of the function suc, or x may be so obtained from y, but not both. We write $x < y$ in the former case. Formally, however, we define $<$ by:

23.1.1 $x < y$ *in* $\mathbb{Z}^+$, *if and only if* $\exists z \in \mathbb{Z}^+$ *such that* $z \neq 0$ *and* $x + z = y$.

Then $x \leqslant y$ if $x < y$ or $x = y$. We may consider for each $n \in \mathbb{Z}^+$, the set of elements x of $\mathbb{Z}^+$ such that $x < n$; call this subset $[\![n]\!]$, so $[\![0]\!] = \varnothing$, $[\![1]\!] = \{0\}$, $[\![2]\!] = \{0, 1\}$, etc. Then we say that a set A has n elements if $A \approx [\![n]\!]$; this allows us to count finite sets. [A commoner way of counting is by taking the subset $0 < x \leqslant n$ of $\mathbb{Z}^+$ as model. It is evident that the two ways lead to the same answer!] There is further discussion of the counting process in Parts I and II.

Exercise 2

(i) Prove that $<$ is a transitive relation but neither reflexive nor symmetric.
(ii) Show that $m \leqslant m' \;.\Rightarrow.\; m + n \leqslant m' + n$.
(iii) Prove the law of trichotomy for $(\mathbb{Z}^+, <)$: given $x, y \in \mathbb{Z}^+$, exactly one of the statements $x = y$, $x < y$, $x > y$ holds. {Fix x, and consider suc (y), using $\mathfrak{P}_2$.}
(iv) Prove that $\mathbb{Z}^+$ is well-ordered (see Chapter 5) by $<$. {Consider the sets $[\![n]\!]$ described following 23.1.1 and use $\mathfrak{P}_2$.}
(v) Show that the equation $m + x = n$ has a solution in $\mathbb{Z}^+$ if and only if $m \leqslant n$; and that the solution, when it exists, is unique.

Next we introduce multiplication by m, defining the function

$$`\times m\text{'}: \mathbb{Z}^+ \to \mathbb{Z}^+ \qquad (m \in \mathbb{Z}^+)$$

recursively:

23.1.2
$$\begin{cases} (\times m)(0) = 0 \\ (\times m)(\operatorname{suc} n) = (\times m)(n) + m; \end{cases}$$

we write $m \times n$ or simply mn for $(\times m)(n)$.

Exercise 3

(For the persistent reader!)
(i) Deduce that multiplication, so defined, is associative and commutative and distributes over addition.
(ii) Show that $m \leqslant m' \Rightarrow mn \leqslant m'n$.
(iii) Show that the equation $mx = n$ where $m, n \neq 0$, has a solution in $\mathbb{Z}^+$ only if $m \leqslant n$; and that the solution if it exists is unique.
(iv) Show that if $mn = m'n$ in $\mathbb{Z}^+$ and $n \neq 0$, then $m = m'$. {See 23.3.10.}
(v) Prove that if $(T, 0', \sigma)$ is a set T with zero $0'$ and successor-function σ satisfying $\mathfrak{P}_1$ and $\mathfrak{P}_2$, then there is an isomorphism $\theta\colon (\mathbb{Z}^+, 0, \text{suc}) \to (T, 0', \sigma)$ such that $\theta(0) = 0'$, $\theta \circ \text{suc} = \sigma \circ \theta$. Thus any system satisfying $\mathfrak{P}_1$ and $\mathfrak{P}_2$ will do for $\mathbb{Z}^+$.

Thus, sketchily and hurriedly and in a way which would surely not satisfy a professional logician, we have rushed to the point we would have wished to have taken as our starting point for the development of the number system. Our interest is primarily in teaching mathematical skill and understanding; because of this—and not for any disrespect for logic and set theory as such—we regard our proper task as that of showing how richer number systems are built out of the natural numbers. This we now proceed to do; the development in this chapter will take us as far as the rationals.

23.2 The System $\mathbb{Z}$

As we know, subtraction is not always possible in $\mathbb{Z}^+$, and we therefore cannot always solve an equation like $a + x = b$ in $\mathbb{Z}^+$. In order to be able to subtract any number from any other it is necessary to pass from the set $\mathbb{Z}^+$ to the larger set $\mathbb{Z}$ of integers, $0, \pm 1, \pm 2, \ldots$; that is, to introduce (or *invent*) the negative numbers. We will now describe in detail the precise mathematical technique for doing this.

The basic notion underlying our procedure is that of regarding an element n of $\mathbb{Z}^+$ as an additive operator converting any number $a \in \mathbb{Z}^+$ into $a + n$. So regarded, n becomes a function $\mathbb{Z}^+ \to \mathbb{Z}^+$ whose graph is a certain subset of $\mathbb{Z}^+ \times \mathbb{Z}^+$, namely the subset consisting of those ordered pairs (a, b) such that $b = a + n$. We see that two pairs (a, b) and (c, d), which are known to belong to two such functions, in fact belong to the same function if and only if $a + d = b + c$ [by cancellation in $\mathbb{Z}^+$; if $b = a + n$ and $d = c + m$, then $a + d = a + c + m$, $b + c = a + c + n$, so $m = n$ if and only if $a + d = b + c$, by Exercise 1(ii)]. We also see that, if we *had* a number $-n$ we would surely wish to associate it with that subset of $\mathbb{Z}^+ \times \mathbb{Z}^+$ consisting of pairs (a, b) such that $a = b + n$.

These observations motivate the following definition. In the set $\mathbb{Z}^+ \times \mathbb{Z}^+$, set up an equivalence relation $(\sim)$ by declaring that $(a, b) \sim (c, d)$ if

$$a + d = b + c.$$

It is evident (formally, we apply the laws of arithmetic referred to above) that this relation is indeed an equivalence relation; we write $[a, b]$ for the equivalence class containing (a, b). We call such a class $[a, b]$ an **integer** and write $\mathbb{Z}$ for the set of integers: that is,

23.2.1 $$\mathbb{Z} = (\mathbb{Z}^+ \times \mathbb{Z}^+)/(\sim).$$

We hasten to reassure the reader that it is only until 23.3 that he is expected to think of an integer as an equivalence class of pairs from $\mathbb{Z}^+ \times \mathbb{Z}^+$. Once the *existence* of a system $\mathbb{Z}$ has been established, we can perform arithmetic according to the rules as we always did before. None of the rules tells us what an integer 'is'.

Using the above set $\mathbb{Z}$ then, we first find within it a copy of $\mathbb{Z}^+$. For, among the elements $[a, b] \in \mathbb{Z}$ there occur in particular the classes $[a, 0]$ containing pairs $(a, 0)$ with second member $0 \in \mathbb{Z}^+$. We therefore define a function

23.2.2 $$f: \mathbb{Z}^+ \to \mathbb{Z}$$

by $f(a) = [a, 0]$; and assert that f is one-one since $(a, 0) \sim (b, 0)$ iff $a = b$. Thus $\mathbb{Z}^+$ is 'embedded' in $\mathbb{Z}$ in the sense that $\mathbb{Z}$ contains a copy $\mathbf{a} = [a, 0]$ of each integer $a \in \mathbb{Z}^+$. These copies are often called 'positive' integers and include $\mathbf{0} = [0, 0]$.

Now given the pair (m, n), such that $m < n$, then by Definition 23.1.1, there exists a non-zero $a \in \mathbb{Z}^+$ such that $m + a = n$; thus $(m, n) \sim (0, a)$. Hence by the law of trichotomy in $\mathbb{Z}^+$ (see Exercise 2(iii)) any class $[c, d]$ in $\mathbb{Z}$ is of one and only one of the forms

$$[a, 0], \qquad [0, 0], \qquad [0, a], \qquad a \neq 0.$$

We agree to write (temporarily) $-\mathbf{a}$ for the integer $[0, a]$ and thus have introduced the **negative integers.** Notice that then the set $\mathbb{Z}$ of integers is the union of the two subsets of positive and negative integers respectively, and that, according to our definition, these two subsets intersect in the set consisting of the positive integer $\mathbf{0}$. Notice, too, that we may interpret $[a, b]$ as $a - b$ if $a \geqslant b$ and as $-(b - a)$ if $b \geqslant a$, (these inequalities in $\mathbb{Z}^+$).

There is next the question of extending the addition and multiplication in the copy $f(\mathbb{Z}^+)$ of $\mathbb{Z}^+$ to the whole of $\mathbb{Z}$. There are two possible approaches, leading of course to the same result.

In the former we first define addition and multiplication in $\mathbb{Z}^+ \times \mathbb{Z}^+$ by equations whose significance will be clear to the reader if he thinks of $[a, b]$ as $a - b$. Thus we define

23.2.3 $$(a, b) + (c, d) = (a + c, b + d),$$
$$(a, b)(c, d) = (ac + bd, ad + bc).$$

We next observe that, with these definitions, if $(a', b') \sim (a, b)$ and $(c', d') \sim (c, d)$, then

$$(a', b') + (c', d') \sim (a, b) + (c, d)$$
$$(a', b')(c', d') \sim (a, b)(c, d).$$

It follows that 23.2.3 determines an addition and multiplication in $\mathbb{Z}$ by the rules (independent of coset representatives):

23.2.4
$$[a, b] + [c, d] = [a + c, b + d],$$
$$[a, b][c, d] = [ac + bd, ad + bc].$$

We next note as immediate consequences of 23.2.4 that

$$[a, 0] + [b, 0] = [a + b, 0], \qquad [a, 0][b, 0] = [ab, 0],$$
$$[a, b] + [0, 0] = [a, b], \qquad [a, b][0, 0] = [0, 0], \qquad [a, b][1, 0] = [a, b].$$

The first two equalities show that the arithmetic of $\mathbb{Z}^+$ is retained when $\mathbb{Z}^+$ is regarded as a subset of $\mathbb{Z}$ by identifying a with $[a, 0]$; that is, the embedding f in 23.2.2 is a homomorphism. Moreover, the last three equalities show that the special numbers **0** and **1** play the same algebraic role in $\mathbb{Z}$ as they do in the copy $f(\mathbb{Z}^+)$ of $\mathbb{Z}^+$ (e.g., **0** is the neutral element for addition in $\mathbb{Z}$ as in $f(\mathbb{Z}^+)$).

As a further consequence of 23.2.4 we observe that the equation

$$[a, b] + x = [c, d]$$

in $\mathbb{Z}$ has the unique solution $x = [b + c, a + d]$, so that we have achieved our objective of being able to carry out any subtraction; and the notation $-\mathbf{a}$ introduced for $[0, a]$ justifies itself since

$$\mathbf{a} + (-\mathbf{a}) = [a, 0] + [0, a] = [a, a] = [0, 0] = \mathbf{0}.$$

Also, $[a, b] = \mathbf{a} + (-\mathbf{b})$.

Finally we check that the laws of arithmetic hold in $\mathbb{Z}$ as in $\mathbb{Z}^+$, so that we have gained much and lost nothing in passing from $\mathbb{Z}^+$ to $\mathbb{Z}$. We therefore revert to writing a for $\mathbf{a}$ when $a \in \mathbb{Z}^+$, since we have shown that $f\colon \mathbb{Z}^+ \to f(\mathbb{Z}^+)$ is an isomorphism (cf. Exercise 3(v)).

Exercise 4

(i) Verify that the rules 23.2.4 are indeed independent of coset representatives. {The cancellation rule (Exercise 1(ii)) must be used.}
(ii) Check in detail that the laws of arithmetic hold in $\mathbb{Z}$ (i.e., that $\mathbb{Z}$ is an integral domain (cf. Chapter 11)).
(iii) Show in a diagram of $\mathbb{Z}^+ \times \mathbb{Z}^+$ the cosets $[n, m]$ and show how this suggests that we could in fact have manufactured $\mathbb{Z}$ from $\mathbb{N}$, thus 'constructing' zero.
(iv) Show that if we go through the whole construction again, using $\mathbb{Z}$ instead of $\mathbb{Z}^+$, then we obtain a system isomorphic to $\mathbb{Z}$.

In the second approach to the introduction of operations in $\mathbb{Z}$, we set out so to define the processes of adding and multiplying by negative integers that the laws of arithmetic remain valid. The discussion of Chapter 9 shows that the laws of arithmetic impose on us *unique* definitions for the addition and multiplication of elements of $\mathbb{Z}$; namely, using f in 23.2.2, where $\mathbf{a} = f(a)$:

23.2.5 $\quad \mathbf{a}\cdot\mathbf{b} = f(ab) = (-\mathbf{a})\cdot(-\mathbf{b}); \qquad \mathbf{a}\cdot(-\mathbf{b}) = -f(ab) = (-\mathbf{a})\cdot\mathbf{b},$

and similarly for addition. One can then verify that each integer $[n, m]$ is $\mathbf{n} + (-\mathbf{m})$, that these operations impose a ring structure on $\mathbb{Z}$, and that the equations 23.2.4 hold. Thus we obtain the same algebraic system as before.

It is very important to emphasize that, whatever approach is adopted, addition and multiplication must be *defined* in $\mathbb{Z}$; the values of these operations in $\mathbb{Z}$ cannot be inferred *in vacuo* from their values in $\mathbb{Z}^+$. The impression is sometimes left with students that they should be able to infer that $(-2)(-3) = +6$ from the meaning of the multiplication of positive integers, but this, of course, is not possible.

Exercise 5

Perform the verifications stated to be possible following 23.2.5.

23.2.6 We now consider how to extend the order relation $\leqslant$, which holds in $\mathbb{Z}^+$ (now regarded as a subset of $\mathbb{Z}$), to a relation still denoted by $\leqslant$ between any two elements of $\mathbb{Z}$: this new relation is said to be **induced** by that in $\mathbb{Z}^+$. It is defined quite simply by saying that if $a, b \in \mathbb{Z}$ then $b \leqslant a$ if $a - b \in \mathbb{Z}^+$. This order relation agrees with that in $\mathbb{Z}^+$, and it is indeed the *only* order relation consistent with that in $\mathbb{Z}^+$ and satisfying the condition (see Exercise 2(ii)):

$$b \leqslant a \;.\!\Rightarrow\!.\; b + c \leqslant a + c, \qquad (a, b, c \in \mathbb{Z}).$$

On the other hand, it is evident that the order relation in $\mathbb{Z}$ is not a well-ordering, in contrast to $\mathbb{Z}^+$; see Exercise 2(iv). The set of negative integers, for example, is a non-empty subset of $\mathbb{Z}$ which does not have a first member (so, of course, is $\mathbb{Z}$ itself). It follows that proofs by induction involving $\mathbb{Z}$ must be handled with great care—it is, for example, sometimes possible to proceed from 0 first in the positive direction and then in the negative direction; it is also sometimes possible to handle the negative integers by some special argument and use induction for the positive integers. The following exercise (which presupposes familiarity with the rationals!) is an example of this second possibility.

Exercise 6

Let $f, g \in \mathbb{Q}[x]$, $g \neq 0$. Prove by induction on $\deg f - \deg g$ that there are polynomials $q, r \in \mathbb{Q}[x]$ with $f = qg + r$ and $\deg r < \deg g$. {See Chapter 12.}

23.3 The System $\mathbb{Q}$

We turn now to the development of the rationals from $\mathbb{Z}$. Just as the negative integers were introduced as a response to the problem of finding a solution of the equation $x + a = b$, so fractions are introduced to solve the equation $ax = b$. Here it should be noted that we exclude the case $a = 0$. For, in any ring, $0x = 0$ for all x (Proposition 9.3.1), so that the equation $0x = b$ could not be solved in any ring if $b \neq 0$, and is satisfied by every x in the ring if $b = 0$. Thus we consider the equation $ax = b$, $a \neq 0$. This motivates the discussion which follows, together with the fact that, for example, 6/4 and 12/8 are names for the one solution of the equation $2x = 3$. The reader should compare the discussion with that of the negative integers in 23.2.

We consider number pairs (b, a), $a, b \in \mathbb{Z}$, $a \neq 0$ and we introduce an equivalence relation $(\sim)$ in the set $E \subseteq \mathbb{Z} \times \mathbb{Z}$ of such pairs by declaring that

23.3.1 $$(b, a) \sim (d, c) \quad \text{if} \quad ad = bc.$$

E.g., $(6, 4) \sim (12, 8)$.

This relation (which has little to do with that in 23.2.1) is plainly reflexive and symmetric; it is also transitive, for if $(b, a) \sim (d, c)$ and $(d, c) \sim (f, e)$, then $ad = bc$, $cf = de$, so $acf = ade = bce$, whence $af = be$, since $c \neq 0$, or $(b, a) \sim (f, e)$. We will, temporarily, write $[\![b, a]\!]$ for the equivalence class mod $(\sim)$ containing (b, a) and write $\mathbb{Q}$ for the set of equivalence classes; such an equivalence class is called a **rational number.** That is, since $E = \mathbb{Z} \times (\mathbb{Z} - \{0\})$, then:

$$\mathbb{Q} = \big(\mathbb{Z} \times (\mathbb{Z} - \{0\})\big)/(\sim) = E/(\sim),$$

and the same reassurance applies here as that following 23.2.1. We now define a ring structure on $\mathbb{Q}$; the reader should notice that these are just the usual rules for adding and multiplying fractions, but expressed in our notation. Recall also that in the highly analogous process of passing from $\mathbb{Z}^+$ to $\mathbb{Z}$ we described two approaches; it is again true here that there are two approaches possible, but we are content to describe only the first.

To define addition and multiplication in $\mathbb{Q}$ we first give rules for adding and multiplying number pairs. Thus we set

23.3.2 $$\begin{cases} (b_1, a_1) + (b_2, a_2) = (b_1a_2 + b_2a_1, a_1a_2) \\ \quad (b_1, a_1)(b_2, a_2) = (b_1b_2, a_1a_2). \end{cases}$$

(Notice that $a_1a_2 \neq 0$ because $a_1 \neq 0$ and $a_2 \neq 0$.) We now wish to show that these laws of addition and multiplication pass to equivalence classes. That is to say, we must show that if $(b_i, a_i) \sim (d_i, c_i)$, $i = 1, 2$, then:

23.3.3 $$\begin{cases} (b_1a_2 + b_2a_1, a_1a_2) \sim (d_1c_2 + d_2c_1, c_1c_2) \\ \quad (b_1b_2, a_1a_2) \sim (d_1d_2, c_1c_2). \end{cases}$$

The second of these involves only a trivial verification, so we are content to prove the first. Thus we are given that $a_id_i = b_ic_i$ $(i = 1, 2)$ and wish to show that $a_1a_2(d_1c_2 + d_2c_1) = (b_1a_2 + b_2a_1)c_1c_2$. But $a_1a_2d_1c_2 = a_1d_1a_2c_2 = b_1c_1a_2c_2 = b_1a_2c_1c_2$ and $a_1a_2d_2c_1 = a_1b_2c_2c_1 = b_2a_1c_1c_2$, whence the desired result is obtained by adding. We may therefore define addition and multiplication in $\mathbb{Q}$ by passing to the cosets $[\![b, a]\!]$, independence of representatives following from 23.3.3:

23.3.4
$$\begin{cases} [\![b_1, a_1]\!] + [\![b_2, a_2]\!] = [\![b_1a_2 + b_2a_1, a_1a_2]\!] \\ [\![b_1, a_1]\!][\![b_2, a_2]\!] = [\![b_1b_2, a_1a_2]\!]. \end{cases}$$

Our early training tells us that the integer b is the same as the fraction $b/1$. We give precise effect to this conviction by defining now a function g from $\mathbb{Z}$ to $\mathbb{Q}$ given by

23.3.5
$$g(b) = [\![b, 1]\!].$$

We note that $g: \mathbb{Z} \to \mathbb{Q}$ is one-one and preserves addition and multiplication (prove these statements!). Thus we may regard $\mathbb{Z}$ as 'embedded', with its ring-structure, in $\mathbb{Q}$; and we temporarily write $\mathbf{b}$ for the copy $[\![b, 1]\!]$ of $b \in \mathbb{Z}$. We next verify axioms $\mathfrak{A}_1$–$\mathfrak{A}_8$ (see Chapter 9) for $\mathbb{Q}$—this is tedious but not difficult. The elements $\mathbf{0}$ and $\mathbf{1}$ in the copy $g(\mathbb{Z})$ perform the same duties in $\mathbb{Q}$. But now $\mathbb{Q}$ is a *field*, that is to say, every non-zero element of $\mathbb{Q}$ has an inverse. For if $[\![b, a]\!] \in \mathbb{Q}$ and $[\![b, a]\!] \neq \mathbf{0}$, then $b \neq 0$ in $\mathbb{Z}$, so $[\![a, b]\!] \in \mathbb{Q}$ and, by 23.3.4, $[\![b, a]\!][\![a, b]\!] = [\![ba, ab]\!] = [\![1, 1]\!]$; therefore

23.3.6
$$[\![b, a]\!]^{-1} = [\![a, b]\!], \qquad (b \neq 0).$$

We have thus achieved our ambition of so enlarging our number system that the equation $\alpha x = \beta$, $(\alpha, \beta \in \mathbb{Q}, \alpha \neq 0)$ has a solution. Indeed $[\![b, a]\!]$ is precisely the solution of the equation $\mathbf{a}x = \mathbf{b}$; more generally, the solution of the equation $[\![q, p]\!]x = [\![s, r]\!]$, $q \neq 0$, is $x = [\![ps, qr]\!]$ since (by 23.3.4):

$$[\![q, p]\!][\![ps, qr]\!] = [\![pqs, pqr]\!] = [\![s, r]\!].$$

Further, if $a \neq 0$ in $\mathbb{Z}$, then by 23.3.6, $\mathbf{a}^{-1} = [\![1, a]\!]$; but by 23.3.3, $[\![b, a]\!] = [\![b, 1]\!][\![1, a]\!]$, whence

$$[\![b, a]\!] = \mathbf{b}\mathbf{a}^{-1}.$$

We may now revert to writing b for $\mathbf{b}$, and then to the traditional notation $1/a$ for a^{-1}, b/a for $[\![b, a]\!]$. This then establishes the arithmetic theory of fractions; we have built it out of $\mathbb{Z}$ which in its turn we built out of $\mathbb{Z}^+$.

Exercise 7

(i) Show on a diagram that the cosets $[\![b, a]\!]$ lie in lines through the origin, in the co-ordinate plane.
(ii) In 23.3.4, check independence of coset representatives in detail.
(iii) Verify that g in 23.3.5 is one-one and a ring-homomorphism.
(iv) Verify that $\mathbb{Q}$ satisfies the axioms $\mathfrak{A}_1$–$\mathfrak{A}_8$ of Chapter 9.
(v) Show that, if we repeat the construction of $\mathbb{Q}$, but replacing $\mathbb{Z}$ throughout by $\mathbb{Q}$, then we merely get an isomorphic copy of $\mathbb{Q}$.
(vi) How could we introduce rationals before negative numbers?

We close this section with some remarks.

23.3.7 REMARK 1. Confusion undoubtedly exists in many students' minds as to what a fraction really is. They are told for example, that the fractions $\frac{1}{3}$ and $\frac{2}{6}$ are equal and that 1 is the numerator of the fraction $\frac{1}{3}$. It thus appears that equal fractions have unequal numerators. It may be better to think of the *fraction* as the number pair (b, a), and the *rational number* as the equivalence class $[\![b, a]\!]$. Thus fractions have numerators and denominators; and the fractions $\frac{1}{3}$ and $\frac{2}{6}$ are *equivalent*, representing the same rational number. This confusion may be eventually eliminated with the increasingly widespread introduction of Cuisenaire rods in schools.

23.3.8 REMARK 2. The description we gave of the passage from $\mathbb{Z}$ to $\mathbb{Q}$ may be applied, word for word, to any integral domain R. If we start from R instead of $\mathbb{Z}$, the field we obtain is called the **quotient field** of R; thus $\mathbb{Q}$ is the quotient field of $\mathbb{Z}$. The reader should verify that the process described does indeed depend only on the fact that $\mathbb{Z}$ is an integral domain.

Exercise 8

(i) Describe the quotient field of the integral domain of polynomials in x with coefficients in $\mathbb{Z}$. Describe the quotient field of a field.
(ii) Let $\mathbb{Z}$ be embedded in the field F as sub-ring. Show how to extend the embedding to give an embedding of $\mathbb{Q}$ in F (e.g., we may have $F = \mathbb{R}$ or

$$F = \{\alpha + \beta\sqrt{2}, (\alpha, \beta \in \mathbb{Q})\}).$$

(iii) Declare $[\![b, a]\!]$ to be *positive* if b, a have the same sign in $\mathbb{Z}$. Show how this gives an order relation in $\mathbb{Q}$ compatible with that in $\mathbb{Z}$. Then introduce a 'modulus', $|\alpha|$ in $\mathbb{Q}$ and verify directly the triangle inequality (cf. Theorem 14.8.1):

$$|\alpha + \beta| \leqslant |\alpha| + |\beta| \qquad (\alpha, \beta \in \mathbb{Q}).$$

23.3.9 REMARK 3. It is noteworthy that the attempt to enlarge the commutative ring R so that the equation $ax = b$ $(a, b \in R,\ a \neq 0)$ can always be solved, *must fail* if R is not an integral domain. Thus, suppose R has divisors of zero, say $cd = 0$, $c \neq 0$, $d \neq 0$; then we cannot enlarge R to a ring S such that in S the equation $dx = 1$ has a solution. For if we could, then $cdx = c$; but $cdx = 0x = 0$, so $cdx = c = 0$, contrary to hypothesis. We saw this already in Chapter 10 with arithmetic mod m, since $\mathbb{Z}_m$ is a commutative ring with unity element and is not an integral domain if m is not prime.

23.3.10 REMARK 4. The diligent and curious reader may well fault us for not having actually proved that $\mathbb{Z}$ is an integral domain! We asserted in 11.1.6 that $\mathbb{Z}$ is an integral domain with Euclidean norm given by the absolute value. However, we never actually *proved* that $\mathbb{Z}$ has no divisors of zero. Since $(-a)b = a(-b) = -ab$ and $(-a)(-b) = ab$, it is sufficient to prove that $\mathbb{Z}^+$ has no divisors of zero. But this may easily be demonstrated by induction using the definition of the function $\times m\colon \mathbb{Z}^+ \to \mathbb{Z}^+$ in 23.1.2.

Chapter 24

THE REAL AND COMPLEX NUMBERS

24.1 The Inadequacy of $\mathbb{Q}$

Although by passing from $\mathbb{Z}$ to $\mathbb{Q}$ we have greatly enriched our number system, we have by no means gone far enough in order to be able to set up the differential calculus. Indeed we have not even gone far enough for a thoroughgoing theory of quadratic equations, since there are certainly quadratic equations with coefficients in $\mathbb{Q}$ without solutions in $\mathbb{Q}$.

One such equation is

24.1.1 $$x^2 - 2 = 0.$$

It was recognized by Pythagoras that if one draws a square of unit side, then the length x of the diagonal must satisfy equation 24.1.1. To their undying glory, the Greeks realized that this fact destroyed completely their picture of how one could represent all geometric quantities in the universe in terms of the rational numbers. [Of course, the Greeks did not admit negative numbers, but this is not crucial to the present argument.] However, they did not know how to set about enriching the number system to make it sufficient for geometry. Tremendous strides towards this superb intellectual achievement were made by Eudoxus and Archimedes; but it was not until the 19th century that the process could be said to be complete.

We first demonstrate that equation 24.1.1 has no solution in $\mathbb{Q}$. In fact we will prove a far more general theorem, which applies to 24.1.1 when $p = 2$.

24.1.2 THEOREM. *Consider the polynomial*

$$f(x) = a_0x^n + a_1x^{n-1} + \cdots + a_{n-1}x + a_n$$

(where each $a_i \in \mathbb{Z}$ and $n > 1$). *Suppose that there is a prime number p such that*

$$p \nmid a_0, \qquad p \mid a_i, \qquad i = 1, \ldots, n; \qquad p^2 \nmid a_n.$$

Then $f(x)$ has no zeros in $\mathbb{Q}$.

Proof. To obtain a contradiction, suppose that $f(x)$ has a zero r/s in $\mathbb{Q}$. We assume, as we may (from the theory of hcf's in $\mathbb{Z}$) that r, s have no common factor other than ± 1. Then

$$a_0r^n + a_1r^{n-1}s + \cdots + a_{n-1}rs^{n-1} + a_ns^n = 0.$$

Since $p|a_i$, $i = 1, \ldots, n$, it follows that $p|a_0r^n$. Since $p\nmid a_0$, it follows that $p|r^n$ and so $p|r$. Since $n > 1$, it now follows that $p^2|a_0r^n, p^2|a_1r^{n-1}s, \ldots, p^2|a_{n-1}rs^{n-1}$. Thus $p^2|a_ns^n$. But $p|r$ and r, s have no common factor. Thus $p\nmid s$, so $p\nmid s^n$, whence $p^2|a_n$. This contradiction with the hypotheses shows that, as asserted, there can be no zero of $f(x)$ in $\mathbb{Q}$. ■

Exercise 1

(i) Deduce that the equation $x^2 - k = 0$, $k \in \mathbb{Z}$, has a solution in $\mathbb{Q}$ if and only if k is the square of some number in $\mathbb{Z}$.
(ii) Show that the polynomial $x^3 - 300x^2 + 60x - 180$ has no zeros in $\mathbb{Q}$.

Theorem 24.1.2 establishes the fact that to solve certain equalities which we feel *should* have solutions (like $x^2 - 2 = 0$) we need to enlarge our number system. However, this is not enough to satisfy our intuition or our geometrical needs, let alone to give sense to the differential calculus. On the geometrical side we meet the mysterious number π, defined as the ratio of the length of the circumference of a circle to the length of its diameter. We are given approximations to π like 22/7, 3.14, 3.14159, 3.14159265, but we are told that none of these is the true value. Although the definition of π given above does not satisfy the strictest canons of logical precision [until, that is, 'circle', 'length', and 'circumference' are defined: see Example 15.5.3], we nevertheless have a clear intuitive idea of this ratio and we find ourselves asking the awesome question 'Well, what *is* the value of π?' It turns out, though the proof is beyond the scope of this book, that indeed π is not the solution of a polynomial equation with integer coefficients (in technical terms, π is not *algebraic* (see Herstein [56]), unlike $\sqrt{2}$, which is algebraic).

However, even at a far more primitive level we feel that there is some incompleteness about our number system. For if we draw a (very long!) line and choose a unit of length, then we feel we can measure the length of a segment of any other line. However, we do not feel confident that, even if we are allowed to subdivide our original unit as a multiple of some much smaller unit, every length can be given as an exact number of units. How then may we ensure that we have 'enough' numbers so that every line segment has a length?

We illustrate the difficulty (hoping to illuminate its solution) by means of the following example. Consider first the sequence of rational numbers

24.1.3 $$0, \frac{1}{2}, \frac{2}{3}, \frac{3}{4}, \ldots, \frac{n}{n+1}, \ldots,$$

where we just continue indefinitely increasing numerator and denominator by 1 at each stage. Plainly the terms of the sequence are getting closer and closer to 1; and we get as close as we please to 1 by going sufficiently far along the sequence. It is also true, as one may readily verify, that not only do the terms of the sequence get closer and closer to 1—they also get closer and closer to each other.

Consider now the sequence of rational numbers

24.1.4 $$1,\ 1+1,\ 1+1+\frac{1}{2!},\ 1+1+\frac{1}{2!}+\frac{1}{3!},\ \ldots.$$

Let us write $a_0, a_1, a_2, a_3, \ldots,$ for this sequence, so that

$$a_n = 1 + 1 + \frac{1}{2!} + \frac{1}{3!} + \cdots + \frac{1}{n!}.$$

It is again true that the terms of the sequence 24.1.4 are getting closer and closer to each other. Obviously $a_{n+1} - a_n = 1/(n+1)!$ and so can be made as small as we like by taking n sufficiently large; thus $a_{n+1} - a_n < \frac{1}{5000}$ if $n \geqslant 6$. However, an even stronger statement is true, namely that if m and n are large enough, then $a_m - a_n$ will be as small as we please. This is the important *precise* meaning to be assigned to the phrase 'the terms of the sequence get closer and closer' and we will now prove it true.

24.1.5 PROPOSITION. *If* $m > n \geqslant 1$, *then*

$$\frac{1}{(n+1)!} + \frac{1}{(n+2)!} + \cdots + \frac{1}{m!} < \frac{1}{n!}.$$

Proof. If $m = n + 1$ the assertion is obvious; assume therefore that $m \geqslant n + 2$. Then

$$\frac{1}{(n+1)!} + \frac{1}{(n+2)!} + \cdots + \frac{1}{m!}$$

$$= \frac{1}{(n+1)!}\left\{1 + \frac{1}{n+2} + \frac{1}{(n+3)(n+2)} + \cdots + \frac{1}{m(m-1)\cdots(n+2)}\right\}$$

We make the expression between the brackets $\{\cdot\}$ bigger if we replace every factor in every denominator by $n + 2$. Thus the left-hand side is less than

$$\frac{1}{(n+1)!}\left\{1 + \frac{1}{n+2} + \frac{1}{(n+2)^2} + \cdots + \frac{1}{(n+2)^{m-n-1}}\right\}$$

$$< \frac{1}{(n+1)!}\left\{1\Big/\left(1 - \frac{1}{n+2}\right)\right\},$$

by summing the geometric progression,

$$= (n+2)/(n+1)(n+1)!$$

$$= (n+2)/(n+1)^2 \cdot 1/n!.$$

But $n + 2 < (n+1)^2$ if $n \geqslant 1$, so the proposition is proved. ■ We hope the reader has not been scared off by this proof—he may, if he prefers, accept the conclusion of the proposition and return to its proof later. However, it is evident that the proposition establishes the fact that $a_m - a_n$ may be made as small as we please. For example, $a_m - a_n < \frac{1}{5000}$ if $m > n \geqslant 6$.

Exercise 2

Show that the terms of the sequence 24.1.3 also get closer and closer together in the precise sense given above.

Now the terms of the sequence 24.1.3, we repeat, not only get closer and closer to each other; they also get closer and closer to the number 1. On the other hand, we now prove that *there is no rational number to which the terms of the sequence* 24.1.4 *get closer and closer.*

Let us first be quite clear what the relevant phrase means. Given any number however small (say, $1/10^6$), we can get to within that distance of 1 by going sufficiently far along the sequence 24.1.3. Obviously, the difference between $(n - 1)/n$ and 1 is less than $1/10^6$ if $n > 10^6$. With the sequence 24.1.4, however, we assert that there is no rational number p/q such that, by taking n sufficiently large, $p/q - a_n$ may be made as small as we like.

We suppose then that there *is* a rational number p/q such that by taking n sufficiently large, $p/q - a_n$ may be made as small as we please; we will derive a contradiction from this hypothesis. First we note that since the terms a_n of the sequence are steadily increasing, p/q must be bigger than any of the numbers a_n. Now consider the positive number

$$q!\,(p/q - a_q).$$

It is plain by looking at a_q that this is an integer k with $k \geqslant 1$. But we showed, in proving Prop. 24.1.5, that $a_n - a_q < 1/q!$ if $n > q$. Also $p/q - a_n = (p/q - a_q) - (a_n - a_q)$. Thus if $k \geqslant 2$, $p/q - a_n$ is at least $1/q!$ for all large n. But q is fixed independently of n; therefore we cannot make $p/q - a_n$ as small as we please. Thus we must have $k = 1$.

It remains to show that even $k = 1$ is impossible. We noted in fact in proving Prop. 24.1.5 that

$$a_n - a_q < \frac{q + 2}{(q + 1)^2}\frac{1}{q!}.$$

Thus even if $k = 1$ we have

$$\frac{p}{q} - a_n > \left(1 - \frac{q + 2}{(q + 1)^2}\right)\frac{1}{q!} = \frac{q^2 + q - 1}{q + 1}\cdot\frac{1}{q!},$$

and so $p/q - a_n$ is never smaller than this positive number (which is fixed, since q is fixed). This proves then that no such p/q can exist. ■

If we imagine the terms of the two sequences 24.1.3 and 24.1.4 marked off on a calibrated line (that is, a line with a scale of measurement provided), they behave very similarly. The points corresponding to the terms increase by ever shorter amounts, these amounts getting as small as we wish if we go far enough along the sequence. But whereas in the case of the sequence 24.1.3 there is a number 1 to which the terms *converge*, there is no such number in the case of the sequence 24.1.4, *unless we enlarge our number system.*

As we said, this example illustrates the problem and suggests the solution. There is a 'hole' in $\mathbb{Q}$, which we would like to 'plug' with a 'number' to which the sequence 24.1.4 could converge. That is, we must *make a number* to which the sequence 24.1.4 is to converge. (In fact, this number, as the more sophisticated reader will have recognized, is the famous number e, the base of natural logarithms.) We now describe the precise algebraic process.

24.2 Sequences

We must refine our language a little and we first make precise our notion of a sequence converging to a limit, as in 24.1.3. Recall that at this stage our number system consists of the elements of $\mathbb{Q}$ only.

24.2.1 DEFINITION. Let S be any set. A **sequence** of elements of the set S is a function $a: \mathbb{N} \to S$. In writing a_n for $a(n)$ it is customary to exhibit the sequence as a linear array of **terms**

24.2.2 $$a_1, a_2, a_3, \ldots, a_n, \ldots.$$

Now let $S = \mathbb{Q}$. Then the sequence 24.2.2 is said to **converge** *to the* **limit** a (in $\mathbb{Q}$) if, given any positive numb r ϵ, however small, there exists $n_0 \in \mathbb{N}$ such that

24.2.3 $$|a_n - a| < \epsilon \quad \textit{provided} \quad n > n_0.$$

(Note that ϵ and $|a_n - a|$ are in $\mathbb{Q}$.) If the sequence happens to converge to 0, then it is called a **null-sequence**. On the other hand, following A. L. Cauchy [(1789–1857); see Bell [10], Ch. 15], the sequence 24.2.2 is called a **fundamental** or **Cauchy sequence** if, given any positive number ϵ, however small, there exists $n_0 \in \mathbb{N}$ such that

24.2.4 $$|a_m - a_n| < \epsilon \quad \textit{provided } m \textit{ and } n \textit{ exceed } n_0.$$

24.2.5 EXAMPLE. The sequence 24.1.3 converges to 1. For, given any $\epsilon > 0$, there exists n_0 such that $\epsilon > 1/n_0$ and then $|1 - (n - 1)/n| < \epsilon$, provided $n > n_0$.

The sequence 24.1.4 is a Cauchy sequence. For, given any $\epsilon > 0$, there exists n_0 such that $1/n_0! < \epsilon$ and then, by Proposition 24.1.5, $|a_m - a_n| < \epsilon$ provided $m, n > n_0$. The sequence 24.1.4 does not converge in $\mathbb{Q}$; this is the statement we have so laboriously proved before giving Definition 24.2.1. [Notice that it was convenient to regard 24.1.4 as a sequence beginning with a_0; that is, as a function $a: \mathbb{Z}^+ \to \mathbb{Q}$. It is very handy to retain this convention in certain cases, and it plainly does not affect the basic theory.]

The sequence $1, \frac{1}{2}, \frac{1}{3}, \frac{1}{4}, \ldots, 1/n, \ldots$ is a null sequence.

24.2.6 PROPOSITION. *A convergent sequence is a Cauchy sequence.*

Proof. Let the sequence 24.2.2 converge to a. Given ϵ we may find n_0 so that $|a_n - a| < \epsilon/2$ if $n > n_0$. But, by the triangle inequality for $\mathbb{Q}$ (see 23 Exercise 8(iii)):

$$|a_m - a_n| \leqslant |a_m - a| + |a - a_n| < \epsilon/2 + \epsilon/2,$$

provided $m, n > n_0$.

Thus $|a_m - a_n| < \epsilon$ if $m, n > n_0$. ■ We may paraphrase this proof by saying 'numbers close to the same number are close to each other'. The reader should draw himself an illustrative picture to help him appreciate the argument.

We have met in Chapter 19 the concept of a *vector space* V *over a field* F, and we next show how the set $\mathbb{Q}^{\mathbb{N}}$ of sequences is given a vector-space structure. [Later, in 24.3.1, we enrich this structure to that of a real algebra.] We *add* two sequences by adding their terms, and to multiply a sequence by a rational number k we multiply each term by k. Thus (when a_n, b_n, k all lie in $\mathbb{Q}$):

24.2.7
$$(a_1, a_2, \ldots, a_n, \ldots) + (b_1, b_2, \ldots, b_n, \ldots) = (a_1 + b_1, a_2 + b_2, \ldots, a_n + b_n, \ldots),$$
$$k(a_1, a_2, \ldots, a_n, \ldots) = (ka_1, ka_2, \ldots, ka_n, \ldots).$$

We easily verify the vector-space axioms (Chapter 19), so that $\mathbb{Q}^{\mathbb{N}}$ has been given the structure of a vector space. The following propositions may now be proved.

24.2.8 PROPOSITION. *If* $(a_1, a_2, \ldots, a_n, \ldots)$ *converges to* a, *and* $(b_1, b_2, \ldots, b_n, \ldots)$ *converges to* b, *then*

(i) $(a_1 + b_1, a_2 + b_2, \ldots, a_n + b_n, \ldots)$ *converges to* $a + b$,
(ii) $(ka_1, ka_2, \ldots, ka_n, \ldots)$ *converges to* ka.

24.2.9 PROPOSITION. *Let* $(a_1, a_2, \ldots, a_n, \ldots)$, $(b_1, b_2, \ldots, b_n, \ldots)$ *be Cauchy sequences. Then*

$$(a_1 + b_1, a_2 + b_2, \ldots, a_n + b_n, \ldots) \quad \textit{and} \quad (ka_1, ka_2, \ldots, ka_n, \ldots)$$

are also Cauchy sequences.

Proof. We will be content to prove Proposition 24.2.8, the proof of Proposition 24.2.9 being very similar. To prove (i), suppose $|a_n - a| < \epsilon/2$ if $n \geqslant n_1$, and $|b_n - b| < \epsilon/2$ if $n \geqslant n_2$. Then, again by the triangle inequality,

$$|a_n + b_n - (a + b)| \leqslant |a_n - a| + |b_n - b| < \epsilon/2 + \epsilon/2 \quad \text{if } n \geqslant n_0$$

where $n_0 = \max(n_1, n_2)$. This proves (i).

To prove (ii), first observe that the assertion is trivial if $k = 0$, so we assume $k \neq 0$. Then given ϵ, there exists n_0 such that $|a_n - a| < \epsilon/|k|$, provided $n > n_0$. This means that $|k|\,|a_n - a| < \epsilon$, or $|ka_n - ka| < \epsilon$ provided $n > n_0$, and so (ii) is proved. ■

At this stage it would be convenient to have some good notation. We shall write† $\{a_n\}$ for the sequence $(a_1, a_2, \ldots, a_n, \ldots)$ and Σ_C, Σ_F, Σ_N for the subsets of $\mathbb{Q}^{\mathbb{N}}$ consisting of the convergent, Cauchy (=fundamental) and null sequences. Then Propositions 24.2.6, 24.2.8, and 24.2.9 together imply that these subsets are in fact sub*spaces* and:

24.2.10 THEOREM. *We have the vector-space inclusions*

$$\Sigma_N \subseteq \Sigma_C \subseteq \Sigma_F \subseteq \mathbb{Q}^{\mathbb{N}}. \quad \blacksquare$$

At the end of 24.1 we indicated our intuitive view of $\mathbb{Q}$ as a row of dots with holes between. Further, a Cauchy sequence α seems to approach either a dot or a hole, and it approaches the same dot or hole as does $\alpha + \beta$, where β is a null sequence. This vague idea leads us to consider the quotient space Σ_F/Σ_N whose vector-space structure is inherited from that of Σ_F. *We define this to be the set of* **real** *numbers* $\mathbb{R}$:

$$\mathbb{R} = \Sigma_F/\Sigma_N$$

which we are therefore viewing as a vector space over $\mathbb{Q}$.

Before considering the details, the reader may still find this rather far removed from his intuitive concept of a real number; granted, however, that intuitive concept, its relation to our precise definition should become clear when we point out that what we are doing is to identify a real number with the set of all sequences of rationals converging to that number. Of course, this last statement must remain imprecise since we have given only the one precise definition of a real number, so we cannot, in *precise fashion*, identify it with an intuitive concept. However, as we proceed with the development, the intuitive concept will become more and more explicit.

Exercise 3

(i) Verify that $\mathbb{Q}^{\mathbb{N}}$ is a vector space over $\mathbb{Q}$.
(ii) Write out a proof of Proposition 24.2.9.
(iii) Denote by Σ_m the subset of $\mathbb{Q}^{\mathbb{N}}$ consisting of all sequences $\{a_n\}$ such that $a_n = 0$ if $n > m$. Prove that Σ_m is a vector subspace of Σ_N of dimension m over $\mathbb{Q}$. Hence prove that dim Σ_N is infinite.
(iv) In (iii) prove that $\Sigma_0 \subseteq \Sigma_1 \subseteq \Sigma_2 \subseteq \cdots \subseteq \Sigma_m \subseteq \cdots$. Show that $\Sigma_\omega = \bigcup_{m=0}^{\infty} \Sigma_m$ is a vector subspace of Σ_N over $\mathbb{Q}$. Investigate Σ_N/Σ_ω. We call Σ_ω the space of **finite** sequences.
(v) Suppose the sequence $\{a_n\}$ converges to l and also to m. Prove that $l = m$. {Use the triangle inequality 23 Exercise 8(iii).}
(vi) Let U denote the set of all 'ultimately constant' sequences $a = \{a_n\}$, i.e., $\exists m$ (depending on a) such that $a_n = a_m$ if $n \geqslant m$. Prove that U is a vector subspace of Σ_C and that $\mathbb{Q} \approx U/\Sigma_\omega$.

† The notation $\{a_n\}$ is traditional, and there is little risk here of confusion with a singleton.

(vii) Define $f: \mathbb{Q} \to \mathbb{Q}^{\mathbb{N}}$ by setting $f(q)$ equal to the sequence with constant value q. Show that f is an embedding of $\mathbb{Q}$ with image contained in U.
(viii) Prove that every Cauchy sequence $\{a_n\}$ is 'bounded', i.e., $\exists k \in \mathbb{Q}$ such that $(\forall n \in \mathbb{N}), |a_n| < k$.
(ix) Negate the propositions '$\{a_n\} \in \Sigma_N$', '$\{a_n\} \in \Sigma_C$', '$\{a_n\} \in \Sigma_F$', obtaining a useful form. {Recall 1.8.7.}

24.3 The Structure of $\mathbb{R}$

So far, $\mathbb{R}$ has only a vector-space structure over $\mathbb{Q}$. However, we point out that $\mathbb{Q}^{\mathbb{N}}$ is actually an **algebra** over $\mathbb{Q}$. This means that, as well as its vector-space structure, $\mathbb{Q}^{\mathbb{N}}$ has the structure of a commutative ring with unity (see Chapter 9) induced by that in $\mathbb{Q}$, and moreover

24.3.1 $$a(\sigma\tau) = (a\sigma)\tau = \sigma(a\tau), \qquad (\sigma, \tau \in \mathbb{Q}^{\mathbb{N}}, a \in \mathbb{Q});$$

of course our definition of the product in $\mathbb{Q}^{\mathbb{N}}$ is the obvious one: if $\sigma = \{b_n\}$ and $\tau = \{c_n\}$, then $\sigma\tau = \{b_n c_n\} = \tau\sigma$.

24.3.2 THEOREM. *Σ_F is a subalgebra of $\mathbb{Q}^{\mathbb{N}}$ and Σ_N is an ideal in Σ_F.*

Proof. Σ_F is already a subspace of $\mathbb{Q}^{\mathbb{N}}$, so we need show only that, if $\{a_n\}, \{b_n\} \in \Sigma_F$, then their product is in Σ_F also. We use the result of Exercise 3(viii) which states that *a Cauchy sequence is bounded.* Leaving the proof of this step to the reader, we proceed. Now

24.3.3 $$\begin{aligned} |a_m b_m - a_n b_n| &= |a_m(b_m - b_n) + b_n(a_m - a_n)| \\ &\leqslant |a_m| \cdot |b_m - b_n| + |b_n| \cdot |a_m - a_n|. \end{aligned}$$

Suppose $\{a_n\}, \{b_n\}$ are Cauchy. Then $|a_n| < K$, $|b_n| < L$ for some constants $K, L \in \mathbb{Q}$. Suppose $|a_m - a_n| < \epsilon/2L$ if $m, n > n_1$ and $|b_m - b_n| < \epsilon/2K$ if $m, n > n_2$, and let $n_0 = \max(n_1, n_2)$. Then 24.3.3 guarantees that, provided $m, n > n_0$,

$$|a_m b_m - a_n b_n| < K \cdot \epsilon/2K + L \cdot \epsilon/2L = \epsilon,$$

so that $\{a_n b_n\}$ is Cauchy. This shows that Σ_F is a subalgebra of $\mathbb{Q}^{\mathbb{N}}$.

Now suppose that $\{a_n\} \in \Sigma_F$ and $\{b_n\} \in \Sigma_N$. Then $|a_n| < K$ for some K and we may find n_0 such that $|b_n| < \epsilon/K$ if $n > n_0$. Thus

$$|a_n b_n| = |a_n| \cdot |b_n| < \epsilon \quad \text{if } n > n_0,$$

whence $\{a_n b_n\}$ is a null sequence. This shows that Σ_N is an ideal in Σ_F. ■

24.3.4 COROLLARY. *$\mathbb{R}$ has a natural structure as an algebra over $\mathbb{Q}$.* ■

For each $a \in \mathbb{Q}$ let $\mathbf{a} \in \mathbb{R}$ be the coset containing the constant sequence $(a, a, \ldots, a, \ldots)$. Then $\mathbf{a}$ contains precisely the sequences of Σ_C converging to a. Moreover, we have the following theorem.

24.3.5 THEOREM. *The function $a \to \mathbf{a}$ embeds $\mathbb{Q}$ as a subalgebra of $\mathbb{R}$.*

Proof. The given function is evidently an algebra homomorphism; it is also injective, since if $(a, a, \ldots, a, \ldots) \in \Sigma_N$, then, given any $\epsilon > 0$, $|a| < \epsilon$, so that $a = 0$. ■

We now have $\mathbb{R}$ presented to us as a ring which contains $\mathbb{Q}$ as a sub-ring, since we will (not surprisingly by now!) identify a with $\mathbf{a}$. It is plain that $\mathbf{1}$ (where $1 \in \mathbb{Q}$) does service as the unity element in $\mathbb{R}$; it remains to show that $\mathbb{R}$ is a field and to extend the order relation from $\mathbb{Q}$ to $\mathbb{R}$.

24.3.6 THEOREM. *$\mathbb{R}$ is a field.*

Proof. Given $x \neq 0$ in $\mathbb{R}$, we have to find some $y \in \mathbb{R}$ with $xy = 1$. A non-zero element $x \in \mathbb{R}$ is a coset mod Σ_N, distinct from Σ_N. Hence, if $\{a_n\} \in x$, then $\{a_n\} \in \Sigma_F - \Sigma_N$.

Since $\{a_n\} \notin \Sigma_N$, $\exists k > 0$ and $n_1 < n_2 < n_3 < \cdots$ such that $|a_{n_r}| \geqslant k$ when $r = 1, 2, 3, \ldots$. Since $\{a_n\} \in \Sigma_F$ there exists n_0 such that $|a_m - a_n| < k/2$ if $m, n > n_0$. Since, for some r, $n_r > n_0$, it follows that, for that r,

$$|a_n - a_{n_r}| < k/2 \qquad \text{if } n \geqslant n_r,$$

whence

$$|a_n| > k/2 \qquad \text{if } n \geqslant n_r. \tag{24.3.7}$$

Form the sequence $\{b_n\}$, where

$$b_n = \begin{cases} 0, & n < n_r. \\ 1/a_n, & n \geqslant n_r. \end{cases}$$

We claim that $\{b_n\} \in \Sigma_F$ and $\{a_n b_n\} \in \mathbf{1}$, so that the coset containing $\{b_n\}$ is the inverse of the coset x containing $\{a_n\}$ and we will have proved that every non-zero element in $\mathbb{R}$ has an inverse; in other words, that $\mathbb{R}$ is a field.

Now $|b_m - b_n| = |1/a_m - 1/a_n| = |a_m - a_n|/|a_m| \cdot |a_n| < 4|a_m - a_n|/k^2$, all this provided $m, n \geqslant n_r$. Since $\{a_n\}$ is Cauchy, it follows that given any ϵ, $\exists m_1$ such that $|a_m - a_n| < \epsilon k^2/4$, provided $m, n > m_1$. Taking $m_0 = \max(n_r, m_1)$, we infer that

$$|b_m - b_n| < \epsilon \quad \text{provided} \quad m, n > m_0,$$

so that $\{b_n\}$ is Cauchy.

We have called a sequence **finite** if all but a finite number of its terms is zero; Σ_ω is the space of finite sequences. Plainly $\Sigma_\omega \subseteq \Sigma_N$. It is also plain that $\{a_n b_n\} - (1, 1, \ldots, 1, \ldots) \in \Sigma_\omega$. Thus $\{a_n b_n\}$ and $\mathbf{1}$ represent the same real number and the proof is complete. ■

Before proceeding to discuss the order relation in $\mathbb{R}$, we extract from the proof above a refined form of 24.3.7 which we will need later.

24.3.8 PROPOSITION. *Let $\{a_n\}$ represent a non-zero real number. Then there exists $k > 0$ ($k \in \mathbb{Q}$) such that either $a_n > k > 0$ for all sufficiently large n, or else $-a_n > k > 0$ for all sufficiently large n.*

Note that we have used the phrase 'sufficiently large' here instead of saying (for example) '$\exists k > 0$ & $\exists n_0$ such that $(\forall n > n_0)\, a_n > k > 0$ if $n > n_0$'. This usage is natural and helpful to avoid clutter. The reader should try recasting the proof of Theorem 24.3.6 using this phrase. [Of course 'k' here plays the role of '$k/2$' in 24.3.7. It is plainly better to simplify notation.] We could even restate the proposition as '*A sequence representing a non-zero real number is eventually bounded away from zero*'.

Exercise 4

(The first three items concern abuses of language.)

(i) The following question is expressed in useful idiomatic language. Make it precise, and then answer it. Given $\{a_n\}$, such that $0 < A < a_n$ when n is sufficiently large, and given $\{b_n\}$ such that $0 < B < b_n$ when n is sufficiently large, show that: $0 < A + B < a_n + b_n$ and $0 < AB < a_n b_n$ when n is sufficiently large.

(ii) Obtain a useful form of negation of the statement 'For all sufficiently large n, $\phi(n)$ holds'.

(iii) After a meeting of physicists, the statement: '1 is approximately 2 if 2 is large', was found on the blackboard. What might they have meant? Similarly assign a reasonable meaning to the statement 'a_n is approximately $2a_n$ if n is sufficiently large'.

(iv) Show that Σ_C is a subalgebra of Σ_F (is it an ideal in Σ_F?) and that $\Sigma_C/\Sigma_N \approx \mathbb{Q}$.

24.4 The Order Relation in ℝ

We now define an order relation in ℝ. First we introduce the notation

$$\{a_n\} \to \alpha$$

to indicate that the Cauchy sequence $\{a_n\}$ represents the real number α; i.e., the coset of $\{a_n\}$ in Σ_F/Σ_N is α, and then α is the image of $\{a_n\}$ under the quotient map $\Sigma_F \to \mathbb{R}$. We notice that this notation (*read*: 'a_n tends to α') is consistent if α is, in fact, rational. We then say that, if $\alpha, \beta \in \mathbb{R}$, *then* $\alpha \geqslant \beta$ (or $\beta \leqslant \alpha$) *if* $\exists \{a_n\} \to \alpha$, $\{b_n\} \to \beta$ *such that* $a_n \geqslant b_n$ *for n sufficiently large.* As usual, '$\alpha > \beta$' means that $\alpha \geqslant \beta$ but $\alpha \neq \beta$. We begin the discussion of the order relation by making the following elementary observation.

24.4.1 PROPOSITION. *$\alpha \geqslant \beta$ iff, given any $\{b_n\} \to \beta$ there exists $\{a_n\} \to \alpha$* with *$a_n \geqslant b_n$, for n sufficiently large. A similar statement holds if $\alpha \leqslant \beta$.*

Proof. Suppose $\alpha \geqslant \beta$. Then $\exists \{a_n'\} \to \alpha$, $\{b_n'\} \to \beta$ with $a_n' \geqslant b_n'$, n sufficiently large. Given any $\{b_n\} \to \beta$, $b_n = b_n' + z_n$ with $\{z_n\} \in \Sigma_N$. Then if $a_n = a_n' + z_n$, $\{a_n\} \to \alpha$ and $a_n \geqslant b_n$, n sufficiently large. A similar argument works if $\alpha \leqslant \beta$. The reverse implication is trivial. ■

Before establishing that $\alpha \geqslant \beta$ really is an order relation we show that it has the proper algebraic properties.

24.4.2 THEOREM. (i) *If* $\alpha \geqslant \beta$ *then* $\alpha + \gamma \geqslant \beta + \gamma$ *for all* $\gamma \in \mathbb{R}$.
(ii) *If* $\alpha \geqslant 0$ *and* $\beta \geqslant 0$, *then* $\alpha\beta \geqslant 0$.

Proof. (i) is easy, since $a_n \geqslant b_n \Rightarrow a_n + c_n \geqslant b_n + c_n$ in $\mathbb{Q}$. As to (ii), Proposition 24.4.1 implies that we may find $\{a_n\} \to \alpha$, $\{b_n\} \to \beta$ with $a_n \geqslant 0$, $b_n \geqslant 0$ for n sufficiently large. But then $a_n b_n \geqslant 0$, n sufficiently large, so that $\alpha\beta \geqslant 0$. ■

Exercise 5

(i) Show that $\alpha \geqslant \beta$ iff $\exists\{a_n\} \to \alpha$, $\{b_n\} \to \beta$ such that $a_n \geqslant b_n$ for all n.
(ii) Show that if $\alpha \geqslant \beta$ and $\gamma \geqslant 0$, then $\alpha\gamma \geqslant \beta\gamma$.

We now prove that the relation $\alpha \geqslant \beta$ is indeed an order relation.

24.4.3 THEOREM. *The relation* $\geqslant$ *is a total ordering of* $\mathbb{R}$, *and* $(\mathbb{R}, \geqslant)$ *is an ordered field.*

Proof. We first show that $\geqslant$ is transitive: $\alpha \geqslant \beta \,\&\, \beta \geqslant \gamma :\Rightarrow \alpha \geqslant \gamma$. Given any $\{c_n\} \to \gamma$, there exists $\{b_n\} \to \beta$ with $b_n \geqslant c_n$, n sufficiently large. Then, given this sequence $\{b_n\}$, there exists $\{a_n\} \to \alpha$ with $a_n \geqslant b_n$, n sufficiently large. Thus $a_n \geqslant c_n$, n sufficiently large, so $\alpha \geqslant \gamma$.

We next show that $\geqslant$ is antisymmetric: $\alpha \geqslant \beta \,\&\, \beta \geqslant \alpha :\Rightarrow \alpha = \beta$. Given $\{a_n\} \to \alpha$, there are sequences $\{b_n\} \to \beta$, $\{b'_n\} \to \beta$ such that $a_n \geqslant b_n$, n sufficiently large, and $a_n \leqslant b'_n$, n sufficiently large. Then, given any $\epsilon > 0$, we have

$$|a_n - b_n| = a_n - b_n \leqslant b'_n - b_n = |b'_n - b_n| < \epsilon,$$

if n is sufficiently large, since $\{b'_n - b_n\} \in \Sigma_N$. This proves that $\{a_n - b_n\} \in \Sigma_N$, so that $\alpha = \beta$.

Finally we show that $\geqslant$ is total: for all $\alpha, \beta \in \mathbb{R}$, $\alpha \leqslant \beta$ or $\beta \leqslant \alpha$. In view of Theorem 24.4.2(i) it is sufficient to show that for all $\alpha \in \mathbb{R}$, $\alpha \geqslant 0$ or $\alpha \leqslant 0$. Let us assume $\alpha \neq 0$, and let $\{a_n\} \to \alpha$. By Proposition 24.3.8, either $a_n > 0$, n sufficiently large, or $a_n < 0$, n sufficiently large. This, however, implies that either $\alpha \geqslant 0$ or $\alpha \leqslant 0$.

That $(\mathbb{R}, \geqslant)$ is an ordered field now follows from Theorem 24.4.2. ■

Moreover, it is plain that $\mathbb{Q}$ is an ordered subfield of $\mathbb{R}$.

We draw attention to the following consequence of a more subtle application of Proposition 24.3.8.

24.4.4 COROLLARY. *Given any* $\alpha > 0$ *in* $\mathbb{R}$, *there exists* $k \in \mathbb{Q}$ *with* $\alpha > k > 0$. ■

This means that we may insert a rational number between any non-zero real number and zero. A generalization of this fact appears in the following set of exercises (see (ii) below).

Exercise 6

(i) Show that $\alpha > \beta$ iff, for all $\{a_n\} \to \alpha$, $\{b_n\} \to \beta$, then $a_n > b_n$, n sufficiently large.
(ii) Show that if $\alpha > \beta$, then there exists $k \in \mathbb{Q}$ with $\alpha > k > \beta$. {Refine (i) by showing that if $\{a_n\} \to \alpha$, $\{b_n\} \to \beta$ and $\alpha > \beta$, then $\exists k \in \mathbb{Q}$ such that $a_n > k > b_n$, n sufficiently large.}
(iii) Show that $(\mathbb{R}, \geqslant)$ is **Archimedean**; that is, that if $\alpha > 0$, $\beta > 0$, then there exists $n \in \mathbb{N}$ with $n\beta > \alpha$. {Use 24.4.4 and the fact that $\mathbb{Q}$ is Archimedean.}
(iv) Extend the definition of $|\alpha|$ to $\mathbb{R}$, so that it agrees with the definition when $\alpha \in \mathbb{Q}$. Prove the triangle inequality (23 Exercise 8(iii)) for this extended modulus-function.

24.5 Decimals

We next introduce one of the most useful practical ways of regarding, and computing with, real numbers, namely the so-called **infinite** (or *non-terminating*) **decimals.** For convenience of discussion, we confine attention to positive decimals, but the reader should have no difficulty in extending the treatment to negative decimals as well.

Let $(a_0, a_1, a_2, \ldots, a_n, \ldots)$ in $\mathbb{Z}^+$ be a sequence of positive integers such that $0 \leqslant a_n \leqslant 9$, all $n > 0$. We may then construct the sequence (of decimals):

24.5.1 $$(a_0, a_0.a_1, a_0.a_1a_2, \ldots, a_0.a_1a_2 \cdots a_n, \ldots).$$

Let $\delta_n = a_0.a_1a_2 \cdots a_n$. It is plain that $\{\delta_n\} \in \Sigma_F$ since $|\delta_m - \delta_n| < 1/10^{n_0}$ if $m, n > n_0$. What we will prove is that every (positive) real number is represented by such an infinite decimal (24.5.1) and that the representation is, almost without exception, unique. Two important lemmas now follow.

24.5.2 LEMMA. *If $\{b_n\} \to \beta$ and $b_n \leqslant \alpha$, all n, then $\beta \leqslant \alpha$.*

Proof. Suppose $\beta > \alpha$ for a contradiction. By Exercise 6(ii), $\exists k \in \mathbb{Q}$ with $\beta > k > \alpha$. Then $b_n < k$, all n, so that $\beta \leqslant k$. We have our contradiction. ∎

24.5.3 LEMMA. *Let $\{b_n\} \in \mathbb{Q}^{\mathbb{N}}$, $\{z_n\} \in \Sigma_N$, $\alpha \in \mathbb{R}$, be such that*

$$b_n \leqslant \alpha \leqslant b_n + z_n, \qquad \text{all } n.$$

Then $\{b_n\} \in \Sigma_F$ and $\{b_n\} \to \alpha$.

Proof. For any m, n we have

$$b_m \leqslant b_n + z_n, \qquad b_n \leqslant b_m + z_m,$$

whence

24.5.4 $$b_m - b_n \leqslant z_n, \qquad b_n - b_m \leqslant z_m.$$

Given any $\epsilon > 0$, $|z_n| < \epsilon$ for n sufficiently large. Thus $|b_m - b_n| < \epsilon$ for m, n sufficiently large, so that $\{b_n\} \in \Sigma_F$. If $\{b_n\} \to \beta$, then Lemma 24.5.2 yields $\beta \leqslant \alpha \leqslant \beta$, whence $\beta = \alpha$. ∎

We now give a construction to represent a given positive real number α by an infinite decimal. Since $\alpha > 0$, there must exist $a_0 \in \mathbb{Z}^+$ such that $a_0 \leqslant \alpha < a_0 + 1$. (This follows from the fact that $\alpha > 0$ and a Cauchy sequence is bounded.) Consider $\alpha_1 = 10(\alpha - a_0)$. Then $0 \leqslant \alpha_1 < 10$, so that there exists $a_1 \in \mathbb{Z}^+$ with $0 \leqslant a_1 \leqslant 9$ and $a_1 \leqslant \alpha_1 < a_1 + 1$. Now consider $\alpha_2 = 10(\alpha_1 - a_1)$. Again $0 \leqslant \alpha_2 < 10$ and we may find $a_2 \in \mathbb{Z}^+$ with $0 \leqslant a_2 \leqslant 9$ and $a_2 \leqslant \alpha_2 < a_2 + 1$. Proceeding in this way, we construct the integers $a_0, a_1, a_2, \ldots, a_n, \ldots$, and hence the infinite decimal (24.5.1).

24.5.5 THEOREM. *Let the decimal $\{\delta_n\}$ be constructed as above from the real number α. Then $\{\delta_n\} \to \alpha$.*

Proof. We prove the key equality

24.5.6 $$\alpha = \delta_n + \alpha_{n+1}/10^{n+1},$$

which is certainly true for $n = 0$, since $\delta_0 = a_0$ and $\alpha = a_0 + \alpha_1/10$. We may thus seek a proof by induction on n, assuming 24.5.6 true for $n = k - 1$. Thus we suppose

$$\alpha = \delta_{k-1} + \alpha_k/10^k.$$

Now by construction, $\alpha_{k+1} = 10(\alpha_k - a_k)$ and $\delta_k = \delta_{k-1} + a_k/10^k$. Thus

$$\delta_{k-1} + \frac{\alpha_k}{10^k} = \delta_{k-1} + \frac{a_k}{10^k} + \frac{\alpha_{k+1}}{10^{k+1}} = \delta_k + \frac{\alpha_{k+1}}{10^{k+1}}.$$

This proves 24.5.6, which moreover implies (since $0 \leqslant \alpha_n < 10$ for all n)

$$\delta_n \leqslant \alpha < \delta_n + 1/10^n.$$

We now apply Lemma 24.5.3 to complete the proof of the theorem. ■

It remains to discuss uniqueness of decimal representation. Let us first exclude from the set of infinite decimals those for which $a_n = 9$ for n sufficiently large; each one such is of the form $x = a_0.a_1a_2 \cdots a_k\dot{9}$ where $a_k < 9$, which, as we shall show, we can rewrite uniquely as $a_0.a_1a_2 \cdots a_{k-1}a_k'$ where $a_k' = a_k + 1$. Thus $x = A/10^k \in \mathbb{Q}$, where $A \in \mathbb{Z}^+$; of course, not every positive rational is a terminating decimal (e.g., $\frac{1}{3}$). Let Δ denote the deleted set of infinite decimals.

24.5.7 PROPOSITION. *Let $\{\delta_n\}, \{\delta_n'\} \in \Delta$ be distinct decimals and let $\{\delta_n\} \to \delta$, $\{\delta_n'\} \to \delta'$. Then $\delta \neq \delta'$.*

Proof. Since $\{\delta_n\} \neq \{\delta_n'\}$ we may suppose, without real loss of generality, that $\exists m$ such that

$$a_n = a_n', \qquad n < m$$
$$a_m > a_m'$$

where $\delta_n = a_0.a_1a_2 \cdots a_n$, $\delta_n' = a_0'.a_1'a_2' \cdots a_n'$. Since $\{\delta_n'\} \in \Delta$, $\exists s > m$ with

$a'_s < 9$. Define (terminating) decimals $\{d_n\}$, $\{d'_n\}$ where $d_n = b_0.b_1b_2\cdots b_n$, and $d'_n = b'_0.b'_1b'_2\cdots b'_n$, by the rules

$$\begin{aligned} b_n &= a_n, & n &\leqslant m,\\ b_n &= 0, & n &> m;\\ b'_n &= a'_n, & n &< s,\\ b'_s &= a'_s + 1 & &\\ b'_n &= 0, & n &> s. \end{aligned}$$

Then plainly $d' < d$ where $\{d_n\} \to d \in \mathbb{Q}$ and $\{d'_n\} \to d' \in \mathbb{Q}$. Moreover $d_n \leqslant \delta_n$ and $\delta'_n \leqslant d'_n$, so that $d \leqslant \delta$, $\delta' \leqslant d'$. Thus $\delta' < \delta$ and the proposition is proved. ■

Finally, we consider the case of an infinite decimal terminating in 9's. To be precise, let $\{\delta_n\}$, $\{\delta'_n\}$ (where $\delta_n = a_0.a_1a_2\cdots a_n$, $\delta'_n = a'_0.a'_1a'_2\cdots a'_n$) be two infinite decimals such that

$$\begin{aligned} a_n &= a'_n, & n &< m,\\ a_m &= a'_m + 1, & &\\ a_n &= 0, & n &> m,\\ a'_n &= 9, & n &> m. \end{aligned}$$

Then it is plain that $\delta_n = \delta'_n + z_n$ where

$$\begin{aligned} z_n &= 0, & n &< m\\ z_n &= \frac{1}{10^n}, & n &\geqslant m. \end{aligned}$$

Plainly $\{z_n\} \in \Sigma_N$, so that $\{\delta_n\}$ and $\{\delta'_n\}$ determine the same real number (which is, of course, *rational*, since δ_n is a terminating decimal).

To sum up then, we have proved that every (positive) real number has a unique infinite decimal in Δ representing it; but, in addition, rational numbers of the form p/q (not necessarily in 'lowest terms') with q a power of 10 have a second representation (outside Δ) by an infinite decimal terminating in 9's.

Exercise 7

(i) Show that, in Lemmas 24.5.2, 24.5.3 we may replace 'all n' by 'for n sufficiently large' and still obtain a valid conclusion.
(ii) A sequence $\{a_n\}$ is said to be *increasing* if $a_{n+1} \geqslant a_n$, all n. Show that if $\{a_n\}$ is an increasing sequence and $\{a_n\} \to \alpha$, then $a_n \leqslant \alpha$, all n.
(iii) Let $\{a_n\} \to \alpha$, $\{b_n\} \to \beta$ and $a_n < b_n$, all n. Can we deduce $\alpha < \beta$?
(iv) Give a non-terminating decimal representation for the following rational numbers

$$\frac{1}{2}, \quad \frac{7}{4}, \quad \frac{99}{100}.$$

(v) (Harder.) Justify the standard process for extracting square roots.

(vi) Define a sequence inductively by the rule

$$a_1 = 1, \qquad a_{n+1} = (a_n^2 + 2)/2a_n.$$

Show that $\{a_n\} \in \Sigma_F$ and that $\{a_n\} \to \sqrt{2}$.

24.6 The Completeness of $\mathbb{R}$

To finish the demonstration that $\mathbb{R}$, as we have defined it, is a good description of our intuitive picture, we need to show that $\mathbb{R}$ has no 'holes'. That is to say, we must show that the reasons which led us to believe that $\mathbb{Q}$ was inadequate do not apply to $\mathbb{R}$; in fact we shall show that every Cauchy sequence in $\mathbb{R}$ actually converges to an element of $\mathbb{R}$. If this were not so, we should (for consistency) have to begin all over again the process laboriously described in the previous sections. There we started with $\mathbb{Q}$ and constructed the vector spaces (over $\mathbb{Q}$):

24.6.1 $$\Sigma_N \subseteq \Sigma_C \subseteq \Sigma_F \subseteq \mathbb{Q}^{\mathbb{N}}$$

(see Theorem 24.2.10). We now have the ordered field $\mathbb{R}$ and may replace $\mathbb{Q}$ by $\mathbb{R}$ throughout the discussion, obtaining

24.6.2 $$\Sigma_N(\mathbb{R}) \subseteq \Sigma_C(\mathbb{R}) \subseteq \Sigma_F(\mathbb{R}) \subseteq \mathbb{R}^{\mathbb{N}}.$$

Here we evidently understand by $\Sigma_F(\mathbb{R})$ the vector space over $\mathbb{R}$ consisting of Cauchy sequences $(\alpha_1, \alpha_2, \ldots, \alpha_n, \ldots)$ where each $\alpha_n \in \mathbb{R}$; here 'Cauchy' is defined as in 24.2.4 but using $|\alpha_m - \alpha_n|$ in the ordered field $\mathbb{R}$ (see Exercise 6(iv)). Similarly, $\Sigma_C(\mathbb{R})$ and $\Sigma_N(\mathbb{R})$ are defined. Now we observed (Theorem 24.3.5 and the remark preceding it) that

24.6.3 $$\Sigma_C/\Sigma_N \approx \mathbb{Q}$$

and that $\mathbb{Q}$ is thus naturally embedded in $\mathbb{R}$. Similarly, we have

24.6.4 $$\Sigma_C(\mathbb{R})/\Sigma_N(\mathbb{R}) \approx \mathbb{R}, \quad \text{and}$$

24.6.5 $$\Sigma_C(\mathbb{R})/\Sigma_N(\mathbb{R}) \subseteq \Sigma_F(\mathbb{R})/\Sigma_N(\mathbb{R}).$$

It would thus appear that we may introduce the field of 'super-reals' $\Sigma_F(\mathbb{R})/\Sigma_N(\mathbb{R})$ and proceed indefinitely in this way. However, 'super-reals' are nothing but reals! The process takes us no further. Precisely, we have the very important theorem:

24.6.6 THEOREM. $$\Sigma_C(\mathbb{R}) = \Sigma_F(\mathbb{R}),$$

i.e., every Cauchy sequence in $\mathbb{R}$ converges to an element of $\mathbb{R}$.

In fact we will prove more (without making the proof any harder!). Given any ordered set U we say that a subset $S \subseteq U$ is **bounded above** if there exists $M \in U$ such that $a \leqslant M$ for all $a \in S$; and M is called an **upper bound** for S. Now suppose M has the property that, given any $M_1 < M$, there exists $a_1 \in S$ with $a_1 > M_1$ (thus M_1 is not an upper bound for S); then M is called the **least** upper bound. 'Lower bounds' and 'greatest lower bounds' are defined similarly; the reader is advised to represent this situation pictorially, on a line.

Two important remarks may be made.

24.6.7 Even when S is bounded above, it may not have a least upper bound. For example, if we take $U = \mathbb{Q}$ and we take for S the set of terms $a_0.a_1a_2 \cdots a_n$ of the infinite decimal expansion of $\sqrt{2}$, then it is easy to see that S is bounded above but lacks a least upper bound *in* $\mathbb{Q}$.

24.6.8 *The least upper bound, if it exists, is certainly unique.* (Proof as exercise.) ■

The crucial theorem about $\mathbb{R}$, which we now state, is the following.

24.6.9 THEOREM. *Every non-empty subset of* $\mathbb{R}$, *which is bounded above, has a least upper bound. Similarly for lower bounds.*

Proof. The idea of this proof is very simple, as the reader will see if he draws his own picture (which of course does not constitute a formal proof). Confining ourselves to *upper* bounds, let $S \subseteq \mathbb{R}$, and suppose that S is non-empty and bounded above. Since $S \neq \varnothing$ it is plain that there exists $k \in \mathbb{Z}$ which is *not* an upper bound for S. It now follows (from the well-ordering of the set of integers $\geqslant k$) that there exists $a_0 \in \mathbb{Z}$ such that a_0 is *not* an upper bound for S but $a_0 + 1$ is an upper bound for S.

We next consider the decimals $a_0.a$, $0 \leqslant a \leqslant 9$. Again it is clear that there exists a_1, $0 \leqslant a_1 \leqslant 9$ such that $a_0.a_1$ is not an upper bound for S but $(a_0.a_1 + \frac{1}{10})$ is an upper bound. Proceeding in this way we generate an infinite decimal:

$$\alpha = a_0.a_1a_2 \cdots a_n \cdots \tag{24.6.10}$$

such that $a_0.a_1a_2 \cdots a_n$ is not an upper bound for S but $(a_0.a_1a_2 \cdots a_n + 1/10^n)$ is an upper bound. (Note that here a_0 may be negative, but this possibility clearly does not affect the theory of infinite decimals which we developed in the preceding section.) We claim that α is the least upper bound for S. First, α *is* an upper bound for S. For, writing α_n for $a_0.a_1a_2 \cdots a_n$ and $\beta_n = \alpha_n + 1/10^n$, we have $\{\beta_n\} \to \alpha$ and $\beta_n \geqslant s$ for every $s \in S$. Thus, by Lemma 24.5.2, $\alpha \geqslant s$. Second, α is the *least* upper bound. For, given any $\epsilon > 0$, there is certainly an n such that $\epsilon > 1/10^n$. Thus

$$\alpha - \epsilon < \beta_n - \frac{1}{10^n} = \alpha_n$$

so that $\alpha - \epsilon$ is not an upper bound, and the theorem is proved. ■ We now use Theorem 24.6.9 to prove Theorem 24.6.6.

Proof of Theorem 24.6.6. Let $\{\gamma_n\} \in \Sigma_F(\mathbb{R})$. Consider the set S of all real numbers α such that $\alpha \leqslant \gamma_k$ for k sufficiently large. Then S is not empty, since $\{\gamma_n\}$, being a Cauchy sequence, is certainly bounded. Thus, by Theorem 24.6.9, S has a least upper bound γ. We prove that $\{\gamma_n\} \to \gamma$.

Now, given any $\epsilon > 0$, it is plain that $\gamma - \epsilon \in S$ (prove this!). Thus $\gamma - \epsilon \leqslant \gamma_n$ for n sufficiently large or

24.6.11 $$\gamma - \gamma_n \leqslant \epsilon \qquad \text{for } n \text{ sufficiently large.}$$

Also $\gamma + \epsilon/2 \notin S$, so that $\gamma + \epsilon/2 > \gamma_k$ for infinitely many values of k. Since $\{\gamma_n\}$ is a Cauchy sequence, $|\gamma_m - \gamma_n| < \epsilon/2$ if m, n are sufficiently large. Choosing m to be sufficiently large and to be such that $\gamma + \epsilon/2 > \gamma_m$, we infer that $\gamma + \epsilon > \gamma_n$, or

24.6.12 $$\gamma_n - \gamma < \epsilon \qquad \text{for } n \text{ sufficiently large.}$$

Putting together 24.6.11 and 24.6.12 we conclude that

$$|\gamma_n - \gamma| \leqslant \epsilon \qquad \text{for } n \text{ sufficiently large,}$$

so that $\{\gamma_n\} \to \gamma$. ■

Exercise 8

(i) Re-prove Theorem 24.6.6, using an argument similar to that used to prove Theorem 24.6.9 (but, of course, without invoking Theorem 24.6.9).
(ii) Give an example of a sequence $\{\alpha_n\} \in \Sigma_C(\mathbb{R})$ with $\alpha_n \in \mathbb{R} - \mathbb{Q}$ but $\{\alpha_n\} \to \alpha \in \mathbb{Q}$.
(iii) Prove the statement above (use the proof of Theorem 24.6.6) that for any $\epsilon > 0$, $\gamma - \epsilon \in S$. Does this property hold for the least upper bound of every set which is bounded above?
(iv) Prove uniqueness of limits in $\mathbb{R}$, analogously to Exercise 3(v).

24.6.13 COMPLETENESS. The fact, presented symbolically in Theorem 24.6.6, that every Cauchy sequence in $\mathbb{R}$ converges, is also expressed by saying that $\mathbb{R}$ is **complete**; we also describe the process of passing from $\mathbb{Q}$ to $\mathbb{R}$ as that of **completing** $\mathbb{Q}$. Indeed the process may be applied to any *metric* space (see 25.5.1) to produce its completion which will, in fact, be complete! Here, of course, we have not only the metric property of $\mathbb{Q}$ (and $\mathbb{R}$) but also the algebraic property of constituting an ordered field. The reader should now be able to assure himself that the ordered field $\mathbb{R}$ has all the properties he has hitherto assigned to it in a somewhat intuitive way. In particular, the fact that $\mathbb{R}$ is complete expresses precisely the idea that the real line has no 'holes' in it—unlike the set of rational points on it.

Moreover, it is no longer necessary to think of the elements of $\mathbb{R}$ as Σ_N-cosets of $\Sigma_C(\mathbb{Q})$, since we now know that there exists a complete ordered field containing $\mathbb{Q}$ as a subalgebra. This remark is consistent with similar ones concerning $\mathbb{Z}$ and $\mathbb{Q}$ in Chapter 23; otherwise we would be forced to think of real numbers as cosets of sequences of cosets of ordered pairs of cosets of ordered pairs of elements from $\mathbb{Z}^+$, an intolerable burden!

Finally, we point out that in Chapter 21 we have already described a way of completing $\mathbb{Q}$, namely by the introduction of Dedekind sections. In fact, the two processes (via Dedekind sections and Cauchy sequences) are quite equivalent and produce isomorphic systems of real numbers. The reader may readily convince himself of this by thinking of the decimal representation of a real number (24.5). For, given such a decimal $\{\delta_n\}$, we may consider the set of all $x \in \mathbb{Q}$ such that $x < \delta_n$ for some n; and this obviously forms a Dedekind section which may be identified with the given decimal. We have preferred to give our detailed attention to the Cauchy process of introducing $\mathbb{R}$ since it seems to us easier to grasp (but not, of course, easy to grasp!) and has the advantage of making use of more standard algebraic notions.

24.7 The Complex Numbers

We now come to the final major extension of the number system to which we wish to direct the reader's attention and which we wish to discuss systematically. The equation $x^2 + 1 = 0$ has no solutions in $\mathbb{R}$ and we wish to enlarge the number system so that it does have a solution. The enlarged number system we shall construct is called the field $\mathbb{C}$ of **complex numbers**. $\mathbb{R}$ is embedded naturally in $\mathbb{C}$ as a sub-field: but $\mathbb{C}$ no longer has the property enjoyed by $\mathbb{R}$ of being an *ordered* field. On the other hand, $\mathbb{C}$ has the very striking and important property, certainly not enjoyed by $\mathbb{R}$, that every polynomial

$$a_0 + a_1x + \cdots + a_nx^n, \qquad (a_n \neq 0, n \geqslant 1)$$

of positive degree in $\mathbb{C}[x]$ has a zero in $\mathbb{C}$ (and hence, by using the remainder theorem (22.3.3), it can be expressed as a product of *linear* factors in $\mathbb{C}[x]$). Thus, in the process of ensuring that the single equation $x^2 + 1 = 0$ should have a solution in $\mathbb{C}$, we have, surprisingly, achieved the effect that *every* polynomial equation in $\mathbb{C}[x]$ has a solution in $\mathbb{C}$. This property is expressed by saying that $\mathbb{C}$ is *algebraically closed*, and Theorem 22.5.2 which asserts this is called *the fundamental theorem of algebra.*

Certainly the quickest way of introducing the complex numbers, to those familiar with the plane $\mathbb{R}^2$ of co-ordinate geometry, is to regard the real numbers as the x-axis and to extend the ring-structure of $\mathbb{R}$ to $\mathbb{R}^2$ by the rules

24.7.1
$$\begin{cases} (x_1, y_1) + (x_2, y_2) = (x_1 + x_2, y_1 + y_2) \\ (x_1, y_1)\cdot(x_2, y_2) = (x_1x_2 - y_1y_2, x_1y_2 + x_2y_1), \end{cases}$$

of which the first is ordinary vector addition in $\mathbb{R}^2$ (which is a vector space over $\mathbb{R}$). But $(x, y) = x\mathbf{i} + y\mathbf{j}$, where $\mathbf{i} = (1, 0)$, $\mathbf{j} = (0, 1)$ are the unit points (see 14.2.1) on the axes and it is easily checked that $(t, 0)\cdot(u, v) = (tu, tv)$ while $\mathbf{j}^2 = -\mathbf{i}$. Therefore, if we identify each $x \in \mathbb{R}$ with $(x, 0)$ (thus identifying $\mathbb{R}$ with the x-axis), then scalar multiplication by t agrees with multiplication by $(t, 0)$; in particular $(1, 0)$ acts as unity element 1, and $\mathbf{j}^2 = -1$. We can therefore write (x, y) as $x + y\mathrm{i}$ in the traditional manner, and $\mathrm{i} = \mathbf{j}$ appears without mystery as $\sqrt{-1}$ since $\mathrm{i}^2 = -1$. It is easily verified that multiplication is associative, commutative, and distributive over addition, while (x, y) (if non-zero) has inverse $(x, -y)/(x^2 + y^2)$.

Thus the triple $(\mathbb{R}^2, +, \cdot)$, consisting of $\mathbb{R}^2$ with the algebraic structure given by 24.7.1, is our required field $\mathbb{C}$ extending $\mathbb{R}$ in which the equation $x^2 + 1 = 0$ has a solution. One property of $\mathbb{R}$ does not extend, however; $\mathbb{C}$ cannot be an ordered field—even in its own right, let alone as an ordered field-extension of $\mathbb{R}$. For if it were, then, since -1 and 1 are both squares in $\mathbb{C}$, we would have the contradiction $-1 > 0$ and $1 > 0$.

When $\mathbb{R}^2$ is thought of in this way, as the algebraic system $\mathbb{C}$, it is called the **Argand diagram**, after its 18th-century inventor (who thought he was merely 'representing' points of $\mathbb{C}$ by those of $\mathbb{R}^2$, whereas he was in fact proving the *existence* of $\mathbb{C}$). We are therefore able to study $\mathbb{C}$ through the intermediary of the geometry of $\mathbb{R}^2$, especially after introducing a metric. Observe that if $x + iy = u + iv$ in $\mathbb{C}$, with x, y, u, v real, then $x = u$ and $y = v$ by definition of the equality $(x, y) = (u, v)$. This process is called 'equating real and imaginary parts', and it is unfortunate that the name 'imaginary' persists, with its psychological overtones; the Arabs called negative numbers 'imaginary' when these were first thought of. Further details about complex numbers are given in Chapter 33.

For a better understanding, however, we now give a second construction, mentioned previously in 22 Exercise 5(iii), which has the advantages (i) of lending itself to generalization in a very evident way and (ii) of reflecting more obviously than do the number pairs the motivation for the introduction of complex numbers through the desire to solve the equation $x^2 + 1 = 0$.

We assume certain relevant concepts from Chapters 11 and 12. Consider then the ring $\mathbb{R}[x]$ of polynomials in the 'indeterminate' x with real coefficients. This ring is a Euclidean domain and $(x^2 + 1)$ is evidently an irreducible element in this domain. We form the ideal I generated by $(x^2 + 1)$ and it is an easy consequence of the theorems we have proved about Euclidean domains that the quotient ring $\mathbb{R}[x]/I$ is a field. We define

$$\mathbb{C}' = \mathbb{R}[x]/I.$$

The composite homomorphism

$$\mathbb{R} \subseteq \mathbb{R}[x] \xrightarrow{\pi} \mathbb{C}',$$

where the arrow represents the canonical projection π of $\mathbb{R}[x]$ onto its quotient ring, is easily verified to be injective and serves to embed $\mathbb{R}$ in $\mathbb{C}'$.

Now every coset of I in $\mathbb{R}[x]$—that is, every element of $\mathbb{C}'$—has a unique representative of the form $a + bx$, $a, b \in \mathbb{R}$. For, given any polynomial $f \in \mathbb{R}[x]$, representing say, $c \in \mathbb{C}'$, then, by the remainder theorem (22.3.3):

$$f = (x^2 + 1)q + r$$

where $\deg r < 2$, so that $r = a + bx$ for some $a, b \in \mathbb{R}$; and f and r represent the same element c of $\mathbb{C}'$. Moreover r (and q) are uniquely determined by f

(and the polynomial $(x^2 + 1)$), and the same polynomial r will be obtained for every f representing c. Thus, if we write i for $\pi(x)$, we obtain a unique expression for an element of $\mathbb{C}'$ in the form $a + bi$ (note that $\pi(t) = t$ if $t \in \mathbb{R}$). If we write the elements of $\mathbb{C}'$ in this way, then we see (by the rules for adding and multiplying polynomials) that

$$\text{24.7.2} \quad \begin{cases} (a_1 + b_1\mathrm{i}) + (a_2 + b_2\mathrm{i}) = (a_1 + a_2) + (b_1 + b_2)\mathrm{i}, \\ (a_1 + b_1\mathrm{i})(a_2 + b_2\mathrm{i}) = (a_1a_2 - b_1b_2) + (a_1b_2 + a_2b_1)\mathrm{i}. \end{cases}$$

For example, to prove the product rule we observe that, in $\mathbb{R}[x]$,

$$(a_1 + b_1x)(a_2 + b_2x) = a_1a_2 + (a_1b_2 + a_2b_1)x + b_1b_2x^2$$

and the rule follows by applying π and observing that $\pi(x^2) = \pi(-1) = -1$. Comparison of 24.7.1 and 24.7.2 now shows that the correspondence $\alpha: \mathbb{C}' \to \mathbb{C}$ given by $a + bi \to (a, b)$ is a ring-homomorphism; while α is obviously a bijection, and $\alpha(\mathbb{R})$ is the x-axis. Hence $\alpha: \mathbb{C}' \approx \mathbb{C}$ is an isomorphism and the two approaches give results which differ only in notation. Henceforth we use the symbol $\mathbb{C}$ rather than $\mathbb{C}'$.

It is evident that we have now reobtained the field of complex numbers as the reader already understands them.

24.8 Completeness of $\mathbb{C}$

The notion of convergence in $\mathbb{R}$, defined in 24.2.3, requires the notion of the 'distance' $|x - y|$ between the points $x, y \in \mathbb{R}$; thus, distance is a *metric* on $\mathbb{R}$, in the sense of 20.1. It may be extended to be a metric on $\mathbb{C}$ by declaring the distance $\rho(z_1, z_2)$ between $z_1 = a_1 + b_1\mathrm{i}$ and $z_2 = a_2 + b_2\mathrm{i}$ to be $\sqrt{[(a_2 - a_1)^2 + (b_2 - b_1)^2]}$; in particular, if $z_1, z_2 \in \mathbb{R} \subseteq \mathbb{C}$, then $\rho(z_1, z_2) = |z_1 - z_2|$, so ρ extends the metric in $\mathbb{R}$. Also, ρ satisfies the axioms $\mathfrak{D}_1$, $\mathfrak{D}_2$, $\mathfrak{D}_3$ of 20.1 for a metric. The definitions of convergent, Cauchy, and null sequence can now be given exactly as in 24.2, except that ρ is used in place of the metric $|x - y|$ of $\mathbb{R}$. (Similarly for any field in which a metric is introduced.) Of course, the extension ρ is not unique; for example, we could have declared the distance between z_1 and z_2 to be $|a_2 - a_1| + |b_2 - b_1|$. This is, indeed, an *equivalent* metric (frequently used by taxi-drivers) and leads to exactly the same notion of convergent sequences as that which we have chosen. Our chosen ρ however, has the distinct advantage that $N(a + bi) = a^2 + b^2 = (\rho(0, a + bi))^2$ is a multiplicative map from $\mathbb{C}$ to $\mathbb{R}^+$.

We are now content to record without proof the (easily proved!) theorem:

24.8.1 THEOREM. *$\mathbb{C}$ is complete with respect to the metric*

$$\rho(a_1 + b_1\mathrm{i}, a_2 + b_2\mathrm{i}) = \sqrt{[(a_2 - a_1)^2 + (b_2 - b_1)^2]}.$$

Moreover, if $c_n = a_n + b_n\mathrm{i}$, ($a_n, b_n \in \mathbb{R}$, $n = 1, 2, 3, \ldots$), then the sequence $\{c_n\} \to c = a + bi$ if and only if $\{a_n\} \to a$ and $\{b_n\} \to b$. □ See also 25.5.4.

Exercise 9

(i) Show that the sequence $\{c_n\}$ in 24.8.1 is Cauchy, if and only if $\{a_n\}$ and $\{b_n\}$ are Cauchy.

(ii) Show that $\mathbb{C}$ is a vector space over $\mathbb{R}$ of dimension 2.

(iii) Show that there is a unique automorphism of $\mathbb{C}$ leaving every element of $\mathbb{R}$ fixed and transforming i into $-$i.

(iv) By using De Moivre's theorem (see Chapter 33), show that every element of $\mathbb{C}$ has n distinct nth roots in $\mathbb{C}$.

(v) Prove Theorem 24.8.1. {Use the remarks preceding 24.8.1, proving them if you are energetic.}

(vi) Discuss in greater detail the relative merits of the two methods (in 24.7) of introducing complex numbers.

(vii) In $\mathbb{R}[x]$, let J denote the ideal generated by $x^2 + 2$. Are the sets $\mathbb{R}[x]/J$ and $\mathbb{C}$ isomorphic as (a) vector spaces over $\mathbb{R}$, (b) fields?

(viii) Show that the set of matrices of the form $\begin{pmatrix} a & b \\ -b & a \end{pmatrix}$, where $a, b \in \mathbb{R}$, is a field under the usual operations of matrix multiplication and addition. Show that this field is isomorphic to $\mathbb{C}$ under the correspondence

$$\begin{pmatrix} a & b \\ -b & a \end{pmatrix} \rightarrow a + bi.$$

24.9 Quaternions and Hypercomplex Numbers

Finally, we discuss briefly the question of further extensions of the number system beyond $\mathbb{C}$. Such a procedure as that used to pass from $\mathbb{R}$ to $\mathbb{C}'$ would, if applied to $\mathbb{C}$ itself, yield nothing; for, the only prime polynomials in $\mathbb{C}[x]$ are linear (by the fundamental theorem of algebra) so that the corresponding quotient field of $\mathbb{C}[x]$ would be just (an isomorph of) $\mathbb{C}$ itself! Moreover, we cannot follow effectively the suggestion contained in Exercise 9(ii) and try to construct a new field which is a *finite-dimensional* vector space over $\mathbb{C}$ since then any element of the new field would be **algebraic** over $\mathbb{C}$, that is, would satisfy a polynomial equation in $\mathbb{C}$, and hence would belong to $\mathbb{C}$, since $\mathbb{C}$ is algebraically closed.

We are thus left with two alternatives: either we must extend $\mathbb{C}$ to a larger field containing elements which are *not* algebraic over $\mathbb{C}$ (such an extension is called **transcendental**) or we must embed $\mathbb{C}$ in a larger algebraic system which is a finite-dimensional vector space over $\mathbb{C}$ but in which not all the field axioms are satisfied.

Either procedure is possible. Thus we may pass to the quotient field (or *field of fractions*: see 23.3.8) of the integral domain $\mathbb{C}[x]$. This field, otherwise known as 'the field of **rational functions** in the indeterminate x over $\mathbb{C}$', clearly contains $\mathbb{C}$ as a sub-field. Moreover, no rational function which is not itself an element of $\mathbb{C}$ can possibly be algebraic over $\mathbb{C}$. The field $\mathbb{C}(x)$ of rational functions is called a 'simple transcendental extension of $\mathbb{C}$'. See, for example Lang [79].

Of perhaps greater immediate interest is the fact, essentially due to the Irish mathematician Hamilton [(1805–65); see Bell [10], Ch. 19], that, if we discard the law of commutativity of multiplication, then we may extend $\mathbb{C}$ to an algebra $\mathbb{H}$ of **quaternions**, which is a two-dimensional vector space over $\mathbb{C}$ and hence of dimension 4 over $\mathbb{R}$. Such an algebra (in which all the field axioms save the commutativity of multiplication are satisfied) is called a **skew** field.

We construct $\mathbb{H}$ by expressing it as a vector space over $\mathbb{R}$ with basis vectors $(1, \mathrm{i}, \mathrm{j}, \mathrm{k})$. $\mathbb{R}$ is embedded in $\mathbb{H}$ by $\alpha \to \alpha 1$, $(\alpha \in \mathbb{R})$. Now, to define a multiplication which shall satisfy the distributive law, on any vector space V over $\mathbb{R}$ with basis vectors $e_1, \ldots, e_n$, it suffices to define all the products $e_r e_s$; for, given $v, v' \in V$, then $v = \sum v_r e_r$, $v' = \sum v'_r e'_r$ $(v_r, v'_r \in \mathbb{R})$ and by distributivity:

24.9.1
$$vv' = \sum_{r,s} v_r v'_s (e_r e_s).$$

Since each $e_r e_s$ is in V,

24.9.2
$$e_r e_s = \sum_t \alpha_{rst} e_t$$

for some $\alpha_{rst} \in \mathbb{R}$ called **structure constants**; and hence we know *every* product vv' when we know the n^3 numbers α_{rst}. Such a vector space V with a multiplication law 24.9.2 is called a **linear algebra over $\mathbb{R}$**, or **real algebra** (for some 'big' examples, see 26.2). In particular, when $(V, e_1, \ldots, e_n)$ is $(\mathbb{H}, 1, \mathrm{i}, \mathrm{j}, \mathrm{k})$, we define the products of basis elements by the rules:

24.9.3
$$\begin{cases} 1\mathrm{i} = \mathrm{i}1 = \mathrm{i}, & 1\mathrm{j} = \mathrm{j}1 = \mathrm{j}, \quad 1\mathrm{k} = \mathrm{k}1 = \mathrm{k}; \\ 1^2 = 1, & \mathrm{i}^2 = \mathrm{j}^2 = \mathrm{k}^2 = -1; \\ \mathrm{ij} = \mathrm{k}, & \mathrm{jk} = \mathrm{i}, \quad \mathrm{ki} = \mathrm{j}; \\ \mathrm{ji} = -\mathrm{k}, & \mathrm{kj} = -\mathrm{i}, \quad \mathrm{ik} = -\mathrm{j}. \end{cases}$$

The reader will readily verify that the associative law holds—a convenience difficult to arrange for a general array (α_{rst}) of structure constants. Moreover the resulting ring has the unity element 1 and the multiplication agrees with that on the embedded copy $\{\alpha 1 + \beta \mathrm{i} | \alpha, \beta \subset \mathbb{R}\}$ of $\mathbb{C}$. Also, the non-zero elements form a group, as follows immediately from the identity

24.9.4
$$(a1 + b\mathrm{i} + c\mathrm{j} + d\mathrm{k})(a1 - b\mathrm{i} - c\mathrm{j} - d\mathrm{k}) = (a^2 + b^2 + c^2 + d^2)1.$$

(We may, of course, suppress the 1 from the notation.) It is plain from 24.9.3, however, that multiplication is not, in general, commutative.

The field $\mathbb{C}$ is embedded in $\mathbb{H}$ as the subspace consisting of vectors $a + b\mathrm{i}$; moreover, $\mathbb{H}$ is also a *complex* vector space, i.e., a space over $\mathbb{C}$ with basis $(1, \mathrm{j})$, since multiplication of j by i (which is now a scalar, being in $\mathbb{C}$) gives $\mathrm{ij} = \mathrm{k}$. It is interesting and important to observe that if we pass to the quotient space of $\mathbb{H}$ by $\mathbb{R}$, embedded in the obvious way, we obtain precisely the usual vector-product structure in $\mathbb{R}^3$ (identifying i, j, k with the unit vectors along the co-ordinate axes).

Exercise 10

(i) Verify that the multiplication 24.9.3 is associative.
(ii) Find the inverses of i, j, k and hence show that the eight elements ± 1, $\pm$i, $\pm$j, $\pm$k form a group with respect to multiplication in $\mathbb{H}$.
(iii) Is $\mathbb{R}$ an ideal in $\mathbb{H}$? Let $h: \mathbb{H} \to \mathbb{H}/\mathbb{R}$ denote the natural map of vector spaces, and recall the remark above that $\mathbb{H}/\mathbb{R} \approx \mathbb{R}^3$. Compare the product $h(vv')$ with the vector product $h(v) \wedge h(v')$ defined in 14.10; thus let

$$X = \{x \in \mathbb{H} | x = b\mathrm{i} + c\mathrm{j} + d\mathrm{k}, (b, c, d \in \mathbb{R})\},$$

and prove that $h|X$ is one-one, while $h(vv') = h(v) \wedge h(v')$ if v, v' and vv' all lie in X. {See 18.10.}
(iv) Let $G = \{g_1, \ldots, g_n\}$ be a group of order n, and let $\mathbb{R}(G)$ denote the set of all functions $f: G \to \mathbb{R}$. Prove that $\mathbb{R}(G)$ is a real vector space of dimension n, with basis elements $\bar{g}_1, \ldots, \bar{g}_n$, where $\bar{g}_r(g_s) = 1$ or 0 according as $r = s$ or not. Use 24.9.1 to show that $\mathbb{R}(G)$ is an associative linear algebra (the '**group algebra**' of G) containing an isomorph of G, when $\mathbb{R}(G)$ is given a certain natural product.

24.9.5 Can we go further in extending the number system in this way (that is, without introducing transcendental elements)? The answer is that we can further embed $\mathbb{H}$ in the so-called 'algebra of Cayley numbers', which is a two-dimensional vector space over $\mathbb{H}$, and hence an eight-dimensional vector space over $\mathbb{R}$ (it is sometimes called the algebra of **octonions**). However, in this algebra we even have to abandon the associative law of multiplication, as well as the commutative law. But we do retain the condition that there shall be no divisors of zero; indeed, the algebra (like $\mathbb{H}$, $\mathbb{C}$, and $\mathbb{R}$) admits a multiplicative function ϕ to $\mathbb{R}^+$ which is zero only on the zero element. It has very recently been proved that the *only* algebras over $\mathbb{R}$ admitting such a function ϕ are vector spaces of dimension 1, 2, 4, or 8 over $\mathbb{R}$, so we certainly cannot go beyond the octonions. But this recent theorem (due to Professor J. F. Adams of Manchester University) required for its proof some of the most sophisticated apparatus of modern algebraic topology; no wonder it had defied the best efforts of mathematicians for many years.

Finite-dimensional vector spaces over $\mathbb{R}$ admitting an *associative* multiplication of the kind 24.9.1, are often called systems of **hypercomplex numbers.** The question arises: what systems of this kind are possible? Apart from the three associative division algebras mentioned above, we know that the sets $\mathscr{M}_n$ ($n = 1, 2, \ldots$) of all real $n \times n$ matrices form real algebras; and then there are the 'pathological' cases such as that where we take any vector space and declare *all* products to be zero. A great step in the development of abstract algebra was Wedderburn's classification (in 1907) of all the possible systems of hypercomplex numbers. Apart from 'pathological' cases, he showed that the possible systems had to be built from Cartesian products of algebras $\mathcal{M}_n$ and division algebras. (For details see, e.g., Albert [1].) Such 'classification' theorems are greatly prized; starting from an apparently limitless family satisfying a set of axioms, they show just how general (or special) the axioms are. Perhaps their greatest appeal is aesthetic, and therefore we have to let them speak for themselves, in their own language.

Exercise 11

(i) Let V be a linear algebra of dimension n over $\mathbb{R}$. Given $\alpha \in V$, define a function $T_\alpha : V \to V$ by $T_\alpha(v) = \alpha v$. Prove that T_α is a linear transformation of V, and that

$$T_{\alpha+\beta} = T_\alpha + T_\beta, \qquad T_{\alpha\beta} = T_\alpha \circ T_\beta \qquad (\alpha, \beta \in V)$$
$$T_{x\alpha} = xT_\alpha (x \in \mathbb{R}), \qquad T_1 = \mathrm{id}_V.$$

Show also that if $T_\alpha = T_\beta$, then $\alpha = \beta$. Thus, the function $\alpha \to T_\alpha$ *embeds* V in the real algebra Hom (V, V) which is itself isomorphic to the algebra of all real $n \times n$ matrices.
(ii) Find matrices corresponding to T_α, when $\alpha = \mathrm{i}, \mathrm{j}, \mathrm{k}$, when V in (i) is $\mathbb{H}$. Hence find a matrix algebra isomorphic to $\mathbb{H}$. (Compare Exercise 9(viii).)
(iii) Let A, B be linear algebras over $\mathbb{R}$. Define addition and multiplication, in their Cartesian product $A \times B$, so that A, B embed as sub-algebras.
(iv) Keeping in mind 24.9.4, consider whether the theory of 15.2 still applies when $\mathbb{R}$ is replaced by $\mathbb{H}$. What happens then to 'geometry' in Chapter 14, and in 17.6?

Chapter 25

TOPOLOGY OF $\mathbb{R}^n$

25.1 Introduction

As indicated at the conclusion of the last chapter, we cannot expect $\mathbb{R}^n$ to behave as a particularly interesting algebraic system for $n \neq 1, 2, 3, 4, 8$. On the other hand, if we look instead at the 'continuity' properties of $\mathbb{R}^n$ (to be precise, its *topology*), we have a very rich situation, which we attempt in this chapter to sketch. More than a sketch is quite impossible, since topology is a vast branch of mathematics, increasing rapidly both in itself and in its applications to mathematics and physics; and the interested reader should turn to the books described in Section 25.14 for detailed information. However, the basic ideas of topology are essential for an educated understanding of elementary mathematics, and they are quite accessible to the reader of this book, as we now show. We begin with some motivating remarks.

25.2 Topology within the Erlanger Program

Broadly speaking, topology is the study of certain transformations, called 'homeomorphisms'†; essentially, these form the extensive class of one-to-one bicontinuous transformations of one subset of $\mathbb{R}^k$ on another. Distinct points are not crushed together, and no tearing is allowed unless the tear is repaired. We ask what properties are left invariant by such transformations, and hope to find enough of these 'invariants' to be able to decide whether or not two given surfaces (for example) can be transformed one into the other in the allowed manner.

This aim is analogous to that in the study of any other branch of geometry. For example, in Euclidean plane geometry, we restrict ourselves to those transformations of figures which are composed of rigid motions (the 'congruence' transformations). A typical theorem here is that two triangles can be transformed into each other by such a congruence if and only if the set S of three side-lengths of one equals the corresponding set for the other; and we say that S is an invariant under the congruence transformations. Of course, other theorems turn up, too, concerning relationships between the invariants, as for example the theorem of Pythagoras which concerns the class of right-angled triangles. Such triangles form a significant class of configurations,

† from *homeo-* = 'same', and *morphe* = 'structure'. Do not confuse with 'homomorphism'.

because the congruence transformations map each element of this class into other elements of the class. A second example is that of plane projective geometry (see 17.10), where the allowed transformations are repeated projection from different centres. Then for instance, four collinear points P, Q, R, S can be transformed into four other collinear points A, B, C, D if and only if the cross-ratios $(PQRS)$, $(ABCD)$ are equal, these numbers being 'projective invariants' of the quadruples of points. But one is led in projective geometry also to theorems about relationships between invariants (e.g., cross-ratio of poles equals cross-ratio of polars). Similarly, in topology, much of the subject-matter concerns mutual relationships between invariants, and methods for working them out. In the language of the Erlanger Program (see 17.11) we are broadening the group of allowed transformations of $\mathbb{R}^k$ from the rigid motions of Euclid, beyond the projectivities of projective geometry, to the homeomorphisms of $\mathbb{R}^k$. Now consider some problems which are of a topological nature.

25.2.1 *The problem of the Koenigsberg Bridges.* This is described in 5 Exercise 1(xiii), and was first solved by Euler in a paper (see Newman [93]), from which topology is usually considered to start.

25.2.2 *The five-colour theorem.* Every map in the plane can be coloured by using no more than five colours. Regions whose boundaries have an arc in common are to have different colours. The same property holds for all sets into which the plane can be transformed by a homeomorphism. It is* an unsolved problem to decide whether or not the number five is best-possible: will four colours always suffice? (Certainly, there are maps for which three colours are inadequate.)

25.2.3 *The Brouwer fixed-point theorem.* Every continuous, not necessarily one-to-one, transformation of a plane circular disc into itself leaves at least one point fixed. The same is true of all sets into which the disc can be transformed by a homeomorphism (e.g., a triangular disc, or a hemisphere).

These results are discussed and proved in Courant–Robbins [25], Ch. V. The first step in the proofs is to start with the simplest possible configuration, e.g., a circular disc for 25.2.3, rather as one 'chooses the triangle of reference' in projective geometry (see Theorem 17.5.4). This comes from the observation that a topological property is being proved, but from then on special methods must be devised, which have been absorbed into topology.

The popular accounts often go on to discuss 'surfaces', and then stop; yet little has been achieved except perhaps to arouse interest. The trouble arises from not stating precisely the meanings of such basic words as 'transformation', 'continuous', and 'one-to-one'. However, the reader of this book

* (Added in present edition) The problem was solved in 1976 by Appel and Haken. The solution relies on a final computer check of several hundred cases to show that, in fact, four colors always suffice.

certainly has the necessary language of sets and functions to be able to go further; in fact, much of the language was designed by topologists for this very purpose! For the moment, we shall take for granted the notion of a continuous function in calculus (as one with an unbroken graph). Precise discussions will be given later, in 25.5.5 and in Chapter 28.

25.3 Some Homeomorphisms

In geometry, the continuous nature of the interesting sets leads us to concentrate on the 'continuous' functions, i.e., those which 'preserve continuity' in some sense. Thus, supposing for the moment that A, B are subsets of $\mathbb{R}^k$, a function $f: A \to B$ is commonly given by specifying the k co-ordinates of $f(a)$:

$$f_1(a), f_2(a), \ldots, f_k(a), a \in \mathbb{R}^k.$$

Each is then an ordinary real-valued function of the co-ordinates of a; we therefore say that f is continuous at a iff each function f_i is continuous in the co-ordinates of a (in the sense of ordinary calculus; see 28.1). And then f is 'continuous on A' iff f is continuous at each $a \in A$. Finally f is a **homeomorphism** of A onto B if f is a bijection, f is continuous on A and f^{-1} on B; and then A and B are **homeomorphic.** All this has been necessary to elucidate the meaning of the first sentence of 25.2 and already reveals a caution there; for as we see in Example 25.3.1 below, a function $g: X \to Y$ can be invertible and continuous, without its inverse being continuous; and in view of the examples of Peano in 7 Exercise 5(vii), continuity is vital if we are to distinguish in a really significant way between sets, for those such as $\mathbb{R}$, $\mathbb{I}$, $\mathbb{I}^2$ clearly are significantly different from each other.

A subset of some $\mathbb{R}^k$ may be called a **(Euclidean) topological space.**

25.3.1 EXAMPLE. Let $\mathbb{S}^1$ denote the unit circle $x^2 + y^2 = 1$ in $\mathbb{R}^2$. On the half-open interval $\langle\!\langle 0, 1\rangle = \{x | 0 \leqslant x < 1\}$, define $f: \langle\!\langle 0, 1\rangle \to \mathbb{S}^1$ by $f(t) = e^{2\pi i t}$. Then f is invertible and continuous; but f^{-1} is not continuous at the point P: $x = 1, y = 0$, as is intuitively clear, since points near P are forced apart, some being sent by f^{-1} near the origin in $\langle\!\langle 0, 1\rangle$ and some near 1. (See Fig. 25.1.)

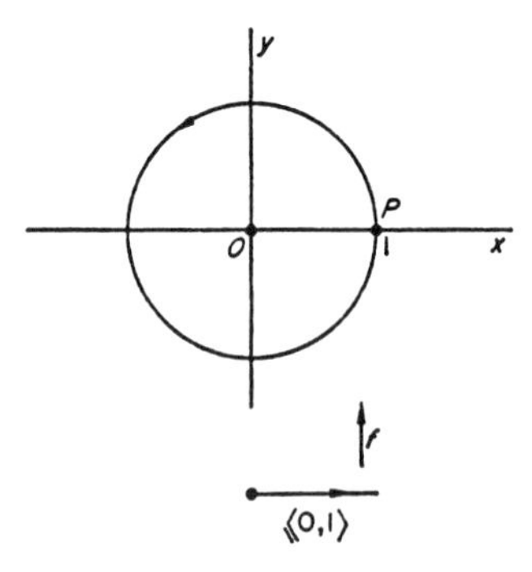

Fig. 25.1

25.3.2 EXAMPLE. In $\mathbb{R}^2$, the ellipse $x^2/a^2 + y^2/b^2 = 1$ is homeomorphic to the circle $\mathbb{S}^1$ because the function which assigns to the point $(a \cos t, b \sin t)$ the point $(\cos t, \sin t)$, is a homeomorphism. Moreover, let B denote the boundary of the unit square $\mathbb{I}^2$. Then $\mathbb{S}^1$ is homeomorphic to B, as we see by projecting it radially onto B (see Fig. 25.2). In formulae, we construct a

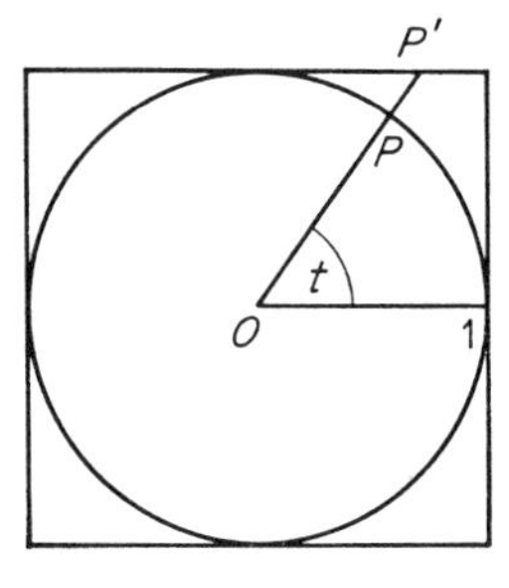

Fig. 25.2

function $g: \mathbb{S}^1 \to B$ given by $g(P) = P'$, the point where the ray OP cuts B; so $g(P)$ is

$$\begin{cases} (\cos t, 1) & \text{if } \pi/4 \leqslant t \leqslant 3\pi/4 \\ (-1, \sin t) & \text{if } 3\pi/4 \leqslant t \leqslant 5\pi/4, \text{ etc.} \end{cases}$$

It is a tedious exercise now to prove in detail that g is in fact a homeomorphism, and we leave the complete account to the reader. {Use your knowledge of sine and cosine to piece some graphs together.} We do not recommend that one always attempt to describe a continuous function by means of an explicit formula!

Similarly, every ellipsoid is homeomorphic to the surface of a cube. Examples such as these should start the reader to acquire skill in recognizing some of the simpler homeomorphic shapes around him; for example, which letters of the alphabet, in which founts of type, are homeomorphic to each other (regarded as one-dimensional sets)? Is the surface of a table homeomorphic to that of a stool, a chair, a cup? Are these items, considered as solids, homeomorphic? Which mammals have homeomorphic surfaces? Is the surface of a given carving by (say) Henry Moore homeomorphic to something familiar? By such exercises can the reader convince himself that topology is (among other things) a science of shape. He may however be a little uneasy that he finds it hard to write out a proof that, say, a solid cup is homeomorphic to a doughnut. A standard procedure is to compare them 'chunk by chunk': the cup handle is homeomorphic to a solid sausage, say by a bijection f: and the solid bowl to another sausage by a bijection g (since a bowl is a very flat sausage with the edges turned up); now form a bijection h from the bowl to the doughnut by making it agree with f on the handle and with g on the bowl, and forming the doughnut out of the two sausages, in the way the handle is stuck on the cup. Since f and g are homeomorphisms, standard theorems ensure that h is also. This illustrates a general technique in which induction is used on the number of chunks, and the last thing we do is try to write down one neat formula.

25.3.3 EXAMPLE. Let $f: \mathbb{S}^1 \to \mathbb{R}$ be a continuous injection of the circle into the plane. The image of f is called a **Jordan curve**, after the French mathematician C. Jordan. It has the essential feature that it is an endless loop which never crosses itself. The 'theorem of Jordan' states that a Jordan curve always splits the plane into two connected pieces, such that any path from one piece to the other cuts the curve; and one of the pieces stretches to infinity (being called the 'outside'), the other being called the 'inside'. This result shows that the intuitive picture of a Jordan curve as an elastic band without crossings is substantially correct. The proof is not easy, even when the curve is a polygon (see 15.6) and of course the general case includes curves with no tangent anywhere; so we cannot just point to pictures. Some idea of the discipline will present itself to the reader if he tries to say what is meant by the 'inside' of a triangle (see 14 Exercise 3(iv)). The Jordan curve theorem was later illuminated by Schoenflies, who proved it by showing that the continuous injection $f: \mathbb{S}^1 \to \mathbb{R}^2$ could be extended to a homeomorphism $g: \mathbb{R}^2 \to \mathbb{R}^2$; that is, g agrees with f on $\mathbb{S}^1$. Hence every topological feature of the figure consisting of a circle and plane is transferred by g to the figure consisting of a Jordan curve and plane; now the circle divides $\mathbb{R}^2$ into exactly two pieces, of which one (the 'inside') is homeomorphic to the open disc $x^2 + y^2 < 1$, so the same must be true of the Jordan curve.

25.3.4 EXAMPLE. A Jordan curve in space is called a **knot.** We can follow Schoenflies and ask if there is a homeomorphism $g: \mathbb{R}^3 \to \mathbb{R}^3$ which carries the given Jordan curve K onto a circle in $\mathbb{R}^2$ (here regarded as the plane $z = 0$). If no such g exists, then K is said to be **knotted**; otherwise it is **unknotted.** If K' is a second knot, we can also ask if there exists a homeomorphism $h: \mathbb{R}^3 \to \mathbb{R}^3$ which carries K onto K'; if h exists, then K and K' belong to the same **knot class.** The enumeration of all possible classes of knots is a difficult problem, still not completely solved, and uses special techniques of group theory (including the algebraic 'free differential calculus'). Observe how we now have the language to frame questions, if not answers; for example, we can give precision to the physical question: can knotted lines of force exist in a potential field? Can knotted vortices exist? Organic chemists are interested in knotted chains of molecules: see [133]. Indeed, the subject arose because of the interest of Maxwell, Tait, and Kelvin in vortices in physics.

25.3.5 EXAMPLE. Now turn from the circle to the sphere $\mathbb{S}^2$ whose equation in $\mathbb{R}^3$ is

$$x^2 + y^2 + z^2 = 1.$$

If $f: \mathbb{S}^2 \to \mathbb{R}^3$ is a continuous injection, the image of f is called a '2-sphere' in $\mathbb{R}^3$. It was proved by the great Dutch mathematician L. E. J. Brouwer, who died at the end of 1966, that every 2-sphere in $\mathbb{R}^3$ separates space, meaning that it splits $\mathbb{R}^3$ into two pieces, as a Jordan curve splits the plane. However, in the 1920s, the American mathematician J. W. Alexander showed that the earlier plane theory of Schoenflies did not apply in $\mathbb{R}^3$, because his 'horned sphere' J was such that there is no homeomorphism of $\mathbb{R}^3$ on itself which extends any continuous injection $\mathbb{S}^2 \to J$. (A beautiful illustration can be found in Hocking–Young [65].) A variant H is shown in Fig. 25.3. In particular, the 'outside' of H is not homeomorphic to the outside of $\mathbb{S}^2$, intuitively because arbitrarily small knots exist near H which cannot be pulled away without breaking, whereas no such knots exist near $\mathbb{S}^2$. A rigorous proof requires group theory. Alexander showed, however, that any *polyhedral* 2-sphere in $\mathbb{R}^3$ (i.e., one whose surface consists of a finite number of flat triangular discs) does have an extending homeomorphism, so that its inside and outside are homeomorphic to those of $\mathbb{S}^2$. The analogous result for an n-dimensional sphere in Euclidean $(n + 1)$-dimensional space was only proved in 1958, as a result of a decisive step made by a nineteen-year-old American, B. Mazur. Similarly for 'smooth' n-spheres, which have a tangent plane everywhere. Alexander's negative result was of great importance as it gave rise to several other 'surprising' examples, which showed just how naïve our untrained spatial perception is. One such example is shown in Fig. 25.4.

To conclude this section, let us prove the assertion in 25.2.3 that the Brouwer fixed-point theorem may be generalized from a circular disc to any homeomorph of a circular disc. Let A be a homeomorph of the circular disc D, and let us show that, given any continuous function $f: A \to A$, there is a fixed point, i.e., a point $x \in A$ such that $f(x) = x$. We have a system of functions as shown, where h is a

$$\begin{array}{ccc} A & \xleftarrow{h} & D \\ {\scriptstyle f}\downarrow & & \downarrow{\scriptstyle g} \\ A & \xrightarrow[h^{-1}]{} & D \end{array}$$

homeomorphism (existing by hypothesis) and g is constructed as $h^{-1} \circ f \circ h$. As a composition of continuous functions, g is continuous [see 25.6.12] and so has a fixed point, by Brouwer's theorem for D. Thus, there exists $y \in D$ such that $g(y) = y$, so applying h to each side gives $h(y) = (f \circ h)(y) = f(h(y))$; so $x = f(x)$ where $x = h(y) \in A$, as required.

The use of diagrams of functions like that above is now standard in topology and algebra; modern texts often look like those on organic chemistry, and have diagrams with dozens of arrows, and beautiful arrangements. The accompanying proofs would be impossible to comprehend if expressed only in words (compare 6.2).

Once we have found a topological property, we can use it to prove the *non-existence* of a homeomorphism. For example, let us prove that the disc D

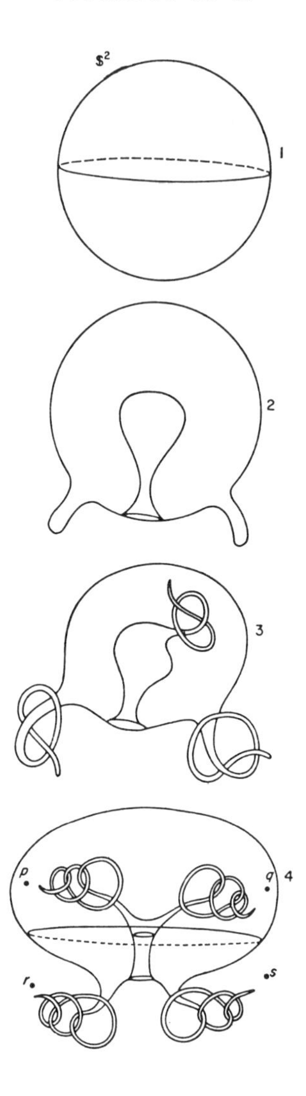

Fig. 25.3 A variant of Alexander's 'horned sphere'. We show four stages in the construction; the horns are intended ultimately to be given an infinity of loops, converging to p, q, r, s. At each stage, we still have a homeomorph of $\mathbb{S}^2$ which can be extended to all of $\mathbb{R}^3$. In the limit we also have a homeomorph H, of $\mathbb{S}^2$, but no extension to all $\mathbb{R}^3$ is possible. That is, no homeomorphism of $\mathbb{R}^3$ will carry H back onto $\mathbb{S}^2$. 1. A 2-sphere in $\mathbb{R}^3$. 2. Grow horns and make a dent. 3. Tie knots in the horns, and push a knotted tunnel into the dent. (The knots must not be pulled tight; otherwise we would not have a homeomorph of $\mathbb{S}^2$.) 4. The horns grow longer and further stitches are 'knitted' into them.

Fig. 25.4 A 'wild' arc α in $\mathbb{R}^3$. (The loops get smaller and smaller, and converge to p, q). There is no homeomorphism $h: \mathbb{R}^3 \to \mathbb{R}^3$ such that $h(\alpha)$ is a straight segment, although α itself is homeomorphic to one.

is not homeomorphic to D', where D' is the result of removing the origin from D; it is a 'punctured' disc. Define $f: D' \to D'$ to be the rotation through 90 degrees about the origin—a continuous function. This rotation leaves only the origin fixed, and so f has no fixed points in D'. Therefore D' cannot be homeomorphic to D, by Brouwer's fixed-point theorem. Similarly, by constructing continuous functions without fixed points, the reader can prove that D is not homeomorphic to a 2-sphere, $\mathbb{R}^2$, an annulus, or a torus. Similarly too, for the Koenigsberg Bridges and the four-colour problem in 25.2.1, 25.2.2.

To date, topologists have been far more successful at showing that homeomorphisms do not exist, than at the reverse problem; and they have discovered a host of very delicate topological properties by which to apply the method suggested above, with fixed points replaced by the appropriate property.

Exercise 1

Let $A \approx B$ mean that A and B are homeomorphic and let $\langle a, b\rangle$, $\langle\!\langle a, b\rangle\!\rangle$ denote the intervals $a < x < b$, $a \leqslant x \leqslant b$ in $\mathbb{R}$.
(i) Prove that $\langle\!\langle a, b\rangle\!\rangle \approx \langle\!\langle c, d\rangle\!\rangle$ for any $a < b$, $c < d$ in $\mathbb{R}$.
(ii) Prove that $\langle a, b\rangle \approx \langle 0, 1\rangle \approx \langle 0, \infty\rangle \approx \mathbb{R}$.
{Consider the log function, and the functions $1/x$, $1/(1 - x)$.} These examples confirm that length is not a topological invariant.
(iii) If A is a subset of Euclidean space, a *path* from x to y in A is defined to be a continuous function $f: \mathbb{I} \to A$, where $\mathbb{I}$ denotes the unit interval in $\mathbb{R}$. (Note: a path is not the image set, as you might expect.) We call A 'pathwise connected' (p.c.) iff there exists a path from any x to any y in A. Prove that if A is p.c., so is every homeomorph of A; and that Euclidean space (of any dimension) is p.c. Prove that a line minus a point is not p.c., whereas a circle (or $\mathbb{R}^2$) minus a point is p.c. Hence prove that

$$\mathbb{R} \not\approx \mathbb{R}^2, \qquad \mathbb{S}^1 \not\approx \mathbb{R}^2, \qquad \mathbb{R} \not\approx \mathbb{S}^1, \qquad \mathbb{R} \not\approx \mathbb{I}, \qquad \mathbb{I} \not\approx \mathbb{S}^1.$$

{A continuous real-valued function, with domain $\mathbb{I}$, which is negative at a and positive at b is zero somewhere; see Theorem 28.5.1.} Show that even a continuous (not necessarily homeomorphic) image of a p.c. topological space is p.c. {See Exercise 2(i) below.}

(iv) An *arc* is a homeomorph of $\mathbb{I}$. Let F_k denote a set formed by joining k arcs at a common end-point, and at no others. Prove that $F_1 \approx F_2 \approx \mathbb{I}$ and $F_3 \approx$ each of the letters 'T', 'Y'. Prove that $F_p \approx F_q$ iff $p = q$, provided $p, q > 2$.

25.4 The Cartesian Product

A common method for studying an object in science is to split it into simpler objects, to study them, and to study the method of assembly back into the original object. The next simplest such method in topology after the use of 'chunks' (see Example 25.3.2) is to use the Cartesian product of spaces.

25.4.1 EXAMPLES. (i) Define $f: \mathbb{R}^3 \to \mathbb{R}^2 \times \mathbb{R}$ by $f(x, y, z) = ((x, y), z)$; clearly f is a bijection. Now $\mathbb{R}^2 \times \mathbb{R}$ is not a subset of any Euclidean space, so we do not yet know what to mean by continuity; but whatever reasonable definition we take (and we give one in 25.5 below), this f and f^{-1} turn out to be continuous.

(ii) We saw in Example 3.1.2 that bijections existed between a cylinder and $\mathbb{S}^1 \times \mathbb{I}$, a torus and $\mathbb{S}^1 \times \mathbb{S}^1$, a solid torus and $\mathbf{D} \times \mathbb{S}^1$, etc. The bijections indicated in Fig. 3.2 are easily seen to be continuous, with continuous inverses, once we have defined continuity in a product. The reader might care to practise by imagining what other Cartesian products look like—e.g., when A, B are shaped like the letters denoting them. {Use the rules 3.2.1 and 3.2.2.}

(iii) If A is a rectangular disc, there is a bijection $f: A \to \mathbb{I} \times F$ where F is an interval; for A is homeomorphic to the rectangle $\{(x, y) | x \in \mathbb{I} \; \& \; 0 \leqslant y \leqslant Y\}$ in $\mathbb{R}^2$ (for any $Y > 0$), so we can take F to be $\langle 0, Y \rangle$. Observe that anything homeomorphic to a rectangle is also homeomorphic to $\mathbb{I} \times F$.

25.5 Metric Spaces

The difficulty we met in Example 25.4.1(i) above, concerning continuity in a Cartesian product, leads us to look more closely at the notion of continuity. We defined it earlier in terms of the continuity of real-valued functions, because we were dealing with a function $f: A \to B$ from one subset of $\mathbb{R}^k$ to another. But many interesting sets do not arise as such subsets, although it is often possible, after investigation, to think of them as if they were. For example, consider the double pendulum, whose 'state' at any time is determined by the two angles s, t of the pendulum, and by the two angular velocities u, v of their ends. The set P of all possible states is then bijective with $\mathbb{S}^1 \times \mathbb{S}^1 \times \mathbb{R} \times \mathbb{R}$ or with $T \times \mathbb{R}^2$ where T is a torus; and $T \times \mathbb{R}^2$ is not a subset of any Euclidean space, although it lies in $\mathbb{R}^3 \times \mathbb{R}^2$ which we feel 'ought' to be $\mathbb{R}^5$. Further, we have a strong feeling that some states of the double pendulum are 'near' to each other, in some sense. What exactly is this sense? We therefore consider now the notion of a 'metric space' defined as follows, examples of which we met, for example, in Chapter 24. We recall from 20.1 the definition:

25.5.1 DEFINITION. A **metric space** X is a set together with a function $d: X \times X \to \mathbb{R}^+$ (called the 'metric') with the properties:

$\mathfrak{D}_1$: *$d(x, y)$ is always positive, and it is zero* iff $x = y$;
$\mathfrak{D}_2$: *$d(x, y) = d(y, x)$ for all $x, y \in X$;*
$\mathfrak{D}_3$: *the 'triangle inequality' holds*, i.e., *for every $x, y, z \in X$,*

$$d(x, y) \leqslant d(x, z) + d(z, y).$$

Strictly speaking we should denote the metric space by (X, d) since there may be more than one metric on X; but if no ambiguity is likely, we speak of 'the metric space X'.

25.5.2 EXAMPLES. (i) For any set X, define $d(x, y)$ to be 0 or 1 according as x is, or is not, equal to y. Conditions $\mathfrak{D}_1$ and $\mathfrak{D}_2$ of Definition 25.5.1 obviously hold, and $\mathfrak{D}_3$ is checked by considering cases. Thus (X, d) is a metric space.

(ii) The real numbers $\mathbb{R}$, with

$$d(x, y) = |x - y|$$

form a metric space; again the conditions $\mathfrak{D}_1$, $\mathfrak{D}_2$ are immediate consequences of the definition of absolute value, while $\mathfrak{D}_3$ is proved in Theorem 14.8.1.

(iii) The Euclidean k-dimensional space $\mathbb{R}^k$ forms a metric space with the Pythagorean metric

$$d((x_1, \ldots, x_k), (y_1, \ldots, y_k)) = \sqrt[+]{\Sigma (x_i - y_i)^2}.$$

Again, the hardest condition to verify is the third, but this follows from the Cauchy–Schwarz inequality (see 20.1.3). Observe that when k is 2 or 3, the condition states that the sum of the lengths of two sides of a triangle is not less then the length of the remaining side. When referring to 'the metric space $\mathbb{R}^k$', we mean it to have this metric.

(iv) If (X, d) is a metric space, so is (A, d) for any subset A of X, where $d(a, a')$ is given by d in X. Hence, for example, the $(k - 1)$-sphere $\mathbb{S}^{k-1}$, given by $x_1^2 + \cdots + x_k^2 = 1$ in $\mathbb{R}^k$, is a metric space. (A, d) is then called a **subspace** of X.

(v) Let C denote the set of all real continuous functions $f: \mathbb{I} \to \mathbb{R}$. Define two metrics on C by

$$d_1(f, g) = \int_0^1 |f(t) - g(t)|\, dt,$$
$$d_2(f, g) = \text{lub}\, |f(t) - g(t)|, \qquad t \in \mathbb{I},$$

where lub means 'least upper bound' as in 24.6.8. It is an exercise in calculus to prove that d_1 and d_2 are metrics.

(vi) Let (A, d), (B, r) be metric spaces. Form a metric space from $A \times B$ by setting

$$p((a, b), (a', b')) = \sqrt[+]{\{d(a, a')^2 + r(b, b')^2\}}.$$

It is an algebraic exercise to verify that p is a metric on $A \times B$, using the fact that d and r are metrics on A, B respectively. We call p the **product metric** on $A \times B$.

Once we have a metric space (X, d) we can talk about convergent sequences, copying the procedure of Section 24.2. Thus if $x_1, x_2, \ldots, x_n, \ldots$ is a sequence of points in X we say that it 'converges' to $x \in X$ iff the sequence $d(x_n, x)$ converges to zero in the usual sense of convergence in $\mathbb{R}$; and we write $\lim x_n = x$.

25.5.3 PROPOSITION. *This limit is unique if it exists.*

Proof. Suppose also that $\lim x_n = y$; if $x \neq y$ we know by $\mathfrak{D}_2$ that $d(x, y) = r > 0$; but for all sufficiently large n, $d(x_n, x) < \frac{1}{2}r$ and $d(x_n, y) < \frac{1}{2}r$. Thus by the triangle inequality $\mathfrak{D}_3$, $d(x, y) < r$, contrary to the fact. Hence $x = y$ and the limit is unique. ■

It is customary to denote the sequence $x_1, x_2, \ldots, x_n, \ldots$ by $\{x_n\}$, and it is rare in practice for this to be confused with the notation for a singleton. Of course the sequence $\{x_n\}$ is a function $s: \mathbb{N} \to X$. If $m: \mathbb{N} \to \mathbb{N}$ is also a function, but such that $m(1) < m(2) < \cdots < m(k) \cdots$, then $s \circ m: \mathbb{N} \to X$ is a sequence commonly denoted by $x_{m(1)}, x_{m(2)}, \ldots, x_{m(k)}, \ldots$ or by $\{x_{m(k)}\}$, and called a sub-sequence of $\{x_n\}$. We invite the reader to check that if s converges to x in X, then any sub-sequence $s \circ m$ also converges to x.

25.5.4 EXAMPLES. In Example 25.5.2(i),

$$\lim x_n = x \text{ iff } x_n = x$$

for all but a finite number of n. In Example 25.5.2(iii),

$$\lim x_n = x$$

iff the ith co-ordinate of x_n converges to that of x, $1 \leqslant i \leqslant k$. In 25.5.2 Example (v) we have

$$\text{(i)}\ d_1(f, g) \leqslant \int_0^1 d_2(f, g)\, dt = d_2(f, g);$$

hence if $\lim f_n = f$ according to d_2, then given $s > 0$ we have for all sufficiently large n,

$$d_2(f_n, f) < s, \quad \text{so} \quad d_1(f_n, f) < s$$

whence $\lim f_n = f$ according to d_1. On the other hand, the sequence $\{\mathrm{id}^n\}$ of functions converges to zero according to d_1 but does not converge to zero according to d_2, for

$$d_1(\mathrm{id}^n, 0) = \int_0^1 |t^n|\, dt = \int_0^1 t^n\, dt = 1/(n+1) \to 0,$$

while $d_2(\mathrm{id}^n, 0) = \mathrm{lub}\, |x^n - 0| = 1$. (Why is this argument sufficient to show that $\{\mathrm{id}^n\}$ does not converge *at all* according to d_2?)

25.5.5 Continuous functions can now be introduced as follows. Let $f: X \to Y$ be a function, where X, Y are metric spaces with metrics d_X, d_Y respectively. We then say that f is **continuous at** $x \in X$ iff for any sequence $\{x_n\}$ converging to x in X, we have $\lim f(x_n) = f(x)$ in Y. Thus $d_Y(fx_n, fx)$ is required to converge to zero whenever $d_X(x_n, x)$ does; fx_n is 'close' to fx if x_n is 'close' to x. We can write

$$fx = f(\lim x_n) = \lim fx_n,$$

or $f(\lim) = \lim (f)$, or 'f commutes with lim'. If f is continuous *at each* $x \in X$, we say that f is **continuous on** X, and call f a **mapping** or **map** of X in Y (using a geographical analogy).

25.5.6 It follows from the second sentence in 25.5.4 about convergence in Euclidean space, that this general notion of continuity agrees with that given in 25.3. As an instructive example, consider the function id: $C \to C$ where C is the set of Example 25.5.2(v). It is a mapping $(C, d_2) \to (C, d_1)$ of metric spaces, by the conclusion from 25.5.4(i); but it is not a mapping $(C, d_1) \to (C, d_2)$, since it does not map the convergent sequence $\{\text{id}^n\}$ to a convergent sequence in (C, d_2), by the last remark in 25.5.4. And, by using the commutativity of lim with mappings, it is easy to prove that 'a map of a map is a map', in the sense that if f: $X \to Y$, g: $Y \to Z$ are continuous, then so is $g \circ f: X \to Z$. As in 25.3, a mapping $f: X \to Y$ is a **homeomorphism** iff it is bijective and f^{-1}: $Y \to X$ is a mapping.

25.5.7 We can now tidy the loose ends in 25.4.1 by taking the metric on each product of metric spaces (A, a), (B, b) to be the product metric given by Example 25.5.2(vi). For example, let us prove that in Example 25.4.1(i) the function f: $\mathbb{R}^3 \to \mathbb{R}^2 \times \mathbb{R}$ is a homeomorphism, where $\mathbb{R}^3$ has its natural metric as in Example 25.5.2(iii), and $\mathbb{R}^2 \times \mathbb{R}$ has the product c of the natural metrics, b, a on $\mathbb{R}^2$, $\mathbb{R}$. We must first prove that f is continuous at a given $p = (x, y, z) \in \mathbb{R}^3$, so we consider any sequence of points (x_n, y_n, z_n) in $\mathbb{R}^3$ converging to p; thus the sequence of numbers

$$\sqrt{((x - x_n)^2 + (y - y_n)^2 + (z - z_n)^2)}$$

converges to zero in the usual sense. But each such number is

$$\sqrt{(b(q_n, q)^2 + a(z_n, z)^2)}, \qquad (q_n = (x_n, y_n),\ q = (x, y) \in \mathbb{R}^2),$$

which is $c((q_n, z_n), (q, z)), = c(f(x_n, y_n, z_n), f(p))$. Therefore the sequence $\{f(x_n, y_n, z_n)\}$ converges to $f(p)$ in $\mathbb{R}^2 \times \mathbb{R}$, so f is continuous at p as required. The continuity of f^{-1} is proved similarly, so f is a homeomorphism. Similarly for the other functions in 25.4.1. We often do not distinguish between a space and a homeomorph, since for topological purposes they are indistinguishable; it is convenient and harmless to write, for example $\mathbb{R}^3 = \mathbb{R}^2 \times \mathbb{R}$, instead of the correct $\mathbb{R}^3 \approx \mathbb{R}^2 \times \mathbb{R}$. It is then possible to say of the double pendulum described prior to 25.5.1 that its set P of states is the Cartesian product of the metric spaces T (the torus) and $\mathbb{R}^2$, where T is regarded as a subspace (in the sense of Example 25.5.2(iv)) of $\mathbb{R}^3$. Thus P is now a metric space, called the 'phase space' of the double pendulum, and we regard it as a subspace of $\mathbb{R}^5$, because of our identification of $\mathbb{R}^5$ with $\mathbb{R}^3 \times \mathbb{R}^2$.

Exercise 2

(i) Prove the statement in 25.5.6 that a map of a map is a map. {Cf. 28.2.1.}
(ii) Prove that f^{-1} in 25.5.7 is a mapping.
(iii) Let r denote the 'taxi-driver's metric' on the set $\mathbb{R}^k$, given by

$$r((x_1, \ldots, x_k), (y_1, \ldots, y_k)) = \sum_{i=1}^{k} |x_i - y_i|.$$

Show that the identity function on $\mathbb{R}^k$ is a homeomorphism $(\mathbb{R}^k, r) \approx (\mathbb{R}^k, d)$ with d as in Example 25.5.1(iii). [Thus, from the point of view of topology (but not from those of a crow or a taxi-driver, or his paying passenger) we can use either metric on $\mathbb{R}^k$.]

25.6 Closed and Open Sets

Let us now take a brief look at the 'pure' topology of a metric space (X, d). When we look for inspiration at $\mathbb{R}$, we see closed and open intervals, i.e., the sets of the forms

25.6.1 $$\{x | a \leqslant x \leqslant b\}, \qquad \{x | a < x < b\},$$

denoted by $\langle\!\langle a, b \rangle\!\rangle$ and $\langle a, b \rangle$ respectively. And when we compare $\langle\!\langle 0, 1 \rangle\!\rangle$ with $\langle 0, 1 \rangle$ we see that the sequence $\frac{1}{2}, \frac{1}{3}, \ldots, 1/n, \ldots$ of points in each interval converges to zero, which lies in $\langle\!\langle 0, 1 \rangle\!\rangle$ but not in $\langle 0, 1 \rangle$. This leads us to the notion of a *closed set* in the metric space X; F is **closed in** X iff F contains all its limit points, i.e., for each sequence $\{x_n\}$ in F which converges to some point x in X, we have $x \in F$. *This notion is relative to* X, since for example the sequence considered above does not converge at all in the space $\langle 0, 1 \rangle$; so although $\langle 0, 1 \rangle$ is closed in itself, it is not closed as a subset of $\langle\!\langle 0, 1 \rangle\!\rangle$. More generally, the whole space X is closed in itself; and if $x \in X$, then the singleton $\{x\}$ is closed in X because the only sequence in $\{x\}$ is the constant one such that $x_n = x$ for all n; and this converges to $x \in \{x\}$. Now let $f: X \to Y$ be a mapping of metric spaces. It need not happen that f maps a closed set of X onto a closed set of Y; for example, take $X = \langle 0, 1 \rangle$, $Y = \langle\!\langle 0, 1 \rangle\!\rangle$ and let f be the inclusion map. Here X is closed in X (as seen above), and f maps X onto itself, a non-closed set in Y. For the general case, however:

25.6.2 THEOREM. *The inverse image of a closed subset of* Y *under a (continuous) mapping* f *is closed in* X. This means: given a mapping $f: X \to Y$ and a *closed* $A \subseteq Y$, then† $f^{-1}(A)$ is closed in X. We omit the (easy) proof, which

† We now use $f^{-1}(A)$ rather than $f^\flat(A)$ because the former is conventional among topologists.

follows from the meaning of continuity. ■ Later on (Theorem 25.6.11) we prove the dual of this theorem.

25.6.3 EXAMPLES. Let $g: \mathbb{R}^2 \to \mathbb{R}$ be given by $g(x, y) = x^2 + y^2$; then for any positive $a \in \mathbb{R}$, the circle $x^2 + y^2 = a$ is closed in $\mathbb{R}^2$. For the circle is $g^{-1}(A)$, where A is the singleton $\{a\}$—a closed subset of $\mathbb{R}$, by a remark above. Similarly the disc $D: x^2 + y^2 \leqslant a^2$ is closed in $\mathbb{R}^2$, since it is $g^{-1}(B)$ where B is the subset of $\mathbb{R}$ consisting of all $x \leqslant a^2$; and to prove that B is closed in $\mathbb{R}$ is a simple exercise on sequences of real numbers. Similarly for parabolae, ellipses and hyperbolae, and indeed for any other subsets of $\mathbb{R}^2$ of the form $h(x, y) = a$ (or $\leqslant a$) where h is a continuous real-valued function. In particular, a line and a half-space (of the forms $ax + by + c = 0$, $ax + by + c \leqslant 0$) are closed subsets of $\mathbb{R}^2$. Similarly, too, for subsets of $\mathbb{R}^k$; and, needless to say, a closed interval $\langle\!\langle a, b \rangle\!\rangle$ is a closed subset of $\mathbb{R}$.

It is also not hard to prove:

25.6.4 THEOREM. *The union of a finite number, and the intersection of any number, of closed sets of X is closed in X.* ■

Exercise 3

(i) Prove that a closed interval $\langle\!\langle a, b \rangle\!\rangle$ in $\mathbb{R}$ is a closed set. Similarly for the sets $\{x | a \leqslant x < \infty\} = \langle\!\langle a, \infty \rangle$ and $\langle -\infty, a \rangle\!\rangle = \{x | -\infty < x \leqslant a\}$ in $\mathbb{R}$.
(ii) Prove Theorems 25.6.2 and 25.6.4.
(iii) Show that a closed subset of $\mathbb{R}^2$ remains closed when $\mathbb{R}^2$ is regarded as embedded in $\mathbb{R}^3$.

25.6.5 A set G in X is said to be **open** in X, iff $X - G$ is closed in X. One should not think that 'open' is the opposite of 'closed', since for example $\langle 0, 1 \rangle\!\rangle$ is neither open nor closed in $\mathbb{R}$. From Theorem 25.6.4 and the rules of complementation in the algebra of sets, it follows that:

25.6.6 THEOREM. *The intersection of a finite number, and the union of any number, of open sets of X is open in X.* ■

Remark. The 'pure' topology of a metric space X is concerned only with the open sets and closed sets of X. Certainly many notions commonly introduced —indeed, introduced by us!—for metric spaces (e.g., continuity of functions, convergence) can be defined without reference to the underlying metric. When we ignore the metric and concern ourselves only with the open sets and closed sets of X, we call X a **topological space** (which is **metrizable** by the given metric). At this point a generalization immediately suggests itself; why not *start* from the concept of a topological space? Thus we could consider a set X and a family of subsets of X which we call the open sets. This family would have to satisfy certain axioms (for example, Theorem 25.6.6 would appear as an axiom, and, indeed, the only other axiom we would require is that X and $\varnothing$ be open). Then we could do topology! This point of view of modern topology finds expression, and some justification, in 25.9.

25.6.7 Definition 25.6.5 above yielded a 'duality' between the open and the closed subsets of X. An alternative, and sometimes more useful, definition of open sets is the following. By an '**r-neighbourhood**' of $x \in X$ (where $r > 0$) we mean the set of all $y \in X$ within a distance r of x; thus $d(x, y) < r$. We denote this neighbourhood by $U_r(x)$. By $\mathfrak{D}_1$, $x \in U_r(x)$. In $\mathbb{R}^2$, $U_r(x)$ is a circular disc minus its boundary, and in $\mathbb{R}$ it is the open interval $\langle x - r, x + r \rangle$. Then:

25.6.8 THEOREM. *A subset G of X is open iff every $x \in G$ has some neighbourhood lying entirely in G.*

Again we omit the easy proof. ■

Remark. The concepts of *closed set*, *open set*, *neighbourhood* are fundamental to the definition of a general topological space. Indeed, Theorem 25.6.8 can then serve as a definition of an open set if the topology of the space is described by giving a system of neighbourhoods. The reader is warned, however, that the definition we have given of a closed set in the metric space X does *not* describe a property enjoyed by closed sets in an arbitrary topological space. On the other hand, if X is regarded as a topological space, then its closed sets as a topological space coincide precisely with its closed sets as a metric space according to our definition. See, e.g., Alexandroff [2] or Newman [94].

25.6.9 EXAMPLES. (i) An open interval $\langle a, b \rangle$ is open in $\mathbb{R}$, for if $x \in \langle a, b \rangle$, then $x - a$ and $b - x$ are both >0; so $U_r(x)$ lies in $\langle a, b \rangle$ for any $r \leqslant \min(x - a, b - x)$. The exterior and interior of the disc D in 25.6.3 are clearly open by the first definition of 'open'; however, we will use Theorem 25.6.8 to show that the interior, $D_1: x^2 + y^2 < a^2$, is open. To see this, we note that each point p in this set has $g(p) < a^2$, $g: \mathbb{R}^2 \to \mathbb{R}$, so we can take $U_r(p)$ with $r < a - \sqrt{g(p)}$; then if $y \in U_r(p)$ we have (since $d(0, y) = \sqrt{g(y)}$):

$$d(0, y) < d(0, p) + d(p, y) \qquad \text{(triangle inequality)}$$

which is $<a$, so $U_r(p)$ lies entirely in D_1.

(ii) Since the last proof used only the properties of a metric, we see that in any metric space X, $U_r(x)$ is open. Also, X is open in itself. Of course, in $\mathbb{R}^k$, a set $U_r(x)$ is a solid ball, but r-neighbourhoods in a general space need not have such a simple structure (see Fig. 25.5). For this reason, it is convenient to define a **neighbourhood** of $x \in X$ as any open set containing x. Observe that we can now use open sets to characterize convergence, as follows: $\lim x_n = x$ *in* X iff *every neighbourhood V of x contains almost all x_n* ('almost all' means 'all but a finite number').

Proof. By definition, $\lim x_n = x \Leftrightarrow d(x_n, x) \to 0$ in $\mathbb{R}$,

$$\Leftrightarrow \forall r > 0, \qquad \text{almost all numbers } d(x_n, x) \text{ lie within } r \text{ of } 0$$

$$\Leftrightarrow \forall r > 0, \qquad \text{almost all } x_n \in U_r(x).$$

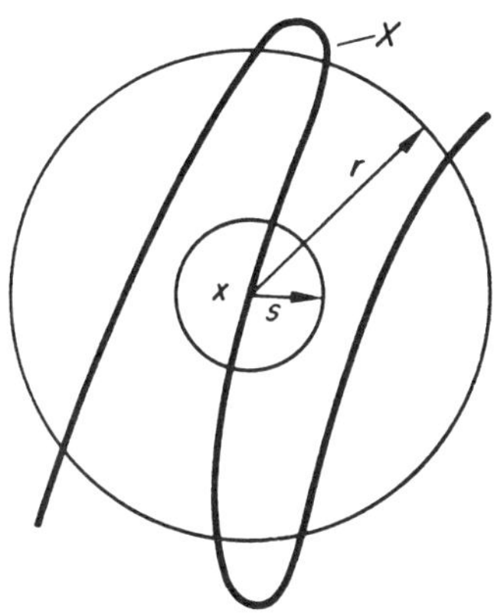

Fig. 25.5 $U(x, r)$ has three components. $U(x, s)$ has just one. $\mathrm{Fr}(U(x, r))$ consists of six points.

But every neighbourhood V of x contains some $U_r(x)$, so if $\lim x_n = x$, then almost all x_n lie in $U_r(x)$ and hence in V. Conversely, if almost all x_n lie in each V, then in particular every $U_r(x)$ contains almost all x_n, whence $\lim x_n = x$. ■

(iii) For brevity we say that a family of neighbourhoods of x is ***arbitrarily small*** iff, given any $U_r(x)$, where r can be taken as small as we like, some member of the family lies in $U_r(x)$. For example, if x lies in the k-sphere $\mathbb{S}^k$ $(k > 0)$ of Example 25.5.2(iv), then x has arbitrarily small neighbourhoods V which are homeomorphic to $\mathbb{R}^k$; for if $y \in \mathbb{S}^k$ be the point diametrically opposite to x, then we can project $\mathbb{S}^k - y$ stereographically, say by $f: \mathbb{S}^k - y \to P$, onto the tangent plane P at x (see Fig. 25.13). Now P is homeomorphic to $\mathbb{R}^k$, so x has arbitrarily small neighbourhoods $U_r(x)$ in P, each homeomorphic to $\mathbb{R}^k$. Then $f^{-1}U_r(x)$ is a neighbourhood of x in $\mathbb{S}^k$ and can be made as small as we like by taking r small, since f^{-1} is a function and continuous.

25.6.10 The **frontier**, $\mathrm{Fr}(A)$, of a set A in X is defined to be the set of all x such that every neighbourhood of x meets both A and $X - A$. For simple sets like a solid ball B in $\mathbb{R}^3$, $\mathrm{Fr}(B)$ is also simple (here it is $\approx \mathbb{S}^2$) but caution is necessary. For example, if A is a curve, say an ellipse in $\mathbb{R}^2$, then $\mathrm{Fr}(A) = A$, and if we delete a radius from the set D_1 in Example 25.6.9(i), the frontier of the resulting set consists of a circle and the radius. And $\mathrm{Fr}(X) = \varnothing$ (in X); for no set meets the empty set $\varnothing$.

The **closure**, $\mathrm{Cl}(A)$ of A in X is defined to be the set of all points lying in A or in $\mathrm{Fr}(A)$ (or both). Thus $x \in \mathrm{Cl}(A)$ iff every neighbourhood of x meets A. It turns out that A is closed iff $A = \mathrm{Cl}(A)$.

Exercise 4

(i) Prove that in X, $\mathrm{Fr}(U_r(x))$ is the set of all $y \in X$ such that $d(x, y) = r$.
(ii) Show that if $x \in \mathbb{Q}$, then $\mathrm{Fr}(U(x))$ is a pair of points or empty, according as r

is rational or irrational; but regarding x as a point of $\mathbb{R}$, then $\mathrm{Fr}(U_r(x))$ is always a pair of points. (In the first case, use the notation $U_r(x, \mathbb{Q})$ to prevent confusion.) Prove that in $\mathbb{R}$, $\mathrm{Fr}(\mathbb{Q}) = \mathbb{R}$, but that in $\mathbb{Q}$, $\mathrm{Fr}(\mathbb{Q}) = \varnothing$.
(iii) Proceed as in (ii) but replace $\mathbb{Q}$ by the set of irrational numbers in $\mathbb{R}$.
(iv) Let A be the graph of the function $y = x \sin 1/x$ $(x > 0)$ in $\mathbb{R}^2$. Show that $\mathrm{Cl}(A)$ is A plus the origin. Discuss the graph B of the function $y = \sin 1/x$ $(x > 0)$. (The set B bristles with strange features, and is commonly used for counter-examples.)
(v) Prove that for any set G open in X,

$$\mathrm{Fr}(G) = \mathrm{Cl}(G) - G.$$

(vi) Prove the statement above that, for any A in X, A is closed iff $\mathrm{Cl}(A) = A$. Show that, for any A, $\mathrm{Cl}(A)$ is closed and that it is, in fact, the intersection of all closed sets containing A.
(vii) Does an open subset of $\mathbb{R}^2$ remain open when $\mathbb{R}^2$ is embedded in $\mathbb{R}^3$ (compare Exercise 3(iii))?

The duality between open and closed sets mentioned in 25.6.7 has probably led the reader to suspect that Theorem 25.6.2 holds with 'closed' replaced by 'open'. This 'dual' is indeed a true theorem. In fact, we now prove that there are four natural properties of a function $f: X \to Y$, each of which characterizes continuity. We particularly draw attention to the form of the following theorem and the framework of its proof, since they are typical of many formulations in modern mathematics.

25.6.11 THEOREM. *Let X, Y be topological spaces and $f: X \to Y$ a function. Then the following four statements are equivalent.*

(i) *$\{fx_n\} \to fx$ whenever $\{x_n\} \to x$;*
(ii) *$f^{-1}F$ is† closed in X whenever F is closed in Y;*
(iii) *$f^{-1}G$ is open in X whenever G is open in Y;*
(iv) *For any $x \in X$ and neighbourhood $V_{f(x)}$ of $f(x)$ in Y, there exists a neighbourhood U_x of x in X with $f(U) \subseteq V$.*

Proof. We show that (i) $\Rightarrow$ (ii) $\Rightarrow$ (iii) $\Rightarrow$ (iv) $\Rightarrow$ (i). This sequence of implications plainly establishes the equivalence of the four assertions.

(i) $\Rightarrow$ (ii). Recall that $X_0 \subseteq X$ is closed iff, for every convergent sequence $\{x_n\}$ with $x_n \in X_0$, then $\lim \{x_n\} \in X_0$. Thus let $\{x_n\} \to x$ with $x_n \in f^{-1}F$. Then $fx_n \in F$ and, by (i), $\{fx_n\} \to fx$. Since F is closed, $fx \in F$, so $x \in f^{-1}F$ and so $f^{-1}F$ is closed.

(ii) $\Rightarrow$ (iii). In fact, obviously (ii) $\Leftrightarrow$ (iii) since $f^{-1}\complement = \complement f^{-1}$.

(iii) $\Rightarrow$ (iv). Now $f^{-1}V$ is, by (iii), an open set of X and certainly contains x. Thus some neighbourhood U_x of x is contained in $f^{-1}V$, so that $f(U_x) \subseteq V$.

(iv) $\Rightarrow$ (i). We must show that every neighbourhood V of fx contains almost all points of the sequence $\{fx_n\}$. Let U be a neighbourhood of x with

† See footnote to 25.6.2.

$f(U) \subseteq V$; such a U exists, by (iv). Since $\lim x_n = x$, then by the characterization of convergence in Example 25.6.9(ii) almost every x_n belongs to U, so that almost every fx_n belongs to $f(U)$ and thus certainly to V. ■

We may now use any of the four statements (i), (ii), (iii), (iv) to characterize a continuous function. In particular, (iv) is the dreaded 'ε–δ' definition of continuity, of the books on analysis, so often the stumbling-block to an understanding of the calculus (see 27.2.7 ff.). Thus, $V_{f(x)}$ is, for some ε, an ε-neighbourhood of $f(x)$, and U_x will then be a δ-neighbourhood of x, for a positive number δ depending on ε. Hence (iv) may be translated as follows: 'Given any ε, there exists δ such that, if x' is within δ of x, then fx' is within ε of fx'. Of course this statement defines **continuity at** x, and f is said to be **continuous** if it is continuous at each $x \in X$.

25.6.12 In Fig. 25.6 we show how statement (ii) of Theorem 25.6.11 may be used to give a vivid proof of an earlier assertion, that a map of a map is a map. The sketches show that for *any* closed F in Z, $(g \circ f)^{-1}(F)$ is closed in X, so $g \circ f$ is continuous. Statement (iv) in Theorem 25.6.11 is useful for Exercise 5(i) below, while statement (i) is useful in 'practical' cases like Exercise 5(ii).

For a detailed treatment, see Newman [94] or Patterson [101].

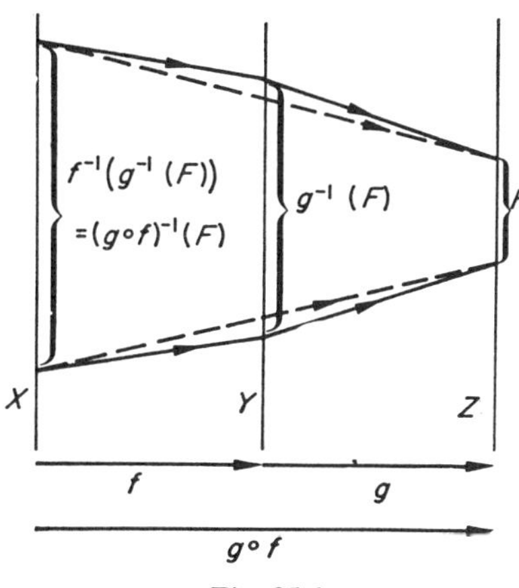

Fig. 25.6

Exercise 5

(i) Let X have metrics d_1, d_2 and let $U_r(x)$, $V_r(x)$ denote r-neighbourhoods in these two metrics, respectively. Prove that $\mathrm{id}_X\colon (X, d_1) \to (X, d_2)$ is continuous iff each $V_r(x)$ contains some $U_s(x)$ (with $s > 0$). (Cf. 25.5.6.) Thus different choices of metric will give homeomorphic spaces (X, d_1), (X, d_2) iff each $V_r(x)$ contains some $U_s(x)$, and each $U_t(x)$ contains some $V_q(x)$. This extends the point made in Exercise 2(iii).

(ii) By the Taylor expansion (see 34.1), $\log(1 + 1/n) = 1/n - A/n^2, n = 1, 2, \ldots,$ where A is a number between 0 and 1, depending on n. Hence use statement (i) of Theorem 25.6.11, with f taken to be the exponential function, to show that $\lim (1 + 1/n)^n = e$.

25.7 Dimension

In any metric space X we can define the notion of dimension. Until the early 20th century, mathematicians had thought that a k-dimensional space should mean 'something with k-parameters', or (in modern parlance) any continuous image of an open set in $\mathbb{R}^k$. But Peano gave examples (the 'space-filling curves', see Fig. 25.7) of mappings of the unit interval onto the unit

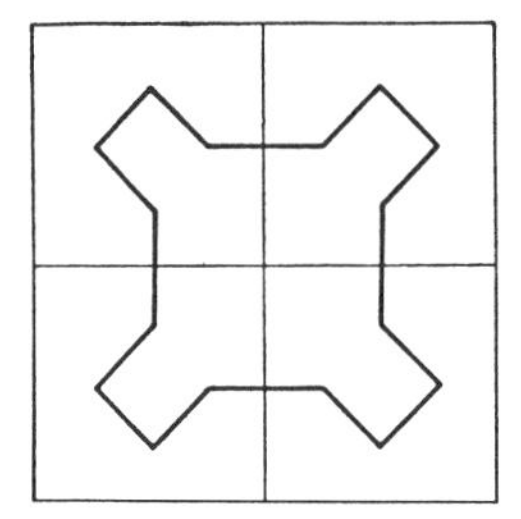
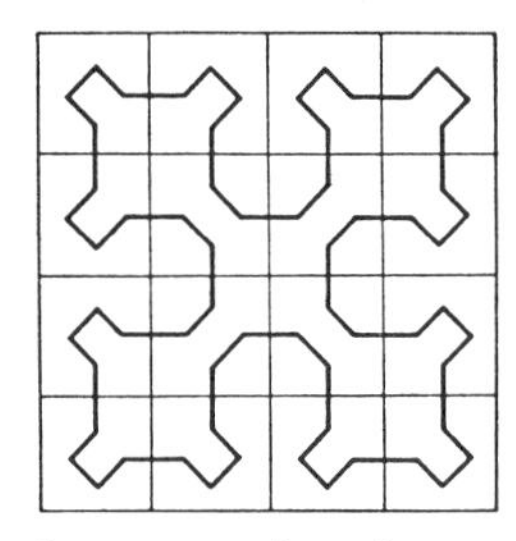

Fig. 25.7 Successive steps in the construction of a space-filling curve. The result after n steps is a Jordan curve which passes within 2^{-n} of each point of the unit square. A limiting process then gives a continuous map of $\mathbb{I}$ onto the whole square; this map is *not* a Jordan curve.

square, so the square would have dimension 1 on the old definition; whereas by any definition a line 'ought' to have dimension 1, a square and $\mathbb{R}^2$ dimension 2, $\mathbb{R}^3$ dimension 3, and so on. The great French mathematician Henri Poincaré (1854–1912), who founded 'algebraic topology' (see Bell [10], Ch. 28; and 25.11 below) gave the first satisfactory definition, and others worked out a theory (see Hurewicz–Wallman [67] for details). Essentially, Poincaré's idea was this: if two fleas A, B are constrained to move only in $\mathbb{R}$, then B can be shielded (see Fig. 25.8) from A by a pair of points, i.e., by a set which 'ought' to be called zero-dimensional. If instead, A and B were in $\mathbb{R}^2$, then the shielding can be supplied by a circle or square, i.e., a set which 'ought' to be called one-dimensional. And in $\mathbb{R}^3$ the shielding is by a room of some shape, i.e., by 'two-dimensional' walls.

25.7.1 Thus, we can use induction; the empty set is declared to be of dimension -1, and X will be said to have dimension $< n + 1$ at $x \in X$ (written $\dim_x X < n + 1$) iff x has arbitrarily small neighbourhoods U such that the frontier of U (see 25.6.10) has dimension $< n$ at each of its points. Finally, $\dim_x X = n$ iff $\dim_x X < n + 1$ and it is false that $\dim_x X < n$.

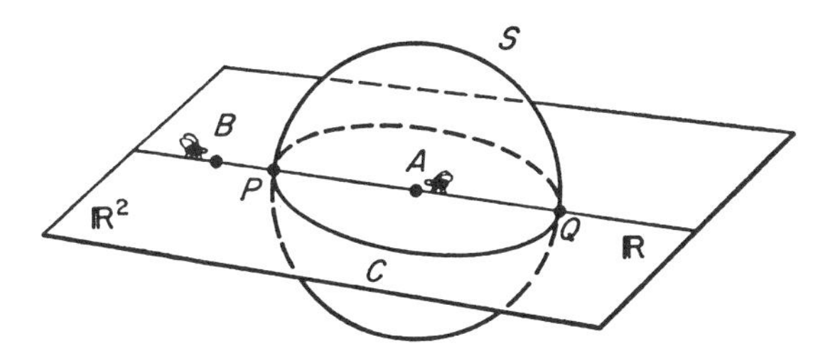

Fig. 25.8 The fleas A and B are shielded from each other on the line $\mathbb{R}$ by the two points P, Q; on the plane $\mathbb{R}^2$ by the circle C; and in space by the sphere S.

[*Caution:* The point of not using equality signs throughout is to exclude the initial possibility that x might otherwise have two sets of arbitrarily small neighbourhoods, one with n-dimensional frontiers and the other with frontiers of dimension $m \neq n$.]

It turns out that $\mathbb{R}^k$ has dimension k at each point (so we write $\dim \mathbb{R}^k = k$), and that all familiar spaces have the 'right' dimension. But the proof is not trivial, because of the possibility in the *Caution* above that besides the solid spherical neighbourhoods $U_r(x)$ (whose frontiers are $(k-1)$-dimensional spheres) there might be queer neighbourhoods whose frontiers have the wrong dimension; and the difficulty is to show that such queer ones do not exist. Observe that dimension is a topological property, because homeomorphisms map arbitrarily small neighbourhoods and frontiers into items of the same kind. Thus, for example, $\mathbb{R}^p \not\approx \mathbb{R}^q$ unless $p = q$. Observe, too, that if $x \in X$ and $y \in Y$ have neighbourhoods U, V respectively such that $U \approx V$ with $x \to y$, then $\dim_x X = \dim_y Y$; for the definition of dimension depends only on small neighbourhoods—it is a 'local' property. Hence it should not be surprising that a spherical surface has dimension 2, although many people seem to feel that it is three-dimensional. The confusion arises because of the difficulty of imagining a sphere separately from our own three-dimensional containing space; and this was an inhibition which the early topologists had to overcome by treating spaces intrinsically, rather than as subsets of something bigger. Such an intrinsic treatment is vital for a study of our own universe, for example. Incidentally, there was also an inhibition about spaces of dimension > 3, but that battle had already been more or less settled by the 19th-century algebraists, who were quite familiar with vector spaces of n 'dimensions' (i.e., 'degrees of freedom').

Other definitions of dimension exist, the most useful depending on the bricklayer's observation that if a wall is covered by sufficiently small tiles, at least three must intersect; and if a volume is packed with sufficiently small boxes, at least four must intersect. Thus X can be defined to be n-dimensional iff in every covering of X by sufficiently small open sets, at least $n + 1$ intersect. It turns out that this definition agrees with the previous one in all important cases.

Finally, there are infinite-dimensional (i.e., not finite-dimensional) spaces. For example, the spaces (C, d_1), (C, d_2) in Example 25.5.2 are not finite-dimensional; roughly speaking, the Fourier coefficients (see 34.5.7) of each function $f \in C$ give f an infinity of co-ordinates.

Exercise 6

(i) Show that $\mathbb{Q}$ and $\mathbb{R} - \mathbb{Q}$ each has dimension zero. [See Exercise 4(ii).]
(ii) Show that each letter of the alphabet has dimension 0, 1, 2, or 3, according to the thickness of the print. [It seems most reasonable to call the letters one-dimensional.]
(iii) Make up intuitive 'proofs' that (a) if A is a subset of a k-dimensional space, then $\dim A \leqslant k$; (b) $\dim (A \times B) = \dim A + \dim B$. [This equation does not hold for certain 'pathological' spaces A, B.]

25.8 Compact Spaces

Among the closed subsets of a space, the most interesting are those which always map into closed sets. This leads to the definition:

25.8.1 DEFINITION. The metric space X is **(sequentially) compact** iff, given any infinite sequence $\{x_n\}$ in X, some subsequence $x_{n(1)}, x_{n(2)}, \ldots,$ of $\{x_n\}$ converges to a limit in X.

We shall omit the word 'sequentially' in this section, but we shall give a second definition of compactness in the next section which has nothing to do with sequences. This second definition is almost universally adopted by topologists today and is equivalent to sequential compactness for metric spaces.

Since everything refers to X, there is no question of compactness depending on a containing space. Examples of *non*-compact spaces are $\mathbb{R}$ (since the set of integers has no limit point: 'infinity' is not in $\mathbb{R}$), $\mathbb{R}^k$, a hyperbola, and in fact any 'unbounded' space Y, i.e., such that there is a sequence $\{y_n\}$ with $d(y_1, y_n) > n$. Examples of compact spaces are not easily proved to be compact until we have quoted some theorems.

First it is shown in books on analysis that:

25.8.2 THEOREM (of Bolzano and Weierstrass).

Every closed interval $\langle\!\langle a, b \rangle\!\rangle$ in $\mathbb{R}$ is compact. □

Secondly, it is easy to prove directly that:

25.8.3 THEOREM. *If X, Y are compact spaces, so is their Cartesian product $X \times Y$.*

Proof. Briefly, if (x_n, y_n) is an infinite sequence in $X \times Y$, then some sub-sequence $x_{n(1)}, x_{n(2)}, \ldots,$ of $\{x_n\}$ converges to $p \in X$, and then some sub-sequence $y_{m(1)}, y_{m(2)}, \ldots,$ of $y_{n(1)}, y_{n(2)}, \ldots,$ converges to $q \in Y$ while $x_{m(1)},$

$x_{m(2)}, \ldots$, also converges to p. Hence $(x_{m(1)}, y_{m(1)}), (x_{m(2)}, y_{m(2)}), \ldots$, is the required convergent sub-sequence of $\{(x_n, y_n)\}$ in $X \times Y$, with limit (p, q). ■

A corollary is that any closed rectangular disc S in $\mathbb{R}^2$ is compact, since S is the Cartesian product of two closed intervals (and so we use Theorems 25.8.2, 25.8.3). Hence by 25.8.3 again, any closed rectangular box B in $\mathbb{R}^3$ is compact, since B is of the form $I \times S$, where I is a closed interval in $\mathbb{R}$; similarly for higher-dimensional boxes in $\mathbb{R}^k$. For a general compact metric space X we quote the following results: proofs are in most of the books listed in 25.14.

25.8.4 THEOREM. *If F is a compact subset of the space X, then F is closed in X.* □

25.8.5 THEOREM. *Every closed subset of a compact space X is itself compact.* □ [(Recall from Example 25.5.2(iv) that the subset is a metric space, using the same metric as in X.] A special corollary for $\mathbb{R}^k$ is:

25.8.6 COROLLARY. *Every closed bounded subset F of $\mathbb{R}^k$ is compact.*

Proof. F then lies in some box B, by definition of 'bounded', and then F, being the intersection of itself and B, is closed in B. We saw above that B is compact, whence, by Theorem 25.8.5, so is F. ■

25.8.7 EXAMPLE. An ellipse E in $\mathbb{R}^2$ is compact, because, by Examples 25.6.3, E is closed, and E is bounded by a square of side twice the major axis. Similarly the $(k-1)$-sphere $\mathbb{S}^{k-1}$ is compact, being closed in $\mathbb{R}^k$ (cf. Examples 25.6.3) and bounded. Hence by Theorem 25.8.3 a torus T is compact, for $T \stackrel{.}{\approx} \mathbb{S}^1 \times \mathbb{S}^1$ (see Examples 25.4.1(ii)); similarly for an annulus ($\stackrel{.}{\approx}$ cylinder) and a doughnut.

It is plain that compactness is a topological property: any space homeomorphic to a compact space is itself compact. More generally one can prove

25.8.8 THEOREM. *Every continuous image of a compact X is itself compact; i.e., if $f: X \to Y$ is continuous and surjective, then Y is compact.* □

We also have a corollary in analysis (called the 'mostest theorem' in 28.5.2): *If $f: X \to \mathbb{R}$ is continuous, and X is compact, then f has a maximum and a minimum in X* (i.e., there exist points $a, b \in X$ such that $f(a)$ is the greatest and $f(b)$ the least value of f). For, the image of f is compact, and therefore a bounded subset of $\mathbb{R}$; so it has a least upper bound and a greatest lower bound.

An important result from dimension theory is the following, which explains the importance of a study of the topology of Euclidean space.

25.8.9 THEOREM. *Every n-dimensional compact metric space X can be embedded in $\mathbb{R}^{2n+1}$, i.e., X is homeomorphic to some (compact) subset of $\mathbb{R}^{2n+1}$.* □ For a proof see Hurewicz–Wallman [67], Ch. V, theorem V.3.

The number $2n + 1$ is 'best possible', because it can be shown that, for example, the set consisting of a pentagon together with five non-intersecting arcs joining pairs of vertices, is one-dimensional, compact and not embeddable in $\mathbb{R}^2$ (see Fig. 25.9); but by the theorem (or directly) we can construct it in $\mathbb{R}^3$, and $3 = 2\cdot 1 + 1$. Of course, for many spaces X, the estimate $2n + 1$ is crude; if X is a torus, the number $3 < 2\cdot 2 + 1$ will do (why can we not embed the torus in $\mathbb{R}^2$?). It is an unsolved problem to give a rule for deciding, for a given space X, the smallest m such that X can be embedded in $\mathbb{R}^m$.

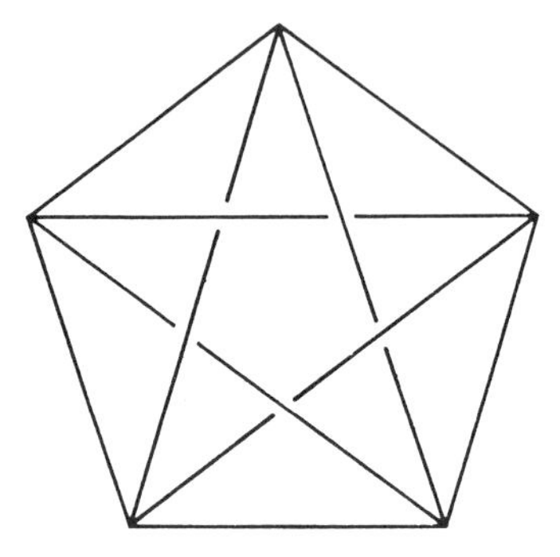

Fig. 25.9 A one-dimensional space which is not embeddable in $\mathbb{R}^2$.

Exercise 7

(i) Prove Theorem 25.8.2, using Theorem 24.6.6. {If A is an infinite subset of $\langle\!\langle a, b\rangle\!\rangle$, then one half, I_1, of $\langle\!\langle a, b\rangle\!\rangle$ contains an infinity A_1 of points of A; one half, I_2, of I_1 contains an infinity of points of A_1, etc. Show that the sequence $I_1 \supseteq I_2 \supseteq \cdots \supseteq I_n \supseteq \cdots$ of intervals contains a single common point, which is a limit point of A.}

(ii) Prove Theorems 25.8.4, 25.8.5, 25.8.8 [they are not difficult].

(iii) Let X be compact and let $f: X \to Y$, $g: X \to Z$ be mappings with f surjective. Show that if there exists a function $h: Y \to Z$ with $hf = g$, then h is a mapping. Show that the conclusion no longer holds if X is not assumed compact.

(iv) Let A denote the Cartesian product of two one-dimensional spaces shaped like the letter Y. Prove that A can be embedded in $\mathbb{R}^4$ and (harder) that A cannot be embedded in $\mathbb{R}^3$.

25.9 Quotient Spaces

Besides studying old spaces, topologists like to invent new ones. One method of construction depends on the following proposition.

25.9.1 THEOREM. *Let F_1, F_2 be compact subspaces of X with no common point. Then F_1, F_2 are a positive distance apart* (i.e., there exists $s > 0$ such that $d(x, y) \geqslant s$ whenever $x \in F_1$, $y \in F_2$). □ [This property still holds if F_1 is compact and F_2 merely closed but fails if F_1, F_2 are merely closed in X;

consider $X = \mathbb{R}^2$, $F_1 = x$-axis, $F_2 =$ one branch of $xy = c^2$.] This theorem leads naturally to the following notion. Let W be a family of compact subsets $F_1, F_2, \ldots$ of X such that no two have a common point and define

$$h(F, F) = 0$$
$$h(F_1, F_2) = \text{largest } s \text{ in } 25.9.1.$$

Then h satisfies properties $\mathfrak{D}_1$ and $\mathfrak{D}_2$ for a metric space and so (W, h) is a candidate for a metric space (we have not yet verified the triangle inequality $\mathfrak{D}_3$); moreover, if F_1, F_2 happen to be single points x, y, then $h(x, y) = d(x, y)$. Thus, W can be thought of as the result of shrinking each F, not already a single point, smaller and smaller until it becomes a point; the single points are left alone. Of course, we have to imagine the shrinking taking place in some larger space, but this is because of our psychology, and is not necessary for the construction. We shall confine ourselves here to those families W of compact subsets of X, satisfying two conditions, $\mathfrak{Q}_1$, $\mathfrak{Q}_2$ as follows.

$\mathfrak{Q}_1$: *each $x \in X$ lies in some $F \in W$.*

Now x can be in only one such F, since no two members of W intersect. If therefore we assign to x this one F, then we have a surjective function $p: X \to W$, the **natural projection.**

Unfortunately, it often happens that (W, h) does not satisfy the triangle inequality $\mathfrak{D}_3$, so (W, h) is *not*, in general, a metric space (contrary to a statement in the pamphlet [45]). Nevertheless, as the examples in 25.9.3 below show, we have a strong feeling that W ought to be a sensible space, sensibly related to X by the projection p. Let us see what we can say about W with the language so far available. We shall find a situation, familiar to working mathematicians, where extra experience forces a change in previously formulated language.

First, we can still discuss limits in W, as before: we say that the sequence $\{F_n\}$ in W converges to F in W provided $\lim h(F_n, F) = 0$ as $n \to \infty$. The limit F is unique, by the triangle inequality in (X, d) (see Exercise 8(i) below). Hence we can still talk about closed subsets of W, as in 25.6, and Theorem 25.6.4 still holds (without change in its proof). The open subsets can then be defined as in 25.6.5, and Theorem 25.6.6 holds. Thus, even if h fails to be a metric, (W, h) is a topological space in the sense of the remark following Theorem 25.6.6.

Second, we can ask if the natural projection $p: X \to W$, defined above, is continuous in the sense of 25.5.5. Thus, we must verify that if $\lim x_n = x$ in X, then $\lim px_n = px$ in W. But, for all $n \in \mathbb{N}$, $0 \leqslant h(px_n, px) \leqslant d(x_n, x)$ by definition of h (because $x \in px$), and therefore

$$\lim h(px_n, px) = \lim d(x_n, x) = 0$$

as $n \to \infty$; hence $\lim px_n = px$, so *the natural projection $p: X \to W$ is continuous.*

Next we suppose that the collection W satisfies the following condition (called 'upper-semicontinuity' in most texts on topology).

$\mathfrak{Q}_2$: *Given $F \in W$ and an open U in X such that $F \subseteq U$, there exists an open V in X, with $F \subseteq V \subseteq U$, such that for any $F' \in W$, if F' meets V, then $F' \subseteq U$.*

(This means roughly that F' does not bend away from F too sharply, once F' hits the V-neighbourhood of F.)

It can be shown—after considerable proof!—that if, for example, X is compact, then there exists a metric q on W such that every sequence which converges in (W, h) also converges to the same limit in (W, q) ,and *vice versa.* This means that (W, h) is homeomorphic to a metric space, or, as we say, that (W, h) is *metrizable.* Thus we were not too far out above, when we called (W, h) a 'candidate' for a metric space, even though h failed to satisfy the triangle inequality; for h gave us the 'right' convergent sequences and hence—what really matters in topology—the 'right' system of closed and open subsets of W.

With this system of closed and open subsets, W is called a **quotient space** (or **decomposition space**) of X, which depends of course on the choice of the particular compact sets F of the family. For *any* choice, we can use Theorem 25.8.8, with the fact proved above that $p: X \to W$ is continuous, to say at once:

25.9.2 THEOREM. *W is compact if X is compact.*

It is a good exercise for the intuition to attempt to recognize W, as the following examples show. [In each case we leave as an exercise to the reader the verification that $\mathfrak{Q}_1$, $\mathfrak{Q}_2$ are satisfied.]

25.9.3 EXAMPLES.

(i) If each $F \in W$ is just a single point, then $f: X \approx W$ because now $(x) = \{x\}$ for each $x \in X$, while

$$h(\{x\}, \{y\}) = d(x, y).$$

(ii) Let X be the unit interval $\langle 0, 1 \rangle$ and let $F \in W$ iff either F is the pair $\{0, 1\}$, or F is a single point x with $0 < x < 1$. We can think of W as the result of gluing the ends of X together, so W is homeomorphic to a circle. [Here, the function h does not satisfy the triangle inequality.]

(iii) Let D be the circular disc in $\mathbb{R}^2$, and let the family W consist of all F such that either F is the boundary S of D, or F is a single point of the interior of D. Then, thinking of W as the result of shrinking S to a point, we see it is homeomorphic to the 2-sphere.

(iv) Let $X = \mathbb{R}^2$, and let W consist of all F such that F is D, or F is a single point of $\mathbb{R}^2 - D$. Then W is the result of shrinking D to a point, so $W \approx \mathbb{R}^2$. If we alter W by exchanging D for the boundary S of D, then W is now homeomorphic to the union of $\mathbb{R}^2$ and a tangent sphere (see Fig. 25.10).

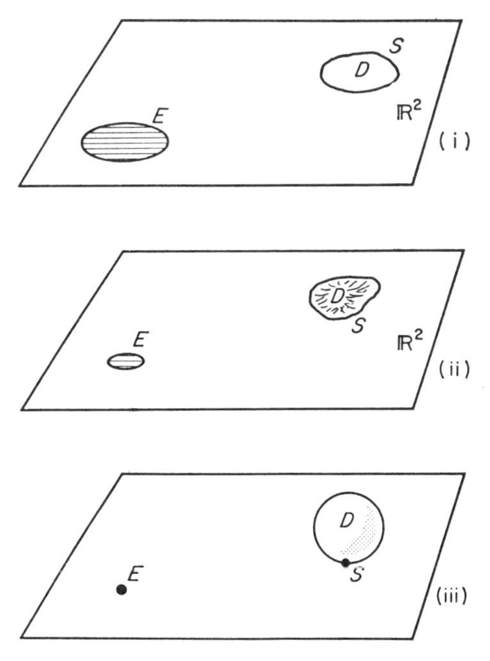

Fig. 25.10 Three stages in shrinking to a point a disc E and a circle S, in $\mathbb{R}^2$. (i) and (ii) are homeomorphic; (iii) gives the quotient space.

(v) Let X be the unit square disc $OABC$. Let W consist of all F such that either F is a point of X not on AB or OC (see Fig. 25.11) or F is a pair of points, of the form $\{(x, 1), (x, 0)\}$. Then W is homeomorphic to a cylinder (obtained by gluing the edges AB, OC together). Now alter W to W_1 by rejecting all F consisting of single points on AO, BC, and replacing them by all F consisting of pairs of the form $\{(0, y), (1, y)\}$. The result is homeomorphic to the torus, obtained by gluing together opposite ends of a cylinder. For brevity, we call 'degenerate' those F in W consisting of single points of X. Form a new family W_2 by replacing the non-degenerate elements in W by all F consisting of pairs $\{(x, 1), (1 - x, 0)\}$. Then the quotient space is the **Möbius band**, obtained by gluing AB, OC together *after* twisting OC through 180°. It has only one boundary curve, and so cannot be homeomorphic to a cylinder. And if we change W_2 to W_3 by using pairs $\{(0, y), (1, 1 - y)\}$ instead of $\{(0, y), (1, y)\}$ we obtain a quotient space called the **Klein bottle**, which cannot be embedded in $\mathbb{R}^3$. For a picture see [25], Ch. V.

(vi) Let X be the 2-sphere $\mathbb{S}^2$ in $\mathbb{R}^3$. Points $P, Q \in \mathbb{S}^2$ are called antipodal iff they are collinear with the origin. Let W consist of all F such that F is a pair of antipodal points on $\mathbb{S}^2$ (here, $h(F_1, F_2)$ is the shortest distance between a point of F_1 and one of F_2); it satisfies the triangle inequality, so (W, h) here

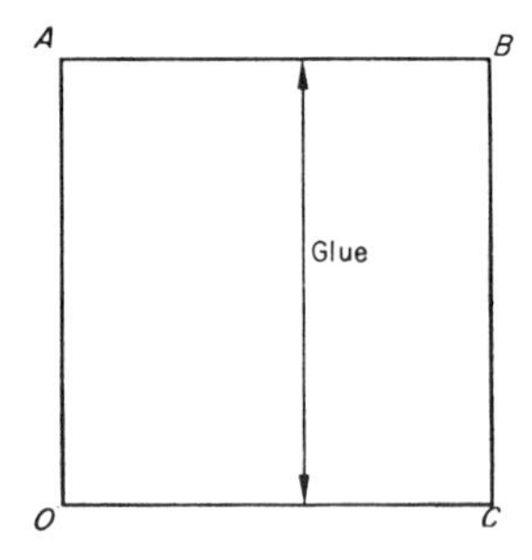

Fig. 25.11(a) A cylinder is formed when AB is glued to OC.

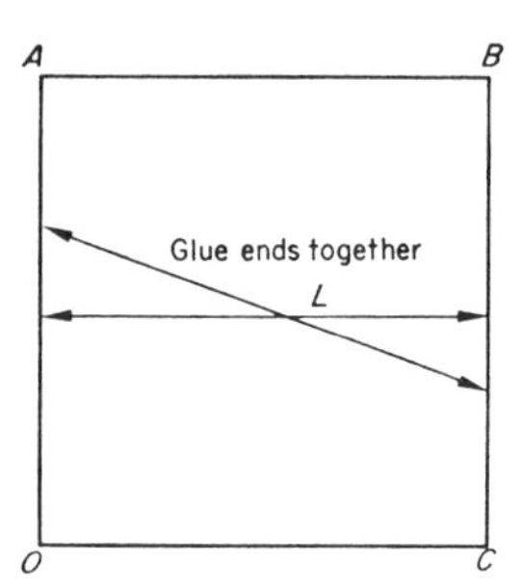

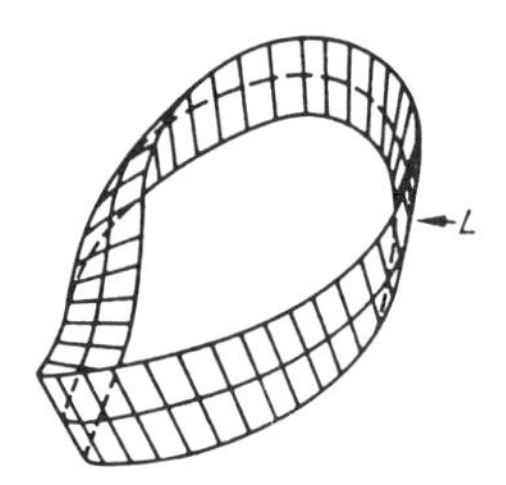

Fig. 25.11(b) Gluing to form a Möbius band. Its edge is formed from AB and OC by gluing B to O and C to A, and is a trefoil knot. The centre line is L.

is a metric space. Then (see 17.8) W is homeomorphic to the real projective plane $\mathbb{RP}^2$ of projective geometry, and by Theorems 25.8.7, 25.8.8 it is compact, since $\mathbb{S}^2$ is.

Exercise 8

(i) Find the lengths of the sides of the triangle of reference in $\mathbb{RP}^2$, using the metric in 25.9.3(vi). Show that each vertex is distant $\sqrt{2}$ from the point $[1, 1, 1]$.
(ii) Let X be compact and $f: X \rightarrow Y$ a surjective mapping. Let W be the family $\{f^{-1}(y), y \in Y\}$. Show that W satisfies $\mathbb{Q}_1, \mathbb{Q}_2$ and that $W \approx Y$. Do you recognize an analogy with a result in group theory?

(iii) Let $f: \mathbb{R}^2 \to \mathbb{R}$ denote the mapping given by $f(x, y) = x$. Show that the family $\{f^{-1}(x) | x \in \mathbb{R}\}$ satisfies $\mathfrak{Q}_1$ but not $\mathfrak{Q}_2$.
(iv) Let W be a family of compact subsets of (X, d) satisfying $\mathfrak{Q}_1$ and $\mathfrak{Q}_2$. If $F \in W$ and F contains more than one point, we call F **non-degenerate.** Suppose that there exists $c > 0$ such that every non-degenerate $F \in W$ has diameter $> c$ (i.e., $\exists x, y \in F$ such that $d(x, y) \geqslant c$). Show that if $\{F_n\}$ converges to F in W, and each F_n is non-degenerate, then also F is non-degenerate. [Observe that the hypotheses are satisfied in all the examples of 25.9.3.]

These examples should suffice to show that the construction of a quotient space is a powerful means of creating new spaces. Also, general problems, still unsolved, are immediately raised: for example, if the sets F in W have certain properties, how are these reflected in the structure of W? Under what conditions, for example, is W homeomorphic to X? These ideas are the basis of some vigorous current research.

25.9.4 Here, as in other branches of topology, the problems require the notion of a 'non-metric space', because the principal role of the metric is to introduce neighbourhoods and open sets in a space; and most proofs in this kind of topology involve the open sets without asking how they happened to be defined. In some problems the metric is a nuisance: in others it cannot be defined at all, although the open sets can. Such spaces are called *non-metrizable*, by contrast with the *metrizable* spaces in which the topology is given by a metric. Notice the distinction between a metric space and a metrizable space; the latter is a topological space whose open sets are precisely those which would be specified by some metric. Remark, however, that different metrics may give rise to the same topological space (see Exercise 2(iii)). For such reasons, approaches to topology have been devised in which metric spaces do not appear for some time; use of a metric presupposes knowledge of real numbers and sequences. However, we have preferred to introduce topological ideas through the metric in order to provide motivation; but the other approach is fundamentally simpler, requiring only basic set theory. For example, if X and Y are topological spaces, a function $f: X \to Y$ can be defined to be continuous iff each set $f^{-1}(A)$ is open in X when A is open in Y (cf. 25.6.11), thus making no mention of sequences, while compactness is generally defined as follows: Let $\mathscr{U} = \{V_\alpha\}_{\alpha \in I}$ be an indexed family of open sets V_α in the space X such that $X = \bigcup_{\alpha \in I} V_\alpha$; suppose that, in fact, for a finite subset $\{V_1, \ldots, V_n\}$ of $\mathscr{U}$, $X = V_1 \cup \cdots \cup V_n$. We then say that the **covering** $\mathscr{U}$ admits a finite subcovering.

If this holds for every such family $\mathscr{U}$ *of open sets, then* X *is called* **compact.** Briefly, 'X is compact if every open covering of X admits a finite subcovering'.

This approach could have been used in the discussion of W prior to 25.9.3; thus we would have used the projection $p: X \to W$ to *define* the open sets A of W to be those sets such that $p^{-1}(A)$ is open in X. Besides supplying W

with its 'right' system of open sets, we have now made p *automatically* continuous in the sense of the definition just mentioned. We can then study W with this 'topology' (as a 'topological space' in the sense of the remark following 25.6.6), without needing the hard theorem that W has a metric if X is compact. For example, it follows easily from these new definitions that W is compact, and more generally that Theorem 25.8.8 holds without using a metric. See also 21 Exercise 13(vii)–(ix).

Exercise 9

(i) Formulate a 'dual' definition of compactness, using only closed sets. Prove that it is equivalent to the last definition.
(ii) Prove that a metric space Y is compact in the sense just defined, iff it is compact in the sense of Definition 25.8.1. {Use (i). If $x_1, \ldots, x_n, \ldots$ is an infinite subset of Y, consider $F_n = \{x_m | m \geqslant n\}$ and show that each F_n is closed in Y. For the converse it suffices to show that if $E_1 \supseteq \cdots \supseteq E_n \supseteq E_{n+1} \supseteq \cdots$ is a sequence of closed subsets of Y, then the E_n's have a common point x. Obtain x as a limit point of a sub-sequence of $\{x_1, \ldots, x_n, \ldots\}$ where $(\forall n)\ x_n \in E_n$.}
§(iii) Let $\mathscr{U} = \{V_\alpha\}_{\alpha \in I}$ be an open covering of the compact metric space X. Show that $\exists c > 0$ such that every set of the form $U_c(x)$ lies in some $V_\alpha \in \mathscr{U}$. {If not, then $\forall n \in \mathbb{N}$, $\exists U_{1/n}(x_n)$ which lies in no V_α. Some sub-sequence $\{x_{n(i)}\}$ converges to some $x \in X$, and x lies in some $V_\beta \in \mathscr{U}$. Now show that $U_{1/n}(x_n) \subseteq V_\beta$ if n is sufficiently large, and obtain a contradiction.}
(iv) Let (X, d) be a metric space, and let $F \subseteq X$ be compact, lying in an open $V \subseteq X$. Show that $\exists c > 0$ such that the set $U_c(F)$, of all $x \in X$ within c of some $y \in F$, lies in V.

25.10 Simply Connected Spaces: Homotopy

To continue the search for topological invariants, we introduce a new property, based on the idea that some spaces have 'holes'—tunnels and chambers—and others not; and that a 'hole' is a topological invariant. Of course, we must be more precise; but how? As a preliminary, we first define a **loop** in a (metric or topological) space X to be a mapping $f: \mathbb{S}^1 \to X$; notice that the loop is *not* the image of f. The image of f may have many self-crossings, and can be thought of as an elastic band. If such a band were to move in X, its position at time t could be described by a mapping $f_t: \mathbb{S}^1 \to X$, and continuity of the motion would be ensured if the point $f_t(s) \in X$ were a continuous function of (s, t) simultaneously; also, $f_0(s) = f(s)$. If X has the property that for any such loop f, there is always a 'motion' f_t with $0 \leqslant t \leqslant 1$ such that the image of f_1 is a single point, then X is said to be **simply connected**. This corresponds in physics to the idea that we should be able to deform the band to a point *without leaving* X; it is fairly clear, for example, that we could not do this with a band surrounding a puncture in a disc, but that we could do it on $\mathbb{S}^2$ or $\mathbb{R}^2$ (see also Fig. 25.12). We show this mathematically below. However, the inherent notion of a deformation can be generalized further as follows.

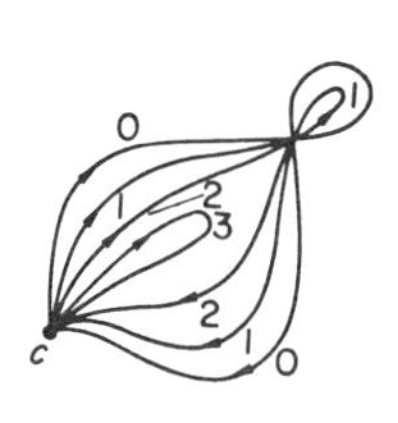

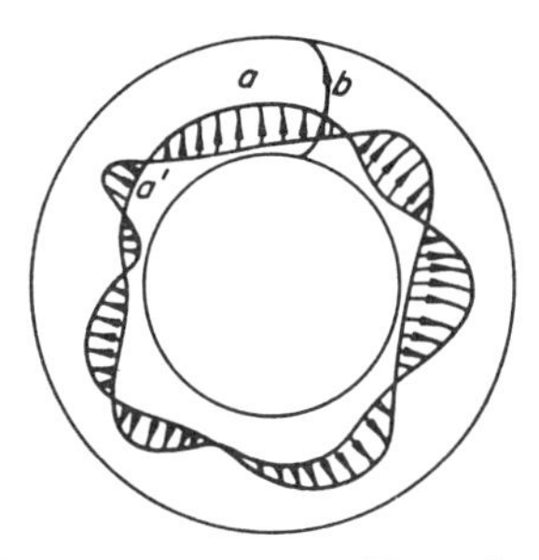

Fig. 25.12 Four stages in a homotopy of a loop ('zero') to a point. (Here, the point c is kept fixed.) The loops a, a' are homotopic on the torus T, and the small arrows give the tracks of the motion. $a \not\simeq b$ on T.

25.10.1 If $f, g: X \to Y$ are two mappings, then we say that they are **homotopic**, or **deformable** into each other (written $f \simeq g$) iff there is a mapping $h: X \times \mathbb{I} \to Y$ (where $\mathbb{I} = \langle\!\langle 0, 1 \rangle\!\rangle$) such that, for each $x \in X$,

$$h(x, 0) = f(x), \qquad h(x, 1) = g(x).$$

Thus, putting $h_t(x) = h(x, t)$, we can think of $h_t: X \to Y$ as a mapping giving the image of X in Y at time t, as this image deforms continuously from its initial position $h_0(X) = f(X)$ to its final position $h_1(X) = g(X)$ in a time interval of one second. Briefly, we write:

25.10.2 $$h_t: X \to Y, \qquad h_0 = f, h_1 = g.$$

By reversing the motion, it follows that $g \simeq f$; keeping it static gives $f \simeq f$; and if also $g \simeq h$, then by deforming f to g in half a second, and g to h in the next half-second, we obtain $f \simeq h$. So homotopy is an *equivalence relation.* In the special case g = constant (i.e., $g(x)$ is the same point $p \in Y$ for all $x \in X$, so $g(X) = p$) we write $f \simeq 0$ in Y, and say that f is *nullhomotopic.* The earlier definition of simple connectivity above requires that any loop f in X be nullhomotopic.

The branch of topology called *homotopy theory* is concerned with the questions of deciding whether given mappings are homotopic, and how many homotopically different mappings there are. The subject becomes especially difficult (see Hilton [62]) when the dimension of X exceeds that of Y; for example, the number p_n of equivalence classes (under the relation $\simeq$) of mappings $\mathbb{S}^{2n+2} \to \mathbb{S}^{n+2}$ has the following peculiar behaviour:

25.10.3

n	0	1	2	3	4	5	6	7	8	9	10	11	12	13	14	15, etc.
p_n	∞	2	2	24	1	1	2	240	4	8	6	504	1	3	4	960

□

We say that X is **contractible** iff the identity mapping id: $X \to X$ is $\simeq 0$ in X. The solid boxes and solid balls in $\mathbb{R}^k$ are all contractible, as well as $\mathbb{R}^k$

itself ($k > 0$); for, define $h_t(P) = Q$, where P, Q are collinear with the origin O of $\mathbb{R}^k$ and Q divides PO in the ratio $t:1 - t$. (Thus at $t = 1$, every point is crushed to O.)

25.10.4 THEOREM. *If X is contractible, then it is simply connected.* (For if everything can be deformed to a point in X, so in particular can any loop.) ■

The converse is false, because $\mathbb{S}^2$ can be proved to be not contractible, whereas it is simply connected. To see the latter, let L be a loop in $\mathbb{S}^2$ and deform it a little into a loop L' which does not meet the North pole N of $\mathbb{S}^2$. Now project $\mathbb{S}^2 - N$ stereographically (see Fig. 25.13) from N onto a plane $\mathbb{R}^2$; the projection of L' is nullhomotopic in $\mathbb{R}^2$ (because $\mathbb{R}^2$ is contractible), so reversing the projection gives in $\mathbb{S}^2$ a deformation of L', hence of L, into a point, so $\mathbb{S}^2$ is simply connected. Of course, contractibility is a topological invariant; for if $h: X \to Y$ is a homeomorphism with X contractible, then there is a homotopy (see 25.10.2) $g_t: X \to X$ such that $g_0 = \mathrm{id}_X$, $g_1 =$ point, whence the homotopy $f_t = h \circ g_t \circ h^{-1}: Y \to Y$ deforms $\mathrm{id}_Y: Y \to Y$ to a point in Y. Similarly, simple connectivity is a topological invariant. ■

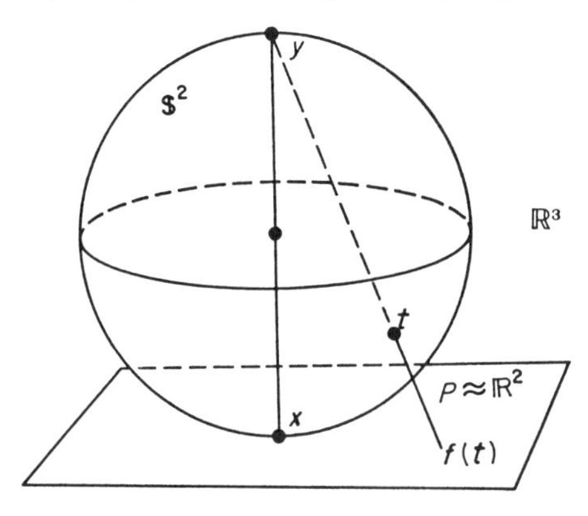

Fig. 25.13 Stereographic projection of a sphere onto a tangent plane P.

Exercise 10

(i) Prove in detail that homotopy is an equivalence relation, by finding the relevant homotopies h_t corresponding to the 'motions' mentioned after 25.10.2.
(ii) Show that any convex set in $\mathbb{R}^k$ is contractible.

Using the Brouwer fixed-point theorem for the disc D, let us now prove:

25.10.5 THEOREM. *$\mathbb{S}^1$ is not simply connected* (and hence not contractible, by Theorem 25.10.4).

Proof. We first need a general result: if $f: \mathbb{S}^1 \to X$ is nullhomotopic in X, then there is a mapping $g: D \to X$ agreeing with f on $\mathbb{S}^1$ (i.e., $g(x) = f(x)$ for every $x \in \mathbb{S}^1$). For, let $f_t\colon \mathbb{S}^1 \to X$ be a homotopy of f to 0; then using polar co-ordinates (r, θ) in D, define g by

$$g(0) = x, \qquad g(r, \theta) = f_{1-r}(1, \theta), \qquad r > 0, \quad \text{where} \quad x = f_1(1, \theta).$$

Each point on $\mathbb{S}^1$ has polar co-ordinates $(1, \theta)$, and $g(1, \theta) = f_0(1, \theta) = f(1, \theta)$, so g agrees with f on $\mathbb{S}^1$ as required. Also g is continuous since f_t and $1 - r$ are.

This fact established, we now show that $\mathbb{S}^1$ is not simply connected by finding a loop in $\mathbb{S}^1$ which is not $\simeq 0$ in $\mathbb{S}^1$. Such a loop is $q\colon \mathbb{S}^1 \to \mathbb{S}^1$ where $q(1, \theta) = (1, \theta + \pi/2)$ (so the loop is a rotation through $\pi/2$ and its image is $\mathbb{S}^1$). Suppose $q \simeq 0$ in $\mathbb{S}^1$. Then there is a mapping $g\colon D \to \mathbb{S}^1$ agreeing with q on $\mathbb{S}^1$, as shown above (taking $X = \mathbb{S}^1$). Since $\mathbb{S}^1$ lies in D, we can apply the Brouwer fixed-point theorem to g, regarded as a map $D \to D$, to say that some $P \in D$ is fixed, i.e., $P = g(P) \in \mathbb{S}^1$. But if $P \in \mathbb{S}^1$, $g(P) = q(P)$ by construction of g, so $P = q(P)$. This is impossible, since q is a rotation through $\pi/2$. From this contradiction it follows that q is not nullhomotopic in $\mathbb{S}^1$, so $\mathbb{S}^1$ is not simply connected. ∎

Notice that, in fact $q \simeq \text{id}$, so $\text{id}\colon \mathbb{S}^1 \to \mathbb{S}^1$ is not nullhomotopic. Our argument, however, required us to consider q rather than id.

25.10.6 The reader should practise eyeing shapes in real life and deciding whether or not they are simply connected. Two useful rules, whose proofs we omit, are as follows (see [62]). If X or Y is not simply connected, neither is the space Z obtained by uniting X and Y at a common point (e.g., take X, Y to be circles and Z a figure eight). If P is deformable to Q, then P is simply connected iff Q is. Hence, for example (see Fig. 25.14), a plane disc

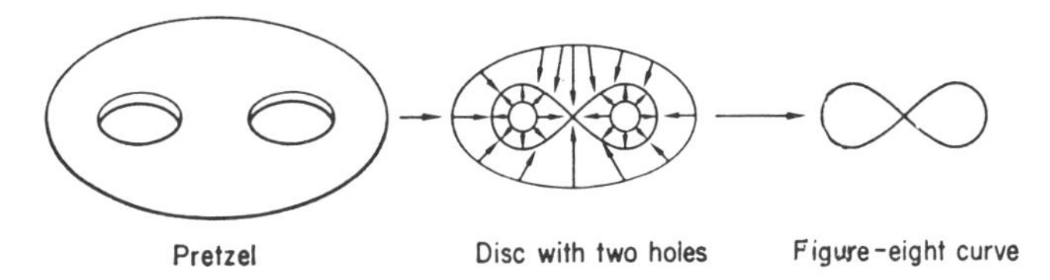

Fig. 25.14 Two deformations: the first flattens the pretzel, the second deforms the disc along the arrows shown.

E with two holes is not simply connected, because it is deformable to a figure-eight curve; and a pretzel (i.e., two-holed doughnut) can be flattened to E by a continuous deformation and is therefore also not simply connected. A third rule is:

25.10.7 THEOREM. *Suppose $X \times Y$ is not empty; then $X \times Y$ is simply connected iff both X and Y are.*

Proof. There are two mappings

(i) $f: X \times Y \to X$, $g: X \times Y \to Y$ given by $f(x, y) = x$, $g(x, y) = y$

(obviously continuous). Observe

(ii) $(x, y) = (f(x, y), g(x, y))$.

If $p: \mathbb{S}^1 \to X$ is a loop in X, then we regard it as a loop p' in $X \times Y$ by setting $p'(t) = (p(t), y_0)$ for some fixed $y_0 \in Y$ (Y is not empty, since $X \times Y$ is not). But, if $X \times Y$ is simply connected, then there is a nullhomotopy $h_t: \mathbb{S}^1 \to X \times Y$ of p'. Hence, using $f: X \times Y \to X$, then $f \circ h_t: \mathbb{S}^1 \to X$ is the required nullhomotopy of p in X; for (recall 25.10.2):

$$\left.\begin{aligned} f \circ h_0(s) &= f \circ p'(s) = f(p(s), y_0) = p(s); \\ f \circ h_1(s) &= f(h_1(s)) = f(\text{constant}) = \text{constant} \end{aligned}\right\} s \in \mathbb{S}^1.$$

Similarly for Y.

Conversely, suppose X, Y are simply connected. To prove $X \times Y$ simply connected, let $p: \mathbb{S}^1 \to X \times Y$ be a loop in $X \times Y$. Then using (i), $f \circ p$, $g \circ p$ are loops in X, Y respectively, so they are deformable to constant maps in X, Y by homotopies $u_t: \mathbb{S}^1 \to X$, $v_t: \mathbb{S}^1 \to Y$. We claim that if $w_t(s) = (u_t(s), v_t(s))$ then $w_t: \mathbb{S}^1 \to X \times Y$ is a nullhomotopy of p in $X \times Y$. For w_t is obviously continuous, while for all $s \in \mathbb{S}^1$,

$$w_0(s) = (u_0(s), v_0(s)) = (f(p(s)), g(p(s))) = p(s) \qquad \text{(by (ii))};$$
$$w_1(s) = (x_1, y_1), \quad \text{say}, \quad = \text{constant},$$

as required. ■

25.10.8 Some corollaries are: a cylinder is not simply connected, since it is homeomorphic to $\mathbb{S}^1 \times \mathbb{I}$; nor is a torus (homeomorphic to $\mathbb{S}^1 \times \mathbb{S}^1$). Again, $\mathbb{R}^2$ minus the origin is not simply connected, being homeomorphic to $\langle 0, \infty \rangle \times \mathbb{S}^1$, since each of its points has polar co-ordinates (r, θ) where $r \in \langle 0, \infty \rangle$ and $\theta \in \mathbb{S}^1$. Hence, without appealing to the theory of dimension, $\mathbb{R}^3 \not\approx \mathbb{R}^2$; for $\mathbb{R}^3$ minus the origin is homeomorphic to $\langle 0, \infty \rangle \times \mathbb{S}^2$, a product of simply connected spaces, hence simply connected. [*Caution:* it is not always true that if $A \times B \approx A \times C$, then $B \approx C$; for, take $A = \mathbb{I}$, and take B, C to be annuli, one with a 'whisker' ($\approx \mathbb{I}$) growing from each edge, the other with two whiskers on one edge: each product is a doughnut with two flat vanes projecting.]

Question: Is the physical universe W simply connected? Very little is known about its shape. Newtonian mechanics assumes $W \approx \mathbb{R}^3$, but in view of relativity theory, this is probably false.

Exercise 11

(i) Prove the statement in 25.10.7, 'if P is deformable to Q, then P is simply connected iff Q is'. [First, what do you consider it to mean?]
(ii) If X is contractible, prove that X is path-connected [see Exercise 1(iii)].
(iii) Use the algebra of sets (3.2.1, 3.2.2) to establish the *Caution* above.

25.11 The Algebraic Approach

An important (and enormous) branch of topology is that known as 'algebraic topology', whose ideas have exerted a great and often unifying influence on other parts of mathematics. The basic idea is that of modelling a geometrical situation by means of algebra, rather as Cartesian geometry expresses Euclidean geometry in algebraic terms. However, the Cartesian situation is ***reversible***, in that the algebra gives complete information about the geometry it describes; but in algebraic topology we deliberately seek to filter out of the geometry only what our algebra is equipped to handle. Otherwise, the algebraic information may swamp us, and in extreme cases (see Exercise 12(vii) below) our algebraic problem would be *exactly* as difficult as the original geometrical one. We shall now consider an important example of this modelling process by associating with each path-connected (metric) space X, and fixed base-point $x_0 \in X$, a group $\pi(X, x_0)$. This group was introduced by Poincaré, and is called the **fundamental group**; among other things, it will give us a measure of the departure of X from being simply connected.

To define $\pi(X, x_0)$, let Ω denote the set of all 'loops at x_0'; each such loop may be regarded as a mapping† $p: \mathbb{I} \to X$ such that $x_0 = p(0) = p(1)$. Two such loops, p, q are '**homotopic rel** x_0' written

25.11.1 $$p \simeq q \qquad \text{rel } x_0$$

if there is a homotopy $h_t: \mathbb{I} \to X$ such that $h_t(0) = h_t(1) = x_0$ for all t; thus p is deformed to q, keeping the ends $p(0)$, $p(1)$ fixed at x_0. As with ordinary homotopy, this is an equivalence relation R, and we define the set $\pi(X, x_0)$ to be Ω/R. Next, we introduce a multiplication first in Ω, then in $\pi(X, x_0)$, by defining the product $p \cdot q: \mathbb{I} \to X$ of loops p, $q \in \Omega$ by

$$p \cdot q(t) = \begin{cases} p(2t), & 0 \leqslant t \leqslant \frac{1}{2}, \\ q(2t-1), & \frac{1}{2} \leqslant t \leqslant 1. \end{cases}$$

Thus $p \cdot q$ may be thought of, intuitively, as the loop p followed by the loop q. It can be shown that if pRp' and qRq', then $p \cdot qRp' \cdot q'$, so we can define multiplication in $\pi(X, x_0)$ by

25.11.2 $$[p][q] = [p \cdot q]$$

† $\mathbb{I}$ is the unit interval, as in Exercise 1(iii); we change here from $\mathbb{S}^1$ to $\mathbb{I}$ for technical reasons. Clearly the geometrical image is unaltered.

independently of coset representatives, where $[p]$ denotes the coset of the loop p. Let c denote the constant loop, for which $c(t) = x_0$ (all $t \in \mathbb{I}$); then for any loop p,

$$c \cdot p \simeq p \simeq p \cdot c \qquad \text{rel } x_0 \text{ in } X$$

whence $[c]$ acts as unit element 1 in $\pi(X, x_0)$. To define inverses, let $\bar{p}$ denote the loop ('reversal of p') given by

$$\bar{p}(t) = p(1 - t) \qquad (t \in \mathbb{I});$$

it can be shown that then $[\bar{p}][p] = 1 = [p][\bar{p}]$, whence we define the inverse $[p]^{-1}$ of $[p]$ by

$$[p]^{-1} = [\bar{p}].$$

Finally, it can be shown that this multiplication in $\pi(X, x_0)$ is associative, and therefore $\pi(X, x_0)$ becomes a *group*. If X is simply connected, then for each loop $p \in \Omega$ we have $p \simeq c$ rel x_0 (watch the base-point!), whence $[p] = [c]$ and $\pi(X, x_0)$ is the trivial group $\{1\}$; and conversely, if $\pi(X, x_0) = \{1\}$, then X is simply connected. Thus, $\pi(X, x_0)$ can be thought of as a measure of X's departure from simple connectivity; also, it need not be commutative.

25.11.3 In particular, we saw in 25.10.5 that $\mathbb{S}^1$ is not simply connected; thus when $X = \mathbb{S}^1$, $\pi(\mathbb{S}^1, x_0) \neq \{1\}$. More precisely, each loop p in $\mathbb{S}$ can be shown to be homotopic, rel x_0, to one which winds round the circle n times in the positive direction or in the negative direction; and the correspondence $p \to \varepsilon n$ (ε = sign of direction) can be shown to yield an isomorphism $\pi(\mathbb{S}^1, x_0) \approx \mathbb{Z}$. Finite algorithms exist for computing $\pi(X, x_0)$, when X lies in a fairly wide class of spaces, including the so-called *compact polyhedra.*

Having associated a group with each based space (X, x_0), we now associate with each mapping $f: X \to Y$, where $f(x_0) = y_0$ (briefly $f: (X, x_0) \to (Y, y_0)$) a group homomorphism $\pi(f) = f_*: \pi(X, x_0) \to \pi(Y, y_0)$ by setting, for each loop p in X,

25.11.4
$$f_*[p] = [f \circ p].$$

This definition is independent of coset representatives; for, if $p \simeq q$ rel x_0 in X, then $f \circ p \simeq f \circ q$ rel y_0 in Y; also

25.11.5
$$f \circ (p \cdot q) = (f \circ p) \cdot (f \circ q)$$

so
$$f_*[p \cdot q] = f_*[p] f_*[q],$$

i.e., *f_* is a homomorphism.* These statements are evident from suitable pictures, and they are easily checked by means of formulae.

From 25.11.4 we have at once the 'covariance' equations

25.11.6
$$(\mathrm{id}_X)_* = \mathrm{id}, \qquad (f \circ g)_* = f_* \circ g_*$$

where id denotes the identity map on $\pi(X, x_0)$ and $g: (Y, y_0) \to (Z, z_0)$ is any mapping. [Compare 6.2.9, and covariant functors in Chapter 38.] In particular, if f is a homeomorphism and g is f^{-1}, then by the algebraic inversion theorem 2.9.8 applied to 25.11.6 we have:

25.11.7 COROLLARY. *If* $f: (X, x_0) \to (Y, y_0)$ *is a homeomorphism, then* $f_*: \pi(X, x_0) \to \pi(Y, y_0)$ *is an isomorphism and* $(f^{-1})_* = (f_*)^{-1}$. ■ In particular then, if $\pi(X, x_0) \not\cong \pi(Y, y_0)$, then X and Y *cannot be homeomorphic.*

Let us now apply this machinery to a geometrical problem (which tells us (on reflection) something about the tightening of drumheads!). The proof of Theorem 25.10.5 yields another proof of this lemma.

25.11.8 LEMMA. *There is no mapping* $f: D \to \mathbb{S}^1$ *of the unit disc into its boundary which leaves the boundary fixed.*

Proof. To get a contradiction, suppose there was such a mapping f. Then we have two diagrams

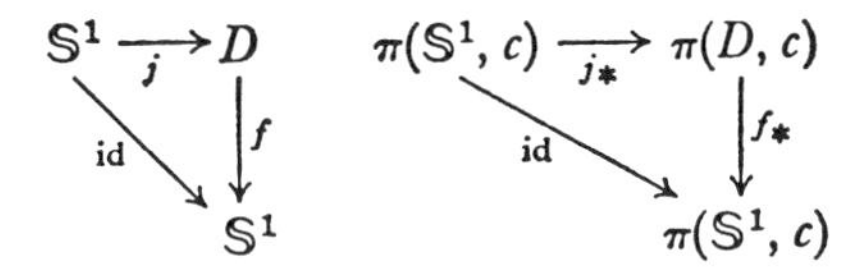

where in the first, j denotes inclusion, while in the second, c is a chosen point on $\mathbb{S}^1$ (hence in D). The second is the 'model', in group theory, of the geometry which the first diagram expresses. Now f leaves $\mathbb{S}^1$ fixed, whence $f \circ j = \text{id}$ and the first diagram is 'commutative'. Therefore, by covariance (25.11.6), $f_* \circ j_* = \text{id}$, so the second diagram is commutative.

Now $\pi(D, c) = \{1\}$, since D is simply connected; while it can be shown that $\pi(\mathbb{S}^1, c) \neq \{1\}$ (see 25.11.3). Hence, $\exists g \in \pi(\mathbb{S}^1, c)$ such that $g \neq 1$, so

$$1 \neq g = \text{id}(g) = f_* \circ j_*(g) = f_*(1) = 1,$$

using 18.6.2(i). This contradiction shows that f cannot exist. ■

Observe that, in the last proof, we cannot retrieve the first diagram from the second, which therefore has filtered out only enough of the geometry to give the contradiction we want. We could have obtained the contradiction from any similar diagram of groups $G(X)$ and homomorphisms $G(f)$ provided $G(\mathbb{S}^1) \neq \{1\}$ and $G(D) = \{1\}$ (so $G(X)$ is sufficiently 'computable') and either the covariance rule 25.11.6 holds or the 'contravariance' rule

25.11.9 $$G(\text{id}) = \text{id}, \qquad G(f \circ g) = G(g) \circ G(f).$$

Observe that we then have an analogue of 25.11.7: $G(X)$ is a *topological invariant.*

25.11.10 Algebraic topology has many such constructions G, where $G(X)$ may have further structure such as that of a ring, or a sequence of groups (for example the sequences $H_n(X)$, $H^n(X)$ of 'homology' and 'cohomology' groups of X) while $G(f)$ preserves this structure also. [A further example is given by the sets counted by p_n in 25.10.3; these sets are in fact 'homotopy groups' of the sphere $\mathbb{S}^{n+2}$.] The choice of a suitable G depends on the problem and may be a delicate matter indeed, as in the solution by Adams (mentioned in 24.9.5) of the problem concerning the possible dimensions of real division algebras. The study of the possible structures $G(X)$ and their mutual relationships forms the core of algebraic topology. It plays a role like that of analytical geometry within geometry, or like that of calculus within physics, with a similar mutual enrichment.

Exercise 12

(i) Prove that the relation R in 25.11.1 is an equivalence relation.
(ii) Prove that if pRp' and qRq', then $p \cdot qRp' \cdot q'$, to establish independence of coset representatives in 25.11.2.
(iii) Prove in detail the steps prior to 25.11.3 showing that $\pi(X, x_0)$ is a group.
(iv) Let $f \simeq g: (X, x_0) \to (Y, y_0)$; thus there is a homotopy from f to g which keeps the base point x_0 in the base point y_0. Show that $f_* = g_*: \pi(X, x_0) \to \pi(Y, y_0)$.
(v) Let x_1 be a second fixed point in X and let λ be a path in X from x_0 to x_1. Using a 'product' rule like that in Ω, associate with each loop p at x_0 a loop $p' = \lambda p \bar{\lambda}$ at x_1, where $\bar{\lambda}$ denotes λ traced in reverse. Prove that the correspondence $p \to p'$ induces an isomorphism $\lambda_\#: \pi(X, x_0) \to \pi(X, x_1)$. If $f: (X, x_0) \to (Y, y_0)$ is a mapping and $y_1 = f(x_1)$, prove that $f'_* \circ \lambda_\# = (f \circ \lambda)_\# \circ f_*: \pi(X, x_0) \to \pi(Y, y_1)$ where $f'_*: \pi(X, x_1) \to \pi(Y, y_1)$ corresponds to f_* with x_1 replacing x_0.
(vi) Adapt the proof of 25.10.7 to show that if X, Y are path-connected metric spaces, and $x_0 \in X$, $y_0 \in Y$, then $\pi(X \times Y, (x_0, y_0))$ is the direct product of $\pi(X, x_0)$ and $\pi(Y, y_0)$. Hence if X is a torus, prove that $\pi(X, x_0) \approx \mathbb{Z} \oplus \mathbb{Z}$, the lattice of points with integer co-ordinates in $\mathbb{R}^2$; while if X is a solid torus, then $\pi(X, x_0) \approx \mathbb{Z}$.
§(vii) Let $C(X)$ denote the real algebra of all continuous functions $f: X \to \mathbb{R}$. Show that a mapping $g: X \to Y$ induces a homomorphism $g^*: C(Y) \to C(X)$ satisfying the contravariance rules 25.11.9. Hence show that g^* is an isomorphism if g is a homeomorphism. (If X is compact, the converse also holds; thus here is a case where the corresponding second diagram in the proof of Lemma 25.11.8 contains *all* the information in the first diagram. It is unknown how to describe $C(X)$ in 'finite' terms, nor how, for example, the contractibility of D implies some algebraic property of $C(D)$.)

25.12 Manifolds

Perhaps the most interesting topological spaces are the manifolds, which are locally like Euclidean space, just as the earth's surface is locally like a plane. We make the formal

25.12.1 DEFINITION. X is an 'n-**manifold**' if each $x \in X$ has a neighbourhood U_x homeomorphic to $\mathbb{R}^n$. [$\mathbb{R}^0$ is the single point $0 \in \mathbb{R}^1 = \mathbb{R}$.] When $n = 2$, X is also called a **surface.**

25.12.2 EXAMPLES.

Clearly $\mathbb{R}^n$ itself is an n-manifold (take U_x to be all of $\mathbb{R}^n$); so is $\mathbb{S}^n$ (see Example 25.6.9(iii)). As a subtler example we show that any locus in $\mathbb{R}^n$ of the form $X = \{x \in \mathbb{R}^n | f(x_1, \ldots, x_n) = 0\}$ is an $(n-1)$-manifold provided $\partial f/\partial x_i \neq 0$ at each $p \in X$ for some i depending on p. In the latter case, an implicit function theorem (see 37.4) tells us that near p we can solve $f = 0$ to give $x_i = q(x_1, \ldots, x_{i-1}, x_{i+1}, \ldots, x_n)$; so, near p, X looks like the graph of a function of $n-1$ variables, and is therefore an $(n-1)$-manifold. The complex projective plane $\mathbb{CP}^2$ is a 4-manifold; for, given P with homogeneous complex co-ordinates $[z_1, z_2, z_3]$ we can suppose z_3 (say) is 1, whence P lies in the set U of all points $(u, v, 1)$ with u, v any complex numbers. Thus $U \mathrel{\dot{\approx}} \mathbb{C} \times \mathbb{C}$; but $\mathbb{C} \mathrel{\dot{\approx}} \mathbb{R}^2$ (Argand diagram), so

$$U \mathrel{\dot{\approx}} \mathbb{R}^2 \times \mathbb{R}^2 \mathrel{\dot{\approx}} \mathbb{R}^4.$$

(We are assuming that $\mathbb{CP}^2$ has a metric so that we can speak of homeomorphisms rather than bijections; such a metric can be defined. But here is a case where we obtain a 'topology' in the non-metrical spirit of 25.9.4, by declaring a set to be 'open' if it intersects each set U in the image of an open subset of $\mathbb{R}^4$ under the correspondence $\mathbb{R}^4 \mathrel{\dot{\approx}} U$.) A complex projective line *is a 2-manifold* (in fact homeomorphic to $\mathbb{S}^2$); for, by 17.5.1, the line PQ in $\mathbb{CP}^2$ is the set of all points of the form $[\lambda P + \mu Q]$ where $\lambda, \mu \in \mathbb{C}$ and distinct points correspond to different ratios $\lambda : \mu$. Now the set of all such ratios is the Gaussian complex 'plane' ($\mathbb{R}^2$ plus the point at infinity), which is topologically a sphere. Similarly, using $\mathbb{R}$ for $\mathbb{C}$, we find the real projective plane is a 2-manifold, and a real projective line is $\mathrel{\dot{\approx}} \mathbb{S}^1$. It is not always easy to prove that a space is an n-manifold; for example, the set of all $n \times n$ orthogonal matrices is a manifold of dimension $\frac{1}{2}n(n-1)$, but the proof is by no means immediate.

Exercise 13

(i) Prove that a non-degenerate conic in $\mathbb{CP}^2$ is a two-dimensional manifold.
(ii) By regarding $\mathbb{R}^6$ as $\mathbb{C}^3$, each point on $\mathbb{S}^5$ may be regarded as a triple (z_1, z_2, z_3) of complex numbers such that $|z_1|^2 + |z_2|^2 + |z_3|^2 = 1$. Let $f\colon \mathbb{S}^5 \to \mathbb{CP}^2$ denote the function such that $f(z_1, z_2, z_3) = [z_1, z_2, z_3]$. Show that $f^{-1}[z_1, z_2, z_3]$ is a Jordan curve in $\mathbb{S}^5$, of the form $\{(z_1 e^{i\theta}, z_2 e^{i\theta}, z_3 e^{i\theta}) | 0 \leqslant \theta \leqslant 2\pi\}$. Now use the method of 25.9 to show that $\mathbb{CP}^2$, regarded as a quotient space of $\mathbb{S}^5$, has a metric.
(iii) Consider similarly the analogous mapping $h\colon \mathbb{S}^2 \to \mathbb{CP}^1$. As indicated in the text, $\mathbb{CP}^1 \mathrel{\dot{\approx}} \mathbb{S}^2$, so we may regard h as a map $\mathbb{S}^3 \to \mathbb{S}^2$ (called the **Hopf map**.) Let E^+, E^- denote the Northern and Southern hemispheres of $\mathbb{S}^2$. Show that $h^{-1}(E^+)$ is a solid torus of the form $\mathbb{S}^1 \times E^+$, and hence that $\mathbb{S}^3$ is the union of the two solid toruses $\mathbb{S}^1 \times E^+$, $\mathbb{S}^1 \times E^-$.

25.12.3 Examples of *non*-manifolds are (i) a one-dimensional space shaped like an X (which is not locally like a line at the crossing point); (ii) the plane

$z = 0$, plus the z-axis in $\mathbb{R}^3$ (not 'nice' at the origin); (iii) three planes meeting in a line (whose points have 'bad' neighbourhoods [see Fig. 15.1(v)]). Of course, in the real world, we never 'see' spaces like these, but fattened-up approximations to them. Each such approximation has a 2-manifold for boundary, provided we disregard its (physical) molecular structure. This leads us to:

25.12.4 DEFINITION. Let X be a space and let B denote the set of $x \in X$ such that x does not possess a neighbourhood $\approx \mathbb{R}^n$. If $B \neq X$ and each $x \in B$ possesses a neighbourhood $\approx$ to the half-space $z_n \geqslant 0$ of $\mathbb{R}^n$, we call X an ***n*-manifold with boundary** B.

It can then be shown that B is an $(n - 1)$-manifold and $X - B$ is an n-manifold whose frontier in X is B.

Clearly an n-manifold satisfies this definition, with B the empty set. B is called the **boundary** of X. If X is compact and B is empty, then X is often called a **closed** manifold.

25.12.5 EXAMPLES.

(i) The unit interval $\langle\!\langle 0, 1\rangle\!\rangle$ is a 1-manifold-with-boundary; here B is the 0-manifold consisting of the pair of points 0 and 1.

(ii) The unit disc D in $\mathbb{R}$ is a 2-manifold-with-boundary $B = \mathbb{S}^1$. The cylinder W in Example 25.9.3(v) is a 2-manifold-with-boundary, where B is a pair of circles; the Möbius strip in the same example is a 2-manifold-with-boundary, and B is a single circle. The torus and Klein bottle in the same example are both closed 2-manifolds.

(iii) A k-dimensional solid box in $\mathbb{R}^k$ is a k-manifold-with-boundary, and then $B \approx \mathbb{S}^{k-1}$, a closed $(k - 1)$-manifold.

25.12.6 THE BIG PROBLEM. For each integer k there is the 'classification problem': list all the topologically different kinds of path-wise connected k-manifolds. When $k = 1$, the list consists only of $\mathbb{S}^1$ and $\mathbb{R}$; and the only 1-manifolds-with-boundary are (homeomorphic to) the intervals $\langle\!\langle 0, 1\rangle\!\rangle$, $\langle 0, 1\rangle\!\rangle$. When $k = 2$, the answer is more complicated and is satisfactory only for compact manifolds; we shall give it in a moment. For higher dimensions, only partial answers are known. In particular, Poincaré conjectured (around 1900) that the only closed, path-connected, simply-connected 3-manifold is (homeomorphic to) $\mathbb{S}^3$. This has never been proved or disproved; showing how little we really understand about three dimensions, in spite of a vast amount of research on 3-manifolds. Curiously, however, the analogous conjecture (that a closed n-manifold with the same homotopy groups as $\mathbb{S}^n$ is homeomorphic to $\mathbb{S}^n$) has been proved when $n \geqslant 5$—by Smale, Zeeman and others around 1960. Thus only the cases $n = 3, 4$ remain open.

25.12.7 Now consider the problem when $k = 2$. First, how can we construct compact 2-manifolds (or surfaces)? Two methods are as follows. One

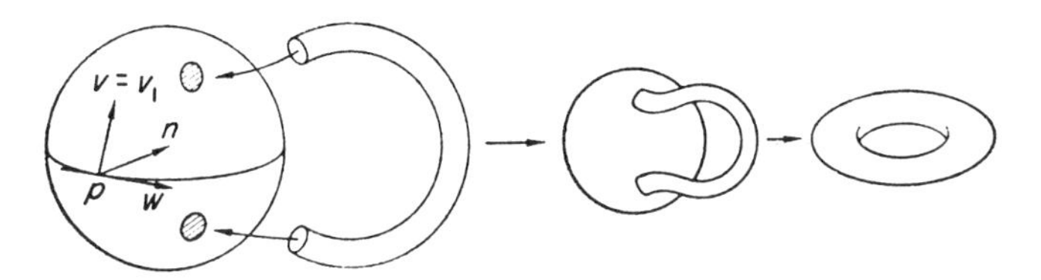

Fig. 25.15 Adding a handle to a sphere to form a torus. (The shaded discs are removed, and their edges joined by a tube.) On the equator of the sphere, we show the three vectors—forwards, left, and inwards—at the start of a journey round the equator (see 25.12.10).

is to take a known surface M, and 'add a handle', i.e., to cut two holes from M, and join their boundaries by a tube (see Fig. 25.15). If $M \approx \mathbb{S}^2$, the result of this 'surgery' is a torus, T; if $M \approx T$, the result is the surface of a pretzel—a 'sphere with two handles'—and after p such steps we obtain a **sphere with p handles** or **orientable surface of genus p**, denoted by M_p. The second method is to start with a known M and 'add a Möbius band', i.e., to cut one hole from M, and to glue its boundary to that of a Möbius band (Example 25.9.3(v)). If $M \approx \mathbb{S}^2$ the result of one such operation is the real projective plane $\mathbb{RP}^2$ (see below), if $M \approx \mathbb{RP}^2$, the result is the Klein bottle, and so on. Let N_p denote the result after p steps; N_p is called a **non-orientable surface with p cross-caps.** It can be proved that all compact surfaces X can be obtained from $\mathbb{S}^2$ by applying one or other method, i.e.,

25.12.8 THEOREM. *Given a compact surface X, not homeomorphic to $\mathbb{S}^2$, then there exists a positive integer p such that either $X \approx M_p$, or $X \approx N_p$ (but not both). Moreover, p is unique; i.e., if $M_p \approx M_q$ or $N_p \approx N_q$, then $p = q$.* □

The simplest proof is too long to give here, but consists essentially in cutting up X into triangles, slitting some edges to flatten X on a plane, and then adding and subtracting triangles to obtain a 'normal form' which gives X as a quotient space of a certain polygonal disc. See, for example, Lefschetz [81].

25.12.9 MODELS. The surfaces M_p are the ones which bound regions in the real world. As a standard model, let B be a solid rectangular box in $\mathbb{R}^3$, and drill p parallel tunnels through B, to leave a solid B_p. Then B_p is a 3-manifold-with-boundary, and (by induction on p) its boundary $b(B_p)$ is homeomorphic to M_p. Let us call any homeomorph of B_p a **solid of genus p.** Then the exercise suggested in 25.3.2 can be refined into that of deciding the genus of 'real-life' solids, still using the 'chunk' technique there described.

Exercise 14

(i) Prove that a spoon, cup, and teapot without lid, have (solid) genus 0, 1, 2, respectively.
(ii) Show that the (solid) genus of a cartwheel with k spokes is $k + 1$ ($k \geqslant 2$). What is the genus of a pair of such wheels on one axle?

(iii) What is the (solid) genus of an unfurnished house with (a) an open door, (b) an open window, (c) a broken window?
(iv) Make suitable simplifying assumptions to determine the genus of an egg-whisk, a bicycle, a piccolo, a violin, the system consisting of a violinist playing a violin (a) *arco*, or (b) *pizzicato*.
(v) Prove that a solid of genus $p > 0$ is not simply connected (use 25.10.6). Prove also that if $p > 1$, the surface of such a solid cannot be a Cartesian product of two 1-manifolds, nor is the solid such a product of two manifolds of dimensions 1 and 2.

25.12.10 ORIENTABILITY. Now consider the basic topological difference between an M_p surface and an N_q. Technically, we say that M_p is **orientable** and N_q **non-orientable** and explain these terms intuitively as follows. Now $M_p \dot{\approx} b(B_p)$, and $b(B_p)$ has a compact 'inside' (viz.: B_p) and a non-compact 'outside' in $\mathbb{R}^3$, so $b(B_p)$ is 'two-sided'. This property is equivalent to the following 'intrinsic' one which depends only on M_p, and not on $\mathbb{R}^3$. Suppose we walk along a smooth loop on M_p, in the direction of increasing θ (where θ is the angular co-ordinate on $\mathbb{S}^1$). Then at any point of the journey we can distinguish between our left and right hand, so let us chalk an arrow $\mathbf{v}$ leftwards as we leave our starting point P along an arrow $\mathbf{w}$. We claim that, when we approach the end of our journey at P (in the direction of $\mathbf{w}$), the chalk direction $\mathbf{v}$ agrees with the direction $\mathbf{v}_1$ we *now* call leftwards. For the journey and arrows have images in $b(B_p)$ under a homeomorphism $h: M_p \dot{\approx} b(B_p)$, while at $h(P)$ we have a further arrow, $\mathbf{n}$, as in Fig. 25.15, pointing along the interior normal (we can smooth $b(B_p)$ to have a tangent plane everywhere). Then the triple of arrows $h(\mathbf{v})$, $h(\mathbf{w})$, $\mathbf{n}$, must form the same kind of left- or right-handed system as the triple $h(\mathbf{v}_1)$, $h(\mathbf{w})$, $\mathbf{n}$ (since we know 'leftwards' uniquely when we know 'downwards' and 'forwards'). Thus $h(\mathbf{v})$ and $h(\mathbf{v}_1)$ agree, whence so do $\mathbf{v}$ and $\mathbf{v}_1$ on M_p. We say that 'orientation is preserved' on all loops of M_p, or briefly 'M_p is orientable'.

We next claim that, if we go down the centre-line L of a Möbius band S, then the corresponding directions $\mathbf{v}$, $\mathbf{v}_1$ are opposed. This is easily deduced from Fig. 25.11; but a real-life walk on a paper model of S is confusing because the paper has thickness, and is really a solid of genus 1. Thus S is not orientable, because we have found an 'orientation-reversing' loop L. Hence, since $N_q(q > 0)$ contains a copy of S by definition (see 25.12.7), N_q is not orientable. Therefore $M_p \dot{\not\approx} N_q$, since an orientable surface cannot be homeomorphic to a non-orientable one. Incidentally we have also proved:

25.12.11 ***N_q cannot be embedded in $\mathbb{R}^3$***,

(otherwise we could use the normal $\mathbf{n}$ to conclude as for M_p that N_q was orientable); this is perhaps why non-orientable surfaces are commonly called 'one-sided'.

Exercise 15

(i) (Unsolved.) Are all loops in the physical universe orientation-preserving?
(ii) (Hard.) Prove that N_{2p+1} is homeomorphic to the result of replacing a disc in M_p by a Möbius band (second method in 25.12.7).

25.12.12 THE EULER CHARACTERISTIC. Suppose a compact surface S is divided into a finite number of triangles $\sigma_1, \ldots, \sigma_n$ such that if any two σ_i, σ_j meet, they have a common vertex only, or a common edge. The σ's form a 'triangulation' Σ of S, and it can be proved that triangulations always exist, as can the analogue (using tetrahedra) for a 3-manifold. Now if S_1, S_2 are triangulated surfaces with triangulations Σ_1, Σ_2 respectively, we know that S_1, S_2 are 'the same' (as topological spaces) if there is a homeomorphism between them; and the question arises whether they are 'the same' as triangulated surfaces, i.e., is $(S_1, \Sigma_1) \approx (S_2, \Sigma_2)$ in some sense?

A reasonable sense, not too narrow for practical purposes, is to ask if it be possible to subdivide the triangles of Σ_1 and Σ_2 respectively in such a way as to obtain triangulations Σ_1', Σ_2' and a homeomorphism k: $S_1 \approxeq S_2$ such that k maps vertices, edges and triangles of Σ_1' to similar items of Σ_1', preserving their incidence relations (e.g., if e is an edge of a triangle in Σ_1', then $k(e)$ has to be an edge of $k(\sigma)$ in Σ_2').

Briefly, (S_1, Σ_1) and (S_2, Σ_2) are then called **combinatorially equivalent**; clearly this notion makes sense in higher dimensions, though a precise description takes some writing out. It was shown by T. Rado in the 1920s that any two homeomorphic compact surfaces are combinatorially equivalent; and the corresponding result for 3-dimensional manifolds was obtained by E. E. Moise in the 1950s, who first demonstrated the difficult theorem that *every* 3-manifold actually possesses a combinatorial structure. In particular, then, any two combinatorial structures on the same n-manifold ($n \leqslant 3$) are unique, in our sense of being 'the same'. In higher dimensions, the first problem is to decide whether an n-manifold ($n > 3$) has a combinatorial structure at all; it is then natural to ask how many. The first important result in this direction was obtained by S. S. Cairns and J. H. C. Whitehead in the 1940s for *differentiable* n-manifolds, i.e., homeomorphs of subsets of $\mathbb{R}^{2n+1}$ possessing a continuously turning tangent plane at each point. They showed that each such n-manifold has exactly one combinatorial structure, which moreover is compatible in a sensible way with its 'differentiable structure'. But then in 1956 John Milnor found two surprising facts to complicate the guessing about the general picture. He showed first that homeomorphic manifolds (in particular homeomorphs P, Q, of the 7-dimensional sphere) could have fundamentally different differentiable structures, so although there is a homeomorphism $h: P \to Q$, we cannot have both h and h^{-1} differentiable. Second, he showed that two homeomorphic spaces A, A' could have *distinct* combinatorial structures. Admittedly A and A' were not manifolds, but they were built fairly simply out of manifolds. However, in 1969, Kirby and Siebenmann, building on work of Sullivan and others, have shown that almost all topological n-manifolds have a unique combinatorial structure if $n \geqslant 5$ (one always expects trouble particularly about the mysterious dimension 4). Indeed, the word 'almost' in our previous sentence may be given a perfectly definite meaning. For, it is possible to state precisely what the obstruction is to a topological n-manifold admitting a compatible combinatorial structure, and what the obstruction is to two homeomorphic combinatorial manifolds being combinatorially equivalent. Actually, an example of a manifold admitting distinct combinatorial structures is the 6-dimensional torus $\mathbb{S} \times \mathbb{S} \times \mathbb{S} \times \mathbb{S} \times \mathbb{S} \times \mathbb{S}$; of course, no such simple description can be given of a topological manifold admitting no combinatorial structure, but there is a 5-dimensional example of this phenomenon.

Returning to the case of a surface S, the Euler characteristic $\chi(S)$ is defined to be the integer

$$\chi(S) = V - E + F$$

where V, E, F denote, respectively, the total number of vertices, edges, and faces (triangles) in the triangulation. Euler showed, at least when $S \stackrel{.}{\approx} \mathbb{S}^2$, that whatever triangulation we use to compute $\chi(S)$ we always get the same integer; a proof by induction on F is given for spheres in Courant–Robbins [25], and can be modified for the other surfaces. The integer $\chi(S)$ is called a 'combinatorial invariant', and once we know its independence of triangulation, we can pick the most convenient one to compute $\chi(S)$. For example, take $S = \mathbb{S}^2$; and take as triangulation the surface of a tetrahedron $ABCD$, consisting of four triangles. Here $V = 4, E = 6, F = 4$, so $\chi(\mathbb{S}^2) = 2$. Similarly, a cylinder and a Möbius band have $\chi = 0$; and if we remove q triangles, but not their boundaries, from a triangulation of a sphere, we reduce F by q and so obtain $\chi = 2 - q$. From this it follows, by adding suitably triangulated handles and Möbius bands as in 25.12.7, that

$$\chi(M_p) = 2 - 2p, \qquad \chi(N_q) = 2 - q,$$

and in particular $\chi(\mathbb{RP}^2) = 1$; while a torus and Klein bottle each have $\chi = 0$. Now, by Theorem 25.12.8, the numbers p, q are topological invariants, and therefore so is $\chi(S)$. This is a commonly occurring phenomenon, that a combinatorial invariant turns out to be a topological invariant; so it can then be computed combinatorially by a finite process. As another example, if we imagine S with mountainous ranges and lake bottoms, then it can be proved that:

25.12.13 *The number $M(S)$, of pits − passes + peaks, is $\chi(S)$.* □ [This result is due to Newton (with S a sphere), and to Kronecker generally.]

Thus $M(S)$ is found by computing $V - E + F$ instead of looking at the detailed topography; for the Earth, regardless of time and geology, $M(S) = \chi(\mathbb{S}^2) = 2$. Generaly, $M(S)$ belongs to calculus, because it refers to the stationary points of a smooth function (the height of the terrain). This is just one of the many encroachments of topology into calculus; in fact, since the precise language of topology became available, enormous strides have been made in the theory of functions of several variables.

Finally 25.12.8 can now be rounded out to the form:

25.12.14 THEOREM. *Two compact surfaces are homeomorphic iff they are both orientable or both non-orientable, with equal Euler characteristics.* □

Exercise 16

Let A, B be two triangulated surfaces-with-boundary. Find conditions under which they can be joined along the boundary to obtain $A \cup B$ as a surface-with-boundary and such that $\chi(A \cup B) = \chi(A) + \chi(B) - \chi(A \cap B)$. Use this to justify the above equations for $\chi(M_p)$, $\chi(N_q)$.

25.13 Applications and Further Outlook

For higher-dimensional triangulable manifolds, the Euler characteristic is defined and turns out to be a topological and combinatorial invariant; but there is no known analogue of 25.12.14. In differentiable manifolds, calculus can be performed, and the formula 25.12.13 holds [see Morse [91]], with appropriate classification of stationary points. From this stems the work of Smale on the generalized Poincaré conjecture, mentioned in 25.12.6. The classical theorems of Stokes and Green also appear in the context of differentiable manifolds, and relate certain tensors to the cohomology groups $H^q(X)$ mentioned in 25.11.10. B. Riemann (1826–66) (see Bell [10], Ch. 26), who had prepared the climate for Poincaré through his great work in geometrical analysis, had generalized 'functions of a complex variable' to a theory about conformal mappings $X \to Y$ of (orientable) surfaces—the 'Riemann surfaces'; and interesting deductions came out of the fact that Cauchy's theorem about contour integrals requires modification on a non-simply-connected surface. Field theories in physics require study of differential equations where the independent variable is a point on a compact manifold (e.g., the physical universe); and topological ideas have been used to give qualitative solutions to non-linear differential equations, for example to show that periodic solutions exist when even a numerical (computer) treatment is impossible. But perhaps the greatest contribution of topology has so far been that it has shown how to pin down elusive and general ideas in mathematical language and supplied methods of proof which have been emulated in other parts of mathematics, e.g., in algebra, logic, and the theory of games and economic behaviour.

25.14 Some Books

There seems to be no one book about topology that beginners enjoy; and none is easy. In the list below, attractive introductions can be found in

Courant–Robbins [25], Rademacher–Toeplitz [105], and Hilbert–Cohn-Vossen [59], with helpful but brief articles in the *Encyclopædia Britannica* (15th ed.) and the current *Chambers.* The analytically minded will like the treatment in Kolmogoroff–Fomin [77]; and Alexandroff [2] gives another Russian approach (in English). Of English writers, Newman [94] is excellent though stern. Patterson [101] has a reasonable account of metric spaces and introductory material on homology groups, while Hilton [62] can be consulted for homotopy theory. For a Texan view, see Bing [14]. Each of the 'introductions' [21, 65, 81] has its supporters, but some still think that [117] has never been beaten; in [118] the same authors give a beautiful introduction to applications of combinatorial topology in analysis. The non-metric treatment mentioned in Section 25.9.4 can be obtained from [17] or [69]. Knot theory now has its introduction in [26], and the reader may care to browse in [37]—especially for Zeeman's topological theory of visual perception. We include also just two [3, 64] of the list of books for professional topologists, and their bibliographies can be consulted. But new books on topology are now appearing frequently, and more suitable ones than those in this list may already be available.

Part VII
CALCULUS

One of the great inventions of human thought is the differential and integral calculus (known as 'the calculus', for short). Its roots go back to the Greeks, but it became a recognized branch of mathematics in the 18th century, after its development by Newton, Leibniz and others. Bold strides were taken by the pioneers, although the logical foundations of the subject were shaky, and their followers in the 19th century laboured hard to make the foundations firm. Their critical scrutiny became itself a branch of mathematics now called *analysis*, which studies the theoretical, rather than the practical aspects, of calculus. Analysis became a substantial part of British university courses in mathematics, especially under the influence of G. H. Hardy's book [52], which was quite unorthodox for its time. Unfortunately, the subject is difficult; it was often used as a criterion for allowing undergraduates to proceed further in mathematics, and it was taught badly, so that many students were put off mathematics simply by failing to learn analysis. In spite of the great emphasis laid upon analysis in universities, the way in which its students later taught calculus in schools was hardly influenced at all. Thus undergraduates continued to arrive from school, primed with an 18th-century spirit of calculus (with all its virtues as well as vices); they still had to make the leap into the quite different ways of thought demanded by their analysis lectures. Fortunately, some university teachers are sympathetic to these students' difficulties, and some improvement is now to be seen in the 'sociological' aspects of the situation.

In this Part of the book, we attempt to expound elementary calculus in a manner which we hope will serve as a half-way house for students, between the early 'intuitive' and the later 'rigorous' approaches (as they are described technically). Thus, we state all the basic theorems, with the precision our set-theoretic language allows us; but we omit those proofs which we feel are difficult and belong to analysis. For example, we sometimes borrow *results* from Chapter 25, without requiring their proofs. The remaining proofs can be given fully, as deductions (often of a purely algebraic form) from the 'hard' results. They will not need proving again—and worse, *unlearning*—as the student matures; only the 'hard' theorems will need proving when his stamina and curiosity are sufficiently vigorous for him to read the standard expositions. Enthusiasts for analysis often seem to believe that every step should be *proved* before proceeding, curiously forgetting that they themselves use all kinds of results (from mathematical logic, for example) which they would be hard-pressed to prove if challenged. We take the view that there will always be gaps, that it is important to recognise them frankly, and that an education in calculus should involve learning to check that hypotheses of theorems are satisfied in applications. Indeed, in their professional careers most graduates are expected to *apply* theorems rather than to prove them.

Our intention in the succeeding chapters of this Part is therefore to consider first limits and their algebra (for calculus is based on the notion of a 'limit'); then the algebra of continuous functions and of differentiable functions; integration and the Fundamental Theorem of Calculus; the 'elementary functions', and functions of several real variables. We also give indications as to how the subject splits into branches and where these branches lead. Some of our discussion will mean more to those who have a 'school' knowledge of calculus as traditionally presented, although such knowledge is not essential.

It is worth pointing out at the start that our treatment depends explicitly on the fact that the set $\mathbb{R}$ of real numbers is a totally ordered field. We do not need to mention the completeness of $\mathbb{R}$ (see 24.6), since this property is needed only for proving (and not for stating) those theorems whose proofs we omit. Questions of uniformity and completeness should surely be raised only on a second appraisal of calculus (when the student knows something worth appraising and cares about it)—unless someone can improve the usual ways of teaching these ideas.

Chapter 26

THE ALGEBRA $\mathbb{R}^I$

26.1 Intervals

We shall be particularly concerned with functions $f: I \to \mathbb{R}$, where $I \subseteq \mathbb{R}$ is a special kind of subset, called an **interval**, and of one of the forms

26.1.1 $$\{x \mid a < x < b\}, \qquad \{x \mid a \leqslant x \leqslant b\},$$

or, less often,

$$\{x \mid a \leqslant x < b\}, \qquad \{x \mid a < x \leqslant b\}.$$

The first form is called **open**, the second **closed**, and the others **half-open**; accordingly we write $I = \langle a, b\rangle$, $I = \langle\!\langle a, b\rangle\!\rangle$, $I = \langle\!\langle a, b\rangle$, and $I = \langle a, b\rangle\!\rangle$, respectively. [Many writers use instead the notations (a, b), $[a, b]$ respectively, and other notations are also current.] Such intervals are **finite** or **bounded.** Sometimes we must consider an 'infinite' or 'unbounded' interval J, for example of the forms

26.1.2 $$\langle -\infty, b\rangle = \{x \mid x < b\}, \qquad \langle\!\langle a, \infty\rangle = \{x \mid x \geqslant a\}.$$

The 'end-points' of I are a, b in 26.1.1, but b, a *respectively* constitute the single end-point of the two intervals in 26.1.2; '∞' is *not* an end-point.

26.2 Algebraic Operations

For theoretical discussions, it is convenient to think of the set of *all* possible functions $f: I \to \mathbb{R}$, with I fixed; and we write $f + g, f \cdot g, \lambda \cdot f$ for those functions $I \to \mathbb{R}$ which are given at each $x \in I$ by

26.2.1 $$\begin{cases} (f + g)(x) = f(x) + g(x); \qquad (f \cdot g)(x) = f(x) \cdot g(x), \\ \qquad\qquad (\lambda \cdot f)(x) = \lambda \cdot f(x), \end{cases}$$

where $f, g: I \to \mathbb{R}$ are any functions and $\lambda \in \mathbb{R}$. The rules 26.2.1 are often described as *pointwise*, since they are defined by specifying their values at each point.

Using an earlier notation, we denote this set of functions by

26.2.2 $$\mathbb{R}^I,$$

but we emphasize that we think of it not 'just' as a set, but as an **algebra**, in that it is equipped with, *and closed under*, the operations 26.2.1 of addition, multiplication, and multiplication by scalars in $\mathbb{R}$. More precisely, we observed in Example 19.1.2(iii) that $\mathbb{R}^I$ is a vector space over $\mathbb{R}$, and in Example 9.2.8, that it is a commutative ring with unity element; moreover, the scalar multiplication and the ring multiplication are bound together by the easily verified rule that

26.2.3 $$(\lambda\cdot f)\cdot g = \lambda\cdot(f\cdot g) \qquad (\lambda \in \mathbb{R}, f, g \in \mathbb{R}^I).$$

Thus $\mathbb{R}^I$ is technically a **commutative algebra over** $\mathbb{R}$ or 'a real algebra'. In calculus, we are much more interested in certain 'sub-algebras' of $\mathbb{R}^I$:

26.2.4 DEFINITION. A subset $X \subseteq \mathbb{R}^I$ is called a **sub-algebra** iff (i) X is a vector subspace of $\mathbb{R}^I$, and (ii) X is a sub-ring of $\mathbb{R}^I$.

(Condition 26.2.3 is automatically satisfied, so X is an algebra according to the previous definition.) Apart from $\mathbb{R}^I$ itself, two simple and important examples of sub-algebras of $\mathbb{R}^I$ are $\mathbb{R}$ and the set $\mathbf{P}(I)$ of real polynomials on I, for the following reasons. The constant function 1 lies in $\mathbb{R}^I$, hence so does each $k \in \mathbb{R}$ since $k = k\cdot 1$; thus $\mathbb{R}$ is naturally embedded in $\mathbb{R}^I$ as a sub*set* and it is now easily verified that this subset is in fact a sub*algebra*. Also, the identity function on I (here denoted by $\mathscr{I}: I \to \mathbb{R}$) is in $\mathbb{R}^I$, hence so is every power (defined inductively by using 26.2.1, but, in fact, $\mathscr{I}^n(x) = x^n$). Therefore, by induction on the number of terms, $\mathbb{R}^I$ contains each polynomial of the form

26.2.5 $$p = \sum_{r=0}^{n} c_r \mathscr{I}^r \qquad (c_r \in \mathbb{R}, 0 \leqslant r \leqslant n).$$

We immediately permit an abuse of language and refer to 'the polynomial $p(x) = \sum_{r=0}^{n} c_r x^r$' regardless of the domain in question. And now, since sums, products, and scalar multiples of polynomials are also polynomials, conditions 26.2.4 are satisfied: so the set $\mathbf{P}(I)$ is a subalgebra of $\mathbb{R}^I$ (it is the smallest subalgebra containing both 1 and $\mathscr{I}$). The point of introducing the concept of real algebra is to help us to state certain theorems succinctly, and possibly to suggest certain questions. We shall certainly not go into a detailed study of algebras, but refer the reader for a first introduction to Bell's remarks in [12], especially p. 252, about Wedderburn, whose pioneer work on the structure of algebras is of enormous importance to mathematics.

Exercise 1

(i) Show that $\mathbb{R}$ is the smallest sub-algebra of $\mathbb{R}^I$, in that $\mathbb{R}$ lies in every subalgebra of $\mathbb{R}^I$.

(ii) Verify in detail that $\mathbb{R}^I$ satisfies every axiom necessary for it to be declared a real algebra. Can you find any interesting subalgebras of $\mathbb{R}^I$ contained in $\mathbf{P}(I)$? Outside $\mathbf{P}(I)$?

(iii) Show that if A is any set, and B any commutative ring with unit, then operations analogous to those of 26.2.1 can be defined in the set B^A of all functions: $A \to B$. Verify that all the vector-space and ring axioms are satisfied in B^A, except that '$\lambda \in \mathbb{R}$' must always be replaced by '$\lambda \in B$'. Thus B^A is by definition an **algebra over** B, with structure **induced** by that of B; and B^A contains a copy of B embedded in a way analogous to that in which $\mathbb{R}^I$ contains $\mathbb{R}$.
(iv) An algebra over a field F is defined to be any set X which is simultaneously a (not necessarily commutative) ring and a vector space over F, and which also satisfies 26.2.3. Show that the set of all real $n \times n$ matrices forms an algebra over $\mathbb{R}$, under the operations of matrix multiplication and addition. If V has dimension n and basis $e_1, \ldots, e_n$, show that the multiplication in V is known as soon as we know the 'multiplication table' of all products $e_i \cdot e_j$. What is the table when V is $\mathbb{C}$, e_1 is 1, and e_2 is i?
(v) An algebra over F in (iv) is called a **division algebra**, provided it has no divisors of zero (i.e., provided $xy = 0 \,.\!\Rightarrow\!.\, x = 0$ or $y = 0$). Prove that the real algebra $\mathbb{R}^I$ is not a division algebra, nor is the matrix algebra of $n \times n$ real matrices when $n > 1$. [See also 24.9.]
(vi) A function $f\colon I \to \mathbb{R}$ is said to be **strictly increasing** iff $f(u) < f(v)$ whenever $u < v$ in I. If, instead, $f(u) > f(v)$, then f is **strictly decreasing**; and in either case, f is **strictly monotonic**. Prove that a strictly monotonic function is one-one. Do the strictly monotonic functions form a subalgebra of $\mathbb{R}^I$?
(vii) Formalize the argument of 26.2.4 as follows. Define a mapping $\theta\colon \mathbb{R} \to \mathbb{R}^I$ by taking $\theta(a)$, for each $a \in \mathbb{R}$, to be the constant function whose value is always a. Prove that θ is injective, that θ is a linear transformation of $\mathbb{R}$ into the *vector space* $\mathbb{R}^I$, and that θ is a homomorphism of $\mathbb{R}$ into the *ring* $\mathbb{R}^I$. (As such, θ is called a **homomorphism of algebras.**) When we 'regard $\mathbb{R}$ as a subalgebra of $\mathbb{R}^I$', we are mentally identifying a with $\theta(a)$, and we speak of 'the function 2' instead of 'the function $\theta(2)$'. No ambiguity arises since θ is one-one, and $\theta(\mathbb{R})$ is a good copy of $\mathbb{R}$ since θ is a homomorphism.
(viii) Let $J \supseteq I$ be an interval. Define a function $\rho\colon \mathbb{R}^J \to \mathbb{R}^I$ by assigning, to each $f \in \mathbb{R}^J$, its restriction $f|I$ (see 2.6.4). Prove that ρ is a homomorphism of algebras, that ρ is onto, but that ρ is not one-one.

26.3 Polynomials

Although we cannot hope to show it in this book, the main importance of the subalgebra $\mathbf{P}(I)$ of all polynomials in $\mathbb{R}^I$ arises in calculus because of a theorem of Weierstrass, which says that if I is closed and finite, then every continuous function in $\mathbb{R}^I$ can be approximated arbitrarily closely by a polynomial in $\mathbf{P}(I)$. There now exist several beautiful proofs using real algebras, within the branch of mathematics called functional analysis; the resulting abstraction enables the theorem to be extended to classes of functions other than $\mathbf{P}(I)$. The classical theory uses a solution of the 'heat equation' of mathematical physics, and it is interesting to see here how much abstract pure mathematics has arisen in connection with the original theorem of Weierstrass, inspired by practical problems of computing (for example).

Exercise 2

(i) Prove that the polynomials in $\mathbf{P} = \mathbf{P}(I)$, of even degree, form a subalgebra of $\mathbf{P}$. Similarly for the polynomials with zero constant term. Show also that the polynomials with rational coefficients do not form a real sub-algebra of $\mathbf{P}$, although they form a sub-ring.
(ii) Prove that $\mathbf{P}$ is a division algebra. {Use the fact that a non-zero polynomial has only a finite number of roots to show that only the zero polynomial represents 0 in $\mathbb{R}^I$.}
(iii) Define a function $D: \mathbf{P} \to \mathbf{P}$ by

$$D\left(\sum_{r=0}^{n} c_r \mathscr{I}^r\right) = \sum_{r=1}^{n} r c_r \mathscr{I}^{r-1}.$$

Prove that D is a linear transformation but not a ring homomorphism. What is the image of D? Show that D is not an injection, and prove that if $Df = Dg$ in $\mathbf{P}$, then $f - g$ is a constant. What replaces the (false) statement $D(fg) = D(f)D(g)$?
(iv) In Exercise 1(viii) prove that the function ρ maps $\mathbf{P}(J)$ into $\mathbf{P}(I)$, and that it is a homomorphism of real algebras. Use the fundamental theorem of algebra (22.5.2) to prove that $\rho' = \rho|\mathbf{P}(J)$ is one-one. Prove also that ρ' is onto (we call it then an **isomorphism** of algebras).
(v) Prove that, if $n, m \in \mathbb{N}$, then $\mathscr{I}^n \cdot \mathscr{I}^m = \mathscr{I}^{n+m}$ and $\mathscr{I}^n \circ \mathscr{I}^m = \mathscr{I}^{nm}$.

26.4 The Reciprocal

Suppose that the interval I does not contain the origin $O \in \mathbb{R}$. Then there is in $\mathbb{R}^I$ the function which assigns, to each $x \in I$, the number $x^{-1} \in \mathbb{R}$; we shall denote this function by

26.4.1 $$1/\mathscr{I}: I \to \mathbb{R}, \qquad (1/\mathscr{I})x = x^{-1}.$$

(there is no standard notation, but we avoid the notation $\mathscr{I}^{-1}$ because of confusion with the *inverse* function (2.9.7)). Of course, the largest (real) domain of '$y = x^{-1}$' is $\mathbb{R} - \{0\}$, and the function in 26.4.1 is the restriction to I of the **reciprocal function**

26.4.2 $$1/\mathscr{I}: \mathbb{R} - \{0\} \to \mathbb{R}.$$

We can, of course, restrict $1/\mathscr{I}$ to any subset of $\mathbb{R} - \{0\}$, not necessarily an interval. With the usual abuse of language (see 2.6.2) we often refer to all such restrictions of the function 26.4.2 as '$1/\mathscr{I}$'.

Exercise 3

(i) Prove that the function $y = \tan x$ is $\sin \cdot ((1/\mathscr{I}) \circ \cos)$, where $\mathscr{I}$ refers to the domain of tan.
(ii) Similarly, express $y = (1 + \tan x)/(\sin x + \cos x)$ in the form $f \cdot ((1/\mathscr{I}) \circ g)$.

The examples of Exercise 3 are special cases of the following fact. Given $f: I \to \mathbb{R}$ such that $f(x) \neq 0$ for any $x \in I$, then $(f(x))^{-1}$ exists in $\mathbb{R}$; and

$$(f(x))^{-1} = (1/\mathscr{I})(f(x)) = ((1/\mathscr{I}) \circ f)(x).$$

We therefore make an abbreviation: define

26.4.3 $$1/f: I \to \mathbb{R}$$

by $$1/f = (1/\mathscr{I}) \circ f,$$

so $$(1/f)(x) = (f(x))^{-1}.$$

Exercise 4

(i) Prove that $(1/f)\cdot(1/g) = 1/fg$, and $(1/f) \circ (1/g) = 1/(f \circ (1/g))$ for any function g for which $1/g$ and the compositions are defined.
ii) If $m, n \in \mathbb{N}$, prove that $1/\mathscr{I}^m = (1/\mathscr{I})^m$, $\mathscr{I}^n \circ 1/\mathscr{I}^m = 1/\mathscr{I}^{nm}$, $(1/\mathscr{I}^m)\cdot(1/\mathscr{I}^m) = 1/\mathscr{I}^{m+n}$, $(1/\mathscr{I}^m) \circ (1/\mathscr{I}^n) = 1/\mathscr{I}^{mn}$.
(iii) If $m, n \in \mathbb{N}$, prove that

$$\mathscr{I}^m \cdot 1/\mathscr{I}^n = \mathscr{I}^{m-n} \text{ if } m > n, \quad \text{or} \quad 1 \text{ if } m = n, \quad \text{or} \quad 1/\mathscr{I}^{n-m} \text{ if } n > m.$$

[Note that we may always write $(\mathscr{I}^m \cdot 1/\mathscr{I}^n)(x) = x^{m-n}$ without confusion; composition applies only to functions, not to elements of $\mathbb{R}$.]

26.5 The Order Relation

The set $\mathbb{R}^I$ inherits from $\mathbb{R}$ an order relation (see 21.1), defined as follows.

26.5.1 DEFINITION. Given $f, g \in \mathbb{R}^I$, we write $f \leqslant g$ (or $g \geqslant f$) whenever, for all $x \in I$, $f(x) \leqslant g(x)$. The relations $f < g$, $g < f$, are defined similarly.

Exercise 5

In $(\mathbb{R}^I, \leqslant)$, ordered as in 26.5.1, prove that
(i) $f \leqslant g$ iff $f - g \leqslant 0$;
(ii) we can have $f \leqslant g$ and $f \neq g$ without having $f < g$;
(iii) $\leqslant$ is reflexive and transitive, and $f = g$ iff $f \leqslant g$ and $g \leqslant f$;
(iv) $<$ is transitive but neither reflexive nor symmetric;
(v) the law of trichotomy (4.4.6) does not hold;
(vi) if $f \leqslant g$, then for any $h \in \mathbb{R}^I$, $h + f \leqslant h + g$, while if also $h \geqslant 0$, then $h \cdot f \leqslant h \cdot g$;
(vii) if $0 < f < g$, then $0 < 1/g < 1/f$.
(viii) Do you approve of our definition of $f < g$ or would you prefer to choose another?
(ix) Show that $\theta: \mathbb{R} \to \mathbb{R}^I$ in Exercise 1(vi) preserves the order (i.e., if $x < y$ in $\mathbb{R}$, then $\theta x < \theta y$ in $\mathbb{R}^I$). Show however that $(\mathbb{R}^I, \leqslant)$ is not Archimedean, in that it is not always true that given $f, g \in \mathbb{R}^I$, and $f < g$, there exists an integer n with $nf > g$.

Chapter 27

LIMITING PROCESSES

27.1 Limits

The fundamental non-algebraic ingredient of calculus is the notion of 'convergence to a limit'. This notion arose from the study of dynamics, when moving objects were seen to approach certain positions in the course of time. Since the object of theoretical dynamics is to make models in pure mathematics of physical situations, and since 'time' and 'moving objects' are not part of pure mathematics, we have to invent suitable language for these models. This process lasted hundreds of years until it was essentially completed in the 19th century.

Let then I be an interval (possibly unbounded) and let $\mathbb{R}^I$ be as in 26.2. Let $f \in \mathbb{R}^I$, $l \in \mathbb{R}$, and let $c \in \mathbb{R}$ either lie in I or be an end-point of I. In dynamics, I will be a model of a time-interval, and $f(t)$ will be (say) the position of a particle at time $t \in I$. We are going to say what we mean by the sentence

27.1.1 *$f(x)$ converges to the limit l on I as x converges to c.*

Informally, we mean that the number $f(x)$ will become (and remain) arbitrarily close to l, provided we take x (in I) sufficiently close to c, but not at c.

The technical reasons for ignoring the value of f at c will appear later; but we emphasize: *c need not lie in I.* A formal definition, which eliminates the vague words like 'become', is as follows:

27.1.2 DEFINITION. $f(x)$ **converges to l on I as x converges to c** provided, given $\varepsilon > 0$, there exists a $\delta > 0$, such that for all $x \in I$,

$$(0 < |x - c| < \delta) \;.\!\Rightarrow\!.\; (|f(x) - l| < \varepsilon).$$

[Note that if c is an end-point, say a, of $I = \langle a, b \rangle$, then $|x - c| = x - a$; similarly if $c = b$, then $|x - c| = b - x$.]

Thus, if we are challenged to prove a claim we may have made, that 27.1.1 holds, then we have to find a δ to match every ε given by our challenger. His ε measures the 'arbitrarily close' part of the informal wording: our δ covers the 'sufficiently close', and need not be particularly 'small'. Straightaway, let us prove:

27.1.3 THEOREM. *If the limit l exists, then it is unique.*

Proof. For suppose, on the contrary, that some number $m \neq l$ satisfies 27.1.1 also. We can assume $m > l$, and so take ε in 27.1.2 to be the positive number $\varepsilon = \frac{1}{3}(m - l) > 0$. Then there exist numbers $\delta, \delta' > 0$ such that 27.1.2 holds for δ while

$$0 < |x - c| < \delta' .\Rightarrow. |f(x) - m| < \varepsilon.$$

Let δ'' be the smaller of δ, δ'. Then since $\delta'' > 0$ and c is in I or an end-point of I, $\exists x \in I$ such that $0 < |x - c| < \delta''$; so we have simultaneously that $|f(x) - m| < \varepsilon$ and $|f(x) - l| < \varepsilon$. Hence, by the triangle inequality (14.8.1), we have

$$|l - m| \leqslant |l - f(x)| + |f(x) - m| < \varepsilon + \varepsilon = 2\varepsilon,$$

hence $m - l = |l - m| \leqslant \frac{2}{3}(m - l)$, contrary to the fact that $m - l \neq 0$. This contradiction arose from assuming the existence of $m \neq l$, so l alone satisfies 27.1.1. ■ We often abbreviate 27.1.1 to

27.1.4 $$\lim_{x \to c} f(x) = l,$$

and write also

$$f(x) \to l \quad \text{as} \quad x \to c,$$

when it is understood that $x \in I$. We do not attempt to say what 'x converges to c' means, since the clause occurs only as part of sentences like 27.1.1; and we have defined the *whole* of 27.1.1.

Not all functions tend to limits, of course. For example, on $\langle 0, 1 \rangle$, $\sin(1/x)$ does not tend to a limit as $x \to 0$, since it never settles to a 'steady state' (as the graph shows in Fig. 27.1(a)).

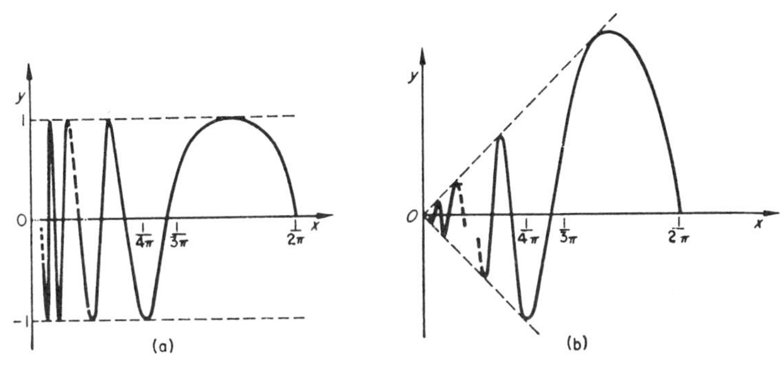

Fig. 27.1 (a) $y = \sin 1/x$ $(0 < x \leqslant 1/\pi)$. (b) $y = x \sin 1/x$ $(0 < x \leqslant 1/\pi)$.

Exercise 1

(i) Prove this last statement formally, using 27.1.2.

(ii) Prove that on $\langle 0, 1 \rangle$, $x \sin(1/x) \to 0$ as $x \to 0$. (See Fig. 27.1(b).)

(iii) In the proof of 27.1.3, write down an explicit number x belonging to I, such that $0 < |x - c| < \delta''$.
(iv) Let $J = \langle u, v \rangle$ be an interval contained in I, and in 27.1.2 let $g = f|J$. If $c \in \langle\!\langle u, v \rangle\!\rangle$, prove that also $g(x) \to l$ as $x \to c$ on J.
(v) Formulate the negation of 27.1.2, using quantifiers. {Recall 1.8.7.}
(vi) Show that, if in 27.1.2 we allow c to lie in $\mathbb{R} - \langle\!\langle a, b \rangle\!\rangle$, then for *any* $l \in \mathbb{R}$, $f(x) \to l$ as $x \to c$ on I! [This is an exercise on logic.]

27.2 The Algebra of Limits

In practice, when deciding whether or not a given function f tends to a limit, it often happens that f is built out of simpler functions; for example, a polynomial is a sum of multiples of powers of $\mathscr{I}$. If these tend to limits, a suitable combination gives the limit of f. Thus to save returning to the definition 27.1.1 again and again in the future it is useful to record algebraic properties of the symbolism 27.1.1 as follows. These properties are proved in books on analysis (see, e.g., Quadling [104]). Suppose then that we are given $f, g: I \to \mathbb{R}$ such that

$$\lim_{x \to c} f(x) = l, \qquad \lim_{x \to c} g(x) = m.$$

Then the following rules hold:

27.2.1 *For any* $\lambda \in \mathbb{R}$, $$\lim_{x \to c} \lambda f(x) = \lambda l;$$

27.2.2 $$\lim_{x \to c} (f(x) + g(x)) = l + m;$$

27.2.3 $$\lim_{x \to c} f(x) \cdot g(x) = l \cdot m;$$

27.2.4 *if* $l \neq 0$, *and* $f(x) \neq 0$ *for any* $x \in I$, *then* $\lim_{x \to c} 1/f(x) = 1/l;$

27.2.5 *if f is the constant function l, then* $\lim_{x \to c} f(x) = l;$

27.2.6 *if f is the identity function $\mathscr{I}$ on I, then* $\lim_{x \to c} \mathscr{I}(x) = c.$

Observe that, for example, 27.2.2 tells us first that $\lim_{x \to c} (f(x) + g(x))$ actually exists, and second that we can compute it by adding $\lim_{x \to c} f(x)$ to $\lim_{x \to c} g(x)$. Similar remarks apply to the other propositions; each backs up a guarantee of existence with a computational rule.

We will be content to prove 27.2.1 and suggest (in an exercise below) that the reader attempt to prove the remaining assertions guided by the model we give.

We first observe that the assertion is trivial if $\lambda = 0$, for obviously $\lim_{x\to c} 0 = 0$. Thus we suppose $\lambda \neq 0$. Our claim is then that, given any $\varepsilon > 0$, there exists $\delta > 0$ such that if $x \in I$ and $0 < |x - c| < \delta$, then $|\lambda f(x) - \lambda l| < \varepsilon$. Now $|\lambda f(x) - \lambda l| = |\lambda|\,|f(x) - l|$. Thus to assert $|\lambda f(x) - \lambda l| < \varepsilon$ is to assert that $|f(x) - l| < \varepsilon/|\lambda|$. But since $\lim_{x\to c} f(x) = l$, we *can* find δ such that if $x \in I$ and $0 < |x - c| < \delta$, then $|f(x) - l| < \varepsilon/|\lambda|$. This δ thus enables us to 'respond to the challenge' and hence shows that $\lim_{x\to c} \lambda f(x) = \lambda l$. ■

It is difficult, and perhaps not especially valuable, to formalize the basic conceptual move in the argument above, but a remark may be helpful. We look for a δ which 'works' for the given ε and the function $\lambda f(x)$; we find that we can use the δ which 'works' for $\varepsilon/|\lambda|$ and the function $f(x)$; in very imprecise but probably useful symbols,

$$\delta(\varepsilon;\, \lambda f(x)) = \delta(\varepsilon/|\lambda|;\, f(x)).$$

The point is that we know that, on the right-hand side above, we may put any positive real number into the first slot so long as we have $f(x)$ in the second. It then turns out that we wish to put $\varepsilon/|\lambda|$ into the first slot in order to have ε in the first slot on the left-hand side. In tackling 27.2.2, the corresponding (imprecise) statement is, as we hope the reader will have no difficulty in discovering,

$$\delta(\varepsilon;\, f(x) + g(x)) = \mathit{min}(\delta(\varepsilon/2;\, f(x)),\ \delta(\varepsilon/2;\, g(x)))$$

where *min* means 'the smaller of'.

Exercise 2

(i) Try proving 27.2.1–27.2.6 directly from 27.1.2.
(ii) In 27.2.4, show that if $l > 0$, then there exists a number $h > 0$, such that $f(x) > 0$ if $x \in \langle c - h, c + h\rangle \subseteq I$. Similarly when $l < 0$.
(iii) Let $g, h \in \mathbb{R}^I$ be functions such that $0 \leqslant g \leqslant h$, and suppose that $h(x) \to 0$ as $x \to c$ on I. Prove that then $g(x) \to 0$ as $x \to c$.
(iv) Use the above rules to show that, if $p(x) = \Sigma a_r x^r$ is a polynomial function in $\mathbb{R}^I$, then $\lim_{x\to c} p(x)$ exists and is $p(c)$ $(= \Sigma a_r c^r)$.

27.2.7 Observe in the last exercise that, if instead one returns to the definition 27.1.1, then direct calculation of the δ required is awkward. For example, suppose that $p(x) = x^2$, and let us prove directly that $p(x) \to c^2$ as $x \to c$. Let $\varepsilon > 0$ be given (by a challenger) and let us find $\delta > 0$, such that $|x^2 - c^2| < \varepsilon$ whenever $0 < |x - c| < \delta$. Now

$$\begin{aligned}|x^2 - c^2| = |x + c|\cdot|x - c| &= |(x - c) + 2c|\cdot|x - c| \\ &\leqslant (|x - c| + 2|c|)\cdot|x - c|,\end{aligned}$$

using the triangle inequality,

$$\leqslant (2|c| + 1)\cdot|x - c|$$

if we make sure first of all that $|x - c| < 1$. Hence, to make $|x^2 - c^2| < \varepsilon$, we need to ensure also that

$$|x - c| < \varepsilon/(2|c| + 1),$$

so we can take our 'δ' to be the lesser of the two numbers 1 and $\varepsilon/(2|c| + 1)$; the challenge has been met.

Note the pattern of this proof. We first play around with the quantity $q = |f(x) - l|$ (here $|x^2 - c^2|$) to get an idea of how small to take $|x - c|$ to make q less than the challenger's ε. Some ingenuity† is usually required, and this is why we prefer to concentrate the ingenuity in proving 27.2.1–27.2.6 once and for all. We then apply algebraic arguments mechanically as needed.

27.3 Infinite limits

It is often necessary to say that $f: I \to \mathbb{R}$ becomes 'large' as $x \to c$; for example when f is $1/\mathscr{I}: \langle 0, 1\rangle \to \mathbb{R}$ and $c = 0$. We therefore make a definition:

27.3.1 DEFINITION. $f: I \to \mathbb{R}$ is said to **tend to infinity as** $x \to c$ (in symbols

$$\lim_{x \to c} f(x) = \infty, \qquad \text{or} \quad f(x) \to \infty \quad \text{as} \quad x \to c),$$

whenever, given $\Delta \in \mathbb{R}$ (however large), there exists $\delta > 0$ such that $\forall x \in I$,

27.3.2 $$0 < |x - c| < \delta \;.\Rightarrow.\; f(x) > \Delta.$$

Observe that we do *not* say what 'infinity' means, but only what '$f(x) \to \infty$ as $x \to c$' means. As an example, let us prove the assertion above that $1/\mathscr{I} \to \infty$ as $x \to 0$ on $\langle 0, 1\rangle$; although it may seem 'obvious', we have to show that it holds by the standards of Definition 27.3.1. Let then $\Delta \in \mathbb{R}$ be given, so that we must find $\delta > 0$ such that (since c here is 0) if $0 < |x| < \delta$ then $(1/\mathscr{I})x > \Delta$. But $(1/\mathscr{I})x = 1/x$, which exceeds Δ whenever $0 < x < 1/\Delta$. Thus we may take δ to be $1/\Delta$; for example, if a challenger gave us $\Delta = 1$ million, then we may take $\delta = 10^{-6}$. Thus the test is passed: we have proved that $1/x \to \infty$ as $x \to 0$ on $\langle 0, 1\rangle$.

27.3.3 We can, as in projective geometry, add a new point labelled '∞' to $\mathbb{R}$, and say that a **neighbourhood** U_a **of** ∞ is any set of the form

$$\{x | x \in \mathbb{R} \;\&\; x > a\}.$$

Then just as 27.1.2 required loosely that $f(x)$ be 'close' to $f(c)$ when x is 'close' to c, the same kind of requirement is stated in 27.3.2: $f(x)$ is 'close' to ∞ (i.e., in U_Δ) when x is 'close' to c. Mathematicians often speak like

† Often called 'epsilonology', 'ε-chasing', or 'making with the ε's'. Learning the technique is a notorious stumbling block to all students of the subject. It is surely helpful in overcoming this block to take some definite values for ε (e.g., $\frac{1}{10}, \frac{1}{100}, \frac{1}{1000}$) and challenge the student to produce suitable values for δ.

this in conversation, but for formal work it is essential to use codifications like 27.1.2 and 27.3.2.

Rules hold similar to those in 27.2, but we need a further definition:

27.3.4 DEFINITION. We write: $\lim f(x) = -\infty$ iff $\lim(-f(x)) = \infty$.

We must then beware of a straight analogue of 27.2.2, since for example $1/x \to \infty$ and $-1/x^2 \to -\infty$, as $x \to 0$, but we cannot say '$1/x + (-1/x^2)$ then tends to $\infty + (-\infty) = 0$', since we have not defined the sum of ∞ and $-\infty$ (even if we introduce them as separate 'points' as suggested in 27.3.3).

Exercise 3

(i) Prove that $1/x - 1/x^2 \to -\infty$ as $x \to 0$.

(ii) Prove analogues of 27.2.1, etc., as $x \to c$:

(a) if $f(x) \to \infty$, then $\lim_{x \to c} \lambda f(x) = \infty$, 0 or $-\infty$, according as λ is positive, zero, or negative;

(b) if in 27.2.2, l is† ∞ but $m \in \mathbb{R}$, then $\lim_{x \to c}(f(x) + g(x)) = \infty$;

(c) if in 27.2.3, l is ∞, then $\lim_{x \to c} f(x)g(x) = X$, where:

(1) if $m \in \mathbb{R}$, then X is ∞, 0, $-\infty$ according as m is positive, zero, or negative, respectively; and

(2) X is ∞ or $-\infty$ if m is ∞ or $-\infty$, respectively; [thus, for example, if l and m are both $-\infty$, then

$$\begin{aligned}\lim_{x \to c} f(x)g(x) &= \lim_{x \to c}(-f(x))(-g(x)) \\ &= \lim_{x \to c}(-f(x)) \lim_{x \to c}(-g(x)) \\ &= \infty \qquad \text{by (2) with } l = m = \infty];\end{aligned}$$

(d) If $f(x) > 0$ on I and $\lim_{x \to c} f(x) = 0$, then $1/f(x) \to \infty$ as $x \to c$; show that $1/x$ on $\mathbb{R} - \{0\}$ does not satisfy the hypothesis and does not tend to any limit as $x \to 0$.

(iii) Formulate 27.3.1 and its negation, using quantifiers.

Given $f: I \to \mathbb{R}$, where I is unbounded and of the form $\langle a, \infty \rangle$, it may happen that the behaviour of $f(x)$ settles down as x becomes large. More precisely then, we make a definition:

27.3.5 DEFINITION. When $l \in \mathbb{R}$, we write $\lim_{x \to \infty} f(x) = l$ (or $f(x) \to l$ as $x \to \infty$), iff given $\varepsilon > 0$, there exists $\Delta \in \mathbb{R}$ such that

$$x > \Delta \;.\Rightarrow.\; |f(x) - l| < \varepsilon.$$

(In the language of 27.3.3, we require that when x is 'close' to ∞, then $f(x)$ is 'close' to l.) Direct analogues of 27.2.1–27.2.5 hold, with c everywhere replaced by ∞.

† Here we do not mean 'l is the number ∞' (there is no such number); we mean 'the symbol l denotes ∞'. We may even *write* '$l = \infty$'.

Exercise 4

(i) Prove the last sentence directly from Definition 27.3.5.
(ii) Formulate a definition of

$$\lim_{x \to -\infty} f(x) = l,$$

and then write out proofs analogous to those in (i) above.
(iii) Formulate 27.3.5, and its negation, using quantifiers.

Finally, how would we say, with the language we now possess, that $f(x) \to \infty$ as $x \to \infty$?

27.3.6 DEFINITION. $\lim_{x \to \infty} f(x) = \infty$ iff given $\Delta \in \mathbb{R}$ (however large), there exists $X \in \mathbb{R}$ such that

$$x > X \,.\!\Rightarrow\!.\, f(x) > \Delta.$$

Exercise 5

(i) Formulate analogously definitions of

$$\lim_{x \to -\infty} f(x) = \infty, \qquad \lim_{x \to \infty} f(x) = -\infty.$$

(ii) Formulate and solve an analogue of Exercise 3, with c replaced first by ∞, then by $-\infty$.
(iii) Formulate 27.3.6 and its negation, using quantifiers.

27.4 Sequences

A **sequence of real numbers** is defined to be a function $a\colon \mathbb{N} \to \mathbb{R}$, but it is customary to denote such a sequence by $a_1, a_2, \ldots, a_n, \ldots$, or by $\{a_n\}$. Sequences are a useful tool in analysis, particularly when they 'converge'; to define convergence we take Definition 27.3.5 as a model and formulate:

27.4.1 DEFINITION. If $l \in \mathbb{R}$, the sequence $a\colon \mathbb{N} \to \mathbb{R}$ **converges to the limit** l (and we write

$$\lim_{n \to \infty} a_n = l, \qquad \text{or} \quad a_n \to l \quad \text{as} \quad n \to \infty)$$

iff given $\varepsilon > 0$, there exists $\Delta \in \mathbb{N}$, such that if $n \in \mathbb{N}$, then

$$n > \Delta \,.\!\Rightarrow\!.\, |a_n - l| < \varepsilon.$$

Direct analogues of 27.2.1–27.2.5 hold, when f and g there are replaced by sequences $a, b\colon \mathbb{N} \to \mathbb{R}$.

Exercise 6

(i) Prove the last sentence.
(ii) Given $f\colon \mathbb{R} \to \mathbb{R}$, let g denote the restriction $f|\mathbb{N}$. If $f(x) \to l$ as $x \to \infty$, prove that $g_n \to l$ as $n \to \infty$. Does the converse hold?

(iii) Prove that the sequence given by

$$a_n = 0 \ (n \text{ even}), \qquad a_n = 1 \ (n \text{ odd})$$

is not convergent (i.e., does not converge to any limit).
(iv) Formulate 27.4.1 and its negation, using quantifiers.
(v) Formulate a definition of '$a_n \to \infty$ as $n \to \infty$', analogous to 27.3.6. Form its negation, using quantifiers. Prove that id: $\mathbb{N} \to \mathbb{R}$ tends to ∞ as $n \to \infty$.
(vi) Prove that the set of all sequences of real numbers forms a real algebra with operations defined as in 26.2.1. Show that a consequence of Exercise 6(i) is that the set of all convergent sequences forms a sub-algebra.
§(vii) Given a sequence $\{a_n\}$ such that $a_n \to 0$ as $n \to \infty$, prove that

$$(a_1 + a_2 + \cdots + a_n)/n \to 0 \quad \text{as} \quad n \to \infty.$$

Does the converse hold?
§(viii) Let $S \subseteq \mathbb{R}$ be a non-empty set with greatest lower bound s (see 24.6.9). Prove that there exists a sequence $\{s_n\}$ of points in S, converging to s.
§(ix) Let Σ denote the set of all bounded sequences $a: \mathbb{N} \to \mathbb{Q}$. Prove that it is an algebra over $\mathbb{Q}$. In Σ, a sequence which converges to zero is called a *null* sequence. Prove that the set R of all null sequences is both an ideal and a sub-algebra of Σ. Define a map η: $\mathbb{Q} \to \Sigma/R$ as follows. Given $q \in \mathbb{Q}$, let q^* denote the constant sequence $q: \mathbb{N} \to \mathbb{Q}$; and then define $\eta(q^*)$ to be the coset of q^* modulo R. Prove that η is one-one and a ring homomorphism. [Thus η 'embeds' a copy of $\mathbb{Q}$ in Σ/R; it can be shown that Σ/R contains a field isomorphic to $\mathbb{R}$, and this is essentially Cantor's construction of $\mathbb{R}$ starting from $\mathbb{Q}$. See Chapter 24.]

Chapter 28

CONTINUOUS FUNCTIONS

28.1 The Algebra $\mathscr{C}(I)$

There are many functions in $\mathbb{R}^I$ which behave in a fashion too random and jumpy for anything significant to be said about them. For example, $\mathbb{R}^I$ contains the function $h: I \to \mathbb{R}$, where

$$h(x) \doteq \begin{cases} x & \text{if } x \text{ is irrational,} \\ 0 & \text{otherwise,} \end{cases}$$

whose graph we cannot possibly draw on paper. We therefore pick out some significant sub-algebras of $\mathbb{R}^I$, and the first of these is $\mathscr{C}(I)$, the subset of all *continuous* functions $f: I \to \mathbb{R}$ where I is now possibly unbounded. A continuous function is described as follows, using the concept of limit defined in 27.1. Let $c \in I$ (so f is defined at the point c; as in 27.1.2, c may be an end-point of I).

28.1.1 DEFINITION. We say that f is **continuous** at c iff $\lim_{x \to c} f(x) = f(c)$ (on I). If f is continuous at every point in I, we say that f is continuous (on I).

The definition was formulated with the intention of picking out those functions whose graph is in one piece, and it can be proved that this intention is justified. The function h, mentioned above, is continuous only at the origin, and its graph is certainly not a curve in one piece, nor even one in 'chunks', as in Fig. 28.1(b). Also, when f is continuous at $p \in I$, and $x \in I$ is near p, then $f(x)$ is near $f(p)$. [But the reader should beware of believing that f is necessarily continuous at x.] In Fig. 28.1(b) we can take x as near as we like to p, but if $x \leqslant p$, then $f(x)$ is no nearer than the distance ST. For a precise discussion, see Quadling [104].

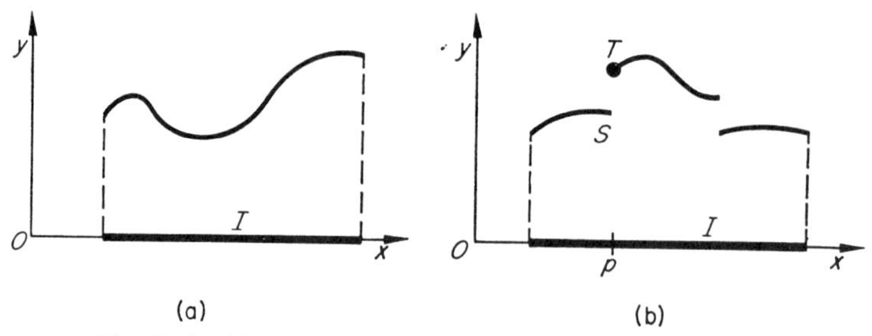

Fig. 28.1 (a) continuous graph. (b) discontinuous graph.

We now state some basic facts about continuous functions. A sample proof follows 28.1.3. For practical purposes, we draw the reader's attention to the criterion in Exercise 2(ii) below.

28.1.2 THEOREM. *$\mathscr{C}(I)$ is an algebra over* $\mathbb{R}$ (in fact a sub-algebra of $\mathbb{R}^I$).

In other words, the sum and product of continuous functions are continuous functions; and a constant times a continuous function is a continuous function.

28.1.3 THEOREM. *A constant function is continuous, and so is the identity function $\mathscr{I}$. If $0 \notin I$, then the reciprocal function* (26.4.1) *$1/\mathscr{I}$ is continuous on I.*

These statements are simple consequences of the algebra of limits 27.2.1–27.2.6. For example, in 28.1.2, it is necessary to show that if $f, g \in \mathscr{C}(I)$, then so does $f + g$. Therefore we must show that at any $c \in I$,

$$\lim_{x \to c} (f + g)(x) = (f + g)(c).$$

But the left-hand side means $\lim_{x \to c} (f(x) + g(x))$

$$
\begin{aligned}
&= \lim_{x \to c} f(x) + \lim_{x \to c} g(x) && \text{(by 27.2.2)} \\
&= f(c) + g(c) && (f, g \text{ continuous at } c) \\
&= (f + g)(c) && (\text{definition of } f + g)
\end{aligned}
$$

as required. Similar algebra supplies proofs of all the statements summarized by 28.1.2 and 28.1.3. No 'epsilon' proofs are needed. ■

Exercise 1

Give these proofs in detail.

A consequence of 28.1.2 is that $\mathscr{I}^2$ is continuous (since $\mathscr{I}$ is); therefore so is $\mathscr{I}^3 = \mathscr{I}^2 \cdot \mathscr{I}$, and by induction on n, $\mathscr{I}^n$ is continuous. Therefore, for every $\lambda \in \mathbb{R}$, $\lambda \cdot \mathscr{I}^n$ is continuous, being a product of two continuous functions. Hence, by induction on the degree,

28.1.4 THEOREM. *Every polynomial function $\sum_{r-0}^{n} c_r \mathscr{I}^r$ is continuous.* ■

This last is a typical exercise in showing that an unfamiliar object (here the polynomial function) is constructed from simpler objects by sums and products, and then applying the rules about such constructions, which are summarized in Theorem 28.1.2. Of course, the reader may well be more familiar with the polynomial function in its traditional guise $\sum_{r=0}^{n} c_r x^r$.

28.2 Composition

Another basic method of constructing new functions from old is to compose them. Thus, we have a further rule (whose proof does not follow immediately from the algebra of 27.2; we have to revert to the definition 27.1.2 (see Exercise 2(ii) below).

28.2.1 THEOREM. *If $f \in \mathscr{C}(I)$, $J \subseteq \mathbb{R}$ is an interval such that $f(I) \subseteq J$, and $g: J \to \mathbb{R}$ is a continuous function, then $g \circ f: I \to \mathbb{R}$ is continuous.*

We shall discuss the proof of this theorem in 28.2.2 below. Meanwhile, for example, once it is known that the cosine function $\cos: \mathbb{R} \to \mathbb{R}$ is continuous, then $\cos(5x^3 + 6x + 1)$ is a continuous function of x on $\mathbb{R}$; for it is the composite $\cos \circ p$, where $p(x) = 5x^3 + 6x + 1$, and $p: \mathbb{R} \to \mathbb{R}$ is a polynomial function, hence (by 28.1.4) continuous.

Exercise 2

(i) If $f \in \mathscr{C}(I)$ and f is nowhere zero on I, prove that $1/f \in \mathscr{C}(I)$. Does this fact make $\mathscr{C}(I)$ a field?

(ii) Using the notion of a sequence (27.4), prove that $f: I \to \mathbb{R}$ is continuous iff the following condition holds. Let $x: \mathbb{N} \to I$ be any sequence with limit c; *then* $f(x_n) \to f(c)$ *as* $n \to \infty$. [Cf. 25.5.5.]

28.2.2 The condition given in the last exercise enables us to express continuity algebraically, by the fact that '*a continuous f commutes with lim*'; for we have just seen that

$$\lim_{n \to \infty} f(x_n) = f(c) = f(\lim_{n \to \infty} x_n)$$

or '$\lim f = f \lim$'. We can now prove algebraically that 28.2.1 holds, as follows.

Proof of 28.2.1. Let $x_n \to c$; then we may suppose $f(x_n) \in J$, so

$$\begin{aligned}\lim (g \circ f)(x_n) &= \lim g(f(x_n)) && \text{(definition of } g \circ f) \\ &= g(\lim f(x_n)) && (g \text{ continuous}) \\ &= g(f(\lim x_n)) && (f \text{ continuous}) \\ &= (g \circ f) \lim x_n,\end{aligned}$$

so $g \circ f$ commutes with lim and is therefore continuous. ■

28.3 The Principle of Preservation of Inequalities

If $f \in \mathbb{R}^I$ converges to l as $x \to c$, we saw in 27 Exercise 2(ii) that if $l > 0$, then $f(x) > 0$ near c, while if $l < 0$, then $f(x) < 0$ near c. From this fact we deduce the following:

28.3.1 **Inertia Principle.** *If I is an open interval, if $c \in I$, if $f \in \mathscr{C}(I)$ and $f(c) > 0$, then there is an open interval $J \subseteq I$, such that $c \in J$ and*† $f > 0$ *on J.*

■ A similar statement holds, of course, when the inequalities are reversed.

† i.e., $f(x) > 0$ for each $x \in J$ (see 26.5.1).

For example, let $f(t)$ denote the velocity at time t, of a particle which is moving continuously. If at some instant t_0 we have $f(t_0) \neq 0$, then the velocity remains non-zero and in the same sense for some positive length of time before and after t_0; one cannot make a car stop dead, nor instantaneously turn off a light.

An extension of the inertia principle is called the **principle of preservation of inequalities**:

28.3.2 *With I, c as in* 28.3.1, *suppose $f, g \in \mathscr{C}(I)$ and $f(c) > g(c)$. Then there is an open interval $J \subseteq I$ such that $c \in J$ and $f > g$ on J.*

Proof. Apply 28.3.1 to $f - g$. ■

28.4 Max and Min

Let $f, g \in \mathscr{C}(I)$. We construct new functions $h, k \in \mathbb{R}^I$ according to the rules

$$h(x) = \max(f(x), g(x))$$
$$k(x) = \min(f(x), g(x))$$

for each $x \in I$. It can be shown (and the proof uses 28.3.2), that *h and k are in fact continuous; so $h, k \in \mathscr{C}(I)$.* This is obvious from a graph. One consequence is that the **absolute value function**

28.4.1 $$\text{abs}: \mathbb{R} \to \langle 0, \infty \rangle,$$

given by $\text{abs}(x) = |x|$, is a continuous function of x on $\mathbb{R}$; for, $|x| = \max(x, -x)$ while $\mathscr{I}$, $-\mathscr{I}$ are continuous by 28.1.3 and 28.1.2. Hence,

$$\text{if} \quad f \in \mathscr{C}(I) \quad \text{then} \quad |f| \in \mathscr{C}(I) \quad [\text{by 28.2.1, since} \quad |f| = \text{abs} \circ f].$$

Alternatively, see Exercise 4(vi), below.

28.5 Two Deeper Theorems

We now state two 'obvious' theorems, whose proofs are much more difficult than any so far, and of a different kind; they depend essentially on the completeness of the number system (24.6).

These proofs are omitted, as they are to be found in any book on analysis. However, each theorem asserts the *existence* of something, which gives the clue that the proof must rely on some axiom of existence in $\mathbb{R}$, such as the 'least upper bound axiom'. [Theorem 24.6.9 states that the system $\mathbb{R}$ constructed in Chapter 24 in fact satisfies the axiom.]

28.5.1 **The intermediate value theorem.** *If $f \in \mathscr{C}(I)$, while $x_1, x_2 \in I$ and α is any number such that*

$$f(x_1) < \alpha < f(x_2),$$

then there exists $z \in I$ between x_1 and x_2 such that $\alpha = f(z)$. □

Geometrically, this is a consequence of saying that the graph of f is in one piece; in Fig. 28.2, the graph is at x_1 below the line $y = \alpha$ and at x_2 above this line; hence the graph hits the line somewhere. If we replace I by the set of *rationals* between a and b, the theorem is false. For suppose J is the set of all rationals between 0 and 2, and $f = \mathscr{I}^2$. Then take $\alpha = 2$, which lies between 0 $(= f(0))$ and 4 $(= f(2))$; there is no $z \in J$ such that $f(z) = 2$, since $\sqrt{2}$ is not rational (see 24 Exercise 1(i)).

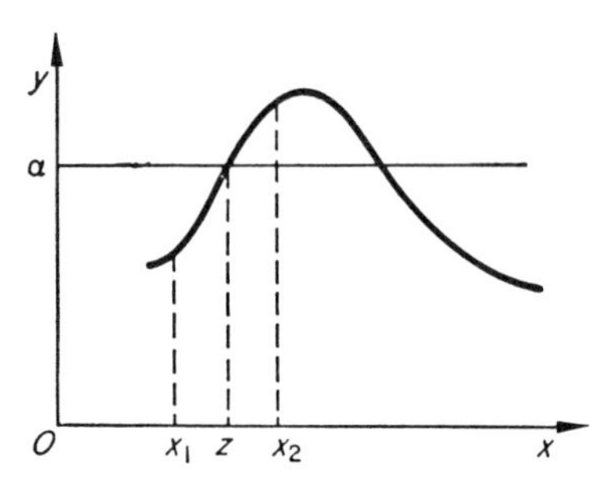

Fig. 28.2

This theorem has some surprising practical consequences: see Courant–Robbins [25], Ch. VI, §6, p. 317. It was, of course, in practical use by mathematicians long before it was proved: for example, to find a root of a polynomial like $p(x) = x^3 + x + 1$; then $p(-1) < 0$, while $p(0) > 0$, so by 28.5.1 a root lies in the interval $\langle\!\langle -1, 0 \rangle\!\rangle$, and one now tries (say) $x = -\frac{1}{2}$, gradually pinning the root in a smaller and smaller interval, ceasing when one has achieved a sufficiently good approximation. Before the theorem could be proved properly (and it eventually was proved by Bolzano in 1850), a precise definition of $\mathbb{R}$ had to be given, and the technique we have outlined above for solving equations gave the clue. (See Chapter 24, and also Bourbaki's *Histoire* [18].)

The second theorem tells us, among other things, that an aeroplane moving continuously for (say) an hour has in that period a greatest and least speed, a greatest and least altitude, and a hottest and coolest spot in its engines. For such reasons, we suggest the informal name 'mostest' for the theorem (which is one of many due to Weierstrass and has no 'official' name). See also Theorem 25.8.8.

28.5.2 **The 'mostest' theorem.** *Let $I = \langle\!\langle a, b \rangle\!\rangle$ be a closed finite interval, and let $f \in \mathscr{C}(I)$. Then* (i) *f is bounded, i.e., there exist $m, M \in \mathbb{R}$ such that for each $x \in I$,*

$$m \leqslant f(x) \leqslant M;$$

(ii) *f achieves its precise bounds, i.e., there exist a greatest such m, and a smallest such M, and there exist $u, v \in I$, such that $f(u) = m, f(v) = M$.* □

As an example when I is not finite, the function $\mathscr{I}: \langle\!\langle 0, \infty\rangle \to \mathbb{R}$ is not bounded above (i.e., no M exists as in (i)); as an example when I is finite but not closed, the function $1/\mathscr{I}: \langle 0, 1\rangle\!\rangle \to \langle\!\langle 1, \infty\rangle$ is not bounded above, since it takes arbitrarily large values x^{-1} as x approaches 0. Also cos: $\langle 0, \pi/2\rangle \to \langle 1, 0\rangle$ is bounded (by 0 and 1), but does not achieve its bounds (since u, v in (ii) would have to be $\pi/2$ and 0 respectively, and these do not lie in the given domain).

If we couple 28.5.2 to the intermediate value theorem 28.5.1, we obtain at once the

28.5.3 **Mapped-interval theorem.** *If $f \in \mathscr{C}(I)$, where I is $\langle\!\langle a, b\rangle\!\rangle$, then $f(I)$ is an interval. Indeed, $f(I) = \langle\!\langle m, M\rangle\!\rangle$, where m, M are as in* 28.5.2(ii). ■

Exercise 3

(i) What can you infer, from the intermediate value theorem and the mostest theorem, about $f(I)$ when $f \in \mathscr{C}(I)$ but I is either not closed or not finite?

(ii) Deduce condition 28.5.2(ii) from 28.5.2(i) as follows. If f did not take the value m in 28.5.2(ii), then $g(x) = 1/(f(x) - m)$ would be a *continuous* function with domain I, and $g(x) > 0$. By condition 28.5.2(i), $g(x) \leqslant K$ for some $K \in \mathbb{R}$, $K > 0$, so $f(x) \geqslant m + K^{-1}$, contrary to the 'greatest' nature of m. From this contradiction, 28.5.2(ii) follows.

28.5.4 If $f \in \mathscr{C}(I)$, then $f(I)$ is an interval J by the mapped-interval theorem. If also $f: I \to J$ is bijective, then $f^{-1}: J \to I$ exists, by the inversion theorem; and since f is continuous, it can be proved, by using the mostest theorem, that *f^{-1} is continuous*, provided always, of course, that I is closed.

Now, to decide whether or not f is bijective, we need only to be able to decide whether or not f is one-one (since f is onto J). For this, we need the notion of a *monotonic function*:

28.5.5 DEFINITION. If $A \subseteq R$ and $g: A \to \mathbb{R}$ is any function, then g is said to be **monotonic increasing** provided

$$x_1 < x_2 \quad \text{on} \quad A \;.\!\Rightarrow\!.\; gx_1 \leqslant gx_2.$$

It is **monotonic decreasing** if, instead, $gx_1 \geqslant gx_2$; and such functions are **monotonic.** They are **strictly monotonic** if strict inequalities are used throughout.

Thus, for example, $\mathscr{I}$ and a constant function are both monotonic increasing, but $\mathscr{I}$ is strictly so; while $-\mathscr{I}^3$ is strictly monotonic decreasing.

Exercise 4

(i) Prove that if g in 28.5.5 is strictly monotonic, then g is one-one.

(ii) If $A = \langle\!\langle a, b\rangle\!\rangle$, while g is continuous and monotonic, prove that $g(a)$ is an end-point of the interval $g(A)$. Show that $g(a) \leqslant g(b)$ if g is increasing. What if g is strictly decreasing?

(iii) Let $m \in \mathbb{N}$. Show that $\mathscr{I}^m$ is strictly monotonic increasing on $\mathbb{R}$ if m is odd, and on $\langle\!\langle 0, \infty\rangle$ if m is even.

§(iv) If $f \in \mathbb{R}^I$, where I is closed and bounded, and if f is both onto and strictly monotonic, prove that f is continuous.
§(v) If, instead, f is continuous and one-one, prove that f is strictly monotonic.
(vi) Define $\mu: \mathbb{R} \to \langle\!\langle 0, \infty \rangle$ by $\mu(x) = \max(x, 0)$. Prove that μ is continuous and hence that *abs* is continuous. [Check that $|x| = 2\mu(x) - x$.] Hence show that h and k in 28.4 are continuous. [$h = \frac{1}{2}(f + g + |f - g|)$].
(vii) Let R be an equivalence relation on the closed interval $I = \langle\!\langle a, b \rangle\!\rangle$, with the property that given $x \in J$, $\exists \varepsilon > 0$ such that for every $y \in J$ with $|x - y| < \varepsilon$, we have xRy. Prove that aRb [Hint: let $Z = \{z \in J \mid aRz\}$ and let w be the least upper bound of Z (see 24.6); show that aRw and then that $w = b$.] This result is the 'Creeping Lemma' of Moss and Roberts, *American Math. Monthly* (1968), Vol. 75, pp. 649–652, and implies several standard theorems of Analysis where the completeness of $\mathbb{R}$ is crucial. In particular, deduce 28.5.2 by taking xRy to mean that f is bounded between x and y; and deduce 28.5.1 by supposing, for a contradiction, that $f(x)$ never equals α, and taking xRy to mean that $(f(x) - \alpha)(f(y) - \alpha) > 0$. These are striking examples of the process of 'casting roles' (see p. 129).

28.6 The Laws of Indices

Let us consider the problem of defining x^a when a is a rational number and $0 \leqslant x \in \mathbb{R}$. From our early days in school, we are brought up to believe that if a is the fraction p/q ($p, q \in \mathbb{N}$), then $x^{p/q}$ is $(x^{1/q})^p$, and $x^{1/q}$ is that number y such that $y^q = x$; and y exists in $\mathbb{R}$ if $x \geqslant 0$. How do we know that such a y exists? The answer comes from the intermediate value theorem—we look at the graph of $\mathscr{I}^q$. More precisely, let $x \geqslant 0$ be given; then one of the powers $2^q, 3^q, \ldots, n^q$ exceeds† x for some *integer* $n > 0$, so $0 \leqslant x \leqslant \mathscr{I}^q(n)$. Therefore $x = \mathscr{I}^q(y)$ for some $y \in \langle\!\langle 0, n \rangle\!\rangle$ by the intermediate value theorem applied to the continuous function $\mathscr{I}^q: \langle\!\langle 0, n \rangle\!\rangle \to \mathbb{R}$.

This proves that, for all $q \in \mathbb{N}$, the function $\mathscr{I}^q: \mathbb{R}^+ \to \mathbb{R}^+$ is onto, where $\mathbb{R}^+ = \langle\!\langle 0, \infty \rangle$. Now by (iii) and (i) of Exercise 4, this function is strictly monotonic and hence one-one on $\mathbb{R}^+$. Therefore

$$\mathscr{I}^q: \mathbb{R}^+ \to \mathbb{R}^+$$

is a bijection. In accord with the traditional notation, we denote its inverse by

28.6.1 $$\mathscr{I}^{1/q}: \mathbb{R}^+ \to \mathbb{R}^+, \qquad \mathscr{I}^{1/q}(x) = x^{1/q}$$

where the last equation now *defines* $x^{1/q}$. By the properties of inverse functions (see 2.9), we have the equations

28.6.2 $$\mathscr{I}^q \circ \mathscr{I}^{1/q} = \mathscr{I} = \mathscr{I}^{1/q} \circ \mathscr{I}^q$$

which respectively confirm the 'school' belief that $(x^{1/q})^q = x = (x^q)^{1/q}$. We now *define* $\mathscr{I}^{p/q}$, by:

28.6.3 $$\mathscr{I}^{p/q}: \mathbb{R}^+ \to \mathbb{R}^+ \quad \textit{is defined to be} \quad \mathscr{I}^p \circ \mathscr{I}^{1/q},$$

† A strict proof of this 'obvious' fact is fairly subtle, and depends on the completeness of the system of real numbers (see Chapter 24).

when p/q is a fraction in its lowest terms, and we write $x^{p/q}$ for $\mathscr{I}^{p/q}(x)$. By a remark in 28.5.4, $\mathscr{I}^{1/q}$ is continuous, and therefore so is $\mathscr{I}^{p/q}$ $(=\mathscr{I}^{p} \circ \mathscr{I}^{1/q})$ by Theorem 28.2.1. We claim

28.6.4 $$\mathscr{I}^{p/q} = \mathscr{I}^{1/q} \circ \mathscr{I}^{p}.$$

Proof. Let f denote $\mathscr{I}^{1/q} \circ \mathscr{I}^{p}: \mathbb{R}^+ \to \mathbb{R}^+$. Then $\mathscr{I}^{q} \circ f = \mathscr{I}^{p}$, by 28.6.2; on the other hand, $\mathscr{I}^{q} \circ \mathscr{I}^{p/q} = \mathscr{I}^{q} \circ (\mathscr{I}^{p} \circ \mathscr{I}^{1/q}) = \mathscr{I}^{p} \circ \mathscr{I}^{q} \circ \mathscr{I}^{1/q} = \mathscr{I}^{p}$, so $\mathscr{I}^{q} \circ f = \mathscr{I}^{q} \circ \mathscr{I}^{p/q}$. Therefore composing each side with $\mathscr{I}^{1/q}$,

$$f = \mathscr{I} \circ f = \mathscr{I}^{1/q} \circ (\mathscr{I}^{q} \circ f) = \mathscr{I}^{1/q} \circ (\mathscr{I}^{q} \circ \mathscr{I}^{p/q}) = \mathscr{I} \circ \mathscr{I}^{p/q} = \mathscr{I}^{p/q}. \quad \blacksquare$$

Now, for any $x \geqslant 0$, $\mathscr{I}^{p/q}(x) = \mathscr{I}^{p} \circ \mathscr{I}^{1/q}(x) = (x^{1/q})^{p}$, and $\mathscr{I}^{1/q} \circ \mathscr{I}^{p}(x) = (x^{p})^{1/q}$, so we have the familiar

$$x^{p/q} = (x^{p})^{1/q} = (x^{1/q})^{p}.$$

We leave now to the reader the task of deriving the traditional laws of indices, by giving suitable hints in the next exercise. (Recall also 26 Exercise 2(v) and 26 Exercise 4.)

Exercise 5

(i) If $k \in \mathbb{N}$, prove that $\mathscr{I}^{kp} \circ \mathscr{I}^{1/kq} = \mathscr{I}^{p} \circ \mathscr{I}^{1/q}$ (so that $\mathscr{I}^{p/q}$ in 28.6.3 is independent of the fraction chosen to represent the number p/q). {The left-hand side is $\mathscr{I}^{k} \circ \mathscr{I}^{p} \circ (\mathscr{I}^{k} \circ \mathscr{I}^{q})^{-1}$, and composition here is commutative.}
(ii) If also $r, s \in \mathbb{N}$, prove that $\mathscr{I}^{p/q} \circ \mathscr{I}^{r/s} = \mathscr{I}^{pr/qs}$ and hence that $(x^{r/s})^{p/q} = x^{pr/qs}$ if $x \in \mathbb{R}^+$. {The left-hand side is $\mathscr{I}^{p} \circ \mathscr{I}^{1/q} \circ \mathscr{I}^{r} \circ \mathscr{I}^{1/s} = \mathscr{I}^{p} \circ \mathscr{I}^{r} \circ (\mathscr{I}^{s} \circ \mathscr{I}^{q})^{-1}$, by 28.6.4.}
(iii) Prove further that $\mathscr{I}^{p/q} . \mathscr{I}^{r/s} = \mathscr{I}^{ps+qr/qs}$, and hence that $x^{p/q} . x^{r/s} = x^{ps+qr/qs}$. {Compose each side with $\mathscr{I}^{qs}$; and note that $(f \cdot g)^{qs} = f^{qs} \cdot g^{qs}$, $(f^{p/q})^{qs} = f^{ps}$.}
(iv) What can you say about $x^{1/q}$ when $x < 0$ and q is odd?
(v) Discuss the continuity of $\mathscr{I}^{p/q}$.

For negative exponents, we proceed similarly but replace $\mathscr{I}: \mathbb{R}^+ \to \mathbb{R}^+$ by $1/\mathscr{I}: \mathbb{R}_+ \to \mathbb{R}_+$, where $\mathbb{R}_+ = \langle 0, \infty \rangle$ to exclude zero. Thus if $p, q \in \mathbb{N}$, we define $\mathscr{I}^{-p/q}$ by

28.6.5 $$\mathscr{I}^{-p/q} = (1/\mathscr{I})^{p/q} = 1/\mathscr{I}^{p/q}: \mathbb{R}_+ \to \mathbb{R}_+.$$

If we restrict all domains to $\mathbb{R}_+$, it is then easy to check that $\mathscr{I}^{p/q} . \mathscr{I}^{-p/q} = 1$; and a long but fairly easy exercise leads us to the same laws of indices as before but with exponents p/q, r/s now arbitrary (positive or negative) rationals. We leave the details to the reader, since we shall give a treatment, in Chapter 31, for arbitrary *real* exponents.

Chapter 29

DIFFERENTIABLE FUNCTIONS

29.1 The Differential Coefficient

Let I denote an interval. We next pick out from $\mathscr{C}(I)$ a subset $\mathscr{D}(I)$ consisting of the 'differentiable' functions, which we will define after some preliminary remarks. Let us first recall the elementary notion of the tangent to a curve. Let $c \in I$, and let P, Q be the points on the graph of $f: I \to \mathbb{R}$ with co-ordinates $(c, f(c))$, $(x, f(x))$ respectively. The **gradient** of the chord PQ is defined to be the tangent of the angle between PQ and Ox; in Fig. 29.1 this angle is QPR (where R is on the ordinate through x, and PR is parallel to Ox). This gradient is therefore

$$\tan QPR = QR/PR = (f(x) - f(c))/(x - c);$$

and if this number should converge to a limit l as $x \in I$ converges to c, the line through P with gradient l is called the *tangent to the graph at c*. Of course, c may be an end-point of I, but in the definitions below (29.1.1, 29.2.1) we will have in mind an open interval I.

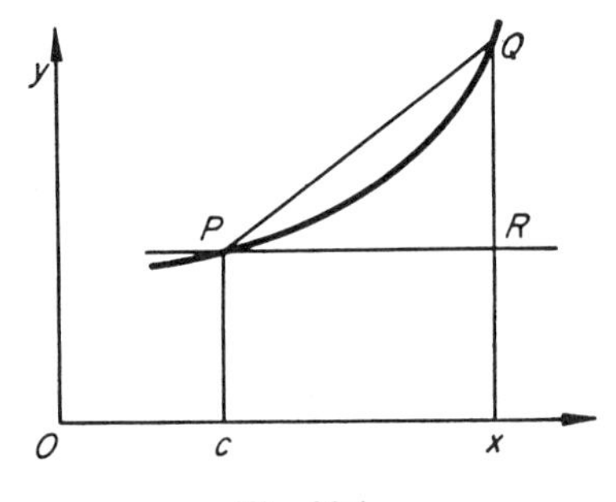

Fig. 29.1

This process of obtaining the number l from the function f and the point c is inspired by geometrical requirements; but we need make no reference to geometry in order to carry it out. We therefore make the following definition.

29.1.1 DEFINITION. The function $f: I \to \mathbb{R}$ has the **differential coefficient** l **at** $c \in I$ iff

$$\lim_{x \to c} (f(x) - f(c))/(x - c) = l.$$

We then say that f is **differentiable at** c.

Observe first that the limit is required to exist (it is then unique, by 27.1.3); second, it is required to equal l; and third, the notion of limit in 27.1.2 required no reference to what happens at $x = c$, a convenience which is essential for this application, since $(f(x) - f(c)/(x - c))$ is not defined at $x = c$.

29.2 The Derivative

We are not very interested in the differentiability of f at a *single* point, but we shall say more about this in 29.5.8, where we show that, in 29.1.1, f must be continuous at c. So we restrict our class of functions still further according to:

29.2.1 DEFINITION. If $f: I \to \mathbb{R}$ is differentiable at each $c \in I$, then f is said to be **differentiable.** The set of all such differentiable functions is denoted by $\mathscr{D}(I)$. If f is differentiable, we can form a new function $g: I \to \mathbb{R}$, where, for each $c \in I$, $g(c)$ is the differential coefficient of f at c. This new function is called the **derivative** of f, and denoted by f', or $\dot{f}$, or Df. In dynamics, $\dot{f}$ is commonly used, being called the 'velocity function' when $f(t)$ represents the distance of a particle from the origin at time t; if instead f is itself the velocity function, then $\dot{f}$ is called the 'acceleration'.

The classical notation (due to Leibniz) is df/dx, arising from the use of 'δf' for $f(x) - f(c)$, and 'δx' for $x - c$; thus the equation in 29.1.1 would be written

29.2.2
$$\lim_{\delta x \to 0} \delta f/\delta x = df/dx,$$

or $df(c)/dx$, to indicate the dependence on c. This notation has virtues, especially in practical work, but it has the same kind of vice as the use of '$f(x)$' for 'the function f'. For example, the above rule for forming δf ought logically to make us write $\delta \mathscr{I}$ for δx.

29.2.3 Historically, the use of these 'infinitesimals' δf was a great source of difficulty and led the subject into criticism by philosophers, since sometimes δf was 'neglected' as if it were zero, and so in 29.2.2 we have apparently a ratio 0/0. These difficulties were not resolved until the concept of limit was clarified and Definition 27.1.2 formulated (using numbers, not infinitesimals).

29.3 The Algebra $\mathscr{D}(I)$

To compute the derivative of a differentiable function, it is best to use where possible the following algebraic rules. In principle, one can always go back to the definition 29.1.1, but this can be awkward (see 27.2.7). As in 29.2.1, let $\mathscr{D}(I) \subseteq \mathbb{R}^I$ consist of all differentiable functions $f: I \to \mathbb{R}$. The operation of forming Df from $f \in \mathscr{D}(I)$ is itself a function $D: \mathscr{D}(I) \to \mathbb{R}^I$ (cf. Example 2.5.8), and the basic facts about D and $\mathscr{D}(I)$ are as follows; full proofs will be found in books on analysis.

29.3.1 THEOREM. *$\mathscr{D}(I)$ is a sub-algebra (over $\mathbb{R}$) of $\mathscr{C}(I)$, containing the constants and the identity function $\mathscr{I}$.* □ This statement implies that every differentiable function is continuous, that the sum and product of a pair of differentiable functions is differentiable, and that a constant times a differentiable function is differentiable. However, we would like to know more: for example, what is the relation (if any) between Df, Dg, and $D(f+g)$? Such questions are dealt with as follows.

29.3.2 THEOREM. *The function $D\colon \mathscr{D}(I) \to \mathbb{R}^I$ is a linear transformation. It is zero on the constants, and $D(\mathscr{I}) = 1$. Moreover, if $f, g \in \mathscr{D}(I)$, then*

29.3.3
$$D(f\cdot g) = Df\cdot g + f\cdot Dg.$$

The proof follows from the algebra of limits; for example, to prove that

$$D(f+g) = Df + Dg$$

(as part of the proof that D is linear), we have, for each $c \in I$, that $D(f+g)$, evaluated at c, is

$$\begin{aligned} D(f+g)|_c &= \lim_{x\to c} \frac{(f+g)(x) - (f+g)(c)}{x-c} \\ &= \lim_{x\to c} \frac{(f(x)+g(x)) - (f(c)+g(c))}{x-c} && \text{(definition of } f+g) \\ &= \lim_{x\to c} \frac{f(x)-f(c)}{x-c} + \lim_{x\to c} \frac{g(x)-g(c)}{x-c} && \text{(by 27.2.2)} \\ &= f'(c) + g'(c) && \text{(since } f, g \in \mathscr{D}(I)) \\ &= (f'+g')(c) && \text{(definition of } f'+g'). \end{aligned}$$

Hence, by definition of the equality of functions,

$$D(f+g) = f' + g' = Df + Dg,$$

as required. ■

Exercise 1

(i) Complete the entire proof of 29.3.3, on the same lines.
(ii) Show that D is not a homomorphism of algebras (26 Exercise 1(vii)) but that $x \to Df|_x$ is a derivation of rings (22 Exercise 8(vi)).

29.3.4 By using the notation of infinitesimals (29.2.3) the last proof might be written

$$\begin{aligned} \lim\,(\delta(f+g)/\delta x) &= \lim\,(\delta f/\delta x + \delta g/\delta x) \\ &= \lim \delta f/\delta x + \lim \delta g/\delta x \\ &= df/dx + dg/dx. \end{aligned}$$

This is the traditional 'school' proof and is very convincing. But in our proof above, we have made explicit the second equality (interchange of lim and sum), using 27.2.2, and it is this step which constitutes the guts of a correct proof.

29.3.5 EXAMPLE. Let us use 29.3.2 to compute the derivative of a polynomial $p = \sum_{r=0}^{n} c_r \mathscr{I}^r$. (The reader is advised to rewrite each step in the traditional 'd/dx' notation to see that we are really doing something quite familiar to him.) First we are given in 29.3.2 that $D(\mathscr{I}) = 1$; and we assert

(i) $D(\mathscr{I}^n) = n\mathscr{I}^{n-1}, \qquad n = 1, 2, 3, \ldots.$

Proof.
$$\begin{aligned} D(\mathscr{I}^{n+1}) &= D(\mathscr{I}^n \cdot \mathscr{I}) \\ &= D(\mathscr{I}^n)\cdot\mathscr{I} + \mathscr{I}^n \cdot D(\mathscr{I}) \end{aligned}$$

by the rule 29.3.3 for products. Thus, making an inductive hypothesis (since $D(\mathscr{I}) = 1$),

$$D(\mathscr{I}^{n+1}) = n\mathscr{I}^{n-1}\cdot\mathscr{I} + \mathscr{I}^n \cdot 1 = (n+1)\mathscr{I}^n,$$

and (i) follows by induction on n. ■

Remark. A common proof of (i) is based on the binomial theorem, and involves all the difficulties (and more) of proving the product rule 29.3.3.

Continuing our computation of $D(p)$, we have

$$\begin{aligned} D(p) &= D\left(\sum_{r=0}^{n} c_r \mathscr{I}^r\right) \\ &= \sum_{r=0}^{n} c_r D(\mathscr{I}^r), \end{aligned}$$

since D is linear; so

(ii) $D(p) = \sum_{r=1}^{n} r c_r \mathscr{I}^{r-1},$

since $D(c_0) = 0$, by 29.3.2. Thus we have proved that every polynomial is differentiable, by appealing to the basic rules, and by giving a formula (namely, equation (ii)) for the derivative. A direct proof from 29.1.1 would be awkward (cf. 27.2.7). These rules were, of course, the motivation behind 26 Exercise 2(iii); but the 'D' there can be used in purely algebraic situations, where no limiting processes are possible. For example, we might be interested in polynomials over a ring, as in Chapter 22. In the theory of knots, this has led to the 'free differential calculus' (see Crowell–Fox [26]).

Another important rule is given by:

29.3.6 THEOREM. *If $f \in \mathscr{D}(I)$ and f is never zero, then $1/f \in \mathscr{D}(I)$ and*

$$D(1/f) = -f'/f^2.$$

Exercise 2

(i) Prove Theorem 29.3.6, using 27.2.4, 27.2.3, and the fact that $\mathscr{D}(I) \subseteq \mathscr{C}(I)$.
(ii) If $f \in \mathscr{D}(I)$ and f is strictly increasing (decreasing) on I (see 26 Exercise 1(vi)), prove that $f' > 0$ (< 0) on I.
(iii) With f as in 29.3.6, let $g \in \mathscr{D}(I)$. Prove (algebraically) the **quotient rule**:

$$D(g/f) = (f \cdot Dg - g \cdot Df)/f^2.$$

(iv) Use 29.3.5(i) and 29.3.6 to prove that if $0 \notin I$, then

$$D(\mathscr{I}^n) = n\mathscr{I}^{n-1}, \qquad n = \pm 1, \pm 2, \ldots.$$

(v) Use 29.3.3 to establish the formula for $D(1/f)$.

29.4 Composition

Coming now to the composition operation, we state the following important rule.

29.4.1 **The chain rule.** *Let $f \in \mathscr{D}(I)$, and suppose $f(I)$ lies in an interval J. Let $g \in \mathscr{D}(J)$. Then $g \circ f \in \mathscr{D}(I)$ and*

(i) $D(g \circ f) = (Dg \circ f) \cdot Df.$

(Here the 'dot' before Df is, of course, the product in $\mathbb{R}^I$.) □ For an outline of the proof, see Exercise 4(iv) below.

In traditional books this rule is misleadingly called the 'function of a function' rule. To recognize the traditional form, write f' for Df, so that (i) takes the briefer form

(ii) $(g \circ f)' = (g' \circ f) \cdot f'.$

Now evaluate this at $x \in I$; the right-hand side is

$$\begin{aligned}((g' \circ f) \cdot f')(x) &= (g' \circ f)(x) \cdot f'(x) \\ &= g'(f(x)) \cdot f'(x),\end{aligned}$$

so in d/dx notation

$$\frac{d}{dx}(g \circ f) = \frac{dg}{df}\frac{df}{dx},$$

using a certain amount of the traditional licence about which function is to be called what!

A crude (but convincing) proof is:

$$\frac{df}{dx} = \lim_{\delta x \to 0} \frac{\delta f}{\delta g} \cdot \frac{\delta g}{\delta x} = \frac{df}{dg} \cdot \frac{dg}{dx} \qquad \text{(by 27.2.3)}.$$

Unfortunately δg might be zero: for a good proof see Exercise 4(iv).

29.4.2 EXAMPLE. In 29.4.1, let $I = \mathbb{R}$, let g be the cosine function, and let $f = \mathscr{I}^2$. (Thus $(g \circ f)(x) = \cos x^2$, and the reader should similarly convert each of our steps into one written in d/dx notation.) Then by 29.4.1,

$$\begin{aligned} D(g \circ f) &= (g \circ f)' = (g' \circ f) \cdot f' \\ &= (\cos' \circ \mathscr{I}^2) \cdot 2\mathscr{I} \qquad \text{(using 29.3.5(i))} \end{aligned}$$

and evaluating at c gives (since $\cos' = -\sin$):

$$\begin{aligned} (\cos \circ \mathscr{I}^2)'|_c &= (\cos' \circ \mathscr{I}^2)(c) \cdot 2\mathscr{I}(c) \\ &= -\sin(c^2) \cdot 2c = -2c \sin c^2. \end{aligned}$$

Exercise 3

(i) In 29.4.2, calculate $D(f \circ g)$ (i.e., $d \cos^2 x/dx$).
(ii) Let $f: \mathbb{R} \to \mathbb{R}$ be given by $f(0) = 0$, $f(x) = x^2 \sin 1/x$ $(x \neq 0)$. Prove that $f \in \mathscr{D}(\mathbb{R})$.

29.5 The Differential $d_c f$

A second formulation of 29.1.1, which is equivalent to the original, is obtained by remarking that on I the difference

29.5.1 $$\rho = \frac{f(x) - f(c)}{x - c} - l$$

must tend to zero as $x \to c$: and if f and c are kept fixed, then ρ is a function of x, with domain $I - \{c\}$. Moreover, $\rho \to 0$ as $x \to c$, so ρ is a *continuous* function at $x = c$ if we define $\rho(c)$ to be zero. Thus we can assert the

29.5.2 THEOREM. *f has differential coefficient l at $c \in I$ iff there exists a function $\rho: I \to \mathbb{R}$ such that*

$$f(x) - f(c) = l \cdot (x - c) + \rho(x) \cdot (x - c)$$

where $\rho(x) \to 0$ as $x \to c$. ■ (Of course the function ρ depends on f and c.)

This second formulation involves no awkward and possibly vanishing denominator. Moreover, since the equation of the tangent to the graph of f at $x = c$ is

29.5.3 $$y - f(c) = l \cdot (x - c),$$

we see from 29.5.1 that the difference between y and $f(x)$ is in modulus $|\rho(x)| \cdot |(x - c)|$. Hence the tangent approximates the graph to within this amount. In infinitesimal notation, $\rho(x)$ is, by 29.5.1,

$$\frac{\delta f}{\delta x} - \frac{df}{dx}.$$

If we put $x = c + t$ in 29.5.2, we have

29.5.4 $$f(c + t) - f(c) = lt + \sigma(t)\cdot t, \qquad (\sigma(t) = \rho(c + t)),$$

and $\sigma(t) \to 0$ as $t \to 0$, with $\sigma(0) = 0$. This says that the left-hand side is approximately that linear function L such that $L(t) = lt$. This linear function is called the **differential** of f at c, and denoted by $d_c f\colon \mathbb{R} \to \mathbb{R}$, since it depends on both f and c. Thus

29.5.5 $$(d_c f)(t) = f'(c)\cdot t.$$

Traditionally, c is ignored and the symbol df is used, being thought of as 'small' yet different from the infinitesimal δf. Sheer confusion consequently reigns! *We therefore emphasize*: $d_c f$ is a linear *function*, and $d_c f(t)$ is small only when t is small; also

29.5.6 $$\delta f = f(c + t) - f(c) = d_c f(t) + \sigma(t)\cdot t$$

where $\sigma(t) \to 0 = \sigma(0)$ as $t \to 0$.

In Chapter 20, it was shown how we may in a natural way identify the set $\mathbb{R}$ with its 'dual', the set of all linear transformations $\mathbb{R} \to \mathbb{R}$. Under this identification, we regard each number $\lambda \in \mathbb{R}$ as also a linear function $(t \to \lambda t)$, and hence we would identify the *number* $f'(c)$ (= differential coefficient of f at c) with the linear function $d_c f$ (= differential of f at c). When we consider functions of several variables, however, such an identification is not possible, because the dual of $\mathbb{R}^n$ $(n > 1)$ is not equivalent in a unique, natural way to $\mathbb{R}^n$. Thus the concept of the differential becomes vital when we come to make this extension of the theory.

Exercise 4

(i) Prove that $d_c\mathscr{I} = \mathscr{I}$. Thus, by 29.5.5,

$$d_c f = f'(c)\cdot\mathscr{I} = f'(c)\cdot d_c\mathscr{I}.$$

[Because of the confusion between x and the function $\mathscr{I}$ in traditional books, the last equation is written: $df = f'(c)\cdot dx$, so, it appears, '$f'(c)$ is the quotient of df by dx'! One might ask: what is dx when $x = 1$? The root of this trouble lies ultimately in not distinguishing between a function f and its value $f(x)$.]

(ii) If f, g are differentiable at c, prove that $d_c(f + g) = d_c f + d_c g$; $d_c(\lambda f) = \lambda\cdot d_c f$ $(\lambda \in \mathbb{R})$; $d_c(f\cdot g) = f(c)\cdot d_c g + g(c)\cdot d_c f$. In what sense is d_c a derivation of rings?

(iii) If, in (ii), we associate functions ρ_f, ρ_g corresponding to ρ in 29.5.1, prove 29.3.2 by means of the equality $\rho_{f+g} = \rho_f + \rho_g$, together with similar ones for $\rho_{\lambda f}$ and ρ_{fg}.

(iv) Use 29.5.2 to prove the chain rule in the form

$$d_c(g \circ f) = (d_{f(c)}g) \circ (d_c f).$$

From 29.5.2 also, we can prove easily that:

29.5.7 THEOREM. *If $f\colon I \to \mathbb{R}$ is differentiable at $x = c$, then also f is continuous at c.*

[Roughly, 29.5.2 says that $\delta f = (l + \rho)\cdot\delta x$, so δf is small if δx is small, and that is essentially what we mean by continuity. A strict proof is as follows.]

Proof of 29.5.7. Given $\varepsilon > 0$, we must show, according to 27.1.2, that there exists $\delta > 0$ such that $|f(x) - f(c)| < \varepsilon$ whenever $|x - c| < \delta$. Now, since $\rho(x) \to 0$ in 29.5.2, then there exists $\delta_1 > 0$ such that $|\rho(x)| < 1$ if $|x - c| < \delta_1$; but then, by 29.5.2, the triangle inequality implies

$$\begin{aligned}|f(x) - f(c)| &\leqslant (|l| + |\rho(x)|)\cdot|x - c|,\\ &< (|l| + 1)\cdot|x - c|,\\ &< \varepsilon \qquad \text{provided } |x - c| < \varepsilon/(|l| + 1).\end{aligned}$$

Hence the δ we require is the smaller of δ_1 and $\varepsilon/(|l| + 1)$. ■

The converse of 29.3.1 is false, as is shown by:

29.5.8 EXAMPLE. $|x|$ is not differentiable at 0 (although it is a continuous function of x, by 28.4.1). Although it is obvious geometrically that the graph of $y = |x|$ has no tangent at $x = 0$ (see Fig. 29.2), we insisted on a non-geometrical definition in 29.1.1 and so should not be content to appeal to

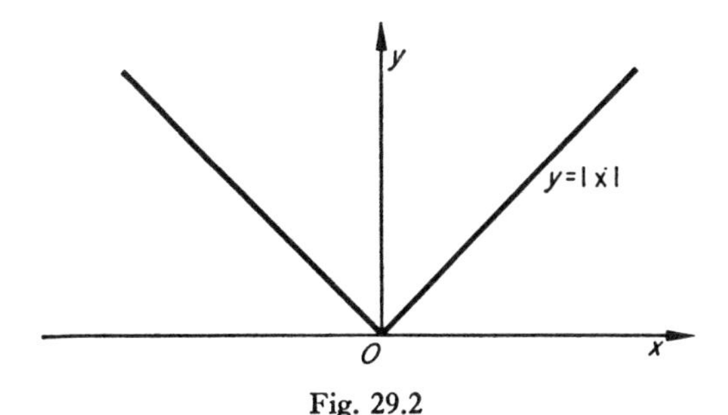

Fig. 29.2

geometry for our answer. By direct calculation we have from 29.1.2—supposing on the contrary that $|x|$ *had* a differential coefficient l at $x = 0$—

$$|x| = x(l + \rho(x)),$$

whence $1 = l + \rho(x)$ if $x > 0$, and $-1 = l + \rho(x)$ if $x < 0$. Now let $x \to 0$, so $\rho(x) \to 0$. Hence $1 = l$ and $-1 = l$, a contradiction. Thus l does not exist. ■

For dealing with this kind of behaviour, there is a theory of 'left and right derivatives'. See, e.g., Olmsted [98], p. 89. These exist, in this example, being, respectively, -1 and $+1$ at $x = 0$. This, of course, is precisely what Fig. 29.2 suggests. But failure to be differentiable can be far more complete than in this simple example. A much more striking (though more difficult) example, given by Weierstrass, consists of a continuous function which is not differentiable at *any* point of the interval constituting its domain! Such a function cannot, of course, be graphed on paper.

29.6 Higher Derivatives

Let $f \in \mathscr{D}(I)$. It might happen that Df is not continuous. On the other hand, Df might also lie in $\mathscr{D}(I)$, in which case we can form $D(Df) = (D \circ D)f$, and say that f is 'twice differentiable'. Similarly, f might be 'three-times differentiable', etc.

We define the nth power D^n of D inductively by

$$D^1 f = Df, \qquad D^{n+1}(f) = D(D^n f);$$

i.e., $D^{n+1}f = (D \circ D^n)f{:}I \to \mathbb{R}$ $(n \geqslant 1)$ and it is natural to write $D^0 f = f$ (='f differentiated no times'). Observe that D^2 means $D \circ D$, whereas in $\mathbb{R}^I$ f^2 means $f \cdot f$; this is because the multiplicative operation of interest for D is composition ($\circ$), while that defined in $\mathbb{R}^I$ is multiplication. The classical notation for $D^n f$ (the 'nth derivative of f') is $d^n f/dx^n$, or $f^{(n)}$.

For many purposes, it is convenient to confine attention to those functions $f \in \mathbb{R}^I$ such that the first n derivatives $f = D^0 f, Df, \ldots, D^n f$ exist and are *continuous*. That is, by induction on n we define the set $\mathscr{C}^n$ of **n-times continuously differentiable functions** by:

29.6.1 $\quad \mathscr{C}^1(I) = \{f \mid f \in \mathscr{D}(I) \,\&\, Df \in \mathscr{C}(I)\}, \qquad \mathscr{C}^{n+1}(I) = \{f \mid Df \in \mathscr{C}^n(I)\};$

thus

$$\mathscr{C}(I) \supseteq \mathscr{C}^1(I) \supseteq \mathscr{C}^2(I) \supseteq \cdots \supseteq \mathscr{C}^n(I) \supseteq \cdots.$$

Exercise 5

(i) Show that $\mathscr{C}^n(I)$ is a sub-algebra, over $\mathbb{R}$, of $\mathscr{C}^i(I)$, $i = 1, 2, \ldots, n$, as also is $\mathscr{C}^\infty(I)$, given by the equation (notation of 4 Exercise 10(iv)):

$$\mathscr{C}^\infty(I) = \bigcap_{n=1}^{\infty} \mathscr{C}^n(I).$$

Show that the constants and polynomials lie in $\mathscr{C}^\infty(I)$.
(ii) Show that $D(\mathscr{C}^{n+1}(I)) \subseteq \mathscr{C}^n(I)$ but that D is not one-one. [In fact, integration theory shows that $D(\mathscr{C}^{n+1}(I)) = \mathscr{C}^n(I)$.] Show that $D(\mathscr{C}^\infty(I)) = \mathscr{C}^\infty(I)$.
(iii) Prove that $D(\mathscr{D}(I)) \nsubseteq \mathscr{D}(I)$, by considering the function f in Exercise 3(ii).

A generalization of 29.3.3 is

29.6.2 **Leibniz' theorem.** *Let $f, g \in \mathscr{C}^n(I)$. Then $f \cdot g \in \mathscr{C}^n(I)$. Moreover,*

$$D^n(f \cdot g) = f \cdot D^n g + n \cdot Df \cdot D^{n-1} g + \cdots + \binom{n}{r} D^r f \cdot D^{n-r} g + \cdots + D^n f \cdot g.$$

For a proof, see 5 Exercise 1(xvi). ■ See also Exercise 5(i) above.

Exercise 6

(i) Given that $y = e^{-x} \log(1 - x)$, prove that

$$(1 - x)y_2 + (1 - 2x)y_1 - xy_0 = 0, \qquad (y_n = d^n y/dx^n,\ y_0 = y).$$

Use Leibniz' theorem to show that, when $x = 0$ and $n \geqslant 1$,

$$y_{n+2} - (n - 1)y_{n+1} - 2ny_n - ny_{n-1} = 0.$$

(ii) Do the same problem for $z = (\tan^{-1} x)^2$, through the equations

$$(1 + x^2)^2 z_2 + 2x(1 + x^2)z_1 - 2 = 0;$$

$$z_{n+2} + 2n^2 z_n + n(n - 1)^2(n - 2)z_{n-2} = 0 \qquad (n \geqslant 2).$$

[These problems are not so formalistic as may first appear; for they give an iterative method of working out the successive differential coefficients at $x = 0$, a useful process when expressing functions in terms of their Taylor expansions (see Chapter 34).]

29.7 The Rolle Conditions

We must mention also two 'qualitative' theorems, which tell us that some geometrical properties, observed in our limited experience of drawing graphs of simple functions, hold more generally. To state the theorems, let $I = \langle\!\langle a, b \rangle\!\rangle$, $I_0 = \langle a, b \rangle$, and let $f \in \mathscr{C}^1(I_0) \cap \mathscr{C}(I)$. Then f is said to satisfy **Rolle's conditions** on $\langle\!\langle a, b \rangle\!\rangle$, or briefly, $f \in \mathscr{R}(I)$, meaning that f is continuous on the closed interval and continuously differentiable on the open interval. The first theorem we state is due to Rolle, and its proof depends on the mostest theorem (28.5.2).

29.7.1 **Rolle's theorem.** *Suppose f satisfies Rolle's conditions on $\langle\!\langle a, b \rangle\!\rangle$, and that $f(a) = f(b)$. Then there exists $c \in \langle a, b \rangle$ such that $f'(c) = 0$.* [Neither the theorem nor its proof gives any method for constructing c.]

Geometrically, the theorem states the 'obvious' fact that a smooth enough curve, joining level end-points, must have a horizontal tangent somewhere between them (see Fig. 29.3). Such a tangent occurs at a maximum or minimum of the function: hence the need to use 28.5.2.

We sketch the proof. If f is constant, the assertion is trivial. If not, we may suppose there is x in $\langle a, b \rangle$ with $f(x) > f(a)$ (otherwise, consider $-f$). Let M be the precise maximum of f in $\langle\!\langle a, b \rangle\!\rangle$, achieved at $x = c$. Then

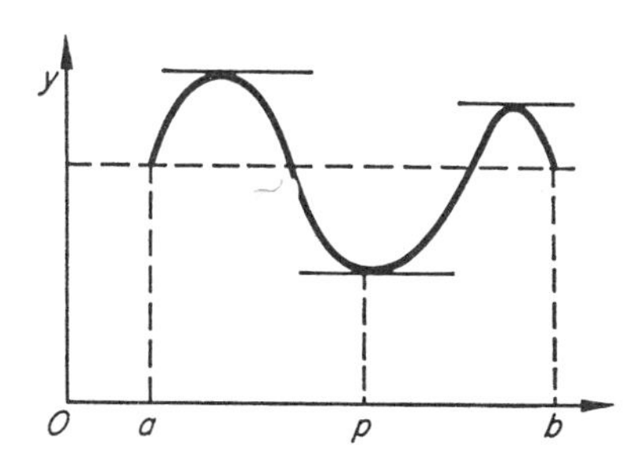

Fig. 29.3

$c \in \langle a, b \rangle$, so $f'(c)$ exists. One readily gets a contradiction (think geometrically!) out of assuming $f'(c) > 0$ or $f'(c) < 0$. ■

29.7.2 COROLLARY. *If f satisfies Rolle's conditions on $\langle\!\langle a, b \rangle\!\rangle$, and if $f' > 0$ $[<0]$, then f is strictly increasing [decreasing] on $\langle\!\langle a, b \rangle\!\rangle$. If $f' \geqslant 0$, then f is increasing, but not necessarily strictly.*

This Corollary gives a converse for Exercise 2(ii), and confirms our experience that a graph with a positive gradient keeps rising. For a proof, see Exercise 8(iii), below.

29.7.3 APPLICATION. With f as in 29.7.2, we know from 28.5.3 that the image of f is an interval $\langle\!\langle m, M \rangle\!\rangle$. Now f is strictly monotonic, and therefore is one-one, by 26 Exercise 1(vi). Thus $f\colon \langle\!\langle a, b \rangle\!\rangle \to \langle\!\langle m, M \rangle\!\rangle$ *is bijective*, so we have a rather simple test to find when a function is bijective. For example, suppose that

$$g(x) = x^3 - 3x + 1, \qquad g\colon \mathbb{R} \to \mathbb{R}.$$

Then g is everywhere differentiable and continuous, being a polynomial, so its restriction to any closed interval I lies in $\mathscr{R}(I)$. Also, $g'(x) = 3x^2 - 3$, so

$$g' < 0 \quad \text{on} \quad \langle -1, 1 \rangle,$$

and

$$g' > 0 \quad \text{on} \quad \langle -\infty, -1 \rangle \cup \langle 1, \infty \rangle.$$

Hence, on $\langle\!\langle -1, 1 \rangle\!\rangle$, g is strictly decreasing, from the value $3 = g(-1)$ to the value $-1 = g(1)$, so $g\colon \langle\!\langle -1, 1 \rangle\!\rangle \to \langle\!\langle -1, 3 \rangle\!\rangle$ is bijective, and 'reverses the sense' (i.e., as t increases from left to right, $g(t)$ moves from right to left). See also 29.9.

Exercise 7

(i) Find $g^{-1}\colon \langle\!\langle -1, 3 \rangle\!\rangle \to \langle\!\langle -1, 1 \rangle\!\rangle$ by a method for solving cubic equations (see, e.g., Birkhoff–MacLane [16]). Similarly, consider g on the intervals $\langle -\infty, -1 \rangle\!\rangle$, $\langle\!\langle 1, \infty \rangle$.
(ii) Discuss similarly $h(x) = x^2 - 4x + 1$, $k(x) = h(x)^2$, on the domain $\mathbb{R}$.
(iii) Prove that the set $\mathscr{R}(I)$ of all functions which satisfy Rolle's conditions on $I = \langle\!\langle a, b \rangle\!\rangle$, is a sub-algebra of $\mathscr{C}(I)$.

The second of the 'qualitative' theorems is the so-called 'mean value theorem'†. For the proof, see Exercise 8(i).

29.7.4 **The mean value theorem.** *Let f satisfy Rolle's conditions on $\langle\!\langle a, b \rangle\!\rangle$. Then there exists $c \in \langle a, b \rangle$, such that*

$$f(b) - f(a) = (b - a) \cdot f'(c).$$

(Observe: if $f(b) = f(a)$, then $f'(c)$ must be zero, since $b \neq a$; thus we have Rolle's theorem again.)

† Not to be confused with the intermediate value theorem (28.5.1).

Geometrically, the mean value theorem tells us that, given the chord PQ (see Fig. 29.4), some tangent to the curve is parallel to PQ. Thus the mean value theorem is 'Rolle's theorem turned through an angle'.

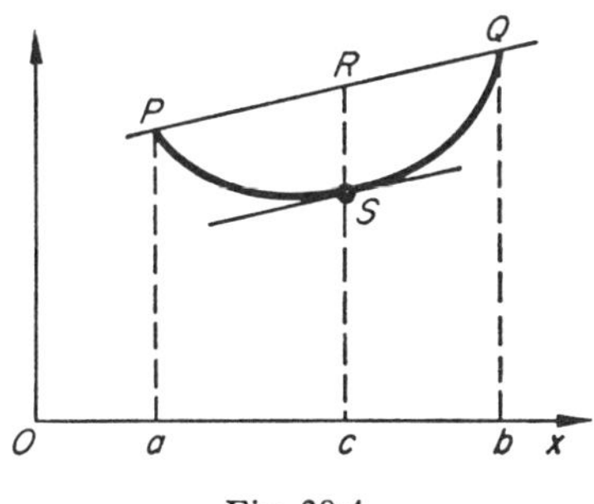

Fig. 29.4

Exercise 8

(i) Write down a formula for $g(c)$, the length of chord RS in Fig. 29.4. Using $\mathscr{R}$ as in Exercise 7(iii), verify that $g \in \mathscr{R}\langle\!\langle a, b\rangle\!\rangle$, and hence prove 29.7.4 from Rolle's theorem.
(ii) $\tan 0 = \tan \pi = 0$, but $\sec^2 x$ does not vanish between 0 and π. Why does this not contradict 29.7.4?
(iii) Deduce 29.7.2 from 29.7.1 by supposing $f' > 0$ and yet (for a contradiction) that $f(u) > f(v)$ for some $u < v$ in $\langle\!\langle a, b\rangle\!\rangle$.

The most important consequence of the mean value theorem is the following. From 29.3.2, we know that a constant function has zero derivative. However, we can deduce the converse. To state the complete result:

29.7.5 THEOREM. *Let the function f satisfy Rolle's conditions on $\langle\!\langle a, b\rangle\!\rangle$. Then f has zero derivative, iff f is constant.*

Proof. If f is constant, then $Df = 0$, remarked above. Conversely, suppose $f \in \mathscr{R}\langle\!\langle a, b\rangle\!\rangle$, and that $f' = 0$. For all $v \in \langle\!\langle a, b\rangle\!\rangle$, $f \in \mathscr{R}\langle\!\langle a, v\rangle\!\rangle$ [why?], so by the mean value theorem, there exists $w \in \langle a, v\rangle$ such that

$$f(v) - f(a) = (v - a)\cdot f'(w).$$

But $f'(w) = 0$ since $f' = 0$, so $f(v) = f(a)$. Hence f is constant. ■

29.7.6 Observe that it is essential, for 29.7.5, that the domain of f be an interval. Otherwise, for example, let

$$A = \langle\!\langle 0, 1\rangle\!\rangle \cup \langle\!\langle 2, 3\rangle\!\rangle \subseteq \mathbb{R}$$

and let $f\colon A \to \mathbb{R}$ be such that $f|\langle\!\langle 0, 1\rangle\!\rangle = 0$, $f|\langle\!\langle 2, 3\rangle\!\rangle = 1$. Then $f' = 0$, but f is not constant.

29.7.7 COROLLARY. *If f, g satisfy Rolle's conditions on $\langle\!\langle a, b\rangle\!\rangle$, and if $f' = g'$, then $f = g$ + constant.*

For $f - g \in \mathscr{R}\langle\!\langle a, b\rangle\!\rangle$ by 28.1.2 and 29.3.1, while $(f - g)' = f' - g'$, by 29.3.2; so $f - g$ = constant by 29.7.5. ■ This gives us an important technique for proving equality of functions. For we have immediately from 29.7.7 a test:

29.7.8 **The Constancy test.** *Suppose $u, v \in \mathscr{R}(I)$. Then $u = v$ iff $u' = v'$ and $\exists a \in I$ with $u(a) = v(a)$.*

Exercise 9

(i) Extend 29.7.5 to intervals of the form $\langle\!\langle a, \infty\rangle$, $\langle -\infty, b\rangle\!\rangle$ and $\mathbb{R}$.
(ii) Suppose I is an interval of the form $\langle -p, p\rangle$, and that $f \in \mathscr{R}\langle -p, p\rangle\!\rangle$. Taking $a = 0$, $b = x \in I$ in the mean value theorem, we have $f(x) = f(0) + xf'(c_1)$ for some $c_1 \in \langle\!\langle 0, x\rangle\!\rangle$. Now suppose also that $f' \in \mathscr{R}\langle\!\langle -p, p\rangle\!\rangle$, so that for some $c_2 \in \langle\!\langle 0, c_1,\rangle\!\rangle$ $f(x) = f(0) + x(f'(0) + c_1 f''(c_2))$.

(1) Prove by induction on n that if $f^{(r)} \in \mathscr{R}\langle\!\langle -p, p\rangle\!\rangle$, $0 \leqslant r \leqslant n + 1$, and if $f^{(r)}(0) = 0$ $(0 \leqslant r \leqslant n)$, then there exist $\lambda, \mu \in \mathbb{R}$ such that $\lambda, \mu \in \langle\!\langle 0, 1\rangle\!\rangle$ and $f(x) = \lambda x^{n+1} f^{(n+1)}(\mu x)$.
(2) Hence prove that, whether or not $f^{(r)}(0) = 0$, there exist $\alpha, \beta \in \langle\!\langle 0, 1\rangle\!\rangle$ such that

$$f(x) - \sum_{r=0}^{n} x^r f^{(r)}(0)/r! = \alpha x^{n+1} f^{(n+1)}(\beta x).$$

(3) Show that α in (2) can be taken to be $1/(n + 1)!$ by applying Rolle's theorem directly to the function of t:

$$g_n(t) - (x - t)^{n+1} g_n(0)/x^{n+1}$$

(with x fixed, $\neq 0$ in I, and $t \in \langle\!\langle 0, x\rangle\!\rangle$), where

$$g_n(t) = f(x) - \sum_{r=0}^{n} (x - t)^r f^{(r)}(t)/r!.$$

[The result is called Taylor's theorem, or 'the general mean value' theorem: see Hardy [52], VII, p. 150, and Chapter 34 below.]

29.8 Example (The Trigonometric Functions)

Elementary trigonometry leads us to infer that if we measure angles in radians, then there exist two functions, sin, cos: $\langle\!\langle 0, \pi/2\rangle\!\rangle \to \mathbb{R}$ such that

$$29.8.1 \quad \begin{cases} \text{(i)} \quad \sin' = \cos, & \cos' = -\sin, \\ \text{(ii)} \ \sin(0) = 0, & \cos(0) = 1. \end{cases}$$

Extending their graphs as in Fig. 29.5 to be periodic with period 2π, we obtain functions now denoted by sin: $\mathbb{R} \to \mathbb{R}$, cos: $\mathbb{R} \to \mathbb{R}$ which still satisfy (i) and (ii). Historically, this method of extension arose from the practical necessities

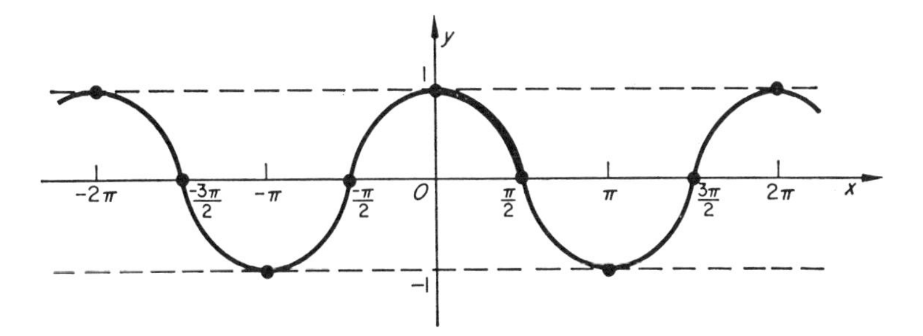

Fig. 29.5 Graph of cosine.

of dynamics: wheels turn through multiples of 2π, whereas trigonometry had grown originally from the 'static' subject of surveying, where angles are relatively small. Also, the method ensures that we still get a differentiable function at the joins (shown by the heavy dots in Fig. 29.5). See also 16.4. This construction is not a *proof*, within pure mathematics, of the existence of such functions, but it is a quick and plausible way of being introduced to them.

Exercise 10

(i) By calculating the points at which the line $y = c$ cuts the graph in Fig. 29.5, write down all the solutions x of the equation $\cos x = c$ (i.e., describe the set $\cos^\flat \{c\}$ for $\cos: \mathbb{R} \to \mathbb{R}$). Do the same for $\sin: \mathbb{R} \to \mathbb{R}$.
(ii) Define $\tan x$ to be $\sin x/\cos x$. What is its domain? Use the graphical method of (i) to work out $\tan^\flat \{c\}$, $c \in \mathbb{R}$.

29.8.2 We shall now sketch a proof of the basic properties of the sin and cos functions, *including their periodicity*, from the basic properties 29.8.1. Later, in 34.5, we shall prove the *existence* of such functions, because by then we shall know what to look for! The steps of the argument are as follows, starting from the general principle that the differential equations 29.8.1 possess solutions $\mathbb{R} \to \mathbb{R}$.

(1) (Uniqueness.) If (s_1, c_1), (s_2, c_2) are pairs of functions in $\mathscr{R}(\mathbb{R})$ such that s_i, c_i satisfy the conditions 29.8.1 on sin and cos respectively ($i = 1, 2$), then $s_1 = s_2$, $c_1 = c_2$.

Proof. Apply 29.7.8 with $u = (s_1 - s_2)^2$, $v = -(c_1 - c_2)^2$, $a = 0$. We deduce that $u = v$, but $u \geqslant 0$, $v \leqslant 0$, so $u = v = 0$. ■

(2) $$\sin^2 + \cos^2 = 1.$$

Proof. Apply the constancy test (29.7.8) with $u = \sin^2 + \cos^2$, $v = 1$, $a = 0$. ■

(3) $$|\sin| \leqslant 1, \qquad |\cos| \leqslant 1.$$

This follows immediately from (2). ■

(4) *Addition formulae.* For all $a, b \in \mathbb{R}$,

$$\sin (a + b) = \sin a \cdot \cos b + \cos a \cdot \sin b$$
$$\cos (a + b) = \cos a \cdot \cos b - \sin a \cdot \sin b.$$

Proof. As functions of x, $\sin x \cdot \cos (c - x) + \cos x \cdot \sin (c - x)$ and $\sin c$ have derivative 0; apply the constancy test with $a = c$. Now put $c = a + b$, $x = a$. Similarly for the second equation. ■

(5) $$\sin 2x = 2 \sin x \cdot \cos x, \qquad \cos 2x = \cos^2 x - \sin^2 x.$$

This follows immediately from (4). ■

(6) sin, cos are *odd and even, respectively*,

i.e., $\sin (-x) = -\sin x$, $\cos (-x) = \cos x$.

Proof. Put $a = x$, $b = -x$ in (4), and regard the resulting pair of equations as linear equations in $\sin (-x)$, $\cos (-x)$. Solve them by using (2). ■

(7) Let $X = \{x \in \mathbb{R} | x > 0 \ \& \ \cos x = 0\}$. *Then* $X \neq \varnothing$.

Proof. By (3) of Exercise 9(ii), with $n = 3$,

$$\cos 2 = 1 - 2^2/2! + (2^4/4!) \cos 2\beta = -1 + r,$$

where, by (3), $|r| \leqslant \frac{2}{3}$; so $\cos 2 < 0$, while $\cos 0 = 1$. Apply the intermediate value theorem to obtain $u \in X$ in $\langle 0, 2 \rangle$. ■

(8) If $x \in X$ in (7), then $\sin 2x = 0$ and $\forall k \in \mathbb{Z}^+$, $(2k + 1)x \in X$.

Proof. By definition of X and (5), $\sin 2x = 0$. Hence by induction on $k \in \mathbb{N}$, and, using (4), we have

$$\cos (2k + 1)x = \cos ((2k - 1)x + 2x) = 0. \ \blacksquare$$

By Theorem 24.6.9 (since $X \neq \varnothing$) there is a greatest lower bound $p \geqslant 0$ of X. We now show that X consists of all odd multiples of p.

(9) $$\cos p = 0 \qquad (\textit{so } p \in X).$$

Proof. By 27 Exercise 6(viii), X contains a sequence x_n convergent to p, so $\cos p = \lim \cos x_n = 0$, by continuity of cos, and the fact that $x_n \in X$. ■

(10) *If* $x \in X$ *and* $x + k \in X$, *then* $k = 0$ *or* $|k| > p$.

Proof. If $x \in X$, then $\cos^2 x = 0$, so $\sin^2 x = 1$ by (2). Hence by (4), $\cos (x + p) = -\sin x \sin p = \pm 1 \neq 0$, so $x + p \notin X$, i.e., $k \neq p$. Hence, if (10) were false, then $0 < |k| < p$. By (4), $0 = \cos (x + k) = -\sin x \cdot \sin k$, so $\sin k = 0$ since $\sin^2 x = 1$. Now apply Rolle's theorem to sin on $\langle\!\langle 0, |k| \rangle\!\rangle$, to obtain a positive zero of cos smaller than p. Contradiction! ■

(11) $$X = \{(2k + 1)p | k \in \mathbb{Z}^+\}.$$

For $\{(2k + 1)p\} \subseteq X$ by (8), and (10) says we can't stick in any more zeros of cos. ■

(12) COROLLARY. *The set of zeros of cos is* $\{(2k + 1)p \mid k \in \mathbb{Z}\}$.
(By (11) and (6).) ■

(13) $$\sin p = 1.$$

Proof. Otherwise $\sin p = -1$, by (9) and (2). But $\sin'(0) = \cos(0) = 1$, so sin increases strictly from the value 0 at $x = 0$. Hence if $\sin p = -1$ the graph of sin cuts the x-axis strictly between 0 and p, say at z. The argument concluding (10), with $k = z$, gives again a contradiction. ■

(14) $\forall x \in \mathbb{R}$, $\cos(p - x) = \sin x$, $\sin(p - x) = \cos x$.

Proof. By (4), using (13). ■

(15)
$$\begin{aligned}
&\sin x = 0 \text{ iff } \exists k \in \mathbb{Z}: x = 2kp;\\
&\cos x = 0 \text{ iff } \exists k \in \mathbb{Z}: x = (2k + 1)p;\\
&\sin x = (-1)^k \text{ iff } \exists j \in \mathbb{Z}: x = (2j + 1)p \quad \text{and} \quad j \equiv k \bmod 2;\\
&\cos x = (-1)^k \text{ iff } \exists j \in \mathbb{Z}: x = 2jp \quad \text{and} \quad j \equiv k \bmod 2.
\end{aligned}$$

Proof. The second statement is (12). Also, $\cos 2p = -1$ by (5) and $\sin 2p = 0$ by (8). Hence by (4), the equations

$$\sin 2kp = 0, \qquad \sin(2k + 1)p = (-1)^k, \qquad \cos 2jp = (-1)^k$$

follow by induction on $k \in \mathbb{N}$, and then by (6) when $k \in \mathbb{Z}$. Conversely, if $\sin x = 0$, then $\cos(p - x) = 0$ by (14), so $x = 2kp$ by (12). If $\sin x = (-1)^k$, then $\cos^2 x = 0$ by (2), so $x = (2j + 1)p$ for some $j \in \mathbb{Z}$, by (14); therefore $j \equiv k \bmod 2$ (already proved). Similarly for the final implication. ■

(16) sin *and* cos *are periodic with period* $4p$.

Proof. A **period** q of a function $f: \mathbb{R} \to \mathbb{R}$ is by definition any positive number such that, $\forall x \in \mathbb{R}$, $f(x + q) = f(x)$. **The** period is the least such period (if any). By (4) and (15), or by (14), $4p$ is *a* period of sin and cos. If there were a smaller period $q(> 0)$ of sin, then in particular $\sin q = \sin(q + 0) = \sin 0 = 0$. Thus $q = 2p$, using (15). But then $\sin(q + p) = 1$, by (13) and definition of period, whereas $\sin 3p = -1$ by (15). Contradiction! Similarly for cos. ■

(17) It is customary to denote p by $\pi/2$. An approximate value of π is 3.14159, which can be found 'arithmetically' by, say, Newton's method of approximation, or 'geometrically' by using integrals to find that the length of a circle is $2\pi \times$ radius. Once this is known, the approximate method of 15.5.3 may be used.

29.9 Inverse Functions

Let $I = \langle\!\langle a, b\rangle\!\rangle$, J be intervals in $\mathbb{R}$, and let $f: I \to J$ be a function. How do we decide whether or not f is bijective? A simple test is the following:

29.9.1 TEST. *If $f \in \mathscr{R}(I)$ and $f' > 0$ on $\langle a, b\rangle$, then the image,* Im f, *is $\langle\!\langle f(a), f(b)\rangle\!\rangle$. If instead, $f' < 0$, then* Im f *is $\langle\!\langle f(b), f(a)\rangle\!\rangle$. In either case, f is one-one.*

Proof. This is merely 29.7.2 coupled with the mapped-interval theorem (28.5.3), since $f \in \mathscr{R}(I) \subseteq \mathscr{C}(I)$. ∎

Suppose then that we have used some test on $f \in \mathscr{C}(I)$, and have found that $f: I \to J$ *is* bijective. As discussed in 28.5.4, there exists a function $f^{-1}: J \to I$ (not to be confused with $1/f$ in 26.4.3) which we know to be continuous, from 28.5.4, since $f \in \mathscr{C}(I)$. Also, f is strictly monotonic, and therefore so is f^{-1}. Hence f and f^{-1} map end-points onto end-points, by 28 Exercise 4(ii). And then, as to differentiability, we have the vital

29.9.2 THEOREM. *Suppose $f \in \mathscr{C}\langle\!\langle a, b\rangle\!\rangle$ and f is a bijection onto an interval J. Then $f^{-1}: J \to \langle\!\langle a, b\rangle\!\rangle$ is continuous. If $c \in \langle a, b\rangle$, then $f(c)$ is not an end-point of J; while if f is differentiable at c and $f'(c) \neq 0$, then f^{-1} is differentiable at $f(c) \in J$.* □

In traditional terms, this adds precision to the saying that 'if y is a differentiable function of x and $dy/dx \neq 0$, then x is a differentiable function of y.' For a proof see Hardy [52]. The theorem allows us to state at once:

29.9.3 COROLLARY (to Theorem 29.9.1). *Under the hypotheses of* 29.9.1, *$f^{-1}: J \to I$ lies in $\mathscr{R}(J)$.* ∎

As a 'practical' application, in the traditional terms mentioned above, y there is often defined on an open interval K on which it possesses a *continuous* derivative. Thus if $y'(c) > 0$ at $c \in K$, then by the inertia principle there is an interval $I_0 = \langle a, b\rangle \subseteq K$ with $a < c < b$, on which $y' > 0$. Then Corollary 29.9.3 implies that x is a differentiable function of y on $y(I_0)$.

Once 29.9.2 is known, we can compute $D(f^{-1})|_{f(c)}$ as follows. The inversion theorem tells us that

$$f \circ f^{-1} = \mathscr{I}_J: J \to J,$$

the identity function on J. Applying the chain rule 29.4.1 we get

$$1 = D(\mathscr{I}_J) = D(f \circ f^{-1}) = (f' \circ f^{-1}) \cdot D(f^{-1});$$

so writing g for $D(f^{-1})$, and evaluating at any point $y = f(c) \in J$, we have

$$1 = f'(f^{-1}(y)) \cdot g(y)$$

by definition of the product of functions. Hence since f' was given to be non-zero on I, we find

29.9.4 *The differential coefficient of f^{-1} at $y \in J$ is*

$$(Df^{-1})(y) = 1/f'(c), \qquad (c = f^{-1}(y)).$$

29.9.5 EXAMPLE. By 29.8.2(15), $\sin' = \cos$ is not zero on $\langle -\pi/2, \pi/2 \rangle$, and since $\cos 0 = 1$ and cos is continuous, then $\cos > 0$ on $\langle -\pi/2, \pi/2 \rangle$. Also, $\sin(-\pi/2) = -1 = -\sin \pi/2$, so, by the test 29.9.1, $\sin: \langle\!\langle -\pi/2, \pi/2 \rangle\!\rangle \to \langle\!\langle -1, 1 \rangle\!\rangle$ is bijective. Its inverse is traditionally denoted by

$$\sin^{-1}: \langle\!\langle -1, 1 \rangle\!\rangle \to \langle\!\langle -\pi/2, \pi/2 \rangle\!\rangle.$$

Since $\cos > 0$, $\sin^{-1}$ is differentiable on $\langle -1, 1 \rangle$ (applying 29.9.2); and then by 29.9.4,

$$(D \sin^{-1})(y) = 1/\sin'(x), \qquad (\text{where } y = \sin x).$$

But $\sin' = \cos$ and $\cos x = \sqrt{(1 - \sin^2 x)} = \sqrt{(1 - y^2)}$, so

$$(D \sin^{-1})(y) = 1/\sqrt{(1 - y^2)}, \qquad y \in \langle -1, 1 \rangle.$$

29.9.6 EXAMPLE. It was shown in 28.6, that for each integer $n > 0$,

$$\mathscr{I}^n: \mathbb{R}^+ \to \mathbb{R}^+ \qquad \text{is bijective, with inverse}$$
$$\mathscr{I}^{1/n}: \mathbb{R}^+ \to \mathbb{R}^+.$$

Now $D\mathscr{I}^n = n\mathscr{I}^{n-1}$ (by 29.3.5), which is never zero on $\mathbb{R}_+$. Hence, by 29.9.2, $\mathscr{I}^{1/n}$ is differentiable (on $\mathbb{R}_+$) with differential coefficient at $z \in \mathbb{R}_+$ given by

$$\begin{aligned} (D\mathscr{I}^{1/n})(z) &= 1/D\mathscr{I}^n(t), \qquad t = \mathscr{I}^{1/n}(z) = z^{1/n} \\ &= 1/n(z^{1/n})^{n-1} \\ &= \frac{1}{n} z^{(1/n)-1} = \frac{1}{n} \mathscr{I}^{(1/n)-1}(z); \end{aligned}$$

hence by definition of equality of functions,

$$D\mathscr{I}^{1/n} = \frac{1}{n} \mathscr{I}^{(1/n)-1}. \tag{29.9.7}$$

This shows why we had to demand in 29.8.1 that f' was never zero in I, since $\mathscr{I}^{(1/n)-1}$ is not defined at $0 \in \mathbb{R}^+$.

Exercise 11

(i) Extend the last discussion to the function $\mathscr{I}^n: \mathbb{R} \to \mathbb{R}$ when n is odd. Obtain the formula of 29.9.7 for all positive integers n. Hence prove that, for all integers m, n with $n \neq 0$,

$$D\mathscr{I}^{m/n} = \frac{m}{n} \mathscr{I}^{(m/n)-1},$$

stating suitable domains for $\mathscr{I}^{m/n}$. {Use the chain rule and the definition (28.6.4) of $\mathscr{I}^{m/n}$ as $\mathscr{I}^m \circ \mathscr{I}^{1/n}$.}

(ii) In Exercise 10(ii), prove that $\tan: \langle -\pi/2, \pi/2\rangle \to \mathbb{R}$ is a bijection. Its inverse is denoted by $\tan^{-1}: \mathbb{R} \to \langle -\pi/2, \pi/2\rangle$. Prove that the latter is differentiable except at $0 \in \mathbb{R}$, and calculate its derivative.

29.9.8 EXAMPLE. Consider the formula written in the traditional notation of mechanics as

(i) $$\ddot{x} = v \cdot dv/dx.$$

The left-hand side is dv/dt, and the chain rule is applied, to give

$$\ddot{x} = \frac{dv}{dt} = \frac{dv}{dx}\cdot\frac{dx}{dt} = \frac{dv}{dx}\cdot v,$$

where v on the right-hand side is regarded as a function of x. Now how do we justify regarding the derivative (v) of the *function* x, as being itself a function of the co-ordinate x? Traditionally one says: 'v is a function of t, and so is x; hence t is a function of x, therefore v is a function of x'. Obviously the inversion theorem lurks here, and we bring it into the open as follows.

By the usual hypotheses of dynamics, we are given a differentiable function $s: T \to \mathbb{R}$, such that $s(t)$ is the distance travelled at time $t \in T$. There is also the 'velocity function' $v: T \to \mathbb{R}$, assumed differentiable. If we suppose T to be a closed interval, then $s(T)$, $v(T)$ are also intervals S, V respectively, by 28.5.3, and we can represent matters by the diagram (a) of Fig. 29.6 where the dotted arrow w denotes the 'velocity considered as a function of x', i.e., the function whose existence we wish to establish.

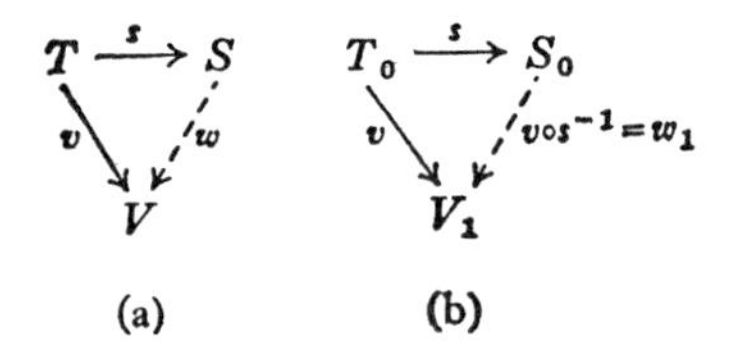

Fig. 29.6

The diagram suggests the solution: let $t_0 \in T$ be a point at which $s' \neq 0$; then there exists an interval $T_1 = \langle\!\langle t_0 - h, t_0 + h\rangle\!\rangle \subseteq T$ on which s' is of constant sign (by the inertia principle (28.3.1) applied to s'—which is continuous by 29.5.7, since the acceleration s'' is supposed to exist). By 28.5.3, $s(T_1)$ and $v(T_1)$ are closed intervals $S_1 = \langle\!\langle p_1, p_2\rangle\!\rangle$, V_1 respectively. Removing their end-points, we let $T_0 = \langle t_0 - h, t_0 + h\rangle$, $S_0 = \langle p_1, p_2\rangle$ and apply 29.9.2 to these open intervals, taking f there to be $s|T_0$. Then we may assert the existence of the differentiable function $s^{-1}: S_0 \to T_0$; and hence we obtain a function $w_1 = v \circ s^{-1}: S_0 \to V_1$ as in Fig. 29.6(b). Thus

$$v = w_1 \circ s: T_0 \to V_1.$$

And now, applying the chain rule, we have

$$v' = (w_1' \circ s) \cdot s' = (w_1' \circ s) \cdot v,$$

since $v = s'$ by definition of velocity. Evaluating at $t \in T_0$ gives $v'(t) = w_1'(s(t)) \cdot v(t)$, which is the traditional formula (i) evaluated at t. [Of course, the function w, required above, is given by $w_1(s(t_0)) = w(s(t_0))$ and thus w is defined, but only on S_0, not necessarily on all of S.]

To summarize:

(1) The formula (i) is correct whenever the velocity is non-zero; and so any solution of a differential equation obtained by its means is valid *a priori* only in these circumstances. However, in most problems, the velocity is non-zero except possibly at the end-points of intervals $I, J, \ldots$ like T_0 above; the set of solutions, one for each interval, can then be pieced together, and the result is a complete, usually differentiable, solution $w: \mathbb{R} \to \mathbb{R}$.

(2) We submit to the reader that our discussion of (i) is what *users* of mathematics should aim to do; namely to say what everything means in standard terms, and to assemble a valid argument by quoting from the stock of proved theorems where necessary. This is quite different from being able to *prove* the theorems, and indeed we have not completely proved (in this book) any of the theorems here quoted. Of course, it is not necessary to do this every time (i) is encountered, but the reader should be aware of the rationale for the traditional procedure.

Exercise 12

Discuss, analogously to 29.9.8, the classical formula '$dy/dx = \dot{y}/\dot{x}$'. {*Hint:* y, x are really functions of t, so that the first y should be $y \circ x^{-1}$ for a suitable inverse x^{-1} of the function x of t.}

Chapter 30

INTEGRATION

30.1 The Problem

Many practical problems raise the question: given $f \in \mathbb{R}^I$, is f in the image of $D: \mathscr{D}(I) \to \mathbb{R}^I$, and if so, what is $D^{-1}(f)$? The most basic of these is that of finding the area under the graph of f, and 18th-century mathematicians argued as follows: Let $I = \langle\!\langle a, b \rangle\!\rangle$; we want the area A bounded by the graph of f, the x-axis (in Fig. 30.1), and the ordinates at a and b. Let $A(x)$ denote

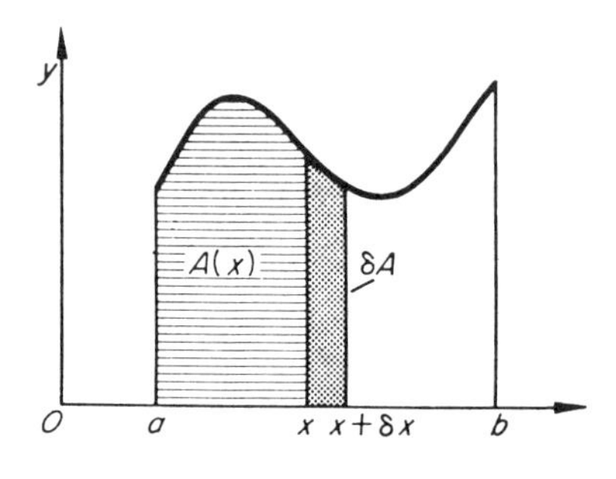

Fig. 30.1

the corresponding area when the ordinate at b is replaced by the ordinate at x, $(x \in I)$. Then, using the classical 'δ' notation, if we increase x to $x + \delta x$, then $A(x)$ increases to $A(x) + \delta A$; and approximately, $\delta A \simeq f(x) \cdot \delta x$, so $\delta A/\delta x \simeq f(x)$. The ratio $\delta A/\delta x$ converges to dA/dx when δx converges to zero, so

$$dA/dx = f(x);$$

i.e., $A(x)$ *is that function of* x *which when differentiated gives* $f(x)$, *and is such that* $A(a) = 0$. Thus we must look for a function $A \in \mathscr{D}(I)$ such that

30.1.1 $$DA = f, \qquad A(a) = 0.$$

By 29.7.8, at most one such function A can satisfy the two conditions 30.1.1. But does any function A *at all* satisfy them? The answer might seem to have been given above: take A to be the area. Two objections arise (the reader should have been prepared for the first by the discussion of area and content in Chapter 15).

(a) What do we mean by 'area'? does it exist?;
(b) Even if it exists, has it a derivative?

A branch of mathematics called measure theory has grown over the years, to answer question (a), and its offspring. (See also 15.7.) However, answers to (a) and (b), which are adequate for the particular problem here, were given by the 'Theory of Riemann Integration' which was built by the 19th century analysts, following ideas of Cauchy and the great German mathematician Bernhard Riemann [1826–1866]. [See Bell [10], Ch. 26.] They proved the following theorem [see Quadling [104], Ch. X, whose proof does not require 'uniform continuity', usual in other proofs; it depends of course on the completeness (24.6) of $\mathbb{R}$.]

30.1.2 (**The fundamental theorem of calculus.**) *Suppose f is a continuous real-valued function on $\langle\!\langle a, b\rangle\!\rangle$. Then there is a function $F\colon \langle\!\langle a, b\rangle\!\rangle \to \mathbb{R}$ such that $DF = f$.* □

The *set* of all such functions F is called the **indefinite integral** of f, and denoted for historical reasons by

30.1.3 $$\int f(t)\, dt.$$

Any member of this set is called a **primitive** for f; and f is the **integrand.** Any two such primitives F, G differ by a constant (by 29.7.5), and the whole set consists of all the functions $F + c$, $c \in \mathbb{R}$. The particular primitive G such that $G(a) = 0$ is called the **definite integral** of f, and denoted by

30.1.4 $$G(x) = \int_a^x f(t)\, dt.$$

Thus, the essential import of the fundamental theorem (30.1.2) is that, first, the number $G(x) = \int_a^x f(t)\, dt$ *exists* for each $x \in \langle\!\langle a, b\rangle\!\rangle$ and, of course,

30.1.5 $$G(a) = \int_a^a f(t)\, dt = 0$$

and, second, that G is a differentiable function of x on $\langle a, b\rangle$, with differential coefficient given by

30.1.6 $$G'(x) = d/dx\left(\int_a^x f(t)\, dt\right) = f(x).$$

However, 30.1.2 gives no simple formula for working out the integral. The 't' which occurs in 30.1.3 and 30.1.4 is quite irrelevant and arises by historical accident; it is a convention of the notation that

30.1.7 $$\int_a^x f(t)\,dt = \int_a^x f(s)\,ds,$$

(and similarly in the indefinite integral), so t and s are called 'dummy' variables, in contrast to x, which is said to be 'free'. Complete confusion can therefore ensue by writing

30.1.8 $$\int_a^x f(x)\,dx,$$

a form to be abhorred—and, consequently, avoided!

The proof of 30.1.2 makes no appeal to the notion of area but only to that of real number; it requires the notion of limit. We can then answer the questions (a) and (b) above, by declaring 'the area $A(x)$' to mean the *value* of the function G in 30.1.4 at x. Certainly 30.1.1 is satisfied, and it turns out that this notion of 'area' behaves in all respects as we would expect it to.

It is also customary, when H is any primitive for f, to write

$$[H(t)]_a^x = \int_a^x f(t)\,dt.$$

As remarked above, $H = G + c$, where c is some constant and G is the definite integral (30.1.4); so $c = H(a)$ since $G(a) = 0$. Hence, by 30.1.4,

30.1.9 $$[H(t)]_a^x = H(x) - H(a).$$

We may well find it useful to assign a meaning to $\int_a^b f(t)\,dt$ even if $b < a$. Equation 30.1.9 suggests the definition

$$\int_a^b f(t)\,dt = -\int_b^a f(t)\,dt \quad \text{if} \quad b < a;$$

for $[H(t)]_x^a =$ (presumably!) $H(a) - H(x) = -[H(t)]_a^x$. See 30.3.3.

Exercise 1

(i) Use the definition of area given after 30.1.8 to prove that the area of a circular quadrant of radius a is $\frac{1}{4}\pi a^2$, with π defined as in 29.8.2(17). Hence, by using strips of width $1/n$, prove that $1/n^2 \sum_{r\geqslant 1} \sqrt{(n^2 - r^2)} \leqslant \pi/4 \leqslant 1/n^2 \sum_{r\geqslant 0} \sqrt{(n^2 - r^2)}$ and calculate π roughly.

(ii) Use elementary geometry to suggest that the 'length' of a smooth curve $y = f(x)$ between a and x is $L(x)$, where $L'(x) = \sqrt{(1 + (y')^2)}$. Hence make a reasonable definition of 'length', without appeal to geometry, and check that it gives the 'correct' answers in simple cases (e.g., for linear segments). For what class of functions f will your definition always give a finite length? [See also 15.4.]

So much for the *problem* of integration. But the *operation* of constructing the definite integral has algebraic properties enabling us to assemble the integral of a complicated function from the integrals of its simpler parts (just as with the operation of differentiation). These 'simpler' parts might however be quite difficult to 'integrate', and two important aids are (i) tables of integrals which record once and for all each integral once worked out; and (ii) numerical approximations to the area measured by the integral. The traditional routine of 'evaluating' an integral chosen and disguised by an examiner to be workable in a given time, is of course almost useless, and even potentially harmful because it may well lead the pupil to think that all integrals are like that; in fact, hardly any are! There would, however, be some value in letting the pupil prepare his own table of all the integrals he has valiantly worked out, to use from the date he started it.

30.2 Rules of Integration

Let $I = \langle a, b \rangle$ as before. We have, then, a function $Q: \mathscr{C}(I) \to \mathscr{D}(I)$, whose value $Q(f)$ at each $f \in \mathscr{C}(I)$ is the definite integral of f; we have met Q before, in 2.5.9. Its properties can all be derived algebraically from 30.1.5 and 30.1.6, together with the properties of continuous and differentiable functions; they are as follows.

30.2.1 THEOREM. *The integral is linear, i.e., given $f, g \in \mathscr{C}(I)$, and given $\lambda, \mu \in \mathbb{R}$, then*

$$\int_a^x (\lambda f(t) + \mu g(t))\, dt = \lambda \int_a^x f(t)\, dt + \mu \int_a^x g(t)\, dt.$$

[Thus Q (defined above) is a *linear transformation.* Note that when using the classical notation we omit the dot from products like $\lambda \cdot f$, except where clarity demands its retention.]

Proof. Since $f, g \in \mathscr{C}(I)$, so does $\lambda f + \mu g$, by 28.1.2. Hence all the integrals involved exist, and can be differentiated, by 30.1.6. But

$$\frac{d}{dx} \int_a^x (\lambda f(t) + \mu g(t))\, dt = \lambda f(x) + \mu g(x) \qquad \text{(by 30.1.6)}$$

and

$$\frac{d}{dx}\left(\lambda \int_a^x f(t)\, dt + \mu \int_a^x g(t)\, dt\right) = \lambda \frac{d}{dx} \int_a^x f(t)\, dt + \mu \frac{d}{dx} \int_a^x g(t)\, dt$$
$$= \lambda f(x) + \mu g(x)$$

by the linearity of differentiation (29.3.2). Thus, the two sides of the equation in 30.2.1 are functions with equal derivatives; but the two functions are each zero at a, and hence, by the constancy theorem 29.7.8, they are equal. ■

Exercise 2

(i) Let p be the polynomial $\Sigma c_r x^r$. Prove that

$$\int_0^x p(t)\,dt = \Sigma\, c_r x^{r+1}/(r+1).$$

(ii) Define a new multiplication in $\mathscr{C}(I)$ by $(f \times g)(x) = \int_a^x f(t)g(t)\,dt$. Prove this multiplication to be distributive (over ordinary addition in $\mathscr{C}(I)$) and commutative, but not associative {e.g., consider $(1 \times \mathscr{I}) \times \mathscr{I}^2$}.

30.2.2 THEOREM. *The integral is order-preserving, i.e., if $f, g \in \mathscr{C}(I)$ and $f(x) \geqslant g(x)$ for all $x \in I$, then*

$$\int_a^x f(t)\,dt \geqslant \int_a^x g(t)\,dt.$$

[In the notations of 26.5 and 30.2, this theorem states: if $f \geqslant g$ in $\mathscr{C}(I)$, then $\int f \geqslant \int g$, where we use $\int f$ instead of Qf; this notation has the advantage of reminding us that we are discussing integrals.]

Proof. Let $h(x) = \int_a^x (f(t) - g(t))\,dt$; then h exists and lies in $\mathscr{C}(I)$ by the fundamental theorem (30.1.2), since $f - g \in \mathscr{C}(I)$. Also, by 30.1.6, $h'(x) = f(x) - g(x) \geqslant 0$ by hypothesis. Hence h has a derivative $\geqslant 0$, so by 29.7.2 h is increasing. Now $h(a) = 0$, so $h(x) \geqslant 0$, $x \in I$. This is the required inequality. ■

Exercise 3

(i) Use the mostest theorem (28.5.2) to show that, if $f > g$ in 30.2.2, then $\forall x \in I$, $\int_a^x f(t)\,dt > \int_a^x g(t)\,dt$. $\{f - g = m + h$, where m is a constant > 0, and $h(x) \geqslant 0.\}$

(ii) Prove that $0 \leqslant \int_0^{\pi/2} \sin(\sin x)\,dx \leqslant \pi/2$.

30.2.3 THEOREM. *If $f \in \mathscr{C}(I)$, then* $\left|\int_a^x f(t)\,dt\right| \leqslant \int_a^x |f(t)|\,dt$. [This is an analogue of the triangle inequality.]

Proof. Recalling 28.4, we define two functions f_+, f_- by

$$f_+ = \max(f, 0), \qquad f_- = \min(f, 0);$$

thus $f_+, f_- \in \mathscr{C}(I)$, by 28.4. Then

30.2.4 $$|f| = f_+ - f_-, \qquad f = f_+ + f_-,$$

so, writing $\int g$ for $\int_a^x g(t)\,dt$—often a better notation—we have

$$\begin{aligned}
\left|\int f\right| &= \left|\int (f_+ - f_-)\right| && \text{(by 30.2.4)}\\
&= \left|\int f_+ - \int f_-\right|, && \text{since } \int \text{ is linear,}\\
&\leqslant \left|\int f_+\right| + \left|\int f_-\right| && \text{(triangle inequality)}\\
&= \int f_+ - \int f_- && \text{(by 30.2.2)}\\
&= \int (f_+ - f_-) = \int |f| && \text{(by 30.2.4).} \quad \blacksquare
\end{aligned}$$

Exercise 4

Using the fact that area is defined by the integral, show from a rough graph that Theorems 30.2.2, 30.2.3 are geometrically obvious.

30.2.5 **The mean value theorem for integrals.** *Suppose $f, g \in \mathscr{C}(I)$ and that $g \geqslant 0$. Then, given $p \leqslant q$ in I, there exists $c \in I$, lying between p and q, such that*

$$\int_p^q f(t)g(t)\,dt = f(c)\int_p^q g(t)\,dt.$$

Proof. By the mostest theorem (28.5.2), there exist numbers m, M, values of f in $\langle p, q\rangle$ such that $m \leqslant f(t) \leqslant M$ for each $t \in \langle p, q\rangle$. Hence, since $g(t) \geqslant 0$,

$$m\cdot g(t) \leqslant f(t)\cdot g(t) \leqslant M\cdot g(t),$$

so by the linearity and order-preserving properties of the integral,

$$m\cdot\int_p^q g(t)\,dt \leqslant \int_p^q f(t)\cdot g(t)\,dt \leqslant M\cdot\int_p^q g(t)\,dt.$$

Now m, M are values of f, so Km, KM are values of the function Kf in $\langle p, q\rangle$, where $K = \int_p^q g(t)\,dt$. Hence, by the intermediate value theorem (28.5.1), the number $\alpha = \int_p^q f(t)g(t)\,dt$ is a value of Kf; i.e., there exists $c \in I$ between p and q such that $\alpha = f(c)\cdot K$. $\blacksquare$

Exercise 5

Prove that $\int_0^{\pi/4} \sqrt{x} \sin x^2 \, dx < \pi^{3/2}/8\sqrt{2}$.

30.2.6 *Integration by parts* If we integrate each side of the equation 29.3.3, we get—when $f, g \in \mathscr{C}^1(I)$ and $x \in I = \langle a, b \rangle$—

$$[f(t)g(t)]_a^x = \int_a^x f'(t)g(t)\,dt + \int_a^x f(t)g'(t)\,dt,$$

for then $f, g \in \mathscr{C}(I)$, by 29.5.7; so $f' \cdot g, f \cdot g'$, and hence $(f \cdot g)', \in \mathscr{C}(I)$ by 28.1.2. Thus all the integrals exist. [Weaker hypotheses are possible; see Olmsted [98], p. 158.]

Exercise 6

(i) If G is *an* integral of g, show that

$$\int_a^x f(t) \cdot g(t)\,dt = [f(t) \cdot G(t)]_a^x - \int_a^x f'(t) \cdot G(t)\,dt,$$
$$(f', g \in \mathscr{C}(I)).$$

Use this formula to evaluate $\int_0^{\pi/2} x \sin x \, dx$.

(ii) For any polynomial f of degree $2n$, let

$$J = \pi \int_0^1 f(x) \sin \pi x \, dx.$$

Prove that

$$J = \sum_{k=0}^{n} (-1)^k \left(\frac{f^{(k)}(0) + f^{(k)}(1)}{\pi^{2k}} \right).$$

{Use integration by parts.}

(iii) Now take $f(x)$ to be $x^n(1 - x)^n/n!$. Prove that $f(x) = f(1 - x)$ and that the derivatives $f^{(k)}(0), f^{(k)}(1)$ are all integers, while $0 < J < \pi/n!$. $\{0 < f(x) < 1/n!.\}$

(iv) Prove that π is irrational, as follows. Suppose, for a contradiction, that π is rational. Then $\pi^2 = p/q$ for some integers p, q. Use (ii) to prove that then $p^n J$ is an integer, and use (iii) to show that $0 < p^n J < 1$ if n is sufficiently large. Hence obtain a contradiction.

[This proof, based on one by I. Niven, is due to W. Scholz, *Math. und Naturwiss. Unterricht*, **20** (1967); a more difficult proof shows that π is not algebraic.]

(v) The following is an exercise on some important integrals called *distributions*. We warn the reader that this is an extremely sophisticated exercise! However, the notions are important for intending physicists, as well as for mathematicians. In 29.6.1, there was defined the set $\mathscr{C}^\infty(I)$ of all functions $f: I \to \mathbb{R}$ which can be differentiated infinitely often. Now let $I = \mathbb{R}$, and let $\mathscr{T}$ denote the subset of all $f \in \mathscr{C}^\infty(\mathbb{R})$ with the property: for some finite interval $\langle a, b \rangle$ (depending on f and called the **support** of f) $f(x)$ is zero if $x < a$ or $x > b$. Thus, $\mathscr{T}$ consists of those infinitely differentiable functions which vanish outside *some* finite interval.

(1) Prove that $\mathscr{T}$ is a sub-algebra of $\mathscr{C}^{\infty}(\mathbb{R})$, not containing $\mathscr{I}$, nor any polynomial other than zero.

(2) Let $\mathscr{T}^*$ denote the set of all linear transformations $T: \mathscr{T} \to \mathbb{R}$. As in 20.4, $\mathscr{T}^*$ is a real vector space, the *dual* of $\mathscr{T}$. Define a function $\eta: \mathscr{C}(\mathbb{R}) \to \mathscr{T}^*$ by defining $\eta(g)$, for each $g \in \mathscr{C}(\mathbb{R})$, to be that linear function $T_g \in \mathscr{T}^*$ whose value at $\phi \in \mathscr{T}$ is

$$T_g(\phi) = \int_{-\infty}^{\infty} \phi(x)g(x)\,dx, \qquad (T_g = \eta(g));$$

we have written $\int_{-\infty}^{\infty}$ instead of $\int_a^b$, to avoid specifying the support (here that of ϕ) of each function which occurs. Prove that η is linear, i.e.,

$$\eta(\lambda g + \mu h) = \lambda \cdot \eta(g) + \mu \cdot \eta(h) \qquad (\lambda, \mu \in \mathbb{R}).$$

(3) Prove that η is one-one. Hence $\mathscr{C}(\mathbb{R})$ can be thought of as being embedded in $\mathscr{T}^*$; by 29.5.7, $\mathscr{T} \subseteq \mathscr{C}(\mathbb{R})$.

(4) If, in (2), g happens to lie in $\mathscr{T}$, use integration by parts (30.2.6) to prove that

$$T_{g'}(\phi) = -T_g(\phi').$$

But by (3), $T_{g'}$ is $\eta(g')$ and is identified with g' under the embedding suggested there. This suggests how we may associate a 'derivative' T' with every $T \in \mathscr{T}^*$; we define T' to be that member of $\mathscr{T}^*$ whose value at each $\phi \in \mathscr{T}$ is

$$T'(\phi) = -T(\phi').$$

Thus we have a function $\Delta: \mathscr{T}^* \to \mathscr{T}^*$, namely, $\Delta(T) = T'$. Prove that Δ is linear. (Hence, we can, by (3), think of each element T of $\mathscr{T}^*$ as a kind of generalized element of $\mathscr{T}$, and this way of thinking is reinforced by the fact that T *has the infinity of derivatives* T, ΔT, $\Delta(\Delta T) = \Delta^2(T), \ldots, \Delta^n(T), \ldots$.)

(5) Let δ be that member of $\mathscr{T}^*$ whose value at each $\phi \in \mathscr{T}$ is

$$\delta(\phi) = \phi(0).$$

Prove that δ is linear (so δ really does lie in $\mathscr{T}^*$) and that the derivative $\Delta\delta$ of δ is given by

$$\delta'(\phi) = -\phi'(0).$$

All this is the initial fragment of the theory of distributions due to L. Schwartz [116] and others; for this reason, the members of $\mathscr{T}^*$ are called 'distributions'. The subject arose because physicists had long used an object called the *Dirac delta-function*. This was a kind of 'function', δ, whose domain was $\mathbb{R}$, whose value was zero everywhere except at 0, where it was infinite, and such that

$$\int_{-\infty}^{\infty} \delta(x)\phi(x)\,dx = \phi(0), \qquad \int_{-\infty}^{\infty} \delta'(x)\phi(x)\,dx = -\phi'(0).$$

No such 'function' $\mathbb{R} \to \mathbb{R}$ can exist, of course; but if it did, and was regarded as in $\mathscr{T}^*$ (by (3)), then these integrals would give the equations in (5). The physicists needed such an object because, for example, it could be regarded as some kind of limit of 'elementary solutions' of the heat equation (see [78], p. 548), and therefore represented an idealized 'distribution' of heat. Their *use* of the 'δ-function' did not lead to errors, because they used only the properties of δ in (5). [Such 'functions' were used by the great British engineer Oliver Heaviside well before Dirac, of course. Readers of his important work *Electromagnetic Theory* must judge for themselves whether he committed any mathematical errors!] Once (5) was proved, physicists merely had to widen their notion of 'infinitely differentiable real-valued function' from $\mathscr{T}$ to $\mathscr{T}^*$; just as mathematicians before them had to widen their notion of rational number through the successive stages of real number, complex number, constant function, and function. Of course, more work must still be done, for example to multiply distributions; and here the reader is referred to Schwartz [116] or Lighthill [82], who calls such distributions *generalized functions*.

30.3 Integration by Substitution

A further useful technique of integration is to substitute a suitable new variable for the existing variable in the integrand, thereby transforming the old integral into a more easily recognizable form. (Examiners frequently use this technique in reverse!) The rule is the following:

30.3.1 *Substitution rule. Let* $J = \langle u, v\rangle$ *and let* $g \in \mathscr{C}^1(J)$ *satisfy* $g(u) = a$, $g(v) = b$. *If* $f \in \mathscr{C}^1(g(J))$, *then*

$$\int_u^v f(g(t))\cdot g'(t)\,dt = \int_a^b f(x)\,dx.$$

Proof. The first integral exists; for $f \circ g$ is continuous by 28.2.1, and g' is continuous by hypothesis, so their product is continuous by 28.1.2, and so has an integral by 30.1.2.

By the fundamental theorem (30.1.2), f has an integral, say F; so $f = F'$. Then

$$\begin{aligned}\int_u^v f(g(t))\cdot g'(t)\,dt &= \int_u^v F'(g(t))\cdot g'(t)\,dt\\ &= \int_u^v (F\circ g)'(t)\,dt \qquad \text{(chain rule 29.4.1)}\\ &= [(F\circ g)(t)]_u^v \qquad \text{(see 30.1.9; } (F\circ g)' \text{ is continuous)}\\ &= F(g(v)) - F(g(u))\\ &= F(b) - F(a) = \int_a^b f(x)\,dx. \quad ■\end{aligned}$$

Exercise 7

(i) If g' is continuous and non-zero in $\langle u, v\rangle$, prove that

$$\int_u^v f(g(t))\,dt = \int_a^b f(x)h'(x)\,dx,$$

where $h = g^{-1}$.

(ii) Discuss the substitution $x = (1 + y^2)/(3 - y^2)$ in the integral

$$\int_a^b dx/x\sqrt{(3x^2 + 2x - 1)},$$

paying special attention to the possible values of the limits a, b.

Notice that in 30.3.1 and Exercise 7, we may well have utilized the convention (following 30.1.9) whereby a meaning is assigned to $\int_a^b f(t)\,dt$, even when $b \leqslant a$; and so we recall the convention. Thus

30.3.2 $$\int_b^a f(t)\,dt = -\int_a^b f(t)\,dt.$$

It is a simple matter of checking cases to prove that, for any a, b, c,

30.3.3 $$\int_a^b f(t)\,dt = \int_a^c f(t)\,dt + \int_c^b f(t)\,dt,$$

assuming that f is defined and continuous on the indicated intervals.

Exercise 8

(i) Let $f \in \mathscr{C}(I)$, $I = \langle a, b \rangle$. Show that for all $k \in \mathbb{R}$

$$k\int_a^x f(kt)\,dt = \int_{ka}^{kx} f(s)\,ds.$$

{Consider $k = 0$ separately.}

(ii) Let $f, g \in \mathscr{C}(\mathbb{R})$. Their **convolution** $f * g\colon \mathbb{R} \to \mathbb{R}$ is the function defined by $(f * g)x = \int_0^x f(t)g(x - t)\,dt$. Prove that $f * g \in \mathscr{C}(\mathbb{R})$, that $*$ is commutative, and that $*$ is distributive over addition. §Prove that $*$ is associative; and calculate $D(f * g)$ when $f, g \in \mathscr{C}^1(\mathbb{R})$.

(iii) Prove that if f is a continuous and odd function on $\langle -1, 1 \rangle$, then

$$\int_{-1}^1 f(t)\,dt = 0.$$

30.4 Convergence of Integrals

It is often necessary to consider integrals of the form $\int_a^x f(t)\,dt$ when x tends to infinity. Thus, we make a definition:

30.4.1 DEFINITION. Suppose that, for each $x \in \langle a, \infty \rangle$, the integral $g(x) = \int_a^x f(t)\,dt$ exists. If $g(x)$ converges to a limit l as $x \to \infty$, then we denote l by

$$l = \int_a^\infty f(t)\,dt = \lim_{x \to \infty} \int_a^x f(t)\,dt.$$

Various tests have been devised to decide when certain such 'infinite integrals' are 'convergent', and they may be found in books on analysis. We shall need only two such tests, those of inspection and comparison. We illustrate these by examples.

30.4.2 EXAMPLE. If $a > 0$, then for each rational $s > 1$,

$$\int_a^\infty t^{-s}\,dt = 1\Big/(s-1)a^{s-1}.$$

For if $x \geqslant a$, then, using 29 Exercise 11, we have

$$g(x) = \int_a^x t^{-s}\,dt = \left[\frac{-t^{-s+1}}{-s+1}\right]_a^x = \frac{x^{-s+1} - a^{-s+1}}{-s+1};$$

now $-s+1 < 0$, so $x^{-s+1} \to 0$ as $x \to \infty$. Hence $\lim_{x\to\infty} g(x)$ exists and is $1/(s-1)a^{s-1}$ as asserted.

30.4.3 EXAMPLE. Given $a \in \mathbb{R}$, $\int_a^\infty \cos t\,dt$ does not exist. For if $g(x) = \int_a^x \cos t\,dt$, then $g(x) = \sin x - \sin a$; and since $\sin x$ does not converge to a limit as $x \to \infty$, then neither does $g(x)$.

30.4.4 §EXAMPLE. For each $n \in \mathbb{N}$, it can be shown that

$$n! = \int_0^\infty t^n e^{-t}\,dt,$$

and also that the integral exists even when n is not an integer but a member of $\mathbb{R}^+$. In the latter case the integral is often denoted by $\Gamma(n+1)$ and called the **gamma function**; it still satisfies the law $\Gamma(n+1) = n\Gamma(n)$, $n \in \mathbb{R}^+$.

Other limits are considered similarly. Thus if $g(x) = \int_a^x f(t)\,dt$ exists for all $x \in \langle\!\langle a, b\rangle$, and $g(x)$ converges to a limit l as $x \to b$, then we say that the (improper) integral $\int_a^b f(t)\,dt$ exists and equals l; and similarly for an interval $\langle b, a\rangle\!\rangle$.

30.4.5 EXAMPLE. $\displaystyle\int_0^1 \frac{dt}{\sqrt{(1-t^2)}} = \lim_{x\to 1}\int_0^x \frac{dt}{\sqrt{(1-t^2)}} = \lim_{x\to 1}\sin^{-1}x = \pi/2,$

since $\sin^{-1}: \langle\!\langle -1, 1\rangle\!\rangle \to \langle\!\langle -\pi/2, \pi/2\rangle\!\rangle$ is continuous at $x = 1$.

§*Exercise 9*

(i) Use a reduction formula to evaluate the improper integral $\int_0^1 x^n(\log x)^r\,dx$, when $0 \leqslant r \leqslant n$ and $r, n \in \mathbb{N}$. (See 34 Exercise 1(iv), (xi).)

(ii) Show that $\int_0^\infty dt/(1+t^2) = \pi/2$.

(iii) Establish the properties of the gamma function stated in 30.4.4.

Part VII continued

ADDITIONAL TOPICS IN THE CALCULUS

Having now assembled some basic machinery of the calculus, we devote the rest of this Part to considering some applications and extensions of it. We first consider briefly the theory of linear differential equations with constant coefficients, and then say a little on Taylor's expansion and infinite series. Finally, we give some of the theory of functions of several real variables, paying particular attention to the defects and difficulties of the classical notation.

Chapter 31

THE LOGARITHM AND THE EXPONENTIAL FUNCTION

31.1 The Logarithm

A good illustration of the power of the fundamental theorem (30.1.2) is shown by the theory of the logarithm. After fruitless attempts at finding a known function with derivative $1/\mathscr{I}$, it was realized by mathematicians that an integral *did* exist (by the fundamental theorem), provided they widened the early point of view that a function had to be 'an expression'; they then had to *make* a new function.

The following definition was made:

31.1.1 $$\log x = \int_1^x (1/t)\,dt, \qquad (x > 0).$$

If $x > 0$, then $1/\mathscr{I}$ is continuous on $\langle\!\langle 1, x \rangle\!\rangle$ or $\langle\!\langle x, 1 \rangle\!\rangle$ (according as $x \geqslant 1$ or $x \leqslant 1$), so the integral exists by 30.1.2; and by 30.1.6,

$$\frac{d}{dx}(\log x) = 1/x.$$

Thus log is a function with domain $\mathbb{R}_+$, range $\mathbb{R}$, and derivative $1/\mathscr{I}$; and the reason for its name is given by the basic algebraic property 31.1.2 below which was previously known to hold for the Napierian logarithms introduced by Napier (1550–1617; see Turnbull [129], Ch. V). Thus we have a proof that Napierian logarithms *exist*. Originally, these were believed to exist, by inference from the laws of indices. People were able to use them, with Napier, in making calculations, before the proof of existence!

31.1.2 THEOREM. *If* $a, b \in \mathbb{R}_+$, *then* $\log ab = \log a + \log b$.

Proof. $$\log ab = \int_1^{ab} \frac{1}{t}\,dt = \int_1^{a} \frac{1}{t}\,dt + \int_a^{ab} \frac{1}{t}\,dt = \log a + \int_a^{ab} \frac{1}{t}\,dt.$$

The substitution $t = as$ transforms the last integral into

$$\int_1^{b} \frac{1}{s}\,ds = \log b. \quad \blacksquare$$

31.1.3 COROLLARY. *If* $a \in \mathbb{R}_+$, *then* $\log 1/a = -\log a$.

Proof. $\log((1/a)\cdot a) = \log 1 = \int_1^1 1/t\,dt = 0;$

but $\log((1/a)\cdot a) = \log 1/a + \log a$ by Theorem 31.1.2. ■

The property 31.1.2 is also enjoyed by the constant function 0. That log is not the zero function is shown by:

31.1.4 PROPOSITION. *If* $a > 1$, *then* $\log a > 0$.

Proof. Apply 30 Exercise 3(i), with $f(t) = 1/t$, $g(t) = 0$. ■

Exercise 1

(i) If $0 < a < 1$, prove that $\log a < 0$.
(ii) If $\alpha \in \mathbb{Q}$ and $\alpha > 0$, then $1/t < 1/t^{\alpha+1}$, when $t > 1$. Hence prove that $\log t < (t^{-\alpha} - 1)/(1 + \alpha)$. Use 27 Exercise 2(iii) to deduce that $\lim_{t\to\infty} t^{-\alpha}\log t = 0$ (see 27.3.5) and then that $\lim_{u\to 0} u^{\alpha}\log u = 0$. [Thus $|\log x|$ is smaller, in a very strong sense, than any power x^{α}, if $\alpha > 0$ and $x > 0$.]
(iii) Evaluate $\int_2^3 1/t\,dt$, $\int_{-2}^{-3} 1/t\,dt$.

31.1.5 THEOREM. *For each integer* $n \in \mathbb{Z}$, *and* $a \in \mathbb{R}_+$,

$$\log a^n = n\log a.$$

Proof. By 31.1.2, and induction on $n \in \mathbb{N}$, and then by 31.1.3. ■

Exercise 2

(i) For each rational $q \in \mathbb{Q}$, prove that $\log a^q = q\log a$. $\{$Let $q = n/m$, $(n, m \in \mathbb{Z})$, and consider first $n = 1$, $n = -1$.$\}$
(ii) Prove that $\log 2 > \frac{1}{2}$.
$\left\{\int_1^2 (1/t)\,dt > \int_1^2 \frac{1}{2}\,dt.\right\}$

Next we prove the very important property

31.1.6 THEOREM. $\log\colon \mathbb{R}_+ \to \mathbb{R}$ *is bijective.*

Proof. We have to prove that log is injective and surjective. Let then $x \neq y$ in $\mathbb{R}_+$; it suffices to prove that log is strictly increasing, to prove it one-one.
We can assume that $x < y$. Then

$$\begin{aligned}\log x &= \int_1^x (1/t)\,dt + \int_x^y (1/t)\,dt \qquad \text{(by 30.3.3)}\\ &= \log x + A, \qquad \text{say,}\end{aligned}$$

where $A = \int_x^y (1/t)\,dt > 0$ by 30 Exercise 3(i). Hence $\log y > \log x$, so $\log y \neq \log x$; i.e., log is injective.

To prove that log is surjective, let $y \in \mathbb{R}$; then we must prove that there exists $a \in \mathbb{R}_+$ such that $y = \log a$. First, if $y = 0$, then $a = 1$. Second, suppose that $y > 0$. Then, since $\log 2$ (in particular) is > 0, (by 31.1.4), there is an integer $n \in \mathbb{N}$, such that $n \log 2 > y$ (by the property of Archimedes, see 24 Exercise 6(iii)). Hence, by 31.1.5,

$$\log 1 = 0 < y < \log 2^n,$$

so y is a number between the two values $\log 1$, $\log 2^n$, of log. By the fundamental theorem (30.1.2), log is continuous, so by the intermediate value theorem (28.5.1), there exists $a \in \langle 1, 2^n \rangle$ such that $y = \log a$; and since $\langle\!\langle 1, 2^n \rangle\!\rangle \subseteq \mathbb{R}_+$, $a \in \mathbb{R}_+$. Third, if $y < 0$, then $-y > 0$, and $-y = \log b$ for some $b \in \mathbb{R}_+$, as we have just seen. Hence, by 31.1.3, $y = \log 1/b$, and $1/b \in \mathbb{R}_+$, since $b \in \mathbb{R}_+$. Therefore log is surjective. ∎

Since $(\log)' = 1/\mathscr{I} > 0$, log is strictly increasing, so its graph looks roughly like that in Fig. 31.1. Every horizontal line $y = \text{const}$ cuts the curve once (since log is surjective) and exactly once (since log is injective).

Caution. It is not really valid to say that the last statement is 'obvious from the graph', because it is only after 31.1.6 is known that we can draw the graph at all.

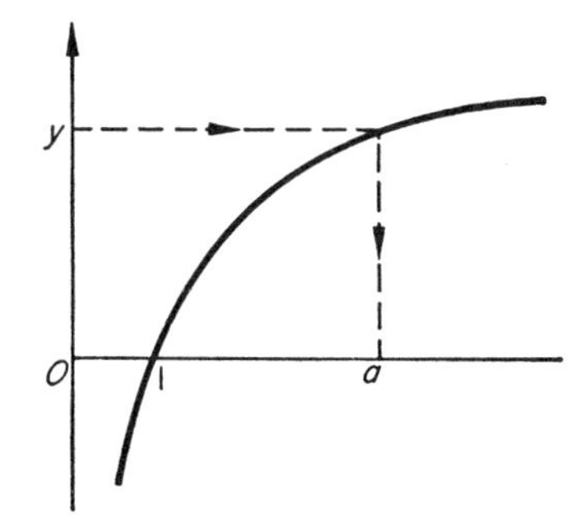

Fig. 31.1

Exercise 3

(i) Show that the groups $(\mathbb{R}_+, \cdot)$, $(\mathbb{R}, +)$ are isomorphic. (Use 31.1.2.)
(ii) Prove that $\log t \to \infty$ as $t \to \infty$, and $\log t \to -\infty$ as $t \to 0$ through positive values.
(iii) Why, from the practical viewpoint of using logarithms in calculation, is 31.1.6 important?

31.2 The Function exp

By 31.1.6 and the inversion theorem, log has an inverse, which is denoted by

$$\exp : \mathbb{R} \to \mathbb{R}_+$$

and called the **exponential function**, for reasons which will appear. Thus

31.2.1 $$\exp \circ \log = \mathscr{I}_1, \qquad \log \circ \exp = \mathscr{I}_2,$$

where $\mathscr{I}_1, \mathscr{I}_2$, are respectively the identity maps on $\mathbb{R}_+, \mathbb{R}$; in other terms, for each $s \in \mathbb{R}_+$, $t \in \mathbb{R}$,

31.2.2 $$\exp(\log s) = s, \qquad \log(\exp(t)) = t.$$

In Fig. 31.1, the arrowed path from y to a shows geometrically how to construct $\exp y$ from y. *Note:* exp *is nowhere zero*, since $0 \notin \mathbb{R}_+$.

Since log has a nowhere-zero derivative, then by Corollary 29.9.3, exp is differentiable, and for each $t \in \mathbb{R}$

$$\begin{aligned}(\exp)'(t) &= 1/(\log)'(s), \qquad s = \exp(t)\\ &= s \qquad ((\log)' = 1/\mathscr{I}_1)\\ &= \exp(t).\end{aligned}$$

Hence we have proved

31.2.3 THEOREM. exp *is its own derivative.* ■

Thus exp is a solution of the differential equation

31.2.4 $$\frac{dy}{dx} = y,$$

with domain $\mathbb{R}$ and initial condition $y(0) = 1$.

31.2.5 THEOREM. *There is no other solution satisfying these conditions.*

Proof. If y is a solution, then $z = y/\exp$ is differentiable, since exp is nowhere zero, and $z(0) = y(0)/\exp(0) = 1$. By the quotient rule,

$$z' = (\exp \cdot y' - y \cdot (\exp)')/(\exp)^2 = 0,$$

since y and exp satisfy 31.2.4. Therefore z is constant, by the constancy test (29.7.8) and this constant is $z(0) = 1$. Hence $y = \exp$. ■

Now, 31.2.3 is the basic *analytic* property of exp. The basic *algebraic* property of exp is:

31.2.6 THEOREM. *If* $u, v \in \mathbb{R}$, *then* $\exp(u + v) = \exp u \cdot \exp v$.

Proof. By 31.1.2,

$$\begin{aligned}\log(\exp u \cdot \exp v) &= \log(\exp u) + \log(\exp v)\\ &= u + v \qquad \text{(by 31.2.2)}.\end{aligned}$$

Apply exp to each side, and use 31.2.2 to obtain

$$\exp u \cdot \exp v = \exp(u + v). \quad ■$$

[This is essentially the same proof as that for h^{-1} in 3.6.9. See also Exercise 4(iv) below.] It is customary to denote the number exp (1) by e, after L. Euler, the great 18th-century Swiss mathematician. [See Bell [10], Ch. 9.] Then, by 31.2.2,

31.2.7 $$\log \mathrm{e} = 1.$$

We can estimate the size of e as follows. We have

$$1 = \int_1^{\mathrm{e}} (1/t)\,dt < \int_1^{\mathrm{e}} (1/t^{3/4})\,dt = [4t^{1/4}]_1^{\mathrm{e}}$$

so $\mathrm{e}^{1/4} > 5/4$, and $\mathrm{e} > 2\frac{113}{256}$; and also

$$1 = \int_1^{\mathrm{e}} (1/t)\,dt > \int_1^{\mathrm{e}} (1/t^{5/4})\,dt = [-4t^{-1/4}]_1^{\mathrm{e}}$$

so $\mathrm{e}^{1/4} < 4/3$, and $\mathrm{e} < 3\frac{13}{81}$.

Exercise 4

(i) By taking different powers of t (and increasing the computational labour), improve these estimates. [In fact, $\mathrm{e} = 2.718 \cdots$.]

(ii) We next remark that

$$\mathrm{e}^2 = \mathrm{e}\cdot\mathrm{e} = \exp(1)\cdot\exp(1) = \exp(1+1) = \exp(2),$$
$$\mathrm{e}^3 = \mathrm{e}^2\cdot\mathrm{e} = \exp(2)\cdot\exp(1) = \exp(3).$$

Prove, by induction on $n \in \mathbb{N}$, that $\exp(n) = \mathrm{e}^n$.

(iii) Prove that: (1) $\exp(-x) = 1/\exp(x)$;

(2) $\exp(n) = \mathrm{e}^n$, $n \in \mathbb{Z}$;

(3) for any rational $q \in \mathbb{Q}$, $\exp(q) = \mathrm{e}^q$.

(iv) Show that the only solution of the equation

$$\frac{dy}{dx} = \lambda x, \qquad y(0) = A \qquad (\lambda, A \in \mathbb{R})$$

is $y = A\exp(\lambda x)$. Hence prove 31.2.6 again by comparing $\exp(u + x)$ and $\exp u\cdot\exp x$.

31.3 The Laws of Indices

Because of Exercise 4(iii) above, it is tempting to suppose tha $\exp(x) = \mathrm{e}^x$ for all $x \in \mathbb{R}$. But what does e^x mean, if x is not rational? If the laws of logarithms are to be obeyed, then for any number a we would require a^x, whatever its meaning, to be such that

31.3.1 $$\log a^x = x \log a,$$

so by 31.2.2,

$$a^x = \exp(\log a^x) = \exp(x \log a).$$

The last term requires that $a \in \mathbb{R}_+$. We therefore take as the *definition* of a^x, for $a \in \mathbb{R}_+$, and $x \in \mathbb{R}$:

31.3.2 DEFINITION. $a^x = \exp(x \log a)$, $a \in \mathbb{R}_+$, $x \in \mathbb{R}$. In particular, since $\log \mathrm{e} = 1$, it follows that

$$\mathrm{e}^x = \exp(x).$$

Also, we can reverse the argument above, and obtain 31.3.1 as a valid equation. Now, when $q \in \mathbb{Q}$, a^q has already been defined, in 28.6.3, but

$$\begin{aligned} \exp(q \log a) &= \exp \log a^q \qquad \text{(Exercise 2(i))} \\ &= a^q \qquad \text{(old definition), by 31.2.2.} \end{aligned}$$

Therefore the two definitions of a^q in 31.3.2 and 28.6.3 agree for $q \in \mathbb{Q}$, so there is no danger of confusion if we use the same notation, a^q, in both. The vague 'feeling', that $\exp x = \mathrm{e}^x$, has now become a precise definition of the right-hand side by means of the left-hand side, for $x \in \mathbb{R}$. It remains only to prove the laws of indices, and these are easily settled; we leave the details for the reader, in the following exercise.

Exercise 5

(i) Verify the laws of indices by showing that for all $x, y \in \mathbb{R}$ and $a \in \mathbb{R}_+$:

$$a^0 = 1,\ a^{x+y} = a^x \cdot a^y,\ a^{-x} = 1/a^x,\ (a^x)^y = a^{xy}.$$

(ii) Using Definition 31.3.2, prove that

$$\frac{d}{dx}(a^x) = a^x \log a \quad (a > 0);$$

and that

(iii)
$$\frac{d}{dx}(x^\alpha) = \alpha x^{\alpha - 1} \qquad (\alpha \text{ constant}).$$

Thus 29.9.7 and its extension in 29 Exercise 11 continue to hold when the exponent is replaced by an arbitrary real number.

(iv) The *hyperbolic functions* are defined by the formulae $\cosh x = \frac{1}{2}(\mathrm{e}^x + \mathrm{e}^{-x})$, $\sinh x = \frac{1}{2}(\mathrm{e}^x - \mathrm{e}^{-x})$, $\tanh x = \sinh x/\cosh x$, $\coth x = \cosh x/\sinh x$, $\operatorname{cosech} x = 1/\sinh x$, $\operatorname{sech} x = 1/\cosh x$.

(1) In $\mathbb{R}$ find the largest domain you can for each of these functions, and work out the corresponding image.
(2) Prove that $(\cosh)' = \sinh$, $(\sinh)' = \cosh$.
(3) Prove that $\cosh^2 - \sinh^2 = 1$.
(4) Given $u, v \in \mathbb{R}$, use 29.7.8 to prove the **addition** formulae:

$$\cosh(u + v) = \cosh u \cdot \cosh v + \sinh u \cdot \sinh v.$$
$$\sinh(u + v) = \cosh u \cdot \sinh v - \sinh u \cdot \cosh v.$$
$$\cosh(-u) = \cosh u; \qquad \sinh(-u) = -\sinh u.$$

(5) Prove that $\exp, \cosh, \sinh, \tanh \in \mathscr{C}^\infty(\mathbb{R})$, while $\log \in \mathscr{C}^\infty(\mathbb{R}_+)$.

(v) Find suitable domains on which to define the inverse functions $\sinh^{-1}$, $\cosh^{-1}$, $\tanh^{-1}$.

Show that the functions

$$y_1 = \sinh^{-1}(\tan x), \qquad y_2 = \tanh^{-1}(\sin x)$$

are defined on $I = \langle -\pi/2, \pi/2 \rangle$, and that they are in fact equal on I. If $x \in I$, calculate dy_1/dx and prove that $y_1 = \log\tan(\frac{1}{2}x + \pi/4)$. (Of course, you may prove that this is the function y_2 if you prefer!)

(vi) Prove that $e^x \to \infty$ as $x \to \infty$ and that $e^x \to 0$ as $x \to -\infty$.

(vii) Extend Exercise 1(ii) to the case $\alpha \in \mathbb{R}$, $(\alpha > 0)$.

(viii) Hence prove that $e^x/x^\alpha \to \infty$ as $x \to \infty$, for all $\alpha \in \mathbb{R}$.

Thus we have the important result that e^x *grows faster than any power of* x, as $x \to \infty$. This rapid **exponential growth** governs many physical and biological processes. Similarly, e^{-x} decays faster than any power of x^{-1} as $x \to \infty$; the mass of a radioactive substance tends to decay exponentially in time, while an unchecked increase of population tends to grow exponentially in time. These qualitative features of the exponential function should be contemplated carefully by the reader, because they tend to be forgotten in the technical manipulations associated with the use of e^x.

Chapter 32

DIFFERENTIAL EQUATIONS

32.1 Linear First-Order Equations

As we saw in Chapter 30, the problem of integration arises in trying to find a solution g to the equation

32.1.1 $$Dg = f,$$

such that $f \in \mathbb{R}^I$, $g \in \mathscr{C}^1(I)$, and with the 'initial condition' $g(a) = 0$. More generally, let $\lambda \in \mathbb{R}$, and form the function $D + \lambda: \mathscr{C}^1(I) \to \mathscr{C}^0(I)$, where for each $p \in \mathscr{C}^1(I)$, $(D + \lambda)(p) = Dp + \lambda \cdot p$. It is easily verified that $D + \lambda$ is *linear*, and for this reason $D + \lambda$ is called a **linear operator.** Then 32.1.1 is included under the problem of finding $g \in \mathscr{C}^1(I)$ such that

32.1.2 $$(D + \lambda)g = f,$$

subject to the *initial condition* that $g(a)$ shall be some number $g_a \in \mathbb{R}$, prescribed in advance. The latter condition gets its name from problems in which I is a time-interval, g a velocity or acceleration, and $a = 0$; more generally, such a condition is called a **boundary condition** on 32.1.2. Then equation 32.1.2, together with the initial condition, is called a linear differential equation of the first order with constant coefficients. (In fact, there is only one 'coefficient', namely λ.) We claim

32.1.3 THEOREM. *There is a unique solution $g \in \mathscr{C}^1(I)$ satisfying* 32.1.2 *and the given initial condition.*

Proof. First, suppose that a solution g does exist. Then multiplying the equation 32.1.2 through by $e^{\lambda t}$ gives

$$e^{\lambda t} \cdot f(t) = \left(\frac{dg}{dt} + \lambda \cdot g(t)\right) \cdot e^{\lambda t} = \frac{d}{dt}(e^{\lambda t} \cdot g(t)) \qquad (t \in I);$$

so, if we integrate each side (as we can, since $\exp, f, g \in \mathscr{C}^0(I)$), then

$$\begin{aligned} \int_a^x e^{\lambda t} f(t)\, dt &= [e^{\lambda t} \cdot g(t)]_a^x \\ &= e^{\lambda x} \cdot g(x) - e^{\lambda a} \cdot g_a, \end{aligned}$$

using the initial condition. Thus

32.1.4 $$g(x) = e^{-\lambda x}\left\{e^{\lambda a}g_a + \int_a^x e^{\lambda t}\cdot f(t)\,dt\right\}$$

—provided a solution exists at all. But the function g defined by the right-hand side of 32.1.4 certainly exists, by 30.1.2. (Admittedly, we may not be able to evaluate the integral in terms of simple functions.) Moreover it has the value g_a at $x = a$, and can be differentiated and found to satisfy the equation 32.1.2. Hence 32.1.2 has at least one solution, and hence it has just one solution, given by 32.1.4. ■

Exercise 1

Carry out the same argument when λ is replaced in 32.1.2 by a continuous function $p: I \to \mathbb{R}$. (We then have a linear differential equation of the **first order** meaning that $D + p$ is both a linear operator and a 'polynomial' of the *first* degree in D.)

32.2 Second-Order Equations

Next, suppose we take a second operator, $D + \mu: \mathscr{C}^1(I) \to \mathscr{C}^0(I)$, like 32.1.2, so that $\lambda, \mu \in \mathbb{R}$. Then we can compose it with $D + \lambda$, at least on the subalgebra $\mathscr{C}^2(I) \subseteq \mathscr{C}^1(I)$, to form a new function:

32.2.1 $$T = (D + \lambda) \circ (D + \mu): \mathscr{C}^2(I) \to \mathscr{C}^0(I).$$

Exercise 2

Prove that T is linear.

Now, for any $h \in \mathscr{C}^2(I)$, we have from 32.2.1

$$\begin{aligned} T(h) = (D + \lambda) \circ (D + \mu)(h) &= (D + \lambda)(D + \mu)(h) \\ &= D(D + \mu)(h) + \lambda\cdot(D + \mu)(h) \\ &= D^2h + (\lambda + \mu)\cdot Dh + \lambda\mu\cdot h \end{aligned}$$

32.2.2 $$= (D^2 + (\lambda + \mu)\,D + \lambda\mu)h$$

in accordance with the definition of addition in the set F of functions from $\mathscr{C}^2(I)$ to $\mathscr{C}^0(I)$. Thus $T = D^2 + (\lambda + \mu)D + \lambda\mu$ factorises into $(D + \lambda)(D + \mu)$ and we drop the symbol for composition since the multiplication in F, inherited from the ring $\mathscr{C}^0(I)$, is not of interest in this particular theory.

We can ask, as in 32.1.2, given $f \in \mathscr{C}^0(I)$ what can be said of the set of functions $g \in \mathscr{C}^2(I)$ such that

32.2.3 $$Tg = (D^2 + (\lambda + \mu)\,D + \lambda\mu)g = f.$$

We suppose a solution g exists, and write

$$\begin{aligned} Tg &= (D+\lambda)(D+\mu)g \\ &= (D+\lambda)u, \quad (=f) \end{aligned}$$

where $u = (D+\mu)g$. Hence, by 32.1.4, we get for any $a \in I$:

32.2.4 $$u(t) = e^{-\lambda t}\left\{e^{\lambda a}u_a + \int_a^t e^{\lambda s}\cdot f(s)\,ds\right\}$$

where $u_a = ((D+\mu)g)(a) = g'(a) + \mu g(a)$. Therefore we need to be told $g'(a), g(a)$ in advance. Now solve the equation $u = (D+\mu)g$ with initial condition $g(a) = g_a$; it has a unique solution, given by 32.1.4:

32.2.5 $$g(x) = e^{-\mu x}\left\{e^{\mu a}g(a) + \int_a^x e^{\mu t}u(t)\,dt\right\},$$

where $u(t)$ is given by 32.2.4. Arguing as for 32.1.4, we see that *if* 32.2.3 has a solution, then that solution is of the form 32.2.5; but 32.2.5 *is* a solution, as can be verified; hence 32.2.5 is *the* solution, subject to the *two* **initial conditions**

32.2.6 $$g(a) = g_1, \qquad g'(a) = g_2 \qquad (\text{so that } u_a = g_2 + \mu g_1)$$

for an arbitrary point $a \in I$. Equation 32.2.3, together with a pair of initial conditions 32.2.6, is called a 'linear differential equation of the second order, with constant coefficients'. One can verify that, in 32.2.5, if $\lambda \neq \mu$, then $g(x)$ is of the form

32.2.7 $$g(x) = Ae^{-\lambda x} + Be^{-\mu x} + p(x)$$

where $A, B \in \mathbb{R}$ and $p(x)$ has the form

32.2.8 $$p(x) = e^{-\mu x}\int_a^x e^{(\mu-\lambda)t}\left(\int_a^t e^{\lambda s}\cdot f(s)\,ds\right)dt.$$

Exercise 3

Suppose that the initial conditions 32.2.6 are replaced by boundary conditions $g(a) = g_1$, $g(b) = g_2$. Find A and B in 32.2.7.

Thus 32.2.7 is often summarized by saying that a linear differential equation, of second order, with constant coefficients, has a solution consisting of the sum of the **complementary function** $Ae^{-\lambda x} + Be^{-\mu x}$, and a **particular integral** $p(x)$; and it contains the two 'arbitrary constants' A and B. However, in practice, correctly formulated differential equations arise only with prescribed boundary conditions, and a solution which does not have A and B evaluated in terms of these is incomplete. Special techniques exist for avoiding the labour of computing the integrals in 32.2.8; they usually involve trial and error, or 'operator' methods, and are dealt with in books on differential equations (see Gaskell [41]). Notice, in particular, that the complementary function is a solution of the differential equation $T(g) = 0$. Indeed it is the general form of any function in the *kernel* of T (see 19.6).

Exercise 4

(i) Show that, if $\lambda = \mu$ in 32.2.3, then $g(t)$ in 32.2.5 is of the form

$$(A + Bt)e^{-\lambda t} + \int_b^t \left(\int_a^s e^{\lambda r} f(r)\, dr\right) ds.$$

Evaluate A, B in terms of g_1, g_2 in 32.2.6, in this case.
(ii) Extend the above method to operators of the form

$$T = (D + \lambda)(D + \mu)(D + \nu), \qquad (\lambda, \mu, \nu \in \mathbb{R}),$$

reducing the problem successively to the solution of three first-order equations. [The corresponding differential equation, $Tg = f$, is then of the third order and requires *three* boundary or initial conditions to be completely formulated. If λ, μ, ν are all distinct, the solution (which exists and is unique) is of the form

$$g(t) = Ae^{-\lambda t} + Be^{-\mu t} + Ce^{-\nu t} + q(t),$$

where $q(t)$ is any function satisfying $Tq = f$ but not necessarily the boundary conditions. Assuming this, one can then choose A, B, C to make the function $g(t)$ fit the boundary conditions, at the cost of having to solve three linear algebraic equations in three unknowns. A similar technique may be applied in principle to nth-order equations with constant coefficients; here, the labour involved in the corresponding problem of solving n linear equations in n unknowns is prohibitive (e.g., when $n = 10$ and $\lambda, \mu, \ldots$ are given to several decimal places, as can happen in engineering or economics). The integrals corresponding to 32.2.8 then provide a much less laborious means of evaluating the solution $g(t)$, especially as computers can cope with integration more easily than with most processes.]
(iii) We start with the equation

$$(D - \lambda)g = f, \qquad \lambda \in \mathbb{R}, \qquad g(0) = A,$$

then $g' = f + \lambda g$, so integrating gives

$$g(t) = A + \int_0^t f(s)\, ds + \lambda \int_0^t g(s)\, ds.$$

Substituting for $g(s)$, we get

$$g(t) = A + \int_0^t f(s)\, ds + \lambda \int_0^t \left(A + \int_0^s f(u)\, du + \lambda \int_0^s g(u)\, du\right) ds.$$

Substitute again for $g(u)$ and obtain an expansion of $g(t)$ involving powers of λ and t, iterated integrals $\int f$, $\iint f$, $\iiint f$, and an 'unknown' remainder, involving g. Apply this process to the equation 31.2.4 to obtain the expansion

$$\exp(x) = 1 + x + x^2/2! + \cdots + x^n/n! + R_{n+1}(x),$$

where $R_{n+1}(x) = \int_0^x R_n(t)\,dt$, $R_1(x) = \int_0^x e^t\,dt$. Carry out a similar process to expand sin and cos, and estimate the size of the remainder integral after n terms. [This method is an example of 'Picard's iteration process', which works for any differential equation of the form $dy/dt = f(y, t)$, $y(0)$ given, when f is (say) differentiable.]

(iv) A function $f: \mathbb{R}^+ \to \mathbb{R}$ is said to be 'exponentially bounded' iff $\exists A, a \in \mathbb{R}$, such that for all sufficiently large $x \in \mathbb{R}$, $|f(x)| < Ae^{ax}$. Let $X_n \subseteq \mathscr{C}^n(\mathbb{R}^+)$ denote the subset of all exponentially bounded functions. Prove that X_n is a subalgebra of $\mathscr{C}^n(\mathbb{R}^+)$.

(v) For each $f \in X_0$ in (iv) above, prove that there is an interval $I_f = \langle s_f, \infty\rangle$ such that $\forall s \in I_f$,

$$F(s) = \int_0^\infty f(t)e^{-st}\,dt$$

exists. Thus $F: I_f \to \mathbb{R}$ is a function called the **Laplace transform** of f, denoted by $L(f)$. Evaluate $L(f)$ when $f(x) = x^n$ ($n \in \mathbb{N}$), 1, $\sin \lambda x$, $\cos \lambda x$, $e^{\lambda x}$ ($\lambda \in \mathbb{R}$). Given $f, g \in X$, $r, t \in \mathbb{R}$, and suitable intervals I_f, I_g prove that we may take $I_{rf+tg} \supseteq I_f \cap I_g$, and that $\forall s \in I_f \cap I_g$,

$$L(rf + tg)|_s = rL(f)|_s + tL(g)|_s.$$

[Laplace (1749–1827) was French. See Bell [10], Ch. 11.]

(vi) Let X_∞ denote $\bigcap_{n=0}^{\infty} X_n$, with X_n as in (iv). (Thus if $f \in X_\infty$, then $\forall n \in \mathbb{N}$, $D^n f \in X_\infty$.) Prove that X_∞ is a subalgebra of each X_n. For each $f \in X_\infty$ use integration by parts to prove that (cf. (v)) $I_{f'} \supseteq I_f \cap \langle a, \infty\rangle$ (where a is such that $|f(x)| < Ae^{ax}$ in (iv)); and that

$$L(f') = \mathscr{I}\cdot L(f) - f(0), \qquad L(f'') = \mathscr{I}^2\cdot L(f) - f'(0) - \mathscr{I}\cdot f(0)$$

on the appropriate domains.

(vii) Solve equation 32.2.3 as follows. Suppose $L(g)$ exists; then there exists an interval $\langle a, \infty\rangle$ such that, taking the Laplace transform of each side of 32.2.3, we have

$$(s^2 + (\lambda + \mu)s + \lambda\mu)G(s) = F(s) + sg'(0) + (\lambda + \mu)g(0),$$

for all $s \in \langle a, \infty\rangle$; where $F = L(f)$, $G = L(g)$. Hence, if $s \neq -\lambda, -\mu$, then

$$G(s) = F(s) + sg'(0) + (\lambda + \mu)g(0))/(s^2 + (\lambda + \mu)s + \lambda\mu)$$

and the right-hand side can often be split into partial fractions. If these are recognizable as $L(p)$, $L(q)$, ..., etc. (using tables of Laplace transforms), then $G = L(p) + L(q) + \cdots = L(p + q + \cdots)$; whence $g = p + q + \cdots$, since it can be proved that L is one-one on the domain X_0. Use the 'table' you made in (v) above to solve

$$(D^2 - 3D + 2)y = t^3 + \sin 4t, \qquad (y(0) = -1, y'(0) = 3)$$

by this method.

(viii) Use Laplace transforms to solve the simultaneous equations

$$x + \dot{y} + \dot{z} = 1$$
$$\dot{x} + y + \dot{z} = 1$$
$$\dot{x} + \dot{y} + z = 1 \qquad (\dot{x} = dx/dt, \text{ etc.})$$

in terms of the values $x(0)$, $y(0)$, $z(0)$.

[Observe that L maps a differential equation into an algebraic equation; and similarly L can map a partial differential equation in two variables, into an ordinary differential equation in one variable. This is an example of the powerful general method in mathematics, of looking for a way of mapping a strange problem into a familiar one. Extensive tables of Laplace transforms exist; and if $L(f) = F$, then f can be expressed as a 'contour integral' involving F, within the theory of functions of a complex variable.]

Besides learning to solve differential equations, it is of great interest and importance to be able to formulate them as well. Well-known examples of this process of mathematical modelling are those associated with the description of radioactive decay, Newton's law of cooling, and population growth. Problems in engineering are to be found in, for example, Gaskell [41]. The biological problem of describing the balance between populations of predators and prey is discussed in Kemeny–Snell [74], and the related descriptions of states of warfare by Richardson and by Lanchester in separate articles in Newman [93]. In the latter works, the modelling has to go through several stages before the mathematics becomes sufficiently tractable. It is instructive to see the physical insight shown by the authors; without it a blind mathematical technique would be useless.

Chapter 33

COMPLEX-VALUED FUNCTIONS

33.1 Differentiation

Imagine a particle moving in the plane $\mathbb{R}^2$, which is at the point z at time t. Its motion gives us a function $\zeta: I \to \mathbb{R}^2$, where I is the time interval of its motion, and $\zeta(t)$ is its position at time t. It is useful to consider $\mathbb{R}^2$ as $\mathbb{C}$, the algebra of complex numbers (24.7), so that $\zeta(t)$ has real and imaginary parts $f(t)$, $g(t)$, and

$$\zeta(t) = f(t) + ig(t):$$

thus $f, g: I \to \mathbb{R}$ are ordinary real-valued functions and we write

33.1.1 $$\zeta = f + \mathrm{i} \cdot g.$$

We therefore widen our algebra of functions to be the set $\mathscr{X}(I)$ of all such functions ζ. It forms an algebra over $\mathbb{C}$ in the obvious way. To discuss, say, the velocity of the particle, we need to be able to differentiate ζ, so we *define* the derivative of ζ to be

33.1.2 $$D\zeta = Df + \mathrm{i} \cdot Dg \qquad \left(\text{or} \quad \frac{d\zeta}{dt} = \frac{df}{dt} + \mathrm{i}\frac{dg}{dt}\right),$$

where we assume of course that $f, g \in \mathscr{D}(I)$. It can be verified that the rules 29.3.2, 29.3.3, 29.3.6 for differentiating sums, products and reciprocals hold for this extended operator, D. Clearly this new operator agrees with the old one on functions of the form $\zeta = f + \mathrm{i}0$, that is, on real-valued functions.

Exercise 1

(i) Perform these verifications in detail. Do the same for integration.
(ii) Extend the constancy test (29.7.8) to functions of the form ζ in 33.1.1.

33.2 The Function cis

A simple and important function of the form 33.1.1 is that which we will temporarily denote by cis: $\mathbb{R} \to \mathbb{C}$, where

33.2.1 $$\text{cis} = \cos + \mathrm{i} \cdot \sin.$$

It is the function ζ (above) for a particle moving on the unit circle† $\mathbb{S}^1$: $|z| = 1$, because, for any $t \in \mathbb{R}$,

33.2.2 $$|\text{cis}\, t| = (\cos^2 t + \sin^2 t)^{1/2} = 1;$$

this shows, incidentally, that cis *is nowhere zero.*

Exercise 2

(i) Prove that the image of cis is the whole of $\mathbb{S}^1$.
(ii) In the plane $\mathbb{R}^2$, let the ray from 0 to z cut $\mathbb{S}^1$ in the point p. Show that $z = |z| \cdot p$, and hence that there is an angle α (measured in radians) with $0 \leqslant \alpha < 2\pi$, such that $z = |z|$ cis α. It is traditional to denote α by arg z, and to measure α counterclockwise, so that

$$\arg 1 = 0, \quad \arg \mathrm{i} = \pi/2, \quad \arg(-1) = \pi, \quad \arg(-\mathrm{i}) = 3\pi/2, \quad \text{etc.}$$

(iii) By drawing a figure as a check, find arg z when

$$z = 1 + \mathrm{i}, \qquad 1 - 2\mathrm{i}, \qquad -1 - \mathrm{i}.$$

(iv) When $z = x + \mathrm{i}y$, use the function $\tan^{-1}: \mathbb{R} \to \langle -\pi/2, \pi/2 \rangle$ (see 29 Exercise 11(ii)) to express arg z in the forms θ, $\pi + \theta$, $2\pi + \theta$ where $\theta = \tan^{-1} y/x$, in the case (a) $x > 0, y \geqslant 0$; (b) $x < 0$; (c) $y \leqslant 0 < x$.

Now, for each $\mu \in \mathbb{R}$, cis μt satisfies the differential equation

33.2.3 $$dv/dt = \mathrm{i}\mu v, \qquad v(0) = 1,$$

which is the analogue of the differential equation

33.2.4 $$du/dt = \lambda u, \qquad u(0) = 1, \qquad (\lambda \in \mathbb{R})$$

satisfied by $\mathrm{e}^{\lambda t}$. This suggests that we therefore *define* $\mathrm{e}^{\mathrm{i}\mu t}$ (since we have given it no meaning previously) by

33.2.5 $$\mathrm{e}^{\mathrm{i}\mu t} = \text{cis}\, \mu t;$$

and then, in the hope of preserving the laws of indices (realized in 33.3.1), we *define* $\mathrm{e}^{(\lambda + \mathrm{i}\mu)t}$, when $\lambda, \mu \in \mathbb{R}$, by

33.2.6 $$\mathrm{e}^{(\lambda + \mathrm{i}\mu)t} = \mathrm{e}^{\lambda t} \cdot \mathrm{e}^{\mathrm{i}\mu t} = \mathrm{e}^{\lambda t} \cdot \text{cis}\, \mu t.$$

Thus, since $\mathrm{e}^{\lambda t}$ and cis μt are nowhere zero, then:

33.2.7 THEOREM. *For any $z = \lambda + \mathrm{i}\mu \in \mathbb{C}$, e^z is never zero; also,*

$$|\mathrm{e}^z| = \mathrm{e}^\lambda \cdot |\mathrm{e}^{\mathrm{i}\mu}| = \mathrm{e}^\lambda \cdot 1 = \mathrm{e}^\lambda. \quad ■$$

Moreover, since

$$\frac{d}{dt}(\mathrm{e}^{(\lambda + \mathrm{i}\mu)t}) = \frac{d}{dt}(\mathrm{e}^{\lambda t} \cos \mu t + \mathrm{i}\mathrm{e}^{\lambda t} \sin \mu t),$$

† Recall that if $z = x + \mathrm{i}y$ $(x, y \in \mathbb{R})$, then $|z|$ is defined to be $\sqrt{(x^2 + y^2)}$; thus $|z| > 0$ unless $z = 0$.

then, using 33.1.2, 33.2.4, and 33.2.5, we see that e^{zt} *satisfies the differential equation*

33.2.8
$$\frac{dw}{dt} = zw, \qquad w(0) = 1.$$

We assert:

33.2.9 THEOREM. *The only solution of equation* 33.2.8 *is* $w(t) = e^{zt}$.

Proof. Since e^{zt} is nowhere zero (by 33.2.7), $q = w/e^{zt}$ is differentiable; by the quotient rule,

$$\frac{dq}{dt} = e^{zt}\frac{\dfrac{dw}{dt} - w\cdot\dfrac{d}{dt}(e^{zt})}{(e^{zt})^2} = \frac{e^{zt}\cdot zw - w\cdot ze^{zt}}{(e^{zt})^2} = 0;$$

so q is constant, by Theorem 29.7.5. But

$$q(0) = w(0)/1 = 1,$$

so q is the constant function 1. Hence $w(t) = e^{zt}$, as asserted. ■

33.3 Algebra of e^z

A corollary of Theorem 33.2.9 is the *law of indices*:

33.3.1 *Given* $u, v \in \mathbb{C}$, $e^{u+v} = e^u \cdot e^v$.

Proof. The reader can verify that each of the functions $e^{(u+v)t}$, and $e^{ut}\cdot e^{vt}$, satisfies the differential equation 33.2.8 with $z = u + v$. Hence by 33.2.9 they are equal for all $t \in \mathbb{R}$. Now put $t = 1$. ■

Hence, we have the analogues of the formulae already proved when u is real, namely,

33.3.2
$$e^{-u} = (e^u)^{-1};$$

33.3.3
$$e^{nu} = (e^u)^n \qquad (n \in \mathbb{N});$$

(meaning of course the nth power of the complex number e^u). We obtain 33.3.2 from 33.3.1 by putting $v = -u$ (since $e^0 = 1$), and 33.3.3 comes from 33.3.1 by induction on $n \in \mathbb{N}$. ■

Taking $u = i\alpha$, $\alpha \in \mathbb{R}$ in 33.3.3, we immediately obtain

33.3.4 **De Moivre's theorem.** $(\text{cis}\ \alpha)^n = \text{cis}\ n\alpha \qquad (\alpha \in \mathbb{R}, n \in \mathbb{N})$.
[*Thus* $(\cos\alpha + i\sin\alpha)^n = \cos n\alpha + i\sin n\alpha$.] ■

If in 33.3.1 we take $u = i\alpha$, $v = i\beta$ $(\alpha, \beta \in \mathbb{R})$, we get $e^{i(\alpha+\beta)} = e^{i\alpha}\cdot e^{i\beta}$; multiplying out and equating real and imaginary parts gives

33.3.5
$$\begin{cases} \cos(\alpha+\beta) = \cos\alpha\cdot\cos\beta - \sin\alpha\cdot\sin\beta \\ \sin(\alpha+\beta) = \sin\alpha\cdot\cos\beta + \cos\alpha\cdot\sin\beta. \end{cases}$$ ■

These are the *addition formulae* of trigonometry (cf. 29.8). Observe that they hold for *all* real α, β and that the only properties of sin and cos that we have used are those in 29.8.1.

From 33.2.5, we have for any $\alpha \in \mathbb{R}$,

33.3.6
$$\begin{cases} e^{i\alpha} = \cos\alpha + i\sin\alpha, \\ e^{-i\alpha} = \cos(-\alpha) + i\sin(-\alpha) = \cos\alpha - i\sin\alpha; \end{cases}$$

thus we are relieved of the necessity of talking about cis α! In particular we have such remarkable formulae as $e^{2\pi i} = 1$, and $e^{\pi i} = -1$. Moreover, by Exercise 2(ii), we may write each $z \in \mathbb{C}$ as

33.3.7
$$z = \rho e^{i\theta}, \qquad \text{where } \rho = |z| \quad \textit{and} \quad \theta = \arg z.$$

Exercise 3

(i) Use 33.3.7 to prove that, for all $z_1, z_2 \in \mathbb{C}$,

$$|z_1 z_2| = |z_1| \cdot |z_2|,$$

and
$$\arg z_1 z_2 = \arg z_1 + \arg z_2 + \lambda,$$

where λ is an integral multiple of 2π. [Thus we write

$$\arg z_1 z_2 = \arg z_1 + \arg z_2 \pmod{2\pi}.]$$

(ii) Use the remark preceding 33.3.7 to show that, for any whole number $n > 1$, and $\forall p, q \in \mathbb{Z}$,

$$e^{2\pi i p/n} = e^{2\pi i q/n} \;.\Leftrightarrow.\; p \equiv q \pmod{n}.$$

(iii) Given $n \in \mathbb{N}$, $n > 1$, use De Moivre's theorem (38.3.4) to show that $e^{2\pi i k/n}$ satisfies the equation $x^n = 1$, when $k = 0, 1, \ldots, n-1$, and that these n numbers are all distinct. Hence show that these are *all* the roots in $\mathbb{C}$ of the equation $x^n = 1$. {See 22.3.5.}
(iv) Hence find all roots in $\mathbb{C}$ of the equations $z^4 = 16$, $z^5 = -1$, and draw a sketch to show their positions in the plane.
(v) Let p denote a prime, and let X denote the set of all 'pth roots of unity' (i.e., all distinct solutions in $\mathbb{C}$ of the equation $x^p = 1$). Show that X is a field under the operations '+', '×', where

$$u \text{ '+' } v = uv, \qquad u \text{ '×' } v = u^r \; (u, v \in \mathbb{C}),$$

where $v = e^{2\pi i r/p}$, and the operation on the right-hand side of each equation is taken in the ordinary sense of $\mathbb{C}$. (Thus 'zero' and the unity element of X are the complex numbers 1, $e^{2\pi i/p}$ respectively.) Do you recognize this field?

The equations 33.3.6 allow us to express $\cos\alpha$ and $\sin\alpha$ in terms of $e^{i\alpha}$ and $e^{-i\alpha}$, namely

33.3.8
$$\cos\alpha = \tfrac{1}{2}(e^{i\alpha} + e^{-i\alpha}), \qquad \sin\alpha = (e^{i\alpha} - e^{-i\alpha})/2i.$$

It is these formulae which suggested the definitions, in 31 Exercise 5(iv) above, of the hyperbolic functions cosh and sinh. Now, since we know from 33.2.6

the meaning of e^{iz} for any $z \in \mathbb{C}$, we continue to know the meaning of the right-hand sides of the equations in 33.3.8 when α is allowed to lie in $\mathbb{C}$ instead of being confined to $\mathbb{R}$. It is then natural to continue to denote these right-hand sides by $\cos \alpha$, $\sin \alpha$, respectively. Thus we *extend* the domain of definition of the functions cos, sin: $\mathbb{R} \to \langle\!\langle -1, 1 \rangle\!\rangle$ to obtain functions: $\mathbb{C} \to \mathbb{C}$ defined by

33.3.9 $$\cos z = \tfrac{1}{2}(e^{iz} + e^{-iz}), \qquad \sin z = (e^{iz} - e^{-iz})/2i$$

for all $z \in \mathbb{C}$.

Exercise 4

(i) Extend the other trigonometrical functions tan, . . . , and the hyperbolic functions, from $\mathbb{R}$ to $\mathbb{C}$.
(ii) Show that cos: $\mathbb{C} \to \mathbb{C}$ is onto, and similarly for sin. Find $\sin^{\flat}\{z\}$ when $z \in \mathbb{C}$.
(iii) Prove the addition formulae 33.3.5 for the extensions in 33.3.9.
(iv) Prove that $\cosh iz = \cos z$, $\sinh iz = i \sin z$.
(v) For any complex numbers z_1, z_2 prove that

$$\cosh(z_1 + z_2) = \cosh z_1 \cdot \cosh z_2 + \sinh z_1 \cdot \sinh z_2.$$

Hence calculate the real and imaginary parts of $\cosh z$, for any $z \in \mathbb{C}$. Show that, as z traces out the line $y = \pi/4$ in $\mathbb{C}$, then $w = \cosh z$ traces out the hyperbola $2u^2 - 2v^2 = 1$, where the real and imaginary parts of w, z are (u, v), (x, y), respectively. [The hyperbola is the same as that given, in the (x, y) plane, by $2x^2 - 2y^2 = 1$, of course; but it often helps to think of w as lying in a separate copy of $\mathbb{C}$, in which co-ordinates are labelled u, v rather than x, y,]
(vi) The function f given by $f(z) = z^{-1}$ $(z \neq 0)$ maps $\mathbb{C} - \{0\}$ into itself. Find the image of f. Let A, B, C denote the points $1 + i$, i, $-1 + i$, respectively; prove that as z traces out the straight segment ABC, then $w = f(z)$ traces out a portion of the circle $u^2 + v^2 + v = 0$. Identify this portion.
(vii) Recall the real algebras $\mathscr{C}(I)$, $\mathscr{D}(I)$, and $\mathscr{C}^n(I)$ of 29.6.1 and let $\mathscr{D}_z(I)$ and $\mathscr{C}^n_z(I)$ denote their analogues in the algebra $\mathscr{Z}(I)$ mentioned after 33.1.1. Show that they are subalgebras over $\mathbb{C}$ of $\mathscr{Z}(I)$, and prove that, as sets

$$\mathscr{Z}(I) \approx \mathscr{C}(I) \times \mathscr{C}(I), \qquad \mathscr{D}_z(I) \approx \mathscr{D}(I) \times \mathscr{D}(I),$$

and $$\mathscr{C}^n_z(I) \approx \mathscr{C}^n(I) \times \mathscr{C}^n(I).$$

(viii) Given an operator $T: \mathscr{C}^n_z(I) \to \mathscr{C}^0_z(I)$ of the form

$$T = D^n + c_1 D^{n-1} + \cdots + c_{n-1} D + c_n \qquad (c_i \in \mathbb{C}),$$

use the Fundamental theorem of algebra (22.5.2) to show that T can be written in the form

$$T = (D + \alpha_1) \cdots (D + \alpha_n),$$

Repeat the theory of Chapter 32 when λ is allowed to be complex, and hence construct a theory of linear differential equations of nth order with constant complex coefficients. As special cases, consider the equations (excluded from consideration in Chapter 32)

$$\left.\begin{aligned} (D^2 + \lambda^2)g &= f, \\ \text{and} \qquad (D^2 + \alpha D + \beta)g &= f, \qquad (\alpha^2 < 4\beta) \end{aligned}\right\} \alpha, \beta, \lambda \in \mathbb{R},$$

subject to *two* boundary conditions. In practical engineering problems, the boundary conditions are usually initial conditions with real constants. For such equations, the nature of the function f often leads to the problem of evaluating integrals of the form

$$\int_0^x t^n e^{\lambda t} \cdot dt, \qquad \int_0^x t^n \sin \mu t \cdot dt, \qquad \int_0^x \sin \mu t \cdot e^{\lambda t}\, dt, \quad \text{etc.} \qquad (n \in \mathbb{N}),$$

all of which are special cases of $\int_0^x t^n e^{(\alpha + i\beta)t}\, dt$. Find an iterative process for evaluating this integral, and then take real and imaginary parts to evaluate

$$\int_0^x t^n \sin \beta t \cdot e^{\alpha t}\, dt, \qquad \int_0^x t^n \cos \beta t \cdot e^{\alpha t}\, dt.$$

(ix) Prove that, if $\zeta = |\zeta| e^{i\theta}$, $\theta = \arg \zeta$, then the solutions $z = x + iy$ of $e^z = \zeta$ are the numbers

$$\log |\zeta| + i(\arg \zeta + 2n\pi), \qquad n \in \mathbb{Z}.$$

This *set*, $\exp^{\flat} \{\zeta\}$, is traditionally denoted by Log ζ; it is customary to describe Log z as a 'many-valued function'. The **principal value** of Log ζ, denoted by $\log \zeta$, is the number $\log |\zeta| + i \arg \zeta$. Thus log is a genuine function and $\log \zeta \in \text{Log}\, \zeta$.

Then $e^{\log \zeta} = \zeta$, but $\log \zeta\zeta'$ will not, in general, be $\log \zeta + \log \zeta'$. One can still define w^z (when w and z are both complex) by $e^{z \log w}$, but the law of indices no longer holds. Show that, in a certain sense, $\text{Log}\, \zeta\zeta' = \text{Log}\, \zeta + \text{Log}\, \zeta'$ and explain how to restore the law of indices by regarding w^z as a many-valued function with a principal value.

What is i^i as a many-valued entity? What is its principal value?

Chapter 34

APPROXIMATION AND ITERATION

34.1 Taylor's Expansion

For practical purposes, it is often necessary to work out a good estimate of the value of a function $f: I \to \mathbb{R}$ at each $x \in I$. One method is to use the **Taylor expansion with integral form of the remainder**, as follows. Let $I = \langle\langle p, q \rangle\rangle$, suppose $f \in \mathscr{C}^{n+1}(I)$ (see 29.6), and $0 \in \langle p, q \rangle$. Given $a \in \langle p, q \rangle$ evaluate

$$\int_0^a f'(a - t)\, dt \qquad (a \in I)$$

in two ways, first (directly) as

$$[-f(a - t)]_0^a = -f(0) + f(a),$$

and second by parts as

$$\int_0^a 1 \cdot f'(a - t)\, dt = [t \cdot f'(a - t)]_0^a - \int_0^a t \cdot (-f''(a - t))\, dt$$

$$= af'(0) + \int_0^a tf''(a - t)\, dt.$$

Proceeding by induction on r, we therefore obtain

34.1.1 $$f(a) = f(0) + af'(0) + \cdots + \frac{a^r f^{(r)}(0)}{r!} + \cdots + \frac{a^n f^{(n)}(0)}{n!} + E_n,$$

where $E_n = \int_0^a t^n f^{(n+1)}(a - t)\, dt/n!$, $f^{(n+1)} = D^{n+1}f$. This is called the Taylor† expansion of f (as far as the term in a^n), and the **remainder** E_n is here expressed in the form of an integral; since $f \in \mathscr{C}^{n+1}(I)$, $D^{n+1}f$ is continuous and the integral exists (by 30.1.2).

It is clearly just as important to estimate $|E_n|$ as to evaluate the successive differential coefficients $f^{(r)}(0)$, in order to arrive at a good estimate of $f(a)$.

† Or Taylor–Maclaurin, after Brook Taylor (1685–1741) and Colin Maclaurin (1698–1746). See Turnbull [129].

34.1.2 EXAMPLE. Let $f = \exp: \mathbb{R} \to \mathbb{R}_+$. Then $(\exp)^{(r)}(0) = \exp(0) = 1$ by 31.2.3. Therefore

$$\exp(a) = 1 + a + \frac{a^2}{2!} + \cdots + \frac{a^n}{n!} + E_n,$$

where $E_n = \left(\int_0^a t^n \cdot e^{a-t}\, dt\right)/n!$. If $a > 0$, then

$$0 < E_n < a^n \int_0^a e^a\, dt/n! = a^{n+1}e^a/n!,$$

while if $a < 0$, then $t \leqslant 0$ (since $a \leqslant t \leqslant 0$), so $e^{a-t} = e^{-(t-a)} < 1$. Therefore, in this case,

$$|E_n| \leqslant 1 \cdot \left|\int_0^a t^n\, dt\right| \Big/ n! = |a|^{n+1}/(n+1)!,$$

so the error in taking $1 + \cdots + a^n/n!$ for $\exp(a)$ is here no worse in modulus than the next term. This is a more useful statement than the error when $a > 0$, since that error involves the (uncomputed) value e^a. We could here compute $e^{-a}(a > 0)$ with error $\leqslant a^{n+1}/(n+1)!$ and then calculate the reciprocal of the number obtained, in order to estimate the error for e^a when $a > 0$.

Exercise 1

(i) When $x > 0$, prove that

$$\int_0^x e^{t^2}\, dt = x + \frac{x^3}{3!} + \cdots + \frac{x^{2n-1}}{(n-1)!\,(2n-1)} + E_n,$$

where $$0 \leqslant E_n \leqslant e^{x^2}x^{2n+1}/n!\,(2n+1).$$

If $x = 1$, how many terms must be taken to make $E_n < 0.01$?

(ii) Use the Taylor expansion to prove the binomial theorem for integral exponent (cf. 5 Exercise 1(xvi).

(iii) Using $|\sin| \leqslant 1$ and $|\cos| \leqslant 1$, show that

(1) $$\sin a = a - \frac{a^3}{3!} + \frac{a^5}{5!} - \cdots + \frac{(-1)^n a^{2n+1}}{(2n+1)!} + S_n,$$

where $|S_n| \leqslant |a|^{2n+3}/(2n+3)!$ and

(2) $$\cos a = 1 - \frac{a^2}{2!} + \frac{a^4}{4!} - \cdots + \frac{(-1)^n a^{2n}}{(2n)!} + C_n$$

where $|C_n| \leqslant |a|^{2n+2}/(2n+2)!$ Compare 32 Exercise 4(iii).

(iv) Let $I_n = \int_0^{\pi/2} \sin^n t\,dt$. Prove that $I_n = (n-1)I_{n-2}/n$ if $n \geqslant 2$. [A formula of this kind is called a 'reduction' formula, for obvious reasons.] Hence prove that, for any $m \in \mathbb{N}$,

$$\int_0^{\pi/2} \sin(\sin t)\,dt = \sum_{r=0}^{m-1} \frac{(-1)^r 2^{2r}(r!)^2}{((2r+1)!)^2} + E_m,$$

where $|E_m| \leqslant 2^{2m}(m!)^2/((2m+1)!)^2$. Treat similarly

$$\int_0^a \sin(\cos t)\,dt, \qquad \int_0^a \exp(\sin t)\,dt.$$

(v) Work out the first six terms of the Taylor expansions for the functions y, z in 29 Exercise 6(i) and (ii). [Here the remainders are not easy to estimate.]
(vi) Use the mean value theorem for integrals (30.2.5) to prove that E_n, in the expansion 34.1.1, is expressible as $a^{n+1}f^{(n+1)}(\theta a)/(n+1)!$, for some θ such that $0 < \theta < 1$. [This is called the **Lagrange form of the remainder**, after Lagrange (Bell [10], Ch. 10). Of course θ depends on f and a and no rule is given for evaluating it. Indeed, it may not even be unique. Nevertheless the Lagrange form of the remainder is often extremely useful for estimation.]
(vii) Hence, show that, if $a \in \mathbb{R}^+$, then

$$\log(1+a) = a - \frac{a^2}{2} + \cdots + (-1)^{n-1}\frac{a^n}{n} + L_n,$$

where

$$L_n = \frac{a^{n+1}(-1)^n}{(n+1)(1+\theta x)^{n+1}},$$

for some θ such that $0 < \theta < 1$.
(viii) Similarly, obtain the **binomial expansion with remainder** for $a, \alpha \in \mathbb{R}$; namely,

$$(1+a)^\alpha = 1 + \alpha a + \frac{\alpha(\alpha-1)a^2}{2!} + \cdots + \frac{\alpha(\alpha-1)\cdots(\alpha-(r-1))a^r}{r!} + A_r$$

where

$$A_r = \frac{\alpha(\alpha-1)\cdots(\alpha-r)}{(r+1)!}a^{r+1}(1+\theta a)^{\alpha-r-1}, \qquad 0 < \theta < 1$$

If $|a| < 1$, prove that $A_r \to 0$ as $r \to \infty$.
(ix) To evaluate a reciprocal $1/X$ numerically may be laborious, and it is easier for a computer to work out a sum of the form $Y(1 + z + z^2 + \cdots + z^n)$ which involves only multiplication and addition. Select suitable Y, z for evaluating X and estimate (in terms of X) how big n need be taken if an accuracy to p decimal places is required. (Compare (viii).) Consider the same problem when X is a non-singular matrix.
(x) Let $f: \mathbb{R} \to \mathbb{R}^+$ be given by

$$f(x) = \exp(-x^2), \qquad (x \neq 0); \qquad f(0) = 0.$$

Prove by induction on n, that $D^n f$ exists for each $n \in \mathbb{N}$; and that $D^n f$ is of the form $D^n f = p_n(1/x)\cdot\exp(-x^2)$ $(x \neq 0)$; $D^n f|_0 = 0$, where p_n is a polynomial. Hence show that, for each $n \in \mathbb{N}$, $f(x)$ is its own remainder, $E_n(x)$.

(xi) Use 30 Exercise 9(i) to show that

$$\int_0^1 x^x\,dx = 1 - \frac{1}{2^2} + \frac{1}{3^3} + \cdots + \frac{(-1)^{n-1}}{n^n} + R_n,$$

where $|R_n| \leqslant 1/(n+1)^{n+1}$. $\{x^x = e^{x \log x} = 1 + x\log x + \cdots + \text{remainder.}\}$

34.2 Maxima and Minima

As a by-product of the Taylor expansion, we can discuss maxima and minima, as follows. Let $f \in \mathscr{C}^n(I)$ as in 29.6, let $a \in I$, and suppose the first $n - 1$ derivatives of f vanish at a; thus $f^{(r)}(a) = 0$, $1 \leqslant r < n$, while $f^{(n)}(a) \neq 0$. Then by Exercise 1(vi) we have, for sufficiently small h,

34.2.1 $$f(a+h) - f(a) = h^n f^{(n)}(a + \theta h)/n!, \qquad 0 < \theta < 1.$$

We say that f has an **isolated maximum** (or **minimum**) at a iff $\exists\delta > 0$ such that the left-hand side is <0 (or >0) whenever $0 < |h| < \delta$. If $f^{(n)}(a) > 0$, then by the inertia principle (28.3.1) applied to $f^{(n)}$, we know that $f^{(n)}(a + \theta h) > 0$ also for θh sufficiently small; similarly when $f^{(n)}(a) < 0$. Thus, from equation 34.2.1 we have at once:

34.2.2 THEOREM. *If n is even, then f has an isolated maximum or minimum at a iff $f^{(n)}(a) < 0$ or > 0 respectively. If n is odd, then f changes sign as h passes through zero.* ■

The function $\mathscr{I}^n$ illustrates this result very well when $n \in \mathbb{N}$, as the reader will see if he also looks at the graph. The point of the word 'isolated' is to eliminate such oddities as a horizontal trough in the graph of a function.

34.2.3 CAUTION. It is important to distinguish between an isolated minimum, and an *absolute* minimum of a function. For example, the function f, given by $f(x) = 2x^3 - 9x^2 + 12x + 1$ with domain $\mathbb{R}$, has an isolated maximum at $x = 1$ but no absolute maximum. If, however, we restrict f to the domain $\langle\!\langle 0, 2\rangle\!\rangle$ (for example), then $x = 1$ is also the absolute maximum of f. The extra check, to see whether (say) an isolated minimum is absolute, must use some criterion like 29.7.2. For example, if $g(x) = x^4$, then by Theorem 34.2.2, g has an isolated minimum at $x = 0$; but applying 29.7.2 to $g'(x) = 4x^3$, we see that g is decreasing if $x \leqslant 0$ and increasing if $x \geqslant 0$. Thus g takes its least value at $x = 0$, and $x = 0$ is also the absolute minimum.

Exercise 2

(i) Find the largest domain D of f in 34.2.3 for which $x = 1$ is the absolute maximum of $f|D$.
Do the same for the isolated minimum of f, and find the largest domain E for which the isolated maxima and minima of f are also the absolute maxima and minima of $f|E$.

(ii) If the length plus girth of a rectangular parcel of square section of side g must not exceed 12 feet, show that the maximum volume V is 16 and occurs when $g = 2$. However, $V = 4g^2(3 - g)$, and this function of g can be arbitrarily large; resolve this paradox.
(iii) The graph of $y = f(x)$ is said to have a **point of inflexion** at $x = a$, iff $f''(a) = 0, f'''(a) \neq 0$. Argue from the Taylor expansion to show that the curve crosses the tangent at $x = a$ (see Fig. 34.1). Show that this result can be false if $f'''(a) = 0$.
(iv) Sketch roughly the graph of $y = x - e^x$, and use it to sketch the graph of $y = \exp(x - e^x)$, showing the behaviour as $x \to \pm\infty$. Prove that the points of inflexion on the second graph are at $x = \pm \log \frac{1}{2}(3 + \sqrt{5})$.

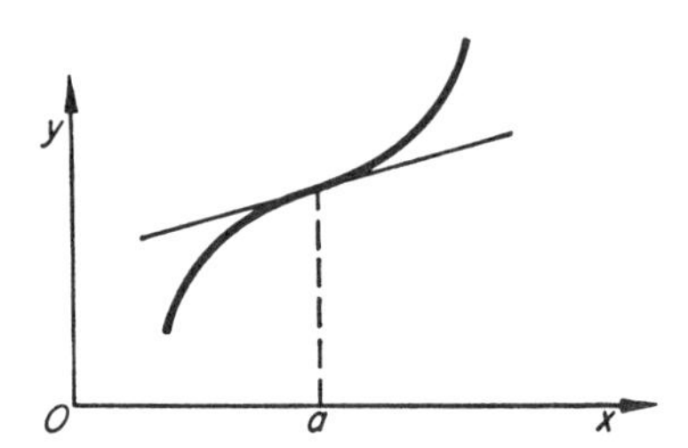

Fig. 34.1 A point of inflexion.

34.3 Newton's Method of Approximation

In calculating the successive derivatives $f^{(n)}(0)$ of a Taylor expansion, or in the inductive steps of Exercise 1(x), the nth formula is calculated by using the preceding ones. Similarly for reduction formulae with integrals (Exercise 1(iii)), Picard interation (32 Exercise 4(iii)), and the use of Leibniz' formula (29 Exercise 6(i), (ii)). The last example, however, has the feature that a sequence of numbers $f(0), \ldots, f^{(n)}(0)$ is being calculated, and the last number obtained is fed back into the formula to get the next one. This is an example of an *iterative process*, since a simple operation is being repeated ('iterated') time after time; hence it is suitable for use with computers. Indeed, it is often quicker, and as accurate when decimals are needed, to provide the computer with an iterative process which uses only additions and multiplications, rather than to offer it an exact but awkward formula which uses more complicated operations. These methods are dealt with in books on numerical methods, e.g., Redish [106] to which we refer the reader for a suitable account of this important field, which is now very lively because of the big computers.

To illustrate iterative processes, we consider a process due to Newton. It is applied to the problem of finding an approximate root of an equation of the form $f(x) = 0$; thus suppose we have guessed—perhaps from drawing a rough graph—that $x = u$ is near a root which we will call v. Write $v = u + r$, and suppose $f \in \mathscr{C}^2(I)$, for some interval $I = \langle a, b \rangle$ containing u and v. Then by the Taylor expansion

$$0 = f(v) = f(u) + rf'(u) + \tfrac{1}{2}r^2 f''(u + s)$$

for some s between 0 and r. Therefore, if $f'(u) \neq 0$,

$$\begin{aligned} r &= -f(u)/f'(u) - \tfrac{1}{2}r^2 f''(u + s)/f'(u) \\ &= r_1 - m, \qquad \text{say.} \end{aligned} \tag{34.3.1}$$

Thus, if we let $u_1 = u + r_1$, then $v - u_1 = -m$, so the new error, in approximating v by u_1 instead of by u, is $-m$; and the worst possible value of $|m|$ can occur only when the numerator is greatest and the denominator least, i.e.,

$$|m| \leqslant \tfrac{1}{2}r^2 \cdot |\max f''/\min f'| = Ar^2 \qquad \text{(say)}, \tag{34.3.2}$$

the max and min referring to the whole interval I.

Hence, if r is small compared with A, the new error, m, will be smaller than the old.

34.3.3 EXAMPLE. Let $f(x) = x^2 - 2$, so $v = \sqrt{2}$. Since $f(1.4) < 0 < f(1.5)$, v lies in the interval $I = \langle 1.4, 1.5 \rangle$ and so, if we let the first approximation be $u = 1.5$, the error r is < 0.1. Then $r_1 = -f(1.5)/f'(1.5) = -\frac{1}{12}$, so $u_1 = \frac{3}{2} - \frac{1}{12} = \frac{17}{12}$. Also, on I, $\min f'$ is attained at $x = 1.4$, while $f'' = 2$; thus in 34.3.2, $A = (\frac{1}{2} \times 2)/(2 \cdot 8)$, and the error in taking $\sqrt{2}$ to be $\frac{17}{12}$ is $\leqslant Ar^2 < \frac{10}{28} \cdot \frac{1}{100} < 0.0036$.

Returning to the general discussion, we can repeat the process by which r_1 was obtained in 34.3.1, with u_1, m playing the roles of u, r respectively, to yield

$$r_2 = -f(u_1)/f'(u_1), \qquad u_2 = u_1 + r_2$$

and

$$|v - u_2| \leqslant Am^2 \leqslant A^3 r^4;$$

and the process can be iterated to obtain a sequence $u, u_1, u_2, \ldots, u_n$ of approximations to v with errors

$$|v - u_n| \leqslant A^{2n-1} r^{2n}.$$

By shrewd choice of r, these errors tend rapidly to zero. As in 34.3.3 a further check is often available, in the case that, say, $f(u_n) < 0 < f(u_{n+1})$; for then, by the intermediate value theorem, v must lie between u_n and u_{n+1}. The geometry behind the method is indicated in Fig. 34.2(a), and it suggests the simplification of Fig. 34.2(b), where the successive tangents at $x = u$ are replaced by lines parallel to the one tangent at $x = u$. This can lead to a saving in computer time, since only the one (expensive) division, $1/f'(u)$, need be calculated, to give successive remainders $-f(u_i)/f'(u)$ by multiplication. We have here a simple example where ingenuity can save expense; this is an important reason why mathematicians are employed by firms which

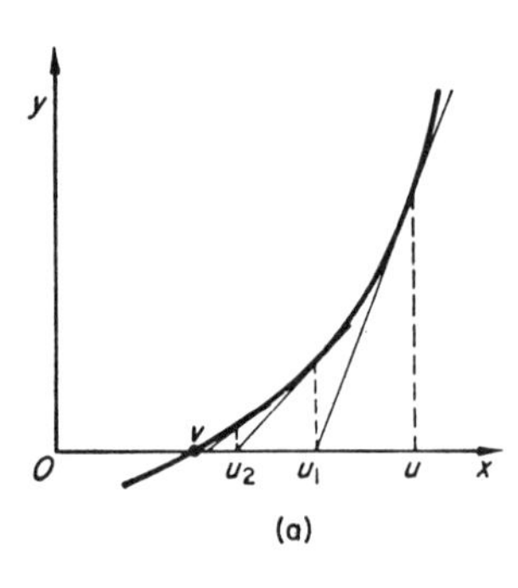

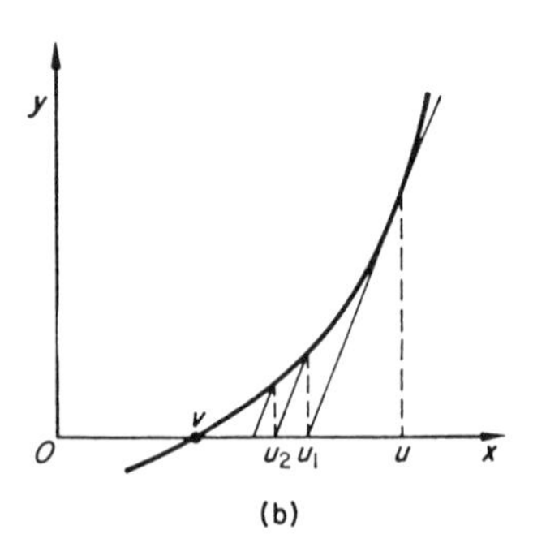

Fig. 34.2

use computers. (Of course, apart from saving money in this rather obvious way, and costing relatively little themselves, mathematicians can make great contributions by the bold and imaginative use of computers in unorthodox situations.)

34.4 Approximate Integration

The connection between areas and definite integrals, as explained in 30.1, suggests methods of approximation. We shall consider three such methods, with their associated error terms; and we confine ourselves to an integral of the form

$$J = \int_a^b f(t)\, dt$$

where $f \in \mathscr{C}^4(I)$, and $f > 0$ on I. In each method, we express J as an easily-computed sum.

We first dissect $\langle a, b \rangle$ by means of the points x_i, where

$$a = x_1 < x_2 < \cdots < x_{n+1} = b,$$

and $x_{i+1} - x_i = h$, $(1 \leqslant i \leqslant n)$. Then the area J under the graph of f is approximately the sum

34.4.1 $$A = \sum_{i=1}^{n} f(x_i)\cdot h$$

of areas of rectangles like those shown in Fig. 34.3. To estimate the error for one typical such rectangle, moved to the origin, we have

$$\int_0^h f(t)\, dt = \int_0^h (f(0) + tf'(\theta t))\, dt$$

(by the mean value theorem)

34.4.2 $$= h\cdot f(0) + E,$$

where $|E| = \left|\int_0^h tf'(\theta t)\, dt\right| \leqslant \frac{1}{2}M_0h^2$, and M_0 is the maximum of f' on

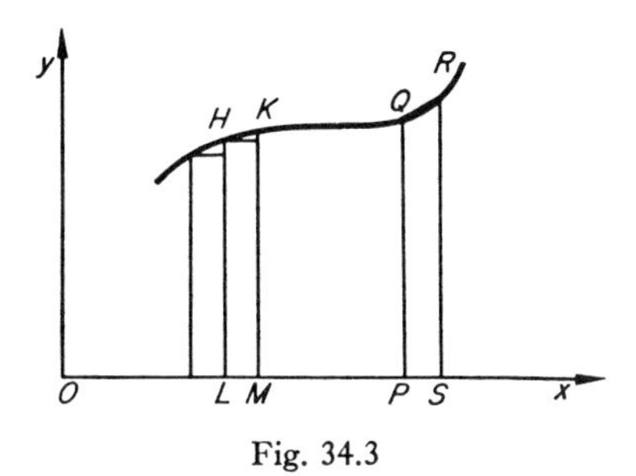

Fig. 34.3

$\langle\!\langle 0, h\rangle\!\rangle$ (which exists if f' is continuous, by the mostest theorem 28.5.2). Applying this to each term in the sum 34.4.1, we get

$$|J - A| \leqslant \tfrac{1}{2}nh^2M$$

where M is the maximum of f' on $\langle\!\langle a, b\rangle\!\rangle$. But $b - a = nh$, whence

$$|J - A| \leqslant \tfrac{1}{2}(b - a)Mh,$$

so the error in taking A for J can be described as 'of order h', written $O(h)$; where for convenience we write

$$O(h^q)$$

for any term X such that $|X| \leqslant Kh^q$, K being a number independent of h.

It is sometimes more convenient to take, instead of the sum A, a sum $S = \sum_{i=1}^{n} f(t_i)\cdot h$, where t_i is chosen from $\langle\!\langle x_i, x_{i+1}\rangle\!\rangle$ $(1 \leqslant i \leqslant n)$. But then

$$|A - S| = \left|\sum_{i=1}^{n} h(t_i - x_i)f'(y_i)\right|$$

for some $y_i \in \langle\!\langle x_i, x_{i+1}\rangle\!\rangle$, so

$$|A - S| \leqslant Mh^2n = M\cdot(b - a)\cdot h;$$

thus the error in taking S for J is still $O(h)$, as before, i.e.,

34.4.3 $$\int_a^b f(t)\,dt = h\sum_{i=1}^{n} f(t_i) + O(h), \qquad (t_i \in \langle\!\langle x_i, x_{i+1}\rangle\!\rangle).$$

A better approximation is obtained from the **trapezium rule,** where we replace the rectangles in Fig. 34.3 by trapezia like $PQRS$. The sum of these areas is then

34.4.4 $$\begin{aligned} B &= \sum_{i=1}^{n} \tfrac{1}{2}(f(x_i) + f(x_{i+1}))h \\ &= h\left\{\tfrac{1}{2}(f(x_1) + f(x_{n+1})) + \sum_{2}^{n} f(x_i)\right\}. \end{aligned}$$

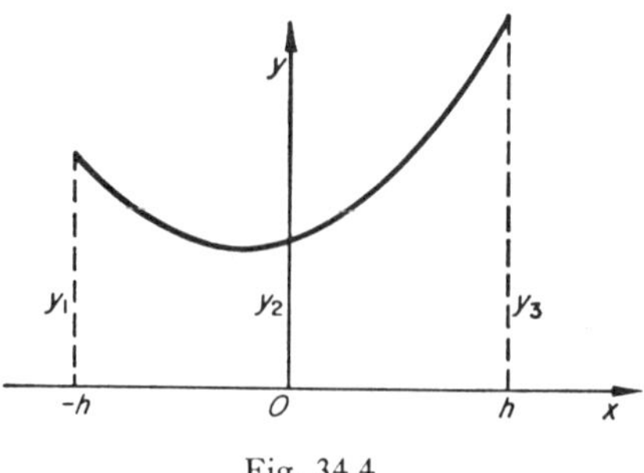

Fig. 34.4

To estimate the error in using a single trapezium we take the Taylor expansion two terms further than in 34.4.2, to give—on moving to the origin—

$$E = \int_0^h f(t)\,dt - \tfrac{1}{2}(f(0) + f(h))\cdot h$$

$$= \int_0^h (f(0) + tf'(0) + \tfrac{1}{2}t^2 f''(0) + O(t^3))\,dt$$

$$- \tfrac{1}{2}h(2f(0) + hf'(0) + \tfrac{1}{2}h^2 f''(0) + O(h^3))$$

since f''' is bounded (being continuous)

$$= -(h^3/12)f''(0) + O(h^4).$$

With each term in 34.4.4, an error of this kind is associated, so from the sum we get

$$J - B = -(h^3/12)\sum_{i=1}^{n} f''(x_i) + O(h^3)$$

(the last term since $nh = b - a$). Now apply 34.4.3 with f replaced by f'' (since f''' is bounded) to get

$$J - B = -(h^2/12)\left(\int_a^b f''(t)\,dt + O(h)\right) + O(h^3)$$

$$= -(h^2/12)(f'(b) - f'(a)) + O(h^3).$$

34.4.5 A better approximation still is that of **Simpson's rule**, where we leave some details to be completed by the reader. We first observe that the parabola $y = ax^2 + bx + c$ through the points $(-h, y_1)$, $(0, y_2)$, (h, y_3) has area (below the curve on $\langle -h, h \rangle$; see Fig. 34.4)

$$C = \tfrac{1}{3}h(y_1 + 4y_2 + y_3).$$

If $y_1 = f(-h)$, $y_2 = f(0)$, $y_3 = f(h)$, then C is an approximation to the area under the graph of $y = f(x)$ on $\langle -h, h \rangle$; and taking the Taylor expansion of f as far as $f^{(\text{iv})}(0)$, we obtain

34.4.6
$$\int_{-h}^{h} f(t)\,dt - C = -h^5 \cdot f^{(\text{iv})}(\xi)/90,$$

for some $\xi \in \langle -h, h \rangle$. Now apply this to the interval $\langle a, b \rangle$, where we take n in Fig. 34.3 to be even, say $n = 2m$; then the portion of curve on $\langle x_{2i-1}, x_{2i+1} \rangle$ is approximated by a parabola as in 34.4.6 to give a total over $\langle a, b \rangle$:

$$S = \tfrac{1}{3}h\left\{f(x_1) + f(x_{2m+1}) + 4\sum_{i=1}^{m} f(x_{2i}) + 2\sum_{i=2}^{m} f(x_{2i-1})\right\}.$$

If we sum the errors, using 34.4.6 and applying 34.4.1 to $f^{(\text{iv})}$, we obtain

34.4.7
$$J - S = -(f'''(b) - f'''(a))h^4/180 + O(h^5).$$

Exercise 3

(i) Prove 34.4.7 in detail.

(ii) Let $J = \int_0^{\pi/2} \sin t\,dt = 1$. Apply the Trapezium rule and Simpson's rule with $h = \pi/8$ to obtain the estimates 0.9976, 1.000 respectively (to four figures).

(iii) Let $J = \int_0^{\pi/2} \sqrt{(1 - t^2)}\,dt\ (= \pi/4)$. Here, the derivatives of $f(t) = \sqrt{(1-t^2)}$ are unbounded at $t = 1$; what modifications would you adopt to estimate π using, say, the Trapezium rule?

(iv) If you use four-place tables, what value of h (in terms of f''') would you decide on for applying Simpson's rule?

(v) Estimate the area under the graph of $y = e^{x^2}$ between $x = 0$ and $x = 1.6$, using Simpson's rule and $h = 0.2$. It is required to find that ordinate $x = k$ which bisects this area; estimate k approximately, estimate the error, and obtain a better approximation to k.

(vi) Estimate the order of magnitude of $n!$ by applying the Trapezium rule to $\int_1^n \log x\,dx$. A better estimate is **Stirling's formula**:

$$P < n! < P \cdot (1 + 1/4n), \qquad P = \sqrt{(2\pi)} \cdot n^{n+1/2} e^{-n}.$$

For a proof see Courant [24].

34.5 Series

In many cases, the remainder term E_n in the Taylor expansion of a function tends to zero as $n \to \infty$. Thus we automatically reach the next stage of calculus, that of the consideration of **power series**, of the form $\sum_{n=0}^{\infty} b_n x^n$. Here, we must discuss the meaning of $\sum_{n=0}^{\infty}$, which must be defined by using limits; thus, if $a\colon \mathbb{Z}^+ \to \mathbb{R}$ is a sequence (see 27.4), we form a new sequence of **partial sums**

$$s_n = \sum_{m=0}^{n} a_m.$$

34.5.1 DEFINITION. If the sequence s converges to a limit l as $n \to \infty$, we say that the series $\sum_{n=0}^{\infty} a_n$ is **convergent** (to the **sum** l). By an abuse of notation, we denote both s and l by the symbol $\sum_{m=0}^{\infty} a_m$. [Notice that n is a dummy index in $\sum_{n=0}^{\infty} a_n$; we could replace it by any other letter. In fact one often writes $\sum_{0}^{\infty} a_n$ for the sake of brevity.] As an example, if in 34.1.1 $E_n \to 0$ as $n \to \infty$, then $S_n = f(a) - E_n$ converges to $f(a) - 0 = f(a)$; so in this case $f(a) = \sum_{0}^{\infty} a^m f^{(m)}(0)/m!$

One then shows that if, for some value of x, $\neq 0$, the series $\sum_{m=0}^{\infty} b_m x^m$ is convergent, then either it converges for all x or there is a largest number $\rho > 0$ such that the series converges for all $x \in \langle -\rho, \rho\rangle$. We accordingly say that the series has **radius of convergence** ∞ or ρ, respectively. The basic theorems are then:

34.5.2 THEOREM. *If the sequence* $|b_{m+1}/b_m|$ *tends to a limit* $\lambda \in \mathbb{R}$ *as* $m \to \infty$, *then* $\sum_{0}^{\infty} b_m x^m$ *has radius of convergence* λ^{-1} *if* $\lambda \neq 0$, *or infinity if* $\lambda = 0$. □

34.5.3 THEOREM. *Suppose the series* $\sum_{0}^{\infty} b_m x^m$ *has radius of convergence* $r > 0$ (perhaps $r = \infty$). *Then the function* $f\colon \langle -r, r\rangle \to \mathbb{R}$, *given by*

$$f(x) = \sum_{0}^{\infty} b_m x^m, \qquad (|x| < r)$$

is differentiable and hence continuous; and

$$f'(x) = \sum_{m=1}^{\infty} m b_m x^{m-1}, \qquad \int_0^x f(t)\,dt = \sum_{m=0}^{\infty} b_m x^{m+1}/(m+1),$$

these new series also having the radius of convergence r. □

The last two series are said to be obtained by **term-by-term** differentiation and integration, respectively, of the original series; and the theorem says that these termwise processes give the derivative and integral of f respectively. In brief,

$$D\sum^{\infty} = \sum^{\infty} D, \qquad \int\sum^{\infty} = \sum^{\infty}\int$$

within the interval $\langle -r, r\rangle$.

34.5.4 COROLLARY. *By iteration of* 34.5.3, *f can be integrated and differentiated term by term, arbitrarily often, on the interval $\langle -r, r\rangle$; the result is always a series with the same radius of convergence r.* ■

These two theorems (which have analogues when $x \in \mathbb{C}$) enable us to clear up the problem of the *existence* of the sin and cos functions. The Taylor expansion (Exercise 1(iii)) forces the following definitions upon us.

34.5.5 DEFINITION.

$$\begin{cases} \sin x = x - \dfrac{x^3}{3!} + \dfrac{x^5}{5!} - \cdots = \displaystyle\sum_{n=0}^{\infty} (-1)^n \frac{x^{2n+1}}{(2n+1)!} \\ \cos x = 1 - \dfrac{x^2}{2!} + \dfrac{x^4}{4!} - \cdots = \displaystyle\sum_{n=0}^{\infty} (-1)^n \frac{x^{2n}}{(2n)!}. \end{cases}$$

By 34.5.2, each has *infinite* radius of convergence; and hence by 34.5.3, each has domain $\mathbb{R}$, and is infinitely differentiable. Hence the conditions 29.8.1 hold, as required.

Exercise 4

(i) In 34.5.3, prove that for each $n = 0, 1, \ldots, m$,

$$b_m = f^{(m)}(0)/m!.$$

(ii) Use 34.1.2 to prove that, if $x \in \mathbb{R}$,

$$e^x = 1 + x + \frac{x^2}{2!} + \frac{x^3}{3!} + \cdots = \sum_{m=0}^{\infty} \frac{x^m}{m!}.$$

(iii) An **analytic** function at $0 \in \mathbb{R}$, with radius a, is of the form $f: \langle -a, a\rangle \to \mathbb{R}$, and such that whenever $|x| < a$, then $f(x) = \sum^{\infty} x^m f^{(m)}(0)/m!$ Prove that the remaindcr, after n terms of the Taylor expansion of f, tends to zero as $n \to \infty$. Hence show that the set of all analytic functions at 0 with radius a forms a real algebra. {Use Leibniz' theorem (29.6.2) and 27 Exercise 6(vii).}
(iv) Show that analytic functions are infinitely differentiable and use Exercise 1(x) to disprove the converse.

For a comprehensive treatment, see any book on analysis (e.g., Osgood [99], Ch. VII). One is first struck in this theory by the riot of 'pretty' results obtained in the 18th century; for example the equations

$$\pi/4 = \sum_{n=0}^{\infty} (-1)^n/(2n+1), \qquad \pi^2/6 = \sum_{n=1}^{\infty} 1/n^2,$$

and 'infinite products' of which an example is

$$\pi^{-1} \sin \pi x = x \prod_{n=1}^{\infty} (1 - x^2/n^2).$$

A typical 18th-century proof of the first equality would be

34.5.6
$$\pi/4 = \int_0^1 (1 + x^2)^{-1}\, dx = \int_0^1 (1 - x^2 + x^4 - \cdots)\, dx$$
$$= \left[x - \frac{x^3}{3} + \frac{x^5}{5} - \cdots\right]_0^1$$

which gives the result. But theorems are needed (and were supplied by the critical 19th-century analysts) to justify (a) the term-by-term integration of the binomial series (valid for $|x| < 1$) and (b) evaluating the result at $x = 1$. This is the point of the critical approach, to be quite sure that the beautiful results *are* right. Of course, in this particular instance, an *ad hoc* proof is easy to give. Thus

$$\int_0^1 (1 - x^2 + x^4 - \cdots)\, dx = \int_0^1 (1 - x^2 + \cdots + (-)^n x^{2n})\, dx + \int_0^1 r_n\, dx$$

where $|r_n| = x^{2n}/(1 + x^2) \leqslant x^{2n}$, so the last integral exists and is $\leqslant 1/(2n + 1)$. Thus

$$\left|\pi/4 - \left(1 - \frac{1}{3} + \cdots + (-1)^n \frac{1}{2n + 1}\right)\right| < \frac{1}{2n + 1},$$

so

$$\pi/4 = \lim_{n \to \infty} \left(1 - \frac{1}{3} + \cdots + (-)^n \frac{1}{2n + 1}\right) = \sum_{n=0}^{\infty} (-1)^n/(2n + 1). \quad ■$$

But we do not wish to have to fall back on ingenuity in particular instances; that is not the way of mathematical progress. That a rigorous general theory is necessary is very clearly shown by the queer results in the allied theory of **Fourier series.** [See G. Fourier in Bell [10], Ch. 12.] These are trigonometrical series of the form

34.5.7
$$f(x) = \sum_{n=0}^{\infty} (a_n \cos nx + b_n \sin nx),$$

and it is true, for example, that every 'reasonable' function $f: \langle -\pi, \pi \rangle \to \mathbb{R}$ can be represented by such a series. But while f may perhaps have a non-periodic extension to a larger domain, the series 34.5.7 must always have period 2π; thus if we work out the Fourier series of the non-periodic function $\mathscr{I}$, we find

$$x = 2\left(\sin x - \frac{\sin 2x}{2} + \frac{\sin 3x}{3} - \cdots\right) \qquad (x \in \langle -\pi, \pi \rangle)$$

and the right-hand side remains convergent outside $\langle -\pi, \pi \rangle$ but it no longer represents there the identity function $\mathscr{I}$. Even worse, a harmless-looking Fourier series can represent a badly discontinuous function. For these and other reasons, the 'manipulative' 18th-century attitude to series had to be supplemented by a 'critical' 19th-century array of theorems delineating what was justifiable and correct and what was not.

Exercise 5

If $y = (1 + x)^2(\alpha \in \mathbb{R}, x > -1)$ prove that y satisfies the differential equation $\alpha y = (1 + x)y'$, as does the binomial series $y = \sum_{r=0}^{\infty} \alpha(\alpha - 1)\cdots(\alpha - r + 1)x^r/r!$ if $|x| < 1$. Use 32 Exercise 1 to conclude that the series then equals $(1 + x)^\alpha$. [This is the *Binomial Theorem* for 'arbitrary index', and the same method can be used in Exercise 4(ii) by considering the equation $y' = y$.]

34.6 Further Outlook

The theory of series also forced generalizations of the ideas of convergence and of integrals; these led to the branches of mathematics called *summability* and *measure theory*. We mentioned the second in 30.1; the first starts with the averaging process of 27 Exercise 6(vii), so that (for example) the obviously non-convergent series $1 - 1 + 1 - 1 \cdots$ is assigned the 'sum' $\frac{1}{2}$ while every convergent series is assigned a 'sum' which turns out to be its sum in the old sense. (See, e.g., Hardy [53].) A further obvious extension of the calculus is to functions with domain *and range* in $\mathbb{C}$, and is expounded in books on *The theory of functions of a complex variable*; and those with an 'applied' point of view tend to give a better perspective, which can later be filled in by the detailed preoccupations of 'purer' books. For example, we recommend Kellogg [71], Ch. XII, and then Osgood [99].

The theory then deepens, as a result of the beautiful contributions of Riemann. The earlier theory considers functions $f: \mathbb{C}_\infty \to \mathbb{C}_\infty$ (where $\mathbb{C}_\infty$ is the 'Gaussian sphere', consisting of $\mathbb{C}$ with an added 'point at infinity'). But Riemann asked: what can be said about functions $f: T \to \mathbb{C}_\infty$ where T is now any smooth surface? Later, $\mathbb{C}_\infty$ here also is changed to a more general surface, e.g., a torus. The result is the theory of 'Riemann surfaces'†, out of which has flowed the modern contemporary theories of analysis on n-dimensional manifolds,‡ with the associated theories of *tensors* and *differential geometry*. Here the flavour is no longer 'arithmetical' (i.e., about $\mathbb{R}$ and $\mathbb{C}$) but 'geometrical', since $\mathbb{R}^n$ has only weak algebraic properties when $n > 2$ (see 24.9).

All this, however, depends on the theory of *functions of several variables*, which was begun in the 18th century for very practical problems in physics. We devote the next chapter to a short survey of this basic and useful theory.

† Klein [75], Weyl [134].
‡ Duff [33], Flanders [36].

Chapter 35

FUNCTIONS OF SEVERAL REAL VARIABLES

35.1 The Problem

In this more difficult stage of calculus, we turn from the functions 'of one variable', which have so far occupied us, to functions 'of several variables'. There are two important reasons for the difficulties: first, the material studied is more complicated, and second, the traditional notation is a marvel of brevity. In the hands of experts, this brevity is all very well, but it is definitely confusing to beginners, and may be a reason for the comparative neglect of the subject by many teachers of analysis. We shall attempt here to show how the language of set-theory can resolve this notational difficulty; but we realize that our notation looks complicated, and will be disliked by those who have mastered the traditional notation. For such experts we naturally recommend no change. However, let us say that nothing is achieved by giving *apparently* simple formulae, at the outset, for complicated matters; and it is better, in introducing a topic, to use symbols in a complicated but self-explanatory formula, rather than to accompany a simple-looking formula with a paragraph of words.

We cannot hope to give here a full treatment of the subject, but we attempt to supply the reader with enough of the set-theoretic point of view to enable him to cope with the classical texts or the book of Nickerson, Spencer, and Steenrod [95] (which is comprehensive and modern). Consequently, we start from the beginning. A summary of the principal formulae of this chapter and Chapter 36 will be found in 36.5.

In practical problems, it is often necessary to consider functions of the form $f\colon A \to \mathbb{R}$, where A is a subset of some n-dimensional space $\mathbb{R}^n$. For example, f might be the temperature in a region $A \subseteq \mathbb{R}^3$, or the height of a roof, above an area $A \subseteq \mathbb{R}^2$. In general, each point $x \in A$ will be of the form

$$x = (x_1, \ldots, x_n),$$

so we write $f(x)$ as $f(x_1, \ldots, x_n)$; it is then customary to call f a **function of n variables.**

When $n = 1$, we have functions $f\colon A \to \mathbb{R}$ of the kind considered before, except that A may be a union of several intervals; and when $n = 2$, the graph of f (in $A \times \mathbb{R} \subseteq \mathbb{R}^3$) will look like a mountainous terrain above A. For larger values of n, we cannot picture the graph, since it lies in $A \times \mathbb{R} \subseteq \mathbb{R}^{n+1}$; so for illustrative purposes we usually take $n = 2$.

A particularly important function is the 'projection of $\mathbb{R}^n$ onto an axis'; thus recall from Chapter 14 the function

35.1.1 $$\pi^i: \mathbb{R}^n \to \mathbb{R}$$

defined by

$$\pi^i(x_1, \ldots, x_n) = x_i.$$

Of course, π^i is a linear transformation.

Exercise 1

Draw the graphs of π^i when $n = 1$, and when $n = 2$; and of the product $\pi^1\pi^2: \mathbb{R}^3 \to \mathbb{R}$.

It is clear then, that functions of several variables abound, and that we need a theory to organize our knowledge of them. As in other situations when studying a new branch of mathematics, it will help to orient the reader if at this point he tries to imagine the form that such a theory might be expected to take. Earlier mathematicians would naturally ask the same kinds of question as they asked when developing the single-variable theory, so we expect each result in that theory to have an analogue in several variables. So, we expect to consider smoothness of functions (continuity and differentiability), approximation by polynomials, and integration; such is the kind of *programme* that guides research and growth in mathematics. Let us then sketch in some details, along this line of approach.

35.2 Continuity

Given a sequence† $\{p_n\}$ of points in $\mathbb{R}^k$, we say that it **converges** to $p \in \mathbb{R}^k$, provided $\pi^i(p_n)$ converges to $\pi^i(p)$ (π^i as in 35.1) for each $i = 1, \ldots, k$. Since the numbers $\pi^i(p_n)$ lie in $\mathbb{R}$, we already know what it means for them to converge. If $\{p_n\}$ converges to p in $\mathbb{R}^k$ we write

35.2.1 $$\lim p_n = p \qquad (\text{or } p_n \to p \text{ as } n \to \infty).$$

As remarked in Examples 25.5.4, $\lim p_n = p$ iff $\lim\|p_n - p\| = 0$ in $\mathbb{R}$. Another characterization is given in (i) of the following exercise. [Compare Example 25.6.9(ii).]

Exercise 2

(i) Using the metric of $\mathbb{R}^k$ (Example 25.5.2(iii)) prove that $\lim p_n = p$ iff given $\varepsilon > 0$, there exists $N \in \mathbb{N}$ such that $\|p - p_n\| < \varepsilon$, whenever $n > N$ (i.e., every ε-neighbourhood of p contains all but a finite number of the points p_n).

† i.e., a function $\mathbb{N} \to \mathbb{R}^k$; cf. 27.4.

(ii) If, together with 35.2.1, a second sequence $\{q_n\}$ in $\mathbb{R}^k$ converges to $q \in \mathbb{R}^k$, prove that $\forall \lambda, \mu \in \mathbb{R}$, the sequence $\{\lambda p_n + \mu q_n\}$ converges to $\lambda p + \mu q$.
(iii) If in $\mathbb{R}$, $\lambda_n \to \lambda$, prove that in 35.2.1, $\lambda_n p_n \to \lambda p$.

By analogy with 28 Exercise 2(ii), we then say that a function $f: A \to \mathbb{R}$ is **continuous at** $p \in A$ whenever $\lim_{n\to\infty} f(p_n) = f(p)$ for every sequence $\{p_n\}$ in A convergent to p. And f is **continuous on** A, iff f is continuous at each $p \in A$. By definition of convergence of $\{p_n\}$ to p, each function π^i in 35.1.1 is continuous at each point. It can be proved that, analogously to Theorems 28.1.2 and 28.1.3:

35.2.2 THEOREM. *The set $\mathscr{C}(A)$ of continuous functions $f: A \to \mathbb{R}$ is an algebra over $\mathbb{R}$. In particular, a constant function is continuous.* □

Moreover, if $\mathscr{C}(I)$ is replaced by $\mathscr{C}(A)$, in 28.1.3, 28.3.1, 28.4, and 28.5.2, then the amended statements are all true. The inertia principle 28.3.1 also holds for each $f \in \mathscr{C}(A)$, except that the interval J there must be replaced by a neighbourhood of a. A function $f: A \to \mathbb{R}^m$ is really nothing other than m functions $f_i: A \to \mathbb{R}$, where $f(p) = (f_1(p), \ldots, f_m(p))$. Then $f_i = \pi^i \circ f$. Moreover, f is said to be continuous (at p) if each f_i is continuous (at p).

Exercise 3

(i) Prove the amended forms of 28.1.3, 28.3.1, 28.4, 28.5.2, in detail.
(ii) Given $A \xrightarrow{f} \mathbb{R}^k \xrightarrow{g} \mathbb{R}^n$, prove that, if f is continuous at $p \in A$, and g is continuous at $f(p)$, then $g \circ f$ is continuous at p. (See 28.2.)

35.3 The Differential

We shall now take up the ideas of 29.5. For this purpose, it is convenient to introduce the 'little O' notation, for estimating remainders. Thus, suppose $g: B \to \mathbb{R}^m$ is a function such that $\|g(x)\|/\|x\| \to 0$ as $x \to 0$. Then we write $o(x)$ instead of $g(x)$ when further information about g is not important. The reader must then be prepared for slightly peculiar but harmless statements like

35.3.1 $$\lambda \cdot o(x) = o(x), \qquad o(x) + o(x) = o(x) = o(x) \cdot o(x)$$

(It is understood throughout that x tends to 0.)
In particular, we can write 29.5.6 as

$$f(c + t) - f(c) = d_c f(t) + o(t)$$

for the function $f: I \to \mathbb{R}$, differentiable at $c \in I$. Since $d_c f: \mathbb{R} \to \mathbb{R}$ is a linear function, this generalizes at once to functions $f: A \to \mathbb{R}$, where A is an open subset of $\mathbb{R}^k$, defined as in Chapter 25 (i.e., recall that then every $a \in A$ lies at the centre of some open ball B, centre a, where $B \subseteq A$). Thus, we say that $f: A \to \mathbb{R}$ is **locally linear** at $p \in A$, iff there exists a linear transformation $T: \mathbb{R}^k \to \mathbb{R}$, such that

35.3.2 $$f(p + x) - f(p) = T(x) + o(x),$$

for all x in some open ball U, centre 0, in $\mathbb{R}^k$ for which $p + x \in A$. (U exists since A is open.) In 36.1.2 below, we shall prove that T is unique, if it exists; and since it depends both on p and on f, we write

$$T = d_p f,$$

calling this the **differential of f at p.** Observe that 35.3.2 says that f is approximately linear on a neighbourhood of p. We can then use our knowledge of linear functions to study f. We shall see in 35.5, that 'local linearity' is the proper definition of 'differentiability'; see 37 Exercise 1(ii) also.

[Traditionally, $T(x)$ in 35.3 is called the 'increment' in f corresponding to the 'increment' x in p, to within second order of small quantities; and $T(x)$ is then written df, (or even δf), and thought of as 'small'. As with 29.5.6, we must emphasize that $d_p f$ is *not* 'small', since it is not a number but a linear transformation. It is true, however, that when x is 'small', then $d_p f(x)$ will usually be small, and can be taken as 'δf' (i.e., $f(p + x) - f(p)$) to within an error of $o(x)$.]

35.3.3 EXAMPLE. Let us compute $d_p \pi^i$ for the function $\pi^i: \mathbb{R}^n \to \mathbb{R}$ of 35.1.1. We have, for each $p \in \mathbb{R}^n$,

$$\pi^i(p + x) - \pi^i(p) = (p_i + x_i) - p_i = x_i = \pi^i(x),$$

so $d_p \pi^i = \pi^i$; and the term $o(x)$ in 35.3.2 is zero.

Exercise 4

(i) Prove 35.3.1; and also, that if $P: \mathbb{R}^m \to \mathbb{R}^n$ is linear, then $P(o(x)) = o(x)$.
(ii) Show that, for any *linear* function $f: \mathbb{R}^n \to \mathbb{R}$,

$$d_p f = f,$$

and for any constant function $c: \mathbb{R}^n \to \mathbb{R}$, $d_p c = 0$.
(iii) In 35.3.2, prove that f is continuous at p.

35.3.4 *Partial Differential Coefficients.* When $k = 1$ in 35.3.2, there is only one projection function, namely π^1 (which is the identity $\mathscr{I}$). By 35.3.3, $\pi^1 = d_p \pi^1$, so by 29 Exercise 4(i):

$$d_p f = f'(p) \cdot \mathscr{I} = f'(p) \cdot \pi^1 = f'(p) \cdot d_p \pi^1.$$

In the more general case, we know that in 35.3.2, $T = d_p f: \mathbb{R}^k \to \mathbb{R}$ is a linear function: thus

$$T(x) = \sum_{i=1}^{k} t_i x_i.$$

The t_i's depend only on p and f, and we now compute them. Using the natural basis $e_1, \ldots, e_k$ of $\mathbb{R}^k$, then, in particular, we have

$$f(p + \lambda e_i) - f(p) = t_i\lambda + o(\lambda)$$

for each $\lambda \in \mathbb{R}$. Hence, since $o(\lambda)/\lambda \to 0$ as $\lambda \to 0$,

35.3.5
$$t_i = \lim_{\lambda \to 0} \frac{f(p + \lambda e_i) - f(p)}{\lambda}.$$

We denote the number t_i by $D_i f|_p$; it is called the ith 'partial differential coefficient' of f at p, and is denoted traditionally by

$$\partial f/\partial x_i|_p, \quad f_{x_i}(p) \quad \text{or} \quad f_i(p).$$

[The first two denotations have the defect that essentially they confuse x_i with π^i in 35.1.1; see the remarks following 29 Exercise 4(i), and the paradox 36.4.2 below.] We then write $T(x)$ in the forms

$$T(x) = d_p f(x) = \sum_{i=1}^{k} D_i f|_p \cdot x_i = \sum_{i=1}^{k} f_i(p) x_i.$$

By 35.3.3, $x_i = \pi^i(x) = d_p\pi^i(x)$; so we obtain an equation about functions:

35.3.6
$$d_p f = \sum_{i=1}^{k} D_i f|_p \cdot d_p\pi^i = \sum_{i=1}^{k} f_i(p) d_p\pi^i.$$

[Traditionally, and with the defects noted above, 35.3.6 is written

$$df = \sum_{i=1}^{n} (\partial f/\partial x_i) dx_i.]$$

35.3.7 COMMENTS. Observe that in 35.3.5, p is kept constant during the evaluation of the limit. Thus if we regard all co-ordinates of p as fixed except the ith, then 35.3.5 is the same as the ordinary formula 29.1.1 for the derivative of a function of one variable. For example, if

$$\begin{aligned} f(x, y) &= x^2 \sin xy^3, \quad \text{then} \\ f_1(u, v) &= 2u \sin uv^3 + u^2v^3 \cos uv^3, \\ f_2(u, v) &= 3u^3v^2 \cos uv^3, \end{aligned}$$

where for f_1 we have worked as if y were constant, and for f_2 we have treated x as constant. We have, too, implicitly used the fact that the rules (29.3.2, 29.3.3) for differentiating sums and products apply also when D is replaced by D_i.

35.3.8 A geometrical interpretation of local linearity, at least when $k = 2$, is the following. By 35.3.2 the difference between $f(p + x)$ and z, where

$$z = d_p f(x) + f(p),$$

is $o(x)$. But the last equation is of the form $z = t_1x + t_2y + \text{const.}$, and so represents a plane in $\mathbb{R}^3$; it is called the **tangent plane** at p to the surface $z = f(x, y)$ and is in a sense the 'best' plane approximation to this surface, the error being $o(r)$ where $r^2 = x^2 + y^2$. (Cf. 29.5.3 when $k = 1$.)

Exercise 5

(i) Find the tangent plane at the point $P = (x_0, y_0, z_0)$, $z_0 > 0$, to the surface $x^2/a^2 + y^2/b^2 + z^2/c^2 = 1$ in $\mathbb{R}^3$, by expressing the surface near P in the form $z = f(x, y)$ (on a suitable domain in $\mathbb{R}^2$).
(ii) Given a function $f: \mathbb{R}^2 \to \mathbb{R}$, let $g(x) = f(x, b)$ (regarding b as fixed). If we think of the graph of f as a mountainous region over the plane $z = 0$ in $\mathbb{R}^3$, show that the graph of g is equivalent to the 'profile' of a section through the mountain, taken along the line $y = b$. Show that $f_1(a, b)$ is the gradient along this profile above the point (a, b).
(iii) In 35.3.5, let $v \in \mathbb{R}^k$, and instead of considering $p + \lambda e_i$ there, evaluate instead the limit (denoted by $\nabla_v f|_p$):

$$\lim_{t \to 0} (f(p + tv) - f(p))/t.$$

This limit is called the '**derivative**' **of** f **at** p **in the direction** v. To what does it correspond in the previous problem? Calculate the derivative of the function in 35.3.7, at the point $(1, 0)$ in the direction $(3, -2)$.
(iv) Prove that, for any λ, $\mu \in \mathbb{R}$ and $w \in \mathbb{R}^k$ in (iii),

$$\nabla_v(\lambda f + \mu g) = \lambda \cdot \nabla_v f + \mu \cdot \nabla_v g$$

$$\nabla_{\lambda v + \mu w} f = \lambda \cdot \nabla_v f + \mu \cdot \nabla_w g,$$

so

$$\nabla_v f = \sum_{i=1}^{k} v_i f_i, \qquad (v_i = \pi^i(v)).$$

See also 36 Exercise 1(iii).
(v) Show that the set of all **differentials** at p (i.e., all linear functions of the form d_pf for some f) form a real vector space with basis $d_p\pi^1, \ldots, d_p\pi^k$. {Use 35.3.6, and the fact that $\pi^1, \ldots, \pi^k$ form a basis for $H(\mathbb{R}^k, \mathbb{R})$ (see 19.4.5).} When is $d_pf = d_pg$?
(vi) Prove $D_i\pi^j = \delta_{ij}$ (Kronecker delta, see 19 Exercise 6(iv)).

35.4 The Formula for Small Errors

Given $f: A \to \mathbb{R}$, where A is open in $\mathbb{R}^k$, it is sometimes necessary to ask for the error in computing f at $p \in A$, when known errors are made in computing the co-ordinates of p. For example, in physics, the co-ordinates of p will be 'observables' which can never be measured exactly, but only to within a known percentage error. Thus, if $p = (p_1, \ldots, p_k)$, and $p' = p + x$ is the 'observed' position of p, then we suppose that $f(p')$ has been calculated.

If we assume f to be locally linear at p, the required value, $f(p)$, is then given by 35.3.2 and 35.3.6:

35.4.1 $$f(p) = f(p') - \sum_{i=1}^{k} f_i(p')\cdot x_i + o(x).$$

Of course, the numbers x_i cannot be calculated, if p is not known; but they can often be estimated. Thus, the sum $\sum_{i=1}^{k} f_i(p')x_i$ gives an estimate of the error in $f(p)$ 'to the second order'.

Exercise 6

(i) Calculate the error in computing $u = \sin i/\sin r$ when errors of $h\%$ are made in computing i and r.
(ii) A triangle ABC is right-angled at B; angle BAC is θ, $AB = x$, $AC = l$. Find the change in θ when x and l change by h, k respectively. For what value of h/k is there no change in θ?

35.5 Differentiability and Derivatives

We saw in 35.3.6 that in particular, if $f: A \to \mathbb{R}$ is locally linear at $p \in A$, and if $A \subseteq \mathbb{R}^k$, then f is 'partially differentiable', in the sense that $f_i(p)$ exists, $1 \leqslant i \leqslant k$. *The converse is false*, however, in that the partial derivatives $f_i(p)$ may exist without f being locally linear at p; see Exercise 7(v). This is scarcely surprising, since knowledge of f_i is knowledge only of a 'slice' of f, as in Exercise 5(ii). It is customary therefore, when $A \subseteq \mathbb{R}^n$, to say that f is **differentiable** at $p \in A$ iff it is locally linear at p; and f is **differentiable on** A, iff f is differentiable at each $p \in A$.

If $f: A \to \mathbb{R}$ is differentiable, then for each $i = 1, \ldots, n$ we can form the ith **partial derivative** $D_i f$, or f_i, of f, defined by

35.5.1 $$(D_i f)(p) = f_i(p).$$

Then we have a function $D_i f: A \to \mathbb{R}$ and if, in addition, $D_i f$ is differentiable, then we can form its partial derivatives $D_j(D_i f): A \to \mathbb{R}$. We write $D_i^2 f$ for $D_i(D_i f)$. These derivatives are denoted traditionally by

$$\partial^2 f/\partial x_j\, \partial x_i \qquad (\partial^2 f/\partial x_i^2 \quad \text{when } i = j)$$

or by f_{ji} and f_{ii}.

It can be proved that if $D_i D_j f$ and $D_j D_i f$ exist and are continuous, then

35.5.2 $$D_i D_j f = D_j D_i f.$$

For this reason, we often consider only the set $\mathscr{C}^2(A)$ of all functions $f: A \to \mathbb{R}$ such that $D_i D_j f$ exists and is continuous at each $p \in A$, $1 \leqslant i, j \leqslant n$. Just as with $\mathscr{C}^2(I)$ in 29.6.1, $\mathscr{C}^2(A)$ is an algebra over $\mathbb{R}$.

Thus, 'most' functions satisfy 35.5.2. There is a proof of 35.5.2 in Hilton [61], p. 49 for such functions, followed by an example of a 'queer' function f which does not satisfy 35.5.2; $f\colon \mathbb{R}^2 \to \mathbb{R}$ is defined by

$$\begin{aligned} f(x, y) &= (x^3 + 2y^3)/2x + y, && \text{if } 2x + y \neq 0, \\ &= 0 && \text{if } 2x + y = 0. \end{aligned}$$

Exercise 7

(i) Verify that if $2x + y \neq 0$, then:

$$f_1(x, y) = (4x^3 + 3x^2y - 4y^3)/(2x + y)^2, \qquad f_1(0, y) = -4y \qquad (y \neq 0);$$

$$f_1(0, 0) = \lim_{x \to 0} \frac{f(x, 0) - f(0, 0)}{x} = 0;$$

so $f_1(0, y) = -4y$ for all y, whence $f_{21}(0, 0) = -4$. Similarly, if $2x + y \neq 0$, then

$$f_2(x, y) = (-x^3 + 12xy^2 + 4y^3)/(2x + y)^2, \qquad f_2(x, 0) = -\tfrac{1}{4}x \qquad (x \neq 0);$$

and $f_2(0, 0) = 0$ whence $f_2(x, 0) = -\frac{1}{4}x$ for all x, so $f_{12}(0, 0) = -\frac{1}{4} \neq f_{21}(0, 0)$. There are points of the line $2x + y = 0$ arbitrarily close to $(0, 0)$ at which neither f_{12} nor f_{21} is defined; hence these functions are not continuous at $(0, 0)$.
(ii) Calculate the image of f, and the set $f^{-1}(\alpha)$, $(\alpha \in \mathbb{R})$ in (i).
(iii) Let $f\colon \mathbb{R}^3 - \{0\} \to \mathbb{R}$ be the function given by

$$f(x, y, z) = r^{-1}, \qquad r^2 = x^2 + y^2 + z^2.$$

Prove that f and all its partial derivatives are differentiable on $\mathbb{R}^3 - \{0\}$ and that f satisfies **Laplace's equation**:

$$(D_1^2 + D_2^2 + D_3^2)f = 0.$$

(Any solution of this equation is called 'harmonic' and has particularly good properties; the operator $D_1^2 + D_2^2 + D_3^2$ is often written Δ or ∇^2 and called 'Laplace's operator'. See, for example Kellogg [71].
(iv) Let $f(x, y) = 2xy/(x^2 + y^2)$ if $xy \neq 0$, and zero otherwise. Prove that f_1, f_2 exist for all $(x, y) \in \mathbb{R}^2$, that $f_1(0, 0) = f_2(0, 0) = 0$, but that f is not even continuous at $0 \in \mathbb{R}^2$.
(v) Let $g \in \mathscr{C}^2(A)$ (see 35.5.2), $u, v \in \mathbb{R}^k$. Prove that $\nabla_u(\nabla_v g)|_p = \sum u_i v_j g_{ij}(p)$, where ∇_u was defined in Exercise 5(iii).

Chapter 36

VECTOR-VALUED FUNCTIONS

36.1 Differentiability

In order to deal with transformations of co-ordinates, we consider, more generally, mappings of the form

$$\mu\colon A \to B, \qquad (A \subseteq \mathbb{R}^n, B \subseteq \mathbb{R}^m),$$

where A, B are usually open sets. Such functions are often said to be 'vector-valued'. Frequently, μ is specified by giving the co-ordinates

$$(\pi^j \circ \mu)(a), \qquad 1 \leqslant j \leqslant m,$$

for each $a \in A$, so here μ can be thought of as a collection of m functions $\mu^j = \pi^j \circ \mu\colon A \to \mathbb{R}$. For example, if $n = 1$, $m = 3$, and A is a (time) interval, then the three co-ordinates $x(t)$, $y(t)$, $z(t)$ give us the position of $\mu(t)$ as it describes a path in space $\mathbb{R}^3$. And if A is an area in $\mathbb{R}^2$, then

$$(\mu^1(x, y), \mu^2(x, y), \mu^3(x, y))$$

will be the point $\mu(x, y)$, and will in general sweep out a portion of a surface as (x, y) runs through B.

For a general $\mu\colon A \to B$, we say (by analogy with 35.3.2) that μ is **differentiable** at $p \in A$ iff there exists a linear transformation $T\colon \mathbb{R}^n \to \mathbb{R}^m$ which approximates μ, i.e., such that

36.1.1 $$\mu(p + x) - \mu(p) = T(x) + o(x)$$

for all x in some ball V, centre O, in $\mathbb{R}^n$, such that $p + V \subseteq B$. This includes 35.3.2, if we take $m = 1$. Let us now prove here what was postponed there.

36.1.2 LEMMA. *If T exists in* 36.1.1, *it is unique.*

Proof. Suppose $S\colon \mathbb{R}^n \to \mathbb{R}^m$ satisfies 36.6.1 also, perhaps with a ball W instead of V. Then there exists a ball U such that $0 \in U \subseteq V \cap W$, and on which $(S - T)(x) = o(x)$ for every $x \in U$. Let $Q = S - T$. Now each point $y \in \mathbb{R}^n$ is expressible as λx with $\|x\|$ arbitrarily small. Since Q is linear,

$$\frac{\|Q(y)\|}{\|y\|} = \frac{\|Q(x)\|}{\|x\|}.$$

Thus letting $x \to 0$, we infer $Q(y) = 0$, so Q is the zero function, whence $S = T$. ■

We denote the unique T satisfying 36.1.1, by

36.1.3 $$T = d_p\mu,$$

calling it the **differential of** μ **at** p. Thus $d_p\mu: \mathbb{R}^n \to \mathbb{R}^m$ is a linear transformation. As before, if μ is already a linear transformation, then $d_p\mu = \mu$.

We now suppose that $B = \mathbb{R}^m$. Then if $\lambda, \mu: A \to \mathbb{R}^m$ are both differentiable at p, we leave the reader to prove that $\lambda + \mu$ and $\alpha\cdot\lambda$ ($\alpha \in \mathbb{R}$), are differentiable at p (they map into $\mathbb{R}^m$, since $\mathbb{R}^m$ is a vector space) and

36.1.4 $$d_p(\lambda + \mu) = d_p\lambda + d_p\mu, \qquad d_p(\alpha\cdot\lambda) = \alpha\cdot d_p\lambda.$$

Also, if $m = 1$,

36.1.5 $$d_p(\lambda\cdot\mu) = \lambda(p)\cdot d_p\mu + \mu(p)\cdot d_p\lambda: \mathbb{R}^n \to \mathbb{R};$$

the last equation follows from the equations (using 36.1.1):

$$\begin{aligned}\lambda\cdot\mu(p + x) - \lambda\cdot\mu(p) &= (\lambda(p) + d_p\lambda(x) + o(x))(\mu(p) + d_p\mu(x) + o(x)) \\ &\qquad - \lambda(p)\cdot\mu(p) \\ &= (\lambda(p)\cdot d_p\mu + \mu(p)\cdot d_p\lambda)(x) + o(x).\end{aligned}$$

Let us denote by $\mathscr{D}_p(A, \mathbb{R}^m)$ the set of all functions $\mu: A \to \mathbb{R}^m$ which are differentiable at p; then we can summarize the last remarks by saying that $\mathscr{D}_p(A, \mathbb{R}^m)$ is a real vector space. Further, using the space $H(U, V)$ of all linear transformations $U \to V$ (see 19.4.5), we see that

36.1.6 $$d_p: \mathscr{D}_p(A, \mathbb{R}^m) \to H(\mathbb{R}^n, \mathbb{R}^m)$$

is a linear transformation (by 36.1.4); if also $m = 1$, then $\mathscr{D}_p(B, \mathbb{R})$ is a real algebra and d_p is a derivation (i.e., its effect on products is given by 36.1.5, the analogue of 22.6.3(i),(ii)).

Exercise 1

(i) Let $\mu: \mathbb{R}^1 \to \mathbb{R}^2$ be the mapping given by $\mu(t) = (\cos t, \sin t)$. Prove that $d_p\mu = j^1: \mathbb{R} \to \mathbb{R}^2$ at $p = 0$, and $j^2: \mathbb{R} \to \mathbb{R}^2$ at $p = \pi/2$, where $j^1(t) = (0, t)$, $j^2(t) = (-t, 0)$.
(ii) Show that the linear transformation $d_p\mu$ in 36.1.3 has matrix

$$J_\mu(p) = (D_i(\pi^j \circ \mu))|_p$$

relative to the natural bases in $\mathbb{R}^m$, $\mathbb{R}^n$. Thus $d_p\mu(e_i) = \sum_{j=1}^{m} D_i\mu^j|_p e_j$, $(i = 1, \ldots, n)$. {Put $x = \lambda e_i$ in 36.1.1, and apply π^i; let $\lambda \to 0$.} [We call $J_\mu(p)$ the **Jacobian matrix** of μ at p, after C. G. J. Jacobi (1804–51), Bell [10], Ch. 18.]
(iii) Let $f \in \mathscr{D}_p(A, \mathbb{R})$ (see 36.1.6). Define the **gradient** of f at p to be the point denoted by $\nabla f|_p$ (*read:* 'grad f at p'):

$$\nabla f|_p = (f_1(p), \ldots, f_n(p)) \in \mathbb{R}^n.$$

Show that the correspondence $f \to \nabla f|_p$ is a linear transformation, and also that if $x \in \mathbb{R}^n$, then (cf. 29.5.5):

$$d_p f(x) = \langle \nabla f|_p, x \rangle \qquad \text{(inner product).}$$

For any $w \in \mathbb{R}^n$, prove that the directional derivative, $\nabla_w f|_p$, is $\langle \nabla f|_p, w \rangle$ (see 35 Exercise 5(iii)).

36.1.7 The statement 36.1.1 and the proof of 36.1.2 would make sense if $\mathbb{R}^m$ were replaced by any real vector space V with a distance (to enable us to define $o(x)$ in V). Thus, given $p \in A \subseteq \mathbb{R}^n$ and, for example, an inner-product space (Chapter 20) V, we may define the **differential** $d_p\mu: \mathbb{R}^n \to V$ by just the same procedure as when V is $\mathbb{R}^m$. If $d_p\mu$ exists for every $p \in A$, we call μ **differentiable on** A; and then we have a new function

$$d\mu: A \to H(\mathbb{R}^n, V), \qquad (d\mu)(p) = d_p\mu: \mathbb{R}^n \to V.$$

If now $H(\mathbb{R}^n, V)$ has also an inner product, then it makes sense to ask whether $d\mu$ is itself differentiable on A, and then to construct

$$d^2\mu = d(d\mu): A \to H(\mathbb{R}^n, H(\mathbb{R}^n, V)),$$

followed by $d^3\mu, \ldots, d^n\mu, \ldots$. We shall here consider only the second differential $d^2 f$ of a real-valued function on A, since it is a sufficiently instructive exercise. To simplify notation, we here write $g \cdot t$ for the value of the function g at t, and omit dots in products; for example $d\mu \cdot p = d_p\mu$.

We look first at $\pi^j: A \to \mathbb{R}$, the projection. Then

$$d\pi^j: A \to H(\mathbb{R}^n, \mathbb{R}^1) = \mathbb{R}^{n*},$$

and the latter space is the *dual* of $\mathbb{R}^n$ with the induced inner product (see 20.4.6). Since $d\pi^j \cdot p = d_p\pi^j = \pi^j$ (by 35.3.3), then $d\pi^j$ is *constant*, its value at each $p \in A$ being the element π^j of $\mathbb{R}^{n*}$; therefore $d^2\pi^j = 0$.

Consider next a product of the form $g d\pi^j: A \to \mathbb{R}^{n*}$ where $g: A \to \mathbb{R}$ is differentiable. Now, the reasoning for 36.1.5 allows us to write (for $\mu: A \to \mathbb{R}^m$, m not necessarily 1)

36.1.8 $$d_p(g\mu) \cdot x = (d_p\mu \cdot x)g(p) + (d_p u \cdot x)g(p) \qquad (x \in \mathbb{R}^m)$$

and the 'x' cannot be omitted to get an equation about functions, since the product $\mu(p)d_p g$ of functions is not defined unless $m = 1$. Hence to calculate $d(g d\pi^j): A \to \mathbb{R}^{n*}$ we have

$$\begin{aligned} d_p(g d\pi^j) \cdot x &= (d_p g \cdot x) d_p \pi^j \\ &= \left(\sum_i g_i(p) x_i\right)\pi^j \qquad \text{by 35.3.6,} \end{aligned}$$

so $d_p(g d\pi^j): \mathbb{R}^n \to \mathbb{R}^{n*}$ is that transformation which assigns to $x \in \mathbb{R}^n$ the function $\left(\sum_i g_i(p)x_i\right)\pi^j: \mathbb{R}^n \to \mathbb{R}$; and the value of the latter at $y \in \mathbb{R}^n$ is

$\sum_{ij} g_i(p)x_iy_j$. We may therefore also regard $d_p(gd\pi^j)$ as a bilinear form: $\beta\colon \mathbb{R}^n \times \mathbb{R}^n \to \mathbb{R}$ (i.e., such that

$$\beta(x, u + v) = \beta(x, u) + \beta(x, v)$$
$$\beta(x + y, u) = \beta(x, u) + \beta(y, u)$$

for all $x, y, u, v \in \mathbb{R}^n$). We can now calculate d^2f as a bilinear form for a differentiable $f\colon A \to \mathbb{R}$, since $d_pf = \sum_i f_i(p)d_p\pi^i$, provided each partial derivative f_i is differentiable on A.

Exercise 2

(i) Complete the last calculation, to show that $d^2f\colon A \to \mathbb{R}^{n*}$ has the property that for each $p \in A$, $(d^2f)\cdot p$ is that bilinear form $\mathbb{R}^n \times \mathbb{R}^n \to \mathbb{R}$ whose value at (x, y) is $\sum_{ij} f_{ij}(p)x_iy_j$. If $\Delta\colon \mathbb{R}^n \to \mathbb{R}^n \times \mathbb{R}^n$ denotes the mapping $x \to (x, x)$, then $((d^2f)\cdot p) \circ \Delta\colon \mathbb{R}^n \to \mathbb{R}$ is a *quadratic form* which we may denote by d_p^2f; its value at $x \in \mathbb{R}^n$ is $(d_p^2f)\cdot x = \sum f_{ij}(p)x_ix_j$, and hence we have, analogously with 35.3.6,

$$d_p^2f = \sum f_{ij}(p)d_p\pi^i d_p\pi^j.$$

[Thus, when $n = 1$, we have $d_p^2f = f''(p)(d_p\pi)^2$, accounting for the classical notation (recall $d_p\pi$ is written dx)

$$f''(p) = d^2f/(dx)^2.]$$

(ii) Calculate d^2f for the function f considered in 35.3.7.
(iii) Discuss d^3f for a real-valued function of n variables.
(iv) For a differentiable function q, let $E(q)$ denote the operation of evaluating $d_p(q)$ at $x \in \mathbb{R}^n$. Does 36.1.8 state that E is a derivation?

36.2 Composition

If $m > 1$, there is no natural multiplication in $\mathscr{D}_p(A, B)$. However, given an open $C \subseteq \mathbb{R}^k$ and points $q \in C$, $p = \sigma(q) \in B$, with functions

$$C \xrightarrow{\sigma} B \xrightarrow{\mu} A,$$

where $\sigma \in \mathscr{D}_q(C, B)$, and $\mu \in \mathscr{D}_p(B, A)$, we can form the composite

$$\mu \circ \sigma\colon C \to A \; (A \subseteq \mathbb{R}^n).$$

Then, generalizing 29 Exercise 4(iv), we have the following rule, which is intuitively obvious inasmuch as it states that the linear approximation to a composite function is the composite of the linear approximations to the components.

36.2.1 LEMMA. $\mu \circ \sigma \in \mathscr{D}_q(C, A)$ *and* $d_q(\mu \circ \sigma) = d_p\mu \circ d_q\sigma\colon \mathbb{R}^k \to \mathbb{R}^n$.

Proof. For all sufficiently small $x \in \mathbb{R}^k$,

$$\begin{aligned}\mu \circ \sigma(q + x) &= \mu(\sigma(q + x)) \\ &= \mu(\sigma(q) + S(x) + o(x))\end{aligned}$$

since σ is differentiable at q, where $S = d_q\sigma$. [It is left as a harder exercise for the reader to work out the size of the ball in $\mathbb{R}^k$, in which the 'sufficiently small' x must lie.] The right-hand side is

$$\mu(\sigma(q)) + M(S(x) + o(x)) + o(y)$$

where $M = d_p\mu$ (since $p = \sigma(q)$) and $y = S(x) + o(x)$; and this, since M is linear, is

$$\mu(\sigma(q)) + (M \circ S)(x) + o(x)$$

since $M(o(x)) + o(y) = o(x)$. Thus $\mu \circ \sigma$ satisfies the requirements for differentiability at $q \in C$, with $d_q(\mu \circ \sigma) = M \circ S$. ■

36.2.2 COROLLARY. *For each $i = 1, \ldots, k$, and $j = 1, \ldots, n$*

$$D_i(\pi^j(\mu \circ \sigma)) = \sum_{r=1}^{n} D_i\mu^r \cdot D_r\sigma^j,$$

where $\mu^r = \pi^r \circ \mu$, and similarly for σ^j.

Proof. By Exercise 1(ii) and the rule (19.4.8) that the matrix of a composite is the product of the matrices of the transformations being composed (see also 36.4.1 below). ■

Exercise 3

(i) Prove that, if $\mu: A \to B$ is differentiable at $p \in A$, then so is each μ^j, and conversely. Say that μ is **continuous** at $p \in A$ provided that, for each sequence $\{p_n\}$ in A convergent to p, $\mu(p_n)$ converges to $\mu(p)$. Prove that μ is continuous at p iff each μ^j is continuous at p.
(ii) Prove that if μ is differentiable at p, then μ is continuous at p.
(iii) Let $A = \langle 0, 2\pi \rangle \subseteq \mathbb{R}$ and let B denote the unit circle $|z| = 1$ in $\mathbb{R}^2$. Let $\mu: A \to B$ be given by $\mu(\theta) = e^{i\theta}$. Prove that μ is one-one, differentiable, and onto, but that $\mu^{-1}: B \to A$ is not continuous at $z = 1$.
(iv) Suppose $I = \langle 0, 1 \rangle$ in $\mathbb{R}$, and let $\mu: I \to \mathbb{R}^k$ be continuous. It is customary to call μ (and not $\mu(I)$) a path in $\mathbb{R}^k$ from $\mu(0)$ to $\mu(1)$. Now let $A \subseteq \mathbb{R}^n$ be such that, $\forall p, q \in A, \exists$ a path in A from p to q. Then A is called **pathwise connected**. Prove that an interval and a plane rectangular region are pathwise connected, while $\mathbb{R} - \{0\}$ is not. Prove also that if A is pathwise connected, then so is $\lambda(A)$, for any continuous $\lambda: A \to \mathbb{R}^m$. [See also 25 Exercise 1(iii).]
(v) Prove the following form of the intermediate value theorem: let $f: A \to \mathbb{R}$ be continuous, where $A \subseteq \mathbb{R}^k$ is pathwise connected. If $p, q \in A$ and α is a number between $f(p)$ and $f(q)$, show that $\exists r \in A$ such that $f(r) = \alpha$. Hence prove that $f(A)$ is an interval or a point, if $A \neq \varnothing$.

36.3 Co-ordinate Systems

Compositions of the form $\mu \circ \sigma$ in 36.2.1 occur particularly when the natural co-ordinate system in $\mathbb{R}^n$ has been changed to some other. For examples of the usefulness of such changes, see, e.g., D'Arcy Thompson [124], Ch. IX (in biology), Milne-Thomson [87] in hydrodynamics. Let us consider first what we mean by 'a co-ordinate system'. We stress that we are not restricting ourselves to Cartesian or even orthogonal systems.

Given $\mu: A \to \mathbb{R}^m$, suppose that μ is one-one and that also $A \subseteq \mathbb{R}^m$. Let B denote the image of μ, so $\mu: A \approx B$ is a set-equivalence; hence $\mu^{-1}: B \to A$ exists. It may occur that μ^{-1} is not continuous (see Exercise 3(iii), above). If, however, μ and μ^{-1} are *both* differentiable at each point of A, B, respectively, then μ is called a 'co-ordinate system' on A.

36.3.1 EXAMPLE. The identity $\mathscr{I}: \mathbb{R}^n \to \mathbb{R}^n$ is a co-ordinate system on $\mathbb{R}^n$; so is any non-singular linear transformation $T: \mathbb{R}^n \to \mathbb{R}^n$.

36.3.2 EXAMPLE. Let $n = 2$, and let σ be the function which assigns to each $P = (x, y)$ the polar co-ordinates (r, θ) of P. Here r is the length OP, which is a real number > 0, and θ is the angle IOP shown in Fig. 36.1. Thus θ is not defined if $P = 0$, and to make θ single-valued, we take $0 \leqslant \theta < 2\pi$. Hence σ has domain $\mathbb{R}^2 - \{0\}$ (it is obviously differentiable and one-one), while its image is $\mathbb{R}_+ \times \langle\!\langle 0, 2\pi \rangle$. Its inverse fails to be continuous on the positive x-axis, for the same reason as μ in Exercise 3(iii) above. However, if A is any subset of $\mathbb{R}^2 - \{x > 0\}$, then σ, restricted to A, is a co-ordinate system on A. In applications, where A is given in advance, we can usually choose the positive x-axis so as to miss the 'interesting' parts of A.

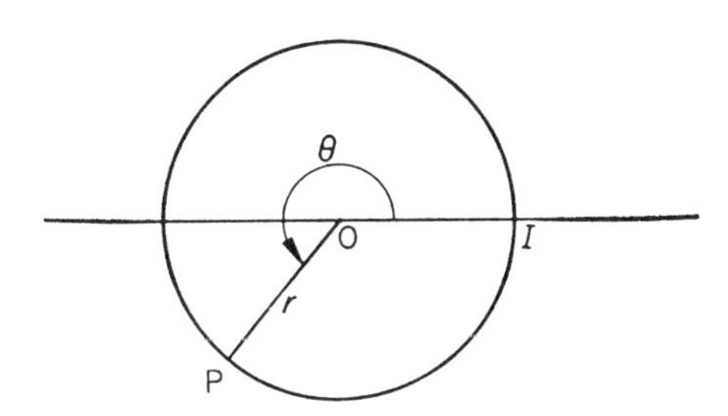

Fig. 36.1

36.3.3 EXAMPLE. (*Graphs in polar co-ordinates.*) A reader who has 'plotted a graph of the form $r = f(\theta)$ in polar co-ordinates' may wonder at the requirement made in the last Example, that $r > 0$; this would cut out certain branches of curves like $r = \cos 3\theta$. However, the standard convention requires that the **graph of** $r = f(\theta)$ is the set S of all points $(x, y) \in \mathbb{R}^2$, such that for some $\theta \in \mathbb{R}$, $x = f(\theta)\cdot\cos\theta$, $y = f(\theta)\cdot\sin\theta$. The set of all points with polar co-ordinates of the form $(f(\theta), \theta)$ is therefore in general only a proper subset of S.

Exercise 4

(i) Find all points on the graph $r = \cos 3\theta$ which do not possess polar co-ordinates of the form $(\cos 3\theta, \theta)$.
(ii) Let $\mathbb{S}^2$ denote the 'unit sphere' $x^2 + y^2 + z^2 = 1$ in $\mathbb{R}^3$, and let N, S denote its North and South poles $(0, 0, 1)$, $(0, 0, -1)$ respectively. Let Π_N, Π_S denote the tangent planes at N, S, respectively. Define a map $\theta: \mathbb{S}^2 - \{S\} \to \Pi_N$ by declaring $\theta(P)$ to be the point where the line SP cuts Π_N. Using analytical geometry show that $\theta(P)$ always exists, and prove that θ is an equivalence. We call θ **stereographic projection onto** Π_N. Similarly, let $\phi: \mathbb{S}^2 - \{N\} \to \Pi_S$ denote stereographic projection onto Π_S; and let $v: \mathbb{R}^3 \to \mathbb{R}^2$ be the projection given by $(x, y, z) \to (x, y, 0)$. Prove that $v \circ \theta = \theta_v: \mathbb{S}^2 - \{N\} \to \mathbb{R}^2$ is a bijection, and similarly for $v \circ \phi = \phi_v$. Prove that $\theta_v \circ \phi_v^{-1}: \mathbb{R}^2 \to \mathbb{R}^2$ is differentiable, with a differentiable inverse.
(iii) Do the same problem for the 'unit n-sphere $\mathbb{S}^n$ in $\mathbb{R}^{n+1}$', where

$$\mathbb{S}^n = \{(x_1, \ldots, x_{n+1}) | x_1^2 + \cdots + x_{n+1}^2 = 1\}.$$

(iv) Do a problem analogous to (ii) when $\mathbb{S}^2$ is replaced by an ellipse in $\mathbb{R}^2$ and by an ellipsoid $x^2/a^2 + y^2/b^2 + z^2/c^2 = 1$ in $\mathbb{R}^3$. §The functions θ, ϕ in (ii), with their analogues in (iii) and (iv), are examples of **a system of co-ordinate charts** on $\mathbb{S}^2$ (or $\mathbb{S}^n$, or an ellipse), since, for example, each point P in $\mathbb{S}^2 - \{S\}$ can be assigned the co-ordinates of $\theta(P)$. Sets which can be covered by such charts are called *differentiable manifolds*, and include the projective plane, projective spaces and (for example) the set of $n \times n$ matrices of determinant 1. See Willmore [137].
(v) Two very useful co-ordinate systems are those in $\mathbb{R}^3$ called 'cylindrical' and 'spherical', and described, for example, in Courant [24]. Decide their domains and images.

36.4 The Chain Rule of Differentiation

Suppose we are given a function $f: A \to \mathbb{R}$ on some domain $A \subseteq \mathbb{R}^m$, and suppose $\mu: A \to \mathbb{R}^m$ is a (differentiable) co-ordinate system on A. It becomes necessary to know the derivatives of $f \circ \mu$ ('the function f considered in the co-ordinates μ') in terms of the derivatives $D_i f$. We therefore use 36.2.2 with $n = 1$ to deduce immediately the so-called **chain rule of differentiation** ($\mathscr{D}_p$ was defined in 36.1.6). In 36.4.1 below, μ need not be one-one.

36.4.1 THEOREM. *Given functions* $A \xrightarrow{\mu} B \xrightarrow{f} \mathbb{R}$ *and points* $p \in A$, $r = \mu(p) \in B$, *where* $B \subseteq \mathbb{R}^m$, $A \subseteq \mathbb{R}^n$, $\mu \in \mathscr{D}_p(A, B)$, $f \in \mathscr{D}_r(B, \mathbb{R})$, *then*

$$\text{(i)}\ D_i(f \circ \mu)|_p = \sum_{j=1}^{m} f_j(r) \cdot D_i(\pi^j \circ \mu)|_p, \qquad (1 \leqslant i \leqslant n). \quad \blacksquare$$

Traditionally, (i) is written

(ii) $\partial f/\partial p_i = \sum (\partial f/\partial r_j)\cdot(\partial r_j/\partial p_i)$,

and the user of the chain-rule in this form must carry in his head, instead of on the page, that

(a) the 'f' on the left-hand side of (ii) is 'f considered as a function of the co-ordinates of p';
(b) the 'f' on the right-hand side of (ii) is 'f considered as a function of $r = \mu(p)$'; and
(c) the various partial derivatives must be evaluated at p, r, and p, respectively.

It is scarcely surprising that students find the subject difficult! The formula (i), on the other hand, contains all the instructions necessary for the computations. Physicists sometimes object to the notation in (i) since they want to think always of f as (say) energy; but f and $f \circ \mu$ are two different descriptions of the same thing, and the difference should be kept explicit. Consider, too the following 'paradox'.

36.4.2 PARADOX. Let $f: \mathbb{R}^3 \to \mathbb{R}$ be given by $f(x, y, z) = x + y + z$, so $\partial f/\partial x = 1$. If, also, $y = x^k$, then $f(x, y, z) = x + x^k + z$, so $\partial f/\partial x = 1 + kx^{k-1}$, a clear contradiction. To resolve it, define $\mu: \mathbb{R}^2 \to \mathbb{R}^3$ by $(x, z) \to (x, x^k, z)$. Then $D_1 f = 1$ (as before), while $1 + kx^{k-1}$ is now $D_1(f \circ \mu)|_{(x,z)}$ and no contradiction arises, since $f \neq f \circ \mu$.

Exercise 5

(i) Recall from Exercise 1(ii), (iii) the Jacobian matrix $J_\mu(p)$ and the gradient operator $\nabla f|_p$. Show that 36.4.1(i) can be written (using matrix multiplication) as

$$\nabla(f \circ \mu)|_p = J_\mu(p)\cdot \nabla f|_{\mu(p)}.$$

(Compare the form of this equation with that of 29.4.1.)
(ii) Choose a new co-ordinate system T of $\mathbb{R}^m$, by taking a new basis $b_1, \ldots, b_m$ where $Te_i = b_i = \sum \beta_{ij} e_j$ and T is linear. Show that the Jacobian of μ at p, relative to this new system, is $J_{\mu \circ T^{-1}} = B^{-1}\cdot J_\mu$, $(B = (\beta_{ij}))$.
(iii) Show that, by removing the variables $p, r = \mu(p)$ from 36.4.1(i), it can be written as an equation about functions:

$$D_i(f \circ \mu) = \sum_j (f_j \circ \mu)\cdot D_i(\pi^j \circ \mu),$$

or, more briefly, as

$$D_i(f \circ \mu) = \sum_j (f_j \circ \mu)\cdot \mu_i^j, \qquad (\mu^j = \pi^j \circ \mu).$$

(iv) Hence, assuming that $f \in \mathscr{C}^2(B)$, $\mu^j \in \mathscr{C}^2(A)$, prove that

$$D_i^2(f \circ \mu) = \sum_j \sum_t (f_{jt} \circ \mu)\cdot \mu_i^t \mu_i^j + \sum_j (f_j \circ \mu)\cdot \mu_{ii}^j.$$

Consider now some applications of the chain rule, which we record here in the brief 'functional form' of Exercise 5(iii):

36.4.3 $$D_i(f \circ \mu) = \sum_{j=1}^{m} (f_j \circ \mu)\cdot \mu_i^j, \qquad (1 \leqslant i \leqslant n).$$

36.4.4 EXAMPLE. Let us derive the product rule 29.3.3. Let $f\colon \mathbb{R}^2 \to \mathbb{R}$ be given by $f(x, y) = xy$, and let $\mu\colon A \to \mathbb{R}^2$ be given by $\mu(s, t) = (u(s, t), v(s, t))$ where $u\colon A \to \mathbb{R}$, $v\colon A \to \mathbb{R}$ are differentiable functions on the open $A \subseteq \mathbb{R}^2$. Then, for each $p = (s, t) \in A$, $f \circ \mu(p) = u(p)\cdot v(p)$, so $f \circ \mu = u \cdot v$. By 36.4.3 then,

$$D_i(u\cdot v) = (f_1 \circ \mu)\cdot D_i(\pi^1 \circ \mu) + (f_2 \circ \mu)\cdot D_i(\pi^2 \circ \mu), \qquad (i = 1, 2);$$

now $f_1(x, y) = y = \pi^2(x, y)$, so $f_1 = \pi^2$, and similarly $f_2 = \pi^1$, so

$$f_1 \circ \mu = \pi^2 \circ \mu = v, \qquad f_2 \circ \mu = u,$$

whence

$$D_i(u\cdot v) = v\cdot D_i u + u \cdot D_i v \qquad (i = 1, 2).$$

This may strike the reader as a round-about proof (indeed, it is!), but we have given it to demonstrate its mechanical, algebraic nature.

36.4.5 EXAMPLE. If $m = 1$, then in 36.4.3, $D_j = D$ and $\pi^j = \mathscr{I}\colon \mathbb{R} \to \mathbb{R}$, so here $\pi^j \circ \mu = \mu$ and

(i) $D_i(f \circ \mu) = (f' \circ \mu)\cdot D_i\mu, \qquad i = 1, \ldots, n.$

Consider, for example, the function g given by $g(x, y) = \cos x$, with $A = \mathbb{R}^2$. Then since $x = \pi^1(x, y)$, g is the function $\cos \circ \pi^1$. Hence, applying (i) with $f = \cos$, $\mu = \pi^1$, we get

(ii) $D_1 g = (-\sin \circ \pi^1)\cdot 1,\ D_2 g = (-\sin \circ \pi^1)\cdot 0,$

since $D_i\pi^j = \delta_{ij}$ by 35 Exercise 5(vi). In traditional notation, (ii) becomes

$$(\partial/\partial x)(\cos x) = -\sin x, \qquad (\partial/\partial y)(\cos x) = 0.$$

36.4.6 EXAMPLE. Suppose A is an open interval in $\mathbb{R}$. Then, changing r to t (for time), we have from 36.4.3 (with $n = 1$)

$$D(f \circ \mu) = \sum_{j=1}^{m} (f_j \circ \mu)\cdot \dot{\mu}^j, \qquad (\dot{\mu}^j = D\mu^j)$$

since there is only one D_i to apply to $f \circ \mu$ and μ, namely $D_1 = D$. In traditional notation, the last equation becomes

$$\frac{df}{dt} = \sum_j \frac{\partial f}{\partial p_j}\cdot\frac{dp_j}{dt}.$$

Exercise 6

(i) In Example 36.4.4, suppose instead that $f(x, y) = x/y$ $(y \neq 0)$. Formulate and prove a 'quotient rule' for evaluating $D_i(u/v)$.
(ii) Extend Example 36.4.4 and your quotient rule to functions of m variables (i.e., when A is open in $\mathbb{R}^m$).
(iii) With the hypotheses of Exercise 5(iv) prove that

$$D^2(f \circ \mu) = \sum_{j,k} (f_j \circ \mu) \cdot \dot{\mu}^j \dot{\mu}^k + \sum_j (f_j \circ \mu) \cdot \ddot{\mu}^j.$$

(iv) In 36.4.1, suppose that $n = 2 = m$, and $f: B \to \mathbb{R}$ is harmonic (i.e., f satisfies **Laplace's equation** $f_{11} + f_{22} = 0: B \to \mathbb{R}$). Suppose that $\mu: A \to \mathbb{R}^2$ is **conformal** (i.e., μ^1, μ^2 satisfy the **Cauchy-Riemann equations** $\mu_1^1 = \mu_2^2, \mu_2^1 = -\mu_1^2$). If also $\mu^1, \mu^2 \in \mathscr{C}^2(A)$, prove that $f \circ \mu$ is harmonic (use Exercise 5(iv)). [This result is of great importance in applied mathematics, for example in hydrodynamics or electricity. It is often necessary to find a harmonic $g: A \to \mathbb{R}$ to fit the physical conditions, and it turns out to be easier to find a conformal μ as above, such that the physical problem can be easily solved for the region $B = \mu(A)$. Then the required g is $f \circ \mu$. For example, B might be the complement in $\mathbb{R}^2$ of a cross-section of an aerofoil, g the 'velocity potential' and A the complement of a circular disc, a much simpler shape.]
(v) For certain applications, it is better to express f in (iv) as a function of the polar co-ordinates (r, θ), so that in traditional notation,

$$\frac{\partial^2 f}{\partial x^2} + \frac{\partial^2 f}{\partial y^2} = \frac{\partial^2 f}{\partial r^2} + \frac{1}{r}\frac{\partial f}{\partial r} + \frac{1}{r^2}\frac{\partial^2 f}{\partial \theta^2}$$

where f is regarded as a function of (x, y), and of (r, θ), on the left-hand and right-hand side, respectively. Use Exercise 5(iv) to prove the last equality, when B is a domain on which the function μ is a co-ordinate system σ (see Example 36.3.2), and $\sigma^1(x, y) = (x^2 + y^2)^{1/2}$, $\sigma^2(x, y) = \tan^{-1}(y/x)$. {On the right-hand side, f must be replaced by $g = f \circ \sigma^{-1}$, so $f = g \circ \sigma$, and Exercise 5(iv) can be applied.}

36.5 Summary of Principal Formulae

For the convenience of the reader, we now summarize the principal formulae and notations. Let $\mu: A \to \mathbb{R}^m$, $f: A \to \mathbb{R}$ be differentiable functions on $A \subseteq \mathbb{R}^n$; let $p \in A$, and let $(e_1, \ldots, e_n)$ denote a basis of $\mathbb{R}^n$, $n = 1, 2, \ldots$. Thus $x \in \mathbb{R}^n$ is of the form $x = \sum_{i=1}^{n} x_i e_i$, abbreviated to $(x_1, \ldots, x_n)$, so $\mu(x) = \mu(x_1, \ldots, x_n) = (\mu^1(x), \ldots, \mu^m(x))$. Then

(1) $d_p\mu: \mathbb{R}^n \to \mathbb{R}^m$ is the unique function such that

$$\mu(p + x) - \mu(p) = d_p\mu(x) + o(x), \; (x \in V)$$

where V is a ball in $\mathbb{R}^n$ such that $p + V \subseteq A$.

(2) The ith partial derivative of $f: A \to \mathbb{R}$ at p is denoted by $D_i f|_p$ or $f_i(p)$ or $\partial f/\partial x_i|_p$ and

$$f_i(p) = d_p f(e_i).$$

Also, the gradient of f at p is the n-tuple

$$\nabla f|_p = (f_1(p), \ldots, f_n(p)) \in \mathbb{R}^n.$$

(3) $d_p\mu(e_i) = \sum_{j=1}^{m} M_{ij}e_j$, where (M_{ij}) is the

Jacobian matrix

$$\begin{pmatrix} D_1\mu^1(p), \ldots, D_1\mu^m(p) \\ \vdots \qquad\qquad \vdots \\ D_n\mu^1(p), \ldots, D_n\mu^m(p) \end{pmatrix}$$

of μ; for,

$$\begin{aligned} M_{ij} = (\pi^j \circ d_p\mu)(e_i) &= d_p\mu^j(e_i) && \text{by (1)} \\ &= D_i\mu^j(p) && \text{by (2).} \end{aligned}$$

Thus (taking $m = 1$), $\nabla f|_p$ is the transpose of the Jacobian matrix of f.

(4) (*Chain rule.*) Given $C \xrightarrow{\sigma} A$, and $q \in C$, then

$$d_q(\mu \circ \sigma) = d_{\sigma(q)}\mu \circ d_q\sigma: C \to \mathbb{R}^m.$$

In particular, writing $p = \sigma(q)$, we have

$$D_i(f \circ \sigma)|_q = \sum_{j=1}^{n} f_j(p) D_i\sigma^j(q),$$

since the left-hand side is

$$\begin{aligned} d_q(f \circ \sigma)(e_i) &= d_p f(d_q\sigma(e_i)) \\ &= d_p f\left(\left(\sum_{j=1}^{n} D_i\sigma^j(q)\right)e_j\right) && \text{(by (3))} \\ &= \sum_j D_i\sigma^j(q) \cdot d_p f(e_j) \end{aligned}$$

because $d_p f$ is linear; by (2) the last sum is the required right-hand side.

Chapter 37

C^r-FUNCTIONS

37.1 The Problem

In this chapter we extend the Taylor expansion (see 34.1) and the theory of maxima and minima ('stationary points') to functions of several variables. We should recall first the setting for this theory, suitably generalized to the case of several variables.

Let $f: A \to \mathbb{R}$ where $A \subseteq \mathbb{R}^n$ is an open set. If $p \in A$, then we say that f is **a C^r-function at p** if all partial derivatives

$$D^{i_1}D^{i_2}\cdots D^{i_m}f\big|_p, \qquad 0 \leqslant i_1 + i_2 + \cdots + i_m \leqslant r,$$

exist and are continuous at p. Thus, for $n = 1$, this asserts that all (ordinary) derivatives of f, up to and including that of order r, exist and are continuous at p. Of course, if they all exist, it follows immediately that all are continuous, with the possible exception of the last. Similarly if $n \geqslant 1$ we may say that f is a C^r-function at p if it is *continuously differentiable* up to order r. Then a C^r-function **on** A is just a function which is a C^r-function at every point p of A. It is easily seen that *the set $\mathscr{C}^r(A)$ of C^r-functions on A is an algebra over* $\mathbb{R}$.

37.2 Taylor's Expansion

We can extend the Taylor expansion 34.1.1 to functions of several variables, as follows. Let $f: A \to \mathbb{R}$ be differentiable on the open $A \subseteq \mathbb{R}^n$, let $p \in A$, and let $v \in \mathbb{R}^n$ be such that $p + tv \in A$ for each $t \in \langle\!\langle 0, 1\rangle\!\rangle = \mathbb{I}$ (such a v exists in each direction since A is open); and as t varies from 0 to 1, then $p + tv$ traces out a segment in A. Let $\mu: \mathbb{I} \to A$ be the path such that $\mu(t) = p + tv$. Then $f(p + tv) = f \circ \mu(t)$, and is a differentiable function of t. Therefore, by Taylor's expansion for $g(t) = f \circ \mu(t)$, we have, supposing $g \in \mathscr{C}^2$—thus if $f \subset \mathscr{C}^2(A)$ and $\mu \in \mathscr{C}^2(\mathbb{I})$:

$$g(t) = g(0) + tg'(0) + \tfrac{1}{2}t^2g''(0) + o(t^2),$$

so using the result of 36 Exercise 6(iii), and putting $t = 1$, we get (since $\dot{\mu}^j = v_j$, $\ddot{\mu}^j = 0$):

37.2.1 $$f(p + v) = f(p) + \sum f_j(p)v_j + \tfrac{1}{2}\sum f_{jk}v_jv_k + o(v^2).$$

When $A \subseteq \mathbb{R}$, this is the ordinary Taylor expansion.

Exercise 1

(i) Use induction on q to compute $f(p + v)$ as far as the term corresponding to t^q in the Taylor expansion of $g(t)$, above.
(ii) Show that f in 37.2.1 is locally linear at p. {See 35.3.2 and use 28.5.2.}

37.3 Critical Points

For several variables, 37.2.1 is particularly useful for examining *stationary* points of f, especially in the dynamics of oscillatory systems with n degrees of freedom. Thus, we call p a **stationary** or **critical** point of f, iff all the partial derivatives f_i are zero at p. The expansion 37.2.1 for $f: A \to \mathbb{R}$ then starts

37.3.1
$$f(p + v) - f(p) = \tfrac{1}{2} \sum f_{jk}(p) v_j v_k + o(v^2);$$

the first sum on the right is called a **quadratic form** in the variables v_j (it is a function $\theta: \mathbb{R}^n \to \mathbb{R}$). If there exists a ball V in $\mathbb{R}^n$, centre O, with $p + v \in A$ when $v \in V$, then p is said to be a (relative) **minimum** or **maximum** of f provided the left-hand side of 37.3.1 is always > 0 or < 0, respectively, for all non-zero $v \in V$.

We call p a **non-degenerate** critical point, iff the **Hessian** matrix $(f_{jk}(p))$ has non-zero determinant. Thus, when $n = 1$, we require simply $f''(p) \neq 0$. Assuming the f_{jk} to be continuous, it can then be shown that p is isolated (i.e., some neighbourhood of p contains no other critical point), and by an orthogonal transformation of co-ordinates the axes can be chosen so that

37.3.2
$$f(p + u) - f(p) = \sum_j \lambda_j u_j^2 + o(u^2),$$

where the λ_j are the eigenvalues [see 19 Exercise 6(x)] of the matrix $(f_{jk}(p))$. Hence if all the λ_j are > 0, then p is a minimum of f; if they are all < 0, then f is a maximum. If exactly q of the λ_j are < 0, then p is called a (non-degenerate) **stationary point of index** q; if $n = 2$ and $q = 1$, then p is called a **saddle-point.** The index q turns out to be independent of the particular co-ordinate system chosen to reduce the quadratic form on the right-hand side of 37.3.1 to the sum of squares on the right-hand side of 37.3.2.

Exercise 2

(i) When $n = 2$, the sets $f =$ constant in A are called **contours** of f (by analogy with the contours of the height function on a geographical map). Find the index of the origin $O \in \mathbb{R}^2$ for the four functions

$$2x^2 + y^2, \qquad x^2 - 4y^2. \qquad 4x^2 - y^2, \qquad -(x^2 + y^2)$$

of x, y, and draw their contours in a neighbourhood of O.
(ii) Show that $ax^2 + bx + c > 0$ for all x if $a > 0$, $c > 0$, and $b^2 < 4ac$. Hence, when $n = 2$ in 37.3.1, prove that a sufficient condition for p there to be a minimum, is that

$$f_{11}(p) > 0, \qquad f_{22}(p) > 0, \quad \text{and} \quad (f_{12}(p))^2 < f_{11}(p) f_{22}(p).$$

Formulate similar sufficient conditions for p to be a maximum and a saddle point. Show that the graph of f in $\mathbb{R}^3$ at such points looks, respectively, like a lake bottom, a mountain top, and a mountain pass. Hence note that the first two inequalities above are appropriate for the 'slices' through the graph, in the x and y directions, to have minima; and similarly in the other cases.
(iii) Points (x_i, y_i), $1 \leqslant i \leqslant n$, are scattered in $\mathbb{R}^2$ and it is required to find a line $ax + by + c = 0$ 'of best fit', that is, such that $\sum (ax_i + by_i + c)^2$ shall be least. Suppose first that $c \neq 0$; we may then take $c = 1$, so we must find $a, b \in \mathbb{R}$ for which

$$f(a, b) = \sum_{i=1}^{n} (ax_i + by_i + 1)^2$$

is least. Prove that a relative minimum of f exists and is unique, using Schwarz's inequality, unless

$$\sum x_i^2 \cdot \sum y_i^2 = \left(\sum x_i y_i\right)^2.$$

In the latter case the required line is

$$x \cdot \sum y_i^2 = y \cdot \sum x_i y_i$$

(with $c = 0$). In the first case, prove that the relative minimum is the least value taken by f. Justify geometrically this choice of a, b as giving 'best fit'.
(iv) In analytic geometry, a common method of finding the centre of the conic $ax^2 + by^2 + 2hxy + 2gx + 2fy + c = 0$ is as follows. The left-hand side defines a function S of (x, y), with domain $\mathbb{R}^2$. The co-ordinates of the centre satisfy the simultaneous equations $S_1 = 0 = S_2$. Justify this method, by relating the symmetry of the conic to the critical point(s) of the function S.
(v) Let l, m be two skew lines in $\mathbb{R}^3$, with vector forms

$$\mathbf{r} = \mathbf{a} + \lambda\mathbf{b}, \qquad \mathbf{r} = \mathbf{c} + \mu\mathbf{d}$$

respectively. Find points $\mathbf{u} \in l$, $\mathbf{v} \in m$ such that $|\mathbf{u} - \mathbf{v}|$ is least, and show that the line joining $\mathbf{u}$ to $\mathbf{v}$ is perpendicular to both l and m. {Regard $(\mathbf{u} - \mathbf{v})\cdot(\mathbf{u} - \mathbf{v})$ as a function f of λ and μ, and use the theory of (ii) above. Observe that the skewness of l and m implies that $(\mathbf{b}\cdot\mathbf{d})^2 < \mathbf{b}^2\mathbf{d}^2$, which inequality implies both that the equations $\partial f/\partial\lambda = 0 = \partial f/\partial\mu$ have a unique solution *and* that this solution is a relative minimum.}

[The general theory of stationary points of index q requires a study of quadratic forms (see, e.g., Mirsky [88]), some topology (see, e.g., Morse [91]), and a study of sets of the form $f^{\flat}(k)$. One of the earliest results of the theory was Theorem 25.12.13. An analogue holds in higher dimensions, for functions on differentiable manifolds; this extension is due to M. Morse and has important consequences in topology.]

37.4 Implicit Functions

Given a function $f\colon A \to \mathbb{R}$ where $A \subseteq \mathbb{R}^n$, we cannot *draw* the graph of f unless $n \leqslant 2$, since we live in a three-dimensional world. But when $n = 3$, we can form a picture of f by studying its **level surfaces**, i.e., the sets of the form $f =$ constant. This is directly analogous to the construction of a contour map for a height function in geography, when $n = 2$ (see Exercise 2(i)). We are therefore confronted with the problem of describing sets of the form $f^\flat(c)$ (which we can picture when $n \leqslant 3$); and for this purpose various theorems have been invented, at least to deal with a differentiable f, and called 'implicit function theorems'. A brief introduction to these follows.

Consider the special case of a function $f: A \to \mathbb{R}$, where A is open in $\mathbb{R}^3$, and suppose $f(p) = c$ for some $p \in A$. Thus the set $S = f^\flat(c)$ is not empty, and our problem is to be able to describe the whole of S. Traditional writers would say that the condition $f(p) = c$ imposes one condition on the variables (p_1, p_2, p_3) so that we can solve for p_3, say, in terms of p_1, p_2; and p_3 is 'defined implicitly' in terms of p_1, p_2. Setting $p_3 = h(p_1, p_2)$, they would then write

37.4.1 $$f(p_1, p_2, h(p_1, p_2)) \equiv c$$

and would assume that h is a differentiable function of (p_1, p_2). They would say, too, that the relation 37.4.1 is an 'identity' in p_1, p_2, meaning in effect that it holds for all (x, y) near (p_1, p_2) and so has zero derivative. With the language now available, we can clarify the matter as follows, in the form of a theorem; for a proof, see Osgood [99].

37.4.2 THEOREM. *Let $f: A \to \mathbb{R}$ be differentiable on the open $A \subseteq \mathbb{R}^3$. If $p \in A$, $f_3(p) \neq 0$ and $f(p) = c$, then there are open neighbourhoods $V \subseteq A$, $U \subseteq \mathbb{R}^2$ such that $(p_1, p_2) \in U$, $p \in V$, and there exists a unique differentiable function $h: U \to \mathbb{R}$ such that*

(i) $h(p_1, p_2) = p_3$

and for each $(x, y) \in U$

(ii) *each point $(x, y, h(x, y)) \in V \subseteq A$,*
(iii) *$f(x, y, h(x, y)) = c$,*
(iv) *if $f(x, y, z) = c$ and $(x, y, z) \in V$, then $z = h(x, y)$.* □

To construct the set $S = f^\flat(c)$, at least near p, let Γ_h denote the graph of h; let $\bar{h}: U \to \Gamma_h$ denote the function $u \to (u, h(u))$. By 3 Exercise 2(iv), $\bar{h}$ is an equivalence; by (i) and (ii), $p \in \Gamma_h \subseteq V$; by (iv), $\Gamma_h = S \cap V$. Thus, the portion of S near p consists exactly of the graph Γ_h, and is therefore a portion of a surface possessing a tangent plane at each point (since h is differentiable). By (iii), $f \circ \bar{h}: U \to \mathbb{R}$ is the constant function c.

Similarly for p_1 or p_2, for functions of more variables, and also for functions of the form $g: A \to \mathbb{R}^m$. In the latter case, if $m > 1$, the condition that some $D_i f$ be non-zero at p, i.e.,

37.4.3 $$\nabla f|_p \neq \mathbf{0},$$

must be replaced by one about a Jacobian matrix (see Osgood [99], IV, 11). Theorem 37.4.2 says essentially that, except possibly at points where $\nabla f = 0$, the set $f^{\flat}(c)$ is a smooth surface; and this surface is like a graph. When $f_3(p) \neq 0$, it *is* the graph Γ_h, but if instead (say) $f_2(p) \neq 0$, then we would have obtained in 37.4.2 a function $j\colon U \to \mathbb{R}$ such that $j(p_1, p_3) = p_2$, with obvious changes in (ii), (iii), and (iv) of 37.4.2. And instead of the graph of j (corresponding to Γ_h, we obtain a surface $\Gamma_j = \{(x, j(x, z), z) | (x, z) \in U\}$ and corresponding to $\bar{h}\colon U \to \Gamma_h$ a map

37.4.4 $$\bar{j}\colon U \to \Gamma_j; \qquad f \circ \bar{j} = c.$$

Similarly for p_1.

When $m = 2$ and we have a differentiable function $g\colon A \to \mathbb{R}$ with A open in $\mathbb{R}^2$, then again except at points where $\nabla g = 0$, the set $g^{\flat}(c)$ is smooth, this time a *curve*. Hence, the set U in 37.4.2 is an interval I, where Γ_h is a curve (see Fig. 37.1). And equation (iii) becomes (assuming $g_2(p) \neq 0$, by 37.4.3):

$$\forall x \in I, \qquad g(x, h(x)) = c,$$

or $g \circ \bar{h} = c$, where (as above)

$$\bar{h}(x) = (x, h(x)) \in \Gamma_h.$$

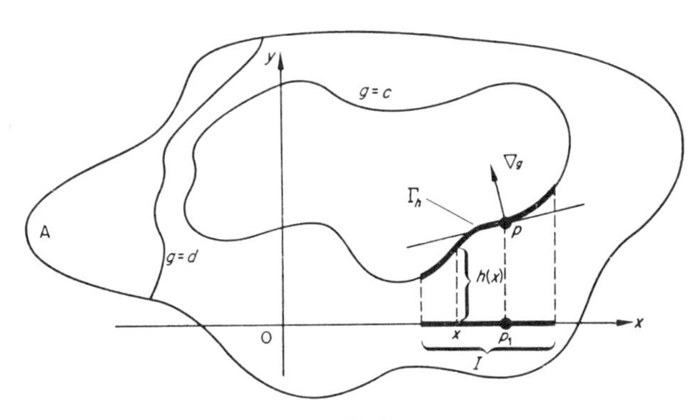

Fig. 37.1

This gives a rapid way of calculating h' as follows. Since $g_2(p) \neq 0$, we may assume by the inertia principle—provided g_2 is continuous on A—that I is so small that g_2 is never zero on $\bar{h}(I)$.

37.4.5 LEMMA. *If g_2 is continuous on A, then for all $x \in I$,*

$$h'(x) = -g_1(x, y)/g_2(x, y) \qquad (y = h(x)),$$

[or in (suggestive) traditional notation, $dy/dx = -(\partial g/\partial x)/(\partial g/\partial y)$].

Proof. We have $g \circ \bar{h} = c: I \to \mathbb{R}$, so for each $x \in I$, Lemma 36.2.1 gives $d_q g \circ d_x \bar{h} = 0$ where $q = \bar{h}(x) \in \bar{h}(I)$. Now for each $v \in \mathbb{R}$, we have, by definition of $\bar{h}$,

$$d_x\bar{h}(v) = (v, d_x h(v)) = (v, h'(x)v),$$

so $0 = (d_q g \circ d_x \bar{h})(v) = d_q g(v, h'(x)v) = g_1(q)\cdot v + g_2(q)\cdot h'(x)\cdot v$; in particular, taking $v = 1$, we get 37.4.5, since $q = \bar{h}(x) = (x, h(x)) \in \bar{h}(I)$ and therefore $g_2(q) \neq 0$, by choice of I explained above. ∎

37.4.6 COROLLARY. *At p on the curve $g = c$, the gradient of g (if non-zero) points along the normal.* ∎

Similarly for more variables (see Exercise 3(v) below).

Exercise 3

(i) Let $f: \mathbb{R}^3 \to \mathbb{R}$ be given by $f(x, y, z) = x + y + z$ and let $p = 0$. Calculate U, V and h in 37.4.2. What are Γ_h and Γ_j in 37.4.4?
(ii) Show that the map $(x, y, z) \to (x, z, y)$ of $\mathbb{R}^3$ onto itself, carries Γ_j in 37.4.4 onto the graph of j in $U \times \mathbb{R}$.
(iii) If $g(x, y) = y^2 - 4ax$, prove that g_2 is zero only on the x-axis. How big is the neighbourhood corresponding to V in 37.4.2? Show that if p is not on the x-axis, then 37.4.5 agrees with h' calculated directly from the solution $y = \pm\sqrt{(4ax + k)}$. How would you apply 37.4.5 if $p_2 = 0$?
(iv) If $f(x, y, z) = ax^2 + by^2 + cz^2 + 2fyz + 2gzx + 2hxy$, prove that ∇f is zero only at $0 \in \mathbb{R}^3$ provided

$$\det\begin{pmatrix} a & h & g \\ h & b & f \\ g & f & c \end{pmatrix} \neq 0.$$

Investigate the structure of the surface $f = c$ in $\mathbb{R}^3$. What do you infer about the centre of a conic in plane projective geometry? (Compare Exercise 2(iv).)
(v) In 37.4.2 prove that $h_1(x, y) = -(f_1/f_3)$, and $h_2(x, y) = -(f_2/f_3)$, each quotient evaluated at $\bar{h}(x, y)$—provided f_3 is non-zero on Γ_h. Hence prove that at p, the gradient of f points along the normal to the tangent plane to Γ_h.
(vi) In 37.4.5, prove that h depends continuously on c. {Use 37.4.2.} Hence clarify the traditional saying that if $u = f(x, y)$, then $x = g(u, y)$ for some function g.

37.5 A Clarification

Let us use this language to make sense of the following problem expressed in traditional notation as follows. 'The equation $f(x, y, z) = 0$ defines each variable as a function of the other two. Establish the formulae:

$$\text{(i)} \begin{cases} \partial z/\partial x = -f_1/f_3; & \partial z/\partial y = -f_2/f_3; & \partial y/\partial x = -f_1/f_2, \\ \partial y/\partial z = -f_3/f_2; & \partial x/\partial y = -f_2/f_1; & \partial x/\partial z = -f_3/f_1; \end{cases}$$

where the variable being differentiated is considered as a function of the other two. In what sense are $\partial z/\partial x$ and $\partial x/\partial z$ reciprocals? Is

$$(\partial x/\partial y)(\partial y/\partial z)(\partial z/\partial x) = -1?\text{'}$$

Solution. In order to apply 37.4.2 we shall suppose that $f: A \to \mathbb{R}$ is given to be differentiable on the open $A \subseteq \mathbb{R}^3$, that $f(p) = 0$ for some $p \in A$, and that each f_i is not zero at p. (This last is required also for the denominators in the formulae (i).) Then by 37.4.2, there exists a function $\alpha: U \to \mathbb{R}$ such that (using 37.4.4) $\bar{\alpha}: U \to \Gamma_\alpha \subseteq A$ and $f \circ \bar{\alpha} = 0$ on U; α is then 'the variable x as a function of y and z'. Let β, V, γ, W denote the analogous objects corresponding to y and z; then γ is h in 37.4.2. Now differentiate the zero function $f \circ \bar{\alpha}$ on the open set U, by the chain rule, to get when $i = 1, 2$:

$$0 = D_i(f \circ \bar{\alpha}) = (f_1 \circ \bar{\alpha})\cdot\bar{\alpha}_i^1 + (f_2 \circ \bar{\alpha})\cdot\bar{\alpha}_i^2 + (f_3 \circ \bar{\alpha})\cdot\bar{\alpha}_i^3.$$

But $\bar{\alpha}(y, z) = (\alpha(y, z), y, z)$, so $\bar{\alpha}_1^1 = D_1\alpha$, $\bar{\alpha}_2^1 = D_2\alpha$ and $\bar{\alpha}_1^2 = \pi_1^1 = 1 = \pi_2^2 = \bar{\alpha}_2^3$, $\bar{\alpha}_2^2 = \pi_2^1 = 0 = \pi_1^2 = \bar{\alpha}_1^3$, where π^1, π^2 are the projections in $\mathbb{R}^2$. Thus, for example,

$$\alpha_2(y, z) = -(f_3/f_1)|_{\bar{\alpha}(y,z)}$$

which is the reciprocal of $\gamma_1(x, y) = -(f_1/f_3)|_{\bar{\gamma}(x,y)}$ (found similarly) at the set X of those points (among others) where $\bar{\gamma}(x, y) = \bar{\alpha}(z, y)$, i.e., where $(x, y, \gamma(x, y)) = (\alpha(y, z), y, z)$; such a point is p, by 37.4.2(i). As to the final question of the problem, the product given must mean $\alpha_1\beta_1\gamma_1$, but it cannot mean a product of *functions*, since α_1, β_1, γ_1 do not necessarily have a common point in their domains; however, on using 37.4.2(i), $\alpha(p_2, p_3) = p_1$, so

$$\alpha_1(p_2, p_3) = -f_2(p)/f_1(p),$$

and similarly for β_1, γ_1. Hence at $p \in A$

$$\alpha_1(p_2, p_3)\cdot\beta_1(p_1, p_3)\cdot\gamma_1(p_1, p_2) = -1.$$

Exercise 4

The following problems are expressed in traditional language. Interpret them along the lines of the text, and solve the resulting problems.

(i) If $w = f(x, y, z)$ and $g(x, y, z) = 0$, then z may be eliminated to get w as a function of x and y. It then has two partial derivatives denoted by $(\partial w/\partial x)_y$, $(\partial w/\partial y)_x$, where the subscripts indicate the variable being kept constant. Similarly we can eliminate y, and we may consider $(\partial w/\partial x)_z$, etc. Prove that

$$\left(\frac{\partial w}{\partial x}\right)_y = \left(\frac{\partial w}{\partial x}\right)_{y,z} + \left(\frac{\partial w}{\partial z}\right)_{x,y}\cdot\left(\frac{\partial z}{\partial x}\right)_y.$$

(ii) Analogously, suppose $z = f(x, y)$, and $y = y(x)$. Prove that

$$\frac{dz}{dx} = \left(\frac{\partial z}{\partial x}\right)_y + \left(\frac{\partial z}{\partial y}\right)_x\cdot\frac{dy}{dx}.$$

(iii) If $x = r\cos\theta$, $y = r\sin\theta$, $r = \sqrt{(x^2 + y^2)}$ and $\theta = \tan^{-1} y/x$, prove that $\partial x/\partial r = \partial r/\partial x$, but $(\partial x/\partial r)_\theta = (\partial r/\partial x)_y$ and $(\partial x/\partial r)_\theta = 1/(\partial r/\partial x)_\theta$.

(iv) If $u = u(x, y)$, $v = v(x, y)$, then the *Jacobian* of the transformation $(x, y) \to (u, v)$ is defined to be the determinant of the Jacobian matrix, that is,

$$\frac{\partial u}{\partial x}\frac{\partial v}{\partial y} - \frac{\partial u}{\partial y}\frac{\partial v}{\partial x}$$

and is denoted by

$$\frac{\partial(u, v)}{\partial(x, y)}.$$

Settle the statement:

$$\text{`If } \partial(u, v)/\partial(x, y) \neq 0, \qquad \text{then } \frac{\partial(x, y)}{\partial(u, v)} = 1 \Big/ \frac{\partial(u, v)}{\partial(x, y)}\text{'}.$$

(v) Make sense of the following two problems, set in University examinations, and solve them.

(a) Prove that if $u = x + y + z$, $u^2 v = y + z$, $u^3 w = z$, then

$$\frac{\partial(u, v, w)}{\partial(x, y, z)} = u^5.$$

(b) If

$$\begin{aligned} u^2 + v^2 + w^2 &= yz + zx + xy, \\ vw + wu + uv &= x^2 + y^2 + z^2, \\ u^3 + v^3 + w^3 &= x^3 + y^3 + z^3, \end{aligned}$$

show that

$$\frac{\partial(u, v, w)}{\partial(x, y, z)} = -\frac{(x + y + z)(x - y)(y - z)(z - x)}{(u + v + w)(u - v)(v - w)(w - u)}.$$

{In each case there is implied a mapping θ: $(x, y, z) \to (u, v, w) \in \mathbb{R}^3$ with Jacobian denoted by $\partial(u, v, w)/\partial(x, y, z)$. In (b) there are two functions F, G: $\mathbb{R}^3 \to \mathbb{R}^3$ such that on a suitable domain $F \circ \theta = G$. Now use 36 Exercise 5(ii).} How do you think the examiner hit on the symmetry in (b)?

(vi) If $u = u(x, y)$, $v = v(x, y)$, and $f(u, v) \equiv 0$, prove that $\partial(u, v)/\partial(x, z) \equiv 0$. Conversely, if the Jacobian vanishes identically near p, while the derivatives of u, v are continuous and not all zero at p, prove that there is a functional relationship $f(u, v)$ near p. {See Hilton [61], p. 33.}

(vii) When v is eliminated between the equations $y = f(x, v)$ and $z = g(x, v)$ the equation $z = \phi(x, y)$ is obtained. Prove that

$$\frac{\partial \phi}{\partial x}\frac{\partial f}{\partial v} = \frac{\partial f}{\partial v}\frac{\partial g}{\partial x} - \frac{\partial f}{\partial x}\frac{\partial g}{\partial v}.$$

Verify this result when

$$\begin{aligned} y &= x \cos v - a \sin v, \\ z &= x \sin v + a \cos v, \end{aligned}$$

a being a constant.

FURTHER OUTLOOK. The reader should now be in a position to understand the standard texts on functions of several variables, and in particular to pursue the question of multiple integrals and their transformation formulae. See for example Courant [24], Vol. 2, Chs. IV and V, and Apostol [4]. The natural extension of that material (apart from applied mathematics) is then the subject of differential geometry: see, for example, Auslander [7] and Auslander–Mackenzie [8].

Part VIII

FOUNDATIONS

This part contains the two final chapters of the book. The first, Categories and Functors (Chapter 38), is an introduction to a new and rapidly growing part of mathematics, which arose from the practice of always asking for the appropriate transformations between newly-defined objects. Thus, the transformations appropriate to groups are homomorphisms, those appropriate to topological spaces are continuous mappings, etc. Such transformations are necessary when we compare a known group (or ring, or Boolean algebra) with an unknown one of the same sort. The family of all groups together with all group homomorphisms is an example of a system called a *category*; another example of a category is the family of all topological spaces and continuous mappings. Such categories themselves have some algebraic structure arising from composition of their transformations. And, following the practice mentioned above, we often compare categories by using transformations called *functors*, which arise when in practice we compare, say, a topological space X (element of one category) with its fundamental group (element of another category). Once the notion is formulated, certain natural questions arise, and a useful framework is created for considering problems in apparently different branches of mathematics, in essentially the same way. This opens up possibilities for generalization, and economy of thought, and the chapter is an introduction to these ideas.

The last chapter, Mathematical Logic (Chapter 39), takes up again in a more formal way such notions as *set* and *proof*, and looks at some of the paradoxes of set-theory. These considerations were already close in Part II, and raise the question whether various parts of mathematics are non-contradictory. We are led once again to the idea of a mathematical model, first in connection with geometry, and then as exploited by Gödel. In particular we discuss three results of Gödel, on the existence of undecidable propositions in formal systems, on consistency of such systems, and on the axiom of choice. The chapter is a mere glimpse of a large and expanding branch of mathematics, which is so often fed by other branches; for example, the theory of categories raises questions of logic which have not yet been finally settled, such as: should categories, rather than sets, be regarded as the fundamental building blocks of mathematics?

set-theory (as it is called), with commentary using illustrations from other, more familiar, parts of mathematics. Thus, Chapter 1 introduces *sets* with the associated ideas of subsets, and inclusion. In Chapter 2 we introduce *functions*, and discuss their composition and inversion, together with the question of counting. Sets and functions are united in Chapter 3, with the construction of the *Cartesian product* of sets. Subsets of a Cartesian product are called *relations*, and two kinds of these are of great importance: equivalence relations (discussed in Chapter 4) and ordering relations (discussed in Chapter 5). Our discussion of the latter includes the method of mathematical induction.

Once armed with the language of this part, the reader can then begin any of the remaining parts, since these are essentially elaborations, in the same spirit, about the special kinds of sets, functions, and relations appropriate to arithmetic, algebra, geometry, and calculus.

Chapter 38

CATEGORIES AND FUNCTORS

Throughout this book we have been concerned with certain domains of mathematical discourse, with their structural properties and with their interrelations. Thus Parts I and II dealt with sets and functions which are set-transformations. Part III dealt with the domain of integers $\mathbb{Z}$ and with similar domains. Part IV dealt with subsets of $\mathbb{R}^3$ and their appropriate transformations. Part V concerned itself with various domains of interest in algebra, and Part VII with domains of interest in analysis. In Chapter 25 (of Part VI) we showed how algebraic topology was concerned with relating domains of interest in geometry to domains of interest in algebra.

Recently mathematicians have found it valuable to formalize these imprecise notions of *domain of discourse* and of *interrelations* between domains, and we believe it may be of value to our readers to give them this foretaste of these important ideas. Though there is a rapidly developing theory accompanying this formalization, we wish to stress that we are mainly concerned here with the development of an appropriate *language* for the expression of mathematical ideas. Thus the spirit of this chapter is very much akin to that which should, in our view, animate the introduction of the notions of sets and functions in the early phases of mathematical education. Then, and now, the emphasis should be on language and not on elaborate theory, since it is inappropriate to study such elaborations before a body of examples is built up. In any case, the elaborations are primarily the responsibility and interest of specialists.

38.1 Categories

Let us recall from Chapter 2 certain essential features of set theory. We are, in that theory, concerned with a class of objects $A, B, C, \ldots$, which we call *sets*, and a class of transformations $f, g, h, \ldots$, which we call *functions*. With each function f is associated a domain A and a range B, and we have written variously

$$f\colon A \longrightarrow B \quad \text{or} \quad A \xrightarrow{f} B$$

to indicate that f has domain A and range B. We have insisted, in defiance of classical tradition, (i) that the *range* is to be distinguished from the *image*, which is the set of values of f and may well be smaller than the range; and (in the same spirit) (ii) that two functions can only be regarded as identical if their domains and ranges coincide. The symmetrical roles played by domain and range in this viewpoint have led some authors to substitute the term 'co-domain' for 'range'.

Further, we know how to compose functions. The functions $f: A \to B_1$, $g: B_2 \to C$ may be composed to yield gf, or $g \circ f$,

$$gf: A \to C$$

if and only if $B_1 = B_2$, that is, if and only if the range of f coincides with the domain of g. This law of composition is *associative*, that is,

$$h(gf) = (hg)f,$$

provided the relevant compositions are defined. Further we may associate with each set A its identity function $1_A: A \to A$; where convenient we drop the suffix and simply write $1: A \to A$. The symbol '1' is particularly suitable in connection with composition since, if we think of composition as a sort of multiplication, then the identity function acts in the way the integer 1 acts in the ring $\mathbb{Z}$, that is,

$$1f = f, \qquad g1 = g,$$

wherever the compositions are defined.

Of course, set theory has more structure than just this. But, just as we found it valuable to abstract certain features of $\mathbb{Z}$ and were led to define the notion of integral domain (say), so we find it valuable here to abstract these features of set theory and define the notion of a *category*. We now come to the precise definition.

38.1.1 DEFINITION. A **category** $\mathcal{C}$ consists of three sets of data:

(1) there is a class of **objects** $A, B, C, \ldots$;

(2) to each ordered pair of objects A, B, there is associated a set

$$M_{\mathcal{C}}(A, B),$$

called the set of **morphisms** (or *maps* or *transformations*) from A to B in $\mathcal{C}$;

(3) to each ordered triple A, B, C, there is a *law of composition* or *composition function*

$$M_{\mathcal{C}}(A, B) \times M_{\mathcal{C}}(B, C) \to M_{\mathcal{C}}(A, C),$$

the image of (f, g) under this function being written† gf or $g \circ f$.

These data satisfy the following axioms:

$\mathfrak{Cat}_1$: *$M_{\mathcal{C}}(A_1, B_1)$ and $M_{\mathcal{C}}(A_2, B_2)$ are **disjoint** unless $A_1 = A_2$, $B_1 = B_2$;*

$\mathfrak{Cat}_2$: *(associative law) $h(gf) = (hg)f$, provided the compositions are defined;*

$\mathfrak{Cat}_3$: *(existence of identities) to each object A in $\mathcal{C}$ there is a morphism $1_A = 1 \in M_{\mathcal{C}}(A, A)$ such that $1f = f$, $g1 = g$, provided the compositions are defined.*

† We would like to write fg for the image of (f, g), but feel compelled to write gf because of our conservative adherence to the convention that functions are written on the *left* of their arguments.

Notice that, just as for groups, integral domains, and so on, what we throw away here is just the specification of the actual nature of the constituent parts of a category. We do *not* assert that $A, B, C, \ldots$ are sets, nor that $f \in M_{\mathcal{C}}(A,B)$ is a function. The law of composition is postulated abstractly and does not have to conform to any familiar concretization of the notion of composition. Nevertheless we naturally conserve notation and terminology, writing

$$f.\ A \longrightarrow B \quad \text{or} \quad A \xrightarrow{f} B$$

as convenient alternatives to $f \in M_{\mathcal{C}}(A, B)$, referring to A, B respectively as the *domain*, *range* (or *codomain*) of f, and to f as a morphism *from* A *to* B. The 'arrow' notation suggests the pictorial representation of $\mathcal{C}$ by means of a linear graph, whose nodes represent the objects of $\mathcal{C}$, and where an edge directed from object A to object B represents an element of $M_{\mathcal{C}}(A, B)$.

The notion of a category may be seen to depend on those of sets and functions for more than just its inspiration. For, the collection of morphisms from A to B forms a *set* $M_{\mathcal{C}}(A, B)$, and composition is a function from the Cartesian product of two sets to a third. [Very recently, F. W. Lawvere has shown how to free the notion of category from even this mild dependence on set theory. We are naturally not concerned with such sophisticated notions here, but we do think it worth mentioning (though the reader should not worry over this) that a category is not itself necessarily a set. The word 'class' which appears in the definition of a category has a technical meaning in many systems of set theory (see Chapter 39), and is used here deliberately so as not to give the impression that the collection of objects of a category may be assumed to be a set. Indeed, when it is a set, the category is said to be **small**].

To prescribe a category, then, one must give all three sets of data. However, in all the examples given below the law of composition is familiar or obvious and will be omitted. We will also leave to the reader the trivial verification of the axioms. Notice that whenever the morphisms are functions (perhaps preserving some structure in the objects) and the law of composition is just the composition of functions, then associativity is automatically guaranteed.

38.1.2 EXAMPLES. The following eleven systems are categories:

(i) Sets and functions;
(ii) Finite sets and functions;
(iii) Sets and injections;
(iv) Sets and surjections;
(v) Groups and homomorphisms;
(vi) Abelian groups and homomorphisms;
(vii) Rings and homomorphisms;
(viii) Vector spaces over a fixed field F, and linear maps;
(ix) Subsets of $\mathbb{R}^3$ and Euclidean movements;

(x) Subsets of $\mathbb{R}^3$ and projective transformations;
(xi) Subsets of $\mathbb{R}^n$ and continuous functions.

(Thus we speak of 'the category of sets and functions', for example.)

The reader should examine exactly what is meant by each example above—and provide many examples of his own. The following two examples are somewhat different in nature and will be treated a little less rapidly—but not because they are more important than the examples above.

38.1.3 EXAMPLE. Let A be any set and let R be the set of endomorphisms η of A (i.e., η is a function $A \to A$). Then we may regard R as a *semigroup* (set with associative multiplication with two-sided unity). Moreover, any semigroup may be realized as a set of endomorphisms. So *any semigroup is a category* with a single object, and a category may be regarded as a generalization of semigroup.

38.1.4 EXAMPLE. Let $(S, \leqslant)$ be an ordered set. We form a category whose objects are the elements of S, and such that for all $a, b \in S$,

$$M_{\mathfrak{C}}(a, b) = \varnothing \quad \text{if} \quad a \nleqslant b,$$

$$M_{\mathfrak{C}}(a, b) \text{ contains the single element } (a, b) \quad \text{if} \quad a \leqslant b.$$

The law of composition simply reflects the transitivity of the order relation; we leave the reader to define it in the obvious way.

Example 38.1.3 suggests that we may be able to transfer to categories such elementary algebraic ideas as those of 3.6. We close this section with one important example of this process.

38.1.5 THEOREM. *The morphisms 1_A are uniquely determined by axiom $\mathfrak{Cat}_3$.*

Proof. Suppose that the morphisms $1'_A$ also satisfy

$$1'f = f, \qquad g1' = g,$$

wherever defined. Then $1'_A 1_A$ is defined, so

$$1_A = 1'_A 1_A = 1'_A. \quad ■$$

The morphisms 1_A enable us to introduce the fundamental concept of an *equivalence*, and *equivalent objects*, in a category. We are motivated by the algebraic inversion theorem 2.9.8. A morphism $f: A \to B$ is called an **equivalence**, or an **isomorphism**, or a **unit**, or **invertible**, if there exists $g: B \to A$ in $\mathfrak{C}$ such that

$$gf = 1_A, \qquad fg = 1_B;$$

and A and B are said to be equivalent if there exists an equivalence $f: A \to B$ in $\mathfrak{C}$. We write $f: A \approx B$, or simply $A \approx B$ in that case. Plainly $A \approx B$ is reflexive, symmetric and transitive (prove the last carefully!). The reader should identify the units in the examples given in 38.1.2.

38.2 Initial, Terminal, Zero Objects

Consider Example 38.1.2(v), the category $\mathcal{G}$ of groups and homomorphisms. If G is a group, then G has a distinguished element, namely the unity element $e = e_G$ of G. This element is 'canonical', in the sense that any group homomorphism maps unity element to unity element. Thus the unity element is emphatically part of the basic structure of a group, and group homomorphisms preserve this structure. We would thus expect the unity element to admit some sort of 'categorical' formulation. In fact what we do below is essentially to generalize the notion of the *trivial group*, whose only element is its unity element, and of the *trivial* or *constant homomorphism* $\phi: G \to H$ which sends the whole of G to the unity element of H.

Thus, we shall call an object I in a category $\mathcal{C}$ an **initial object** if $M_{\mathcal{C}}(I, A)$ is a singleton for every object A in $\mathcal{C}$. Plainly any two initial objects in $\mathcal{C}$ are equivalent (indeed, they are 'canonically' equivalent, in the sense that the equivalence required is uniquely determined). Similarly an object T in $\mathcal{C}$ is called a **terminal object** if $M_{\mathcal{C}}(A, T)$ is a singleton for every object A in $\mathcal{C}$, and again any two terminal objects are canonically equivalent. If there is an object P which is both initial and terminal, then P is called a **zero object**, or a **point**, and every initial or terminal object in $\mathcal{C}$ is a point if $\mathcal{C}$ contains a point.

38.2.1 EXAMPLES. (i) In the category $\mathcal{S}$ of sets and functions, the empty set is an initial object (by 3.4.3) and every singleton is a terminal object. There are no zero objects.

(ii) In the category of sets and injections, the empty set is an initial object and there is no terminal object.

(iii) In the category $\mathcal{G}$ of groups and homomorphisms, and in the category $\mathcal{A}b$ of abelian groups and homomorphisms, the singletons (or trivial groups) are the zero objects.

Now let $\mathcal{C}$ be a category with zero object P. Then, for all A, B in $\mathcal{C}$, $M_{\mathcal{C}}(A, B)$ is non-empty; for $M_{\mathcal{C}}(A, B)$ certainly contains the morphism $A \to P \to B$. Moreover, this morphism is independent of the choice of zero object P. For if Q is any other zero object, then we have a commutative diagram

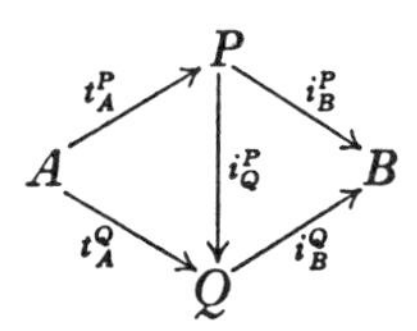

and so $i_B^P \circ t_A^P = i_B^Q \circ t_A^Q$. Call this last morphism the **zero morphism** $0 = 0_{AB}: A \to B$. Then evidently

38.2.2 $$0f = 0, \qquad g0 = 0,$$

provided the compositions are defined (and we give appropriate suffixes to the zeros). Conversely, we say that $\mathcal{C}$ is a **category with zero morphisms** if for each ordered pair of objects A, B of $\mathcal{C}$ there exists a morphism

$$0 = 0_{AB}: A \to B$$

satisfying 38.2.2. Thus a category with zero object is a category with zero morphisms. The converse is evidently false. For if we throw all the trivial groups out of $\mathcal{G}$, the resulting category still has zero morphisms (the constant homomorphisms), but no zero objects [prove this!]. On the other hand, as this example suggests, the situation where zero morphisms are present in $\mathcal{C}$, but $\mathcal{C}$ lacks zero objects, is artificial; we may, indeed, always restore zero objects to a category with zero morphisms. This we do by the following device.

Let $\mathcal{C}$ be a category with zero morphisms. Enlarge $\mathcal{C}$ to $\overline{\mathcal{C}}$ by (i) adjoining a new object P, (ii) adjoining a morphism t_A^P, constituting $M_{\overline{\mathcal{C}}}(A, P)$, for each A in $\mathcal{C}$, (iii) adjoining a morphism i_A^P, constituting $M_{\overline{\mathcal{C}}}(P, A)$, for each A in $\mathcal{C}$, (iv) adjoining 1_P constituting $M_{\overline{\mathcal{C}}}(P, P)$, (v) extending the law of composition in $\mathcal{C}$ to $\overline{\mathcal{C}}$ by the rules:

$$\begin{aligned} t_B^P \circ f &= t_A^P, && \text{for every } f: A \to B \text{ in } \mathcal{C}, \\ g \circ i_C^P &= i_D^P, && \text{for every } g: C \to D \text{ in } \mathcal{C}, \\ 1_P \circ t_A^P &= t_A^P, && \text{for every } A \text{ in } \mathcal{C}, \\ i_A^P \circ 1_P &= i_A^P, && \text{for every } A \text{ in } \mathcal{C}, \\ i_B^P \circ t_A^P &= 0_{AB}, && \text{for every } A, B \text{ in } \mathcal{C}. \end{aligned}$$

38.2.3 THEOREM. *$\overline{\mathcal{C}}$ is a category, containing $\mathcal{C}$ as a subcategory. P is a zero object in $\overline{\mathcal{C}}$ and the zero morphisms of $\overline{\mathcal{C}}$ are the zero morphisms of $\mathcal{C}$, together with the morphisms t_B^P, i_A^P, 1_P.* ∎

We have sneaked in here the notion of a *subcategory*: it is surely evident what is meant by this term. In fact, a subcategory $\mathcal{D}$ of $\mathcal{C}$ is said to be **full** if

$$M_{\mathcal{D}}(A, B) = M_{\mathcal{C}}(A, B),$$

whenever $A, B \in \mathcal{D}$; and, of course, $\mathcal{C}$ is a full subcategory of $\overline{\mathcal{C}}$ above. The reader should study Examples 38.1.2 with a view to detecting occurrences of subcategories of categories, and should observe when the subcategories are full. Naturally, he should also make up his own examples.

In view of Theorem 38.2.3 we will take no interest in those categories which have zero morphisms but lack zero objects. On the other hand the example of the category of sets shows that we may not confine attention simply to categories with zero objects.

Exercise 1

(i) Develop, along lines similar to the treatment of zero morphisms, the notions of initial and terminal morphisms.
(ii) Show, using 38.2.2, that zero morphisms, where they exist, are unique.
(iii) Show that the category of sets cannot be embedded as a full subcategory of a category with zero objects. Generalize.
(iv) Interpret the device used above to construct $\overline{\mathcal{C}}$, when $\mathcal{C}$ is represented by a linear graph (as observed after 38.1.1), and so prove Theorem 38.2.3.

38.3 Functors

One of the most important methods in mathematics today, and one of the most important points of view, may be described informally as follows; we gave an example of it in Chapter 25. Finding ourselves faced with some question about certain objects and morphisms of a category $\mathcal{C}$, we attempt to answer this question by transforming it into a question about objects and morphisms of a category $\mathcal{D}$. In the course of the passage from $\mathcal{C}$ to $\mathcal{D}$ certain information may have been thrown away, so the question in $\mathcal{D}$ is by no means a perfect replica of that in $\mathcal{C}$; but, by way of compensation, we hope to have simplified the situation sufficiently to have a good chance of answering the question and hence of obtaining information about our original problem. The passage from $\mathcal{C}$ to $\mathcal{D}$ may be likened to that of a homomorphic mapping from one group to another—indeed, such a homomorphism is a very special case of a *functor* from one category to another, whose definition we give below. We give one example of the technique described informally above before passing to the precise definition.

38.3.1 EXAMPLE. Groups may be given by generators and relations (see Ledermann [80]). That is, we may specify a group by saying that its elements may be expressed as words in certain symbols $a, b, c, \ldots$, and their inverses: and that the generators are *related* by the requirement that certain words

$$R_1(a, b, c, \ldots), \ldots, R_n(a, b, c, \ldots), \ldots$$

are equal to the unity element. Precisely, let F be the free group freely generated by $a, b, c, \ldots$; and let R be the smallest normal subgroup of F containing the elements $R_1, \ldots, R_n, \ldots$. Then the group generated by $a, b, c, \ldots$, subject to the relations $R_1, \ldots, R_n, \ldots$, is by definition the quotient group F/R. For example, the group of symmetries of the square is the group generated by a (rotation through a right angle) and b (reflection in a line joining mid-points of opposite sides), subject to the relations $a^4 = 1$, $b^2 = 1$, $aba = b$.

Now suppose G is the group generated by a, b with relation $a^2b^5 = b^3a^4$, while H is the group generated by c, d with the relation $c^3d^5 = d^3c^8$; and suppose we wish to know whether the groups G and H are isomorphic. As a first attempt to answer this question (which, in this case, succeeds) we pass from the category of groups to the category of abelian groups. That is, we 'abelianize' each group by adding the relations $xy = yx$ (for all x, y) so that its elements now all commute. From G we obtain in this way a commutative group $\bar{G}$, which we write additively. $\bar{G}$ is generated by α, β, and subject to the relation $2\alpha + 5\beta = 3\beta + 4\alpha$, or, since $\bar{G}$ is abelian,

$$2\alpha = 2\beta.$$

From H we obtain $\bar{H}$, generated by γ, δ, subject to the relation

$$3\gamma + 5\delta = 3\delta + 8\gamma, \quad \text{or} \quad 2\delta = 5\gamma.$$

It is now not at all difficult to see that $\bar{G}$ and $\bar{H}$ are *not isomorphic*; for $\bar{G}$ contains an element $\alpha - \beta$ of order 2, whereas $\bar{H}$ is cyclic infinite (generated by $2\gamma - \delta$). Since $\bar{G}$ and $\bar{H}$ are not isomorphic, it is certain that G and H are not isomorphic (why?), so our question is answered. Of course, had $\bar{G}$ and $\bar{H}$ turned out to be isomorphic, we could not have inferred that G and H were isomorphic—and a probe sharper than the ***abelianizing functor*** would have been needed. For plainly we can have two non-isomorphic groups whose abelianizations are isomorphic—just take a non-abelian group and its abelianization!

We now proceed to our main definition.

38.3.2 DEFINITION. Let $\mathcal{C}$ and $\mathcal{D}$ be two categories. Then a (**covariant**) **functor** F from $\mathcal{C}$ to $\mathcal{D}$ is a rule assigning

(i) to each object A in $\mathcal{C}$ an object FA in $\mathcal{D}$;
(ii) to each morphism $f\colon A \to B$ in $\mathcal{C}$ a morphism $Ff\colon FA \to FB$ in $\mathcal{D}$,
subject to the conditions

(iii) $F(gf) = (Fg)(Ff)$;
(iv) $F(1_A) = 1_{FA}$.

Before giving examples, we make some comments. The reader will note how closely this definition resembles that of a homomorphism of groups. It is true that, in defining a group homomorphism, we do not demand that it map unity element to unity element—but that is only because, in the case of groups, we get this property free, simply by demanding that a homomorphism respect the group operations. Notice that in (iii) above, we have suppressed the phrase 'where defined'; because (plainly) the right-hand side is defined if the left-hand side is. Next, we explain the qualification *covariant* in the title of F. This is explained, in fact, by reference to (ii) and (iii) above. The reader will notice that, in (ii), the direction of the arrow is preserved by F; that is, Ff goes from the F-image of the domain of f to the F-image of the range of F. This, in turn, makes sense of (iii). (Compare 6.2.9.) On the other hand, examples abound in modern mathematics where one wishes to reverse the arrow, that is, to consider functors F such that (i) holds, but

(ii)′ to each morphism $f\colon A \to B$ in $\mathcal{C}$ a morphism $Ff\colon FB \to FA$ in $\mathcal{D}$ is assigned.

Then (iv) is preserved, but (iii) is replaced by

(iii)′ $F(gf) = F(f)F(g)$.

A functor satisfying this variant of the definition is said to be **contravariant.** We now give examples of each type; in (i) and (ii) below the effect of the functor in question on *morphisms* is explained in the reference given.

38.3.3 EXAMPLES. (i) Let $\mathcal{S}$ be the category of sets and functions. Then if we assign to each set A its family $\mathfrak{p}A$ of subsets, we obtain a functor $\mathfrak{p}: \mathcal{S} \to \mathcal{S}$ which is *contravariant*, by 2 Exercise 7(i).

(ii) Let $\mathcal{S}'$, $\mathcal{G}'$ denote respectively the categories of sets and bijections, and of groups and isomorphisms. To each set A we may assign the group Perm A, so Perm: $\mathcal{S}' \to \mathcal{G}'$ is a functor which is *covariant* by 6 Exercise 7(ii).

(iii) Let $\mathcal{G}$ be the category of groups and homomorphisms and $\mathcal{S}$ the category of sets and functions. Let $F: \mathcal{G} \to \mathcal{S}$ be the **forgetful** (or **underlying) functor**, which associates with every group its underlying set and with every homomorphism the associated function of the underlying sets. Evidently F is a (covariant) functor which ignores, or forgets, the group structure. The reader may think it absurdly fussy to emphasize that F is a functor, and in most cases one does not bother to stress its presence by appropriate notational distinctions. But in fact certain fundamental ideas have been clarified by making this notion of a forgetful functor explicit. We will not go into details, but will content ourselves by remarking that what we frequently have to do in group theory may be expressed as follows: G and H are two groups, and we have some function f from FG to FH. We want to know if f is a homomorphism i.e., is $f = Ff'$, with $f': G \to H$ in $\mathcal{G}$? Since F is (clearly) one-one on morphisms, we naturally identify f with f' if $f = Ff'$. Thus, with slight but quite admissible inaccuracy, we re-express the question by asking whether f is itself actually a homomorphism. Question: is F one-one on objects? (Of course, this group-theoretical remark barely scratches the surface of significance of underlying functors!)

(iv) Let $\mathcal{Ab}$ be the category of abelian groups and let $\mathcal{T}$ be the category of abelian groups all of whose elements are of finite order. Then $\mathcal{T}$ is a full subcategory of $\mathcal{Ab}$ in the following sense. As a subcategory, it gives rise to an embedding functor $E: \mathcal{T} \to \mathcal{Ab}$, which is one-one on objects and maps; also, E induces a surjection from $M_{\mathcal{T}}(A, B)$ to $M_{\mathcal{Ab}}(EA, EB)$. This is the characteristic property of a full subcategory. There is also an important functor $T: \mathcal{Ab} \to \mathcal{T}$, which associates with every abelian group its *torsion subgroup* (subgroup of elements of finite order) and with every homomorphism $\phi: A \to B$ in $\mathcal{Ab}$ the restriction $T\phi: TA \to TB$. Here the crucial observation is that any homomorphism $\phi: A \to B$ maps the torsion subgroup of A to the torsion subgroup of B. Plainly T *retracts* $\mathcal{Ab}$ onto $\mathcal{T}$, that is, the *composite functor* TE (definition obvious) is the identity functor from $\mathcal{T}$ to $\mathcal{T}$.

$\mathcal{A}b_{fg}$, the category of finitely generated abelian groups, is a full subcategory of $\mathcal{A}b$; while $\mathcal{F}ab$, the category of finite abelian groups, is a full subcategory of $\mathcal{A}b_{fg}$ and of $\mathcal{T}$. We therefore have the associated embedding functors. We make two observations about these embeddings: (i) $\mathcal{F}ab$ may be regarded as the *intersection* of $\mathcal{A}b_{fg}$ and $\mathcal{T}$ in $\mathcal{A}b$; and (ii) the retracting functor K retracts $\mathcal{A}b_{fg}$ onto $\mathcal{F}ab$.

(v) Let $\mathcal{G}$ be the category of groups and $\mathcal{A}b$ the category of abelian groups. Then, as in Example 38.3.1, the abelianizing functor *Abel*: $\mathcal{G} \to \mathcal{A}b$ associates with every group G the abelianized group $\bar{G}$, which is the quotient group of G by the commutator subgroup $[G, G]$. As to the effect on maps, a homomorphism $\phi: G \to H$ maps a commutator $g_1^{-1}g_2^{-1}g_1g_2$ in G to the commutator $(\phi g_1)^{-1}(\phi g_2)^{-1}\phi g_1 \phi g_2$ in H, and hence maps $[G, G]$ to $[H, H]$; it follows that ϕ induces $\bar{\phi}: \bar{G} \to \bar{H}$. We therefore set $\bar{\phi} = Abel(\phi)$ and this rule defines a functor *Abel*. This is the precise form of the functor of Example 38.3.1. Of course there is an embedding of $\mathcal{A}b$ as a full subcategory of $\mathcal{G}$, and *Abel* is a retraction with respect to this embedding.

(vi) In the course of talking about groups in Example 38.3.1, we quietly inserted the notion of a free group on a certain set of generators. We claim that this is a functor $Fr: \mathcal{S} \to \mathcal{G}$, the free group functor. For, given any function $f: S \to T$, there is plainly an induced homomorphism of the free group $Fr(S)$ on S to the free group $Fr(T)$ on T; and Fr is clearly functorial. The relation of Fr to the forgetful functor F of Example 38.3.3(iii) is very interesting and important. Given a set S and a group G, there is a natural one-one correspondence between $M_{\mathcal{G}}(Fr(S), G)$ and $M_{\mathcal{S}}(S, FG)$. The reader is strongly advised to examine the validity of this statement, to describe the correspondence explicitly, and even to attempt to give explicit precision to the term 'natural' in this context.

(vii) The following is a slightly more sophisticated example. Let X be an open interval on the real line and let $x_0 \in X$. We refer to (X, x_0) as a *based interval*. By a differentiable map of based intervals $f: X, x_0 \to Y, y_0$, we mean a differentiable function $f: X \to Y$ such that $f(x_0) = y_0$. The differential $d_{x_0}f: \mathbb{R} \to \mathbb{R}$ is then a (linear) function. Thus, defining D on objects by $D(X, x_0) = \mathbb{R}$, and on maps by $Df = d_{x_0}f$, D is a candidate to be a functor from the category of based intervals and differentiable maps to Hom $(\mathbb{R}, \mathbb{R})$ (regarded as a category with only one object, $\mathbb{R}$). To check that D is indeed a functor, we need to recall that the statement

$$D(gf) = (Dg)(Df)$$

is just the celebrated *chain rule* (Chapter 29).

(viii) We come now to an example of a contravariant functor. Let $\mathcal{V}$ be the category of vector spaces over a given field F. With any vector space V we may associate as in Chapter 20 the *dual* vector space V^*, consisting of linear transformations from V to F. This is a contravariant functor from $\mathcal{V}$ to itself. For a linear transformation $\phi: U \to V$ induces a linear transformation $\phi^*: V^* \to U^*$ by the rule:

$$\phi^*(\xi) = \xi \circ \phi, \qquad \xi \in V^*,$$

and plainly 38.3.2(iii)′ is satisfied. Notice that $**: \mathcal{V} \to \mathcal{V}$ is, of course, a covariant functor.

(ix) Two functors, one covariant, the other contravariant, are embedded in the very definition of a category. Thus, given a category $\mathcal{C}$, let Q be a fixed object of $\mathcal{C}$. Then we have a covariant functor to the category $\mathcal{S}$ of sets, given by

$$M_{\mathcal{C}}(Q, \cdot): \mathcal{C} \to \mathcal{S}$$

and a contravariant functor

$$M_{\mathcal{C}}(\cdot, Q): \mathcal{C} \to \mathcal{S};$$

here $M_{\mathcal{C}}(Q, \cdot)(X)$ is, of course, $M_{\mathcal{C}}(Q, X)$ (with the obvious effect on morphisms), and $M_{\mathcal{C}}(\cdot, Q)(X) = M_{\mathcal{C}}(X, Q)$. We can indeed regard $M_{\mathcal{C}}$ as a *functor* of *two variables*. Thus there is an evident notion of the *Cartesian product* $\mathcal{C} \times \mathcal{D}$ of two categories $\mathcal{C}$, $\mathcal{D}$ (supply details!); and $M_{\mathcal{C}}$ is then a functor from $\mathcal{C} \times \mathcal{C}$ to S, contravariant in the first variable and covariant in the second. The reader should supply other examples of functors of several variables.

There is one more basic notion in the theory of categories. This is the notion of a *natural transformation* of functors. Indeed the whole language and apparatus of categories and functors were developed in order to give precision to the intuitive concept of a natural transformation.

Let $F, G: \mathcal{C} \to \mathcal{D}$ be two (covariant) functors. Then a **natural transformation** T, from F to G, is a rule assigning to each object A of $\mathcal{C}$ a morphism T_A in the set $M_{\mathcal{D}}(FA, GA)$, subject to the condition that the diagram

38.3.4
$$\begin{array}{ccc} FA & \xrightarrow{T_A} & GA \\ \downarrow{\scriptstyle Ff} & & \downarrow{\scriptstyle Gf} \\ FB & \xrightarrow{T_B} & GB \end{array}$$

be commutative for every $f: A \to B$ in $\mathcal{C}$.

If each T_A is invertible in $\mathcal{D}$, we say that T is a **natural equivalence.** Plainly, if T is a natural equivalence from F to G, then T^{-1}, given by

$$(T^{-1})_A = T_A^{-1},$$

is a natural equivalence from G to F.

If T is a natural transformation from F to G, and S is a natural transformation from G to H, then a natural transformation ST from F to H is given by

38.3.5 $$(ST)_A = S_A T_A;$$

it is then clear that T, from F to G, is a natural equivalence if and only if there exists S, from G to F, such that $ST = \text{Id}$, $TS = \text{Id}$, where Id stands for the identity transformation of functors.

We regard the concepts of *category*, *functor*, and *natural transformation* as essential to mathematical literacy and by no means the property of specialists in category theory. (In the same way, the concepts of *group* and *homomorphism* are necessary for all who wish to understand and do mathematics, not simply for group-theorists.) Category theory (like group theory) has developed now as an independent mathematical discipline, but we do not discuss such further refinements.

We saw in 38.3.5 how to compose natural transformations. The reader should observe that, effectively, what we did was to structure the family of functors from $\mathcal{C}$ to $\mathcal{D}$, together with their natural transformations, into a category, which we may call the *functor category* and write $\mathcal{D}^{\mathcal{C}}$; there are, we admit, certain sophisticated set-theoretical difficulties about this formulation (which would certainly be eliminated by insisting that $\mathcal{C}$ be a small category), but these need not detain us. In this formulation the natural equivalences are just the units of $\mathcal{D}^{\mathcal{C}}$.

This formulation suggests that a certain procedure from elementary set theory, namely that of 6.2, should work in the context of categories, namely:

38.3.6 PROPOSITION. (i) *A functor* $U: \mathcal{D}_1 \to \mathcal{D}_2$ *induces a functor*

$$U^{\mathcal{C}}: \mathcal{D}_1^{\mathcal{C}} \to \mathcal{D}_2^{\mathcal{C}}.$$

(ii) *A functor* $V: \mathcal{C}_1 \to \mathcal{C}_2$ *induces a functor* $\mathcal{D}^V: \mathcal{D}_2^{\mathcal{C}} \to \mathcal{D}_1^{\mathcal{C}}$.

Proof. (i) Let $F: \mathcal{C} \to \mathcal{D}_1$ be a (covariant) functor. Then $UF: \mathcal{C} \to \mathcal{D}_2$ is a functor and we set $U^{\mathcal{C}}(F) = UF$. This defines $U^{\mathcal{C}}$ on the objects of $\mathcal{D}_1^{\mathcal{C}}$. Now let T, from F to G (write $T: F \to G$) be a natural transformation. Define the natural transformation $U^{\mathcal{C}}(T): UF \to UG$ by the rule

$$U^{\mathcal{C}}(T)_A = U(T_A): UF(A) \to UG(A), \qquad A \in \mathcal{C}.$$

We check that $U^{\mathcal{C}}(T)$ is a natural transformation. To this end, let $f: A \to B$ in $\mathcal{C}$. Then we apply the functor U to 38.3.4. Since U is a functor, the commutativity of the square is preserved, so that $U^{\mathcal{C}}(T)$ is indeed a natural transformation.

Next we check that $U^{\mathcal{C}}$ is a functor. Plainly if $T: F \to F$ is the identity natural equivalence (i.e., $T_A: FA \to FA$ is just 1_{FA}), then $U(T_A) = 1_{UFA}$, all

A, so that $U^{\mathcal{C}}(T): UF \to UF$ is the identity natural equivalence. Secondly, if $T: F \to G$ and $S: G \to H$, then

$$\begin{aligned} U^{\mathcal{C}}(ST)_A &= U((ST)_A), && \text{by definition of } U^{\mathcal{C}},\\ &= U(S_A T_A), && \text{by 38.3.5}\\ &= U(S_A)U(T_A), && \text{since } U \text{ is a functor,}\\ &= U^{\mathcal{C}}(S)_A U^{\mathcal{C}}(T)_A, && \text{by definition of } U^{\mathcal{C}},\\ &= (U^{\mathcal{C}}(S)U^{\mathcal{C}}(T))_A, && \text{by 38.3.5.} \end{aligned}$$

This completes the proof of (i).

We leave the proof of (ii) to the reader (who must therefore also supply the explicit definition of $\mathfrak{D}^V$). ■ We also leave to the reader the formulation of the dependence of $U^{\mathcal{C}}$ on U and of $\mathfrak{D}^V$ on V, along the lines of the discussion in 6.2.

As a final charge to the reader we ask him to look at the notion of natural transformation where *contravariant* functors are involved. We ourselves turn to examples.

38.3.7 EXAMPLES. (i) Consider the (covariant) functor $**: \mathcal{V} \to \mathcal{V}$ of 38.3.3(viii). We define a natural transformation T: Id $\to **$ by the rule (for $T_V: V \to V^{**}$)

$$T_V(v)(\phi) = \phi(v), \qquad V \in \mathcal{V}, \quad v \in V, \quad \phi \in V^*.$$

It is easy to verify that T is a natural transformation. It is also not difficult to see that T_V is always a monomorphism of V in V^{**}. However, if V is not finitely generated, then T_V is not onto V^{**}, whereas if V is finitely generated, then T_V is indeed onto V^{**} (this follows from the fact that V and V^*, and hence V^{**}, then have the same dimension). Thus if we restrict attention to the full subcategory $\mathcal{V}_{fg}$ of $\mathcal{V}$, whose objects are the finite-dimensional vector spaces over F, then $**$ may be regarded as a functor from $\mathcal{V}_{fg}$ to itself and T is then a natural equivalence.

(ii) We introduced in 38.3.3(iv) the torsion subgroup functor $T: \mathcal{Ab} \to \mathcal{T}$ and the embedding functor $E: \mathcal{T} \to \mathcal{Ab}$, and remarked that TE is the identity functor on $\mathcal{T}$. On the other hand there is an evident natural transformation S, from ET to the identity functor on $\mathcal{Ab}$; namely, if A is any abelian group and $ET(A)$ is its torsion subgroup, then

$$S_A: ET(A) \to A$$

is just the embedding of the torsion subgroup of A in A.

(iii) We take up again the relation between the forgetful functor $F: \mathcal{G} \to \mathcal{S}$ of 38.3.3(iii) and the free group functor $Fr: \mathcal{S} \to \mathcal{G}$ of 38.3.3(vi). Consider first the composite functor $F \circ Fr: \mathcal{S} \to \mathcal{S}$. This functor associates with each set S the set of elements in the free group generated by S. Since the elements of S certainly figure among the elements of $F(Fr\ S)$, there is a natural embedding of S into $F(Fr\ S)$ which yields a natural transformation

$$T: \mathrm{Id} \to F \circ Fr.$$

Now consider the composite functor $Fr \circ F: \mathcal{G} \to \mathcal{G}$. This functor associates with a group G the free group generated by the set of elements of G. For the sake of clarity let us put a bar over an element of G when it is to be regarded as a free generator of $Fr\ (FG)$. Now if S is a set and H is a group, then any function from S to H (more strictly, from S to FH) has a unique extension to a homomorphism from $Fr\ S$ to H (see 38.3.3(vi)). But we *have* a particular function from FG to G, which maps $\bar{g}$ to g! This function gives rise to a homomorphism $Fr\ (FG) \to G$ and hence to a natural transformation

$$U: Fr \circ F \to \mathrm{Id}.$$

The natural transformations T and U are intimately connected. We invite the reader to verify that

$$FG \xrightarrow{T_{FG}} F\ Fr\ FG \xrightarrow{F(U_G)} FG, \qquad F(U_G) \circ T_{FG} = 1_{FG},$$

and

$$Fr\ S \xrightarrow{Fr\,(T_S)} Fr\ F\ Fr\ S \xrightarrow{U_{Fr\,S}} Fr\ S, \qquad U_{Fr\,S} \circ Fr\,(T_S) = 1_{Fr\,S}.$$

(iv) Consider the functor $Abel: \mathcal{G} \to \mathcal{A}b$ of Example 31.3.3(v), together with the embedding functor $E: \mathcal{A}b \to \mathcal{G}$. Then $(Abel)E = \mathrm{Id}: \mathcal{A}b \to \mathcal{A}b$. We consider $E(Abel): \mathcal{G} \to \mathcal{G}$. The natural projection $G \to \bar{G} = G/[G, G]$ then determines a natural transformation from the identity functor to $E(Abel)$.

Finally we remark that all we have said carries over to categories with zero morphisms; and our examples can be adapted to that context. If $\mathcal{C}$ and $\mathcal{D}$ are categories with zero morphisms, then it is natural to demand of a (covariant) functor $F: \mathcal{C} \to \mathcal{D}$ that

$$F(0_{AB}) = 0_{FA,FB}.$$

Exercise 2

(i) Adapt Example 38.3.7(iii) to the context of categories with zero morphisms. {Define $\mathcal{S}_0$ to be the category whose objects are pairs (X, x_0), where X is a set and x_0 is a distinguished point of X, and whose morphisms $X, x_0 \to Y, y_0$ are functions from X to Y mapping x_0 to y_0.}

(ii) Show that *Comm* = commutator-subgroup is a functor from the category of groups to itself, and exhibit a natural transformation from *Comm* to the identity. Define the **lower central series** $G_0, G_1, G_2, \ldots,$ of a group G by the rule

$$G_0 = G, \qquad G_n = [G, G_{n-1}], \qquad n \geqslant 1,$$

where, if A, B are subsets of G, $[A, B]$ is the subgroup generated by commutators $a^{-1}b^{-1}ab$, $a \in A$, $b \in B$. Show that $G_n \subseteq G_{n-1}$, and G_n is normal in G. Exhibit a natural transformation from the *functor* G_m to the functor G_n, $m \geqslant n$ (having shown that G_n *is* a functor!).

(iii) Let $\mathcal{S}$ be the category of sets in a given universe of discourse. Exhibit union and intersection as functors $\mathcal{S} \times \mathcal{S} \to \mathcal{S}$.
(iv) Describe the category of sets and relations and the embedding functor in the category of sets and functions.

38.4 Standard Notions in the Theory of Categories

In this final section we discuss a few examples of one of the most fundamental processes in mathematics, namely that of generalization, as it relates to the theory of categories. We may express the question this way: suppose we have a notion which plays a fundamental role in set theory, that is, in the category $\mathcal{S}$—can we define a notion which is meaningful for any category $\mathcal{C}$ and specializes to the given notion when we take $\mathcal{C} = \mathcal{S}$? Of course, certain such notions (e.g., domain, range) are already present in $\mathcal{C}$ through the very definition of a category. For others it may or may not prove possible to provide the requisite generalization; and even where it does prove possible there is not necessarily a *unique* generalization. Indeed we would obtain a *theorem* of set theory by demonstrating that two notions of category theory which do not coincide for all categories do in fact coincide for $\mathcal{S}$. We also remark that what we have said for the category $\mathcal{S}$ holds also for other 'concrete' categories (e.g., for the category $\mathcal{G}$ of groups).

A category really consists essentially of its morphisms and their composition law, since each object A may, theoretically, be replaced by the identity morphism 1_A. Thus any definition in category theory must involve just morphisms, and experience shows that the canonical method of definition is by reference to *universal mapping properties.* Rather than give a precise definition of such a property, we prefer to let the idea emerge from the examples we give.

We propose first to generalize the fundamental notions of *one-one*, or monomorphic, and *onto*, or epimorphic. Let $\mathcal{C}$ be a category and let $f: A \to B$ be a morphism of $\mathcal{C}$. Then we say that f is **monic** if, for all X and all $u, v: X \to A$,

38.4.1 $$(fu = fv) \Rightarrow (u = v).$$

Similarly we say that f is **epic** if, for all X and all $u, v: B \to X$,

38.4.2 $$(uf = vf) \Rightarrow (u = v).$$

Notice first that these definitions require us, at least in principle, to test f by reference to the entire content of the category. Thus 'monicity' and 'epicity' are 'universal' properties. It is thus surprising to find, as we shall below, that in the category of sets, 'monic' means 'one-one' and 'epic' means 'onto'. For to find out whether a *function* f is one-one, or onto, it is blatantly superfluous to take into consideration any function but f itself. On the other hand, the fact that 'monic' means 'one-one' in the category of sets must be due to certain features of the objects and morphisms of the category $\mathcal{S}$.

38.4.3 THEOREM. *In the category $\mathcal{S}$ of sets, 'monic' means 'one-one' and 'epic' means 'onto'.*

Proof. Let $f: A \to B$ be a function from the set A to the set B. Then if f is one-one it is monic. For if $fu = fv: X \to B$, then $\forall x \in X$, $fu(x) = fv(x)$, so that, f being one-one, $\forall x \in X$, $u(x) = v(x)$; thus $u = v$. Conversely, let f be monic, and let $f(a_1) = f(a_2)$; we want to prove that $a_1 = a_2$. To this end, let X be the singleton set $\{x\}$ and let $u, v: X \to A$ be given by $u(x) = a_1$, $v(x) = a_2$. Then $fu = fv: X \to B$ whence, f being monic, $u = v: X \to A$, or $a_1 = a_2$.

Now let f be onto B; we show f epic. For if $uf = vf: A \to X$, then $\forall a \in A$, $uf(a) = vf(a)$. But every $b \in B$ is expressible as $f(a)$, so that $\forall b \in B$, $u(b) = v(b)$; thus $u = v$. Conversely, let f be epic and let $b_0 \in B$; we wish to show that $b_0 \in fA$. Suppose not; let X be a doubleton $\{x_0, x_1\}$ and define $u, v: B \to X$ by $uB = x_1$ and $vb_0 = x_0$, $vb = x_1$, $b \neq b_0$. Then if $b_0 \notin fA$, $uf = vf$, whence $u = v$. This is a contradiction, so that $b_0 \in fA$ and f is onto. ■

We have shown that 'monic' and 'epic' are indeed generalizations of 'one-one' and 'onto'. These are certainly not the only possible generalizations. For example we might have demanded that a morphism f in the category $\mathcal{C}$ have a left (right†) inverse. It is easy to prove that the function f has a left (right) inverse if and only if it is one-one (onto); it is also easy to see that if the morphism f in $\mathcal{C}$ has a left (right) inverse, then it is monic (epic). Thus this definition, though providing a valid generalization of one-one and onto, leads to a more restricted class of morphisms and is on those grounds less desirable. It would also have the effect of *not* yielding monomorphisms and epimorphisms in the category of groups, since not every group-monomorphism (epimorphism) has a left (right) inverse. [Why?]

We know that in the category of sets a function is invertible if and only if it is one-one and onto. It turns out that this is also true in the category of groups. But in general it is not true that a morphism in $\mathcal{C}$ is invertible if and only if it is monic and epic. To show this the following example will suffice—at least for those familiar with certain basic notions in topology. It turns out that, in the category of topological spaces and continuous maps, 'monic' means 'one-one' and 'epic' means 'onto'. However, a continuous map may be one-one and onto without being invertible (consider, for example, the map which wraps the half-open interval $\langle 0, 1\rangle$ round the unit circle (see 25.3.1)). The reader should prove as an exercise that a morphism in $\mathcal{C}$ is invertible if and only if it is monic and has a right inverse (why is this criterion unsymmetrical?).

† For brevity the parentheses are used, so that all bracketed adjectives should be read together.

We now take up, by way of illustration, a second fundamental concept drawn from set theory, that of the Cartesian product of two (or more) sets. This is defined as a set of ordered pairs of elements and thus does not lend itself to immediate generalization. Thus in order to generalize the notion we must seek a characteristic property of the Cartesian product which is expressed entirely in terms of the morphisms of the category of sets, that is, of functions. In the spirit of 3 Exercise 1(viii), we enunciate the following result.

38.4.4 PROPOSITION. *Let A_1, A_2 be sets. Then if p_i: $A_1 \times A_2 \to A_i$ $(i = 1, 2)$ is the projection, the triple $(A_1 \times A_2; p_1, p_2)$ has the following property: given any set X and any functions f_i: $X \to A_i$ $(i = 1, 2)$, there is a unique function f: $X \to A_1 \times A_2$ such that $p_i f = f_i$, $i = 1, 2$.*

Of course, f is given by $f(x) = (f_1(x), f_2(x))$. ■ There is no doubt that $A_1 \times A_2$ enjoys the property enunciated, but we have not shown that this property is characteristic of the Cartesian product. However, this demonstration is superfluous, since we will show below that the property is characteristic in any category. The important thing for us is that the property is expressed entirely in category-theoretical language. Thus we are led to the following definition.

38.4.5 DEFINITION. Let A_1, A_2 be two objects in a category $\mathcal{C}$. Then a **product** of A_1 and A_2 in $\mathcal{C}$ (if it exists) is a triple $(A; p_1, p_2)$ consisting of an object A of $\mathcal{C}$ and two morphisms p_i: $A \to A_i$, $(i = 1, 2)$, with the property: given any object X and any morphisms f_i: $X \to A_i$, $(i = 1, 2)$, there is a unique morphism f: $X \to A$ such that $p_i f = f_i$, $(i = 1, 2)$.

We now enunciate the uniqueness of the product.

38.4.6 THEOREM. *Suppose $(A; p_1, p_2)$ and $(A'; p_1', p_2')$ are both products of A_1 and A_2. Then there is a unique unit u: $A \to A'$ such that $p_i' u = p_i$, $(i = 1, 2)$. Conversely, if $(A; p_1, p_2)$ is a product, and u: $A \to A'$ is a unit, then*

$$(A'; p_1 u^{-1}, p_2 u^{-1})$$

is also a product.

Proof. Since $(A; p_1, p_2)$ is a product and p_i': $A' \to A_i$ are morphisms, there is a unique morphism u': $A' \to A$ such that

$$p_i u' = p_i' \qquad (i = 1, 2).$$

Similarly there is a unique morphism u: $A \to A'$ such that

$$p_i' u = p_i, \qquad (i = 1, 2).$$

Then $p_i u' u = p_i$, $p_i' u u' = p_i'$, $(i = 1, 2)$. Now the uniqueness part of the definition asserts that f is completely determined by $p_1 f$ and $p_2 f$. Since $p_1(u'u) = p_1 1$ and $p_2(u'u) = p_2 1$, it follows that $u'u = 1$; similarly $uu' = 1$, so that u is a unit.

The converse is evident: if f is the unique morphism $X \to A$ such that $p_i f = f_i$, $(i = 1, 2)$, then $uf: X \to A'$ is the unique morphism such that $p_i u^{-1}(uf) = f_i$, $(i = 1, 2)$. ∎

The perspicacious reader may object at this point that we have not proved the uniqueness of the product—all we have proved is that any two products of A_1 and A_2 are equivalent. First, we reply that we have proved more than that, since not only are any two products equivalent but they are equivalent under a uniquely determined equivalence. Second, definitions by universal mapping properties do inevitably lead to this perhaps somewhat broad notion of uniqueness—if two objects are genuinely equivalent we must expect they can be substituted for each other. There is a close analogy here with certain formulations used in the development of the number concept. For example, when we extend from the integers to the rational numbers we say that the integer n can be identified with the rational number $n/1$—is, then, n the same as or different from $n/1$? Again, complex numbers may be defined by means of pairs of real numbers or by means of polynomial cosets—do we get the same complex numbers in the two cases? We are used to regarding such a question as a little pointless, because we know the mathematics is not affected by our 'choice' of complex numbers. So it is with the product in a category—we get the same properties whatever choice of product we make. With this explanation we will feel free to talk of *the* product of A_1 and A_2.

Of course, the product may not exist. In the category of cyclic groups, or in the category of sets with no more than 10 elements, there are not always products. Thus $\mathbb{Z}_6$ is the product of $\mathbb{Z}_2$ and $\mathbb{Z}_3$, but $\mathbb{Z}_2$ and $\mathbb{Z}_4$ have no product in the category of cyclic groups. The reader should prove these statements for himself, noting, however, that the fact that the direct product $\mathbb{Z}_2 \times \mathbb{Z}_4$ is not cyclic is *not* the same fact as the fact that $\mathbb{Z}_2$ and $\mathbb{Z}_4$ do not have a product $\mathbb{Z}_n$ (for some n) in the category of cyclic groups. (Why not?)

Having defined the product of two objects in a category $\mathcal{C}$, we have no difficulty in extending the definition to any number of objects (not necessarily finite). Thus, given an indexed family of objects A_i, their product (if it exists) consists of an object A together with morphisms $p_i: A \to A_i$ having the following property: given an object X and morphisms $f_i: X \to A_i$, there exists a unique morphism $f: X \to A$ such that $p_i f = f_i$, all i. We call the morphisms f_i the **components** of f and we may write

$$A = \prod A_i, \qquad f = \{f_i\};$$

if the indexing set is finite, say $i = 1, 2, \ldots, n$, then we may write

$$A = A_1 \times A_2 \times \cdots \times A_n, \qquad f = \{f_1, f_2, \ldots, f_n\}.$$

38.4.7 THEOREM. *If any two objects of $\mathcal{C}$ have a product, then any finite number of objects of $\mathcal{C}$ have a product.*

Proof. We argue by induction on n, the number of objects whose product we wish to form. Then products exist by hypothesis if $n = 2$, so we assume that collections of $(n-1)$ objects have products and prove that sets of n objects have a product.

Thus, let $A_1, A_2, \ldots, A_n$ be objects of $\mathcal{C}$ and let $(A'; p', \ldots, p'_{n-1})$ be the product of $A_1, \ldots, A_{n-1}$. Now let $(A; p''_1, p''_2)$ be the product of A' and A_n. Then we assert that $(A; p_1, p_2, \ldots, p_n)$ is the product of $A_1, A_2, \ldots, A_n$, where

$$p_i = p'_i p''_1, \qquad 1 \leqslant i \leqslant n-1,$$
$$p_n = p''_2.$$

We leave the proof of this assertion to the reader. ■

The reader should note (by examples) that it may well happen that any two objects of a category have a product, but some enumerable collection of objects in the category may lack a product.

Let us consider in particular the case of three objects. The proof of Theorem 38.4.7 shows that (with suitable projections to A_1, A_2, A_3) both $A_1 \times (A_2 \times A_3)$ and $(A_1 \times A_2) \times A_3$ are products of A_1, A_2, A_3; thus there exists a unique equivalence†

38.4.8 $$a: A_1 \times (A_2 \times A_3) \rightarrow (A_1 \times A_2) \times A_3$$

transforming the projections. The equivalence 38.4.8 expresses the *associativity* of the product construction. One may ask about the commutativity of the product—is there a canonical equivalence between $A_2 \times A_1$ and $A_1 \times A_2$? The answer is, trivially, affirmative; for either is just a way of *writing* the product of the pair of objects A_1, A_2. Some subtlety only enters into the discussion when we consider the product of two copies of the *same* object A. There is then a 'twisting' equivalence

$$t: A \times A \rightarrow A \times A$$

given by $t = \{p_2, p_1\}$; in the category of sets, t is given by $t(a_1, a_2) = (a_2, a_1)$.

Great importance attaches to *product-preserving* functors. Let $\mathcal{C}$ be a category in which products exist, and let $F: \mathcal{C} \rightarrow \mathcal{D}$ be a functor. Then we say that F is **product-preserving** if, for any indexed family of objects A_i, with product $(A; p_i)$, the collection $(FA; Fp_i)$ is the product of the family FA_i. We might say that F **preserves finite products** if the above condition is satisfied whenever the indexing set is finite. In the exercises below we study the question of whether functors are product-preserving.

We take up two further questions which naturally suggest themselves, when we come to consider products and exploit our experience of the Cartesian product of sets. First, are the projections always epic? It turns out to be possible to construct crazy examples in which the projections are not epic. However we easily prove the following.

38.4.9 THEOREM. *Let $(A; p_i)$ be the product of the indexed family of objects A_i in the category $\mathcal{C}$ with zero morphisms. Then each p_i is epic.*

† Theorem 38.4.6 obviously extends to products of any number of objects.

Proof. Our proof will, in fact, prove much more. Let $u_i: A_i \to A$ be the morphism given by

$$(u_i)_j = \delta_{ij}, \qquad \text{the Kronecker delta;}$$

that is to say, the *i*th component of u_i is $1: A_i \to A_i$ and, if $j \neq i$, the *j*th component of u_i is $0: A_i \to A_j$. Then, by very definition of u_i, we have

$$p_i u_i = 1 \qquad (\text{and } p_j u_i = 0, \quad j \neq i).$$

Thus p_i has a right inverse and so is certainly epic. In fact it is a projection onto a *retract*; that is, if u_i is used to embed A_i in A, then p_i projects (or **retracts**) A back onto A_i, acting as the identity on the sub-object A_i. ∎
The reader will note that this proof only requires that the morphism sets $M^{\mathcal{C}}(A_i, A_j)$ be non-empty. Thus to construct a counterexample to the epicity of p_i, one must suppose (at least) that $A = A_1 \times A_2$ and $M^{\mathcal{C}}(A_1, A_2)$ is empty.

Next, consider the functors $M^{\mathcal{C}}(\cdot, A)$ and $M^{\mathcal{C}}(\cdot, A_1) \times M^{\mathcal{C}}(\cdot, A_2)$, where $A = A_1 \times A_2$. There is a natural equivalence between these two functors, given by the transformation

$$\{f_1, f_2\} \to (f_1, f_2).$$

Conversely, suppose that, given A_1 and A_2, there is an object A and a natural equivalence η (of contravariant functors from $\mathcal{C}$ to $\mathcal{S}$)

$$\eta: M^{\mathcal{C}}(\cdot, A) \to M^{\mathcal{C}}(\cdot, A_1) \times M^{\mathcal{C}}(\cdot, A_2).$$

Let $\eta(1_A) = (p_1, p_2)$, so that $p_i: A \to A_i$ $(i = 1, 2)$. Then we assert that *$(A; p_1, p_2)$ is the product of A_1 and A_2.* For, given $f_i: X \to A_i$ $(i = 1, 2)$, let $f: X \to A$ be given by $\eta(f) = (f_1, f_2)$. Then, by naturality,

$$(f_1, f_2) = \eta(f) = \eta(1 \circ f) = (p_1 f, p_2 f),$$

so that $f_i = p_i f$ $(i = 1, 2)$; and, if $f_i = p_i f'$ $(i = 1, 2)$, then, reversing the steps in the line above,

$$\eta(f) = \eta(f'),$$

whence, η being an equivalence, $f = f'$.

There is, of course, much more to be said about products (for a very few such things, see the Exercises at the end of this section). However, in order not to lengthen this chapter unduly, we will leave products here and pass to one more typical example of a generalization.

Here, however, we will generalize not from the category of sets but from the category of groups. A basic idea in group theory is that of the kernel of a homomorphism. Thus if $\phi: G \to H$ is a homomorphism, the kernel of ϕ is the subgroup of G consisting of those elements of G which are mapped by ϕ to the unity element of H. At first sight this notion looks unpromising for generalization to general categories, since it depends so heavily on a discussion of the elements in a group. Thus, in order to generalize, we must characterize the kernel entirely by means of the morphisms of the category, that is, of group-homomorphisms. Since $\mathcal{G}$ is, of course, a category with zero morphisms, it is reasonable that our generalization should be to arbitrary categories with zero morphisms. The vital clue to the generalization is provided by the following theorem.

38.4.10 THEOREM. *Let $\phi: G \to H$ be a group-homomorphism and let $\kappa: K \to G$ be a monomorphism such that*† $\phi\kappa = 0: K \to H$, *and with the following property: given any group X and any homomorphism $\xi: X \to G$ such that $\phi\xi = 0: X \to H$, there exists a homomorphism $\eta: X \to K$ with $\kappa\eta = \xi: X \to G$. Then κ maps K isomorphically onto the kernel of ϕ.*

Proof. The various homomorphisms are displayed in Fig. 38.1. Since $\phi\kappa = 0$, κ certainly maps K into the kernel of ϕ. To show that κ is onto the kernel of ϕ, let g belong to the kernel of ϕ; thus $g \in G$ and $\phi(g) = e$. Choose X to be cyclic infinite, generated by x, and let $\xi: X \to G$ be the unique homomorphism given by $\xi(x) = g$. Then $\phi\xi = 0$ since $\phi\xi(x) = e$. Thus, by the given property of κ, there exists $\eta: X \to K$ with $\kappa\eta = \xi$. Thus $\kappa(\eta(x)) = g$, so that κ is onto the kernel of ϕ and the theorem is proved. ■ We note that if K_0 is the kernel of G and $\kappa_0: K_0 \to G$, then, of course, $\phi\kappa_0 = 0$, and moreover κ_0 does have the property ascribed to κ in the theorem; for such a ξ (with $\phi\xi = 0$) has its image contained in K_0 and so η is just ξ itself, except that the range is now to be regarded as K_0. We are therefore led to the following definition.

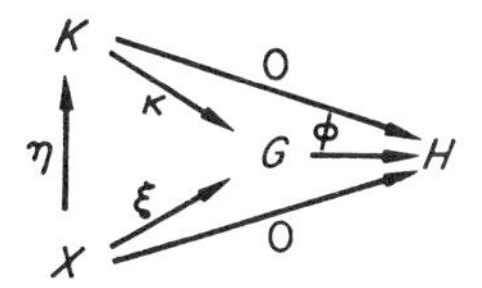

Fig. 38.1

38.4.11 DEFINITION. Let $f: A \to B$ be a morphism in the category $\mathcal{C}$ with zero morphisms. Then the pair $(K; \kappa)$, where $\kappa: K \rightarrowtail A$ is monic, is called a **kernel** of f if $f\kappa = 0$ and also κ has the following property: given any object X and any morphism $\xi: X \to A$ such that $f\xi = 0$, there exists a morphism $\eta: X \to K$ with $\kappa\eta = \xi$.

One readily proves the uniqueness of $(K; \kappa)$:

38.4.12 PROPOSITION. *Let $(K_1; \kappa_1)$ and $(K_2; \kappa_2)$ both be kernels of f. Then there exists a unique unit $\omega: K_1 \to K_2$ such that $\kappa_2\omega = \kappa_1$. Moreover, if $(K; \kappa)$ is a kernel of f and $\omega: K' \to K$ is a unit, then $(K'; \kappa\omega)$ is also a kernel.* ■

† Recall that, if A and B are groups, then $0: A \to B$ is just the constant homomorphism which sends every element of A to the unity element e of B.

Thus, just as for products, we may talk of *the* kernel. Notice that, in an arbitrary category with zero morphisms, the kernel is really a (monic) *morphism* and not an *object*. If we wish to speak of just K, we may refer to the **kernel-object**. Notice also that we do not insist that every morphism in a category have a kernel.

We draw attention to some properties of the kernel, generalizing familiar or evident properties of kernels in the category of groups.

38.4.13 PROPOSITION. *If μ is monic, then f and μf have the same kernel.* ■

38.4.14 PROPOSITION. *Let $\mathcal{C}$ be a category with zero morphisms and products. If $f_i: A_i \to B_i$ has kernel $\kappa_i: K_i \to A_i$, then $f_1 \times f_2: A_1 \times A_2 \to B_1 \times B_2$ has kernel $\kappa_1 \times \kappa_2: K_1 \times K_2 \to A_1 \times A_2$.*

Proof. First, $\kappa_1 \times \kappa_2$ is monic. For let $\xi^i = \{\xi_1^i, \xi_2^i\}: X \to K_1 \times K_2$ $(i = 1, 2)$ with $(\kappa_1 \times \kappa_2)\xi^1 = (\kappa_1 \times \kappa_2)\xi^2$. But

$$(\kappa_1 \times \kappa_2)\xi^i = (\kappa_1 \times \kappa_2)\{\xi_1^i, \xi_2^i\} = \{\kappa_1\xi_1^i, \kappa_2\xi_2^i\}.$$

(Prove this last equality!) Thus

$$\{\kappa_1\xi_1^1, \kappa_2\xi_2^1\} = \{\kappa_1\xi_1^2, \kappa_2\xi_2^2\}, \quad \text{or}$$
$$\kappa_1\xi_1^1 = \kappa_1\xi_1^2,$$
$$\kappa_2\xi_2^1 = \kappa_2\xi_2^2.$$

Now κ_1, κ_2 are monic, so $\xi_1^1 = \xi_1^2$, $\xi_2^1 = \xi_2^2$. Thus $\xi^1 = \xi^2$, so $\kappa_1 \times \kappa_2$ is monic. Next $(f_1 \times f_2)(\kappa_1 \times \kappa_2) = f_1\kappa_1 \times f_2\kappa_2 = 0 \times 0 = 0$. Finally, suppose $\xi = \{\xi_1, \xi_2\}: X \to A_1 \times A_2$, and $(f_1 \times f_2)\xi = 0$. Then $\{f_1\xi_1, f_2\xi_2\} = 0$, so $f_1\xi_1 = 0$, $f_2\xi_2 = 0$, $\xi_1 = \kappa_1\eta_1$, $\xi_2 = \kappa_2\eta_2$, and

$$\xi = \{\kappa_1\eta_1, \kappa_2\eta_2\} = (\kappa_1 \times \kappa_2)\{\eta_1, \eta_2\}. \quad ■$$

Exercise 3

(i) Identify the monics and epics in the following categories:

(a) based sets and based functions;
(b) groups and homomorphisms;
(c) fields and homomorphisms;
(d) the category with two objects A and B, and three morphisms, 1_A, 1_B and $f: A \to B$.

What are the units in these categories?
(ii) Construct a subcategory of the category of sets, which possesses a monic which is *not* a one-one function.
(iii) Show that, in the categories $\mathcal{S}$, $\mathcal{G}$, and $\mathcal{A}b$, every morphism can be expressed as an epic followed by a monic. Are such expressions unique? Can every morphism be expressed as a monic followed by an epic? {Cf. 4 Exercise 10(i).}
(iv) Show that if $\mathcal{C}$, $\mathcal{C}'$ are categories with products, so are $\mathcal{C} \times \mathcal{C}'$, $\mathcal{C}^{\mathcal{D}}$.

(v) Show how to embed $\mathcal{C}$ in the category of contravariant functors from $\mathcal{C}$ to $\mathcal{S}$, which we write $\mathcal{S}^{\mathcal{C}^*}$. If $\mathcal{C}$ admits products, does this embedding preserve products?
(vi) Let $F: \mathcal{G} \to \mathcal{G}$ be the functor which associates with every group its commutator-subgroup. Is F product-preserving?
(vii) Let $\mathcal{G}^{\text{epic}}$ be the category of groups and 'onto' homomorphisms. Does $\mathcal{G}^{\text{epic}}$ admit products? Show that the *centre* of a group yields a functor $\mathcal{G}^{\text{epic}} \to \mathcal{G}$. Is the centre-functor $\mathcal{G}^{\text{epic}} \to \mathcal{G}$ product-preserving?
(viii) Express u_i (see Theorem 38.4.9) as a natural transformation.
(ix) Identify the kernel-functor in the categories of Exercise (i) above.
(x) In the notation of Theorem 38.4.9, show that $u_2: A_2 \to A_1 \times A_2$ is the kernel of $p_1: A_1 \times A_2 \to A_1$.
(xi) Given a category $\mathcal{C}$, construct from it a new category $\mathcal{C}^*$ (called the **dual** of $\mathcal{C}$) as follows. The objects of $\mathcal{C}^*$ are just those of $\mathcal{C}$, but for every pair A, B of objects, we define $M_{\mathcal{C}^*}(A, B) = M_{\mathcal{C}}(B, A)$. Show that a morphism f in $\mathcal{C}$ is monic if f is epic in $\mathcal{C}^*$. Deduce *from this* that if g is an epic morphism in $\mathcal{C}$, then g is monic in $\mathcal{C}^*$. {Use the fact that $\mathcal{C} = (\mathcal{C}^*)^*$.}
(xii) Let $T: \mathcal{C} \to \mathcal{D}$ be a covariant functor. Show that it may be regarded as a contravariant functor: $\mathcal{C} \to \mathcal{D}^*$, a contravariant functor $\mathcal{C}^* \to \mathcal{D}$, and as a covariant functor $\mathcal{C}^* \to \mathcal{D}^*$. [The formal notion of duality here should be compared with duality in projective geometry (Chapter 17). Despite its simplicity, it has proved a useful concept in category theory.]
(xiii) Generalize the notion of the *disjoint union* of sets (along the lines of the generalization of Cartesian product) to yield a notion meaningful in any category. In fact this notion is called the *coproduct*, since the coproduct in $\mathcal{C}$ is just the product in $\mathcal{C}^*$. Identify the *coproduct* in (a) the category of based sets, (b) the category of abelian groups, (c) the category of groups. {Regard groups as given by generators and relations, as in 38.3.1.}

For further information about categories, see MacLane–Birkhoff [85], Hilton [63], MacLane [84], Mitchell [89].

Chapter 39

MATHEMATICAL LOGIC

In this final chapter we shall take up several points of logic about which we have been silent in previous chapters. The notions of axiom and proof need clarification, and we need to say more about the notion of set, which was left extremely vague in Part I. We shall discuss these points first at an informal practical level in sections 39.1 and 39.2 and then conclude with a more formal approach, still far from that of a professional logician, but one which we hope may arouse the reader's interest in reading the specialized texts.

39.1 Axioms

The discussion in Chapter 16 on the 'logic of geometry' showed how axioms were historically first distinguished, in Euclidean geometry, as a collection of statements which nobody would deny. The remaining statements of the geometry (variously called 'propositions', 'lemmas', 'theorems') must then be deduced from the axioms, by a process with specified rules. In historical fact, as we remarked, the axioms of Euclid were not formulated with sufficient precision to deduce all the theorems according to the rules, but Hilbert eventually did formulate a suitable system. Thus, in a sense, Hilbert's system contained *all* the information about Euclidean geometry: only time, energy and ingenuity were needed to draw out from his axioms all the remaining 'facts' of the geometry. There was also the big difference that nobody was expected to agree to their truth, since truth was not here in question. The axioms were merely statements of the mutual relationships between the objects of the particular system of geometry, and the axioms do not state what the objects are. The only requirement is the logical consistency of the axioms. It is this compact portmanteau quality of axiom systems which explains the modern preoccupation with them. For example, the set of axioms $\mathfrak{A}_1$–$\mathfrak{A}_9$ (for a commutative ring with unity) tell us all the rules for doing arithmetic in such rings; a similar statement holds for groups. The versatility of these axiom systems is important because, by the method of casting rôles which we have explained before, we have theorems valid in all rings or all groups. Such versatility is not possessed by all sets of axioms; for example, we saw in Chapter 14 that for the scalar product in $\mathbb{R}^3$ the axioms $\mathfrak{S}\mathfrak{P}_1$–$\mathfrak{S}\mathfrak{P}_4$ characterize this product completely.

There is then in mathematics a general philosophy of separating our work into four kinds of activity, once we have decided to consider a set S of axioms.

In practice, of course, the axioms were probably written down with some special system in mind; otherwise they are likely to be barren. (We intend this remark as a serious warning to prospective mathematicians!) The four kinds of activity may be described as follows:

(1) We investigate the properties which the statements of S contain implicitly.

(2) We prove an *existence theorem* to show that some system actually satisfies the axioms. [Both (1) and (2) are inspired by the special system we started from, and which led us to write down the statements of S at the outset.] Of course our interest in the set S largely derives from the richness of the source of examples satisfying the axioms.

(3) We next look for other systems satisfying the axioms, so that we can apply to such systems (as well as to the special one we perhaps started from) the theorems which we have deduced under (1). Thus we look for an *application* of our work by means of suitable interpretations of the axioms of S.

(4) We seek to *classify* all such systems by using a notion of isomorphism to say when we would regard two such systems A and A' satisfying S as being 'the same'. If this classification turns out to be not too narrow, it may also be interesting to consider mappings between such systems, and hence to consider the category formed by them; thus we are looking for the units among the morphisms of this category. The attempt to say when things are 'the same', that is, to provide effective criteria for the isomorphism of two systems, is a vital and fruitful activity, characteristic of modern mathematics and rare in classical mathematics.

We should add that different mathematicians lay greater stress on some of these activities than on others, depending on interest and taste. Applied mathematicians are rarely interested in activity (4) (except that their attempts to say when things are 'the same' become an important discussion about degrees of approximation), but the first three activities are very much their concern; their 'special system' will be some situation in the physical or human sciences.

39.1.1 EXAMPLE. When S is the system $\mathfrak{A}_1$–$\mathfrak{A}_9$, activity (1) of the work is to prove such theorems as $(-1) \times (-1) = +1$. For activity (2) there is the kind of existence theorem which we met in Chapter 23, where we constructed $\mathbb{Z}$ from the system $\mathbb{N}$ of whole numbers; or the work we met in Chapter 10, when we constructed the system of residues of $\mathbb{Z}$. We make applications of these elementary theorems, under activity (3), so often and so implicitly that we are hardly aware of them. Under activity (4) we have such theorems as that of 22.8.2 in which we are able to classify all finite integral domains. Such a theorem is something of a luxury, since we certainly cannot make such a classification of infinite integral domains (for instance).

39.1.2 EXAMPLE. When S is the set of axioms for a vector space, the activity (1) is the general theory of vector spaces, while (2) is the checking that the well-known examples are vector spaces. Activity (3) is the interpretation of the theorems as, for example, the theorems about linear differential equations in Chapter 19. Activity (4) corresponds (at least for finite-dimensional spaces) to the statement of Corollary 19.4.11 that every n-dimensional vector space over a field F is (isomorphic to) the space F^n of all n-tuples from F.

39.1.3 EXAMPLE. In applied mathematics, activity (1) is often already carried out within pure mathematics; but see for example the theory of voting (Steiner [122]), and the axioms of probability theory. Activity (3) has often been very satisfying, as the many applications of potential theory, or of the wave equation, or of the calculus of variations, testify. Many other such examples can be cited

39.1.4 EXAMPLE. If S is the system of axioms for a complete ordered field, then the theory of Chapter 24 shows us that such a system is unique in the sense that there is only one field satisfying the axioms, namely the set of real numbers. This comes under activity (4), so that activity (3) is here redundant. Activity (2) is the construction of the real numbers either by Cantor's process (24.2) or by that of Dedekind (21.3), and the activity (1) is really the subject called 'real variable analysis'.

Other sets of axioms which we have met in the book, such as those for Boolean algebra, or for a category, or for a projective geometry, should have given the reader ample illustration of the activities (1)–(4) recorded above. Observe that the construction of a set with the required structure as in activity (2) may be quite complicated, as, for example, our construction of the real numbers in Chapter 24. It then becomes a positive advantage to be able to work with the axioms directly, so as to be able to forget about the possibly complex nature of the elements; we may then simply work with their mutual inter-relationships as expressed by the axioms. Also, the formulation of the axioms may act as a *specification* of what the system is that one wishes to construct. An outstanding contemporary example is the axiomatic formulation of homology theory in algebraic topology by Eilenberg and Steenrod (see [35]), and subsequent work using the same technique.

The classification problem in activity (4) may be extremely difficult, if not impossible, as we remarked concerning groups in Chapter 18 or topological manifolds in Chapters 25.

The impression is often given in books, and by natural scientists, that pure mathematicians consider just any old system of axioms and thereby have a freedom (perhaps to evade difficulties) which is not allowed to such people as applied mathematicians. This is likely to convey a false impression. Most mathematicians when considering a system of axioms have in mind a particular application for those axioms, and they require the axioms to be appropriate to the application they want to pursue. Once such a system—suggested by a problem in physics, geometry or economics, say—has been investigated, it may be interesting to modify the axioms a little as was done with the parallel postulate (see p. 600) in Euclidean geometry to obtain the systems of non-Euclidean geometry. Here the motivation may be curiosity but again it may be because we have another problem in mind. The point we stress is that the selection of axiom systems is not arbitrary but is done for good reasons. These reasons and their effect on our selection of axioms constitute an essential part of the art of mathematics. In any case, moreover, the system must be logically consistent, and experience shows that it is quite difficult to be both arbitrary *and* logically consistent in the selection of axioms. Occasionally mathematicians have tried to be somewhat arbitrary, but on the whole their systems have led to barren theories because there has been no motivation. We dwell on this point because some enthusiasts for teaching modern mathematics give the impression that a subject like group theory (for example) is concerned only with fiddling with the wording of the axioms to get equivalent ones. This is sometimes good to do as a sideline, for reasons of economy, but it is not a primary aim in practice.

From these considerations of the *practical* side of the axiomatic method, we move to the logical considerations; and first we need to discuss the notion of a 'set'.

39.2 Sets

In the years from 1874 Cantor developed his theory of sets very rapidly, his interest being originally aroused by questions of analysis and number theory. As he developed his theory of the transfinite arithmetic of cardinals and ordinals, the boldness of his approach alarmed some of his contemporaries. Eventually there began to appear some flaws, technically described as paradoxes. Here we use the term 'paradoxes' in the technical sense that they implied contradictions in mathematics. This is in contrast with the usage in ordinary language where a paradox tends to mean 'that which would surprise a person steeped in convention'—for example, statements like '$\aleph_0 + 1 = \aleph_0$', or 'In country X charladies earn more than doctors'. The most easily accessible paradox, in the technical sense, is that discovered by Bertrand Russell at the turn of the century. It runs as follows:

Let us call a class 'extraordinary' if it is a member of itself—for example, the class of all classes which are definable in fewer than 100 words is an extraordinary class. Let us call all other classes 'ordinary', and let C denote the class of all ordinary classes. Now consider the question: Is C extraordinary? If the answer is 'yes', then $C \notin C$ by definition, so C is ordinary and therefore $C \in C$ by definition of C. If the answer is 'no', then C is ordinary, so $C \in C$; that is, C is extraordinary. In either case we have inferred the genuine paradox that

39.2.1 $$C \in C \textit{ and } C \notin C \text{ (or } C \in C \text{ and } \neg(C \in C)),$$

which has the standard contradictory form (p and $\neg p$) of a technical paradox.

Observe how closely the argument resembles that for Cantor's theorem (6.3.3), that $2^a > a$; there we assumed we had a surjective function $\theta: A \to \mathfrak{p}A$, and obtained a set $T \subseteq A$, with $T = \theta(t)$, $t \in A$, such that $t \in T$ and $t \notin T$. Thus we inferred that θ did not exist. This suggests then that the only way out of Russell's paradox (39.2.1) is to *deny that C is properly defined*, thus forcing us to inquire more closely into the definition of sets and classes. The result of such an enquiry should, we hope, allow us to retain Cantor's theorem because it seems to our instinct to be a genuinely good theorem of mathematics. There is also the technical point (which we shall see in Theorem 39.5.8 later) that if mathematics contains a paradox, then all statements become provable, and this is contrary to our feeling about the rightness of the world and the utility of mathematics. (These 'instincts' and 'feelings' are of course not part of written mathematics; they are there, because mathematics is a human activity.)

Again, consider the paradox of Burali-Forti: let W be the set of everything there is, and let a denote its cardinal. Clearly then for any cardinal b, $b \leqslant a$ by definition of $\leqslant$ in Chapter 7. In particular take $b = 2^a$, so $2^a \leqslant a$, contrary to Cantor's theorem itself.

Most mathematicians feel intuitively that the ideas of the proof of Cantor's theorem are good mathematics and somehow different from these strange big classes like C or W. In their definitions, words are used somewhat vaguely—what do we mean by 'everything there is?', 'definable in a finite number of words', etc.? Such usage is reminiscent of well-known classical paradoxes like that concerning the Cretan who says 'all Cretans are liars'. Is he telling the truth or not? Such paradoxes were called *semantic*, because they seemed to depend merely on the use of language and to be remote from mathematics. But just how is such a paradox different from the others? Clearly, the situation was disturbing to mathematicians.

Several ways round the difficulties have been suggested. For example, Russell invented his elaborate 'theory of types' whose complexity led to its neglect. Eventually in the 1930's a system of rules was gradually worked out by a series of logicians including von Neumann and Gödel, and these are stated precisely in Gödel's book [44]. We shall here simply describe the spirit of them, warning the reader that recent work in the theory of categories (as mentioned in Chapter 38) is beginning to show that even this system is not adequate when mathematicians wish to work with categories. It is, however, reasonably adequate for most of mathematics as known at the present time.

First, the rules distinguish between 'sets' and 'classes' in such a way that any reasonable candidate X may be called a class but we are not allowed to call X a set without knowing more about it. If X is admitted as a set, then we say that $X \in \mathbf{M}$ ($\mathbf{M}$ for the German 'Menge'). Thus, for example, the empty class is declared to be a set, i.e., $\varnothing \in \mathbf{M}$; and then if A, B are known to be sets, we have some rules including the following:

Rule 1: $A \cup B$, $A \cap B$, $A \times B$, $\mathfrak{p}A$ are declared to be sets.

Rule 2: A sub-class of a set A is a set.

Rule 3: If A is a set and $\{B_\alpha(\alpha \in A)\}$ is an indexed family of sets, then $\bigcup_{\alpha \in A} B_\alpha$ is a set.

For example, therefore, the finite cardinals $0 = \varnothing$, $1 = \{\varnothing\}$, etc., are sets, by successive applications of rules 1 and 2, because if $A \in \mathbf{M}$, then the family of subsets of A is also in $\mathbf{M}$.

Rule 4: We may write $A \in B$ *only* when A is known to be a set.

The previous deduction of Russell's paradox is now not legitimate, since the rules force us to cast it in the following mould. In order to be extraordinary, a class X must satisfy $X \in X$; therefore $X \in \mathbf{M}$ by rule 4; and hence extraordinary classes must be *sets*. Now form the class C of all ordinary classes. If C were a set, then we could argue as above and obtain $C \in C$ and $C \notin C$; whence we can escape a paradox by concluding that C *is not a set*. This does not of course exclude the possibility that the Russell paradox can still be deduced, perhaps with much greater ingenuity; but neither it nor any other has up to date been (correctly) deduced. Moreover, the above rules are not excessively restrictive when we want to do ordinary mathematics. For example, by scrutinizing the proof of Cantor's theorem carefully in Chapter 6 we see that we can allow the theorem in the form that when A is a *set* there is no surjection: $A \to \mathfrak{p}A$. By contrast, other suggested cures like Russell's theory of types would prohibit a great deal of mathematics. Another workable scheme is that due to Quine, who simplified Russell's theory of types (see Rosser [108]). Whereas this system does not distinguish between sets and classes, it cannot allow for all classes A that A shall be bijective with the class $\{\{a\} | a \in A\}$ of singletons (so 2 Exercise 5(ii) will not always hold in Quine's system).

Exercise 1

(i) Prove that if $A, B \in \mathbf{M}$, then $\mathbf{M}$ contains every relation with domain A and range B. Does B^A lie in $\mathbf{M}$?
(ii) Let $A \in \mathbf{M}$ and let R be an equivalence relation on A. Prove that $A/R \in \mathbf{M}$.
(iii) Assuming that $\mathbb{N} \in \mathbf{M}$, use the theory of Chapters 23, 24 to prove that $\mathbb{R} \in \mathbf{M}$ and $\mathbb{R}^n \in \mathbf{M}$ $(n \in \mathbb{N})$.

The reader, however, may wonder about the status of the theory of categories, his suspicions having been aroused by remarks in Chapter 38 about 'small' categories. Is the category of all groups a set, in the Gödel formulation? Such questions are still the subject of controversy; and meanwhile the theory of categories is being pursued vigorously, in the same way as Cantor pursued his theory of sets without troubling too much about logical problems. This is because the interest lies, for many, in questions which seem safe enough and unfraught with semantic troubles. Meanwhile Lawvere has proposed a new foundation of mathematics independent of set theory; see his contribution to *Proceedings of the Conference on Categorical Algebra*, Springer, Berlin (1966).

39.3 Consistency

By now we have probably raised some doubts in the reader's mind concerning logic. Supposing the known paradoxes have been excluded by suitably adjusting the rules, how do we know other paradoxes cannot arise? Technically speaking, how do we know when a part of mathematics is *consistent*? This is a very practical question since every proof by *reductio ad absurdum* (see 8 Exercise 11(iii)) assumes consistency, such proofs being of the following form: We assume that a proposition P is true and then deduce a contradiction Q and $\neg Q$, for some Q; hence P must have been false, since the assumption is the only weak link in the argument. When discussing consistency, sets of axioms again become important tools, enabling us to trace a fault (in an argument) of which these axioms are at the source. Such a set S of axioms is said to be **consistent** if there is no proposition P such that we can deduce P and $\neg P$ from S (thus S implies no paradox). We therefore allow proofs by *reductio ad absurdum* because we believe that the axioms of mathematics are consistent. This belief is not something we can *prove*, as we shall see below, but it is a belief about the natural world.

The question of consistency first presented itself urgently at the beginning of the 19th century when Bolyai and Lobachevsky investigated non-Euclidean geometry. Recall that one of Euclid's postulates was the so-called **parallel postulate** $\mathfrak{P}$, which says that if l is a line in a plane Π and Q is a point of Π not on l, then there is exactly one line in Π through Q and parallel to l. Geometers from the Greeks onwards felt that this postulate was different from the others, and they made many efforts to prove it as a theorem from the remaining axioms. As an attempt in this direction Bolyai and Lobachevsky (as well as earlier workers) tried to obtain a proof by *reductio ad absurdum*, starting with the assumption that $\mathfrak{P}$ was false. In particular they supposed that at least two lines existed through Q which were parallel to l (technically their geometry was *hyperbolic geometry*). A long chain of deductions was obtained, some of which were strange by the standards of ordinary geometry (these were 'paradoxes' in the sense of ordinary speech); but Bolyai and Lobachevsky obtained no logical contradiction. There was not even a conflict with Nature as far as experimental error could detect. Eventually Beltrami showed that this hyperbolic geometry was consistent with Euclidean geometry as follows.

Consider first the system $\mathfrak{Euc}$ of plane Euclidean geometry itself. Its axioms and rules tell us relationships between the entities of interest—points, lines, distances, and angles—without specifying what these objects are. True, Euclid tried to specify, but as we observed in Chapter 16 his attempt was flawed and also unnecessary for doing his geometry. If now, with Descartes, we use the real numbers $\mathbb{R}$ to make a *model* of this geometry, we have a correspondence between the entities of $\mathfrak{Euc}$ and notions of $\mathbb{R}^2$ as shown by the following list:

$\mathfrak{Euc}$	$\mathbb{R}^2$
point P	ordered pair (x, y)
line l	set L of zeros of linear function
P on l	$(x, y) \in L$
distance (P_1, P_2)	$\sqrt{[(x_1 - x_2)^2 + (y_1 - y_2)^2]}$

... etc. ...

As we know, the axioms of $\mathfrak{Euc}$ hold in $\mathbb{R}^2$, but now they hold as *theorems about* $\mathbb{R}$; for example, $\mathfrak{P}$ is no longer self-evident in $\mathbb{R}^2$ as it might have been in $\mathfrak{Euc}$ to the Greeks, but we can check its validity by using the rules of arithmetic in $\mathbb{R}^2$. Thus, to Q and l respectively correspond (q_1, q_2) and $ax + by + c = 0$ in $\mathbb{R}^2$, and the unique line required has equation

$$a(x - q_1) + b(y - q_1) = 0$$

by the algebraic theory of $\mathbb{R}$. Having thus modelled $\mathfrak{Euc}$ in $\mathbb{R}^2$, we infer that every argument in $\mathfrak{Euc}$ can be decoded into an argument in $\mathbb{R}^2$ using the above list. [The decoded argument may not be as efficient as a direct argument using the techniques of co-ordinate geometry, but that is not the point here.] Hence if $\mathfrak{Euc}$ contains an argument which leads to a logical paradox, that argument can be copied into co-ordinate geometry; and hence a paradox would lie in the system $\mathbb{R}$ of real numbers. Similarly we saw, in the case of the $\mathbb{R}^3$-model of plane projective geometry in Chapter 17, that if the axioms of the projective geometry were inconsistent, then the geometry of $\mathbb{R}^3$ would be inconsistent also.

In fact, every time in this book that we have laid out a set of axioms, (say for groups, categories, ...) we have followed it with one or more 'examples' to show the reader what kind of system the axioms were intended to describe; but each such example may also be regarded as a 'model' for the axiom system. Since the integers $\mathbb{Z}$ form such a model for the group axioms, these axioms are consistent provided the theory of $\mathbb{Z}$ is consistent.

Beltrami worked similarly with hyperbolic geometry. His model for hyperbolic geometry consisted of the points $A, B, \ldots$ of a certain unbounded surface in $\mathbb{R}^3$ which had the technical property of possessing constant negative curvature. But as the model of the line through A and B he took the curve of shortest length (a so-called *geodesic*) joining A and B but lying in the surface. The proof that such a line exists is a theorem of differential geometry; hence it is a problem about differential equations and hence ultimately a theorem about $\mathbb{R}$. Thus, if a contradiction could be deduced from the axioms of hyperbolic geometry the deduction could be decoded into an argument about the real numbers, and hence the system $\mathbb{R}$ would be inconsistent. This demonstrates, then, the *relative consistency* of hyperbolic geometry. It is, so to speak, 'as consistent as' $\mathbb{R}$, and to deny its consistency implies denying that of $\mathbb{R}$.

As to the consistency of the real-number system, we showed in Chapters 23 and 24 how to build $\mathbb{R}$ via the rationals and the integers from the system $\mathbb{Z}^+$ of numbers $0, 1, \ldots, n, \ldots$. The latter system is specified in Peano's axioms. Let us then denote by $\mathfrak{Z}_0$ this system of Peano axioms, together with the rules for manipulating sets which allow us to form the Cartesian products, equivalence classes, infinite sequences, and so on, that we used in Chapters 23 and 24. Then since we have modelled $\mathbb{R}$ within $\mathfrak{Z}_0$ we see that the theory of $\mathbb{R}$ is consistent with that of $\mathbb{Z}$; a contradiction deducible about $\mathbb{R}$ could be copied to give a contradiction within $\mathfrak{Z}_0$. It therefore becomes of prime importance to investigate the consistency of $\mathfrak{Z}_0$, and we discuss this in 39.6 below.

Meanwhile, however, we pause to remark that this technique of demonstrating relative consistency is clearly of wide applicability. Moreover, it can be used to show that a particular axiom A in a system X is **independent**, i.e., A is not deducible from the other axioms in X. This is not so important as consistency, but it does lead to economy. For example, let us denote by Y the set consisting of all the axioms of $\mathfrak{Euc}$ except $\mathfrak{P}$. Then Beltrami's model of hyperbolic geometry is a system S satisfying Y, while S also satisfies $\neg\mathfrak{P}$. Therefore, $\mathfrak{P}$ cannot be deducible from the set Y unless we admit the contradiction $\mathfrak{P}$ & $\neg\mathfrak{P}$ in the theory of $\mathbb{R}^3$—that is, unless $\mathbb{R}$ is not consistent. With this proviso, $\neg\mathfrak{P}$ is *also* independent of Y, because the geometry of $\mathbb{R}^2$ satisfies $\mathfrak{P}$ as well as all the axioms of Y; and we assume that $\mathfrak{P}$ & $\neg\mathfrak{P}$ cannot both be valid in $\mathbb{R}$.

The consistency of hyperbolic geometry both with $\mathbb{R}$ and with the facts of Nature showed that there was no special reason for singling out $\mathfrak{Euc}$ as something special. Nor could the statements of $\mathfrak{Euc}$ any longer be regarded as true or false—they were merely consistent or inconsistent with the axioms. This led to a powerful change of outlook, in that pure mathematics could now be regarded as a supplier of a wide variety of models by which to study the universe. It showed, too, that serious errors can arise if we identify the natural object studied with our model of it. In fact, the latter should be expected to be at best an approximation, and there might be advantages in choosing a model which is deliberately crude in some respects. Also, several models may be available, some better for some purposes than others. We have discussed examples concerning algebraic topology in Chapter 25, and the example of truth and the propositional calculus in Chapter 8; see also Tuller [126]. We shall shortly consider a very remarkable example of mathematical model-making due to Gödel, who has exploited the method to get several profound and influential results. But let us conclude this section with the general remark that words themselves are material from which we make models of our thoughts, and the symbols of mathematics may often refine these models by being more explicit. On the other hand, in poetry or drama, for example, an author may deliberately make understatements, thus making a cruder model (in one sense); but if he is skilful, such understatements may actually heighten the effect and convey a much better impression of his thoughts!

39.4 Formal Systems

The spur for Gödel's work was Hilbert's concern with the consistency of mathematics. Hilbert felt that there was a 'safe part', which would use only finite processes of proof (for example, of the kind used in arithmetic: as in Part III of this book). He thought that this safe kind of reasoning might be used to extend the system to show that more spectacular reasoning was in fact still safe. To make sure that a theorem really is deduced by logic from a set of axioms, he saw that these axioms must be taken 'devoid of meaning'. That is, it must be shown that the theorem is obtainable by manipulating the axioms according to the rules, but without making any appeal to what the axioms mean. Recall that the Greeks appealed to meaning when they took for granted the statement that the diagonals of a parallelogram intersect inside the parallelogram; such a statement when analysed is not deducible from their axioms, as we saw in Chapter 16. Of course, this is not to say that in practice a mathematician does not privately assign meaning to the axioms. In fact this assignment helps him to decide which statements are likely to be things he can find proofs for. But according to Hilbert the finished, printed, proofs should appeal only to the axioms and the rules of logic, not to the meaning privately assigned. In this strict sense then, the symbols are manipulated as in a game where the rules of the game are prescribed. In chess, for example, the relationships between the King and Bishop are prescribed only by the rules of chess, and the interest lies in the consequences of the rules, not in possible analogies with a real-life society. Similarly for Hilbert's attitude to mathematics (in its formal sense; not, of course, when he was working in some technical branch like integral equations or number theory).

In particular, Hilbert hoped to formulate part or all of mathematics as a game of this kind. Then he hoped to be able to show that there was consistency by using the 'safe part' mentioned above. Now if $\mathfrak{T}$ was a part under discussion, he pointed out that we must distinguish between statements *in* $\mathfrak{T}$ and statements *about* $\mathfrak{T}$. The statements in $\mathfrak{T}$ form what he calls the **object-language**, whereas those about $\mathfrak{T}$ form the **meta-language.** Deductions in the meta-language must use the safe part of mathematics mentioned above, and the result is *meta-mathematics relative to* $\mathfrak{T}$. Hilbert's hope was shattered by Gödel, who showed that Hilbert's plan is too ambitious; and we now give a sketch of his method.

In Chapter 8, where we looked at the propositional calculus, we have already had one example of formalizing a part of our thought-processes. This calculus is too weak as a model of mathematics, and we must make it more complicated by using further formulae corresponding to 'predicates' $\phi(x)$. Such a predicate ϕ describes a property which x may or may not have, and if

x is replaced by something specific like 3, then we obtain a *proposition* $\phi(3)$. For example, if $\phi(x)$ is the predicate '$x < 2$', then $\phi(3)$ is the proposition $3 < 2$ (which happens to be false). It is assumed that ϕ, like a function, has a domain $\mathscr{U}$ in which the variable x lies; thus $\phi(x)$ is always either true or false when $x \in \mathscr{U}$. We constantly use predicates in mathematics—very often with the existential quantifiers $\exists$, $\forall$—and then $(\exists x)\phi(x)$, $(\forall x)\phi(x)$ are propositions. Further, we can substitute one variable for another; for example, if x is $y^2 + 1$ and $\phi(x)$ holds, then $\phi(y^2 + 1)$ holds. Thus any formal model of mathematics must contain some formulae corresponding to predicates, with rules of quantification and substitution. Further, to talk about *sets* in the model, the variables $x, y, \ldots$, may be regarded as elements of the universe $\mathscr{U}$, and predicates $\phi, \psi, \ldots$ can be used to pick out (and hence talk about) the subsets of $\mathscr{U}$; ϕ determines the subset $\{x|\phi(x)\}$. To describe such a working model in full detail is a lengthy task, and the interested reader may consult Rosser [108]. We content ourselves with giving a sketch to indicate how the full project would go, because we want to get on to Gödel's theorems. Any formal model $\mathfrak{F}$ will have certain general features which we now describe before we specify a particular model. In particular, $\mathfrak{F}$ is to be a game (using the word in a non-frivolous sense) consisting of objects which are to be manipulated according to prescribed rules. We emphasize that the language used to state the rules is not part of $\mathfrak{F}$; it is part of the meta-language. Nevertheless, it must be precise and unambiguous; and it is helpful, when setting it up, to imagine we are explaining to a computing machine how to check a proof in $\mathfrak{F}$. [This approach was exploited by Turing [128] with great success and with great influence on technical logic and the theory of automata.]

The objects of $\mathfrak{F}$ are to be formulae $\mathfrak{A}$, $\mathfrak{B}$, $\mathfrak{B}(x), \ldots$, built from alphabets of symbols, of three (or more) kinds:

39.4.1(1)
$$\begin{cases} X, Y, \ldots & \text{(for propositions)} \\ \phi, \psi, \ldots, & \text{(for predicates)} \\ x, y, \ldots & \text{(for variables);} \end{cases}$$

the formulae of $\mathfrak{F}$ also include the logical symbols (using '$\vee$' for 'or'):

(2) $\&$, $\vee$, $\neg$, $\rightarrow$, $\leftrightarrow$, $\exists$, $\forall$,

as well as symbols for punctuation:

(3) (,), ,, ., :, #,

(where # is here used to denote a space between two formulae, and all the commas except the third belong to our meta-language, and not to $\mathfrak{F}$). Moreover, $\mathfrak{F}$ will have some mathematical symbols

(4) $+$, $\times$, 0, $'$(successor), etc.

Any finite row of these symbols is called a *formula*, but certain formulae are picked out (by rules 39.5.2 given below) and called 'wffs' (well-formed formulae). A finite set S of wffs of $\mathfrak{F}$ is selected arbitrarily and called the $\mathfrak{F}$-set of axioms, and to these we add the axioms of logic 39.4.2, 39.4.3 below. Thus for all wffs $\mathfrak{A}$, $\mathfrak{B}$, $\mathfrak{B}(x)$, ..., of† $\mathfrak{F}$, the following formulae, together with those of S, constitute the *axioms of* $\mathfrak{F}$:

39.4.2
$$\begin{cases} \mathfrak{A} \vee \mathfrak{A} \to \mathfrak{A}, \quad \mathfrak{A} \to \mathfrak{A} \vee \mathfrak{B}, \quad \mathfrak{A} \vee \mathfrak{B} \to \mathfrak{B} \vee \mathfrak{A} \\ (\mathfrak{A} \to \mathfrak{B}) \to (\mathfrak{C} \vee \mathfrak{A} \to \mathfrak{C} \vee \mathfrak{B}). \end{cases}$$

39.4.3
$$\begin{cases} \mathfrak{B}(x) \to (\exists y)\mathfrak{B}(y) \\ (\exists x)\mathfrak{B}(x) \to \neg((\forall x) \neg \mathfrak{B}(x)). \end{cases}$$

[Observe that the formulae 39.4.2 are all tautologies in the propositional calculus (see 8.5), while those in 39.4.3 correspond to our normal practical usage of the words 'for all', 'there exist' and 'implies'. These axioms ensure that, to some extent, $\mathfrak{F}$ is a model of our thought processes, however arbitrary the chosen $\mathfrak{F}$-set S may be.] The rules for picking out wffs will ensure that the axioms 39.4.2 and 39.4.3 are all wffs. Moreover, the logical symbols are related to each other, in the sense that we really need only two of them (say $\vee$ and $\neg$) and then the symbols &, $\to$, and $\leftrightarrow$ are introduced as abbreviations —based on our experience in Chapter 8—by agreeing that $\mathfrak{A}$ & $\mathfrak{B}$, $\mathfrak{A} \to \mathfrak{B}$, and $\mathfrak{A} \leftrightarrow \mathfrak{B}$ are abbreviations for the formulae $\neg(\neg\mathfrak{A} \vee \neg\mathfrak{B})$, $(\neg\mathfrak{A} \vee \mathfrak{B})$, $(\mathfrak{A} \to \mathfrak{B}$ & $\mathfrak{B} \to \mathfrak{A})$.

39.4.4 To play the 'game' of $\mathfrak{F}$ is to increase its stock of provable wffs, in the following sense (where we model our practical notion of a proof). The axioms of $\mathfrak{F}$ are declared to be provable wffs at the start; and if $\mathfrak{A} \to \mathfrak{B}$ and $\mathfrak{A}$ are both known to be provable wffs, then $\mathfrak{B}$ is added to the stock of provable wffs. This is the 'rule of Modus Ponens' (see 8.5.3).

Such is the specification (omitting some details) of a general formal system $\mathfrak{F}$; clearly it is devoid of meaning, in the sense that a computer could blindly grind out provable wffs according to the rules of the game. In fact, a computer in 1958 re-proved the first 220 or so theorems of *Principia Mathematica* in a few minutes: see Wang [132].

39.4.5 EXAMPLE. Based on a suggestion of T. Rowland, we give a purely formal example of a system E which is too weak (and arbitrary) to serve as a model of our deductive processes, but where, for that very reason, the 'proof game' of E is more clearly seen. The only symbols of E are declared to be 0 and 1, and the wffs are non-empty finite rows of 0's and 1's. Thus $0^3 1^5 0 1^2 0$ is a wff, where 1^5 stands for a row of five 1's, and similarly for 0^3, etc. The only axiom of E is the wff 0, and the set $\mathscr{P}$ of provable wffs is given by the rules

(i) if $\mathfrak{A}$ is a wff in $\mathscr{P}$, so is $\mathfrak{A}1$;
(ii) if $\mathfrak{A}$ and $\mathfrak{A}\mathfrak{B} \in \mathscr{P}$, then $\mathfrak{B}0 \in \mathscr{P}$,

† Here $\mathfrak{B}(x)$ is a 'one-place predicate', i.e., a wff in which the variable x is 'free' in the sense of the rules 39.5.2 below.

(and $0 \in \mathscr{P}$ because 0 is an axiom). Then it is easy to verify that the wffs of $\mathscr{P}$ are all of the forms

$$0, 01^n, 1^n0, \quad \text{and} \quad 1^n01^m, \qquad (n, m \in \mathbb{N}).$$

Suppose we change the second rule to 'then $\mathfrak{B}0$ and $0\mathfrak{A}$ are in $\mathscr{P}$'. We invite the reader to describe the set of provable wffs now. He should also try other rules, suggested by himself.

39.5 Examples of the 'Proof Game'

Before going into greater detail, let us now illustrate the 'proof game', by deriving some theorems from the axioms 39.4.2 (so we are doing another version of the propositional calculus considered in Chapter 8). To give the formulae a simpler appearance, we shall here use $p, q, \ldots$ instead of $\mathfrak{A}, \mathfrak{B}, \ldots$; and we denote $\neg p$ by p'. To indicate that p is provable by the rules given above, we shall write $\vdash: p$ (*read:* 'p is formally provable'). For brevity we display the axioms of 39.4.2 as

I $\vdash: p \vee p \rightarrow p$;
II $\vdash: p \rightarrow p \vee q$;
III $\vdash: p \vee q \rightarrow q \vee p$;
IV $\vdash: (p \rightarrow q) \rightarrow [(r \vee p) \rightarrow (r \vee q)]$.

Here, as below, such phrases as 'for all wffs p' are to be understood; and all the formulae will be wffs according to the detailed rules given later in 39.6.

First, $\rightarrow$ is transitive:

39.5.1 THEOREM. *If* $\vdash: p \rightarrow q$ *and* $\vdash: q \rightarrow r$, *then* $\vdash: p \rightarrow r$.

Proof. If in axiom IV we replace p, q, r by q, r, p', respectively, we get

$$\vdash: (q \rightarrow r) \rightarrow [(p \rightarrow q) \rightarrow (p \rightarrow r)],$$

since $p \rightarrow q$ is $p' \vee q$, as we agreed above. Since $\vdash: q \rightarrow r$, then, by the Modus Ponens rule, $\vdash: [(p \rightarrow q) \rightarrow (p \rightarrow r)]$; but $\vdash: p \rightarrow q$, so applying the rule again we conclude $\vdash: p \rightarrow r$. ■

39.5.2 COROLLARY. $\vdash: p \rightarrow p$.

Proof. By axiom II with $p = q$, $\vdash: p \rightarrow p \vee p$; by axiom I, $\vdash: p \vee p \rightarrow p$, so by transitivity $\vdash: p \rightarrow p$. ■

Since $p \rightarrow p$ is (an abbreviation of) $p' \vee p$, we obtain at once:

39.5.3 COROLLARY. **(Law of excluded middle.)** $\vdash: p' \vee p$. ■

Another consequence of axiom IV is that $\vee$ distributes over $\rightarrow$:

39.5.4 THEOREM. *If* $\vdash: p \rightarrow q$, *then* $\vdash: p \vee r \rightarrow q \vee r$.

Proof. By axiom IV and Modus Ponens, $\vdash: r \vee p \to r \vee q$. By axiom III, $\vdash: p \vee r \to r \vee p$, so by transitivity $\vdash: p \vee r \to r \vee q$. A second application of axiom III completes the proof. ■ (Observe that the proof shows how axiom III enables us to replace $p \vee r$ by $r \vee p$ in any implication.)

Corresponding to the rule of double negation ($8.1.2_p$), we have

39.5.5 THEOREM. $\vdash: p \leftrightarrow p''$.

Proof. In 39.5.3, we put $p = q'$ to get $\vdash: q'' \vee q'$. Then

$$\vdash: q' \vee q'' \qquad \text{(by axiom III and Modus Ponens)},$$

i.e.,

$$\vdash: q \to q''.$$

By definition of $\mathfrak{A} \leftrightarrow \mathfrak{B}$ above, it now remains to prove that $\vdash: q'' \to q$. In axiom IV, replace p, q, r by p', p''', p, respectively, to get

$$\vdash: (p' \to p''') \to [p \vee p' \to p \vee p'''];$$

but we proved $\vdash: q \to q''$ for all q, so putting $q = p'$ and applying Modus Ponens, we get $\vdash: p \vee p' \to p \vee p'''$. Hence by axiom II,

$$\vdash: (p \to p) \to (p'' \to p).$$

But $\vdash: p \to p$ (by 39.5.2), so $\vdash: p'' \to p$ by Modus Ponens. Since $\vdash: p \to p''$ (already proved), we have $p \leftrightarrow p''$. ■

39.5.6 COROLLARY. *If* $\vdash: p \to q$, *then* $\vdash: q' \to p'$.

Proof. Since $\to$ is transitive, $\vdash: p \to q''$, i.e., $\vdash: p' \vee q''$; whence

$$\vdash: q'' \vee p', \qquad \text{i.e.,} \quad \vdash: q' \to p'. \quad ■$$

This result enables us to go from results about $\vee$ to those about & (recall that $\mathfrak{A}$ & $\mathfrak{B}$ is $\neg(\neg\mathfrak{A} \vee \neg\mathfrak{B})$); for example,

39.5.7 THEOREM. *If* $\vdash: p \to q$, *then* $\vdash: p \,\&\, r \to q \,\&\, r$.

Proof. By hypothesis and 39.5.6, $\vdash: q' \to p'$, whence by 39.5.4,

$$\vdash: q' \vee r' \to p' \vee r'.$$

By 39.5.6 again, the theorem follows, since $p \,\&\, r$ is $(p' \vee r')'$. ■

Just as axioms I and II yield $\vdash: p \vee p \to p \vee q$, we now leave the reader to prove some analogues as an exercise.

Exercise 2

(i) Prove the statements:

$$\vdash: p \to p \,\&\, p, \qquad \vdash: p \,\&\, q \to p, \qquad \vdash: p \,\&\, q \to p \,\&\, p.$$

{Use 39.5.5.}

(ii) Hence prove that *if* $\vdash: p \,\&\, q$, *then* $\vdash: p$ *and* $\vdash: q$. [This corresponds to a fact of practical deduction; so also does the converse, but that is harder {see Hilbert–Ackermann [58], p. 36, Th. 18}.]

The following theorem is vital; it shows that if a paradox exists in the system $\mathfrak{F}$, then all formulae are provable.

39.5.8 THEOREM. *If, for some p, $\vdash: p$ and $\vdash: p'$, then for all q, $\vdash: q$.*

Proof. By axiom II, $\vdash: p \to (p' \to q)$; but $\vdash: p$ (given), so $\vdash: p' \to q$ by Modus Ponens. Also $\vdash: p'$ is given, whence application of Modus Ponens to $\vdash: p' \to q$ yields $\vdash: q$. ■

Many other theorems (such as the associativity of $\vee$, for example) can be proved; see Hilbert–Ackermann [58]. We stress again, however, that all the results 39.5.1–39.5.8 were proved according to the rules, with no appeal to the possible meaning of the formulae. Other systems of axioms could be used instead, for example the intuitively obvious statements we took as basic in 8.1.

As to the 'existential' axioms 39.4.3, we remark that they are weaker than those normally used in practical mathematics, in that the rule

39.5.9 $$(\forall x)\mathfrak{B}(x) \to (\exists x)\mathfrak{B}(x)$$

is not present. This omission arises because 39.5.9 is not used in the proof of Gödel's theorems below; if it *had* been used, some schools of mathematical philosophy would have preferred to reject 39.5.9 rather than accept the theorems. Such philosophies object to 39.5.9 on the following grounds: knowledge that $\mathfrak{B}(x)$ holds for all x does not of itself tell us how to *construct* a single x with $\mathfrak{B}(x)$ true. But this objection is based on an assignment of *meaning* to the symbols $\forall$, and $\exists$, even if we define one in terms of the other by

$$(\forall x)\mathfrak{B}(x) = \neg(\exists x)(\neg\mathfrak{B}(x)).$$

For an account of such philosophies, see Wilder [136].

39.6 Gödel's Theorems

Consider now the case when $\mathfrak{P}$ is Hilbert's 'safe' part of mathematics, the system $\mathfrak{Z}$ of elementary number theory. Then $\mathfrak{Z}$ requires only the two ('constant') predicates, '$x = y$', and 'y is the successor of x', written $y = x'$ (and not to be confused with 'not x'); and only the alphabet of variables $x, y, \ldots, \phi, \psi, \ldots$ in 39.4.1(1). It uses also the symbols of 39.4.1(2)–(4), quantifying only over variables $x, y, \ldots$, but not over predicates $\phi, \psi, \ldots$ (i.e., it allows $\exists x$ and $\forall x$ in formulae, but not $\exists\phi$, $\forall\phi$).

These suffice to state the Peano axioms in the following form, in contrast to the form we chose in 23.1 where we used the informal notions of set and function; such general notions are not necessary for the relatively limited purpose we have in mind in discussing $\mathfrak{Z}$ (which is therefore more restricted than $\mathfrak{Z}_0$, the system mentioned in 39.3, because $\mathfrak{Z}_0$ allows more complicated operations of set-theory). The Peano axioms become:

39.6.1 $$(\forall x)(\exists y)(y = x' \,\&\, (\forall z)((z = x' \rightarrow z = y) \,\&\, (z' = y \rightarrow z = x)),$$

corresponding to 'every x has a unique successor y and the successor function is one-one';

39.6.2 $$\neg(\exists x)(0 = x'),$$

corresponding to '0 is not the successor of any x';

39.6.3 $$(\phi(0) \,\&\, (\forall x)(\phi(x) \rightarrow \phi(x')) \rightarrow (\forall x)\phi(x),$$

corresponding to the statement of the principle of induction (5.1.4). Notice that 39.6.3 holds for all ϕ, a fact ensured by the substitution rule (8) in 39.6.5 below; thus we do not need, *within the formal system*, to quantify over ϕ. The usual axioms of equality are also to be included (saying that $x = y$ is an equivalence relation between variables).

We therefore think of $\mathfrak{Z}$ as the algebra of logic (i.e., a system satisfying 39.4.1–39.4.3) together with Peano's axioms, the axioms of equality, and the rules given in 39.6.5 below for constructing formulae. It is now possible to define the notion of a *recursive* function $f(x_1, \ldots, x_k)$ (with arguments x_i, and values in† $\mathbb{Z}^+$) as one built from the constants and the successor function, using only finite combinations of sums, products, compositions, and restriction. Thus if f, g are recursive, so are the functions h, j, k where

39.6.4 $$h(n, m) = f(n, g(m, n)), \qquad j(n) = f(n, n), \qquad k(n) = f(n, a)$$

where $a \in \mathbb{Z}^+$ is fixed; a precise formulation can be given by a procedure with rules analogous to the rules (2)–(8) for wffs, below.

We define the wffs of $\mathfrak{Z}$ by the following rules. First note that we do not need an infinite alphabet for the variables, since we can manufacture an infinite number from two symbols in the form x, x_1, x_{11}, x_{111}, etc. Then we define a 'term' by:

39.6.5

(1) any variable standing alone is a term;
(2) 0 is a term;
(3) if $\mathbf{a}$ is a row of symbols and is a term, then $\mathbf{a}'$ is a term;
(4) if $\mathbf{a}$ and $\mathbf{b}$ are terms, so are $(\mathbf{a} + \mathbf{b})$ and $(\mathbf{a} \times \mathbf{b})$;
(5) only a formula shown to conform to the last four rules is a term.

A wff is now defined by:

(6) if $\mathbf{a}$ and $\mathbf{b}$ are terms, then $(\mathbf{a} = \mathbf{b})$ is a wff;
(7) if $\mathfrak{A}$ and $\mathfrak{B}$ are wffs, then so are

$$(\neg\mathfrak{A}), \qquad (\mathfrak{A} \vee \mathfrak{B}), \qquad (\mathfrak{A} \,\&\, \mathfrak{B}), \qquad (\mathfrak{A} \rightarrow \mathfrak{B}), \qquad (\mathfrak{A} \leftrightarrow \mathfrak{B}).$$

† $\mathbb{Z}^+$ is not a notion in $\mathfrak{Z}$; it is a notion of our metamathematics, 'the set of variables $x, y, \ldots$'.

In (6) any variable $a_{11}, \ldots,$ in $(\mathbf{a} = \mathbf{b})$ is said to be *free*. In (7) any free variable $a_{11}, \ldots,$ in $\mathfrak{A}$ or $\mathfrak{B}$ is *free* in the new wff.

(8) (i) If x is free in a wff $\mathfrak{A}$ and x does not occur in a term $\mathbf{b}$, then the result $\mathfrak{C}$ of replacing x by $\mathbf{b}$ throughout $\mathfrak{A}$ is also a wff; moreover
(ii) $(\exists x)\mathfrak{A}$ and $(\forall x)\mathfrak{A}$ are wffs and x is now *bound*. Further, if $\mathbf{b}$ in (i) happens to be a variable y, $(\exists x)\mathfrak{A} \leftrightarrow (\exists y)\mathfrak{C}$ and $(\forall x)\mathfrak{A} \leftrightarrow (\forall y)\mathfrak{C}$.

[The reader should here compare the similar situation in calculus; in the formula $\int_a^b f(x)\,dx$, a and b are free, while x is bound; also

$$\int_a^b f(x)\,dx = \int_a^b f(y)\,dy$$

analogously with (ii).]

The axioms 39.4.2, 39.4.3 and the Peano axioms 39.6.1–39.6.3 are therefore wffs.

39.6.6 A *proof* (in any formal system $\mathfrak{F}$) is defined to be a finite series of wffs containing no free variables, and related to each other as follows. Each wff is an axiom, or related to the axioms by the rules of procedure (e.g., by Modus Ponens (39.4.4) or substitution). The proof is said to be a proof of the last wff in its series. The word 'proof' will from now on be understood in this sense. To avoid ambiguity, a wff $\mathfrak{A}$ will be described as *$\mathfrak{F}$-provable* if there exists in $\mathfrak{F}$ a proof of which $\mathfrak{A}$ is the last row. To show whether the system is consistent, Theorem 39.5.8 tells us to prove:

39.6.7 *There is no wff $\mathfrak{A}$ in $\mathfrak{F}$ such that both $\mathfrak{A}$ and $\neg\mathfrak{A}$ are $\mathfrak{F}$-provable.*

Hilbert wanted to prove three things about the system $\mathfrak{Z}$ of elementary number theory:

(a) its *consistency*: $\mathfrak{A}$ and $\neg\mathfrak{A}$ are not simultaneously $\mathfrak{Z}$-provable;
(b) its *completeness*: given $\mathfrak{A}$, either $\mathfrak{A}$ or $\neg\mathfrak{A}$ is $\mathfrak{Z}$-provable;
(c) the *Entscheidungsproblem:* to invent a *general procedure* to find out whether $\mathfrak{A}$ is $\mathfrak{Z}$-provable or $(\neg\mathfrak{A})$ is $\mathfrak{Z}$-provable.

Having proved these, the next part of Hilbert's programme would be to make the axioms for real numbers and some more powerful logic. For example, he would then allow quantifiers $\exists\phi$, $\forall\phi$ where ϕ is now a predicate, and not just a variable x. In 1936, the consistency of $\mathfrak{Z}$ was proved by Gentzen, but by methods not allowed by Hilbert; i.e., they go outside Hilbert's 'safe' framework, because Gentzen uses transfinite arithmetic.

The completeness of $\mathfrak{Z}$ (in (b)) was proved false by Gödel (1931), which meant that (c) must be modified to
(c′) Given $\mathfrak{A}$, is there a general procedure to decide whether $\mathfrak{A}$ is (within $\mathfrak{Z}$) true, false, or undecidable?

This was proved, in the negative, by Turing (1936). Most important of all, Kurt Gödel proved two theorems, of which the first asserts:

39.6.8 THEOREM. *Let $\mathfrak{F}$ be a system with a finite number of symbols, rules of procedure, and axiom schemes. Further, suppose $\mathfrak{F}$ contains $\mathfrak{Z}$, and is consistent. Then, in such a system, rules can be given for finding a formula $\mathfrak{K}$, such that neither $\mathfrak{K}$ nor $\neg\mathfrak{K}$ is $\mathfrak{F}$-provable.*

Note that the system of algebraic logic (with $\exists$ and $\forall$) is too small to contain $\mathfrak{Z}$, and indeed it can be shown to be complete (by a method akin to truth tables).

Gödel proved a second theorem:

39.6.9 THEOREM. *For such a system $\mathfrak{F}$ (as in Theorem* 39.6.8), *it is not possible to formulate in $\mathfrak{F}$ a proof of the statement*

$$\mathfrak{C}\text{: '}\mathfrak{F} \textit{ is not contradictory'},$$

although we can formulate the statement $\mathfrak{C}$ in $\mathfrak{F}$. If, on the other hand, $\mathfrak{F}$ were not consistent, the proof could be formulated.

This result shows why Gentzen had to use results not permitted by Hilbert; $\mathfrak{F}$ cannot be *proved within itself* to be consistent.

We now sketch the proof of Gödel's first theorem; his second theorem uses this proof and involves more technicalities.

39.7 Gödel's Proofs

39.7.1 In our metamathematical arguments about the system $\mathfrak{F}$, we shall use the system $\mathbb{Z}^+$ of natural numbers in the usual mathematical way (but using only the finite type of process envisaged as 'safe' by Hilbert). We then refer to the 'informal system $\mathbb{Z}^+$'. In particular, we need the notion of a recursive function, i.e., a function $f\colon \mathbb{Z}^+ \times \mathbb{Z}^+ \to \mathbb{Z}^+$ as in 39.6.4. Further, if A is a relation on $\mathbb{Z}^+$, then we call A *recursive* if there exists a recursive function f such that for all $m, n \in \mathbb{Z}^+$, $f(m, n) = 1$ if mAn holds, while $f(m, n) = 0$ if mAn does not hold. Here 'holds' means 'is true in the usual (unformalized) sense of elementary mathematics'; it is a notion in our metamathematics. For convenience we shall write $A(m, n)$ instead of mAn, so that when g is a recursive function it should be clear what $A(m, g(u, v))$ means. Further, if A is also recursive, then it can be proved that $A(m, g(n, n))$ is a recursive relation between m and n.

39.7.2 The important idea behind the proofs of Gödel's theorems consists in a process for labelling all the formulae of $\mathfrak{F}$ with integers of the informal system $\mathbb{Z}^+$; this metamathematical process cannot be described in $\mathfrak{F}$, but since $\mathfrak{F}$ contains the formal system $\mathfrak{Z}$, the labelling can be copied into $\mathfrak{F}$.

Thus $\mathfrak{F}$ can be made to talk about itself, so to speak, and to state something like the Cretan paradox mentioned above. The resulting statement is *not* a paradox, however, and turns out to be the undecidable formula $\mathfrak{R}$ whose existence is asserted in 39.6.8. The proof of the second theorem (39.6.9) develops in the same way.

39.7.3 More precisely then, we start by translating the formulae of $\mathfrak{Z}$ into the informal number system $\mathbb{Z}^+$. Our symbols are

$$\vee, \quad \neg, \quad \exists, \quad \forall, \quad (, \quad), \quad =, \quad +, \quad \times, \quad ', \quad 0, \quad \#,$$

together with variables $x, y, z, \ldots$. To these we assign the integers

$$3, \quad 5, \quad 7, \quad 9, \quad 11, \quad 13, \quad 15, \quad 17, \quad 19, \quad 21, \quad 23, \quad 25,$$

together with $2, 2^2, 2^3, \ldots$, respectively. Then to a row **a** of symbols, say $\vee x \exists$, with assigned numbers 3, 2, 7, we associate the single number $G(\mathbf{a}) = 2^3 \cdot 3^2 \cdot 5^7$; that is, we take the first three prime numbers, raise each to an exponent equal to the assigned integer, and multiply the results; for longer rows we use more prime numbers in the obvious way. Hence, to every row of symbols corresponds a positive integer $G(\mathbf{a})$, called the **Gödel number** of **a**. Conversely, by splitting a positive integer n into prime factors, clearly we either arrive back at a unique **a** of symbols such that $n = G(\mathbf{a})$, or we can conclude that n is not the Gödel number of any row of symbols. All this (including the prime factorization theorem) is carried out in the informal system $\mathbb{Z}^+$, and this labelling process cannot be expressed formally in $\mathfrak{F}$.

39.7.4 However, we can make the following rules of arithmetic in $\mathfrak{F}$:

If x, y are integers, expressible in the form

$$x = 2^{m_1} \cdot 3^{m_2} \cdot 5^{m_3} \cdot \cdots \cdot p_k^{m_k}; \qquad y = 2^{n_1} \cdot 3^{n_2} \cdot \cdots \cdot p_j^{n_j},$$

then we define $x * y$ to be the number

$$x * y = 2^{m_1} \cdot 3^{m_2} \cdots p_k^{m_k} \cdot p_{k+1}^{n_1} \cdots p_{k+j}^{n_j}$$

where p_r denotes the rth prime. Notice that $*$ is associative. Now define the following functions of x (where $x \in \mathbb{N}$):

$$\begin{aligned}
\text{neg}(x) &= 2^{11} * 2^5 * x * 2^{13},\\
\text{dis}(x, y) &= 2^{11} * x * 2^3 * y * 2^{13},\\
\text{pl}(x, y) &= 2^{11} * x * 2^{17} * y * 2^{13},\\
\text{mul}(x, y) &= 2^{11} * x * 2^{19} * y * 2^{13},\\
\text{seq}(x) &= x * 2^{21}.
\end{aligned}$$

From the point of view of $\mathfrak{Z}$, these are legitimate (if mystifying) functions; from the point of view of the metamathematics outside $\mathfrak{Z}$, these functions represent Gödel numbers of certain formulae provided x and y do.

Exercise 3

(i) Show that $(\forall x)(x = x)$ has Gödel number $2^{11}3^{9}5^{2}7^{13}11^{11}13^{2}17^{15}19^{2}23^{13}$. Estimate its magnitude.
(ii) Is the number 123,456,789,101,112,131,415 the Gödel number of any formula?
(iii) Show that with the functions neg, ..., defined in 39.7.4, then if x, y are Gödel numbers of formulae $\mathfrak{X}$, $\mathfrak{Y}$, the values of the functions listed represent the Gödel numbers of

$$(\neg\mathfrak{X}), \quad (\mathfrak{X} \vee \mathfrak{Y}), \quad (\mathfrak{X} + \mathfrak{Y}), \quad (\mathfrak{X} \times \mathfrak{Y}), \quad \text{and } \mathfrak{X}' \text{ respectively.}$$

Recall from 39.6.6 the notion of a proof in $\mathfrak{F}$. Hence, if z is an integer and

$$z = x_1 * 2^{25} * x_2 * 2^{25} * \cdots * x_k,$$

then z represents a proof, whose lines have Gödel numbers $x_1, x_2, \ldots, x_k$; that is to say, z is the Gödel number of a proof of which the last line has Gödel number x_k. This suggests that we define a relation Q on $\mathbb{Z}^+$ by writing

39.7.5 $Q(x, y)$ means: x is the Gödel number of a proof of which the last line has Gödel number y.

It can be proved that Q is a *recursive* relation in the sense described in 39.7.1.

We now use the fact that $\mathfrak{F}$ contains $\mathfrak{Z}$, to copy into $\mathfrak{Z}$ certain integers and relations of $\mathbb{Z}^+$. In particular, each integer n in our metamathematics corresponds to a unique integer in $\mathfrak{Z}$ which we shall denote by $\mathbf{n}$ (so $\mathbf{0}$ denotes the 0 of $\mathfrak{Z}$, $\mathbf{1}$ denotes $0'$, $\mathbf{2}$ denotes $0''$, etc.). The following result can then be proved (and it is not surprising on reflection):

39.7.6 LEMMA. *Let $A(x, y)$ be a recursive relation on $\mathbb{Z}^+$. There is a formula $\mathfrak{A}(\mathbf{x}, \mathbf{y})$ in the system $\mathfrak{F}$, such that given any numbers $\mathbf{m}, \mathbf{n}$ in $\mathfrak{F}$, then $\mathfrak{A}(\mathbf{m}, \mathbf{n})$ is $\mathfrak{F}$-provable if $A(m, n)$ holds and $\neg\mathfrak{A}(\mathbf{m}, \mathbf{n})$ is $\mathfrak{F}$-provable if $A(m, n)$ does not hold.*

We call $\mathfrak{A}$ the *$\mathfrak{F}$-analogue* of A.

Suppose we have a formula $\mathfrak{A}(\mathbf{x})$ in $\mathfrak{F}$ such that

$$G[\mathfrak{A}(\mathbf{x})] = p.$$

Then $G[\mathfrak{A}(\mathbf{0})] = a$, $G[\mathfrak{A}(\mathbf{1})] = b, \ldots, G[\mathfrak{A}(\mathbf{r})] = q$, say, so that q is the Gödel number of a one-place predicate with Gödel number p, when the variable is replaced by $\mathbf{r}$. Let us write

39.7.7 $$q = s(p, r).$$

If p does not correspond to a formula with a free variable, define $s(p, r)$ to be 0. Then we have defined a function $s\colon \mathbb{Z}^+ \times \mathbb{Z}^+ \to \mathbb{Z}^+$; and it turns out that s is a recursive function, though the proof of this is rather elaborate. With this apparatus, we can now prove the first of Gödel's theorems. Consider the relation Q of 39.7.5, and let $R(x, y)$ denote the relation $Q(y, s(x, x))$, with s as in 39.7.7. Then R is recursive by the remark concluding 39.7.1. Hence by Lemma 39.7.6, there is a formula $\mathfrak{R}(\mathbf{x}, \mathbf{y})$ in $\mathfrak{F}$, the $\mathfrak{F}$-analogue of R. Let

$\mathfrak{P}(\mathbf{x})$ be the formula $(\exists \mathbf{y})\mathfrak{R}(\mathbf{x}, \mathbf{y})$ and let $G[\neg\mathfrak{P}(\mathbf{x})] = a$. We claim (and this clearly will prove Theorem 39.6.8):

39.7.8 *Neither* $\neg\mathfrak{P}(\mathbf{a})$ *nor* $\mathfrak{P}(\mathbf{a})$ *is* $\mathfrak{F}$*-provable.*

For, suppose $\neg\mathfrak{P}(\mathbf{a})$ is provable. Then there is in $\mathfrak{F}$ a proof $\mathfrak{B}$, of $\neg\mathfrak{P}(\mathbf{a})$. Now by 39.7.7,

$$G[\neg\mathfrak{P}(\mathbf{a})] = s(a, a).$$

Let $G[\mathfrak{B}] = b$. Hence, in our metamathematics the relation $Q(b, s(a, a))$ holds. By Lemma 39.7.6, therefore, the formula $\mathfrak{R}(\mathbf{a}, \mathbf{b})$ is provable in $\mathfrak{F}$. Hence by the first axiom in 39.4.3, together with Modus Ponens, $(\exists \mathbf{y})\mathfrak{R}(\mathbf{a}, \mathbf{y})$ is $\mathfrak{F}$-provable, i.e., $\mathfrak{P}(\mathbf{a})$ is provable in $\mathfrak{F}$. Thus we have shown that if $\neg\mathfrak{P}(\mathbf{a})$ is provable, then $\mathfrak{P}(\mathbf{a})$ is provable (in $\mathfrak{F}$). But $\mathfrak{F}$ is consistent by hypothesis, so this cannot be (by 39.6.7). Hence $\neg\mathfrak{P}(\mathbf{a})$ is *not* $\mathfrak{F}$-provable.

For the second half of the assertion 39.7.8, Gödel had to assume the 'ω-consistency' of $\mathfrak{F}$, that is, *if the formulae* $\mathfrak{A}(\mathbf{0}), \mathfrak{A}(\mathbf{1}), \mathfrak{A}(\mathbf{2}), \ldots, \mathfrak{A}(\mathbf{n}), \ldots$ *are all* $\mathfrak{F}$*-provable, then the formula* $\neg(\forall \mathbf{x})\mathfrak{A}(\mathbf{x})$ *is not* $\mathfrak{F}$*-provable, simultaneously.* [Even if the latter formula was provable, there would still be no contradiction, for no proof arises that, for a given $\mathbf{n}$, $\neg\mathfrak{A}(\mathbf{n})$ is provable.] The condition of ω-consistency was later shown to be unnecessary by Rosser, but we shall follow Gödel's argument.

To prove the second half of 39.7.8, suppose that $\mathfrak{P}(\mathbf{a})$ is provable. We may assume, by the first part, that given any formula $\mathfrak{N}$, then $\mathfrak{N}$ is not a proof of $\neg\mathfrak{P}(\mathbf{a})$. Hence, if n is any positive integer, then $Q(n, s(a, a))$ does not hold in the informal system $\mathbb{Z}^+$. Therefore by Lemma 39.7.6, $\neg\mathfrak{R}(\mathbf{a}, \mathbf{n})$ is provable in $\mathfrak{F}$. Thus if $\mathfrak{A}(\mathbf{n})$ denotes $\neg\mathfrak{R}(\mathbf{a}, \mathbf{n})$, then

(i) $\mathfrak{A}(\mathbf{0}), \mathfrak{A}(\mathbf{1}), \ldots, \mathfrak{A}(\mathbf{n}), \ldots$ are all $\mathfrak{F}$-provable.

Now $\mathfrak{P}(\mathbf{x})$ means $(\exists \mathbf{y})\mathfrak{R}(\mathbf{x}, \mathbf{y})$, so by the second axiom in 39.4.3,

$$\mathfrak{P}(\mathbf{a}) \rightarrow \neg((\forall \mathbf{y}) \neg \mathfrak{R}(\mathbf{a}, \mathbf{y})).$$

Since also $\mathfrak{P}(\mathbf{a})$ is $\mathfrak{F}$-provable by assumption, then by the Modus Ponens rule 39.4.4, $\neg((\forall \mathbf{y}) \neg \mathfrak{R}(\mathbf{a}, \mathbf{y}))$ is $\mathfrak{F}$-provable, i.e.,

(ii) $\neg((\forall \mathbf{y})\mathfrak{A}(\mathbf{y}))$ is $\mathfrak{F}$-provable.

Hence, (i) and (ii) together contradict the ω-consistency of $\mathfrak{F}$, and it follows that $\mathfrak{P}(\mathbf{a})$ is not $\mathfrak{F}$-provable. This proves Theorem 39.6.8. ■

Observe that $\mathfrak{P}(\mathbf{a})$ is $(\exists \mathbf{y})(\mathfrak{R}(\mathbf{a}, \mathbf{y}))$, so that if we were to identify each A in Lemma 39.7.6 with its $\mathfrak{F}$-analogue $\mathfrak{A}$, then $\mathfrak{P}(\mathbf{a})$ would mean $(\exists y)Q(y, s(a, a))$, i.e., 'the formula with Gödel number $s(a, a)$ has a proof in $\mathfrak{F}$ (with Gödel number y)', which is to say 'the formula $\neg\mathfrak{P}(\mathbf{a})$ is $\mathfrak{F}$-provable'. Thus (outside $\mathfrak{F}$) $\mathfrak{P}(\mathbf{a})$ would *assert its own falsity*, like the Cretan paradox. This is probably what motivated Gödel, but by taking $\mathfrak{F}$ 'devoid of meaning' he avoids the confusion inherent in the Cretan paradox, and relies on a 'diagonal argument' whose germ goes back to Cantor.

We can now enlarge $\mathfrak{F}$ by adding either $\mathfrak{P}(\mathbf{a})$ or $\neg\mathfrak{P}(\mathbf{a})$ as an extra axiom, thus making two quite distinct systems $\mathfrak{F}_+$, $\mathfrak{F}_-$. If $\mathfrak{F}$ is to be interpreted as a model of our thought-processes, then, as we have seen, $\mathfrak{P}(\mathbf{a})$ has an interpretation we know to be untrue, whereas the same interpretation gives $\neg\mathfrak{P}(\mathbf{a})$ as "$\neg\mathfrak{P}(\mathbf{a})$ is not provable in $\mathfrak{F}$'—a statement which fits the facts. Thus $\mathfrak{F}_-$ is likely to be a more applicable model than $\mathfrak{F}_+$. In either case, there will of course be a new formula $\mathfrak{P}_+$ undecidable in $\mathfrak{F}_+$, and a formula $\mathfrak{P}_-$ undecidable in $\mathfrak{F}_-$; $\mathfrak{P}_+$ depends on the construction of $\mathfrak{F}_+$-analogues (see Lemma 39.7.6), which ultimately depends on the axioms of $\mathfrak{F}_+$—and similarly for $\mathfrak{P}_-$ in $\mathfrak{F}_-$. These interpretations of $\mathfrak{P}(\mathbf{a})$ and $\neg\mathfrak{P}(\mathbf{a})$ given above may be regarded as demonstrating that any attempt to describe our thought-processes by a finite model such as $\mathfrak{F}$ will be inadequate; for our brains can decide the acceptability of $\mathfrak{P}(\mathbf{a})$, whereas $\mathfrak{F}$ cannot.

39.8 The Axiom of Choice, and the Continuum Hypothesis

In Chapter 7, we mentioned two difficulties, in the foundations of mathematics, namely the axiom of choice and the continuum hypothesis. Following his earlier achievements, Gödel dealt with these in his monograph *The Consistency of the Continuum Hypothesis* [44]. There he gave a system Σ of axioms for set-theory (which are adequate for doing virtually all mathematics). Using the rules of Σ, he constructed within Σ another mathematical system Δ which is also a model of set-theory just as Σ is. But Δ turns out to be 'richer' than Σ, because all the axioms of Σ hold in Δ, while the axiom of choice and the continuum hypothesis both hold also in Δ. Let us denote the axiom of choice and the continuum hypothesis by A, C, respectively. Then Gödel shows that these statements about Δ, including the statements of A and of C, are provable theorems of Σ, carried out using the rules of Σ. Hence, we are no more likely to meet a contradiction if we use A and C in practice than if we forswear their use; for we may choose to use Δ (rather than Σ) as our model of set-theory on which to build mathematics, and we cannot thereby introduce new errors. If Σ is consistent, so is Δ, while if Σ is not consistent, use of Δ might show a paradox rather more quickly than otherwise (but the paradox would already be inherent in Σ).

The possibility remained, however, that A and C for Σ itself might in fact be deducible within Σ. But around 1963, Paul Cohen showed that A and C (and a natural generalization of C to higher cardinals) are *independent* of the axioms of Σ. This shows why Gödel had to pass to the model Δ: he could not have deduced that A and C held in Σ itself. New speculations therefore arise about different possible systems of mathematics (each of which contain the same elementary mathematics, of course), but we cannot pursue them here. Suffice it to say that all such systems would be presumably equally valid as mathematics, but perhaps some would make better models for investigating problems than others. We should always welcome new models in mathematics because of their great usefulness for increasing our understanding, not only of logic, but of many other activities. For a very clear technical account, see Cohen's expository notes [22].

BIBLIOGRAPHY

In the following list of books we have not attempted to be comprehensive. New books are appearing rapidly, and the reader may very easily find some among them which he prefers. Also there are now several series of surveys on mathematics, sometimes translated from foreign languages and issued in English, particularly by American and British publishers. We have in mind the series by Random House and by the Mathematical Association of America, together with some excellent elementary texts from Russia. Also there are now several journals catering for the teaching of mathematics, such as *Mathematics Teaching* and the *Mathematical Gazette*, published in Britain, and the *American Mathematical Monthly*, published in the United States. These contain many interesting articles relating to small points in mathematics which have been found to lead to interesting teaching situations. At the same time they review the flood of relevant modern literature.

Perhaps we should draw attention also to books by mathematicians which might help to reveal the thought-processes of a mathematician; in particular, we mention the books by Hardy [51] and Littlewood [83], those of Bell [10, 12] and several articles in Newman [93].

[1] ALBERT, A. A. *Structure of Algebras*, American Mathematical Society Colloquium Publications no. 24 (1939).

[2] ALEXANDROFF, P. *Elementary Concepts of Topology*, Dover, New York (1961).

[3] ALEXANDROFF, P. and HOPF, H. *Topologie*, Vol. I (*Grundlehr. math. Wiss.* **5**) Springer, Berlin (1935).

[4] APOSTOL, T. M. *Mathematical Analysis*, Addison-Wesley, Reading, Mass. (1960).

[5] ARNOLD, B. H. *Introduction to Concepts of Elementary Topology*, Prentice-Hall, Englewood Cliffs, N.J. (1964).

[6] ARTIN, E. *Galois Theory*, 2nd edn, Notre Dame Mathematical Lectures (1948).

[7] AUSLANDER, L. *Differential Geometry*, Harper and Row, New York (1967).

[8] AUSLANDER, L. and MACKENZIE, R. E. *Introduction to Differentiable Manifolds*, McGraw-Hill, New York (1963).

[9] BELL, A. W. and FLETCHER, T. J. *Symmetry Groups*, pamphlet no. 12 of *Mathematics Teaching*, Association of Teachers of Mathematics (1964).

[10] BELL, E. T. *Men of Mathematics*, Penguin Books, Harmondsworth, Middx (1953).

[11] BELL, E. T. *Mathematics, Queen and Servant of Science*, Bell, London (1952).

[12] BELL, E. T. *The Development of Mathematics*, 2nd edn, McGraw-Hill, New York (1945).

[13] BERGE, C. *The Theory of Graphs*, Methuen, London (1962).

[14] BING, R. H. *Elementary Point-Set Theory*, pamphlet no. 8 of *American Mathematical Monthly* (1960).

[15] BIRKHOFF, G. *Lattice Theory*, 3rd edn, American Mathematical Society Colloquium Publications no. 25 (1966).
[16] BIRKHOFF, G. and MACLANE, S. *A Survey of Modern Algebra*, revised edn, Macmillan, London (1965).
[17] BOURBAKI, N. *Éléments de mathématique*, Hermann, Paris (1939 on); the parts *General Topology*, Parts 1 and 2, and *Theory of Sets* are available in English: Addison-Wesley, Reading, Mass. (1967, 1968).
[18] BOURBAKI, N. *Éléments d'histoire des mathématiques*, Hermann, Paris.
[19] BOYER, C. B. *A History of Mathematics*, Wiley, New York (1968).
[20] BUSACKER, R. G. and SAATY, T. L. *Finite Graphs and Networks*, McGraw-Hill, New York (1965).
[21] CAIRNS, S. S. *Introductory Topology*, Ronald, New York (1961).
[22] COHEN, P. J. *Set Theory and the Continuum Hypothesis*, Benjamin, New York (1966).
[23] COHN, P. M. *Linear Algebra*, Routledge and Kegan Paul, London (1959).
[24] COURANT, R. *Differential and Integral Calculus*, 2 vols, Blackie, London (1952).
[25] COURANT, R. and ROBBINS, H. *What is Mathematics?*, Oxford University Press, London (1941).
[26] CROWELL, R. H. and FOX, R. H. *Introduction to Knot Theory*, Ginn, Waltham, Mass. (1963).
[27] CUNDY, M. and ROLLETT, A. P. *Mathematical Models*, 2nd edn, Clarendon Press, Oxford (1961).
[28] DAVENPORT, H. *The Higher Arithmetic*, Hutchinson, London (1952).
[29] DAVIS, M. *Computability and Solvability*, McGraw-Hill, New York (1958).
[30] DIEUDONNÉ, J. *Foundations of Modern Analysis*, Academic Press, New York (1960).
[31] DODGSON, Rev. C. L. (Lewis Carroll) *Symbolic Logic*, reprint, Dover, New York (1958).
[32] DORN, W. S. and GREENBERG, H. J. *Mathematics and Computing*, Wiley, New York (1967).
[33] DUFF, G. F. D. *Partial Differential Equations*, Clarendon Press, Oxford (1956).
[34] DURELL, C. V. *Projective Geometry*, Macmillan, London (1931).
[35] EILENBERG, S. and STEENROD, N. *Foundations of Algebraic Topology* Princeton University Press, Princeton, N.J. (1952).
[36] FLANDERS, H. *Differential Forms with Applications to the Physical Sciences*, Academic Press, New York (1963).
[36a] FLETCHER, T. J. (editor) *Some Lessons in Mathematics: A Handbook on the Teaching of 'Modern Mathematics'* by members of the Association of Teachers of Mathematics, Cambridge University Press, London (1964).
[37] FORT, M. K (editor) *Topology of 3-Manifolds and Related Topics*, Prentice-Hall, Englewood Cliffs, N.J. (1962).
[38] GARDNER, M. *Scientific American Book of Puzzles and Diversions*, Simon and Schuster, New York (1959).
[39] GARDNER, M. *The Second Scientific American Book of Puzzles and Diversions*, Simon and Schuster, New York (1961).
[40] GARDNER, M. *New Mathematical Diversions from Scientific American*, Simon and Schuster, New York (1966).

[41] Gaskell, R. E. *Engineering Mathematics*, Staples, London (1960).
[42] Gelfand, I. M., Raikov, D., and Shilov, G. *Commutative Normed Rings*, Chelsea, New York (1964).
[43] Gödel, K. *On Formally Undecidable Propositions* (English trans.), Oliver and Boyd, Edinburgh (1962).
[44] Gödel, K. 'The consistency of the continuum hypothesis' *Ann. Math. Stud.* **3** (1940).
[45] Griffiths, H. B. *Topology*, reprint from *Mathematics Teaching*, Association of Teachers of Mathematics (1967).
[46] Hadwiger, H., Debrunner, H. and Klee, V. *Combinatorial Geometry in the Plane*, Holt and Rinehart, New York (1964).
[47] Hall, D. W. and Spencer, G. L. *Elementary Topology*, Chapman and Hall, London (1955).
[48] Hall, M. *Theory of Groups*, Macmillan, London (1959).
[49] Halmos, P. R. *Finite-Dimensional Vector Spaces*, Van Nostrand, Princeton, N.J. (1958).
[49a] Halmos, P. R. *Measure Theory*, Van Nostrand, Princeton, N.J. (1950).
[50] Halmos, P. R. *Naïve Set Theory*, Van Nostrand, Princeton, N.J. (1960).
[51] Hardy, G. H. *A Mathematician's Apology*, reprint of 1st edn with foreword by C. P. Snow, Cambridge University Press, London (1967).
[52] Hardy, G. H. *Pure Mathematics*, Cambridge University Press, London (1965).
[53] Hardy, G. H. *Divergent Series*, Clarendon Press, Oxford (1949).
[54] Hardy, G. H. and Wright, E. M. *An Introduction to the Theory of Numbers*, Clarendon Press, Oxford (1959).
[55] Hausdorff, F. *Mengenlehre* (1927); English trans., *Set Theory*, Dover, New York (1957).
[56] Herstein, I. N. *Topics in Algebra*, Blaisdell, Waltham, Mass. (1964).
[57] Hilbert, D. *The Foundations of Geometry*, English trans. by E. J. Townsend, Open Court, Chicago (1902); reprinted 1962.
[58] Hilbert, D. and Ackermann, W. *Principles of Mathematical Logic*, Chelsea, New York (1950).
[59] Hilbert, D. and Cohn-Vossen, S. *Geometry and the Imagination*, Chelsea, New York (1952).
[60] Hilton, P. J. *Differential Calculus*, Routledge and Kegan Paul, London (1958).
[61] Hilton, P. J. *Partial Differentiation*, Routledge and Kegan Paul, London (1960).
[62] Hilton, P. J. *An Introduction to Homotopy Theory*, Cambridge Tracts no. 43, Cambridge University Press, London (1953).
[63] Hilton, P. J. *Homotopy Theory and Duality*, Nelson, London (1967).
[64] Hilton, P. J. and Wylie, S. *Homology Theory*, Cambridge University Press, London (1960).
[65] Hocking, J. G. and Young, G. S. *Topology*, Addison-Wesley, Reading, Mass. (1961).
[66] Hodge, W. V. D. and Pedoe, D. *Methods of Algebraic Geometry*, Cambridge University Press, London (1947–54).
[67] Hurewicz, W. and Wallman, H. *Dimension Theory*, Princeton University Press, Princeton, N.J. (1961).

[68] JEFFREYS, B. S. and JEFFREYS, H. *Methods of Mathematical Physics*, 3rd edn, Cambridge University Press, London (1956).
[69] KELLEY, J. L. *General Topology*, Van Nostrand, Princeton, N.J. (1955).
[70] KELLEY, J. L. *Introduction to Modern Algebra*, Van Nostrand, Princeton, N.J. (1960).
[71] KELLOGG, O. D. *Potential Theory*, Chelsea, New York (1947).
[72] KEMENY, J. G., SNELL, J. C., and THOMPSON, G. L. *Introduction to Finite Mathematics*, Prentice-Hall, Englewood Cliffs, N.J. (1959).
[73] KEMENY, J. G., SNELL, J. C., THOMPSON, G. L., and MIRKIL, H. *Finite Mathematical Structures*, Prentice-Hall, Englewood Cliffs, N.J. (1959).
[74] KEMENY, J. G. and SNELL, L. *Mathematical Models in the Social Sciences*, Blaisdell, Waltham, Mass. (1962).
[74a] KINGMAN, J. F. C. and TAYLOR, S. J. *Introduction to Measure and Probability*, Cambridge University Press, London (1966).
[75] KLEIN, F. *On Riemann's Theory of Algebraic Functions and Their Integrals*, Chelsea, New York (1963).
[76] KLEIN, F. *Elementary Mathematics from the Advanced Standpoint*, Dover, New York (1939).
[77] KOLMOGOROFF, A. N. and FOMIN, S. V. *Elements of the Theory of Functions and Functional Analysis*, Graylock, New York (1957).
[78] KREYSZIG, E. *Advanced Engineering Mathematics*, Wiley, New York (1962).
[79] LANG, S. *Algebra*, Addison-Wesley, Reading, Mass. (1967).
[80] LEDERMANN, W. *Introduction to the Theory of Finite Groups*, 4th edn, Oliver and Boyd, Edinburgh (1961).
[81] LEFSCHETZ, S. *Introduction to Topology*, Princeton University Press, Princeton, N.J. (1949).
[82] LIGHTHILL, M. J. *Introduction to Fourier Analysis and Generalised Functions*, Cambridge University Press, London (1958).
[83] LITTLEWOOD, J. E. *A Mathematician's Miscellany*, Methuen, London (1953).
[84] MACLANE, S. *Homology*, Springer, New York (1963).
[85] MACLANE, S. and BIRKHOFF, G. *Algebra*, Macmillan, London (1967).
[86] MICHIE, D. 'Trial and error' *Penguin Science Surveys*, Penguin Books, Harmondsworth, Middx (1961).
[87] MILNE-THOMSON, L. M. *Theoretical Hydrodynamics*, 4th edn, Macmillan, London (1962).
[88] MIRSKY, L. *An Introduction to Linear Algebra*, Clarendon Press, Oxford (1955).
[89] MITCHELL, B. *Theory of Categories*, Academic Press, New York (1965).
[90] MOISE, E. E. *Elementary Geometry from an Advanced Standpoint*, Addison-Wesley, Reading, Mass. (1963).
[91] MORSE, M. *Calculus of Variations in the Large*, American Mathematical Society Colloquium Publications no. 18 (1934).
[92] NAGEL, E. and NEWMAN, J. R. *Gödel's Proof*, Routledge and Kegan Paul, London (1959).
[93] NEWMAN, J. R. (editor) *The World of Mathematics*, Simon and Schuster, New York (1956).
[94] NEWMAN, M. H. A. *Elements of the Topology of Plane Sets of Points*, Cambridge University Press, London (1954).

[95] NICKERSON, H. K, SPENCER, D. C., and STEENROD, N. *Advanced Calculus*, Van Nostrand, Princeton, N.J. (1959).
[96] O'HARA, C. W. and WARD, D. R. *An Introduction to Projective Geometry*, Clarendon Press, Oxford (1937).
[97] OLDS, C. D. *Continued Fractions*, Random House, New York (1963).
[98] OLMSTED, J. M. H. *Intermediate Analysis*, Appleton-Century-Crofts, New York (1956).
[99] OSGOOD, W. F. *Functions of Real Variables*, Chelsea, New York (1935).
[100] PAPY, G. *Groups*, Macmillan, London (1964).
[101] PATTERSON, E. M. *Topology*, Oliver and Boyd, Edinburgh (1959).
[102] PATTERSON, E. M. *Vector Algebra*, Oliver and Boyd, Edinburgh (1968).
[103] POSTNIKOV, M. M. *Foundations of Galois Theory*, Clarendon Press, Oxford (1962).
[104] QUADLING, D. A. *Introduction to Analysis*, Clarendon Press, Oxford (1959).
[105] RADEMACHER, H. and TOEPLITZ, O. *The Enjoyment of Mathematics*, Princeton University Press, Princeton, N.J. (1957).
[106] REDISH, K. A. *An Introduction to Computational Methods*, E.U.P., London (1961).
[107] REUTER, G. E. H. *An Introduction to Differential Equations, and Linear Operators*, Routledge and Kegan Paul, London (1958).
[108] ROSSER, J. B. *Logic for Mathematicians*, McGraw-Hill, New York (1953).
[109] RUDIN, W. *Principles of Mathematical Analysis*, 2nd edn, McGraw-Hill, New York (1964).
[110] RUDIN, W. *Real and Complex Analysis*, McGraw-Hill, New York (1966).
[111] RUSSELL, B. *History of Western Philosophy*, Allen and Unwin, London (1957).
[112] RUSSELL, B. and WHITEHEAD, A. N. *Principia Mathematica*, 3 vols, Cambridge University Press, London (1910–13).
[113] RYSER, H. J. *Combinatorial Mathematics*, Wiley, New York (1963).
[114] SAATY, T. L. *Lectures on Modern Mathematics*, Vols I, II, III, Wiley, New York (1963, 1964, 1966).
[115] SAWYER, W. W. *Prelude to Mathematics*, Penguin Books, Harmondsworth, Middx (1955).
[116] SCHWARTZ, L. *Introduction to the Theory of Distributions*, Toronto University Press, Toronto (1952).
[117] SEIFERT, H. and THRELFALL, W. *Lehrbuch der Topologie*, Chelsea, New York (1947).
[118] SEIFERT, H. and THRELFALL, W. *Variationsrechnung in Grossen*, Chelsea, New York (1948).
[119] SIERPINSKI, W. *Leçons sur l'hypothèse du Continu*, Warsaw–Lwow (1934).
[120] SNOW, C. P. *Science and Government*, Oxford University Press, London (1961).
[121] SOMMERVILLE, D. M. Y. *Analytical Conics*, Bell, London (1937).
[122] STEINER, H. G. 'Mathematisierung und Axiomatisierung einer politischen Struktur' *Der Mathematikunterricht* **3** (1966).
[123] STOLL, R. R. *Sets, Logic and Axiomatic Theories*, Freeman, San Francisco (1962).
[124] THOMPSON, D'ARCY W. *On Growth and Form*, abridged and edited by J. T. Bonner, Cambridge University Press, London (1961).

[125] THURSTON, H. A. *The Number System*, Interscience, New York (1959).
[126] TULLER, A. *A Modern Introduction to Geometries*, Van Nostrand, Princeton, N.J. (1967).
[127] TURING, A. M. 'Computing machinery and intelligence' *Mind* **59**, 433–460 (1950).
[128] TURING, A. M. 'On computable numbers...' *Proc. Lond. math. Soc.* **42**, 230–265 (1937).
[129] TURNBULL, H. W. *The Great Mathematicians*, Methuen, London (1966).
[130] VEBLEN, O. and YOUNG, J. W. *Projective Geometry*, 2 vols, reprint, Blaisdell, Waltham, Mass. (1960).
[131] VAN DER WAERDEN, B. L. *Modern Algebra*, Ungar, New York (1953).
[132] WANG, HAO *IBM Jl. Res. Dev.* **4** (1960).
[133] WASSERMAN, E. 'Chemical topology' *Scient. Am.* (Nov. 1962).
[134] WEYL, H. *Symmetry*, Princeton University Press, Princeton, N.J. (1952).
[135] WHITESITT, J. E. *Boolean Algebra and its Applications*, Addison-Wesley, Reading, Mass. (1961).
[135a] WIGNER, E. P. *Group Theory and its Applications to the Quantum Mechanics of Atomic Spectra*, Academic Press, New York (1959).
[136] WILDER, R. L. *Introduction to the Foundations of Mathematics*, Wiley, New York (1965).
[137] WILLMORE, T. J. *An Introduction to Differential Geometry*, Clarendon Press, Oxford (1959).
[138] ZASSENHAUS, H. *The Theory of Groups*, Chelsea, New York (1958).
[139] ZIPPIN, L. *Uses of Infinity*, Vol. 7, Random House, New York (1962).

INDEX

INDEX OF SPECIAL SYMBOLS

This index contains those symbols and notations which have a special meaning. We do not include notation (like $f(x)$) which is both standard and familiar to the reader. We have included very brief descriptions of the meanings of the symbols and a reference to a page number for those readers requiring further detail. Since some symbols defy alphabetical classification, we cannot avoid asking the reader occasionally to skim through several sections to find the symbol required. We draw particular attention to *the last two items*!

Encounter with Mathematics

By **L. Garding**
1977. ix, 270p. 82 illus. cloth

The purpose of this text is to provide an historical, scientific, and cultural frame for the basic parts of mathematics encountered in college.

Nine chapters cover the topics of Number Theory, Geometry and Linear Algebra, Limiting Processes of Analysis and Topology, Differentiation and Integration, Series and Probability, and Applications. Each of these chapters moves from an historical introduction into a basic factual account, and finally into a presentation of the present state of the subject including, wherever possible, most recent research. Most end with passages from historical mathematical papers, as well as references to additional literature. Three remaining chapters deal with models and reality, the sociology, psychology, and teaching of mathematics, and the mathematics of the seventeenth century, providing a fuller historical background to infinitesimal calculus. Intended for beginning undergraduates, the text assumes background in high school or some college mathematics.

Undergraduate Texts in Mathematics

Apostol: Introduction to Analytic Number Theory.
1976. xii, 334 pages. 24 illus.

Childs: A Concrete Introduction to Higher Algebra.
1979. Approx. 336 pages. Approx. 7 illus.

Chung: Elementary Probability Theory with Stochastic Processes.
1975. x, 325 pages. 36 illus.

Croom: Basic Concepts of Algebraic Topology.
1978. x, 177 pages. 46 illus.

Fleming: Functions of Several Variables. Second edition.
1977. xi, 411 pages. 96 illus.

Halmos: Finite-Dimensional Vector Spaces. Second edition.
1974. viii, 200 pages.

Halmos: Naive Set Theory.
1974. vii, 104 pages.

Hewitt: Numbers, Series, and Integrals.
1979. Approx. 450 pages.

Kemeny/Snell: Finite Markov Chains.
1976. ix, 210 pages.

Lax/Burstein/Lax: Calculus with Applications and Computing, Volume 1.
1976. xi, 513 pages. 170 illus.

LeCuyer: College Mathematics with A Programming Language.
1978. xii, 420 pages. 126 illus. 64 diagrams.

Malitz: Introduction to Mathematical Logic.
Set Theory - Computable Functions - Model Theory.
1979. Approx. 250 pages. Approx. 2 illus.

Prenowitz/Jantosciak: The Theory of Join Spaces.
A Contemporary Approach to Convex Sets and Linear Geometry.
1979. Approx. 350 pages. Approx. 400 illus.

Priestley: Calculus: An Historical Approach.
1979. Approx. 409 pages. Approx. 269 illus.

Protter/Morrey: A First Course in Real Analysis.
1977. xii, 507 pages. 135 illus.

Sigler: Algebra.
1976. xii, 419 pages. 32 illus.

Singer/Thorpe: Lecture Notes on Elementary Topology and Geometry.
1976. viii, 232 pages. 109 illus.

Smith: Linear Algebra
1977. vii, 280 pages. 21 illus.

Thorpe: Elementary Topics in Differential Geometry.
1979. Approx. 250 pages. Approx. 111 illus.

Wilson: Much Ado About Calculus.
A Modern Treatment with Applications Prepared for Use with the Computer.
1979. Approx. 500 pages. Approx. 145 illus.

Wyburn/Duda: Dynamic Topology.
1979. Approx. 175 pages. Approx. 20 illus.